For the last century, economic analysis has been wedded to the idea of equilibrium, in spite of the evident fact that most economic relationships are in flux. The theory of transformational growth in this work replaces equilibrium with history. The role of the market is not to allocate resources, but to generate innovations, which are "selected" by competition in an evolutionary process. These innovations in turn change the way markets work and how they adjust, thus creating new problems and new kinds of pressures to innovate. Different historical periods can be distinguished, with a new one perhaps on the horizon. In each period a distinctive style of technology prevails, associated with corresponding institutions and patterns of market behavior.

To analyze market behavior requires theory. This rests on a core of basic value relationships between prices and profits, growth and quantities, centering on the cross-sectoral wage-bill capital requirements condition, which can be shown to reflect the Golden Rule, and to imply a linear wage-profit tradeoff. The core relationships provide the foundations for a theory of monetary circulation, which makes possible a revised Keynesian approach, based on Classical foundations.

THE GENERAL THEORY OF
TRANSFORMATIONAL GROWTH

The general theory of
transformational growth
Keynes after Sraffa

EDWARD J. NELL

The New School for Social Research

CAMBRIDGE
UNIVERSITY PRESS

PUBLISHED BY THE PRESS SYNDICATE OF THE UNIVERSITY OF CAMBRIDGE
The Pitt Building, Trumpington Street, Cambridge CB2 1RP, United Kingdom

CAMBRIDGE UNIVERSITY PRESS
The Edinburgh Building, Cambridge, CB2 2RU, United Kingdom
40 West 20th Street, New York, NY 10011-4211, USA
10 Stamford Road, Oakleigh, Melbourne 3166, Australia

© Edward J. Nell 1998

First published 1998

Printed in the United States of America

Typeset in Times Roman

Library of Congress Cataloging-in-Publication Data is available

*A catalog record for this book is available from
the British Library*

ISBN 0 521 59006 X hardback

In memory of Marcella
With gratitude to Marsha
And hope of better times
For Adam and Jacob, Miranda and Gwen

CONTENTS

EXTENDED TABLE OF CONTENTS

Contents   xi

Contents

Contents

Contents   xix

PREFACE

Economics is a difficult subject at the best of times, but taking an unconventional approach makes it doubly so. It is hard enough to keep up with current research as it is, so there must be important reasons for asking the reader to take the trouble to look at the world in a different way.

Moreover, some efforts at a new approach may give one cause to be skeptical. A good deal of the best work is chiefly critical (which is emphatically *not* the case here). Such work is widely ignored, or cited only in obscure footnotes. It is generally acknowledged that neo-Classical theory has significant flaws – but it is equally widely believed to be the only useful and relevant scientific approach. Neo-Ricardian theory, for example, is just as abstract, calls for a counter-intuitive interpretation of Keynesian macroeconomics, and has little or no behavioral theory. Its approach to policy and empirical work is underdeveloped, to say the least. The post-Keynesian school has a better record on policy and empirical work, and begins from a reasonable position in macroeconomics, but has an eclectic approach to pricing, and, indeed, to theory in general. Some would say it is more an attitude than a doctrine.

By contrast, neo-Classical theory is not only fully articulated in virtually every area of interest, but is able to find room for almost any position on any issue. It has become the language of economics, and like a language it has a grammar. But the grammar does not commit one to any particular explanation or ideology over another, any more than English or French grammar does. However, it does have implications, as we shall see.

Nevertheless the alternative schools have focused heavily on criticism, and this has not proved useful. Firing critical salvoes at neo-Classical theory can be considered anything but constructive. For one thing, neo-Classical thinkers themselves have already done the job. In a practical sense general equilibrium theory has all but self-destructed. What can we learn about the economy from it? It is unrealistically abstract, yet nevertheless requires strong assumptions with little or no economic justification. And even so, it still leaves us with the possibility of multiple equilibria, some or even all of which may be unstable. Moreover, the theory cannot easily accommodate a general rate of profit on capital. How then can it be a theory of capitalism? What has happened to the profit motive if business and finance are not interested in obtaining the highest returns on their investments? Even worse – if possible – conventional general equilibrium theory can incorporate only the most rudimentary concept of money, and it handles that badly.

The reply is surely that general equilibrium theory is not supposed to be practical – it was designed to illuminate market interdependence. Practical theory operates at a different level, the partial equilibrium theory of Chicago, or Solow's "middlebrow" aggregate theory. Of course, these approaches are legitimate targets for "capital theory" criticisms. But again, what use is such criticism? How will it help anyone to understand the causes of unemployment or inflation or the productivity slowdown, or whether corporate pricing policies will pass along the effects of changes in exchange rates? Flawed or not, neo-Classical models can be used to address the questions of the day. If the flaw is important, account can presumably be taken of it in drawing conclusions from the model. Or the model can be recast to avoid relying too heavily on the flawed bits.

The critics try to show that neo-Classical theory rests on one or another fatal error of logic, for example, in distribution theory, or in theory construction in regard to effective demand. Yet this sometimes seems almost arrogant. Is it plausible that so many, for so long, have been deceived by so few – the founders of neo-Classical theory? Is the last century of mainstream economics nothing but a vast sea of errors and mistakes? Did neo-Classical theory develop and become dominant for no good reason at all? Can all this be explained by scientifically irrelevant factors, for example, that it serves as an ideological cover story for modern capitalism? This hardly seems convincing.

On the other hand, the defenders of neo-Classical thinking do not seem on firm ground, either. Virtually every trained economist has his or her own pet peeve about accepted professional thinking – an utterly indefensible, but basic, assumption here, an outrageous oversimplification there, in some other place a conclusion wholly at variance with well-established facts. These are not isolated cases; the aura of unrealism is absolutely pervasive. Everyone who has worked with neo-Classical models has felt the sense of being in an imaginary land – a market-driven Oz with an all-too-fallible Wizard half-concealed behind the veil of assumptions, working the optimizing machinery.

Critics and defenders alike tend to present the profession with a choice: the theory, or parts of it, are right or wrong. It passes or fails. If it passes it can continue to be accepted; if it fails it must be replaced by a correct, or more nearly correct, theory. The Bohr model of the atom was accepted as correct; it was then shown to be flawed, and replaced by a model based on quantum mechanics. The critics contend that the "capital theory" arguments have shown the flaw in marginal productivity theory, and therefore in the whole edifice of costs and supply, as well as distribution theory. Defenders maintain that the flaw is truly marginal, and that, in any case, no quantum mechanics are in sight. "You can't beat Something with Nothing." Both sides treat the question as an issue of the timeless truth of a scientific theory.

Atoms have no history. They do not behave at one time, in one era, in one way, and then differently at another time or place. Markets, however, may do just that. They are social institutions, and institutions develop and change, historically. If markets do change, then a theory describing their working may be true for one era, for one time and place, and not true for another. Classical theory, or theories, for example, might describe the eighteenth and early nineteenth centuries, neo-Classical approaches the later nineteenth and early twentieth, and variants of Keynesian theory, the Great Depression and after. This is not to underrate the importance of getting theory right; a model with a logical flaw describes nothing at all. But it is to suggest that different approaches to theory may have developed for good historical reasons, and may reflect different institutional arrangements – differences that may still be relevant today.

But there is no room for such an idea in the dominant methodology of economics. The language of the subject is the vocabulary of rational choice, its grammar, the algebra of optimizing. Rationality is the same in all eras, all times, and all places. The setting and constraints may differ, but the problems and the methods of solution are the same. Markets at bottom reflect rationality, i.e. optimal choices, which, in competitive conditions, will emerge in equilibrium. The method of analysis is to compare equilibria; dynamics, insofar as it is necessary or useful, is equilibrium dynamics, or concerns the stability of equilibria.

It is not what this points to that is objectionable; optimizing surely has a role in any sensible economic analysis. But the neo-Classical approach uses optimizing to determine equilibrium, usually understood to be the norm toward which ordinary behavior will gravitate. In actual practice, optimizing – for instance, operations research – is generally used to find a way to *improve on* the current norms of ordinary behavior. Normal behavior – the empirical counterpart to equilibrium – is likely to reflect rules of thumb and practical compromises; optimizing will lead to changes in this behavior – changes which may be quite disruptive.

The problem is that in focusing on allocation the neo-Classical approach has directed our attention *away* from vastly more important problems – the nonallocational effects of markets, their role in *creative destruction*. Competition may or may not have strictly allocational effects.

If it does they are surely minor compared to its role in bringing about innovation and productivity growth, causing uneven development, leading to stagnation, inflation, booms and busts, and all the rest.

The pernicious effect of the presumption that economic analysis must be equilibrium analysis can be illustrated by considering the theory of steady growth. This began from two sources in the 1930s – the von Neumann model, and Harrod's famous essay, amplified later by Domar. During the 1950s and 60s an explosion of models developed the Harrod-Domar analysis, the neo-Classical counterapproach, a neo-Keynesian version, and two-sector variants of both. These were used to explore problems in distribution theory and the monetary role in growth, and to develop the theory of optimal growth, among other things. Multisector models were developed to examine the conditions for steady growth, the nature of the turnpike, the optimal choice of technique, and many other topics.

During this time, while economists were developing the theory of *steady* growth,

- most advanced capitalist economies (ACEs) operated with a substantial and varying margin of excess capacity
- most Soviet-style economies operated under varying conditions of shortage
- in all ACEs, the share of agriculture in labor and output steadily declined
- in all ACEs, the share of manufacturing first rose, then leveled off; in some it then began to decline
- in all ACEs, the share of services rose, and changed character
- in all ACEs, the share of white-collar work rose relative to blue-collar
- toward the end of the period, "high-tech" industries began to increase, while steel and other traditional Mass Production industries began to decline.
- in the United States recent "downsizing" has brought a sharp loss in high-paying manufacturing jobs and in white-collar middle management

Except for the last, these have been persistent trends, forming a long-term pattern. Arguably – it will be argued in this book – this was the way the economy grew; specifically, this was how it worked to increase productivity. But in developing growth theory this pattern has been ignored – in the "new" growth theory as much as in the old. While it is perfectly reasonable to disregard short-term fluctuations in conducting a long-term analysis, it is not reasonable to ignore persistent long-term changes. There must be something wrong with a discipline that devotes its scarcest and most valuable resource – high-powered analytical skill – to the analysis of steady, balanced expansion in the face of a pattern of persistent trends leading over a century to fundamental changes in the proportions and structure of the economy. In general, economists have allocated a massive amount of theoretical time and energy to the study of equilibrium, in spite of the all-too-obvious fact that it is seldom, if ever, encountered.

There may well be good reasons to study equilibrium. Certainly a large and complex array of theoretical tools has been devised, for which any serious student must be grateful. Yet it can easily be seen that the price – the opportunity cost – has been high: we know very little about the way the ACEs actually grow, particularly about how they generate productivity growth. Indeed, our theoretical understanding of the causes and effects of productivity growth is, in general, remarkably slim. And part of the reason must be that economists, concentrating on equilibrium, have paid comparatively little attention to historical dynamics and the causes of structural change.

The theory of Transformational Growth is an effort to make a fresh start, to begin all over again, from History rather than from Equilibrium, from structures and institutions rather than from rational choice. In particular, markets will be approached as competitive monetary exchange systems, rather than implicit barter systems. This will require a new treatment of the theory of money, one in which circulation, rather than asset-holding, assumes pride of place, and in which circulation velocity is tied to productivity.

In emphasizing History we shall nevertheless find a place for Equilibrium; in emphasizing structure and institutions, we will find a place – many places – for optimizing and rational choice, so long as the relative roles are properly understood. Traditional microeconomics, in its Marshallian configuration, will turn out to have a major assignment, as a theory of market adjustment under certain conditions. Even equations determining vectors of barter exchanges will find a place in the sun – though only at certain times of day. The traditional tool-box will find many tasks. Much of the vocabulary of economics will remain, but there will be significant changes in grammar. So the overall balance will be very different, and the subdued roles assigned to many conventional stars may not prove satisfactory to some audiences. Nevertheless, this appears to be the price which must be paid to convert our subject into one capable of addressing the transformations of the economy.

ACKNOWLEDGMENTS

In a work that has taken as long as this to mature, giving adequate acknowledgement to all who have influenced it is bound to be a problem. My apologies at the outset to my friends and associates – I have borrowed freely from all of you. I only hope the use I have made of your ideas is constructive.

Nevertheless a number of conversations and contributions do stand out. First, and most profound, there is my debt to my wife, Marsha Nell. Without her support and encouragement – more than encouragement, her hard work, love and devotion – this simply would never have been finished.

Then there are my teachers and elders: Joan Robinson, Piero Sraffa, Nicky Kaldor, Maurice Dobb, Richard Goodwin, and Adolph Lowe. And perhaps John Hicks (though *not* J. R. Hicks, with whom I had differences), and Robert Solow and Paul Samuelson. From each of these I learned something indispensable – though not necessarily what they wished to teach me!.

Next, my friends and colleagues: Over the years I have had innumerable illuminating conversations with Piero Garegnani, Heinz Kurz, Geoff Harcourt, and Bertram Schefold. I have also learned at various times from discussions with Luigi Pasinetti, Paolo Sylos-Labini, Harald Hagemann, Uli Krause, Fernando Vianello, Ghislain Deleplace, Will Milberg, Sergio Parrinello, Fabio Petri, and the late Krishna Bharadwaj. I have benefited from co-teaching with Willi Semmler. Others who should be mentioned include Paul Davidson, Alfred Eichner, Robert Heilbroner, Jan Kregel, Hy Minsky, and Anwar Shaikh. Many of these conversations and friendships stemmed from the International Summer School at Trieste; others developed out of the programs at the New School.

Over the years, Ross Thomson has helped to educate me in Economic History; Martin Hollis and Onora O'Neill have kept me in touch with Philosophy. Many years ago, at Nuffield College, conversations with Peter Browne and Laszlo Peter developed some of the foundations for the point of view developed here.

Many of my students and former students have aided this project in various ways. Gary Mongiovi assisted with numerical calculations;

Enrique Delamonica with some of the mathematics and an econometric study of money demand and the mark-up; Stephanie Thomas also worked on the econometrics of money, as well as preparing the charts in chapter six; Susan Pashkoff and George Argyrous read sections; Tom Phillips, Thorsten Block, Stephanie Thomas, Ray Majewski and David Kucera provided studies of the 'old' and 'new' business cycle; and Gang Gong provided simulations for the appendix to Chapter Thirteen. Others have helped in various ways: Louis-Phillipe Rochon, Garrett Bekker, and my sons Adam Nell and Jacob Nell. In earlier stages Chidem Kurdas, Mat Forstater, Steve Pressman, Scott Moss and Peter Anthony also provided assistance.

I am grateful to Heinz Kurz for inviting me to lecture on these matters at Graz, and to Bertram Schefold for a similar invitation to Frankfurt.

Stephanie Thomas transcribed the diagrams into a computer program, rationalized the numbering of the equations, and prepared the text for copy editing. Stephanie Thomas, Ray Majewski and Thorsten Block helped with proofreading.

Tom Phillips assisted in the preparation of the bibliography.

Judy Fischetti of the Woodstock Library, drawing on the Mid-Hudson Library System, provided invaluable help, with good cheer and remarkable efficiency.

Two recent books of mine complement this one. *Transformational Growth and the Business Cycle* provides empirical support for many of the positions developed here, and *Making Sense of a Changing Economy* sketches the larger picture and draws some of the policy conclusions. Finally, I should note that I have written on many of the subjects covered here on previous occasions. Much of what I have said in those earlier efforts is superseded by the present work.

PART I

History or equilibrium?

For the last century, economic analysis has been wedded to the idea of equilibrium, in spite of the evident fact that most economic relationships are in flux. The theory of steady growth was developed over a quarter century during which, in the advanced countries, the share of agriculture fell dramatically, manufacturing first rose, then leveled off, and then began to fall, while services and especially government rose, and changed character. Growth was never even approximately steady. Nor did it simply fluctuate. Instead, the composition and character of the economy changed according to a definite pattern, which rested, in turn, on a core structure that remained largely unchanged. This core consisted of – and is today – the set of basic value relationships between wages, prices, and profits, determining the value of capital. This is what provides the continuity within the processes of change.

The study of such patterns of directed change, and the forces that determine them, is the subject of the theory of transformational growth. The first chapter argues that theory should assign the market the role, not of allocating resources, but of generating forces that bring about innovations. These innovations, in turn, change the way markets work – how they adjust – creating new problems, and thus new kinds of pressures to innovate. Evolutionary forces can be seen at work. Three historical periods can be distinguished – with a fourth perhaps on the horizon. In each a distinctive style of technology prevails, associated with corresponding institutions and patterns of market behavior.

The second chapter compares and contrasts market behavior and economic structure in the two periods for which adequate data exists – the late nineteenth century, roughly 1870–1914, and the post – World War II era of fully developed Mass Production. Two patterns of market adjustment are identified, which appear to work according to different principles. Understanding these systems – how they work, how and why they differ, and how the later evolved out of the earlier – will be our chief project here.

CHAPTER 1

The idea of Transformational Growth

An equilibrium approach to the economy necessarily underplays change. The conventional wisdom looks for equilibrium configurations and so has missed a highly visible pattern of change that has characterized virtually all major capitalist economies. According to this pattern, agricultural productivity rises faster than demand, so that labor shifts out of agriculture, moving first to the cities, then to the suburbs. Manufacturing increases its share of employment, then levels off, and eventually finds its employment share declining. Within manufacturing the capital goods industries increase in importance relative to consumer goods. Labor employed in services increases, but at a certain point the services change character, shifting from household and personal to business and financial. Similar changes take place in the proportions of output. The share of government in output and employment at all levels rises, then levels off. Within government the share of transfers and welfare rises. Nonprofit acitivities rise relative to the rest of the economy. All these changes take place slowly but steadily, and can be observed whether the economy is booming or slumping. But they are incompatible with the idea of steady growth and cannot be properly understood if we think in such terms.

Nor are these the only changes. At least as important are the changes in the way markets work, in their pattern and manner of adjustment. As we shall see in detail later, prior to World War I markets adjusted primarily through price changes; following World War II, the dominant adjustments took place through changes in output and employment. Prior to World War I the role of the state was limited; markets operated autonomously and the state could plausibly be described as a "Night Watchman" – although most night watchmen don't hand out free land and subsidies! After World War II government policy played a major, sometimes dominant, role in determining the overall level of output, employment, and inflation. The Night Watchman became the Welfare State, provided we understand that in the United States much of the welfare went to military contractors – "Warfare State" might be almost as appropriate at times. Prior to World War I financial markets were closely linked to both booms and slumps; following World War II the cycle is only loosely tied to financial markets – although their importance in other respects is at least as significant, and grows throughout the period. Before the First World War the majority of businesses were organized as family firms, which sought to build to an optimal size, and thereafter tended to operate at that level, distributing their profits as dividends. After the Second World War businesses, organized as modern corporations, retained their earnings and reinvested them to grow at an optimal rate, given the conditions of their markets. Before the First World War family ties were stronger, and many households tended to include parts of the extended family, sometimes taking in members from several generations; even if not living under the same roof, the extended family formed a household production system, bartering services. After the Second World War households came to be based on the nuclear family, household production declined, and as divorce rates rose, single-parent families became common.

Emphasizing these changes takes us away from the traditional approach. In the view of the conventional wisdom, markets are a reflection of people's rational choices; hence the way markets work will be the same in all times and places, since rationality is universal and timeless (Walras, 1954; Robbins, 1936; Samuelson, 1989). If economic analysis can be extended beyond the marketplace, it is because people make rational choices in all aspects of life.

The claim for the universality of rational choice has to be stated carefully, however. Market outcomes, of course, will differ in different times and places, since people's tastes and technologies will be different. Market institutions will also differ, but, although they introduce frictions and imperfections into market processes – transaction costs, in the idiom of the day – they are ultimately not determining. They are simply the vehicles for rational choices; they facilitate or sometimes inhibit the process of reaching positions of constrained optimization. Markets, in short, may look very different in different times and places, but the way they work will always be to compare alternative possibilities by assigning prices to them, and then solving the appropriate constrained maximizing problems. Choices and outcomes will then be determined through market clearing. Nonmarket decisions can often be treated by similar methods.

But if markets in one era adjust through price changes, and in another through quantity variations, if in one era firms find their optimal size, while in another they seek their optimal rate of expansion, then it seems that more is involved in the contrast than mere appearances. It is not just how things look, it is how they work that differs.

Contrary to the conventional wisdom, then, the theory of Transformational Growth takes changes in the way markets work as part of its subject, and is designed to explain and help to analyze the causes and effects of these changes, including an understanding of their persistence.

The nature of markets

The textbooks and the conventional wisdom see markets as social devices – institutions – designed to resolve conflicts by finding the optimal allocation of scarce resources, working through the price mechanism. Scarcity is a fact of life, a condition imposed on human society by nature.[1] There simply isn't, wasn't, and can never be enough to go around, that is, to satisfy everyone's preferences to the fullest extent desired. Hence some preferences of some people will go unsatisfied. But the question, whose preferences, and which of them, will go unsatisfied to what extent, is potentially explosive. The social function of the market is to prevent an outburst by finding the optimal solution, a solution which will be subject, of course, to a number of constraints, such as the ownership of property, which themselves may not be optimal at all. Nevertheless, tensions will be reduced by the simple fact that there exists an institutional method of resolving disputes over jobs and pay and other forms of distribution, which also provides a way of deciding what will be produced for whose consumption.

What markets do, in short, is to find the equilibrium which will balance competing interests and bring everyone to the highest degree of satisfaction possible, given all the other demands, in light of the constraints imposed by the nature-determined fact of scarcity. Markets are harmonizers; they seek out equilibrium, and they enable us to deal rationally with the niggardliness of nature (Samuelson, 1947). Indeed they impose rationality on us, since those who fail to make rational choices will tend to lose out in competition to those who do. Market-clearing through price adjustments establishes the equilibrium; everyone is in the best position possible – for if they were not, the market would not clear (Hahn, 1973). Once reached, this situation will be maintained indefinitely, until the "givens" change.

This understanding of markets cannot be accepted. For one thing it is utterly static. Technology, tastes, and resources are treated as exogenous, as given, and changes in these are outside the scope of the analysis. They are, in fact, the fundamental data used to derive equilibrium positions. Yet it is obvious to common sense that, all through recent history, market pressures have driven the advance of technology and have forced changes

---

[1] The discussion here is not concerned specifically with what is known as the "non-satiation" postulate. That postulate is indeed basic to neo-Classical theory – without it solutions might fail to exist – but the point in the text concerns the general approach of analyzing market behavior, and interpreting market outcomes, on the basis of models of constrained maximizing. "Scarcity" in this sense is a philosophical doctrine – a point Walras made quite clearly.

in tastes. Advances in technology, in turn, have led to redefinitions of resources. Besides being static, this treatment of markets elevates equilibrium to a central place. Yet, in practice, many markets seldom clear, and even those that do don't appear to be in a state that could be called "equilibrium."[2] Again this is obvious in the case of the labor market, but it is also apparent in many other markets. To be sure, markets and competition do impose a kind of rationality on anyone who participates in them. Calculation of advantage is forced on one; anyone who fails to watch their pocketbook will lose it. And there is a sense in which the market system brings competing interests into an arena in which their struggles will, in some ways, benefit rather than injure society. Certainly conflicts in the marketplace are generally less destructive than on the battlefield.

But that is not what markets do. Struggles don't harmonize interests, except by conquest. "They made a desert and called it 'peace,' " it was said of the war party in the Roman Senate. Outcomes, in reality, are not in any obvious sense optimal; indeed, they are often quite evidently wasteful. And the resulting allocation of resources is frequently anything but efficient. Of course, it will be argued that in this imperfect world, markets, too, are imperfect. We should not expect too much; the scientific point is the understanding of the tendencies – what the system is trying to do, not what the actual outcomes are.

Perhaps – the story sounds good and is hard to refute. But it also looks evasive. If markets do not appear to be allocating resources efficiently, it is an act of faith to believe that, nevertheless, they "really" are. Many economists and observers of the business and political world have been willing to make that leap of faith, simply because there is no credible alternative account of markets. Markets are obviously doing something; and without question they are one of the major institutions of the modern world. In some ways, perhaps, the most important. Indeed, it will be argued here that markets are largely responsible for having *created* the modern world. But not because they allocate resources efficiently. The belief that markets "allocate" has stood in the way of understanding how markets actually work, and what they actually do. Once we look clearly at the historical record, especially at the interaction of markets and technology, and the resulting impact on the organization of production, that leap of faith, hopefully, should no longer be necessary.

[2] Market-clearing implies that all offers to buy and to sell were matched – no unsatisfied demands and no unsold goods. Equilibrium requires in addition that a single uniform price should have been established. Market-clearing is consistent with multiple and even with fluctuating prices.

For what markets do – to state a grand generalization in the manner of the textbooks – is to generate the pressures that increase productivity. As we shall show throughout the rest of this book, these pressures bring about innovations, organizational innovations as well as new technologies, which markets then diffuse throughout the system by the force of competition. Market adjustment – the price system – mobilizes the profits to underwrite investment in these innovations, making the diffusion possible. This means that markets pick winners and losers, which is, indeed, a rough sort of allocation. But it is a far cry from the smooth harmony of the conventional version.

In some respects, this is a return to the views of the great Classical economists, Smith, Ricardo, and Marx. The price mechanism cannot realistically be pictured as a sensitive selector of optima; markets create and destroy, reward and punish. Competitive behavior is managing resources in what amounts to economic warfare. Efficiency is rewarded because it confers advantages in the struggle. But equally or more, so does successful innovation. Faced with a shortage, the winning entrepreneur sees an opportunity. When a resource is scarce and badly needed – rubber at the beginning of World War II – innovators will find a way to replace it. Scarcity will drive up prices, but the high price, bringing high profits, together with an assured market, will stimulate innovation – synthetic rubber, for example. In the long run, scarcity is *incompatible* with equilibrium. New products and new processes result from the innovation stimulated by scarcities, and these disrupt any settled arrangements. "Given conditions" mean given problems, but the price mechanism rewards solutions, and punishes those who simply accept problems as given. The equilibrium approach is on the wrong track altogether.

Schumpeter, of course, tried to combine the two (Schumpeter, 1934). He argued that businesses normally operate according to familiar patterns, following tried-and-true paths. The result is an equilibrium of conventional behavior, basically describing a system stuck in a rut. Existing businesses will normally be run by those who find success and satisfaction in living up to the standards of bourgeois life. Nothing new and little worthwhile could ever come from this. By contrast, entrepreneurs, imaginative and daring, innovate, and thereby lift the system out of the mud of everyday life. The entrepreneurs will develop new products and new technologies, found new firms, overthrow the old order, and transform the markets. Entrepreneurs unleash the whirlwind of creative destruction; their new processes and products may wipe out whole sectors or long-established businesses. But entrepreneurs are the exception, singular beings who come along once a generation, and they form a self-selected elite, a natural aristocracy. Their activities cannot be regulated, and their coming cannot be predicted. Prediction and scientific study must concern itself with the mundane level of ordinary business and normal life. Economics therefore studies the equilibrium system, but everything that matters, the innovations that revolutionize the world, come from the entrepreneurs.[3]

As descriptive sociology this won't do. Too many major innovations have poured out of the scientific labs of the great universities and government agencies, to be developed and tested and thrown into the market by scientific bureaucrats. Even more seriously, however, marrying the two views of the market faces a logical hurdle, which threatens to block the effort to combine equilibrium ideas with the view that the market engenders innovation.

Neo-Classical market-clearing equilibrium depends on a balance between the patterns of behavior of suppliers, on the one hand, and buyers, on the other – each agent in each group presumably deciding what to do by solving an appropriate constrained maximizing problem for the various likely hypothetical prices. The solutions to these maximizing problems, of course, depend upon the givens – resources, technology, and preferences. Market-clearing, in addition, depends on agents being able to form reliable expectations as to the reactions of others, especially when there is no "auctioneer." Such expectations must be rationally formed. Entrepreneurs, however, will upset the givens. They introduce new products and new technologies, they create or discover new resources, and they open new horizons, changing what people want and expect out of life. Furthermore, they do this repeatedly, if at unpredictable times. Rational agents will therefore learn to expect the unexpected from the entrepreneurs, and will look for signs that will indicate which resources will change in value, what new products or processes will create new conditions, where "creative destruction" will strike next, and so on. Since success in innovation cannot ever be guaranteed, there can be no certainty, and different agents will form different views of the future. The givens will no

---

[3] In Schumpeter's original vision (1934) industry is characterized by a large number of small firms and easy conditions of entry. Entrepreneurs create new firms which move in and displace the former industry leaders (Nelson and Winter, 1982). But in a later work (1942), he recognizes the importance of large, oligopolistic firms and barriers to entry. But he argues that to maintain and protect their oligopolistic position, large firms will be obliged to develop research laboratories and pursue a strategy of innovation based on R&D. The former is sometimes characterized as "widening" innovation – new firms enter with new products or processes – while the latter is termed "deepening," as existing firms develop the implications of their accumulated knowledge (Malerba and Orsenigo, 1995). But in both cases innovations are the work of entrepreneurs.

longer be given, as far as agents' plans and expectations are concerned. Nor will agents be able to form reliable expectations as to what others will do in various market situations – for some may be acting entrepreneurially, while others are behaving conventionally. Expectations will therefore be tentative and liable to sudden shifts. In short, if entrepreneurs are active, equilibrium will be difficult to determine, let alone to reach, on the basis of the data normally deemed sufficient by neo-Classical theory.[4]

Harmonious equilibrium is not a struggle that engenders pressures for innovation and creative destruction. Markets cannot easily be portrayed as both equilibrating and creative. In one sense neo-Classical thinking has rightly backed away from Schumpeter, for in doing so it has maintained its consistency. But at a terrrible price. Equilibrium theory is largely useless for practical purposes. It is unable to explain the ordinary behavior of business and the working of markets, let alone speculate on the economic forces that helped to create the modern world, and now perhaps threaten to destroy it – or at least to undermine its prosperity.[5]

The proper course, then, is to abandon the view that markets allocate resources through optimizing and market-clearing adjustments, in favor of a vision that sees them as forcing development through a competitive struggle that creates incentives to innovate and to spread innovations rapidly and widely. With this understanding of the market, we can begin to appreciate the role that it has played in the development of the modern world.

From natural order to progress

The recent slowdown in the rate of increase in productivity, and the consequent stagnation in living standards, has caused great alarm. A regular rise in the standard of living has come to be the norm among the advanced nations. It is worth noting, however, that even two centuries ago a change in the standard of living would have been the exception, and would most likely have been attributed to natural causes, like the weather and its effect on crops. The Physiocrats, following Boisguilbert, understood that misguided policies could lead a (largely) market-based economy

---

[4] In the language of economic theory, there are not enough givens to determine the existence of equilibrium, and even if an equilibrium could be shown to exist, the market would not reliably move toward it, because of the instability of expectations.

[5] A simple example: equilibrium comparative static analysis suggests that Prohibition and tough law enforcement will raise prices enough to reduce demand and stamp out the use of recreational addictive drugs. Dynamic analysis tells us, on the contrary, that Prohibition and law enforcement may actually tend to *expand* the market (Nell, 1994).

into ruinous difficulties. But the idea that living standards could be expected to rise *regularly* because of activities in the marketplace, or as a consequence of policies of the state, or both together, only emerged in the writings of Adam Smith, to be developed by Jeremy Bentham and the other members of the Political Economy Club. Two centuries ago it had hardly penetrated beyond that circle; as yet it played no role in politics. Similarly, the idea of Progress had only just taken root on the continent; it had yet to flower. And prior to the Enlightenment, no one thought of regular and successive improvement in material well-being as something to be expected, let alone as an aim of normal politics.

Consider the contrasts between two general characterizations of the social order, one corresponding to the traditional forms of society, whether "river valley" systems (Egypt, Mesopotamia), slave-based (Athens, Rome), or feudal (Europe in the Middle Ages), the other describing the West since the Industrial Revolution. The first is essentially static, not in the sense that nothing changes, but rather that whatever happens to bring about changes comes from the outside. The system itself is self-reproducing, and when impacted by outside causes will adjust so as far as possible to restore the earlier equilibrium. Traditional society is indeed an equilibrium system, although not a market-based one, since in it the role of the market is strictly limited. It is a system of "natural order."

By contrast, there is a second form of social arrangements, evident among the advanced nations, the actively – though not always success-fully – developing countries, plus the nations of the Socialist Bloc both before and after the collapse of the Berlin Wall. It is essentially driven by an internal dynamic, leading to innovation, social change, and institutional development. So fundamental is this dynamic that a failure of material prosperity to grow creates a sense of crisis. It is a system of "regular progress."

The condition of natural order

From the earliest times of recorded history, through the civilizations of the Ancient World to the Middle Ages and even into the Renaissance, the vast majority of humanity lived in an unchanging world, a stable cycle of revolving time, days running from sunrise to sunset, months passing through the phases of the moon, seasons rolling across the year, with the lives of people flowing in the same way through phases similar to the seasons. People are born, just as life bursts forth anew in the spring, grow and flourish in their summer, mature in the autumn years, and then wither and die in the cold winter of old age. Just as everything in nature has its proper place and exists in a harmonious, if sometimes cruel, balance, so everyone in the social system has an appointed role, likewise balanced with the rest. This social order renews itself in a manner like that of nature.[6] New people are born to step into the shoes of the aged and dying; new leaders replace old, sons replace fathers, daughters become new mothers. Crafts and trades are passed on from father to son,[7] domestic skills from mother to daughter, as generation replaces generation, in an unchanging cycle, just as the seasons roll. Status and position are handed down through the family. Lives in any one generation echo the corresponding lives in all previous generations. And the whole society will first be young and vibrant, then will mature and finally age, perhaps even collapsing from decay; but then spring will come again in a new age, and the kingdom will be reborn. Plus ça change, plus cèst la même chose. The water flows, the river stays, though it will rise and fall, flood in the spring, run dry in the summer (Laslett, 1984).

The condition of regular progress

Systems in this condition are organized around the market, and progress is defined in market terms. Normal functioning for any sector requires success in the market; no matter how steeped in tradition, an activity that fails the market test must change, perhaps drastically. But to function normally, and to successfully maintain and reproduce its basic institutions, the market system must *expand*. This need to expand is driven by competition; for competition compels cost-cutting and parsimony. In conditions of early capitalism, these apply not only to the firm, but also to the household, for the two are not fully separate. But if competition is expanding productivity on the one hand, and enforcing cost reduction and parsimony on the other, there has to be investment and expansion to absorb the savings. If this fails to take place, some firms or even some sectors will be unable to make ends meet.[8]

---

[6] Different cultures have different concepts of nature, and so are likely to attach different meanings to the idea of the "natural order" in society. The point, however, is not that all traditional societies share a common natural order, but that all relate their – quite different – static orders to their – likewise very different – concepts of nature.

[7] "It is in accordance with the fate decreed by Enlil for man that a son follow in the work of his father." So records a clay tablet of nearly 2000 B.C.E., translated by S. N. Kramer, quoted in his *History Begins at Sumer* (New York: Doubleday, 1959). Also cf. Weisberg (1967, ch. 6), analyzing the guild structure of ancient Mesopotamia, with particular reference to fathers and sons.

[8] As we shall see, in the earlier periods of capitalism, a failure of investment to keep pace with rising productivity and savings resulted in a collapse of prices; later, it manifested itself in a decline of output and employment.

Expansion, however, should not be understood as simply more of the same; an even stronger competitive position can be attained through successful innovation, offering new or improved products, or more efficient methods of production, particularly where economies of scale exist. Successful innovations are those selected by the market. These come to be expected, and plans – even government policies – are based on anticipated increases in productivity and rising standards of living. Education is undertaken by the state, in view of the fact that skills can no longer be passed on from father to son and mother to daughter, because the technologies are changing so rapidly that the technical skills of one generation will no longer be useful in the next. But this implies that the jobs and the organization of work will differ from generation to generation. If jobs have changed, the system of authority is also likely to change. Certainly it will have to be flexible enough to accommodate innovation. Hence it will be difficult to hand positions of authority down from generation to generation – this is one of the points of connection between capitalism and democracy. Further, if jobs and organization and authority positions all differ, it is likely that places of work will also change; hence to be close to their work, when grown, a new generation may move to different living places than those of their parents.

A society in a condition of natural order, then, functions quite differently from one in a condition of regular progress. The first is characterized by replacement, repetition, stability, unchanging order, and stable authority, and the passing on of skills, knowledge, position, and property through the family. The second is driven by an internal dynamic, and exhibits expansion, innovation, instability, structural development, and shifting patterns of authority. In particular, the classification, selection, and preparation of the new generation shifts away from kinship and comes to take place through the institutions of the State.

How does an economy change from functioning in a condition of natural order, to one of regular progress? The short answer is that it enters on a path of transformational growth. The long answer – the project of this book – and the only way to make the short one intelligible, has to explain how this happens. So first, we must spell out exactly what is involved in transformational growth.

Defining Transformational Growth

A society that is successfully functioning in a condition of natural order does not one day start to swell up equiproportionally. In its natural condition it reproduces itself; it is stable and static. Nothing changes. An exceptional surplus will be put aside and stored against bad times – but saving or investment for growth is simply unknown. Steady growth is not possible in these conditions. Exogenous factors may lead to expansion, of course. Exceptionally good weather could lead to bountiful harvests over a number of years, resulting in a better diet, improved health, and a population increase. New lands would then be cleared and settled. Or a wave of immigrants might descend on the system. But adjustments to one-time exogenous causes are not at issue. Steady growth means a regular, normal pattern of expansion, and this will not result from a one-time exogenous shock.

In fact, cases of steady growth are rare – if they exist at all.[9] Pre-capitalist societies grew by fits and starts, and often went long periods without growth, even declining at times. Capitalist economies have never been stationary, nor have they declined, except for short periods of crisis; they have always grown, but their growth has been irregular and unbalanced. It is hard to see how this can be analyzed in terms of steady growth or deviations from steady growth.

Rejecting "steady growth" as an analytical approach is one thing; but it may be thought quite another to jump to a concept of transformational

---

[9] Steady growth in the neo-Classical mode is not only a misleading idea; it is virtually impossible in practice. To see just how implausible it is, consider that in any society there are likely to be rich and poor, powerful and weak. Suppose for simplicity that they divide into only two classes, neither pure rentiers, so that both work for earned income. The first group in early capitalism, might be owner-operators (who receive "wages of superintendence" as well as profits), in a later era, professional managers, who own stock; the other will be workers. Both classes save and own capital, and thus obtain interest (indistinguishable from dividends), even the poorest. Let the rate of interest be the same on capital, no matter who owns it. But their respective possession of assets can be expected to affect the ability of the two classes to acquire the skills or influence to command earned income. The wealthier class will earn more in proportion to its wealth; that is, salaries will be higher than wages, as we might expect. Suppose, in addition, as is quite reasonable, that the wealthy save a higher proportion of their income. Under these very simple and rather plausible circumstances, the wealth of the richer class will grow faster than the wealth of the poorer, thereby ensuring that the gap between the salaries of the professionals and managers and the wages of the workers also expands. Given that the consumption patterns of rich and poor must differ, producers and markets will specialize, and the markets serving the rich will be expanding faster than those serving the poor. The argument is presented in detail in chapter 10 of *Transformational Growth and Effective Demand* (1992).

growth.[10] For this is not just "unsteady" or nonproportional growth; it is a mixture of such growth and structural development. Not only does this way of thinking picture the economy as growing unevenly, it also suggests that institutions and technologies regularly undergo radical changes.

Yet, on reflection, this should not seem so surprising. That it does is perhaps a measure of the extent to which our thinking has been shaped by equilibrium ideas. But, as we have seen, growth is fundamentally disruptive. Think of the growth of a person, or kittens, or a flower. Nothing complex grows equiproportionally. Shapes change, proportions change, even functions change. The ratio between volume and surface area necessarily changes – an important determinant of shape and structure in plants that depend on photosynthesis. Perhaps the analogy raises too many questions. Think instead of the growth of a company or a corporation. As it adds new products and new divisions, it has to change its management structure. Even the culture and ethos will change. As a university expands, adding new programs, it has to reorganize. Priorities shift, the conditions for tenure change, the day-to-day experience of both students and faculty alters. Two-and-a-half decades ago, major league baseball began adding the expansion teams; as any fan can tell you, the game has not been the same since.

Growth, it seems, then, generally tends to be non-proportional; steady growth should be thought of as an exception, rather than as the norm. Of course, conventional economics does not claim that the economy actually grows along a steady path – it is rather supposed that the actual path is a deviation from the steady path, caused by market imperfections and government intervention, abetted by various exogenous shocks. Market forces tend to push the economy toward the steady path, whereas the various interferences tend to deflect it. The actual path will be the resultant of these two conflicting tendencies. Just as real economies are never in equilibrium, no actual cases of steady growth are ever found, even though equilibrium and steady growth are the outcomes toward which market forces are tending.

On reflection, this explanation has to be dismissed as implausible; a good try, perhaps, but it just can't carry the burden. The economy is

[10] The "new growth theory," of course, might be considered the mainstream response. Unfortunately it addresses none of the major issues. What it does do is determine the growth rate endogenously, allowing preferences to play a role, and sometime providing a market account of innovation. But this is accomplished, in most of the models, by allowing the profit rate to be determined exogenously, by technology (the neo-Classical production function), or in a separate subset of the model (Kurz and Salvadori, 1994; Grossman and Helpman, 1994).

really heading for steady growth, but are there too many bad influences? Our skepticism is based on more than suspicion, however; it follows from our basic approach. Is it really plausible to suppose that market forces tend to push the economy to steady growth? The *market* is going to press the economy to *steady, balanced* growth? Not if the principal function of the market is to generate pressures to innovate and, through competition, to finance and spread innovations. Not if the nature of the market is to provide a field for competitive struggle, the outcome of which will enrich some and impoverish others. The market is the battle-field on which the forces of creative destruction wreck their havoc – and build their empires. What could be steady or balanced in such a drama?

So let's go right to the beginning. How can we explain the change from a traditional society in a stable condition of natural order to a growing society operating in a regularly progressive mode? This change actually occurred in Western Europe during the seventeenth and eighteenth centuries, so it is not an unreasonable question. And it has occurred, or is taking place now, among some of the developing countries of the Third World. But the question here is not concerned with historical details or specific events. It is a matter of theory, and the issue is causality – what are the differences between the ways the two systems work, and what brought about the change?[11]

Broadly speaking, the answer is that it was the development of the market, its generalized spread through all sectors of society, that brought about the change. For technology to develop there have to be pressures to improve the ways things are made and done, there has to be a general willingness to accept such improvements, and there has to be a selection process, choosing which of various possible improvements will actually be taken up.[12] When all production has come to be destined for the market, with the object of realizing a profit, everything produced can be measured in money, and everything is subject to competitive pressure. The

---

[11] It is not clear that traditional economics has an answer, because it does not deal with historical issues. It tends to assume that rationality is the same anywhere, at any time, and since markets reflect rational behavior, it treats markets as essentially the same in all times and places. Philosophers might well disagree – and a Transformational Growth perspective would indeed argue that rationality itself is more highly developed now than, for example, when Walras and Jevons wrote.

[12] Basalla (1994), carefully lays out an evolutionary perspective and sees clearly that alternative paths of technological development were possible – canals vs. railroads, external vs. internal combustion – but seems largely oblivious to the role of the market.

corollary of profit is loss. Improvements will strengthen market position; their cost can be estimated in money, as can the potential benefits. And so can the dangers of not doing anything, given that competitors might act. Under these circumstances there will be general pressures to innovate as a competitive strategy.

This suggests that a universal competitive market system causes the economy to grow.[13] In a sense this is correct, but there are two important qualifications. First, the market economy must be a capitalist economy. Goods must not only be marketed for profit, they must also be produced by wage labor. If goods were marketed for profit, but produced by a cooperative of workers (as in the guilds, or later in conditions of market socialism), there would not be the same intensive pressure for growth and innovation. Capitalists consider the wages of their workers to be a cost, something to be minimized. Workers are to be pushed as hard as may be expedient; the objective is to get as much work out of them as possible during the hours for which they are being paid. By contrast, for a workers cooperative, work is part of a way of life. It is not necessarily something to be minimized, or to be overdone at the expense of health, well-being, or pleasure. On the contrary, in a cooperative enterprise part of the goal is likely to be to structure work to make it satisfying and fulfilling. In general, this means that work will be adjusted to the needs and desires of workers. In the same way, wages are family income; they are hardly an appropriate target to be minimized. Under these conditions, innovations will not tend to cut labor costs or dramatically restructure work.

---

[13] But traditional economics does explain savings and investment, and this account comes closest to an explanation of the emergence of regularly progressive economies. Savings means postponing consumption, but present consumption will always be preferred to future (by assumption). Hence postponement will have to be compensated, and *interest* is the compensation. The more consumption is postponed the higher the compensation will have to be. Investment can afford to pay the compensation, because "more roundabout," i.e., more capital-intensive, methods of production are more productive. However, additional "roundaboutness" adds progressively smaller increments to productivity. Thus higher and higher savings calls for higher and higher interest, while larger and larger investments have a diminishing ability to pay higher interest. At a certain point the two will be equal, and this will be the equilbrium rate of interest, determining an equilibrium level of investment, and therefore a rate of growth. The "Capital Controversy," however, has shown that the concept of "round-aboutness," or capital-intensity, cannot play the role assigned to it in this story (Harcourt, 1972; Steedman, 1986; Laibman and Nell, 1978). There are also difficulties in the notion of "time preference." In any case, the story provides no answer to the question why saving and investment, and growth, emerged when they did.

The second major qualification is that the markets must be expanding. Stagnant or depressed markets offer much less scope for innovation, and obviously little room for economies of scale. Why should anyone invest, and build more capacity, if there is no additional demand expected? One reason might be that a producer believes that he can make a better product, or produce it more cheaply with the new plant and equipment. But this is very risky – the new process may not work, or may have "bugs," the new product may not be as attractive as hoped. If the market is not expanding, room for the new producer can only be found by displacing existing firms – whose customers may be loyal. However, if the market were expanding, this would not be a problem. New firms could still find customers even if their product were no better and their costs no less than their competitors'. Clearly expanding markets are necessary to encourage innovation.

Yet this brings us face-to-face with a problem: how can markets be expanding in a static society? To put it another way, if markets are expanding, how can the society be static? If we try to explain the shift from a traditional to a regularly progressive condition by appealing to expanding markets, are we not begging the question?

The answer is no, and the reasons will give us some insight into the nature of transformational growth. It is usual to think of markets expanding in one of two ways: either more customers are added, for example, through population growth, or existing customers find themselves with larger incomes and more money to spend. Either of these may be compatible with steady growth, so these are the forms of market expansion that have received attention. But there is another, more interesting way in which markets may expand, and this will give us our clue. Markets may take over functions that were formerly carried out through non-market procedures. In traditional society many branches of production, and most services, were handled in the domestic system, abetted by local barter. The household made much of its own furniture, its clothing, put up its own preserves, grew much of its food, and looked after its young and elderly. The local community provided more specialized services – medical, smithwork, needlework, carpentry, construction – and more complex goods, often in barter (Larkin, 1988; Laslett, 1986). Money was used for goods obtained in trade from outside the neighborhood or local community. But these hitherto domestically produced goods could be more cheaply produced, and the corresponding services provided, by organized industry working through the market. To take over these functions would provide an enormous field for enterprise – and, in fact, that is precisely what happened all during the nineteenth century, in both

England and America. The story of the market in the last century is chiefly about its conquest of domestic production.

Nevertheless it is still necessary to explain how this invasion of the domestic sphere by the market began, and what forces kept it going. The answer is important and illuminating: it began because of pressures that brought about an imbalance in the economy, forcing labor to migrate from the countryside to the cities in search of work – and a place to live. Initially – as a matter of historical fact – the change came about because of a shift from grain and cattle to sheep, which required far fewer hands. Enclosures then forced families off the land. People protested and complained of "sheep eating men," but to no avail. Sheep were more profitable; woolens did well in international trade, and mutton was valued locally.[14] Later enclosures in Great Britain took place for other reasons, but the effect was the same – to generate a migration of labor to the cities, and out of agriculture into industry. In the United States hired farm labor was forced off the land largely because of high and rising productivity, not necessarily due to innovation, but to the clearing of better land. Not only laborers were driven out. Family farms failed, too. High productivity produced bumper crops, and drove prices down, thereby forcing the high-cost farms out of business.

The migration of families to the cities in search of work and places to live created a potential market. Families moved to cities because there were economies of scale in living. Of course, if those driven off the land ended up with no incomes, no market would materialize. But once they found work, they would have to purchase many goods and services that they and their relatives and neighbors formerly provided for themselves. Even if their real incomes stayed the same or decreased somewhat, the size of the market would increase. This, then, opened the way for competition and innovation.

But how were they to find work? What work was there to be found? Initially, there may well have been none. But the arrival of large numbers of individuals and families in search of work created both a problem and an opportunity. Their presence would drive wages down, in their desperation they would take any job – and if nothing materialized, they would pose a problem for the authorities. The situation called out for creative solutions, that is, for innovation. Jobs had to be *created*, not found.

---

[14] The earliest enclosures of this kind were in Norfolk in the fifteenth century, running on into the sixteenth. The major movement came in the late seventeenth and eighteenth centuries; but the later movement was significantly augmented by the increasing productivity of labor in agriculture. Fewer farmhands were needed.

Moreover, urban congestion increased contact between the classes, and the contact spurred envy and imitation. The styles and patterns of consumption of the rich could be seen by the middle classes and the poor, increasing their demand for goods of higher quality, and making them willing to work harder or in new ways to earn the means to satisfy their new wants (Gilboy, 1967).

These pressures were strengthened by another important effect of the migration to the cities. The crafts and trades that serviced families and their needs had formerly been spread out spatially. Each village or town had its blacksmith shop, its wheelwright, its seamstresses, and so on. Urban migration concentrates population, driving up land values, so that space will be at a premium. It will be worthwhile to figure out how to concentrate production to serve a large number of families in a small space – quite apart from *economies* of scale. Simply to work out how to produce on a large scale will be advantageous, because it will reduce the rent for space. But that cannot be done so long as production relies on human or animal power. Nor can it depend on wind or water power, for they are as unreliable as the weather, and changeable as the seasons. Anyone who could harness mechanical power for production would be in a position to make a great deal of money. And, of course, they did.

The double nature of Transformational Growth

For growth to start up, then, the economy must become unbalanced.[15] As the pearl is the oyster's response to the irritation caused by the grain of sand, so growth, at least initially, is the response of the economy to the irritation of structural imbalance. But this is not a one-time, exogenously caused imbalance; it is an imbalance which results from an ongoing process, an imbalance which will be reproduced if corrected. The outflow of labor from agriculture, which set off the Industrial Revolution, itself also became a product of that revolution. As innovation developed in the cities, the new methods and machines found use in agriculture, raising productivity there, thus further adding to the displacement of labor. The movement generates a response which, in turn, further intensifies the forces that brought about the initial movement (Nell, 1992, ch. 15).

This, then is one aspect of Transformational Growth: a structural imbalance is regularly reproduced, so that it becomes a trend, a long-term

---

[15] There is an echo here of Albert Hirschman's famous thesis (Hirschman, 1959) that development occurred in response to imbalances, so that international aid and development planning should not be aimed at fostering balanced growth, but instead should look for and support constructive responses to imbalances.

process of change in the relationships between the sectors of the economy.

But there is a second aspect. The effects of the flow into the cities provide both an incentive and an opportunity for innovation *of a specific type*. The incentives derive from social pressures – work has to be found for the displaced labor; while the opportunity calls up visions of profit – for wages are forced down. Both are well-known. But the specificity has been less noted: just as congestion creates a large market in a compact area, so it forces up the price of urban land, and these two factors together afford a premium to production on a large scale in a small space. Even if a new large-scale method is more expensive per unit in terms of direct inputs into production, it may pay, nevertheless, because of the saving on space. Thus the costs created by congestion provide a margin for experimentation.

When the new methods are introduced they will change the pattern of costs, and the structure of industry. Initially this will happen in a few industries in a few places, but these will become centers of innovation and investment (Pacey, 1985). Over time the innovations will spread, gradually being adapted to other uses, often in sectors wholly unrelated to those in which they first occurred. As large-scale methods spread, the institutions of production change. What were formerly artisan shops, often located in the home, and staffed by members of the craftsman's immediate family, become family-based firms, with production centered in a separate building – a factory – operated by hired labor. When this has taken place in a sufficient number of sectors, it will be found that the character of the economy has changed.

Stages in transformational development

Identifying stages of a system's development is no easy task. To begin with, we must be clear about the "system." A social order has many aspects. Political and social factors are relevant, but the central issues here are economic – concerned with the development of the capitalist system. Capitalism means that production is organized as "capital," that is, that the means of production are operated for a profit by capitalist owners, who compete with one another and hire labor to perform the necessary work. Capitalism is characterized by stable long-term relationships between prices, wages, and the rate of profits on capital. But institutions – families and firms – the working of competition and patterns of market adjustments, all may change, within a framework of stable value relationships. In the early stages, for example, capitalists form a class; that is, ownership of funds and productive facilities is passed down from generation to generation within families. Workers likewise form a class; skills are passed along within families. In later stages, we shall see, neither capital nor skills are passed along through families, and "classes" are more complex and harder to identify.

Usually in referring to a "system," we mean a stable social formation; that is, a social-economic system works in a way that makes it possible for it to continue more or less indefinitely. As its members grow old and die, new ones are born and appropriately trained to replace them. As it uses up resources, new ones are found or produced to replace them. As its tools and buildings wear out, new ones are produced. The system consists of institutions that make sure these replacement functions are carried out.

Capitalism, however, is unusual among social-economic formations in that it systematically develops. That is, while it does train and socialize replacements who will take over the functions of aging members, and similarly it does reproduce resources and tools as they are used up and wear out, it also generates pressures that lead to changes and improvements, sometimes leading to vast upheavals. The system contains features that generate innovations, and also has a process of selection. Moreover, unless these changes and upheavals take place, the system is likely to falter or stagnate, even collapse.

A "stage," or an era, can be defined as a particular configuration of production processes, education and training, resources use, and institutions of control that makes such reproduction possible. Economists have not always paid attention to the details of production; they will be all-important here.

Stages here will be defined by the methods of production, meaning the way tools, equipment, and energy are organized to turn materials into useful artifacts, including how skilled workers and personnel perform services. Starting with the artisan, it will be shown that skills are progressively rationalized and sequenced for efficiency, so that they can be embodied in equipment powered by energy from various sources. As changes accumulate and spread, costs are affected. As the composition of costs changes, competitive strategies change, and the markets come to work differently.

Stages are not easy to date. History flows continuously – with breaks, to be sure, but most, like the French Revolution, tend to be social and political rather than economic. The Industrial Revolution is a major economic dividing line – but when, exactly, did it take place? And where, for that matter? Stages will, in a sense, ultimately be defined by technology, but they will be distinguished in practice here by the changes in the way markets adjust. We will find a distinctive pattern of market behavior corresponding to each major division in the character of technology.

This can be illustrated with general sketches drawn chiefly from the history of the advanced countries of Western Europe and North America, which, however, with some modifications, also apply to Japan and to the patterns of development of Third World countries in the period since World War II.

The period of early industry: artisan shops, traditional agriculture (AS/TA)

Industrialization, as we know it, began in Western Europe in the late seventeenth and early eighteenth centuries. At that time, production was organized largely around the household. Artisans operated workshops and places of business that were attached to their homes, employing family members and apprentices or servants as labor. Apprentices and servants working in the enterprise, who were not related, would normally live as part of the family. Journeymen would take meals with the family though they might sleep elsewhere. Even on the farm workspace and living space tended to be joined physically; the barn would be an extension of the house, or the buildings would be attached to one another in some way.

Craft methods were employed in production, requiring traditional skills. These were passed on from father to son, as household and domestic skills were passed on from mother to daughter. Such household-run shops made many of their own tools. Energy came from human effort, animal power, wind and water. Operations were small-scale, single-batch affairs, in which the division of labor was limited, with the organization of work catering to skilled craftsmen. These last actually carried out the processes; the others assisted them, setting up work, handling and carrying tools and materials, holding objects in place while an operation was performed, providing strength when required and general support services at other times (Hunter, 1979; Hounshell, 1984).

A work team had to function as a unit. Work was coordinated as it was carried out. Very little was programmed in advance; designs were either traditional, or developed and modified for particular customers. Processes were worked out in practice and depended on the particular characteristics of the work team. Morale counted heavily, as did leadership.

There were a few large-scale shops during this period, the Armory of Venice, for example, or the Meissen Works, or the silk manufacturing in Lyons. But these were exceptions; and more important, they did not engage in large-scale production so much as concentrate large numbers of small-scale processes in one place. This permitted a greater division of labor and also made it possible to adjust tasks to levels of skills better. (Craftsmen did not have to do their own carrying or setups, for instance.)

Plus shipping and storage could be centralized, while repairs, maintenance, and administrative costs could be spread over larger numbers. But the actual processes of production were carried out in very much the same way as in the small shops of local craftsmen.

Agriculture, for the most part, was conducted in this period as it had been for centuries. There were improvements and innovations, in seeds, for example, and in drainage systems. The three-field system had already replaced the two in most places. Ploughs were improved, and the substitution of the horse for oxen allowed deeper ploughing and larger fields. Nevertheless, farming continued to be based on human and animal power, operating the same kinds of tools – even if improved – in the same ways as past generations. In most of Europe and North America farming was a family affair; even the large manors were organized and run by a family, with many dependent families assigned to various specialized tasks – cooking, gardening, washing, baking, sewing, etc. The plantations of the Caribbean and the American South differed chiefly in that they were more market-oriented, depending largely on a single cash crop, sugar or cotton. But in other respects they were similar, in being structured around family relationships, and a unity of work and living space.

Marketing was carried out largely by the producers. In many cases the place of production was also the place of sale, as with bakeries, butchers, blacksmith shops, carriage-makers, drapers, saddlers, and other crafts. The word "shop" still carries this ambiguity, with its dual meaning of workshop or retail outlet. The building trades of course plied their crafts on the spot. However, a number of activities, especially in textiles, were organized through the "putting-out" system. Merchants would bring materials to homes, where women, for the most part, would work them up, whereupon the finished goods would be collected (Mantoux, 1961).

Farmers sold in local markets and often paid their rents in kind. A small amount of produce might be sold on the farm itself or in roadside stands; but the bulk would be taken to local markets by the farmers themselves. Grain and produce would typically be stored in barns and farm buildings. Animals would often be slaughtered on the farm, then carted off to the butcher, although they might also be driven to market by farmboys. Production and sale were localized, and carried on together, as part of the same pattern of activity.

At the risk of oversimplifying, then, AS / TA can be characterized by a set of unities. Human activities were part of the world of nature. The weather, daylight, the seasons – all affected the ability to carry out production and to perform the functions of the family. Buildings provided shelter, and firewood heat. Limited forms of lighting were available. But human society, while not at the mercy of nature, was far from independent of it. Communications depended on transport; a message could be sent no faster than a person could travel – and transporation depended on the seasons and the weather. Then there were the unities of workplace and living space, and place of production and place of sale. Here "place" should be understood not simply as geographical location, but also, and more importantly, as a focal point in social space. When work and living space coincide, this means that work activities and family life overlap and intermingle. In the same way, the activities of production and marketing are mixed together, both in shops and on the farm, and cannot easily be separated.

Craft-based factories and mechanized agriculture (CBF/MA)

It was well known that production could be speeded up and performed on a larger scale,[16] using better and stronger materials, if human and animal energy could be supplemented or replaced by mechanical energy. Wind and water power had long been used for milling and similar operations. But the scale had remained small; wooden wheels and driving mechanisms turned millstones. The strength of metal was needed to make it possible to develop the equipment that would work on a larger scale.

Mining and metallurgy improved markedly during the century prior to the First Industrial Revolution – the transition from AS/TA to CBF/MA. Waterwheels could be strengthened with metal bolts and clamps: later the entire wheel could be fabricated of first wrought iron, then steel. As size increased, design improved also. The turbine wheel, later to be used in steam, internal combustion, and jet engines, was invented by French hydraulic engineers in the 1830s and improved in the 1840s. Almost

---

[16] In Adam Smith's pin factory each man specializes in a particular operation, drawing the wire, straightening it, cutting, pointing, grinding, putting the head on, whitening or polishing, etc., amounting to about eighteen distinct operations, which can be divided up in several different ways. Clearly they have to be coordinated, and each worker has to complete his operation in a manner that eases the way for the next. The description makes it clear that the division of labor itself requires practice; the advantages cannot be reaped without effort and skill. Moreover, much of the machinery and improvements "were originally the inventions of common workmen, who · · · naturally turned their thoughts towards finding out easier and readier methods" (1961, p. 9). Marshall (1961) argues that while machinery displaces purely manual skill, it increases the demand for judgment and general intelligence. Again his examples illustrate the need for coordination of the workforce and for high morale (cf. pp. 255–64).

immediately it was imported to the United States and adapted and further improved (Hunter, 1979).

But greater power was of limited use unless it could be adapted to the tasks at hand. This required transmission, belting and gearing. During the development of more powerful waterworks, transmission systems improved, with gearing, the use of cams, connecting rods, crankshafts, and tappets all becoming more sophisticated. The American system, in particular, allowed for the most flexible use of power, based on lightweight, rapidly spinning line shafts, sometimes running as much as a thousand feet or more. These also had to be balanced with heavy flywheels, to counter the "whipping" effects of major changes in machine loads (Hunter, 1979; Hounshell, 1984).

Steam power developed both for moving and stationary uses. The railroad made it possible to travel and move goods independently of the weather, and far more rapidly than ever before. The telegraph for the first time broke the link between travel and communications. Hitherto, to get a message somewhere it had to be borne by a messenger.[17] This meant among other things, that no message could tell anyone when the train was coming, since the train was the fastest form of travel on earth! The telegraph solved that problem and made it possible to send messages without messengers, to buy and sell, transfer money, place orders, and generally transmit news without anyone having to leave home. This facility was vastly extended by the telephone and later the radio (Hunter, 1985).

Greater power called for longer worktime. Lighting prolonged the day; electric lights, when they came, made multiple-shift working possible, since this required not only light in the workplace but also street lighting to enable safe travel to and from work at night (Hunter, 1991).

Better transportation and communications greatly expanded the market; steam power and then electricity and internal combustion made large-scale production possible. The unities that characterized AS/TA are broken apart in CBF/MA. Production is no longer dependent on nature; communications is independent of transportation; the workplace and the home have been separated, and marketing is conducted apart from production. Company towns – like Lowell, Massachusetts – remain and even flourish for a time, then gradually die out. Cities grow up around large-scale operations. But the tools and the organization of production are still essentially the same, even though the tools are now driven by new and larger sources of power. As a consequence the system still behaved

---

[17] Carrier pigeons were not reliable; smoke signals and mirrors reflecting light depended on the weather, etc.

like a Craft economy; indeed, in a sense it still was – the Craft tools were driven by power on a large scale, but organized in craft fashion. Hence employment and output were still relatively inflexible, and markets adjusted through prices. But now larger investments were at stake, since fixed capital had become more important.

In AS/TA price fluctuations were not so dangerous. But now the inflexibility of Craft-based systems created problems. Since the new factories were essentially Craft operations writ large,[18] the price fluctuations characteristic of craft-based systems endangered the large capital investments. These fluctuations exhibited a clear pattern; prices in general were more flexible in both directions than money wages, implying that profits tended to move procyclically and real wages countercyclically, creating serious pressures on businesses. Moreover, keeping the labor force on during times of weak demand was manifestly wasteful; yet it was hard to manage any production without all the crafts and skills being present.

The solution came in the form of "continuous throughput" Mass Production (Chandler, 1977).

Mass Production and corporate agriculture (MP/CA)

The key to Mass Production is not just scale, although with its advent scale was enlarged greatly. But economies of scale had already been achieved in CBF. Mass Production adapts the tools to the power sources, and vice versa. It develops flexible, adjustable power, making it possible to adapt production quickly and easily to what is needed. It reorganizes work to take advantage of the possibilities inherent in the technology, restructuring and rescheduling until the most efficient arrangements have been found. The result is that Mass Production processes are *flexible* in the sense that output and employment, and so running costs, can be varied as needed. It is not necessary to operate with all hands working, in order to produce a small amount. Plants can be run at various fractions of capacity, or on short time. They can be closed down for short periods and started up again at very little extra cost. Unit running costs

---

[18] With characteristic acumen Marx observed, "On a closer examination of the working machine proper, we find in it, as a general rule, though often, no doubt, under very altered forms, the apparatus and tools used by the handicraftsman or manufacturing workman; with this difference, that instead of being human implements, they are the implements of a mechanism, or mechanical implements. Either the entire machine is only a more or less altered mechanical edition of the old handicraft tool, as, for instance, the powerloom, or the working parts fitted in the frame of the machine are old acquaintances, as spindles are in a mule, needles in a stocking loom" (Marx, 1967, Vol. 1, p. 373).

thus tend to be constant over large ranges of output. Employment can be varied, while productivity remains constant (Hounshell, 1984).

Mass Production need not rely so heavily on worker morale because the machinery itself carries out large parts of the skilled work. The skills, in effect, are built into the machinery. Workers will have to be semiskilled, but they chiefly perform repetitive tasks, as the production line itself moves, operating equipment that is partially preprogrammed. The equipment itself does the cutting, shaping, processing, and so on; the worker guides and positions the equipment, and begins and ends the operation. But the energy that drives the tool, and action of the tool itself, are normally independent of the worker.

An important consequence follows: projects started by one group of workers can be carried on and finished by another. Alternatively, membership in the working group can be varied easily – workers do not have to learn each other's idiosyncracies. Under craft conditions, the detail of the work will vary with the skills and pace of individual workers, so they must adapt to each other. Jobs started by one worker or one team must be finished by them. It will be difficult "to change horses in midstream" – a craft metaphor.[19] Once production is independent of particular groups of workers, shift work is possible. At the end of a shift workers put down their tools; the next shift simply picks them up and starts where the first left off. In a boom extra shifts can be added; in a slump shifts can be laid off – or put on part time. Given that start-up and shut-down costs are minimal, production can easily be adjusted to sales (Nell, 1993; Chandler, 1977).

An important reason that Mass Production can adapt to variations in demand is that it draws on new and more flexible sources of power. Power transmission has become cheaper, and capable of covering great distances. Electrical power can be turned on with the flick of a switch; electricity can be conveyed in large or small quantities to any workspace, anywhere in a plant. It can be transmitted from large central power stations to all points in a city or geographical region. Internal combustion engines provide great power in a small space, with rapid acceleration and deceleration. Moreover they are mobile. Steam could only be used with rails, or on water; the internal combustion engine, because of its great

---

[19] Hitching a new team to sinking wagons in midcrossing is the stuff of heroic Westerns; hooking up one tractor to pull out another in trouble is doing what they're designed for. The point is that Mass Production equipment has been designed to overcome the characteristic problems of the Craft economy – especially problems that made production dependent on the skills, attention and effort of the workforce.

ability to deliver accelerating power, was quickly adapted to road travel, revolutionizing transportation (Hunter, 1985, 1991).

As a result, under Mass Production transportation costs fall, as movement becomes more rapid, both for people and goods. Urbanization thus gives way to suburb creation; workers do not have to live within walking or cycling distance of the factory. They can commute by car. Similarly goods can be shipped rapidly and cheaply – as was already possible on the railroad in CBF. But now trucking greatly increases the flexilibity and range of shipping. Storage becomes cheap and reliable, creating greater scope for inventory management.

Mass production also institutionalizes research and development. Processes and products are improved continuously. Jobs are progressively redefined, as tasks are broken up and skills built into the equipment. Production becomes faster and cheaper; products are targeted for specific markets. Agricultural productivity rises steadily; crops become increasingly independent of the weather, and resistant to pests. Firms must continually improve their plant and equipment, and continually buy new vintages of equipment, in order to remain competitive.

The character of labor changes, too. Blue-collar work is increasingly mechanized and deskilled. White-collar work, managerial and sales work, rises relative to blue-collar production work, creating a characteristic source of tension. Blue-collar productivity rises steadily, and blue-collar wages tend to rise in proportion, as a consequence of productivity bargaining. But white-collar jobs are higher in status and must maintain a traditional differential. Hence white-collar pay will rise, even though productivity has not, indeed, may not even be accurately measurable. The result is a steady inflationary pressure.

Computerized production and bio-tech agriculture (CP/BA)

Mass Production tends to divide markets into large segments, so as to produce the same product for everyone, with some simple variations for specialized buyers. But a large part of the public wants customized products, and that is where the computer revolution comes in. By computerizing production equipment, possibilities for changes in product design can be built into the system, preserving the cost advantages of long production runs, while still allowing for product variation.

Products, however, appear to be changing in character. Instead of an *item*, high-tech consumers (and, increasingly, others) now purchase a *system*, or components of a system. The good comes with a warranty, a service contract, slots for adding on further components, etc., and the possibility of a discount for trading it in on a new model.

Computerized control over production allows processes to be split up, without losing the economies of scale characteristic of Mass Production. This makes it possible for business to locate activities all over the world, choosing the most advantageous spots in terms of costs or marketing. Economies of scale and scope no longer require a single centralized location.

Computerized production carries the deskilling of labor even further than in the past. Blue-collar work is reduced both in amount and in importance. High-skill technicians and engineers are needed to monitor, maintain, and repair the equipment. Computerized high-tech communications monitoring makes it possible for low-skill work to be done in low-wage areas, still meeting quality standards. Cheap rapid transport makes it worthwhile to ship the intermediate products back and forth.

The emerging system has a new cost structure. The reduction of blue-collar work and the reduction in direct production costs generally means that variable costs are relatively low, whereas capital costs are high. Also high are development costs, sunk costs for research and trial and error testing. Risks are very great, particularly as the new technologies are being tested, leading to calls for government participation.

The reduction in the importance of labor appears to have significant political implications. The voice of organized labor carries much less weight than in the recent past.

The behavior of markets: stylized facts in each era

Both AS/TA and CBF/MA exhibit characteristic patterns of price and quantity movements. We will explore this in detail in the next chapter. Prices of all kinds, including money wages, fluctuate in both directions. Primary prices show the greatest flexibility, measured as deviations from the trend (if any). Next are prices of manufactures, while money wages are the slowest to move and have the lowest range of variance. Output is more flexible in both directions than employment. But both kinds of prices are more flexible than output, and so also than employment, while money wages show about the same degree of variability as employment. These patterns imply that real wages will tend to move inversely to output and employment, that is, real wages will move countercyclically. But productivity also moved procyclically in the short run, contrary, for example, to the implications of marginal productivity theory. Finally, given somewhat inflexible employment and countercyclical real wages, aggregate consumption will tend to vary inversely with investment.

During the century preceding World War I, the general trend of prices was downwards, that of money wages flat, until the very last years. This suggests that the gains from productivity were transmitted to the economy through lower prices. Prices also moved closely with interest rates. The monetary system was based on gold or gold and silver. Paper money tended to be convertible. Financial crises were frequent and tended to destabilize the economy, though recoveries were usually rapid.

Firms tended to be small and family-based, typically aiming to reach an optimal size and remain at that level. Ownership and control were united in the family. Profits were distributed rather than retained.

Following World War II, however, the behavior of prices, money wages, output, and employment was very different. Prices only rose; except for brief and special instances they never turned down. Money wages also rose, and rose faster than prices. The benefits of productivity increase were thus transmitted to the economy by rising wages, not by falling prices. Output and employment, however, fluctuated markedly in both directions. Variations in employment were accomplished through an organized system of layoffs, often following an agreement reached through collective bargaining. Real wages – product wages – moved procyclically. Productivity increases were regular; innovation resulted from research and development, hence new equipment tended to be better than old. Firms therefore had to invest regularly in order to stay competitive. Growth thus took place in the form of expansion of existing firms, which came to be organized as corporations, with a separation of ownership and control. The share of government at all levels in GNP rose, and continued to rise.

Two patterns of growth

We can distinguish two idealized systems of market adjustment, based on historical conditions, but defined as pure types – a stylized "Craft" economy, in which production is carried out on a small scale by groups of specialized, skilled workers, assisted by animal, water, or mechanical (steam) power, and a stylized "industrial" economy in which mass production is carried out by workers operating along an assembly line or similar system that sets the pace of work. ("Craft" conditions thus characterize much primary sector production, and "industrial" conditions correspond to manufacturing.) Each can be considered a complete economy, capitalist, described by the Classical equations and capable of growth. But the way investment brings about growth will be different in each, and so will the behavior of costs and prices when demand fluctuates (Leijonhufvud, 1985).

A Craft economy can either operate in stationary equilibrium or it can grow. If it grows it will characteristically expand through the lending of household savings to new firms, which set up shops that replicate existing ones, but serve new customers. But – at least under certain conditions – it is under less pressure to grow, and its prices do not depend on growth. By contrast, an industrial economy cannot ever settle into stationary conditions; in general it must expand or fall into recession, and its prices do reflect its growth. The reason for this lies in the different relationship of technology to competition in the two cases.

In the Craft economy success in competition comes through the development of the skills and morale of workers. The successful firm has the better product, and has more reliable delivery times and quicker production times (with unit costs therefore lower), etc. – all of which depend on worker skills and their ability to function together as a team. Such characteristics are personal and intangible; improving them does not depend on rebuilding factories or reequipping shops. But they are also unreliable; sickness or disaffection of key workers could undermine the whole year's effort.

In the industrial economy competitive success likewise depends on cost-cutting and improved product design, but the difference is that these have been built into the production process, literally into the equipment, and no longer depend on intangible personal characteristics. They are therefore reliable, but limited by the characteristics of the technology. Improvements require retooling or rebuilding when they are "embodied"; they require investment. Plants have to be shut down and renovated or scrapped and rebuilt. Even "disembodied" technical change, however, requires redesigning the work flow and the organizational chart.

An improvement provides a competitive advantage and must therefore be matched. So there will be need and incentive for the economy to contain a sector that specializes in supplying the means of production, and large enough to meet the demand for rebuilding entire industries. Investment will be more or less continuous, and productivity will regularly rise, though not necessarily in step with demand.

Competitive improvements (cost-cutting, product quality) thus shift from depending on personal characteristics of artisans to being embodied in equipment and job design; as a result, the nature of labor changes. In a Craft economy a high proportion of labor is fixed. The entire work team must function continuously in order for there to be any production at all. The firm has only two choices: to produce or to shut down. The only way to vary production is to vary the productivity of labor. As a consequence, the ability of firms to respond to fluctuations in demand by adjusting output and costs will be very limited. (This is the clue to understanding what we shall call the "Marshallian" short-run production function.) By contrast, in an industrial economy productivity is built into the equipment, and labor is highly simplified. Tasks are broken down into their components, and jobs are deskilled as far as possible. The flow of product runs only one way; goods are never returned to earlier stages in the production process – as they typically are in Craft processing (for example, to adjust tolerances). The work crew is not important, and productivity no longer depends on workers. Output can be quickly adapted to the level of sales; labor can be laid off or put on short time; it becomes a variable cost, and its productivity is fixed.

The shift from Fixed Employment to Flexible Employment technology also changes the nature of the saving-investment process. When the artisan economy expands, household savings will be advanced to set up new firms. The system of small establishments will replicate itself; growth will not normally be undertaken by adding to the capacity of existing firms, for there are few economies of scale in traditional crafts, while adding to the investment under one management increases risk. Industrial systems, on the other hand, do provide economies of scale, both in the design of equipment and in the organization of work. Even more important, however, technological competition between suppliers of capital goods means that new equipment is likely to be better or cheaper than old.[20] New investors will have an edge; existing firms cannot afford to remain satisfied with their present scale of operations, leaving growth to new entrants, for new firms will be able to undercut them in their own markets.

But existing firms do not necessarily have to scrap and rebuild every time there is a significant innovation. This would be wasteful, both socially and privately. Instead they can adopt the innovation in building new capacity to meet growing demand, carefully building just enough – at an appropriate price – to prevent newcomers from entering. The industry will then consist of a number of firms each having both new and old plants, rather than of older firms with outdated plants and newer ones with superior equipment.

To achieve this, however, growth must proceed differently than in the artisan economy. In the artisan economy firms move to an optimal size at which they tend to remain; growth takes place through the lending of

---

[20] New equipment may be different – a new product – or imply a new or modified technique. Firms did not previously purchase or use such equipment because it had not yet been developed. It is therefore not possible, in the case of an industrial economy, to start with the assumption that the list of products and the set of techniques is given at the outset, with optimal selections being made each period. On the contrary, firms will compete to produce new and improved products and cheaper techniques. To assume a fixed list, however large, implies ruling out the form of industrial competition that determines the nature of both labor and the investment process. Or to put it another way, the assumption of a fixed list is an essential underpinning of the idea that prices and quantities are determined simultaneously by market-clearing.

savings to create new firms.[21] To avoid competing for household savings, firms in a mass production economy will retain their earnings and invest them directly. So long as their investments are judged to be wise, the value of their equity will rise in proportion – which means that shareholders desiring funds can obtain them by selling off an appropriate part of their holdings at the higher price.

Growth in an industrial system thus differs fundamentally from growth in a Craft-based economy. Competition requires regular investment, financed by retained earnings, with a consequent rising price of equity. Labor becomes a variable cost, and output and employment vary together in line with sales, while productivity, fixed by technology, stays constant. Capacity capable of meeting the maximum likely demand can be installed, thereby ensuring that there will be no room for newcomers, without any risk of having to meet the labor cost of that capacity when it is not in full use.

But perhaps the most relevant differences, for the present discussion, concern costs and prices. In both, normal (or long-run) prices will reflect normal costs.[22] But the character of these costs differs. In a Craft-based economy, or in the primary sector of a modern economy, labor costs are largely fixed; variable costs are therefore confined to materials. Start-up and shut-down costs are large; variations in demand cannot be matched by variations in running costs. Output can only be adjusted by varying productivity. So when demand falls, for example, profits will be hit the hardest. And, since output cannot be readily contracted, excess inventories will have to be "dumped" for whatever they will bring, forcing prices down. Market prices will therefore reflect the requirements of market clearing. In an industrial economy, or in the manufacturing sector, by contrast, labor and other costs will be variable, so that operating costs and output can be adjusted to variations in demand, keeping productivity

---

[21] In the traditional theory, growth is understood to mean movement toward the optimal size. For example, in his classic *Structure of Competitive Industry* (1931), E. A. G. Robinson devotes chapters to discussing optimal technical size and managerial size, the optimum financial unit and the optimal marketing unit, thus necessitating a separate chapter on the "Reconciliation of Differing Optima."

[22] "Normal cost" equations show that the price of one unit equals its wage cost plus the input costs plus the profit. Breaking this down shows the coefficients of the various inputs into each process, each multiplied by the amounts employed and by its price, aggregated to get total input cost. This figure will be multiplied by the gross rate of profit – the net rate plus the rate of depreciation – and then added to the labor inputs times the wage rate to get total costs. This results in the familiar "Sraffian" or Classical matrix expression for prices and the rate of profit, given the wage, on the assumption that competition renders wages, prices, and profit rates uniform. (See Chapter 7.)

constant, and shifting part of the burden of adjustment from profits to wages.

In the Craft-based world, growth simply replicates existing stationary relationships and market prices reflect the current balance of *spending* and output. Since spending is a reflection, in turn, of effective demand, we will postpone consideration of Fixed Employment prices until we are ready to analyze aggregate demand.

By contrast, in the Mass Production system growth is a major agent of innovation and change and is central to the normal working of markets; for example, it must occur for potential profits to be realized. Of even more relevance here, it is part of the competitive process. If markets currently clear, but there is an imbalance between rates of growth of supply and demand, they will fall into disequilibrium in the future. If they do not clear now, but their rates of growth are in balance, they will eventually even out. Prices are obviously both affected and influential – at low prices new customers can adopt a new good, so demand can expand; but low prices mean low profit margins, so little finance for construction of new capacity. This suggests that a price might be found that would just balance the rates of growth of supply and demand.[23] We will need, however, to take a closer look at demand in a modern industrial economy.

The long-run growth of demand is governed by the development of markets. Markets, in turn, both reflect and stimulate the progressive shift in the composition of output of modern economies from a largely agricultural system, with growing manufacturing and small services, to a predominantly industrial system, with a small agriculture, stable or even slightly declining manufacturing, and a large service sector, in which professional services are growing relative to personal and, more recently, to an increasingly information-oriented economy with a tiny but increasingly "corporate-industrial" agriculture, a stagnant manufacturing sector, and a growing new high-tech sector, with expanding services, which, in turn, are increasingly business-oriented.

Implications for economic theory

Broadly speaking in Craft conditions – AS/TA and CBF/MA – prices are flexible, reacting to changes in demand pressures. In both, the price mechanism tends to stabilize the system of production, distribution, and expenditure, that is, aggregate demand and supply. In AS/TA,

---

[23] The relation between pricing and growth has received official recognition. During the late 1960s and the 1970s the British National Board for Prices and Incomes used "the effect which a particular level of profit has on the firm's ability to finance future investment," as one of the criteria for approving a higher price (Pickering, 1971, p. 232). (See Chapter 10.)

however, price adjustments follow the "gravitation" model, meaning that changes in output require the shifting of capital in response to changes in demand. This is the Classical model, in which market prices driven by supply and demand gravitate around natural prices, which are determined by the rate of profit. By contrast, CBF/MA exhibits something more like Marshallian adjustment, in which output and employment can vary even though plant and equipment remain fixed. Price flexibility in this case is associated with changes in outputs and employment. Employment moves inversely to real wages; marginal productivity conditions seem broadly plausible. Investment and consumption spending appear to move inversely to one another. But the monetary and financial system is not stable, which tends to undermine the stability provided by the working of the price system.

By contrast, under MP/CA, production, distribution, and expenditure are unstable, in one or more senses of the term; left alone, that is, the system will tend to move cumulatively in a given direction. Consumption spending tends to vary directly with investment spending, exacerbating rather than dampening any fluctuations. Movements may or may not converge to a definite position, but even if they do, the position certainly need not be one that anyone desires. However, this tendency to move cumulatively is partially kept in check, and even reversed, by the monetary and financial system, which therefore provides a weak stabilizing force. (It will be shown in Chapter 13 that interaction between the two can generate cycles.) Unfortunately, the financial markets also produce separate destabilizing tendencies of their own.

"Price theory" or microeconomics, as it appears in the textbooks, is thus the theory of markets in Craft conditions. Of course it is presented as a fully *general* theory, resting on universal principles deriving from the choices of rational agents. As will be shown in Part II, to derive descriptive universality from rationality in this way is not valid. But if rational agents are excised from the theory and replaced, in a more or less Marshallian fashion, with agents representing the general forms of actual institutions, then the price-flexibility theory does describe the working of markets in a Craft-based economy. This corresponds to the models of competitive pricing.

As technology became more flexible, permitting variations of employment and output as demand changed, prices and employment moved positively with demand, so that real wages moved inversely to employment, implying rising marginal costs. But with further technological development, output and employment became able to adjust fully to demand changes, so that prices no longer varied significantly. This led to the development, first, of the theory of imperfect competition, then of theories of full cost pricing.

Modern macroeconomics, on the other hand, is the theory of market adjustments under Mass Production. In the earlier era, the Quantity Theory had provided an account of the value of money, that is, the general level of prices, and this served as a kind of macroeconomics. But the approach depended on fixed employment and output. With the advent of Mass Production, variations of demand lead to changes in output and employment that create further change in the same direction. Prices are largely independent of current levels of activity, being planned together with investment in accordance with expectations about the growth of markets, in marked contrast to the flexible prices of the Craft economy. Price determination must be based on the institution of the corporation, including an account of its financial structure.

The behavioral assumptions of macroeconomics, for example in Keynes and Kalecki, were originally based on institutional requirements and realities. Only later did economists demand "rational foundations," as they sought to develop a synthesis of macro and micro. But no synthesis is possible – or necessary.

The attempt of mainstream economics to develop a universal theory has not been achieved. Indeed, in its most general form, the theory has largely self-destructed. But the various branches of neo-Classical theory all are likely to have something important to tell us. The question, then, is, What are they telling us, and when did it apply, where, in what circumstances?

Conclusions

Much of the confusion in economics is generated by trying to apply models and approaches suited to one era, and its institutions, to another, with different institutions. Science is supposed to provide universal laws, and under this misapprehension, economics has sought to develop universal principles of market behavior. Transformational growth takes the opposite view: markets in each era adjust in ways that reflect the technology and institutions of that time. What we now call "microeconomics" began as the theory of the working of the flexible price system of the Craft-based economy (although it then got caught up in rational choice theory, which took it in a different direction). "Macroeconomics," in turn, explains the adjustment of the Mass Production system. Markets can, however, be given a broad and general characterization: they generate innovations competitively and, through price and output adjustments, they generate the profits to finance investment in those innovations.

CHAPTER 2

The stylized facts of the old business cycle and the new

Macroeconomics makes an occasional bow to history and institutions, noting their importance in understanding policy, particularly in connection with exogenous "shocks." But the models advanced in most mainstream work, as well as those in contemporary "alternative" schools of thought, tend to be perfectly general. They are not considered specific to any one historical period. Economic behavior and the working of markets are treated as universal, essentially the same, aside from "imperfections," in all times and places. Economics is grounded in rational choice, expressed most fully in general equilibrium price theory. Macroeconomics can then be derived from this by specifying any of a large number of imperfections or institutional barriers to the smooth adjustment of markets.

General equilibrium theory is abstract and nonempirical. But price theory does not have to take this form.[1] Ordinary textbook micro-economics – supply and demand – offers a strikingly detailed picture of the way markets work, from which a number of plausible empirical propositions can be derived. For example, this picture suggests that when demand fluctuates, prices will fluctuate in the same direction, and the fluctuations will be greater the more inelastic the supply is. Movements in

---

[1] Indeed, it can be argued that the theory of flexible price adjustment *should not* be cast in such a mold. Neo-Classical general equilibrium theory, based on rational choice, provides few, if any, empirically testable insights, while at the same time, generating serious theoretical difficulties. It cannot rule out multiple equilibria, some or all of which may be unstable (Ingrao, 1992). It cannot find a plausible rationale for the use of money (Hahn, 1985). It cannot make room for capital, earning a rate of return which competitively tends to uniformity (Garegnani, 1990; Schefold, 1997). A sensible price theory should give us plausible criteria for uniqueness and stability, a reasonable account of the usefulness of money, and a coherent understanding of the rate of return on capital. Early neo-Classical theory, as developed by Wicksell (1967), Walras (1954), Clark (1895), Marshall (1960), and Pigou (1907), for example, aimed to supply all of these. Their constructions were defective, as later studies have shown, but the answer is surely not to abandon their goals, in order to preserve their approach, but to adjust the approach in order to achieve the goals. Applied economics needs a plausible account of flex-price adjustment.

prices will therefore be positively correlated with movements in output. Real wages will be inversely related to employment and output. Increases in productivity will lead to lower prices. The picture has institutional implications as well: firms will grow to an optimal size and operate at that level indefinitely.

These empirical implications of ordinary price theory are seldom stressed, perhaps because they do not appear to be true of today's economy. Instead, price theory tends to be developed axiomatically and is presented as an offshoot of a more general theory of rational choice. But this is to do microeconomics a disservice. It was originally formulated as a theory of the working of markets, and it deserves to be taken seriously as just that.

Old-fashioned macroeconomics, as presented in the textbooks of the 1950s, like the simplest models today, took prices as fixed, and examined quantity adjustments. These reflected the multiplier and the accelerator, or "capital stock adjustment" principle, and provided an account of market adjustment, which could be and was examined empirically in extensive studies.[2] These models also yielded policy implications.

Macroeconomic data suggest a great difference between the working of the economy in the era in which price theory was founded and its behavior later. In the late nineteenth century when price theory developed, production was organized largely through family firms and family farms, and steam power was used to operate processes that still reflected traditional crafts. In the Keynesian era, production came to be organized by giant modern corporations running modern technologies on electric power and internal combustion (Tylecote, 1991; Perez, 1983, 1985; Solomou, 1986). The technologies are different, and so are the institutions. As a result, it will be argued, so is the way the market works.

In the earlier period the market appeared to function in some respects, as would be expected from neo-Classical theory, at least in Marshallian form. In the later period aspects of its working appear to be Keynesian, and the neo-Classical elements have largely disappeared. In the earlier period there is some evidence to suggest that the market and the price mechanism responded in a stabilizing manner. Financial markets and the

---

[2] Macroeconomics is often presented as Aggregate Supply and Demand, a sibling to microeconomic theory, expressed in the same format, with downward sloping demand, and rising supply curves. The general price *level* functions analogously to the microeconomic price variable. The analogy is flawed. Market prices are paid and received; no one pays or receives the price *level*. Market quantities are measurable in their own units; the macro analogues are revenues, which have the units, price times quantity. Worst of all, in many formulations movements along the Aggregate Supply curve will cause the Aggregate Demand curve to shift – the two functions are not independent of one another (Nell, 1992, 1995).

monetary system, however, tended to be unstable. The turning points of the business cycle appear to have been endogenous. In the later period, however, the stabilizing aspects of market adjustment appear to have vanished. Indeed, market responses appear to exacerbate fluctuations, as would be expected from Keynesian theory and from early Keynesian accounts of the business cycle. The government, however, perhaps in conjunction with the financial system, has tended to provide a stabilizing influence.

An explanation for the differences between the eras can be suggested, which is supported by the record, namely, that the prevalent technology in the earlier period prevented easy adjustment of output and employment. This, in effect, imposed a form of price flexibility, which can be shown to have had a moderately stabilizing influence. But technological innovation greatly increased the adaptability of production processes, so that by the later period, output and employment could be adjusted easily, and the resulting system can be shown to be unstable in a Keynesian sense.

Stylized facts

Many extraneous influences affect economic variables. So it is difficult to make general claims about the economy – there will always be exceptions. Moreover few relationships in economics are fully stable; they tend to be affected by external and arguably irrelevant forces. To deal with this Kaldor suggested the use of "stylized facts."[3] These pick out central and defining features and present them with the rough edges smoothed over, highlighted, so they can be seen with clarity.

"Stylized facts" are stated in general propositions; they present observable, repeatable relationships between measurable variables. They state that two or more variables move together in some definite pattern; or that two or more variables are independent of one another, or that certain relationships, for example, certain ratios, can be expressed by constants (Klein and Kosobud, 1961). These facts are said to be valid over some considerable range of times and places, and can be verified or supported by different bodies of data.

"Stylizing" facts means to remove noise, to remove the influence of irrelevant variables, to cut away random or extraneous factors, so as to present the central relationship in pure – and often, in simplified – form. If

---

[3] Kaldor observes that "in the social sciences, unlike the natural sciences, it is impossible to establish facts that are precise and at the same time suggestive and intriguing in their implications, and that admit to no exceptions.... [W]e do not imply that any of these 'facts' are invariably true in every conceivable instance but that they are true in the broad majority of observed cases – in a sufficient number of cases to call for an explanation" (Kaldor, 1985, pp. 8–9).

the relationship is complex or awkward, it may be "rounded off," or reduced to a more manageable format. What is irrelevant, or extraneous, however, may not be obvious. It will always call for judgment; it may also be a matter of theory.

In particular, many relationships involve variations that take place in the context of a *trend*, so that the relationship cannot be seen, or seen clearly, until the trend has been removed. Detrending, however, requires identifying the trend, which may well depend on deciding which factors determine the trend, and which the variations around it. Different detrending procedures are likely to result in different patterns of fluctuation around the trend (Canova, 1991).

Stylized facts can be considered at two levels. There are the individual facts, each of which tells us something about a particular area of the economy, and then there is the pattern or configuration that can be seen in a group of such facts.[4] If the stylized facts encompass the main features of the economy this will give us a picture of the system as a whole. To make that judgment, of course, requires a theory that defines the main features of the economy.

A different kind of judgment is needed to determine the range of times and places for which these facts should be expected to hold. Are economic relationships timeless, that is, expected to hold always and everywhere? If they are derived from rational choice, perhaps they should. But if such a notion of rationality is unrealistic, or inconsistent with other aspects of human thought and culture, as philosophers have suggested (Hollis, 1995; Hargreaves-Heap, 1989; Hollis and Nell, 1975), economic relationships may be *historical*, in the sense that they hold for particular periods of history, and not otherwise.

This is the perspective adopted here. Fifteen general propositions have been established about the trade cycle at different times. These will be grouped under six headings, with representative sources cited. The claims will be presented separately, under the same headings, for an earlier and a later historical period. In each case, taken together the propositions provide an approximately accurate picture over most of the period. The two pictures present a striking contrast. Moreover, the subject matter is central to economic analysis; prices, money wages, employment, produc-

---

[4] A perhaps extreme case is the construction of "reference cycles," the method developed by Burns and Mitchell (1946) for the study of business cycles. The full cycle is divided into eight segments, and the behavior of the key variables as they move through the cycle is presented by averaging their values in these periods over a set of actual cycles.

tivity, expenditure, trade, investment, and money are at issue. Institutions – government and the firm – are also portrayed. Sources and brief explanations will be given, but no attempt will be made here to justify the claims in detail. Nor is it claimed that the list of proposed "facts" is complete – only that it is sufficient to suggest two different coherent pictures. The first group of propositions presents a portrait of the Old Trade Cycle of the nineteenth century, running roughly from the Napoleonic Wars to World War I, although respectable data only exists after about 1860 – and even then much is questionable. The second covers the post-World War II era.

The old trade cycle

Business units tended to be small, operating relatively inflexible methods of production, meaning that the factory or shop could either be operated or shut down, but could not easily be adjusted to variable levels of output. Prices, on the other hand, were flexible in both directions, as were money wages. The price mechanism appeared to operate. The cycle could be seen in price data.

Prices and money wages

1. The trend of prices was downwards over the whole period. By contrast, the trend of money wages was more or less flat in the first half century, then moderately rising.[5]

Sources: Sylos-Labini (1989), esp. tables 1, 2 (1992), esp. table 1, appendix 1. Pigou (1929), esp. charts 3, 11, 14, 15, 16. Phelps Brown (1981), chs. 7, 8. Phelps Brown and Hopkins, in Carus-Wilson (1962), vol. 2, esp. fig. 1, p. 183. There was an upturn in prices in the 1860s, and a smaller one just before World War I, but the trend is dominant. The latter half of the nineteenth century shows a slight upward trend in money wages, becoming more pronounced after 1900.[6]

---

[5] Measuring trends presents notorious problems. In most of the earlier discussions variations were calculated by simple differencing. In more recent work segmented linear trends have been fitted. In the work done at the New School many approaches have been tried; the results reported have survived different methods of detrending (Canova, 1992), and have been supported in studies using "error-correction" methods (Nell, 1997).

[6] These trends are consistent with the patterns of the previous centuries. In the latter part of the eighteenth century prices rose dramatically as a result of the French Revolution and the Napoleonic Wars. From 1600 to 1775, however, prices of consumables in Southern England were more or less flat, with a slight downward trend from 1650 on, broken only by sudden upturns due to wars. Builders' wages rose very moderately, staying flat for long periods, from 1600 to 1775, then rose steeply up to 1815, then went flat until the last quarter of the nineteenth century (Phelps Brown and Hopkins, 1962, p. 170).

2. Both prices and money wages changed in both directions. Changes in raw materials prices (deviations from the trend) were greater in both directions than changes in manufacturing prices, which in turn were greater than changes in money wages.[7]

Sources: As above, plus Pedersen and Petersen (1938), who focus on the contrast between flexible and relatively inflexible prices. Most of their most flexible prices were raw materials. It is noticeable, however, that even their "inflexible" prices (prices that remain unchanged for more than one year, a number of times over the century) exhibit a downward trend (p. 222). See also Zarnowitz (1992), pp. 150–1.

Employment, output, and real wages

3. Changes in unemployment (proxy for output) were less than the changes in prices; changes in unemployment were "small." Although direct measurements of output are hard to come by, output and employment varied together, with output variations being larger. Prices and output varied together, with price fluctuations being somewhat greater than those of output (deviations from trend). Changes in investment and net exports are often associated with opposite variations in consumption; they certainly do not lead to variations in the same direction, as the multiplier would require.

Sources: As above. Double-digit unemployment was rare, cf. Pigou (1927), charts 18, 19. Hoffmann (1959) provides an output index based on forty-three series, which Phelps Brown adapts for 1861–1913. Pigou uses unemployment as a proxy for output. Sylos-Labini (1986) compares changes in prices, wages and output. Nell and Phillips (1995) found evidence inconsistent with a multiplier in Canadian data for 1870–1914. Block and Kucera (1997) have confirmed the correlation between prices and output for Germany and Japan respectively.

4. Putting these together, it can be seen that real wages, or more particularly product wages, moved countercyclically. That is to say, real wages varied directly with unemployment.

Sources: Pigou (1927), esp. charts 16, 18, 20; Michie (1987). Michie recalculates the work of Dunlop (1938) and Tarshis (1939), and finds that product wages moved countercyclically before World War I (ch. 8). U.S. figures are problematical, but a weak countercyclical pattern is evident in the late nineteenth

---

[7] Since money wages are less flexible than money prices, it is implied that real wages are flexible. Since primary products are more flexible than manufacturing goods, it is implied that the real price of primaries in terms of manufactures changes. Other real price variations can be calculated from the data.

century.[8] (This will be a major point of contrast with the postwar era, although Michie contends that international comparisons are so difficult that it is hard to generalize. But in the later period, some patterns of procyclical movement can be detected.) Nell and Phillips (1995) find evidence tending to confirm an inverse relationship between real wages and employment in Canadian data; Block, and Kucera (1997), respectively, confirm the inverse relationship for Germany and Japan.

Productivity and output

5. Output as a function of labor, both for individual plants and for the economy as a whole, was believed by virtually all contemporary – and later – economists to exhibit diminishing returns. Actual evidence, however, is weak, although, as will be explained later, a good case can be made for a version of diminishing returns. Productivity, however, is closely correlated with short run variations in output in many industries, and positively correlated in general, and varies in both directions more than employment.

Sources: Pigou (1927), ch. 1, pp. 9–10; Aftalion (1913). Calculations made from Hoffmann's data on nineteenth-century Germany show the strong correlations between productivity and output in the short run, and the greater variation of productivity compared to employment.[9]

6. Long-run productivity growth (measured in moving averages) was irregular and unpredictable, and lower than in later periods, although significant. It was transmitted to the economy through falling prices, with stable money wages. The rise in long-run real wages is closely correlated to productivity growth.

Sources: Pigou (1927); Phelps Brown (1962); Sylos-Labini (1989, 1993).

---

[8] Ray Majewski has examined these figures in a dissertation, using the *Shipping and Commercial List*, and the *Aldrich Report* for pre-1890 prices, as a check on Warren and Pearson's Wholesale Price Index. The BLS provides a historic Index of Wages per Hour, and Angus Maddison has developed (1982) an index of real GNP, based on Gallman's (1966) study, which in turn is based on Kuznets (1961). Kuznets has been criticized by Romer (1989) and defended and revised by Balke and Gordon. Neither the Romer nor the Balke and Gordon data change the result that there is a countercyclical pattern evident in the U.S. data.

[9] If the relationship between output and employment were described by a well-behaved neo-Classical production function, then output and productivity should be related *inversely*, rather than directly. The widespread evidence of a positive relationship makes it clear that the countercyclical movement of product wages does not "confirm" traditional marginal productivity theory. Nevertheless, the traditional theory appears to have been "on to something," and the object here is to find out what that was.

Money and interest

7. The (nominal) Quantity of Money was correlated with both output and prices. Changes in the Quantity of Money appeared to affect prices. Income velocity fluctuated somewhat, but showed no trend. In some respects the system behaved as if money were fixed exogenously. This requires some explanation.

Sources: Pigou (1927), p. 132, et passim, pp. 166–72; Snyder (1924). By midcentury the economies of Europe had shifted to the gold standard, prior to which they had operated on bi-metallist principles. It is generally agreed that the gold standard behaved as if the economy relied on "outside" money, that is, on an exogenous money supply (Patinkin, 1965). To be sure bank checking deposits were beginning, and note issue by country banks was not closely bound by reserves, either in the United States or the United Kingdom. But in a loosely organized banking system, without clearly defined policies governing the lender of last resort, prudent financial management required tightening reserves and raising the discount rate in the face of expansion and rising prices, and vice versa in times of falling prices. Central banks followed the "rules of the game" (Pigou, 1927, p. 279; Eichengreen, 1985).[10] Money may not have been strictly exogenous, but prudent management required the banking system to behave as if it were.

8. Investment booms were accompanied by over-eager financial expansion, leading to crises and crashes; these precipitated investment slumps and financial contraction. Variations in employment and prices closely matched expansions and contractions of credit.

Sources: Hicks (1989), ch. 11; Mill (1848), book 3, ch. 12; Pigou (1927), Kindleberger (1978), esp. 3, 4, 6, 8, and appendix. Interest rates and prices rose together in the upswing and fell together in the downswing. The financial crash was usually the signal for the expansion to collapse.

9. The average level of the long-term rate of interest was fairly stable, from the mid-nineteenth century until World War I, and after the war continued to be moderately stable until the 1930s. Interest rates in the United States, however, fluctuated more than those of the United Kingdom, but nominal rates tended to fall as prices also fell. What Keynes termed "Gibson's Paradox" held

---

[10] Some studies, notably Nurkse (1944), for the interwar years, and Bloomfield (1959), for 1880–1914, have shown that Central Banks widely violated the rules. It is now generally recognized that Central Banks had considerable discretion, and that the stability of the gold standard system (which held only for the countries at the center – the periphery suffered frequent convertibility crises, devaluations, and internal credit cruches) reflected successful management (Sayers, 1957), especially by the Bank of England. It was as much a sterling reserve system as a gold standard system.

during more than a century – levels and changes in the nominal interest rate were closely correlated with levels and changes of the wholesale price index, and the long rate was more closely correlated than the short rate. (Hence the nominal interest rate and the nominal quantity of money were correlated.) Both contrast markedly with the postwar era.

Sources: Kalecki calculates deviations from a nine-year (cycle-long) moving average of U.K. consols, and shows that they are very small; Osiatinski (1990), p. 297, table 16. Kalecki considers this sufficient justification to treat the long rate as a constant in developing models of the business cycle. B. Friedman (1986), p. 404, fig 7.1, shows stable interest rates on commercial paper from 1890 to 1914. Sidney Homer, p. 289, shows that money rates fluctuated around a declining average the decade following 1870 and then were steady the next two decades, rising after 1900 to the outbreak of the war. Prussian and German imperial bonds were steady from 1850 to the end of the century; Homer and Sylla (1991), p. 258. Keynes (1931), vol. 2, discusses "Gibson's Paradox."

Besides these strictly economic trends and relationships there are a number of important institutional facts that have changed dramatically. These, of course, are more difficult to substantiate with hard data. Nevertheless the historical record seems to support a set of generalizations – with the caveat that there may be many exceptions.

Business organization, finance, and the state

10. Business was organized and operated by family firms. Firms invested to achieve an optimum size, at which they would then remain, varying their output around the least cost level.

Sources: Pigou (1927). Chandler (1977, 1989) examines the rise of large-scale corporations, beginning in the late nineteenth century. These early corporations are clearly the exceptions. Firms grew to their optimum size and remained at that level thereafter (Robinson, 1931). Nell and Phillips, drawing on Urquhart, found marked changes in firm size and organization for Canada.

11. Once firms reached their optimum size, they did not retain earnings for investment; profits were distributed, saved (or spent) and then loaned for investment by new firms. Finance for investment was thus predominantly mobilized savings, raised through issuing bonds (railroads), borrowing and finding active or "silent" partners.

Sources: As above. The bulk of investment represented borrowed savings and was carried out by new firms Clark (1895). See also Urquhart (1986) on Canada.

12. Governments tended to play a passive role in economic affairs; the "Night Watchman State" intervened little and planned less. Most intervention took the form of subsidizing development.

Government spending and transfers together normally amounted to less than 10 percent of GNP, in some cases near 5 percent, and showed no trend until just before World War I.

Sources: A. Maddison (1984, 1991), esp. table 1 (1991); Hoffmann (1985), Urquhart (1986).

World trade and investment

The period of the old business cycle saw the first great global expansion of trade and investment. By 1914 the ratio of trade to GNP for Great Britain rose to a level that was not reached again until the 1960s. Foreign investment also rose to levels comparable to those of today. But the composition and nature of that era's trade and investment differed importantly from today's.

13. Trade between advanced countries and colonies or less developed regions tended to exceed the volume of trade between advanced countries and other advanced countries. Intrafirm trade was low, and there was very little cross-border manufacturing.

Sources: Michie and Smith (1995); Nayyar (1995); Krugman (1995); World Bank (1994).

14. In composition, trade in the old period was made up of primary goods, coming from the colonies and less developed regions, exchanged against final products from the central countries. The ratio of primary and final products to the total was high. The list of goods involved in trade was stable; the volume of services in trade was low in comparison to goods.

Sources: as above.

15. In 1913 the ratio of the stock of foreign direct investment to world GNP was 9 percent; today it is 7.2 percent. The stock of foreign investment in developing countries in 1914 was $179 billion, almost double the stock in 1980, which stood at $96 billion. In the old period 55 percent of foreign investment went to mining and other primary sector activities, 30 percent to transport, trade and distribution, with only 10 percent in manufacturing, much of which was concentrated in North America and Europe.

Sources: as above, esp. Nayyar (1995).

The character of the cycle

In general, the old business cycle was long, ranging from eight to eleven years, with a long slow buildup to a rapid boom culminating in a crisis with a sharp short fall to the bottom, followed by a slump of variable duration. Then recovery would come, gradually slowing down to a long period of normalcy, followed by another boom. The shape has been compared to a "saw-tooth." Two versions are illustrated. The cycle fluctuated around a moderate growth rate (Boyd and Blatt, 1985).

Now consider the same categories in the post–World War II era.

The new trade cycle

The family firm has been superseded by the modern corporation, operating mass production technology, in which it is able to lay off labor and adapt output and employment easily to changing sales.[11] The price mechanism is no longer in evidence. The cycle is more evident in relations between quantities than in price data.[12]

Prices and money wages

1. The trend of prices was upward the whole period, and the trend of money wages rose even more steeply. Neither prices nor money wages (as measured by general indices) turned down, though rates of inflation sometimes slowed in recessions.

Sources: As above. Also, for comparisons of prices and outputs with the earlier period, Taylor (1986). For wages and prices, Gordon (1983). For a brief discussion of "stylized facts" regarding wholesale and retail prices, the price level,

---

[11] Several studies of the contrasts between the "old" and the "new" business cycle are now available. Nell and Phillips (1995), studying Canada, find good evidence for an inverse relationship between product wages and employment in the old period, and a direct relationship in the new. They also find significant differences in the sizes and characteristics of firms, and in the nature of government between the periods. There is little evidence of a multiplier in the earlier period. Kucera (1997), studying Japan, found similar results for product wages and output – with certain qualifications – and also found a weakly negative relationship between Consumption and Investment in the earlier period, in contrast to a strongly positive relationship in the later. Block (1997), studying Germany, found strongly contrasting relationships between product wages and output in the two periods, as did Thomas (1997) in an examination of British data for the two periods. These studies are gathered together in Nell (ed) 1997, which also contains material from the present chapter.

[12] The period from World War II to the present breaks somewhere in the early 1970s. The first part has been termed the "Golden Age" of modern capitalism: growth rates of output and productivity were high, inflation and unemployment low, and the amplitude of the cycle was moderate. By contrast in the later period, growth of output and productivity became erratic and fell, inflation and unemployment became severe, and the cycle intensified. However, examining this change is not our purpose here. The era prior to World War I can also be subdivided into periods.

monetary aggregates, short and long interest rates, investment, and the timing of indicators, Zarnowitz (1985).

2. Raw materials prices fluctuated more than manufacturing prices, and occasionally fell, though less (in proportion) than in the old trade cycle. Money wage changes were proportionally greater than price changes. Real prices showed great stability, changing only with changes in productivity.

Sources: As above, plus Nield (1963) and Coutts, Godley, and Nordhaus (1987). Ochoa (1986, 1988), demonstrated the strong stability of real prices using 86 × 86 input-output tables. See also Carter (1970), and Leontief (1986).

Employment, output, and real wages

3. Changes in unemployment and output were greater proportionally than changes in prices or money wages; changes in unemployment were large.[13] Output varied in both directions, while prices only rose; the correlation between the two was weak, although price rates of change – inflation – weakly correlate with output in the first half of the era. Output variations exhibit a multiplier relationship: autonomous fluctuations in investment and net exports are magnified by a factor estimated at a little less than 2.

Sources: As above, esp. Sylos-Labini (1989). Evans (1969) surveys the estimates of the value of the multiplier as of that date.

4. Changes in real wages (product wages) tended either to be mildly procyclical, or not to exhibit a distinct pattern. For the United States a weak procyclical pattern has been "largely confirmed" according to Blanchard and Fischer (1989), p.17.

Sources: Michie (1986), chs. 4, 5, 6; Blanchard and Fischer (1989), ch. 1, pp. 17–19; Zarnowitz (1992), pp. 146–50

Productivity and output

5. Output as a function of employment tended to exhibit constant or increasing returns, according to Okun's Law, supported by Kaldor's Laws.

Sources: Lowe (1970), esp. ch. 10.

6. Productivity growth was transmitted to households through money wages rising more rapidly than prices. It tended to move procyclically and was the major source of increasing per capita

---

[13] That is, even *with* countercyclical government intervention, the variations in employment are large in the postwar cycle. They would be far greater in the absence of such policies.

income; the trend over the cycle was stable until the seventies; its decline since then has led to stagnant real incomes.

Sources: Michie (1987), Okun (1981).

Money and interest

7. The supply of money was endogenous, responding to demand pressures. The quantity of money for transactions (M_1) was correlated with *nominal* income, but is not closely related to either output or prices. Changes in the Quantity of Money appear to affect interest rates. Income velocity for M_1 shows a strong upward trend.

Source: Moore (1988); Wray (1990); Nell (1992). "Endogenous money" has many meanings, but the point is that the money supply is not a constraint on real expansion.

8. Over time financial booms and crises became more loosely linked with the movement of prices, unemployment, and output. Real booms generated financial expansion, but financial expansion proved possible even in sluggish and slumping conditions. Credit crunches sometimes, but not always, appeared to slow inflation, and sometimes, but not always, slowed expansion. Crashes no longer led to immediate slumps.

Sources: Hicks (1989), ch. 11; Wolfson (1986); Wray (1990).

9. The long-term rate of interest varied substantially in the postwar era. From the early 1950s to the early 1960s, the real long-term rate rose from near zero in both the United States and Great Britain; it then fell to nearly zero – below by some accounts – in 1975, then rose steeply to over 7.5 percent in 1985, and fell again thereafter. Thus it fell during the inflation of the 1970s, and rose during the early 1980s, as inflation declined. But the *nominal* long-term rate closely tracked the rate of inflation, with interest close to inflation in the 1950s, lying above it in the 1960s, then falling below in the mid-1970s, and rising above again in the 1980s. The correlation is high, and the turning points match closely.

Sources: Calculated from Citibase. The long rate was calculated as a five-year moving average (the average length of the postwar cycle) from 1950–90 and then plotted. The prime rate and triple A bond rates were regressed on the GNP deflator. For a similar relation between nominal short rates and inflation, see Mishkin (1981, 1992).

Business organization, finance, and the state

10. The modern multidivisional corporation has replaced the family firm as the organizing institution through which most of GNP is

created. Growth was carried out largely by existing firms. Under conditions of Mass Production there are economies of scale, and technological progress accompanies investment. Firms must invest continually just to keep up. It is no longer possible to define an optimal size for firms; the question has become their optimal rate of growth.

Sources: Eichner (1976); Wood (1976); Penrose (1954, 1974); Herman (1981); Williamson (1980).

11. Finance for investment has come to be largely internal, raised through retained earnings, for expansion projects carried out by existing firms.

Sources: As above. In the 1960s the ratio of corporate debt to assets rose, then fell in the 1970s, but rose again very steeply in the 1980s (Semmler and Franke, 1993). Gross investment is largely financed by retained earnings, but it could be argued that a large part of net investment is financed by borrowing. Gross investment is the relevant figure for growth, however, since replacements incorporate technical innovations. Moreover, much of the growth of corporate debt in the 1980s was connected with takeovers and mergers (Caskey and Fazzari, 1992).

12. Government intervention and planning became a regular feature of the postwar economic scene. Government expenditures plus transfers had risen to over a third of GNP after the war, and continued to rise as a percentage of GNP throughout the period, faltering only in the 1980s.[14]

Sources: OECD (1986); Nell (1988).

World trade and investment

13. The volume of trade among advanced countries exceeded the trade between advanced countries and developing nations. Intrafirm trade was high, making up to 40 percent of the total of world trade, and as much as 60 percent of United States trade. Cross-border manufacturing has become extensive, with estimates suggesting that one-third of world manufacturing involves global outsourcing by transnational corporations.

Sources: Michie and Smith (1995); Nayyar (1995).

14. In trade in the postwar world the ratio of intermediate goods and capital goods to the total has been high. The list of goods traded

---

[14] In addition, it should be noted that state expenditure in relation to GNP was high and rising during the period, in both military and civilian categories. It rose or remained high, in spite of explicit and politically inspired attempts to cut it back.

has changed rapidly, and the ratio of services to goods in trade has been high.

Sources: Michie and Smith (1995); Nayyar (1995).

15. The stock of direct foreign investment in the modern era, having fallen substantially in relation to world GNP, has now risen to a level comparable to that just before the First World War. In 1913 it was 9.0%; in 1960 it was only 4.4%, in 1980 4.7%, but by 1991 it had risen to 7.2% (Kozul-Wright, 1995, table 6.9, p. 158). The flow of world FDI rose from 1.1% of world gross fixed capital formation in 1960 to 2.9% in 1991. However, its composition is very different from the FDI of pre–World War I. Today only 10% of direct foreign investment goes to primary activities, whereas 40% goes to manufacturing and 50% is in services.

The character of the cycle

The new business cycle has generally been shorter, with less precipitous collapses but longer slumps and slower recoveries than the old. The boom is not so sweeping. The shape of the cycle is broadly a succession of hills and valleys, with a slow climb at the start, then up the steep slope, easing off toward the peak, turning down slowly, and then acclerating down, gradually easing off and sliding into the bottom.

The pattern of growth is significantly different. In the period of the old business cycle population growth was an important determinant of economic growth. In the new period this is less important for the United States, and population plays no role at all in driving the growth of the major European states. In general the cycle has tended to fluctuate around a much higher rate of growth and, in particular, a far higher rate of productivity growth. The latter has been closely associated with increases in the capital-labor ratio, and both are positively related to the level of the rate of growth. The overall rate of growth and the average level of the general rate of profit are closely associated (to put it another way: profits are highly correlated with investment). The labor force participation rate is positively associated with the rate of growth.

Summary

To summarize the "stylized" differences between the two periods:

The agents have changed in size and character, from family firms to modern corporations

Markets have changed their patterns of adjustments, from one in which prices move procyclically and real wages countercyclically, to one in which prices only rise, and real wages move erratically or procyclically

A macro economy in which consumption varies inversely to investment and net exports changes to one in which the multiplier is prominent (variations in investment and net exports set off similar variations in consumption)

The system in which productivity increases are transmitted through falling prices changes to one in which the transmission comes through rising money wages

A financial system in which interest rates are procyclical changes to one in which they appear to behave erratically

A money supply that is, or behaves as if it were, exogenous, changes to one that is endogenous

And, finally, a nonintervening "Night Watchman" state changes to an interventionist Keynesian state.

Structural differences

The preceding points concern the nature of firms and the way markets work. Besides these differences there are others which describe the changes in the structure of the economy between the two eras. Two are particularly noticeable: the size relationships between sectors changed, and so did the character of costs.

All through the period of the "old trade cycle" labor flowed out of agriculture and primary products into manufacturing and services. Output in the latter two grew more rapidly. As labor moved out of the primary sector it settled in large towns and cities, which grew rapidly. In the period of the "new trade cycle" labor continued to leave agriculture, but manufacturing ceased to grow, while services changed character and became the fastest expanding sector. Urbanization ceased, the cities stagnated and even declined. But the suburbs expanded, as did the large metropolitan areas.

The following table shows the approximate range of sizes of sectors as proportions of GDP in the two periods:[15]

Sectors	Craft-based factories	Mass Production
Agriculture	40–50%	5–10%
Services	35–50%	40–60%
	Mostly personal	Mostly business
	Low-tech, unproductive	Increasingly high-tech
Manufacturing	10–15%	35–50%
	Increasing	Stable or decreasing

[15] Because these are ranges the percentages do not add up.

The above table includes government under services. Separating it out is revealing.

Government	10% or lower	40–55%
	Included in services	Spending on services and manufacturing; transfers; government production
	Stable	Rising

Labor costs have fallen in all sectors as a proportion of total costs; they were higher in the earlier era in every sector, but they have fallen as fast in agriculture as in manufacturing:

Agriculture	2/3–3/4	1/5–1/4
Services	3/4	1/3–2/3
Manufacturing	2/3	1/5–1/4

In the earlier period blue-collar labor costs made up between half and two-thirds of all labor costs. In the era of Mass Production blue-collar work has fallen to much less than half of total labor costs.

In the earlier period plant was designed to produce a certain level of output; varying production was costly and difficult. In the later period, employment and output could be varied more easily, so that average variable cost curves contained a long flat stretch (Hansen, 1948; Mansfield, 1978; Lavoie, 1992, pp. 118–28).

Price versus quantity adjustments

In the earlier era markets evidently adjusted through price changes; in the later, however, prices no longer seem to be changing in relevant ways. Instead, employment and output are adjusted when demand fluctuates. These two patterns of market response are significantly different. The first is broadly stabilizing; the second, however, is not.[16]

---

[16] This discussion concerns the "normal" behavior of "stylized" actual market agents, for example, family firms or modern corporations, working class or middle class households. It is not concerned with the "idealized rational agents," conceived independently of social context, that populate the models of much contemporary economic theory. The normal behavior of stylized actual agents may well involve maximizing subject to constraints, and will generally be "rational" in some sense. But it is shaped and determined by context and institutions (Nell and Semmler, 1991, Introduction; see also chs. 3 and 4).

Market adjustment in the pre–World War I era

In the earlier era, when production was carried out with an inflexible technology, a decline in autonomous components of aggregate demand – investment or net exports – would lead prices to decline; since output could not easily be adjusted, it would have to be thrown on the market for whatever it would fetch. For similar reasons employment could not easily be cut back; hence there would be little or no downward pressure on money wages in the short run. As a consequence, when the current levels of the autonomous components of aggregate demand fall, real wages rise, in conditions in which employment remains generally unchanged. Hence – to put it compactly – when investment declines, consumption spending rises. Investment and consumption move inversely to one another.

For *relatively small* variations in autonomous demand this is a stabilizing pattern of market adjustment. For *large* – and prolonged – collapses of demand, however, the relative inflexibility of output and employment can lead to disaster. Unable to cut current costs, or unable to cut them in proportion, and facing declining prices, firms will eventually have to shut down. When prices fall to the breakeven point, all their employees will be out of work. With no revenue, the firm will have to meet its fixed charges out of reserves, and when these are exhausted, it will face bankruptcy. Shut-downs, of course, reduce consumption and are destabilizing.

Similarly, a rise in the autonomous components of demand lead to a bidding up of prices, but not, initially, of money wage rates. Hence the real wage falls. With employment fixed, consumption declines in real terms. Again, consumption and investment spending move inversely. In addition, the fall in the real wage makes it possible for employers to absorb the costs of reorganizing work and thus, in the longer term, to hire additional employees. But so long as the proportional increase in employment is less than the proportion decline in the real wage, consumption will fall.

Such a fall in consumption following a rise in investment can be expected to exert a dampening influence on investment. Similarly the rise in consumption following a decline in investment activity can be expected to provide a stimulus.

These stabilizing influences are reinforced by the behavior of interest rates. When demand falls, prices fall, and interest rates follow suit. We saw that according to "Gibson's Paradox" interest rates were highly correlated with the wholesale price index. Hence a decline in investment will be followed by a fall in interest rates, just as consumption spending picks up. The effect will be to provide a stimulus. By contrast, in a boom, interest rates will rise, just as consumption spending turns down.

Of course, the impact of these countervailing tendencies will be reduced by bankruptcies and capacity shrinkage in the slump and by the formation of new firms and the expansion of capacity in the boom. When demand falls sharply and closures and bankruptcies reduce the number of firms, output shrinks, and the pressure on prices might seem to be reduced. But bankruptcies and closures reduce employment, and therefore consumption demand. So demand declines further, and prices continue their downward course, pulling interest rates down with them. Falling prices and low interest rates make replacement investment attractive. At some point it will be worthwhile shifting replacement forward in time. This could then start an upswing. In the same way, capacity expansion will tend to inhibit the rise in prices in the boom – but building new capacity itself increases demand, which will feed the pressure on prices. Interest rates will continue to rise; at some point interest and prices will be sufficiently above normal that it will seem worthwhile to postpone replacement. This could then prove the start of the downturn.

In short, the pattern of market adjustment provides endogenous mechanisms that could bring a boom to a close, and lead to recovery from a slump. The system is self-adjusting, and capable of generating an endogenous cycle around a normal trend. The three internal processes just described contribute to this – real wages, and therefore consumption, move countercyclically, replacement investment moves countercyclically, and the interest rate moves procyclically. These combine to provide pressure on net new investment to eventually turn against the cycle, perhaps – or probably – with a variable lag that depends on circumstances. Whether such a cycle actually manifests itself, and what its characteristics, amplitude, etc., will be, of course, will depend on the current parameters of the system, and on historical conditions.

Market adjustment in the post – World War II era

The mechanism of market adjustment in the earlier era rested on the countercyclical movement of real wages, coupled with the procyclical movement of interest rates. Neither of these patterns is observable in the postwar era. The mechanism just does not exist.[17]

---

[17] The shift to reduced stability was remarked by Duesenberry (1958), p. 285, "the historical changes in the structure of the American economy which occurred during the first quarter of the twentieth century tended to reduce the stability of the system." However, Duesenberry did not offer a clear explanation. He suggested that there was a tendency for changes in investment to be offset by opposite changes in consumption in the era between the Civil War and World War I (p. 287), and he developed a multiplier-accelerator model, which, however, was defective (Pasinetti, 1974). But his approach outlined a loose general framework that would apply universally, allowing for changes in parameters that would allow each cycle to be different. He did not suggest a systematic change from a stabilizing market mechanism to an essentially unstable one.

In this period prices no longer vary with demand; instead prices are driven by inflationary pressures, partly generated by the new process of transmitting productivity gains through increases in money wage rates. This tends to upset socially important income relativities. If these are restored as a result of social pressures, costs will be increased without corresponding gains in productivity, thereby leading to price rises, setting off a wage-price spiral.

But the system does respond to variations in autonomous demand. Mass production processes can easily be adjusted to changes in the level of sales. Employment and output will vary directly with sales. Hence when investment rises or falls, employment (including extra shifts and overtime for those already on the job) will also rise or fall, while prices and money wages remain unchanged. Since production can be adjusted, there is no necessity to "dump" when demand falls, nor will there be shortages when demand rises.[18] In the simplest case, consumption depends on the real wage and employment; as a result consumption will vary directly, rather than inversely, with investment. This is a version of the multiplier (Nell, 1978, 1992).

Multiplier expansions and contractions of demand, if substantial or prolonged, will tend to induce further variations in investment in the same direction. This is the accelerator, or capital stock adjustment principle.[19]

Early in the postwar era many Keynesian trade cycle theorists argued that the endogenous processes of the modern economy were fundamentally unstable.[20] The plausible range of values for the multiplier and accelerator seemed to imply either exponential expansion and contraction,

---

[18] The comparative "downward rigidity" of prices is thus explained by the technological shift (which also implies a change in business organization), rather than by the growth of monopoly. Zarnowitz (1992), p. 152, summarizing the literature, concludes, "there is no convincing evidence that greater monopoly power is what actually distinguishes the last 40 years from the earlier era."

[19] Evans (1969), chs. 19 and 20, calculates a variety of multipliers, including "multipliers" with induced investment, on various assumptions, and presents numerical estimates for the postwar United States.

[20] Hicks (1950) examined the plausible ranges of values of the multiplier and capital-output ratios, and concluded that, empirically, the system had to be either unstable or generate antidamped cycles. Matthews (1959) reviews the literature and appears to regard models with an unstable endogenous mechanism, running up against buffers, as the most reasonable. A related school of thought argued that advanced capitalist economies had an inbuilt propensity to stagnate, which would have to be offset by government expenditure (Kalecki, 1972; Steindl, 1976), possibly abetted by various kinds of private "unproductive" expenditure (Baran and Sweezy, 1966). In this case, the instability is seen to hold in one direction only – or chiefly – namely downwards. But it is denied that there are any "self-correcting" adjustment mechanisms.

or, if a lag were introduced, antidamped cycles. To develop a theory of the business cycle, it was necessary to postulate "floors" and "ceilings," which these movements run up against. The floor was set by gross investment; it could not fall below zero, and arguably it could not fall to zero, since existing capital had to be maintained, which required replacement. Full employment and supply bottlenecks of all kinds provided ceilings. Once the explosive movement was halted, various factors were supposed to lead to turnarounds (which might be endogenous in the case where the multiplier-accelerator generates antidamped movements). Thus the business cycle was seen to be made up of three parts – an unstable endogenous mechanism, which runs up against external buffers, slowing movement down or bringing it to a halt, at which point various ad hoc factors come into play, leading to a turnaround and unstable movement again but in the opposite direction. In short, a mixture of endogenous and exogenous.

The floors and ceilings, however, in practice have seemed too elastic to explain the turning points; depressions could keep sinking, and full employment did not reliably stop booms.[21] Nor was it clear why, when an expansion or contraction hit a ceiling or floor, it should turn around. Even at full employment, demand in monetary terms could keep rising; even when net investment hits zero, replacements could be postponed – and even when replacements have fallen off, consumption might be curtailed. Moreover, even if expansion or contraction stops, will the accelerator actually turn the movement around? The argument is more plausible for the upper turning point. But, in fact, in the postwar era most upper turning points appear to have occurred before the economy pressed against full capacity or full employment, while the economy has normally turned up before net investment had settled definitively at zero. Many suggestions have been offered to account for these anomalies, yet no single explanation, or combination of accounts, has generally appeared convincing. Some authors even contended that different cycles might rest on different factors (Duesenberry, 1958). Yet however unsatisfactory the theory as a whole might have been, the argument that the endogenous mechanism had become unstable appears to be sound.

However, it has been argued that the financial system might stabilize an otherwise unstable economy. A multiplier-accelerator boom would raise incomes, increasing the transactions demand for money. Such a rise in demand for money will increase interest rates, which, in turn, will act as a drag on investment, bringing the boom to a halt. The multiplier-

---

[21] Duesenberry (1958) judges that ceiling theories cannot explain the upper turning point (p. 278).

accelerator then goes into reverse, throwing the economy into a downswing, but the falling level of income will bring down the transactions demand, thereby pulling interest rates down. The lower interest rates will then stimulate investment, starting the upswing, setting off the multiplier-accelerator.

Recent estimates of the "multiplier" (Bryant et al., 1988) take these relationships into account. Most econometric models try to introduce and estimate all relevant factors (Fair, 1984); hence they likewise include interest rate effects, and perhaps other factors as well.[22] This may be a mistake. Both the simple multiplier and the capital stock adjustment principle are based on solid relationships, which are structurally based and economically motivated. When spending in one sector increases, it sets off repercussions in other sectors, leading to further increases in spending. When demand increases, pushing producers against capacity, it makes economic sense for them to increase their capacity. By contrast, when income increases, while the need for a circulating medium increases, it is not at all obvious that an "increased demand for money" pushes up against a given supply, driving up interest rates. Quite the contrary: as will be argued in Part III, in such cases credit expands, near monies arise, and/or velocity increases – all without any effect on interest rates. The chief determinant of interest rates in the postwar era appears to be central bank monetary policy. Moreover, even when interest increases, its effect on investment is unreliable. It may take a very steep rise in interest, kept in place for a long period, to bring a boom to a halt. As is evident from the early 1990s, a fall in interest rates by no means leads to expansion.

---

[22] Fair, for example, holds that the "word 'multiplier' should be interpreted in a very general way ... [as showing] how the predicted values of the endogenous variables change when one or more exogenous variables are changed" (1984, p. 301). First the model is estimated, then the initial value(s) of the exogenous parameters are set, and the value(s) of the endogenous variables are calculated. The exogenous parameters are then changed, and the new value(s) of the endogenous variables are found. The difference between the two sets of values shows the impact of the change; if only a single parameter is changed, then the value of a single endogenous variable can be divided by that change to calculate a "multiplier." The advantage of this approach is its generality; the disadvantage is that it incorporates into the same calculation of the impact of a change, processes that rest on foundations that differ greatly in reliability. The "passing along" of expenditure and of costs is measurable and reliable, but the response of financial variables to other changes is less so, and the response of real variables to financial variables is notoriously unstable. Multipliers calculated by Fair's method are unlikely to be worth the trouble of estimating them. By contrast, Keynes-Kalecki expenditure multipliers are based on reliable relationships, as are the *long-run* "multipliers" calculated from input-output data. These latter, however, do not tell us anything about the effects of changes in spending.

Rather than floors and ceilings, or the working of the financial system, it can be argued that politics has chiefly provided the turning points. Booms led to balance of payments crises or to inflationary wage-price spirals. Pressure from business interests would lead to an induced recession. Full employment also threatened – or was perceived to threaten – work discipline. On the other hand, slumps threatened governments at the ballot box. The actual business cycle of the postwar era has had an irregular and distinctly political character – although the ability to control the economy may well have eroded over time.

However, the turning points do not coincide that neatly with political interests, and in several cases, it is evident that policy did not produce the desired effects. Yet the cycle is still apparent, suggesting that there is room for an endogenous theory.

Changes in technology

There are no doubt many ways to approach explaining the variance between the two periods. However, the technological contrasts between the two eras are so marked that it seems reasonable to turn to the economic implications of such differences for a possible explanation (Nell, 1992, chs. 16, 17; 1993).[23] The main features of the old trade cycle are all related, directly or indirectly, to the characteristics of the technology of the period. As we saw earlier, until comparatively recently technology was developed by and for small-scale operations, run largely by households or groups of households. These evolved into family firms. The first Industrial Revolution brought the shift from small Craft operations to factories, which, however, were based on essentially the same technologies. Even though, at the end of the nineteenth century, great advances were made as steam power and steel were brought into widespread use, enabling substantial expansions in the size of plants, reaping economies of scale, the technologies still largely operated on the principles of "batch" production, rather than continuous throughput. In many cases the use of steam power simply permitted a large number of workstations, each organized according to older principles, to run at the same time off a central power source. The power, in turn, ran essentially the same tools that had previously been operated by hand. Operatives had to be present at all workstations in order for any production to take

---

[23] These technological changes did not just happen; they were themselves the product of market incentives and pressures, brought about by the problems and opportunities faced by firms in their everyday business. This is the subject of the theory developed in later chapters, especially 9 and 11.

place. Even where continuous throughput developed, start-up and shut-down costs were high.

These limitations had economic consequences. The economy faces continuous shocks from the outside world. Of particular importance are exogenous fluctuations in sales. Firms could not easily vary output to match changes in sales – a firm could either produce or shut down. Craft technologies were inflexible in terms of adapting output and employment (and so costs) to changes in the rate of sales. As a consequence, when demand rose (fell) output could only be increased (decreased) by varying productivity, that is, work effort. The technology required team effort among workers, generally performing on a small scale, so that changes in output could only come with changes in effort – or by reorganizing the work team. But neither labor nor capital was willing to change work norms, except temporarily. Hence the level of employment would have to change, but this in turn would be costly in terms of disruption, and would take place only if compensated by higher prices, at least for a time. Thus a rise in demand would drive up prices, lowering the real wage, thereby leading to an expansion of employment. Inflexibility thus can help to explain the characteristic patterns of variations in prices, output, wages, and employment (Hicks, 1989; Nell, 1992).

Family firms operating Craft technologies do not require extensive government oversight or intervention. A private enterprise financial system may serve this kind of economy well, except in hard times, when it will prove unstable.

By contrast, Mass Production technology permits easy adaptation of employment and output to changes in sales, while leaving productivity unaffected. Variable costs will thus be constant over a large range (Hansen, 1949; Lavoie, 1992). Prices will therefore tend not to vary with changes in demand. Mass Production technology also permits expansion to reap economies of scale, leading to larger firms, differently organized, and motivated to grow. Under Mass Production productivity will tend to grow regularly and will be reflected in wage bargains. Rising wages for production workers will create tensions with other social groups, leading to pressure to raise their incomes, creating inflationary pressures. Large growing corporations cannot tolerate a financial system prone to crisis; Mass Production requires government oversight and intervention in many related dimensions. As a consequence the new trade cycle differs in every one of the above respects.[24]

---

[24] For further elaboration, cf. Nell (1992), chs. 16, 17; Nell (1993) and the references cited there; and Nell (1997). Argyrous (1992), provides a case study of the aircraft industry. Howell (1992), proposes a similar classification of time periods.

How can this be reflected in elementary economic theory? The production function has traditionally been the basic analytic tool of neo-Classical theory in regard to pricing, employment, and output. High theory interprets each point on the production function as representing a different choice of technique – but this was not how Marshall and Pigou understood it (Marshall, 1961, p. 374; Pigou, 1944, pp. 51–2). For them the production function showed output as a function of current employment and the available plant and equipment. This discussion suggests that changes in technology are a primary cause of the changes in the behavior of economic variables from the old to the new trade cycle. Such a shift in technology can perhaps be represented as a change in a Marshallian production function from one with a pronounced curvature, so that the slope declines as employment increases, to one that is a straight line with a constant slope (Nell, 1992).[25]

Some implications for current debates

As between the two eras, the pattern of the cycle is different, the structure of the economy has changed, and the state has developed from a Night Watchman to the guarantor of welfare. Given the different patterns of wage, price, output, and productivity movements, it may seem unlikely that the same models will apply to both. Yet many contemporary discussions appear to be predicated on the belief that the basic explanatory models should be universal, implying that the market mechanism, apart from imperfections, would be the same in the two periods. Different results will be the consequence of institutions, interventions, or imperfections.

Contemporary schools of thought

First, consider two groups that stress "microfoundations," and hold that macro-phenomena are to be explained by theories of rational choice under constraints, where these constraints may include various

---

[25] Assuming for simplicity that all and only wages are consumed, the crucial dividing line occurs when the curvature of the production function is such as to give rise to a marginal product curve of unitary elasticity. At this point the proportional decline in the real wage is exactly offset by the proportional rise in employment. If the curvature were greater, the rise in employment would not offset the fall in the real wage – so that consumption would *decline* as a result of a rise in investment demand that led to a bidding up of prices. If the curvature were less, then a rise in investment, bidding up prices, would lead to such a large rise in employment that consumption would *increase*. This is the multiplier relationship. See Chapters 9 and 11 below.

kinds of imperfections, and interferences by government or other institutions, and where rationality itself may be limited. In each case we will see that the way markets and the cycle have developed is the opposite of what might be expected from the theory.

The New Keynesians

Keynesian theory was developed at the beginning of the second period, in which output adjustment is comparatively rapid, while price fluctuations are slight. But Keynesian theory is not consistent with full neo-Classical equilibrium. To justify Keynesian theory, while still accepting the basic premises of the neo-Classical approach, "New Keynesians" have proposed a variety of mechanisms that purport to explain the "rigidity" of real or nominal prices and/or of real or nominal wages. Being rigid means that the price or wage in question will not adjust to clear the corresponding market, and this failure is then shown to result in Keynesian consequences.

Many ingenious suggestions have been offered: nominal wages may be rigid because actual or implicit labor contracts are cast in nominal terms (Gordon, 1986). Such contracts, however, are often clearly suboptimal; moreover, if prices are flexible, they imply a countercyclical movement of the real wage. Recent work has instead focused on rigidities in nominal prices, attributed to "menu costs" (Mankiw, 1990). Such nominal rigidities may interact with real rigidities – it may cost more to change prices than the expected gain, because of difficulties in disseminating information (Ball and Romer, 1990). Real wages may be sticky because of fears that a variable, market-driven real wage will result in productivity losses – the idea underlying the "efficiency" wage, first noted by Adam Smith (Michl, 1993). Or there may be "coordination failures," resulting from the resistance of firms to lowering prices. In such cases there may be multiple positions of "normal" output. Capital and labor markets may fail to adjust readily because of asymmetric information or risk aversion (Greenwald and Stiglitz, 1993). Similarly, small effects may be magnified, because of risk aversion. Firms may take decisions in these circumstances in the light of "near rationality," rather than full rationality (Akerlof and Yellen, 1985). That is, they may decide it is not worth the trouble to recalculate continually (Romer, 1984; Greenwald and Stiglitz, 1993; Mankiw, 1990; Gordon, 1990).

Broadly speaking, the New Keynesians focus on one or another realistic aspect of market interaction, which would be ruled out by assumption in a world of "perfect markets," and then develop models of maximizing behavior showing how such "imperfections" prevent prices or wages from adjusting to clear the relevant markets. There is thus no single dominant explanation for "Keynesian" results; rather there is a whole class of possible explanations, each applicable to appropriate circumstances.[26]

Almost without exception, however, the imperfections cited in these models were more serious in the period of the old trade cycle, when prices and money wages were flexible, than in the postwar era, when they were not. The costs of changing prices, for example, were greater when printing costs were larger, mail slower, and faxes nonexistent. Asymmetric information must have been more serious before the existence of databanks and computers. Informational problems must have been greater in the days before telecommunications. (If information costs were not greater then, it would not have been worthwhile to invent and introduce the new methods of communication.) Insider-outsider relations must have been more important before the development of standardized tools and equipment, for then training had to be done on the job, and workers had to learn to cooperate together under unique circumstances. No shop would be exactly like any other. And so on. The conditions that the New Keynesians have identified as causing prices to fail to adjust were more prominent in the period when prices *did* adjust.

In the same way, the claim that small "shocks" can have large effects, either because of the risk aversion of firms (Greenwald and Stiglitz, 1993), or because of limited rationality and real rigidities (Mankiw, 1990; Akerlof and Yellin, 1985), also appears likely to be more true of the older period, in which, in fact, prices and wages, both nominal and real, were relatively more flexible. Surely risk aversion would be greater when uncertainty was greater, communication poorer, and information harder to come by. Full rationality would be less likely under these conditions, and it would be true both that there were more "frictions" and that the real costs of each would be larger, since adjustment would be slower.

In short, while the New Keynesian approach directs attention to important aspects of markets, it cannot explain the change from relatively flexible money prices and wages, with an inverse relationship between real wages and employment, to downwardly inflexible, upwardly drifting nominal wages and prices, exhibiting a mildly procyclical real wage-employment pattern.

---

[26] "The challenge is to choose between the myriad of ways in which markets can be imperfect, and to decide on the central questions and puzzles to be explained" (Greenwald and Stiglitz, 1993, p. 25). Zarnowitz (1992), ch. 4, carefully presents a number of important ways in which these models of imperfection fail to explain the observed behavior of prices, wages, employment, and costs.

The New Classicals

New Classicals consider that the price mechanism works to bring about market-clearing in all sectors, impeded only by market imperfections or government interference. The latter only works when market agents do not expect it, or during the time it takes for them to learn how to adapt their behavior to compensate. Since market imperfections prevent optimality, it will pay those in suboptimal positions to remove the imperfections, and it will be in no one's long-run interest to preserve them – the gains from a change would outweigh the losses, so the losers could be compensated. Hence over time imperfections will be eliminated. We should expect, therefore, to see market processes improve their operation over time; market-clearing and market adjustments should be more efficient as time goes on (Hoover, 1990).

Price flexibility appears only in the first period, at a time when market imperfections must be considered more serious than later. Communications were less developed, transportation was slower and more costly, credit was more difficult to check, and calculation was harder and slower. Yet in this period, in spite of the imperfections, the price mechanism appeared to play a role, and the real wage behaved in accordance with marginal productivity theory. But it is only in this period that we see evidence in time series statistics of the price mechanism at work. Market adjustment through prices and market-clearing is less in evidence as time passes.

(Of course, the market did not always clear in the early period, but crises and periods of unemployment went with *falling*, never with stable or rising prices, and money wages, as happened later.)

"Rational expectations," at least when combined with market-clearing, imply results that also fit the earlier period better, although the formation of such expectations only makes sense in the later one. It is relatively plausible to assume rational expectations in the modern period, when firms have computers, modern telecommunications, access to extensive databanks, and employ trained statisticians and economists – although it must also be assumed that agents know what the relevant variables are, something economists cannot agree on! But having "rational expectations," as the phrase is usually understood, does not make much sense in an era of family firms, little education, pencil and paper calculation, poor communications – and when economic theory was so little developed that it would not be possible to identify the relevant variables in many situations. Yet it is in the earlier era that we see evidence of price movements that suggest a tendency toward market-clearing, and where market adjustments appear to accord with marginal productivity theory.

Next consider two groups that explain macro-phenomena by reference to institutions, including competition, economic and technological structure, and the conditions of the world.[27]

The post-Keynesians

Rather than adapt Keynes to neo-Classical microfoundations, post-Keynesians have sought to defend and develop Keynesian thinking, building foundations on a realistic account of institutions (Davidson, 1988, 1994). Lexicographical and need-based theories of household choice, together with markup accounts of corporate pricing, provide an appropriate setting for the theory of effective demand (Nell, 1992; Lavoie, 1992; Eichner, 1989). Labor markets respond chiefly to demand pressures (Nell, 1976, 1988; Lavoie, 1992). Money is seen as adapting endogenously to demand (Moore, 1988). Financial institutions are treated as simply another form of profit-seeking firm responding to market incentives (Minsky, 1986). Uncertainty is ubiquitous, and money and monetary contracts are understood as institutions designed to provide a way of managing practical affairs in the face of our inability to predict the future (Davidson, 1988). Uncertainty is so pervasive that the economy cannot be expected to gravitate towards equilibrium; as Keynes remarked, "equilibrium is blither." Not all post-Keynesians agree (Davidson, 1994). The concept may be useful at times for organizing our thoughts. But many post-Keynesians would argue that it can play no practical role, and should be replaced with the study of dynamics. Investment, in particular, will be volatile, and through the multiplier this will cause fluctuations throughout the economy. These will be exacerbated by financial markets, in which instability is endemic, since financial fragility tends to grow during boom periods (Minsky, 1978, 1986). Conflicting claims during booms give rise to built-in inflation, which is not corrected during the slump, since money wages are not flexible downwards (Rowthorn, 1982, Lavoie, 1992).

Uncertainty, however, must have been much more serious and pervasive in the economic conditions prior to World War I. Communications were

---

[27] It is not implied that these models eschew maximizing behavior; on the contrary, virtually all draw on some form of maximizing, or profit-seeking behavior, under some circumstances. But both the goals and the means are shaped by the institutions and social conditions. What both deny is that there could exist an abstract individual with well-ordered preferences, endowments, etc., able to act in a similarly abstract market. Agents in the market, if persons, are themselves *products* of training and education. That is how they acquired their skills and knowledge. Agents which are institutions – corporations – have to be modeled as institutions, since they typically make decisions in different ways than individuals. (See Chapters 3 and 4 below.)

poorer, databases were less developed, calculation was slower, and the basic economic relationships were less understood.[28] There was far less control over the natural environment, and methods of storage and preservation were still backward. Yet in this period talk of equilibrium was not altogether blither; the economy had built-in stabilizing influences. Conflicts, especially class conflicts, were more intense and less civilized in this period; but there was no inflation at all. Prior to World War I financial and real crises were strongly linked. Each, it seemed, was capable of precipitating the other, and certainly each exacerbated the other. But in the postwar world, for the developed economies, the linkage is much weaker. A financial crisis, as in 1987, may do no significant damage to the real economy. A serious recession, as in the early 1990s, may do no harm to the stock and bond markets. The characteristics of the postwar world cannot be adequately understood in terms of post-Keynesian uncertainty and its effects on financial markets.

The neo-Ricardians

Taking its cue from Sraffa (1960), neo-Ricardian theory builds on given technology, given size and composition of output and a given real wage. From these givens the set of relative prices that will support a uniform rate of profit in a "long-period position" can be found (Garegnani, 1976). Alternative real wages will be associated with different rates of profit and prices; the wage-profit rate trade-off can be defined, and its properties examined. Choices of technique can be analyzed. A devastating critique of the marginal productivity theory of distribution follows from this; in addition, Walrasian general equilibrium theory can be shown to be characteristically overdetermined, or unable to accommodate a uniform rate of profit (Pasinetti, 1976, 1984). The "dual" consumption-growth rate trade-off can be examined in relation to the relative sizes of sectors. Paths of steady growth can be examined, and the effect of alternative wages or techniques explored (Kurz, 1991; Kurz and Salvadori, 1995).

The neo-Ricardian method is to compare alternative "long-period positions." The economy is assumed to gravitate toward those positions, or revolve around them. Actual positions of the economy will not

---

[28] At least two senses of "uncertainty" can usefully be distinguished – "natural uncertainty" meaning that the world is nonergodic and that in general the future cannot be predicted from study of the past, and "market uncertainty" which arises from the fact that agents do not know each other's intentions, and/or how the various strategies will work out when played. Neither can be reduced to calculable risk. Davidson (1996), for example, stresses the former; Graziani (1996), Nell (1996), and Cartelier (1996) the latter. The former is compatible with endogenous stability, the latter is not.

normally, perhaps not ever, be fully adjusted long-period positions. The latter refers to a *theoretical ideal*. But although an ideal, it is the goal toward which the economy is moving, under the pressure of competitive forces. Capital will be shifted about until prices and industry sizes are correct. (This same perspective is taken by many who work in the newly developing fields of nonlinear dynamic analysis and chaos theory.)

But in the modern era, technological change is regular and widespread; it *results* from economic activity – "learning by doing" – and from organized research and development. Innovation is a part of competitive investment strategy. The coefficients are changing continually, and investment plans are subject to constant revision. Movement toward a long-period position – or for that matter, in any direction – is quite likely to change the data on which that position is based. In addition, for many purposes neo-Ricardian theory takes the size and composition of output to be given; but in the modern era the composition of output, and the structure of the economy generally, are continually changing.

Moreover, in the modern world market forces are often destabilizing, which means that there is no process of gravitation. Even if a long-period position could be defined, the forces of competition would not direct the economy toward it.

By contrast, in the era of Craft-based production, it may well have made sense to approach the economy on the assumption that at any time it was tending toward a long-period position. Technological change was irregular; firms distributed profits, and entrepreneurs borrowed them to invest. Market forces were stabilizing. Processes of structural change were slower. Under these conditions the "long-period method" could provide insights. But it is not an appropriate method for studying the postwar economies of Mass Production.[29]

The "cyclical amplitudes" debate

In a different vein a dispute has arisen over the relative amplitude of fluctuations prior to World War I compared to after World War II. Christina Romer (1986, 1987) has argued, in a series of papers, that the fluctuations in unemployment (and in output) in the pre–World War I

---

[29] This does not imply that the Sraffa equations and related models (von Neumann, 1945; Morishima Pasinetti, 1980) are inapplicable; only that they cannot be applied by way of the "long-period approach." But if the coefficients are interpreted as weighted averages of the vintages in use (rather than as "best-practice") the equations will exhibit a picture of the position of the economy at a particular moment. This will change only slowly, as the capital stock changes, and it represents the starting point of dynamic adjustment processes (Roncaglia, 1988; Nell, 1993). (see Chapters 7 and 8.)

economy have been overstated. Her argument begins with a critique of Lebergott (1964) on whose painstaking work most estimates rely. She notes that he had to interpolate extensively to construct his series, but argues that in doing so, he relied on assumptions that magnified the actual fluctuations. She advances similar objections to Kuznets's series. Her recalculations reduce the fluctuations considerably (though they are still greater than those of the postwar era); but her methodology requires assuming that rela-tionships, such as Okun's Law, which characterize the postwar economy, also apply to the pre–World War I economy (Lebergott, 1986; Zarnowitz, 1992). This is unlikely; moreover she ignores the extensive contemporary commentaries on economic events which Lebergott, especially, used to corroborate his work (Sheffrin, 1989). Other writers, for example, Balke and Gordon (1989) and Altman (1992), reexamining the question, find much larger differences than she did. Taylor (1986) found that wages and prices were more flexible in the earlier period, but that fluctuations were also more severe, findings echoed by Zarnowitz (1992). But if pre–World War I fluctuations turn out to be smaller than hitherto believed, the distinctive patterns outlined above will only be enhanced.

From the perspective suggested here, then, the debate over business cycle volatility is on the wrong track. The issue appears to be whether post–World War II Keynesian policies helped to stabilize the economy, with Keynesian supporters arguing that such policies made a difference, while critics hold that little or no benefit is evident.[30] The method has been to compare the amplitude of postwar fluctuations with those of an era in which there was no government intervention, namely, the period prior to World War I. But the character of the cycle in the two periods is not comparable – prices, wages and employment behaved differently. So did money and interest. And the size and nature of government spending differed dramatically. Focusing on the amplitude of fluctuations in employment and output simply misses the more significant changes, which occur, for example, in the relations between the fluctuations in

[30] Since the late 1970s Western governments have adopted austerity policies and have tried to cut back on the growth of state expenditure. These efforts have tended to slow growth and raise unemployment. In addition, world trade has grown faster than world output, without a corresponding development of credit to ease balance of payments problems. Keynesians tend to argue that many of the economic difficulties of the last two decades stem from mistaken policies. However, the perspective here would suggest looking at developments in technology as well. Are new technologies leading to changes in patterns of cost, and in methods of organizing production? If so – and surely they are – what effects are they having on the responsiveness of markets to policy? (See Chapter 13 below.)

prices, wages, and employment. Instead of comparing the time-series of a variable from one period directly with that from the other, more would be revealed by comparing the *patterns* made by the relationships between the time-series variables in one period, with the patterns revealed among those relationships in the other.

Conclusions

A review of "stylized facts" reveals dramatic differences between periods, regarding both the structure and institutions of the economy and the way markets work. A plausible explanation for the changes can be found in the development of technology, as it evolved from what can be termed "Craft-based factory production" to Mass Production. The reason is that this change affects the nature of costs, turning fixed current costs into variable costs, which, in turn, affects the way markets adjust. In the earlier period markets adjusted through stabilizing price and real wage changes, often upset by unstable financial markets; in the later, the market mechanism, working through output and employment adjustments, is unstable, although financial markets are no longer so problematic. Floors and ceilings help to prevent excessive fluctuation, but government policy is the key to the economy's behavior, although there may be an endogenous cyclical mechanism.

Between the two periods considered here lie the interwar years. In this period Mass Production had not yet developed fully, and governments had not learned to cope with the evolving instability of markets. From the perspective suggested here we should expect this period to be the most unstable of all, as indeed it was.

The implication for applied economic analysis is that no "microfoundations" for "macro" are possible or relevant. "Micro" concerns adjustment through flexible prices and applies to the earlier period – and to developing countries with a large sector of Craft-based production – whereas "macro" applies to Mass Production economies. Each describes a distinct pattern of adjustment and neither is more fundamental than the other.

PART II

Method and approach: the active mind

The methodology of "scientific economics" adopted the traditional empiricist's view of the mind as the passive recipient of sense impressions, organized by definitions and analytic truths. Sense data provided the basis of our understanding of the external world, the building blocks out of which the edifice of knowledge was constructed. These were classified and manipulated by means of analytic truths, such as those of mathematics, forming the building blocks into patterns and structures which "pictured" the world, that is, were isomorphic to it. Sense data were passively recorded; the structures were built to conform to external reality – the structure of knowledge, even the logical structure of propositions, mimicked the structure of the world. Knowledge was *recorded*, it was not created. Analytic truths, in turn, were taken to be simply given to the mind and to utterly lack content. Facing the world, the individual mind received sense data as stimuli, and responded by forming them into a picture, using analytical formulae. In this account of knowledge, the mind of the observing agent played no active role.

In the same way, in the picture of the economy sketched by neo-Classical theory, the minds of economic agents play no role. They are likewise the recipients of stimuli, to which they respond automatically, applying analytic formulae. The formulae follow from the axioms of rationality – the axioms, in turn, are taken as given. The values and motives of the agents are likewise given; faced with stimuli, for example, in the form of prices, they respond with appropriate actions, making choices that involve quantities. The stimuli are received and accepted passively; the actions result from applying algorithms to calculate the optimal responses, and, given the data, the correct choices follow directly. The same stimulus, in the same conditions, will always elicit the same response – *ceteris paribus*, of course.

This vision of the passive mind, however, is no longer acceptable philosophically. The underlying theory of perception has been shown to be inadequate. Sensory impressions, especially those arising in social contexts, have to be actively interpreted. Interpretation, in turn, is guided by conceptual truths which themselves arise from the activity of the mind. As a consequence empirical observation cannot be neatly separated from reasoning; to put it another way, truths of reason may sometimes convey important information about the world. And the world may intrude upon the activity of the mind. In economics, in particular, truths of reason provide us with a map of the relationships between agents and the material world – in economic terms, rational choice and production.

But by the same token, to understand and sometimes even to discover these truths of reason, it is necessary to investigate the world, and especially, perhaps, to investigate investigating. Conceptual truths – those that are not merely analytic – are truths that exhibit or assert the preconditions for the mind's activity and relationships to the world. They are true because the act of denying them would itself, directly or indirectly, presuppose them. Such truths, however, are not fixed and immutable, given once and for all. They can be presented well or badly, understood deeply or in a shallow way, developed in more or less detail. Like mathematics, they are capable of infinite elaboration. In economics such truths provide a framework, a set of guidelines, telling us how to construct theory to picture the world adequately.

In particular, conceptual truths provide a basic framework for understanding the structure of human social systems. Such structure, in turn, provides the setting in which behavior takes place, a setting which limits and conditions behavior. Finally, rationality guides behavior, but rationality works through, and must be understood in terms of, conceptual truths.

Such truths can be identified through philosophical reasoning, but their real import can only be appreciated by observing them in practice. This requires empirical study of a special kind, namely fieldwork, an approach that has not been much cultivated in economics. But fieldwork can focus on institutions and structure or on subjective attitudes and behavior. Corresponding to this we find two kinds of models in economics, of structure and of behavior. Each, however, is deficient and requires the other. Regular, planned behavior – such as optimizing – is not possible except in a well-understood context, which, in turn, depends on institutions, rules, customs, and technology, and these are maintained and supported in the ways explained in structural models.

Institutions, etc., however, are intangible. They are manifested in the words and deeds of agents. But to be so manifested, institutions, rules, customs, technology, etc., must first be understood and interpreted; they must be adapted to the circumstances at hand, applied to the situation. And this is a project for the *active* mind. For the active mind, however, "givens" are always open to question.

The paradigmatic models of neo-Classical economics are almost exclusively behavioral, but they adopt a particular form in which to model behavior: that of stimulus and response, giving rise to the characteristic problem of "inexact" laws or generalizations. But the stimulus-response approach is appropriate only for describing agents who are understood as having *given* motivations and values. The agents must also be understood as having *given* knowledge of the world; they do not learn or innovate, nor do they experiment with interpreting the stimuli they receive. The neo-Classical approach therefore adopts the passive picture of the mind. Yet such models also rest on an assumed but largely unexamined structure, the context in which stimulus and response take place. But the structure of an economy implies the presence of agents who must be understood as active minds, for it requires active minds to interpret and apply abstract rules in concrete situations.

Structural models show how the economy maintains and reproduces itself. But it will not do so in exactly the same way every time – agents with active minds will see to that. Market adjustment will confront agents with characteristic problems. Whoever solves these problems will be rewarded – at the expense of those who don't. Competition in the market will judge the innovations and reward the improvers, while discarding the failures and punishing the losers and laggards. Over time this will lead to changes in the way the market works; as they adapt to their altered market environment, the agents – households and firms – will take on new characteristics. The system evolves.

The stimulus-response characterization of agents must therefore be rejected in favor of a "problem-solving" picture, which, in turn, implies that innovation must always be a possibility. Problems can only be solved in a context, which itself must be accounted for. This is the job of "structural analysis." Innovation, in turn, feeds back into the structure, changing the environment within which agents act.

Behavior takes place within this structure and may be examined descriptively – what is likely to happen? How will the various strategies of the agents interact? Or it may be considered prescriptively, asking how can agents innovate and improve their circumstances? The first will give prominence to self-interested behavior, with given skills and information, taking place in given circumstances. Optimization will be practical and limited to choosing the best among available options – and it may be mixed with or even superseded by rule-following. It is in the second that optimization comes into its own, in a more general and abstract approach to problem-solving. But the reason it can be abstract and general is that it is prescriptive. The result is a shift in the role of optimization from determining equilibrium – normal behavior – to determining the most advantageous deviations from the norm. Equilibrium methods must therefore be downplayed and linked to dynamics. Indeed, the idea of equilibrium has to be considered in three separate contexts. A notion must be developed appropriate for structural analysis – reproduction – another for descriptive behavioral modeling – demand equilibrium, the balance of leakages and injections – and a third for prescriptive argument – optimization subject to constraints, implying scarcity values. But in whatever sense equilibrium is used, its role in analysis will be subordinate to the study of the pressures for change.

CHAPTER 3

Conceptual truths and empirical observations

Introduction

Many contemporary economists appear to have drunk deeply from a concoction best described as a pragmatist approach to methodology, although in its "rhetorical" form it borders on the postmodern. It seems to have been considered satisfying because it apparently supports and explains conventional practices, and, as well, helps to defend at least some aspects of neo-Classical theory against competing theories and critics. But the recipe for the pragmatist brew retains important residues from earlier empiricist distillations, and may not be as digestible as it seems at first. A more satisfactory philosophical blend, while mixing in aspects of pragmatism, raises important issues for theories of market behavior.

The most convenient potion mixes Popperian falsification with Lakatos's sociological account of knowledge. This brew leads to a defense of conventional theory on the ground that it works, that is, it guides policy, and seems to be empirically satisfactory in a broad way, so that the "hard core" deserves to be protected. On the other hand, competing theories are required to meet the falsification test, since not being established, they cannot claim exemption for their "hard cores." This they tend to fail. (The approach, however, leaves general equilibrium theory unprotected – it has no empirical content and generates no falsifiable propositions.)

A variant of this approach, likewise pragmatist, but operating with a weaker criterion of falsification, and rejecting Lakatos's sociology, holds, with John Stuart Mill, that economics is a "separate and inexact science" (Hausman, 1992). It is separate because it can identify and study the chief causes of the principal phenomena that interest it (broadly, wealth), but it is inexact because its generalizations and laws are subject to a long list of *ceteris paribus* clauses. Yet it is a science, because the (as yet poorly understood) variables alluded to in these clauses are, in principle, identifiable, reliable, and capable of refinement. Nevertheless, this approach ultimately rests on the hypothetico-deductive model of explanation, which requires a criterion for accepting/rejecting generalizations, yet it adopts a pragmatist stance toward the problem of induction (namely, that the problem can't be serious because science works).

But as philosophy, the methodology of Popper-Lakatos and related approaches can be shown to be flawed; it both draws on and at a crucial point denies the concept of the active mind. The rules and maxims of the active mind must be self-justifying; but the Popper-Lakatos approach cannot justify itself, since it rejects the only kind of conceptual analysis that could provide a justification. Yet conceptual analysis derived from the active mind is exactly what is needed in developing economic theory.

In particular, such analysis allows us to understand the relationships between agents, institutions, and the material world in an economic system, providing an account of structure. Structure, in turn, is the setting for behavior; behavior has to be seen in the context which defines not only opportunities and limitations, but also commitments and expectations. With these in place, the role of rationality for the individual agent can be addressed. One aspect is instrumental; the rational agent seeks to choose the most advantageous option among those available. But another is procedural; the rational agent carries out his or her commitments in the most appropriate way. And finally, rationality can be both critical and imaginative with respect to ends and objectives.

Conceptual analysis also can provide guidance in adapting these general points to particular cases through empirical fieldwork (*not* library studies). The resulting view of the economy gives rise to an account of value, competition and markets that differs from the mainstream. Moreover, it supports the view that history cannot be properly studied by equilibrium methods, and that economic analysis is likely to be different in different historical eras.

Methodology and falsification

It is popular today among economists to adopt a modified pragmatist position, accepting the Popperian falsification criterion as a line of demarcation, but also accepting Lakatos's view of the development of scientific knowledge (Blaug, 1990; Caldwell, 1991). This view holds that the criterion of falsifiability should be applied to economic theories; those that do not engender falsifiable predictions should be considered defective, specifically including, for example, general equilibrium theories. An exception may be made, though, for "standard" neo-Classical theory, on the grounds that it is and has been the basis for virtually all mainstream research. Since such research continues to be acceptably successful, its foundation deserves protection. This blend of Popper and Lakatos is open to objections that undermine any variant of empiricism (Hollis and Nell, 1975) – but it also has some problems peculiar to itself.[1]

Approaches to falsification

The trouble with verification is that no matter how many times a general statement is confirmed, we have no good reason to suppose that it will still hold the next time. For some general statements, this seems reasonable. There was no problem in revising "All swans are white" when black swans were discovered in Australia. But it would be a different story if we met a problem with "Solid objects fall to earth when unsupported." This statement seems to be "lawlike," to be more than a simple generalization from limited experience. Yet it has not so far proved possible to find a solid basis for distinguishing the two.[2] Both depend on experience, and past experience provides no sure guide to the future. This is known as the "problem of induction"; the fact that things have always been a certain way doesn't mean they will continue that way. We may have simply met with a biased sample. Popper strongly rejects any idea of "confirmation"; no matter how many cases accord with a general rule, the rule gains no support from them. General statements are conjectures; they cannot be verified.

This poses a problem for empiricists. Explanations are supposed to be the same in social and natural science. An event or a feature of something is explained by showing that it falls under a verified "covering law," that is, under a true general statement of "lawlike" form. But if "lawlike" general statements cannot be distinguished from mere generalizations, and if neither can be conclusively verified, what explanatory force can a "covering law" have?

Popper and his followers among economists accept the covering law approach. To cope with the difficulty, Popper introduced the Principle of Falsification (PF). Rather than trying, impossibly, to verify general

---

[1] Other recent approaches – e.g., Hausman (1992) – differ superficially, but also end up with a naturalist/empiricist position that is very close to the pragmatists'. (Economics does not differ significantly from the natural sciences; propositions are known to be true through some set of verification tests, either empirical or rational or some mixture.) D. Wade Hands has plausibly objected that Popper and Lakatos do not mix very well, and that economists should find Lakatos more congenial (Hands, 1993).

[2] After careful consideration, a distinguished commentator concluded that a universal conditional is lawlike "just to the degree that it is inductively supported or confirmed" (Pap, 1959, p. 301). This is no help unless there is a solution to the problems of induction and confirmation.

statements, he argued, the implications of theories should, instead, be subjected to the test of attempting to falsify them. A general statement or a theory cannot be verified, but a falsifiable implication of such a statement or theory can be confronted with the evidence. Theories or statements that pass such tests should be provisionally accepted; those that fail should be modified accordingly. This is the method of "conjecture and refutation." Perhaps most important, theories that do not result in falsifiable implications should be dismissed as "non-scientific"; such theories are metaphysics or ideology.[3]

"Conjecture and refutation" implies that all knowledge is provisional;[4] we *decide* what to accept or reject; we devise not only the tests, but also the criteria that determine what shall be counted "pass" or "fail." And we also decide in each case, whether the application has been valid. The contrast with the passive role of the mind in traditional empiricism could hardly be greater.

A problem, however, undermines the simple clarity of this view: as Duhem and Quine pointed out long ago (and Hollis and Nell emphasized) tests cannot be decisive, because the statements derived from theories also rest on *ceteris paribus* clauses, conditions specifying the appropriateness of the test, the suitability of proxy variables, and many other factors, while the validity of these tests themselves depends on the skill of the testers, the accuracy of the equipment, the reliability of the reporting, the freedom of the test environment from extraneous influences, and so on. An indefinitely large number of subordinate conditions has to be

---

[3] In fact, the problem really cannot be evaded this way. It is true that general affirmative statements cannot be conclusively verified, because we cannot check all cases; whereas they can be falsified by a single counterexample. But this does not provide a way out of the problem of induction. For by the same token, affirmative existential statements cannot be falsified in principle, although they can be verified by a single example. "There is a needle in this haystack" cannot be conclusively falsified, however hard and long one looks, since one might simply have missed it. But finding the needle conclusively verifies it. But for the general statement, "no haystacks contain needles" to be falsifiable, it is necessary to be able to come up with a counterexample, such as "this haystack contains a needle," which can be verified, but cannot be falsified. "All haystacks contain needles" cannot be verified, but to be falsified we would have to confirm "this haystack has no needle," which cannot be done, though it could be falsified. The Principle of Falsification cannot function without relying on verification, and even the two together cannot evade the problem of induction.

[4] Including, of course, the claim that all knowledge is provisional. But this did not bother Popper – it was a position that should be adopted, provisionally, because it was fruitful. Pragmatism is OK as long as it works!

accounted for, any one of which could conceivably account for a negative result.[5]

Faced with this problem Popper added a further wrinkle to his argument: for a theory to be scientific, it had not only to issue falsifiable implications or predictions, but it had also to eschew "immunizing strategies"; that is, it had to say in advance what would be a decisive test. It could not simply dismiss negative results, saying that the test conditions were unsuitable, or the test procedures unreliable. The side conditions had to be enumerated and specified as well.[6]

But current writings go a step beyond Popper. They adopt Lakatos as well, to claim that valid scientific practice may distinguish between a "hard core" of central propositions that will not be given up whatever the empirical results, and a "protective belt" of working hypotheses that will be adjusted in the light of testing. The Falsification criterion applies to the propositions of the "protective belt," though in practice many mainstream methodologists seem to want to subject the hard core of nonmainstream theories to it as well (Hands, 1993).

Method, individuals, and wholes

There is, however, one important respect in which many of today's methodologists differ from Popper. Blaug, for example, agrees with Popper that there is only one scientific method, valid for both social

---

[5] Hausman (1992), ch. 8, offers a defence: putative laws with *ceteris paribus* clauses are inexact, vague generalizations which are justified if four conditions can be met: the statement must be lawlike, reliable, refineable, and "excusable," meaning we know and can identify the other things that must be *paribus*. Rosenberg (1992), pp. 112–17, points out that, in fact, for neo-Classical economics these are simply not met. He argues further that it is unreasonable to expect them to be met, that economics is therefore not a "separate" science in Mills's terms, and that the "inexactness" cannot be eliminated or even reduced. Moreover economists have not tried to reduce it; instead they have weakened the claims of rational choice theory to the point of vacuity. As for empirical work, Summers (1991) offers a disturbing critique of recent macroeconometric papers, pointing out that virtually all sophisticated studies fail to establish or support any general position – that is, they fail to convince. Econometric studies almost never replicate results, and new work seldom or never builds on accepted earlier findings – unlike empirical work in the natural sciences (Mirowski, 1992). By contrast the empirical work that *is* convincing to the profession is looser, less precise, and more historical.

[6] There is a logical problem, however. To dismiss a theory because it has been falsified is justified only if it is believed that it would fail the same test again, in similar conditions. That is to say, it must be a lawlike statement that the generalization at issue will fail a certain test. But this is itself an inductive generalization, based on a single instance! Even if the test is repeated a number of times, the problem of induction remains.

and natural science. This is the method of covering-law explanations based on generalizations that have passed the test of Falsification.

But while social scientists (for example, Blaug, 1992) tend to accept the unity of method, many macroeconomists disagree that in the social sciences such unity must, in turn, be based on "methodological individualism." This last is the doctrine that all social actions or events are constructs out of actions, decisions, or events done by individuals, roughly as molecules are constructed out of atoms. Blaug, for example, rejects this; it would imply that Keynesian theory would only be valid if it could be reduced to microeconomic decisions and actions. "Methodological holism," in his view, may be a perfectly valid approach for certain questions; this need not cast doubt on the unity of method, since the covering-law approach still applies.

But there may nevertheless be important differences. In the individualistic approach, the Principle of Rationality clearly can be invoked, and Popper argues that it must be. Even though individuals may fail to act rationally, it is the appropriate assumption to make (though he does not justify this claim!). But in analyzing the behavior – or the structure – of wholes, rationality and optimizing may have no place. Simple Keynesian models, for example, may contain no optimizing behavior, nor do some models of the business cycle. But these models do contain behavioral rules; how are these to be verified? And what does "verification" mean – that the rule exists, or that it is followed? Since there is no optimizing a narrow concept of rationality cannot be invoked. But behavior breaking a rule does not "falsify" the rule. So the Principle of Falsification does not appear to be appropriate either. What, then, could we construe as a "covering law"? (Later we shall argue that fieldwork, in the anthropologist's sense provides the understanding and the data to verify – and interpret – such rules. But they are not covering laws.) This is not the only problem. In other holistic models there may be no agents at all, so no behavior, optimizing or otherwise. (An example is part I of Sraffa or Chapter 7 below). How are we to interpret such a model? Before the question of falsification can be raised we have to know what the variables and equations mean.[7]

---

[7] Of course it could be argued that what appears to be nonoptimizing behavior is actually a disguised form of rational choice. It could also be claimed that actual behavior only approximates the ideal type, namely, rational choice. But these moves only define the problem away. For the covering law model is saved only by abandoning the holistic approach – i.e., insisting that the behavior in question is rational choice by agents after all. Macroeconomics must be explained as micro; macro relationships show the optimizing choices of "representative agent" (Kirwan, 1993).

So it seems that permitting dualism in ontology may create a problem for monism in method. In this respect Popper himself is more consistent – at the price of ruling invalid a great deal of social and economic theory.[8] A high price indeed, for we shall see that individual behavior cannot be explained apart from rules and insititutions, on the one hand, and the agent's relation to the material world – technology – on the other. And these call for holistic explanations.

A critique of the Principle of Falsification

The centerpiece of the Method of Falsification is the Principle, namely that a theory is to be accepted as scientifically valid if and only if it generates falsifiable propositions which are not found to be false. Popper considers this to be the "line of demarcation," between science and non-science – the latter comprising metaphysics, literary criticism, rhetoric, ideology, etc. But it should be remembered that he presented the idea in the course of debates with logical positivists, who sought, using the Principle of Verification, to discriminate between what could be true or false, and what could be dismissed as ultimately meaningless, because it was not possible to determine its truth or falsity. Popper's line of demarcation is much the same as that of the positivists, however much he may seek to distance himself from them, although he will allow that unscientific thinking need not be meaningless. The problems which follow confront any criterion for the validity of generalizations – inexact or otherwise – used in hypothetico-deductive explanations.

It has been claimed – with rhetorical fireworks! (Feyerabend, 1975; McCloskey, 1985) – that no foundations are needed, that there are no general criteria of validity or principles of method. Validity is determined entirely within a discipline or subject, and differences must be settled by persuasion. Rhetoric is the ultimate arbiter. This puts astrologers on a par

---

[8] In other works Popper developed a special version of the covering-law model which he called "situational analysis," and claimed this to be the appropriate, and only valid, method of explanation in social sciences. The significant point was that the covering law or laws appropriate to a given situation would be those based on the Principle of Rationality. However, he was never able to provide a satisfactory account of the status of the Principle of Rationality. It is certainly not empirically true that people are always rational; on the other hand it cannot be a tautology that people behave rationally. Yet rationality is a powerful idea. Models of rational behavior are widely used to criticize what people or firms actually do – if a firm wants to maximize profit, and linear programming, using correct data, suggests a certain choice of inputs, then the firm has made a mistake if it chooses otherwise. The model predicted incorrectly, but the error was the firm's, not the model-builder's. We are inclined to accept this, but why? What is the logical status of such models?

with astronomers; indeed, to judge from numbers of followers they are a good deal more persuasive. More to the point, it puts all schools of economics on a par; there is no way *in principle* to discriminate between two well-worked-out, but contradictory, approaches, each of which passes the crucial tests imposed by its own internal standards. This is the relevant problem in economics today. The Method of Falsification and its variants address the issue; the anti-foundationalists and rhetoricians do not and cannot.

Falsification has significant implications. It has been argued that it rules out whole areas of economic theory as "unscientific," and that it provides grounds to dismiss the pretensions of various political ideologies. But these implications are only as sound as the principle itself. If we are to dismiss whole branches of economics and political theory we must have solid grounds. There must be good reasons to accept the principle, that is, to hold that it is true that the Principle of Falsification demarcates scientific knowledge from other kinds of thinking.

Broadly speaking such general statements are accepted either because they are supported by facts, or because they follow from reasoned argument. An important philosophical tool of analysis, central to modern empiricism, the "analytic-synthetic distinction" (ASD), tried to make this more precise.[9] This distinction was used by logical positivists to consign metaphysics, theology, speculative philosophy, and social theory to the wastebasket. Analytic propositions are true by definition, and are ultimately trivial – the predicate is in a manner of speaking "contained" in the subject: "All bachelors are unmarried." Synthetic propositions are ones in which the predicate adds to the subject, so the truth cannot be determined by simply unpacking the meaning of the subject. "All bachelors are young and handsome." Analytic propositions are true or false necessarily, synthetic ones are contingent. All well-formed declarative propositions, it is contended, are one or the other. (Clearly nothing can be both. But if a proposition is not analytic, then the predicate "adds" to the subject, so it must be synthetic. Of course, evaluative propositions or recommendations cannot be true or false, so are in a different category.) With this weapon in hand, then, the positivists attacked: Are propositions about God or History or Human Nature analytic? No? Then they must be

---

[9] "Anti-foundationalists" will object, saying no such distinction can finally be drawn, or defended, since – extreme cases apart – all truths exhibit both analytic and synthetic aspects and cannot finally be classified as one or the other. We will consider this in a moment.

synthetic; so how are they to be verified? If they cannot be verified, then they must be meaningless – for meaning is the method of verification.

Now consider the Principle of Falsification (PF): is it analytic or synthetic? Clearly it is not analytic; if it were it would be trivial, a mere consequence of definitions. Since it is a major and controversial philosophical position, it is neither trivial nor a simple deduction from definitions. But is it synthetic, then? If synthetic means "empirical" this is difficult to understand. How could a criterion demarcating scientifically valid knowledge from nonscientific claims be empirical? Separating two types of theories and propositions involves distinguishing their logical status; logical differences are not detectable in laboratories or by sensory means. Further, if PF were empirical, it would confront the problem of induction, which implies that even if it had worked in the past, we could not know that it would continue to correctly demarcate scientifically valid propositions. Indeed, PF was introduced in part to overcome the problem of induction, since that problem undermined the Verification Principle. If PF is now found to be subject to that problem, there has been no advance.

"Synthetic," however, need not be taken to mean empirical; perhaps it means "contingent." At first glance this might seem to make it possible to bypass the problem of induction. But contingent general propositions have to be confirmed, so we still face the question of how – on what grounds – we could know PF to be true? It cannot be as a logical consequence of theoretical or philosophical argument, for then it would be a necessary truth. But it cannot be empirical, since then it would be subject to the problem of induction. What is left?

Perhaps we should apply PF to itself? Could it be self-validating? As a strategy this has the advantage of allowing us to consider PF contingent, while avoiding the problem of induction – since PF itself has been advanced as a way around that problem. The trouble is, what falsifiable conclusions can be derived from PF? The obvious candidate would be the claim that scientifically valid theories have falsifiable implications. But is this falsifiable? How could one test it? For a theory that did not have such implications would not be scientific – unless, of course, there is another, wholly independent definition of "scientific." If there is, then what is the point of PF? But if PF does not generate falsifiable implications, then it is not itself scientifically valid; it is metaphysics, rhetoric, or ideology.

At this point it might be objected that many philosophers no longer accept the ASD (Quine, 1953; Davidson, 1984; Rorty, 1991). On the contrary, they hold that no sharp line can be drawn between truths of reason and truths of experience. Theoretical propositions, in particular, depend on a mixture of both; moreover, such propositions cannot be tested independently of one another. They face the tribunal of experience together, jointly. Negative tests do not falsify theories or even propositions – they lead to revisions and reconsiderations. Some propositions are more deeply embedded than others, and we will be more reluctant to revise them. What we revise in the face of experience depends on our interests and purposes; we decide what to keep as true.

This would be acceptable to many economists, for example, Blaug, for it corresponds well with Lakatos. The propositions of the "protective belt" are subject to revision and reconsideration; those of the "hard core" will be defended to the end. The distinction between the two depends on our interests and preferences as researchers – although who "we" are, and how "we" arrive at agreement in such matters is nowhere spelled out. It should be noted, however, for future reference, that this implies a very active role for the mind. We *decide* what to revise in the face of experience.

Very well, suppose, then, that we reject the ASD. Where does that leave our Popperian methodology? What grounds can be given to accept PF? It is still the case that reason and experience are the only grounds for accepting or rejecting propositions, even if no sharp distinction can be drawn between the two kinds of grounds. Since neither provides support for PF, combining them will provide no help either. PF still cannot be based on experience, for the problem of induction has not been dissolved; and reason still gives us no compelling reason to accept it. On Quinean and pragmatic grounds, our interests and purposes may be allowed to help us decide what to revise in the light of experience. But what tells us that PF should not be revised?

The pragmatist will answer that we should accept PF if it works, that is, if it helps us to distinguish scientific approaches from nonscientific ones. But what are nonscientific approaches? Surely those that fail to generate falsifiable propositions. This way leads only in a circle. Even worse, the pragmatist must tell us how to avoid the Quine-Duhem/Hollis-Nell trap, before he or she can claim that PF "works" (Hollis and Nell 1975, chs. 4, 5).

Perhaps it accords well with our interests and purposes? That depends on who "we" are. What reference group is being cited? Which of their interests and purposes are at issue? Pragmatists have, of course, been far from silent on these questions, but it is difficult to justify any one answer. On one influential view, "we" are the educated and concerned, who understand the issues, "acting at our best." It is hard to see how can this be justified, without a circular appeal back to "our" interests or judgments.

In any case, under closer scrutiny PF does not fit so easily into the Pragmatist mode. If all parts of a theory face the tribunal of experience together, and if we may decide what to revise and what to hold on to, then there cannot be any decisive tests. For example, how can a Popperian defend the Lakatosian refusal to develop potentially falsifying tests for the propositions of the "hard core"? Could such tests be defined in principle – or could reasons always be found to reject falsification? If the latter, then surely PF has been supplanted. For the point of PF was to define a criterion based on decisive tests. "Immunizing strategems" were to be avoided, even abolished, and auxiliary propositions which might invalidate a test were to be identified in advance. Now it seems that decisive tests are a will-o'-the-wisp, a figment of positivist imagination.

PF as a prescription

There is another route. Perhaps we should abandon altogether the idea of justifying PF, and cease to consider it as a proposition capable of truth or falsity. Instead it should be seen as a recommendation, a proposal for developing research strategies, and classifying knowledge usefully. It is valid, not because it is true, but because it accords well with our interests and purposes in research and scientific activity. On this view, then, PF would fall outside the realm of scientifically valid statements. (Popper appears to accept this view of the status of PF, although it is less clear that he considers the assumption of rationality merely a recommendation, since he seems to argue that it cannot be escaped; Popper, 1985.)

This is a coherent view, but it reduces PF (and perhaps situational logic) to a mere matter of our preferences and interests. Notoriously these vary, as in practice do research strategies and methodologies; they vary from time to time, from subject to subject and from person to person, and perhaps also from institution or class to institution or class. PF should be followed, say its proponents, because it works for us, it is useful. In short, they like it. They recommend it, presumably on the grounds that what works for them will also work for us. But it might not. Or even if it does, we might not like it. Or we might think or hope that another strategy will work better, or, perhaps, do the same job for less money, time, or effort.

If PF has the status of a prescription, then what is involved is rational choice of method on the part of researchers. Think back to what we know about the theory of rational choice: In normal circumstances, and making the usual assumptions, if preferences differ, choices will differ. If preferences are all identical, but constraints differ, choices will differ. If preferences and constraints are the same, but prices differ, choices will differ. If preferences, constraints, and prices are the same, but incomes differ (or endowments, distinguishing them from other constraints), choices will differ. And this is just the formal approach to rational choice.

If we allow for different concepts of rationality, even more variation in choice will be possible. In short, there can be no presumption that everyone will or should follow this approach. The choice of scientific method becomes arbitrary, even though it may be "rationally" chosen (Hollis and Nell, 1975, ch. 6, appendix). No one can be criticized for following another strategy. On this view it is impossible to hold the doctrine of methodological monism. There can be as many methodological strategies as there are researchers. It all depends on tastes, for which, as the saying goes, there is no accounting.

In addition, there is something "logically odd" about treating PF as a recommendation. PF is supposed to provide a line of demarcation between scientific and nonscientific statements and theories. The *only* rational ground for recommending it, then, is that it works in a specific way: namely that it picks out all and only the valid scientific theories. If it does not do the job, or if it does it badly, then it is inadequate, and cannot rationally be recommended; another criterion must be sought. No other grounds could reasonably outweigh this and provide a rationale for recommending it. If it can be shown to do the job, no other reason to support it will be needed – or indeed, would be relevant. (If two different criteria both worked equally well, but one were simpler or more elegant, these secondary characteristics would count. But first the criteria must be shown to work.) However, showing that PF does the job for which it is designed is to justify it; that is, it is to show that it is, indeed, a criterion demarcating scientific knowledge. But as we have seen, falsification, by itself, runs into problems; even with verification added, there is no evading the problem of induction, and if deciding what to revise is left "up to us, in the light of our interests," the idea of decisive tests collapses, and nothing can be demarcated. If PF is a recommendation, it is a poor one.

Summing up

In short, the methodology widely favored today is irredeemably flawed. It cannot be justified: if the ASD is accepted, the Principle of Falsification cannot be validated as either analytic or synthetic. If the ASD is rejected, a justification still cannot be found for PF, since neither reason nor experience nor any combination of the two supports it. Nor can PF be considered a plausible "rational choice" by researchers. For preferences, constraints, costs, and endowments differ among researchers, so choices will, too. Nor, finally, does it make sense to consider it as a "recommendation."

But the pragmatist approach does emphasize the active role of the mind in establishing our knowledge of the world. So, too, do the arguments advanced against PF, for they are precisely arguments about what argument can establish, and how. They are applications of reason to the project of reasoning. The mind is self-reflective; it considers its own powers and methods in relation to its projects, and in doing so develops guidelines for establishing knowledge.[10]

Conceptual analysis and method

Truths of reason – a priori truths – which result from reflection on the processes of understanding, can tell us about the world.[11] This is a strong claim which many might view with suspicion. But it is not so implausible, as a sneak preview of our argument – in nutshell form – will show. Consider the opposite statement, "truths of reason can tell us nothing about the world." This is certainly not an empirical generalization; if true, it must be a truth of reason. But it is an informative one. It tells us that a long tradition in philosophy and certain contemporary research programs in economics (for example, Austrians following von Mises) are quite wrong. If true, therefore, it tells us something about the world; so it is false. Since the negation is false, the original statement must be true.

Truths of reason do not tell us about the world in the same way that empirical propositions do, but they are both informative and indispensable. They are not independent units whose truth or falsity is determined separately from other propositions. Truths of reason are embedded in our conceptual framework; they are part of our theory of the world. They tell us how to look at the world, and what to look for, by telling us in general terms what *must* be there, and what *cannot* be there. They are guidelines, or maps, not detailed maps, but outlines. They give us the framework of theory, by laying out the meanings of the basic terms.

Fundamental philosophical propositions do tell us something – to take a negative example, the positivist claim that "all general propositions are

---

[10] It is worth noting why this must be true. Consider a claim to the contrary: "the mind is not self-reflective, reason cannot apply to reason." Such a claim instantiates what it denies.

[11] In Kantian terms this would be to claim that a priori truths can be synthetic. But too much should not be made of this traditional formulation. The argument here is not about the history of philosophy, nor is it about particular puzzles in reasoning about reasoning. It is rather that a stronger conception of the role of reason may help both to clarify philosophical issues in economics, and underpin a better methodology.

either analytic and a priori or synthetic and empirical, or meaningless" is neither analytic nor empirical, nor meaningless. If it were valid it would have to be a conceptual truth, able to tell us something about the nature of thought. It is not valid, however, and so the statement that it is not is an informative conceptual truth.

The same point can be seen in regard to the problem of induction and to the problems besetting PF. In one sense the pragmatist is correct: scientists have nothing to fear from the problem of induction. But that does not justify a casual dismissal. It follows logically from the premises of empiricism, so the fact that it is not a practical problem shows that empiricism is flawed – and *that* is a conceptual truth. In the same way, the reason underlying the failure to find a justification for PF is that such a justification would have to be a conceptual truth in the strong sense that empiricist philosophy wishes to reject. But if there is what amounts to "a priori knowledge" in this sense, then PF is not, as it stands, valid. For in this case there must be some knowledge which is scientific – tells us about the world – but not falsifiable by empirical tests, so PF, alone, does not do its assigned job. The realm of such scientifically valid knowledge may well be marked out by a set of conditions that include among them a version of PF, but this set of conditions will itself be developed by philosophical reasoning, and so will be a priori, even though it will have to take account of the actual practices of the various sciences.

The pragmatist's reply

Empiricists and pragmatists will reply with Quine that "no statement is immune to revision"; any truth can be revised, including the laws of physics, even the laws of logic, if we are prepared to make enough changes in our system of thought (Quine, 1965; Nell, 1976). The image is that of a web or network, that covers experience, rather than corresponding to it. The problem of induction vanishes, because general statements are not verified or falsified – we *decide*, on the balance of the evidence and in the light of our interests, when to accept them and what status to give them. When a problem arises, for example, when some statement appears to be falsified, it is "up to us," in the light of our interests, to *decide* what changes to make. We can continue to hold the statement in question to be true, for example, and make other adjustments in the system. We are not compelled to reject the apparently falsified statement. And just as we are not compelled to reject statements falsified by events, we are not compelled to accept statements verified by them, either. *Not even statements verified analytically* have to be accepted. Of course, to reject an analytic truth would require extensive conceptual revision, but, according to the pragmatist perspective, such a revision is, in principle, possible. In short, there are no conceptual truths.[12]

One aspect of this view is worth special attention here. Although there are no conceptual or a priori truths on this view, there are certainly degrees of difference. There are statements that are primarily to be examined conceptually, and others that are obviously primarily empirical. But no statement is purely one or the other. All statements, all arguments, are at least a little of both. That is what it means to say that our statements are all part of the network of knowledge, and all face the "tribunal of experience" together. Thus the pragmatist has no problem with the claim that seemingly conceptual truths tell us about the world; that will be granted. The problem comes in holding them to be necessary.

Yet even here the pragmatist is obliged to agree that to reject or revise a "conceptual truth" – humans are (potentially) rational animals,[13] humans have free will – or a law of physics – action equals reaction – will be more difficult and call for more extensive reworking of the rest of the system than maintaining, say, "all swans are white," in the face of the discovery of black swans. Black swans will have to be considered a different sub-species; this will require revising the criteria for belonging to a species. Besides the ability to interbreed, it will now have to include color. It might take some work to make sense of this; perhaps the easiest course would be to reserve the word "swan" for the special case of white members of the species. "Making revisions" in our systems of thought is not easy; we seldom see such projects outside of the development of science.

To revise "humans are (potentially) rational animals," however, would be an undertaking of a wholly different order of magnitude. Suppose we

[12] If true, of course, this would have to be a conceptual truth. It clearly is not an empirical claim; how could it be confirmed empirically? It would be necessary to canvass *all* truths, to see if any were conceptual. But the set of *all* truths would have to contain the proposition, "this is the set of all the truths there are," the truth of which cannot be known until all members of the set have been determined to be true. Since it is itself a potential member of the set, its truth must forever be undetermined, and the set of all truths can never be completed. The proposition "there are no conceptual truths" could be falsified, of course, by finding a conceptual truth, but that is precisely the point at issue. There are plenty of candidates; the claim is that what appear to be conceptual truths are actually not. What is inescapable, however, is that if the argument succeeds, it generates a conceptual truth of precisely the kind it denies. For a similar problem in the arguments of Quine and Morton White, see Nell (1976).

[13] "Potentially" is added here to Aristotle's proposition – and it will be argued that the need to add this qualification is itself a conceptual truth. One cannot *be* rational without learning language, problem-solving, and social skills; the potential is ingrained, but it will not become actual except as a consequence of social processes.

proposed to include Anne Rice's vampires as humans, on the grounds that they were derived from humans – as some of her characters have suggested. We would have to revise our biology: Now not all humans can interbreed. Not all humans are mortal; not all humans need water and vegetables. Sunlight and garlic are harmful to some humans. Some humans have telepathic and psycho-kinetic powers. This begins to call for revision of physics as it applies to humans. Some humans can fly, unaided. It raises questions of politics: should vampires be allowed to vote, to have citizenship? The ethical question is central – are some humans intrinsically evil? Or amoral – and how does "amoral" relate to "evil"? As for rationality, it is evident that vampires are partly rational – we can talk to them – but their behavior makes it clear that their rationality extends only to choice of means. Their ends are nonrational, even irrational, and while they acknowledge, they do not accept, the force of reason.

Still, they are mythological, fictitious. Take real cases where rationality is limited or unattainable. Do we consider the insane as human? Do we accept those with Down's syndrome, and other mental shortfalls? It is ambiguous; we do not treat them as legal persons. They cannot vote or manage property. We can love them, even interbreed with them, but they cannot participate fully in human life. We accept them, but only so far. And we draw the line at those who are brain dead; they are only "vegetables." Nor will we accept talking chimps; not at all.

As even these simple and partly whimsical examples show, the criteria for "being human" are various and complex – an indefinite list of definite descriptions, to use John Searle's terms. But rationality and animal nature are central; when violated even in part, the ascription of humanity is put in question, or reduced in degree. To change this – to ascribe humanity, for example, in the absence of rationality – would require changing our conceptual framework in many different areas. Exactly what aspects of rationality are excised? What will be the implications for human speech, for government, for law, property and contract, for family life? And each of these changes might, in turn, plausibly require further changes in other areas. The pragmatist assumes that these changes can be successfully made, and that the further revisions called for can also be successfully made – and that any still further revisions will be increasingly minor and eventually peter out. In other words the chain of revisions will be finite; it will come to an end within a reasonable period.

(As a thought experiment, imagine modifying our conception of humanity to include a group, in all other respects the same as we, but whose behavior is purely and wholly *instinctual*. They never reason, never weigh alternatives, never argue, think, or plan. They act only on instinct. Now – including them – describe what human beings are, what human institutions are, the basis of ethics and morals, the chief determining factors in human history, the relations between the sexes, and so on. In each area the revisions will surely set up consequences for related areas.)

It does not seem plausible in the case of a major revision that the chain of required changes will be finite; certainly pragmatists have not argued the case in detail. What if, instead of coming to a stop, the chain of revisions continued endlessly? At each step, the proposed change could be "saved" by making revisions – that is the pragmatist's claim – but each such set of revisions would call for further revisions, in other fields.[14] These in turn could be made, but then there would be still further revisions in fields even more distant, without end. Consider also: perhaps eventually the chain will require further revisions to statements already revised once. So there are two problems: first, the chain might go on forever, and second, it might double back on itself, requiring further revisions to those already made. Either of these possibilities would suggest that the original statement was not revisable in the way attempted.

However, the fact that pragmatism allows that different kinds of revisions differ in degrees of difficulty does make a sort of compromise possible. To put our project in terms acceptable to pragmatism: a conceptual truth could be understood as one that could not be revised without upsetting a vast range of subjects, generating an infinitely long, and possibly backward-folding, chain of revisions. Such truths will apply to the world, and are the ones that it will be most difficult and complicated to revise. Instead of "necessary" and "contingent" statements, we would have a gradation by degrees of difficulty in revising. The most difficult, requiring the longest chains of revisions, would correspond to necessary truths, the easiest, to contingent. In between, however, there would be many grades, from those whose revision, while not utterly unthinkable, presents daunting complexities, through those whose seemingly easy revision turns out to involve unexpected awkardness, to those that are in fact simple to revise.

Such a gradation may in fact be rather plausible; the blunt distinction between necessary and contingent may well be an oversimplification. Some statements and concepts do seem more "deeply embedded" in our framework of thought than others. This approach captures that. It also allows for an interpretation of conceptual truths as the most deeply embedded, ones that could not be revised without an endless chain of further revisions.

---

[14] In order for this argument to be non-trivial, there must be a criterion marking off different "fields."

(This approach would account for the status of PF; if justifiable, it would be such a deeply embedded conceptual truth. To justify it would require showing both that it worked in practice, and that to revise it would require an endless chain of further revisions. But it doesn't work, and it doesn't require such further revisions, as we have seen.)

Conceptual truths in economics

In the same way, the boundaries of economics and the conditions for meaningful, and separately, the conditions for valid, economic statements, will be marked out by philosphical reasoning. Theory is not merely a process of unpacking the implications of arbitrary or "useful" assumptions. Even theoretical propositions in economics, as close cousins to "truths of reason," can be derived from nonarbitrary assumptions, in a process that is not unlike philosophical investigation.

Such basic philosophical statements will be about economics, and how it applies to the world; hence, indirectly, they will be truths about the world, even though a priori. However, "a priori" in relation to economics should not be understood in the same way as in philosophy. In the latter case it is fully general; an a priori proposition must, in some sense, be "self-evident," or rest on presuppositions which cannot be denied, and therefore must be presumed.[15] Rationality in economic behavior is an obvious example.

Economics is about agents choosing to take jobs, to invest, to produce certain lines of goods or services or purchase various products. Choices must be made by weighing the costs of alternative strategies in the light of the ends to be achieved; this is the traditional province of rationality in economics. The format of rational choice expresses a conceptual truth – the optimizing procedure determines the best choice. But while this is the most readily apparent case of conceptual truths, certain others will prove more central to our argument. To understand them we must remember: economics applies to the activities of human agents in the world; it therefore presupposes the general principles of the physics of material objects and the biology of the human race.

So there will be other important basic truths. Human reproduction requires two partners; so if Robinson Crusoe is to provide a paradigm for economic analysis he will need Ms. Friday. The isolated individual cannot be the basis for economic analysis. More generally, human life is built around social relations; and economics, in particular, depends on property, for exchange is the exchange of ownership rights. Rights, in turn, depend

---

[15] Hollis and Nell offer a summary account of such propositions. See also Nell (1974) and Hollis (1995).

on contracts, that is, promises which give rise to obligations. But this depends on authority; contracts must be enforced, which requires judgments and settlements of disputes. Hence there must be rules and institutions. Agents in turn interact with the world; they are able to affect and alter material objects in various ways – which is production. How they do this is the province of technology. Thus the economic activity of agents rests on two foundations: their relations with each other – institutions – and their relations with the material world – technology.

These activities take place regularly, organized through institutions. The forms according to which agents behave depend on the institutions that define everyday roles – family, household, job. These provide grounds for agents' expectations about each other, in terms of their respective *roles*, as determined by birth, education, or appointment. Understanding, discriminating, forming expectations, and making informed choices, all require skills and training, which are passed along from one generation to another through institutions. Rationality itself is *learned*; only the potential for it is inbred. Such activities also require current material support, which comes from interaction with others. Institutions develop and channel human skills, but the limits to what people can do, and what they need as support, ultimately rest on human biology.

The activities of investing and producing require engineering skills, based on technology, which sums up the various ways agents can affect the material world and bend it to their purposes. But in doing so, agents, being themselves material objects, must also alter themselves and their relationships. Technology is not simply a set of ways agents can affect the world; to affect the world agents must also alter themselves and their institutions.

Technology ultimately rests on principles elaborated by physics and chemistry, just as institutions reflect the limits and possibilites implied by biology. "Ultimately" is important here; neither physics nor biology determines any economic relationships. But the portrayal of economic relationships must be *consistent* with the basic principles of each. For example, it may be convenient and, for some purposes, reasonable to ignore intermediate products and the depreciation of capital goods. Many models do.[16] But no *general* principles can be derived from such analyses. Action

---

[16] Including famous and influential ones, like Hicks's *Value and Capital* (1939). Intermediate products are treated, following Pigou, as a "lake," fed by current factor services and drained by current output (p. 118), while depreciation is mentioned briefly in the discussion of income on p. 187, and again on p. 196. The need for replacement and the implications – that replacement depends on production and implies limitations on the pattern of exchange – are not explored at all.

equals reaction; any materials, anything manufactured, will eventually be used up or wear out, and have to be replaced. Energy is used up, materials wear out, no supplies last forever; the Second Law of Thermodynamics ensures that energy will always be costly (Georgescu-Roegen, 1971). Similarly, for some limited purposes, agents may be assumed to be indefinitely long-lived. But nothing essential can rest on this assumption; human agents are at least as mortal as they are rational, and in practice more certainly so.

General equilibrium theory provides another important example. Equilibrium is defined without reference to the level of real consumption necessary to provide agents with the support and training to enable them to carry on and reproduce. Hence an equilibrium position may be one in which some or all agents could not survive and reproduce (Rizvi, 1991). For some limited purposes, again, this may be a convenient simplification. But nothing *general* can be concluded. Moreover, adding an account of the consumption necessary for agents to function creates problems for the theory – no equilibria may exist, and if any do they may not be stable. If a model cannot account for the training, support, and replacement of its agents, the approach is surely flawed.

Conceptual truths in economics, then, trace the general forms of the relationships holding between economic agents, on the one hand, that is, economic institutions, such as firms and households, and between agents and the material world, that is, technology, bearing in mind that agents themselves are part of the material world.

Interpreting conceptual truths

To claim that there can be a priori knowledge of the world does not imply that we can sit in our armchairs and figure out the ways African markets differ from those in Latin America. Such specific matters are never a priori. Truths of reason provide direction to research; they tell us where to look and what kinds of things to look for. They tell us about the shape of the world; they don't give us facts – they outline the possibilities and the limits.

To understand this better, consider the opposite position. Conceptual truths, according to empiricism, are conceived to be analytic, that is, trivial. They cannot provide information about the world and are little more than a system of classification, resulting from the manipulation of stipulated definitions. However, on this view, definitions, in turn, must themselves be arbitrary, since if they were not, there would be true or necessary definitions, which would not be empirical, but would not be analytic either (W. E. Johnson, 1933). (Think of Socrates' arguments over the correct definition of justice in *The Republic.*) Thus it can be argued that analytic truths turn out to be the consequences of decisions, which are ultimately undetermined stipulations. In the end, there is no a priori knowledge, since conceptual analysis is simply drawing out the consequences of arbitrary classifications. However, the problem of induction stands in the way of empirical general knowledge. Since analytical knowledge and empirical knowledge are the only two possibilities, if the first is trivial and the second unattainable, empiricism is in danger of collapsing into universal skepticism.

By contrast, an approach built around conceptual truths provides a role for reason in the formation of theories.[17]

Conceptual truths and "armchair empiricism"

Reason is not confined to the manipulation of arbitrary definitions; it can be concerned with establishing the correct definitions, definitions that capture the essential characteristics of the objects under study. "Essential," in turn, is not a matter of mysterious essences; essential characteristics can be understood in two ways: on the one hand,

---

[17] Most textbooks reject the idea that there might be truths of reason and embrace empiricism or pragmatism. The reason may lie in a belief that anything known "a priori" must be fixed and immutable. Moreover, because such truths are necessary and we know them, we impose them on others. Since we have found the truth, we cannot in good conscience permit others to remain in the dark shadows of error and ignorance. It is our duty to enlighten them, with bullets if need be. This is wholly absurd. Conceptual truths, like any others, can be understood and stated fully or partially; they can be known in depth or only approximately. They can also be misstated, or understood incorrectly. They can be developed, as the implications and connections between their terms are drawn out. Particular versions of a conceptual truth may be approximations which can be improved by further analysis, just as particular theorems in mathematics can be deepened and improved. A simple example: $2 + 2 = 4$. This can be understood at very different levels, depending on the conceptualization of "number"; moreover the proposition has developed significantly in the last half century, as number theory has developed. The proposition is true; but the meaning of "number" has changed! In the same way, Aristotle provided philosophical arguments for "man is a rational animal." The truth of this has not changed, but our understanding of both rationality (e.g., game theory), and of what it is to be an animal (e.g., genetics), has deepened greatly. A priori knowledge of the world requires examining the world, too. Just because knowledge is a priori, does not mean that anyone has privileged access to it, or that the conclusions cannot be criticized, disputed, and revised.

as related to the ability of something to persist, to stay in existence, to remain "the same," that is, itself, during processes of change. The essential characteristics of something, on this view, are those characteristics which must be implied when we make reference to it. These characteristics are implied, because if they were not present, the thing would not exist, or would not be able to maintain itself in existence.[18] On the other hand, essential characteristics can be understood as those necessary for a thing to be "re-identified" by an observer as the "same thing" at a different time or place. Again, these are characteristics which must be implied when we make reference to the object (Wiggins, 1980).

These two perspectives do not differ, for our purposes, when material objects are concerned; but when the objects of study are themselves *agents*, then the distinction matters. For the essential characteristics are the agent's identity.

It's easy to see the significance of this. A widespread problem in the economics profession is "armchair empiricism," the idea that empirical work can be done sitting in a room with a computer, messing around with a database. The empirical economist doesn't have to know anything about the world, about the way things are actually done, "know," that is, in the sense of having direct, intimate acquaintance. Labor market economists don't have to experience job lineups, get laid off, do temp work, or work on shop floors – or even interview those who do. Monetary economists don't have to process mortgages, or car loans, or handle portfolios, or manage banks. Price theorists never have to work in sales, or do the shopping. Anyone can become an authority; no experience is necessary. But if theory is based on real definitions, that is, on essential characteristics expressed in conceptual truths about the world, then to develop and understand the foundations of theory, the theorist would have to know the world in precisely that intimate direct way. Conceptual theorizing must be based on and embody empirical work, which will tell us the identifying characteristics of the objects under study. The common belief that conceptual truths are supposed to make it possible to understand the world by just thinking about it has the true relationship exactly back-

---

[18] Conceptual truths are particularly important in the analysis of social systems, since they regulate the thinking of the agents as well as the theory building of the observers. The problem of "rationality" for Popper is that the ascription of rationality to human agents is a conceptual truth, of exactly the kind for which he can find no place in his philosophy. It is a priori that human agents are rational animals, and also a priori that they have "free will," and hence can choose an irrational course of action.

wards. On the contrary, to do pure thinking, to theorize about the world, it is *also* necessary to investigate the world.

By contrast, pragmatism leads to armchair empirical work. There is no need to distinguish the essential characteristics of an institution from its accidental properties, because there are no essential characteristics. No such distinction can be drawn. There is no need to investigate the inner workings of a system, because "inner" and "outer" are just a matter of the observer's position, an accident of perspective. You don't have to try to understand what "really" happened – in Dallas, or in 1929 – because nothing "really" happened; it's all a matter of what explanation works best, for us, now, in the light of our present needs. And we can revise the story later. What the numbers are numbers of is simply a matter of what we choose them to be of. It's Humpty Dumpty's theory of meaning.

Method in economics

"Humans are (potentially) rational animals," "animals are mortal," "humans have free will," "action equals reaction." These are all rather traditional "a priori" propositions. The first and third are true because they are "undeniable," which is to say that an attempt to deny them would end up instantiating either them or propositions which imply them. (To argue against the first would be to exhibit rationality; if the third were false, the argument would not matter, since what everyone believes would be determined.) The second and fourth are examples of "natural necessity," and follow from fundamental features of the natural world. If they were contravened our ordinary notions of material objects and biological life would require extensive, arguably endless, revision.

It may be admitted that in some broad sense they are true, yet denied that they can be of any use. All are so general, so independent of context, that they describe nothing. But that is the point. They are not scientific laws, or generalizations. They are *guides* to thinking, tools for developing theory.

All four are stated here very loosely. When amplified by related conceptions, they are capable of providing a basic framework for economic theory. That is, they provide guidance, a way of formulating the subject. Consider each, noting the implications for economics.

The minimal sense of rationality is instrumental, finding the best means to given ends. This provides a basis for economic calculation, which is certainly prescriptive, and can under appropriate conditions be used descriptively. But the concept will be broader than that – the notion of rationality that is "undeniable" implies much more. (*Much* more – an extensive and Kantian line of argument holds that rationality implies the obligation to tell the truth, clearly fundamental to contract, and therefore to exchange, and also to the formation of expectations in economics; Hollis, 1995; O'Neill, 1992; Nell, 1995.) At the very least it implies an ability to reflect on, weigh, refine, judge, and choose among ends, which, in turn, implies the ability to develop criteria by means of which to discriminate among ends. In other words, rationality is more than the mechanical application of criteria and algorithms to questions of choice; it is *active*. The rational mind defines problems and creates the tools for solving them.

In conjunction with our animal nature, this has an important implication. Rational agents are mortal; they are born and will die and have to be replaced. When born, however, they are unable to function rationally. They not only need support; they must learn to think and to speak. Babies are not born with language. Nor can they acquire language on their own. There is no such thing as a (fully) private language. But without language there is no rationality.[19] The need to learn to think and to behave is a consequence of being rational, in the full-fledged sense that rationality will guide action. Consider: to the extent that actions are governed by instinct, a newborn animal does not have to learn, or needs to learn less. But if actions are instinctual, they are not governed by reason. Hence if rationality is to govern action, action cannot be governed by instinct, and will therefore have to be learned. Thus human rationality *presupposes* (minimal) social relationships.

To this mix now add free will. This implies that predicting actions, even rational ones, cannot be the sole basis on which theories stand or fall. Active choice, reflection, and innovation are always possible. Situations can be reinterpreted; new forms of behavior can be invented. Any motivation to choose the best among the givens is also an invitation to find or invent something still better.

Since human agents are animals – in fact, mammals – material support will be needed in order for their rationality to function. Being an animal implies the need for subsistence; the agents of an economic system must be fed and otherwise supported. Material support, in turn, must be provided by processes governed by natural necessity – the elementary laws of physics. That action equals reaction implies that when anything is made by material effort or processes, other things will be used up. Cutting dulls the knife; sawing wears down the sawteeth; tools wear out and must be replaced. To be sustained over time, a production process requires re-

---

[19] That there are no private languages, and that rationality requires language are conceptual truths, which have been extensively explored – and disputed – in contemporary philosophy (Wittgenstein, 1956; Hollis, 1995; Winch, 1960; Strawson, 1959).

placements, which in general will have to be produced by other processes (Hollis and Nell, 1975).[20]

What does this tell us about the correct method for economics? First, that it must begin with conceptual analysis, and the starting point must be to establish the form of an economic system, namely, the relations between agents, and between agents and the material world – institutions and technology and their interaction. Then we must ask, how a system can continue to exist, that is, how it can support itself. We cannot refer to an economic system in a general manner, unless its existence is stable and continuous. That is to say, it must be able to support itself, and continue to function. This will provide the concept of the system, that is, that which will remain the same while its properties and characteristics change. To put it in terms of Popper's "situational logic": before the logic of what to do can be studied, we have to understand what the situation is, that is, what its continued existence depends on, and what can and cannot change without it becoming a different situation.

Such an approach implies, for example, a critique of the idea of an isolated human agent. Such an agent – Robinson Crusoe – could not reproduce himself.[21] Secondly, it implies material activity, again an a priori truth – a priori at least with reference to economics; a social system uses up energy and material goods, so it must replace or reproduce them in order to continue. Any "natural endowments" will be used up, and will have to be replaced. It therefore presupposes the general principles of physics and chemistry as these apply to the ordinary practices of life

[20] A good deal of effort has gone into trying to find or define processes of which this is not true, that is, processes which are self-sustaining and produce their own replacements. Ricardo suggested corn – which is its own seed, and provides the support for the labor that grows and harvests it. Knight offered the parable of the "Crusonia" plant, which likewise supported its tenders and reseeded itself. Notably both are *biological*, rather than mechanical processes; the growth is brought about by internal causes that are unrelated to the economy. Both are admittedly fictional. If there is specialization and division of labor – if agents specialize in what they are relatively best at doing – then every production process will have to be resupplied by other processes. Hence production implies some form of transfer of products, or rudimentary exchange.

[21] Robinson was a shipwrecked slave-trader! The wreckage washed up on shore, and he was able to retrieve dried food, tools, knives, firearms, gunpowder, and many other things. So he was not thrown on his own resources – he had the tools and equipment of European civilization at his disposal. Using these he was able to make Friday his servant – a relationship of domination, not exchange between equals. Economic calculation certainly served him well, but he was a well-trained, not an abstract, individual, and both his training and his endowments were products of a complex economy; cf. Hymer (1980).

according to which we make and use material objects, tools, furniture, machinery, etc.

A social system is based on human agents; these are born, grow, mature, learn the skills of living, etc., and finally die. Human agents are not born like Venus, fully developed; there are no independent, mature babies.[22] Economics presupposes the general principles of biology as they apply to the human race – and also in regard to farm animals, crops, etc. For a social order to continue to function it must replace those who have aged, or who are no longer able to carry out their duties. As new people are born, they must be cared for, socialized, and prepared for their roles in later life. To feed the active members of the system as well as the children who will replace them, the society must manage crops, domesticated animals, etc. A social system is based on specialization of function and division of labor; hence production implies exchange (Hollis and Nell, 1975, ch. 9).

Notoriously, economic agents are assumed to function rationally: they adopt means to achieve ends, they form expectations of the future on the basis of present evidence, they order their preferences consistently, and in general, they try to make the best of their circumstances. Moreover, in the struggle for existence, those who fail to make the best of things tend to lose out to those who succeed. Different theoretical approaches will present these points differently, but some form of active rational behavior appears in every account. But pragmatism and, even more strongly, the philosphical position suggested here have implications for understanding rationality and its role in economic behavior. We have seen that concepts and categories are "theory-laden," meaning that events and perceptions are not passively recorded, but must be actively interpreted. Conceptual truths direct inquiry; but we must decide how to follow the directions and how to treat the results, what to accept, revise, or reject. Nothing is finally settled; new results may upset old truths at any time – but exactly what is upset will depend on our interpretation, guided by conceptual truths – which we also have to interpret and apply. Rationality, therefore, must be open-ended, able to learn and to innovate; it can never be mechanical, the calculated response to a foreseen stimulus. Purely instrumental rationality – as in neo-Classical economics – sits uneasily in the company of the active mind. This will prove important.

---

[22] As a thought experiment imagine that babies were born fully developed. The result would be a society without education, training, apprenticeship, or socialization. What kind of learning would be necessary, or possible? How could there be innovation? How could people adapt to changes? Anthills and beehives do not innovate.

The purpose of conceptual analysis here is to spell out the priorities and map the logical geography of the relationships. The object is to identify the forms that any human social system must display, and to classify, provisionally, the different types of system.[23] Defining the form means showing how the system can support and maintain itself, how goods are produced and distributed, how roles and duties are assigned and authority is determined, and what is likely to happen if these relationships break down. It establishes the nature of the rational mind and outlines the place of rationality in economic activity.

The next step is to move from this very general level, to the study of a particular society and economy. This jump cannot be made by collecting some statistics and trying to fill in the general categories developed by conceptual analysis. First the general categories have to be adapted to the particular case; but that has been done by the people of the particular society themselves! We, the observers, have to discover how this adaptation has taken place, in the history and development of the society. This requires what anthropologists call "fieldwork."

Fieldwork

Fieldwork means finding out what people actually do, how they actually think and behave, and what they mean when they say something. Fieldwork has not been prominent in economics. There are exceptions: the Institutionalists did it; much Industrial Organization is based on fieldwork, as is much Labor Economics (J. R. Commons, 1968; P. Sargent Florence, 1972; P. Andrews, 1951; Edwards, 1979). Recent work on the "informal economy" provides a good contemporary example (Portes, Castells and Benton, 1989). Surveys of consumer confidence (Survey Research Center, Conference Board) reflect fieldwork, but most so-called empirical work today is based on number crunching.

Fieldwork calls for *participation*; to know the meaning of a social practice, it is necessary to experience it in some way. It may be possible to gain an understanding imaginatively, or through discussions with participants, and it is certainly not necessary to participate in every aspect. But participation ensures that the observer directly experiences the social practice, and can check the meaning and appreciate the nuances by asking other participants. The object is to get beneath the surface, to contrast

---

[23] We have already argued that the existence of "synthetic" truths of reason implies that we will have to know something about the world, in order to develop basic theory. "High theory" cannot be done in isolation, relying on abstract postulates and mathematics. Even more upsetting to the conventional wisdom, "armchair empiricism" will not suffice either, for empirical studies will have to inform conceptual ones.

actual behavior with the "official" view, and to relate language and description to behavior (McCloskey, 1985). It draws on the method of "*verstehen*," a method that economists tend to regard with suspicion, although it was central to the work of the German Historical School. Indeed, this suspicion seems unwarranted; there is a widespread appreciation for realism among economists – at least those who reject Friedman's extreme position. Even Blaug refers with approval to "realism," for example, in his comments on Hicks, who regarded it as central.[24] Yet "realism" can only be verified by fieldwork.

In economics, fieldwork is necessary, for example, to tell us the real relations in a corporation, as opposed to what the Table of Organization says; it is needed to tell us what really motivates people – as opposed to what they say motivates them, or what we – or the corporations! – think should motivate them. It can tell us how prices are actually fixed, and what was paid as opposed to what was reported; what things are really costs, as opposed to ways of concealing profits; what is the difference between income and income defined for tax purposes; what inputs are really necessary, what is really work, as opposed to sophisticated shirking; what consumers really want, as opposed to what they have been induced to want – or whether such a distinction can be drawn. (An example: Is the recent 1985–95 epidemic of chronic pain, especially lower back pain, the result of technological changes in the workplace, or increases in stress due to economic uncertainty, or is it due to the increased willingness of juries to award large damages and disability pay?) Fieldwork can give us a picture of markets in operation, of the institutions that organize production and sales, and of the way work is structured – as seen from the "inside," and balanced against the "official" picture, for both – and the contrasts – will be part of the truth. Without fieldwork our numbers and therefore our statistical analyses will give us a distorted picture of the

---

[24] Given his approving stance towards realism, one might expect Blaug to be rather well-disposed to the proposed "Cambridge Revolution." He approves of the Classical Economists and disapproves of attempts to treat them as precursors to modern marginalism. He strongly supports Keynesian economics and regards it as highly practical. He is antagonistic to mainstream general equilibrium theory, regarding it, on the one hand, as too abstract to be of any practical use, but on the other as providing a misleading understanding of competition. All these points have been made at one time or another by supporters of the post-Keynesian approach. Yet Blaug rejects the Sraffian treatment of the Classical economists, and wrote a furious, and some would say unfair, critique of *The New Palgrave*, which provided a modest forum for some neo-Ricardian views. Why? Perhaps because he regards marginalism as a serious and at least partly successful attempt to understand markets – a view that finds support here, subject to the proviso that the markets in question are historically specific.

world. Without fieldwork we cannot know the operating rules in our economic institutions, or the true motivations of agents.[25]

Fieldwork does not result in scientific theories, let alone covering-law explanations (if there are any such!). As we shall see, two types of fieldwork can be distinguished. One kind can give us a carefully drawn picture of institutions and practices, general in that it applies to all activities of a certain kind in a particular society or social setting, but specialized to that society or setting. Although institutions and practices are intangible, such a picture will be objective, a matter of fact, independent of the state of mind of the particular agents reported on. Approaching the economy from a different angle, another kind of fieldwork can give us a picture of the state of mind of economic agents – their true motivations, their beliefs, state of knowledge, expectations, their preferences and values. These results will also be matters of fact, but they will be records of the subjective states of the agents reported on – their feelings, attitudes, beliefs, preferences, and values. Fieldwork is reporting, but it is at times an exceptionally sophisticated reporting, because it requires the observer to penetrate the disguises of key roles in society and the economy. This requires careful judgment, since the mask will usually display a partial truth.

[25] Mayer (1993) gives the example of "time inconsistency" theory, in which a game theoretic analysis demonstrates the case for a rule-based rather than a discretionary monetary policy. In this approach the central bank is assumed to generate inflation in order to trick agents into overestimating their real wages and therefore supplying greater work effort. As Mayer points out (pp. 64–5), the statistical evidence suggests strongly that Fed policy has been anti-inflationary during most of its existence. The only exceptions were during wartime. This could be supported even more strongly by reading the records of meetings of the Board of Governors and the Open Market Committee. Further, even if the Fed had an inflationary bias, the reason for this bias might be quite different than that assumed by time inconsistency theory. That theory rests on an attribution of intentions to an institution, the Fed, an attribution made without considering the available evidence, or doing the fieldwork necessary to gather and evaluate new or better evidence. A different but even more extreme example is provided by Lucas's (in)famous claim that "Involuntary unemployment is not a fact or phenomenon which it is the task of theorists to explain. It is ... a theoretical construct which Keynes introduced in the hope that it would be helpful in discovering a correct explanation for a genuine phenomenon: large-scale fluctuations in measured, total employment" (Lucas, 1978, p. 354; see also the commentary in Rosenberg, 1992, pp. 77–8. Even minimal fieldwork will establish that "involuntary unemployment," in the normal sense of the terms, is a fact, and, moreover, one in need of explanation. Further (historical) fieldwork will show that the *character* of employment in leading industrial countries changed from before 1914 to after 1945. The legal, regulatory, and institutional arrangements changed.

The standards and practices of fieldwork itself are designed to produce valid information, which is to say, the information should be reliable; it must have passed an appropriate test of verification, and it should be publicly verifiable. Another fieldworker should be able to repeat the study and confirm the results. But neither falsifiability nor situational logic is involved. Fieldwork has its own rules (Nadel, 1959; Harner, 1970; 1973; Tambiah, 1990), Geertz, 1973).

Fieldwork has to understand the society or sector being studied in its own terms, and then translate those terms into the observer's language. What Geertz calls "experience-near" concepts must be interpreted in terms of "experience-distant" ideas. This is a problem in hermeneutics; translation may well be difficult (Bernstein, 1978). No simple one-to-one mappings may exist. Interpretations may involve suggestion, images, literary devices and "putting oneself into another's shoes." This may sound foreign to the economist's conception of science, but as anthropologists have shown, if done well it yields reliable information – although doing it well makes rigorous demands on the investigator. It is not easy resisting the temptation to simplify and yield to the familiar.

Conceptual analysis of fieldwork can then put together the real patterns of behavior and motivation, in the context of the available and actually operating technology, ways of working, making, and doing things. Such conceptual analysis may be concerned with "deconstruction," a literary analysis taking apart the reported picture, discovering concealed meanings and hidden agendas, both on the part of observers and the observed. Or – the program of economics – it may accept the picture, and set out to construct models that will show how the system works in various ways, including how it may fail to work and break down. Conceptual analysis based on fieldwork will provide the essential assumptions and definitions on which model-building should be based. In order to construct the kind of models that will enable economists to understand the way the system works, we need to start from conceptual truths, fleshed out by understanding "from the inside," and then to develop "stylized facts" by interpreting statistics in the light of the fieldwork.

These can then be further developed on the basis of published statistics (adjusted in the light of information uncovered in fieldwork), and the models can be tested, revised, and so forth. Verification and falsification have a place here, not a privileged place, but a role to play nevertheless. They are not decisive, but they are useful.

Conclusions

Conceptual analysis, understood as a flexible search for conceptual truths, interacting with fieldwork, provides a method by which to approach economic issues. This approach is superior to that advocated by Pragmatism, which cannot give a coherent account of theoretical concepts, especially in relation to empirical work. But fieldwork may be directed toward economic structure or toward behavior. These raise distinct issues and have been treated differently by different economic traditions. We will explore the different problems fieldwork faces with respect to structure and behavior.

But then we must confront the fact that the neo-Classical tradition has little interest in fieldwork, or in structure. Its approach is to identify an agent, termed an individual, and characterized by preferences that possess certain very general qualities. The agent can be virtually any kind of economic actor, for example, a household or a firm, a borrower or a lender, a worker or an employer, because the specifics don't matter. Given the preferences, the method is to construct an optimizing model to predict what that agent will do in various assumed circumstances, responding to market stimuli. As the market stimuli vary, the results of the optimization will vary, giving rise to functional relationships. These are then said to hold generally, *ceteris paribus*.

Besides the well-known *ceteris paribus* problems, there are two issues here: first, agents have not been properly related to the structure in which they are assumed to act, and second, an optimizing model yields *prescriptions*, not descriptions. When combined with the *active* portrayal of the mind, this provides good reasons for rejecting the conventional approach and developing an alternative. Such a move leads directly to rejecting the view that "markets efficiently allocate scarce resources" in favor of the idea that markets generate and finance innovations – the basis of transformational growth.

CHAPTER 4

Rationality, structure, and behavior

Behavior takes place in a social context, some elements of which will be transitory, others permanent. Regular or repeated behavior will depend chiefly on the latter. But such permanent features of the social setting must themselves be reproduced physically, if material, or must be reproduced in the actions and behavior of agents, if, like rules and customs, they are intangible. This suggests that inquiry must proceed along two related fronts, delving into institutions and into technology. But it also calls for two different kinds of fieldwork, one exploring structure, the enduring features of the social context, whether material or institutional, the other looking into behavior itself, whether reflecting institutional imperatives or individual choice.

Fieldwork

On the one hand, then, there is the investigation of institutions and normal practices, as they relate to economic questions – family and kinship on the one side, jobs, offices and the workplace on the other, and the market in between. In each case the aim will be to clearly identify the relevant roles, duties, social conventions, jobs, and offices. On the other hand, fieldwork will be needed to uncover the true motivations, beliefs, expectations, and subjective intentions of those currently occupying roles, jobs, and offices, carrying out duties, and (supposedly) observing conventions. The first type of investigation is objective, in the sense that what it finds will be factual (although intangible), and so independent of the states of mind of those reporting on it, but the second investigation precisely concerns the states of mind of those reporting, as they interact with others (though it has to be ascertained that the respondents have reported accurately and truthfully – not everyone really knows, or is willing to face, their own state of mind!) The second is therefore subjective, but it is meant to be an accurate account of the current state of the agents' subjectivity.

Individuals and institutions

It has sometimes been argued that these two are not really different. More precisely, it is claimed that institutions, roles and rules, norms, and the like are simply *constructs* out of the beliefs, preferences,

106

attitudes, and other mental states of individual people. What appear to be objective social entities really have no independent existence. They are simply a convenient shorthand for describing the choices and actions of individuals.[1] In particular, individual agents choose to enter into agreements and contracts with each other, thereby creating institutions, which are thus simply shorthand descriptions for cooperating individuals. Churches are nothing but agreements of people to come together from time to time to worship, as schools are agreed-upon places for learning. Even nation-states ultimately are merely the complex outgrowths of social contracts. More to the point, households are groupings of rational agents for convenience in consuming, while firms are groupings designed to minimize transactions costs in the processes of production and marketing.

But – it will be argued here – institutions are prior to, and do not depend on – let alone exist as a construct out of – the beliefs, preferences, etc., of individuals. First, priority: human "rational individuals" are *created* by means of social actions in a social context. First, they are conceived and born as a result of interaction between (at least) two people. Since babies are not independent, and do not know how to act when first born, they must be raised and socialized, before they can even attempt to act or to choose. Second, an individual's potential rationality becomes actual only as the result of learning language and how to act, especially how to perform speech acts. Since the notion of a "private language" is not coherent (Wittgenstein, 1956), learning a language implies participation in a rule-guided, convention-based activity, just what is supposed to be accounted for in terms of individual choices and preferences.

Next, independence. The claim here is that institutions do not depend on the subjective beliefs or preferences of those currently involved with them. Consider performative acts: two people are married if they went through a legitimate, authorized ceremony properly. This is a matter of fact. It does not matter whether anyone currently believes it. A bill becomes the law of the land, if it has been passed by the legislature following correct procedures, and is signed by the chief executive. "Ignorance of the law is no excuse," implies precisely that the law is independent of our knowledge of it, or beliefs about it. The responsibilities of a particular job, the operations to be performed by

[1] Mrs Thatcher, in an interview while prime minister, is reported to have summed it up succinctly: "There is no such thing as society; there are only individuals" (*Women's Own*, 1982, Hugo Grant, *One of Us*).

the jobholder, are defined by the company engineers; any given worker, or even all the workers at a particular time, may be mistaken about them.

Someone must know, it might be argued, the engineers, for example. Yet each engineer, privately, might have a different opinion of how the operations are defined. The definitions were arrived at by a committee of engineers, where each member took responsibility for a different section, and all finally signed the final document without reading the other sections carefully. So none of the engineers, in fact, know. But in the case of a dispute they have a document, computer programs, and a *procedure* for reaching a settlement. This procedure is a set of rules and norms.

Consider the appellate judge who pronounces a law unconstitutional, following the ruling of the Supreme Court, although, privately, he believes that the Court was wrong in its interpretation of the Constitution. Neither the status of the law nor the meaning of the Constitution depend on his beliefs – or anyone else's; they depend on public *acts*, properly carried out by constituted authorities – for example, a judge rendering a decision, or the Constitution being ratified. To be sure, some properly positioned and suitably trained agents must verify that the acts have been properly performed. But such judgments are supposed to be *objective*, that is, are supposed to be made independently of the agents' private feelings, and are supposed to be reached by following appropriate procedures. Such acts must be properly understood, for the validity of institutions depends on them, and the normal behavior of agents must be undertaken in the context of institutions, for example, the law (Austin, 1960).

Institutional descriptions tell us what *ought* to happen, or what ought to be done, not (necessarily) in the moral sense, but in the sense of what anyone has a right to expect. Arguably, this is generally true, but here we will concentrate on economic institutions. A bank ought to make loans, a firm goods, a salesperson ought to attend customers, a barber cuts hair, a policeman enforces the law, a teacher teaches. When someone holds a job or an office they have responsibilities, and they are empowered to act in certain ways.

Why should anyone do what they ought to do, in this sense? Because they get rewarded if they do, and fired if they don't. And how do we know this will happen? Because the appropriate supervisors are *obliged* to do it; hiring and firing, rewarding and punishing is their job. People are appointed to jobs, generally after appropriate training, and frequently in competition with other candidates. Appointments, to be valid, must observe the correct form. They can be terminated for nonperformance, again following proper procedures. Institutions, such as firms or banks, are incorporated following legal procedures, and must obey regulations and observe the law.[2]

Any account of norms and institutions must be able to capture this fundamental sense of obligation. A norm expresses what one ought to do in appropriate circumstances. An office carries duties as well as privileges; a job calls for performance; a role generates expectations. In each case the implied "ought" may be moral or conventional, strong or weak. But it is deontic, not descriptive. To account for it is the challenge which faces any attempt to reconstruct institutional descriptions in terms of individual choices, based on beliefs and preferences.

It is a challenge which is hard to meet. How can duties be explained in terms of rational choices?[3] Two parties choose to work together, for example, to reduce transaction costs. But at a certain point a parameter changes and the project ceases to be useful to one party. Why should that party keep the agreement, if it has ceased to be advantageous? Circumstances have changed, so the outcome of the optimizing process will be different. Instrumental rationality dictates a change in behavior.

[2] But, to repeat, why *should* anyone observe the law? In one sense, because it *is* the law. We are obligated to observe it because in accepting our positions, offices, roles, we have accepted the responsibilities inherent in them. In another sense, because we will be punished if we don't, and it is the responsibility of our superiors to see that this happens. They will do their duty, in turn, because it is the responsibility of *their* superiors to see that they do, and so on. In some cases the chain of responsibility goes on up to the king or emperor (and then, through the high priests, to the gods). In other social systems, it doubles back, so to speak, and those below have the duty to judge the performance of their superiors, and remove them if they fail to uphold their office. A limited failure to observe the law or carry out duties can always be handled; a widespread or general failure will mean the breakdown of order – the "war of all against all" – since no one can form reliable expectations of the conduct of others, which requires each to be wary, and "the life of man will be solitary, poor, nasty, brutish, and short."

[3] Suppose we "construct" the concept of the firm out of cooperation between workers, managers, and capitalists. Each has a comparative advantage and preferences. Workers like to work and prefer to be supervised and paid regularly; managers prefer to supervise and manage projects, for regular pay; capitalists prefer to pay others to manage and work, but like to take risks. On this basis we could surely "account for" firms in terms of cooperation based on preferences. But "workers," "managers," and "capitalists" are not terms designating *individuals*. They are institutional terms, designating *roles*. Each implies a social position, which has been reached as a result of training, appointment, or inheritance, that is, as a result of a social process. (An individual may simultaneously occupy several roles; a manager may also be a capitalist. A capitalist may take a job even though he doesn't need it. Workers may have savings.) So this approach would "explain" one institution in terms of other institutions. To meet the challenge, the building blocks of the construction have to be purely individual and subjective.

But in the ordinary world – in moral philosophy and in business both – one who makes an agreement is honor-bound to keep it. Agreements are based on the institution of promising, which, in turn, gives rise to the laws of contract. (And promises are the foundation of credit.)

In short, it seems reasonable to conclude that these two spheres – institutions, roles, and obligations, on the one hand, and subjective attitudes, preferences, and beliefs, on the other – are sufficiently separate to warrant different investigations, given that both are evidently significant in economics.

Fieldwork and structure: the context of behavior

Structural fieldwork investigates the economy by looking at relationships in production, exchange, and distribution – such as the linkages between sectors or agents, for example, technological and legal interdependences (input-output relationships, interest on capital, wage or salary contracts), or relationships of status and authority, as in comparing the positions of property or wealth-owners and the property-less in various sectors. Fieldwork establishes the linkages between these features of the system and ranks them in importance; it is concerned with gathering and interpreting statistics, but also with the character of technology, with job titles and descriptions, contracts, chains of command, responsibilities, and so on. Objects of study will include roles – producers and consumers, suppliers of labor or of savings and wealth, and institutions – firms and households.

The study of households will raise the question of the position of families and the kinship system as holders and transmitters of wealth, as well as consuming units and suppliers of labor: does this make them also the unit of social classes? Can we usefully distinguish classes, that is, classes of families, by their holding or not holding income-earning wealth?

But before thinking about classes, it will prove important to distinguish two kinds of institutions – one kind can be termed those that run the world, the other, those that prepare people to hold positions in the institutions that run the world (Nell, 1996). That there must be these two follows from the fact that humans are mortal; if institutions are to continue to function, properly prepared people must be available to succeed those who currently hold the positions. Hence there must be institutions that prepare them – families, schools, churches, training programs, apprenticeship systems, and the like. People to replace those currently running the world must be born, raised, socialized, educated, and trained for their roles in later life. In a broad and metaphorical sense, the two kinds of institutions represent a demand and a supply of suitably prepared individuals – the institutions that make up the world of practical affairs demand replacement personnel, while the educating and socializing institutions – families in the first instance, then schools and training programs – provide the supply (including the supply of those to replace the present managers of socialization).

Studying this second kind of institution leads naturally to a study of the products of socialization, the different social types and personality profiles that the system turns out. Warriors, priests, shopkeepers, bureaucrats, engineers, explorers, farmers, rabbis, all differ in attitudes as well as in skills. Many of these attitudes are learned, the results of nurture; however, some, of course, are inborn, deriving from nature. But even some of these latter may be the consequence of social processes, deriving from the gene pool established by the rules and customs governing partner selection in the marriage system. But the study of the types produced by the system's socialization and educational processes is not the same thing as examining the states of mind of people in their day-to-day lives. As we shall see, this is a separate inquiry, a different kind of fieldwork.

To obtain a picture of the whole, the two kinds of institutions must be put together. It must be shown, first, how each works, and then it must be shown how the two fit together. The roles and activities that we called "running the world" are interdependent; they produce goods and services for each other, in the process consuming the very goods and services they produce. These must be shown to make sense taken together; the different aspects of society mesh, join to make up a culture. Economics will contribute by exploring whether the linkages and connections are mathematically consistent, and stable. Along with this it must be seen whether the socialization institutions produce appropriately prepared replacements for those occupying the positions of society. A particular form of this is the "class society," in which those occupying the leading roles form families that produce a new generation prepared to take over those leading roles (and who will, in turn, contract marriages that will produce the following generation), while those in the lower roles likewise form families that will produce future occupants of the same lower-level roles.

But all this does not take place according to a plan, nor on the other hand does it happen by accident. The running of capitalist society, its production, exchange, and distribution, and the filling of its jobs and positions are all coordinated by the market. But the actions that people take in the market are carried out in pursuit of self-interest; each agent is free to choose among a variety of possibilities, and does so in the light of material advantage. Workers may choose jobs, employers their workers, consumers their goods, producers their target markets. Adam Smith spoke of a "system of perfect liberty." (Ideally, that is; in reality they all face constraints.)

In this system, market outcomes will not in general be those intended by the market participants. Some will be winners, others losers, and there will be many who are disappointed at least in part. And while the market coordinates activities, balancing supplies and demands, no one has specifically acted with the intent to bring about such coordination. It comes about as an unintended consequence. Sometimes, in fact, the market fails, and rather than coordination, it brings about a breakdown – depression or inflation. To understand this requires putting all the pictures together. In a sense, the final objective of fieldwork in economics is to give us a practical picture of the working of the market.

Yet this raises a question about the market, as the institution that holds the society together. For while the market acts to coordinate activities, on the face of it, this is not a function for which it is well suited. *Pace* neo-Classical theory, in actual practice competition is notoriously wasteful – activities and facilities are duplicated, time and effort are put into defeating competitors and misleading consumers rather than producing useful goods and services, and so on. Competitive strategy often dictates concealing information, when social well-being would be furthered by its dissemination. Advertising and sales campaigns spread false and misleading ideas. Market systems are frequently unstable, generating booms and busts. These matters are swept under the rug by the assumptions of neo-Classical theory – which are required for reasons we will explain shortly – but, in fact, are an essential part of the actual working of markets.

Surely, then, coordination would be better handled by a system of planning, administered by a well-organized bureaucracy? Bureaucracy has a bad name today, but Weber, for example, considered it the expression of rationality in human affairs (Weber, 1949). Indeed, when effective coordination becomes urgent, as in wartime, governments all over the world typically suspend the market and set up administrative systems. Why, then, do we normally rely on market coordination? Or is there a more important function of the market that justifies it, in spite of its wastefulness? We will return to this point.

Fieldwork and behavior
    The second kind of fieldwork concerns motivation, attitudes, preferences, and other subjective influences on behavior, *given* the context – laws, customs, technology, etc. It is an exploration and mapping of the chief features of the states of mind of the agents, picturing such states as they are likely to affect behavior. It is not, however, personal biography; the issues concern the subjective influences on economic behavior, *typical* economic behavior. Personal histories may well be illuminating, but they

are relevant only insofar as they shed light on economic decisions and actions.

These studies can be complicated by the fact that people are not always truthful about their states of mind, and, worse, even if they try to be, they may fail because they are unaware of their own motivations or attitudes, or are subject to self-deception. (To take examples in regard to economic questions: where preferences reflect officially discouraged prejudices, for example, the true preferences may not be acknowledged. Also, people frequently understate the extent to which they are motivated by money, and often hold false beliefs about their own and other's wealth, sometimes stubbornly clinging to expectations they know will never be fulfilled.)

To map the actual states of mind of agents is to study people, who are social products and have been prepared for certain roles, acting in the roles which they have assumed or to which they have been appointed (which may or may not be the ones for which they were prepared). There will be mixed loyalties, conflicts, uncertainties, and very often contradictory and unreliable reports will have to be reconciled.

What such a mapping will show is how agents see the world, how they value its various aspects, and how they plan strategy and tactics in regard to economic activities. In particular, it will show their understanding and motivation in regard to the market.

Modeling behavior and structure

These two aspects of the economy, roughly its structure and the typical motivations and behavior of its agents, give rise to two lines of analysis. The first will show the linkages and connections between economic institutions, making it possible to calculate various relationships. The second will examine motivation and strategy in various contexts, showing how these can explain behavior. There is an obvious sense in which each needs the other as a complement: structure without behavior is lifeless, behavior without structure has neither basis nor focus. This will be argued later; first we must examine the models of contemporary economics.

Model building

A "model" can be said to have two aspects, or to be composed of two kinds of elements. On the one hand, there is the purely formal part, and on the other, there is the interpretation which clothes the formal skeleton with meaning. The formal part of a model consists of an algorithm in some formal calculus. Two algorithms commonly used in economics are, first, maximizing a function of many variables in the differential calculus (usually subject to some constraints), and secondly, determining the existence of a solution to a set of linear equations, namely that the rank of the augmented matrix equal the rank of the coefficient matrix. Each of these formal models is purely abstract and must be given an interpretation; it must be applied to a subject matter. This requires making its variables and relations represent certain concepts. Thus one variable will stand for "price," another for "quantity demanded," and so on. By this route the maximization algorithm, for example, can be made the basis of the model of demand theory, in one interpretation, and supply theory in another. The formal side of the model thus provides the method for the determination of the unknowns in terms of the given conditions, while the substantive interpretation applies this method to the problem at hand.[4]

When behavior is the object of study the existence, the characteristics and the positions of those whose behavior it is must be taken as given. It is here, in connecting behavioral functions to agents as they are assumed to exist, that the "subjectivity" of the approach lies. It has rightly been pointed out, in answer to the charge of subjectivity, that the variables of behavioral models in economics refer to publicly observable acts. Hardly anything could be more objective. The theory of demand has choices and market prices as its variables; both are observable, open and publicly verifiable; the theory of supply refers to inputs, prices, and outputs, all likewise public and observable.

Nevertheless this misses the point, which is that none of these is observable except in connection with some *actual* agent; but actual or observable agents are rarely similar to the ideal types postulated by the model. The real significance of "subjectivity" lies here, in the fact that acts, however public they may be, are always someone's acts, that is, they belong to a subject. The identity of the action – what exactly was done on a given occasion – depends on the intention of the agent (Anscombe, 1958; Wiggins, 1980). A theory of behavior must therefore always predicate its behavioral functions of some agents or kind of agents. For acts done in the real world to correspond to the actions of theory, the agents of the real world must correspond in all essentials to the agents postulated by theory. But behavioral theory tends to concentrate attention on the way agents with assumed knowledge, abilities, and desires make decisions and affect one another's actions, neglecting the question of what these agents are and how they and their characteristics are brought into being and maintained.

---

[4] Many other mathematical algorithms can be, and are, used, especially in dynamic theory. But maximizing has a special place in neo-Classical economics.

Stimulus-response models

Neo-Classical models analyze behavior in specific ways. Instead of drawing on fieldwork, to define motivation and set the problems of choice in well-described institutional context, agents are considered abstractly and presumed to be rational and to choose freely. This, then, leads to models which exhibit a particular kind of market behavior, which we can call a "stimulus-response" pattern.

These models are strongly behavioral, paying little attention to structure. The context of action is abstract; the questions concern what an agent, usually a "household" or a "firm," would normally do, acting under the influence of an assumed motivation, and calculating rationally, when presented with various stimuli. It is assumed that the actions in response to stimuli are successful – a harmless assumption, when it is households making purchases, but question-begging, when it is investors introducing a new technology. Given the behavioral assumptions, reaction patterns to such hypothetical stimuli are constructed, and from these sets market functions are aggregated. Equilibrium market positions are then determined by solving the market equations on the hypothesis that behavior will be adjusted as stimuli move, until the markets are cleared.

Such a procedure takes a kind of agent and a social context as given and assumes them to persist. But the structural question has been evaded: what conditions must be met for the social context (and, a fortiori, for any behavior which depends on this context) to continue to exist? Further, what limitations does maintaining the structure put on the possibilities of behavior? The answers to these questions are independent of all specific behavioral assumptions, since they are questions about what institutions require in order to continue to operate, and not about how persons are likely or inclined to act in various social contexts.

Because the neo-Classical approach is behavioral in the stimulus-response sense there is a necessary connection between the patterns of behavior and the characteristics assumed to hold of the agents. These patterns consist of a series of choices made under the influence of the desire to maximize private gain (utility or profit) in response to a variety of hypothetical economic situations. Economic behavior is assumed to take the form of rational responses to clearly perceived stimuli, calculated on the basis of given preferences, technology, and endowments. The behavior called forth by the stimuli will always be to make a selection of the choice variables – subject to *ceteris paribus*.

But even given the narrow concept of rationality, this is not acceptable. The maximizing procedures that determine the best selection of the choice variables *also* set valuations on the constraints, indicating which would make the most difference if shifted.[5] Why should agents not redefine the problem and attack the "givens"? Why not invent new methods of production, new products, or new strategies? Why don't they *learn* that the same response will leave them no better off than last time, so if they really want to do better they will have to try something new? For that matter, why not redefine the stimuli, or reorder their preferences? If the agents are rational, they have free will. They can choose to reexamine their situation.

By saying that a pattern of behavior has the stimulus-response form we mean nothing more than that when an agent perceives an external stimulus he or she performs some particular and predetermined act in response to it. The act is making a choice from a predefined response set. The optimizing model is applied in one way only, to determine this choice – even though it implies other possibilities.[6]

Moreover, the complexity of action is overlooked, as are the possibilities for learning and innovation implied by such complexity. The actions in response to stimuli are presumed to be always successful; there is no sequence of efforts, building to a climax, a point at which success or failure will be determined – will we make the sale? Will the production run be completed on time? Will the new technology work? What can we learn from the problems encountered along the way? Any action that makes a difference to others or has an effect on the material world can (necessarily!) be done either well or badly, and thus can be

[5] Consider, for example, what is implied by the "Theorem of the Alternative" in linear programming. A verbal statement runs: "If a commodity is produced in the optimal solution of the primal, then the value of the resources needed to produce a unit amount cannot exceed its unit price; and if a positive valuation is assigned to a resource in the dual optimal solution, then that resource must be used fully." Since the solution of the primal equals the solution of the dual, taking the derivative of the solution with respect to a constraint tells us how much the solution would be improved by shifting the constraint; this derivative is the shadow price. The solution to the *choice of output* problem implies values for binding constraints, and these values tell us how worthwhile it would be to *shift* these givens.

[6] An illustration may help. A firm can produce two goods, and is subject to two constraints, as shown in the figure. Its profit function is indicated by dotted lines; the feasible region is shaded. Initially it produces only one good, at point A. It hires a team of management consultants, who tell it that point B is the optimum. They also observe that to reach point B, truck production must be reduced and car production brought to a substantial level from scratch. Both require shutting down the plant and reorganizing. To shift constraint I to position I′, however, would require some rebuilding and might lead to unexpected troubles, but the cost might not be much greater than that of reorganizing. It would then be easy to reach A′, a point only a little less profitable than B. The firm could then move to C, which is more profitable than B, at its convenience.

(continued)

successfully completed, or can fail.[7] If done badly, doing it better can be learned. And the success can either be confidently expected, or regarded as uncertain – a point which will prove to be important later. The advantage of the stimulus-response pattern lies in its easy translation into the calculus, since the stimuli can be treated as the arguments of a function and the response as its value for a selected argument. But there is no attempt to learn, to develop a *new* response, or to define and solve a problem. The advantage is bought at the price of oversimplifying action and restricting the activity of the mind. And this implies a number of other, related, rigidities, some more important than others.

For one thing it is normal to assume that both stimuli and responses may occur in any order. To price p_1 there will correspond demand q_1 regardless of whether the preceding price was higher (and quantity lower) or lower (quantity higher). There is no difficulty in *moving* from any one pattern to any other; neither the direction nor the magnitude of movement matters. If time enters into the analysis at all, it enters as a separate variable to which there corresponds a definite pattern of response, or which leads to a shift in the pattern of responses to other variables. In addition, the time needed to bring forth and complete a particular response to a particular stimulus is normally defined as part of that response, and is both independent of what went before or will come after, and of any other variables in the system (unless some specific dependence is explicitly shown, of course). Finally, the behavioral responses of other

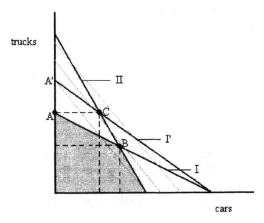

---

[7] If it has an effect on others or on the world, it either has the intended effect or another effect. If it is not the intended effect, the action has failed to achieve its objective. It is a consequence of nondeterminism that not everything we intend is what we actually accomplish. Free will implies, as well, that we can always change our minds.

agents are assumed to be either independent of the behavior of any given agent, or to function as stimuli to which he responds in some definite way.

A second problem arises from the fact that it is customary in neo-Classical writing to represent the stimulus-response patterns by functions of real variables, implying that the behavioral patterns are capable of continuous variation. (It is admitted, of course, that this is unrealistic).

A third and widely discussed problem concerns the characteristically subjective aspects of behavior: perceptions, knowledge, motivation, learning, and the like. In actual people and even more in actual institutional (and therefore political) situations, these are highly complex, and more often than not, fraught with contradcitions. Information is ambiguous, perceptions are uncertain, and expectations volatile, while motivations (and the loyalty of subordinates) may be undercut by rivalry or self-destructive impulses, and the whole may be undergoing continuous transformation through learning. Under such conditions no equilibrium patterns or behavior could possibly be defined. So assumptions are made to rule all of this out of court, with the result that the model takes on an artificial and unrealistic air. The "behavior" it purportedly describes is often not recognizable.

One reason is that this approach tries to incorporate structural features into the attitudes, beliefs, and so on, of agents, rather than presenting them directly. Institutions are thus treated simply as ways that people prefer to interact. This misses an important point about obligation, promises, and the rule of law. For the stimulus-response/rational choice approach has tacitly assumed a world where there is no conflict – no conflict of duties, no conflict between agents, no self-deception.

A simple example will illustrate the point. Two agents meet, each with goods that the other desires. For conventional economics, this is the starting point for the theory of exchange. However, this depends on a hidden assumption, namely, that each party holds property rights over the respective bundles of goods, recognized by law, and sanctioned by police with enforcement powers. Otherwise the stronger could simply appropriate the goods of the weaker, rather than paying for them. Indeed, the stronger could force the weaker to work for him, as Robinson Crusoe did to Friday. If it is not clear who is the stronger there could well be a war. This is not so fanciful; there are plenty of examples all through history. For exchange to take place, rather than a battle, requires the existence of a recognized authority.[8]

---

[8] Even if the two (or more) agents are of equal prowess, if the one who strikes first has an advantage, then attacking will be the best strategy. Hence there will be a Prisoner's Dilemma: for both to exchange would be better for both, but both will nevertheless choose to fight, in the absence of an authority (Schelling, 1960).

In short, to analyze behavior, it is necessary to characterize the context in which it takes place.

Models of structure

The aim of a structural model is to examine the very features of social life that the behavioral models have taken for granted, and to determine under precisely what conditions, and with what consequences, it is appropriate to take these matters for granted.

Instead of describing the behavior of agents, a structural model shows the rules governing behavior, the methods and procedures of production, the legal and property relationships. These may describe systems such as capitalism (Sraffa, 1960; Bharadwaj and Schefold, 1992) or feudalism (Nell, 1968; 1992, chs. 12, 13); or they may describe a particular institution, as in a flow chart showing the working of a production process, or an organizational table for a firm. What is shown is not tangible, perceptible, or "objective" in the same way. But this does not imply that these rules are any the less real or objective. The intangibility of structure does not imply its subjectivity. The duties of, for example, the president of the United States are not a matter of subjective preference. A proposition stating them is not a "value judgment" or an "expression of feeling." It is a proposition stating a fact, although one of a different kind than those the natural sciences examine.

Structural models represent intangible, immaterial relationships. They show rules and formulae; methods of production, like the entries in cookbooks, are recipes, the rules of distribution are just that – rules. Organizational tables show the structure of a company, its accounts show its balance of profit and loss. Structural models show relations between agents, and between agents and the world, but what they give us is a blueprint, an outline, a pattern which has to be instantiated. And this may be done well or badly, rightly or wrongly.

There is an important difference in focus here compared to neo-Classical thinking. Both are concerned with intangibles, but the latter's concern is with states of mind that are properly ascribed to individuals, whereas structural models relate to features of institutions. As we saw earlier, this calls for a focus on roles, duties, and norms rather than preferences, wants, and desires.

A structural model of production and distribution – a model setting out the framework of the economy – determines actions that must be taken to maintain or reproduce the system, that is, given the initial position and how it is changed by the processes of the economy, it shows what must be done to restore the initial position, or perhaps some expanded or altered version of it. The "equilibrium condition" is therefore that this position should be restored in the original or altered form. Since the model does not ascribe actions to agents, there is no attempt to predict what will actually happen; the model merely shows what has to be done if reproduction is to take place. There is no balancing of contending forces here; the calculation shows the necessary conditions for keeping the economy going, given the maxims and rules governing and defining the courses of action available in the system.

Lowe (1965, pp. 37–40, 265–75) argues that since a structural determination of the conditions for reproduction – what he calls a *homeostatic model* – cannot be described without reference to goals, it must be considered *prescriptive*, "instrumental" in his terms. It has to be understood as determining the conditions for the realization of the goals of restoring the initial (or expanded) conditions consistently with a certain distribution of the surplus. He concludes that such models are neither descriptive nor predictive. The point is correct, but the conclusion does not follow. Such models show what *ought* to happen, in given conditions, in the sense of how things could be expected to work out if everyone did their jobs properly, carried out their duties, and acted normally, given the circumstances. This *is* descriptive, although in a sense which must be understood carefully. Such a model describes and analyzes actions in conformity with the norms of the system; it presents the coordinated patterns of action that correspond to the norms. A different form of description would focus on the actions that fit the historical record. That would be the province of a *predictive* model.

Certainly Lowe is correct to consider the structural model to be "instrumental" in the sense that it works out the conditions necessary to realize the goals of the system – but they *are* the goals inherent in the economy. So if the economy is going to keep running, it has to (more or less) achieve these goals. A *prescriptive* analysis would show how circumstances could be changed so that the patterns of action could be coordinated better, or how the goals could be achieved more cheaply or to a higher degree. Structural models – and some other models, too – have a normative element, but they are descriptive, all the same because the norms *are* the norms of the system under analysis. This might be called *normative description.*

A structural analysis thus considers the logical relations holding between given institutions, through examining the extent and nature of their mutual dependence. The interdependence may be revealed in the market place, but it is based on the structure of the possibilities of action. Before one can choose to buy or sell anything the goods must be brought into existence, and since doing this entails using up other goods, which, in turn, are produced using still other goods, linkages and dependencies are established which are independent of what happens in the market place on any particular occasion. Such structural relationships are, in the first place, technological: Are the means of production sufficient? Is there enough to go around, and if so, what exchanges will allocate it? Secondly, structural relationships involve legal matters: exchanges are contracts, raising questions of equity. Ownership of means of production gives rise to ownership of the product. If there is a surplus, who has the right to claim how much of it? It is of no importance to such an inquiry whether the market will actually settle on the prices determined by such considerations, though of course, if such prices are not realized there may be structural repercussions. The important point is that subjective considerations are directly relevant only to a market analysis, never to a structural one.

A comparison of neo-Classical behavioral and Classical structural models

Neo-Classical analysis not only emphasizes behavior, it takes a rigorous stimulus-response approach. Classical theory, by contrast, tends to analyze structure, adding along the way some, often ad hoc, behavioral assumptions, which may or may not involve optimizing. Each approach has strengths and weaknesses.

First, the stimulus-response approach describes how an agent acts under various assumed conditions. For example, in a behavioral model, a Marshallian utility function is a predicate which might or might not describe a particular consumer. A consumer cannot acquire such a function, either through purchase or by inheritance. It is "his" or "hers" if and only if it correctly *describes* their preferences. By contrast, an agent could purchase the design or otherwise adopt a method of production such as is represented in input-output studies. Unlike behavioral models, structural models are predicated on social relationships into which agents can enter, such as positions of ownership, or jobs. The terms of a behavioral model describe the behavior or will of agents, given their social positions, whereas the terms of a structural model are possible objects of an agent's will.

Secondly, to employ a stimulus-response approach in concrete circumstances requires listing the particular agents to whom the behavioral functions are to apply. The prediction of behavior is necessarily the prediction of someone's behavior, hence the agents must be specified even if only in the most abstract terms. By contrast a structural model need never specify particular agents, times, or places. It makes reference only to the rules, methods, and procedures employed by actual institutions; there is, indeed, a sense in which it is they who refer to the model and not the other way around. The model is, so to speak, inherent in the institutions; it shows what they are trying to achieve.

Thirdly, the consequences of falsity are very different for the two kinds of models. If it is shown that in a given context a behavioral model (known to be applicable to that context) is false, that is, gives wrong predictions, then all that follows is that the postulated behavior pattern is not true of the agent or agents in question. No doubt is thereby cast upon their existence or continued existence. But if it is shown that a structural model is false relative to the institutions which it supposedly represents, for example, that their continuous existence can be achieved only with other values of the unknowns than those calculated, or perhaps through reliance on a different set of factors, then it is implied that the actually existing institutions were not correctly represented by the model, or, to put it another way, the model in fact shows a different set of institutions from those actually present. The applicability of a behavioral model must be decided prior to and independently of the determination of the truth or falsity of the predication of the behavior patterns. But in the case of structural models the question of applicability and the question of truth are one and the same.

Fourthly, a behavioral model implies a definite set of actions which will occur if the model is truly predictive of the agents. (The set of actions need not, of course, be unique; multiple equilibria are possible.) A structural model, however, implies a definite set of results which must be achieved if the social context is to be preserved; but the set of actions which will bring these results about under particular circumstances is not specified. Hence, there is always room for a behavioral model to decide this further question in the context of any structural model.

Structural models must be "stylized," but don't carry *ceteris paribus* clauses. The only question is whether the structure has been fully identified – in all *essentials* – and modeled accurately, given the agreed upon degree of "stylization." Stylizing covers two different adjustments. First, the level of analysis must be settled. This means aggregating groups of industries and groups of payments (for example, wages and salaries, corporate and unincorporated business earnings), averaging different methods of production for the same goods, abstracting from failed or failing operations (that is, industries or processes that are being phased out), among other things. Second, the data must be adjusted in the light of the purposes of the study, which calls for smoothing over irregularities, filling in and correcting incomplete data, and so on. If the model is accurate, then the calculations *must* be true. By contrast, the implications of behavioral models are always subject to an indefinite list of *ceteris paribus* clauses.

"Maximizing" poses a central issue. Optimizing behavior is central to the neo-Classical project, in which it functions *descriptively*. But it could be treated in a different way, "instrumentally," rather than in terms of stimulus and response (Lowe, 1965; Hollis and Nell, 1975). Instead of being a description of what rational agents *will* do, optimizing, suitably embedded in a structural context (e.g., a classical model of production), could be taken as a representation of what they *should* do. The traditional – stimulus-response – approach postulates motivation and tries to predict what will happen, on the assumptions that agents behave "rationally," confine their actions to the set of choice variables, and that all actions undertaken automatically succeed. The suggested alternative postulates a goal, presumably implied by the structure (institutions, etc.), and then inquires what would have to be done to achieve that goal. The failure of a maximizing model to accord with experience, when interpreted this way, leads to criticism, not of the model, but of the agents. The model shows that they could have done better. On this interpretation the model functions as a *tool* for the agent. Rationality could be instrumental rather than descriptive. This point will come up again.

Finally, the meaning of the equilibrium conditions in the two cases is altogether different. In a behavioral model the equilibrium condition implies the balancing of two kinds of forces impinging on an agent's decisions to act, a balancing of desire and effort, or of benefits with costs. In any case, the basis of the equilibrium is to be found in the motivations of the agents of the system. By contrast, the equilibrium condition of a structural model makes no reference at all to agents or to their motives, and consists purely and simply in the condition that a certain pattern of material allocation be restored, enabling the system to reproduce itself. The equilibrium condition, in short, is that existence be preserved. In the one case, therefore, the equilibrium condition determines the existence of the system; in the other case it presupposes it.

So what a behavioral model predicts about its subject does not explain that subject's continued existence, but on the contrary presupposes it: whereas what a structural model tells us about its subject is precisely what its continued existence depends on. The agents of a behavioral model are assumed to be self-perpetuating; the question of how they continue to exist is never raised. Yet this is exactly the question which a structural model is designed to answer – and it cannot be assumed away, for the conditions of continued existence set limits on behavior.

It follows that the behavioral concept of equilibrium is deficient in an important way. For it is possible to conceive of a case where desires and efforts are in a neat balance in each market (taken individually and assuming other things to be equal), yet the "equilibrium" prices might not be such as to permit reproduction (Koopmans, 1955; Rizvi, 1990). Nowhere in the neo-Classical system are there any equations which show precisely what is required to support the institutions which provide the context for the agents whose behavior is being determined. Nowhere are the values, customs, laws, or normal rules explained. The context of behavior is assumed, but not described – and the fact that that context must be supported and maintained, at a significant material cost, is nowhere acknowledged.

So far structural models may seem to do rather well in the comparison. But they have one grave deficiency – they tell us very little (Hahn, 1988). In a sense they are nothing but accounting – given certain information, properties of the system can be calculated. These properties tell us the conditions for success in some activities and give us the limits of others. But nothing in all this tells us what is likely to happen, or why.

The dependence of behavior on structure

In order to act agents must have the power to do so; they must be in an appropriate position, possessing the authority as well as the skills. This means they must occupy a place in the structure that is suitable or propitious for the kinds of actions they propose. A firm must possess the rights to produce a product, must own capital, must have the right to hire labor and to borrow money; a household must be able to enter into contracts, to handle money, to sell labor, to buy goods. These all require various abilities, which have to be learned, but in addition, they require that the agents be *authorized* to perform these actions. If they are not, then the actions, contracts, agreements, etc., are void; they have no standing and cannot be enforced.

Agents acquire standing in two ways. Most fundamentally, it comes as a birthright, which has two aspects – being born into a family, and as a citizen of a nation-state. These provide the basis for an agent's general ability to enter into contracts and transactions. Further and more specific standing comes as a consequence of achievement or various actions leading to qualifications, which state that one can perform various tasks. Such qualifications then lead to appointments that confer specific powers and privileges and require specific duties.

Agents must have appropriate standing to perform economic activities. An agent unable to contract cannot take part in the market; an agent cannot sell what he does not own, cannot borrow if she has no collateral, cannot lend if she has no assets. Agents, in short, cannot be taken as abstract; they must be agents with standing, and their standing comes as a result of holding places in a social structure, the continued existence of which must be ensured by their activities.

Neo-Classical rational agents are not properly related to structure; indeed, neo-Classical models generally lack an adequate account of structure. Since neo-Classical agents are abstract, they also lack "standing," or rather, their standing is left undefined. This poses a problem. In any society-wide market model there are two sets of agents – households and firms. But in society as a whole there is only *one* set of persons, each of whom is supposedly both a member of a household and an employee or owner of (supplier of savings to) a firm.[9] To whom are preferences to be attributed – to persons, or to households and firms? If to persons, then a business decision based on preferences could be affected by household related matters, and vice versa. Moreover, in constructing their preference scales persons face the diffcult matter of resolving the often, and in some areas perhaps necessarily, conflicting claims of the two. Since people grow, learn, and age, such resolutions, when achieved, cannot reasonably by supposed to hold once-and-for-all. Hence preferences may be liable to sudden shifts. But if preferences are attributed to households and firms – to *institutions* – satisfying them cannot be equated with enhancing individual welfare, nor can providing opportunities to satisfy them be judged to advance individual freedom. Even more problematically, a prediction derived from models based on preferences attributed to households and firms, rather than persons, would be descriptively normative, that is, its validity would be based on the presumption that agents will act as they are supposed to, that is, they will carry out their duties.[10]

In addition, these "agents" can also each be identified with a commodity. Two of the "commodities" for which there are supply and demand functions, showing offers to buy and sell at various prices, are labor and capital. The support of working-class households at least is obviously the same thing as the support of labor, and the maintenance of a firm is clearly impossible unless its capital is kept intact, which means keeping intact not only its plant and equipment, but also the circulating fund out of which it pays wages, salaries and other running costs. Failure

[9] Some persons may belong to households, but own nothing, and be unable to find a job. By leaving the question of standing undefined, neo-Classical theory in effect assumes that all agents have the *same* standing. A fieldwork approach will show that this is not true: in virtually any society there will be citizens and non-citizens, employed and unemployed, property owners and non-owners. There may be other distinctions as well, for example, to take a famous case: free and unfree.

[10] It might be argued that "firms" and "households" refer to nothing really existing. They are just ways of describing individuals *acting in certain capacities*. But where do those capacities come from? How are they supported and maintained? How are they passed along from individual to individual, as one retires or dies and is replaced? Moreover, what is the exact relation of the person to the "capacity" or role? Does the person have to do what the firm wants, or are the firms' goals simply constructs out of those of the employees and suppliers of savings? "Employee" also appears to designate a role or capacity, for which one must be trained and to which one must be appointed.

to keep the firm's capital intact means a loss of its competitive position and ultimately results in bankruptcy. Clearly the support of capital and labor need to be expressed in structural terms, drawing on information gathered in fieldwork.

No actual agents can therefore be fitted into the neo-Classical mold. Nevertheless, neo-Classical models have a great hold over modern economic thinking, for, as we shall see, abstract rational calculation has great *prescriptive* power – and this may be the secret of neo-Classical economics: it has appropriated the prescriptive power of rationality to "describe" the (alleged) allocational efficiency of free markets.

Types of models

At this point an important distinction must be made between *three* kinds of economic models (Hollis and Nell, 1975, ch. 5). The most basic models are structural and analyze reproduction, the way the system can maintain itself or expand. These must be based on a conceptual analysis of the institutions of the economic system as determined in fieldwork. Input-output models, for example, show the basic exchanges that have to be made to keep the system running; they exhibit the possibilities of investment, consumption, export, etc., and they show the structure of interdependence, the wage-profit trade-off, and so on. The analysis can be formal and abstract; institutions can be simplified, but the essential features cannot be distorted.

But such models say rather little about behavior and tell us almost nothing about what is likely to happen in various markets. A second type of model, therefore, can be designed to explain and predict behavior. But it cannot be based on the stimulus-response framework, for the reasons just presented. This second type of model must situate behavior in structure and must draw on the results of fieldwork to make assumptions about motivation, rule-governed behavior, and decision procedures, including optimizing. Economic agents may plausibly be assumed to pursue their self-interest, although not to the exclusion of all other motives. Such models will certainly assume rationality – humans are rational animals – but the concept of the rationality will be broad, and will include propensities to learn and to innovate. Agents have active minds. They draw up plans, and then execute them. Planning does not *entail* implementation; commitment does not entail fulfillment. Implementation occurs in stages, and actions reach climaxes at which they may succeed or fail. At any point actions can be reexamined; commitments can be revised – and at any point things can go awry, simply because human action can always fail. An irreducible residue of uncertainty resides here.

As noted, such models will have to draw on simplified results from structural analysis, to present the context in which behavior takes place. Agents must be properly situated, and have access to the means to act. They can be assumed to pursue their self-interest in markets. But self-interested behavior does *not* mean generalized "rational choice." Agents pursuing their self-interest must be considered in their actual circumstances, facing the options that exist for them. To add to those options something *abstractly* possible, but not *actually* a present option, is to introduce an irrelevancy in a predictive model. Such matters should be studied in programming models. Moreover, the "self" in question is the product of a family/kinship/educational system, now acting in a role in a production organization, and also holding a position (breadwinner?) in a family household, which will produce the next generation of agents. The actual motivation of an actor will develop out of the interaction of these various components of the self.

The choice set facing an agent is composed of those options between which the agent *should* choose, given the responsibilities of his or her position. That is, these are the choices they are supposed to make, are empowered to make, and have the skills and information to decide on. The agent can be expected to do a good job choosing among these options; to consider others might be beyond the agent's capabilities or powers. We can assume an agent will do the best job possible, that is, will maximize among routine options, when doing so is implied by his responsibilities, meaning it is normal and expected of him, and not to do so would waste resources needed for other activities. This can be described as "self-interested behavior," or "role-based maximizing," and is a proper foundation for models of economic behavior. (A powerful solvent of hypocrisy, too. We are rightly skeptical when people claim to have no economic motivation: as Deep Throat said, "follow the money.")

But such routine maximizing is not "rational choice." There is a division of labor issue here: the agent is an *actor*, that is, the agent makes choices and *carries out the resulting actions*. Actors are trained to act; households not only buy, they consume. Firms choose factors and inputs; but their main activity is producing. Management consultants and interior designers, on the other hand, are specialists in *choosing and evaluating*. They have studied the opportunities, know the possibilities and circumstances in detail, and have mastered the methods – and pitfalls – of making optimal choices under various kinds of constraints. Their *job* is to examine all the options, assess the constraints, rethink the possibilities, and then optimize. Or at any rate, lay out and rank the best courses of action, spelling out the likely implications of each. By contrast, the actor's job is to make *routine* choices and get on with the program.[11]

Once agents are properly understood, and placed in their appropriate circumstances, important questions can be addressed. What will happen in the business cycle? Will inflation intensify, will unemployment rise or fall? Such models can also be developed for particular markets or sectors. The actual practice of the economy in question must be known if it is to be modeled accurately. Fieldwork is therefore essential to success, and the lack of attention to systematic fieldwork by economists may help to account for the generally poor record in predictive econometrics.

Finally, there is a third kind of model, quite different from the others, with a different conceptual foundation. These were called "programming" models in Hollis and Nell, and termed "instrumental" by Lowe. They are not predictive; rather they determine what the best course of action would be for given agents in given circumstances, and moreover, they allow for a reconsideration of the "givens," since they make it possible to determine how much would be gained by shifting the constraints. Such models provide the natural and proper home for the narrow "means-ends" concept of rationality. If the agents do not do what the model calls for, it is the agents who are to be criticized, not the model, assuming that the model correctly represents the circumstances, and the goals and motivation of the agents (including the possibility that the agents might innovate or otherwise change their circumstances or goals, in line with the "shadow prices" determined by the model). The conclusions of a rational choice model have an extraordinary power. They represent what *ought* to be done in the given conditions, not what should be done morally, but rationally. The model tells us the right, proper, sensible, best thing to do in the circumstances. Agents in the given conditions who do not act in accordance with the model may be considered foolish.

---

[11] Strictly speaking, when maximizing requires specialized skills and absorbs resources, to make choices wholly and only in accordance with rational optimizing leads to paradox. For if optimizing requires resources, and will be done better as more resources are devoted to it, then the optimal amount of resources to be devoted to optimizing must be decided. This requires balancing costs against expected gains – the gain from optimizing as opposed to using a rule of thumb compared to the costs of obtaining the information and carrying out the calculations. But to know the gains accurately requires *solving* the original optimizing problem! Yet spending the time solving it might well turn out not to be optimal. Moreover, there is an infinite regress. How much effort should be put into deciding how much effort should go into optimizing? This is itself an optimizing problem, as is the problem of deciding how much effort to devote to solving it, and so on, ad infinitum.

These models embody the optimizing interpretation of rationality, and thus generate the concepts of scarcity and opportunity costs. But they are *prescriptive* rather than descriptive. Such models are *tools* for the active mind. And, like any tools, they can be used in more than one way. The narrow, calculating concept of rationality may determine the best choice from a given set, but it is also a means by which an agent may analyze how to improve on the "givens" in the situation. Moreover, calculating rationality does not necessarily characterize agents or their behavior; the actions which follow from a model of rational choice will only be *descriptive* if and when the agents *accept* the results prescriptively.

"Rationality" is not a dispositional predicate, like "nervous" or "stolid." Behavior is rational in the required narrow sense only if it embodies or rests on an appropriate relation between means and ends. But this cannot be adequately judged only from "outside"; agents may rethink their goals, may reorder their ideas of short run and long, may wish simply to try something new on the chance that it might work better, and so on. For behavior to be judged "rational" – carrying out the rational choice model – the model's relationship between means and ends must be that intended by the agent. The agent must accept the logic of the model, for if the agent rejects the model for good reasons (or even for bad ones, so long as they are *reasons*) it will not be descriptive or predictive.

There are thus deep-rooted problems in the assumption that rational choice will govern behavior according to the stimulus-response approach.[12] Agents will only "behave rationally," that is, act in accordance with the dictates of a rational calculation, if they accept that calculation, which means that they must agree that the problem posed is the one they, in fact, face, and that it is posed in a manner that will yield the results of most use to them. The choice must be within their powers – or they must have commissioned the study! They must accept the choice variables and the constraints and agree to decline to try to shift the constraints. Otherwise, they are entitled to dismiss the calculation as irrelevant. Hence a rational choice model must be based on realistic assumptions. To be descriptive, maximizing models must be closely tied to the roles and circumstances of agents.

[12] It is sometimes argued that it is reasonable to assume that all agents are rational, since rational agents would win out in competition over nonrational ones. Enough is known about this sort of dynamics that we can dismiss such an argument out of hand (Arthur, 1990). (For example, the nonrational agents could do well by accident at the outset and become "locked in," on analogy with inferior technologies; the costs and time involved in calculation could slow the decision-making of the rational agents, allowing nonrational followers of rules of thumb to move faster in the markets, etc.)

However, this implies that outcomes will be specific and sensitive to the choice of assumptions, as in linear programming and operations research. But the neo-Classical model seeks to use rationality as the foundation for making *universal* claims; as a result the models are highly abstract and decidedly unrealistic. It would be reasonable, therefore, for agents to reject such models and insist on developing calculations that are closely based on their immediate conditions.

This leads once again to the *prescriptive* power of maximizing, which now becomes a reason to reject descriptive and universal models based on maximizing. Given a model specific enough that agents would accept it, its conclusions might be used to show the agents how those conditions might be altered in their favor. In other words, the model might be used as a guide to innovation. Far from yielding universal results, then, rational calculation might give very particular answers, and some of these, at least, could become part of an effort to change the givens of the problem.[13] Maximizing models are not a good foundation for equilibrium behavior; they are just as likely to suggest changes and deviations from the norm!

On this basis some conclusions with respect to method can be provisionally sketched. The basic idea is to develop Structural Analyses, and then to consider the Behavioral Options (SA/BO). First, it is necessary to develop a picture of the basic structure of the system. That requires understanding the technologies in use, the organizations that use them and how they are controlled, and the way the human population is supported and enabled to reproduce. This means gathering the relevant information and devising structural models at various levels of abstraction. Reproduction models provide the foundations; they show what the system is, and how it works. They provide the blueprint, so to speak.

Once these are in place, behavioral questions can be considered – very little can be said about what will actually happen until behavior is

---

[13] A simple example of the kind mentioned earlier: consider finding the optimal mix of activities to achieve a given end to the highest degree, subject to various constraints, say, in linear programming form. Given the solution, agents might choose to use the calculation, not to determine their activity mix, but to evaluate the potential gains from shifting the various constraints, given the costs of doing so. That they might do this could be ruled out empirically in a given situation, but it cannot be ruled out a priori. A general equilibrium model based on constrained maximizing must be considered essentially incomplete. The model calculates the gain from shifting the constraints, thereby defining a demand for constraint-shifting. Why is there no supply? The implicit answer is that it is too costly, or no technology is yet known, but, of course, this is simply an invitation to innovate. Maximizing models, of course, will be appropriate descriptions of the behavior of agents in roles in which choosing the best mix or process is part of the job. But even here an operations research analysis might come up with a better approach or a new algorithm.

specified. Two general types can be considered. On the one hand, predictive models can be set up, for the system as a whole, or for various subsectors, down to individual agents. These models must be based on well-grounded assumptions about the circumstances and motivations of the actual agents. This depends on a good account of the technology, rules, and institutions, including the situation and motivations of agents, which must come from fieldwork. Agents must be in a *position* to act, which means they must occupy an appropriate place in the structure. Given the results of fieldwork, behavioral patterns can be developed and the course of the economy through time can be projected.

On the other hand, rather than descriptive, *prescriptive* models can be devised to consider the ways the actual performance of the system can be improved, from various (possibly conflicting) points of view. Again, these can be developed for the system as a whole, or for subsectors, down to individual agents. If aspects of economic behavior could be improved or better results obtained, programming models will indicate where and how this might happen. Programming models show possibilities for innovation and learning. They may also indicate, for particular agents, how competitive strategies could be improved. In turn such changes may affect the basic reproduction/expansion conditions, leading, quite possibly, to changes in rules and institutions, which would have to be ascertained by more fieldwork.

All three levels, conceptual analysis, fieldwork, and model-building, interact; each can help to extend and develop the others. No single criterion governs all. Each draws on precepts and practical maxims peculiar to itself, but each provides assistance to the others, and in some measure each is necessary to the others.

The nature and function of the market

At this point we can return to a question posed earlier: if the market, in actual fact, is an inferior coordinating mechanism, why has it been assigned that function? If it is not a coordinating system, then what is it supposed to do?

The view that the market efficiently coordinates is based on the stimulus-response behavioral approach. Scarcity underlies optimizing; faced with scarcity, agents must make the best of what they have. Competitive conditions are those that allow each agent full flexibility; resources are mobile, information prevalent, and no institutional barriers exist. Such conditions are among the givens and are assumed not to be affected by what agents do. Hence agents will be enabled to optimize fully, subject to the actions and strategies of others. Competition is therefore the setting for optimizing behavior, and it is supposed to lead in some sense to the best outcome for everyone.

This set of ideas – scarcity, competition, individual optimizing, equilibrium in supply and demand – may be considered the central core of the neo-Classical vision (Hausman, 1992; Rosenberg, 1992). Whether the analysis pertains to one market or many, or even all markets – partial or general equilibrium – is less fundamental. Early neo-Classical accounts focused on one or a few markets, and emphasized long-period equilibrium, following the Classics. Walras took the interaction of all markets as his subject, but his analysis rested on scarcity, competition, and optimizing. Like the others he focused chiefly on equilibrium; the *process* was not analyzed, because the outcome was what counted, and stability or gravitation was presumed. (The *tatonnement* is more metaphor than model; Edgeworth referred to Walras's "noisy and unconvincing dynamics.") More recent work has questioned this presumption, and the neo-Classical approach has found itself unable to demonstrate stability in general equilibrium. This makes it difficult to claim, even subject to various qualifications, that private optimizing leads to a social optimum.[14]

The structural approach is concerned with reproduction and the rules governing the distribution of the surplus. It does not analyze competition, but tends to adopt the Classical view which tends to see competition as a struggle over shares of the surplus. The picture is one of contained conflict, by no means exclusively class conflict. Sectors are also in contention, and each capital competes with other capitals, workers with workers. A competitive society is one in which conflict is endemic, but controlled and channeled. Moreover, the competitive struggle leads to an expansion of output. The struggle over the pie makes the pie larger! However, conflict has winners and losers; competition is the race toward the

---

[14] The Classics emphasized structure, and the neo-Classics have focused on behavior, assuming it to take the stimulus-response form. Contrary to a view expressed in recent discussions (Garegnani, 1976; Eatwell, 1983; Milgate, 1986) the marginalist revolution was not simply a change in theory while maintaining adherence to a common, long-period method. It was a change from Classical structural analysis to an altogether different kind of theory – neo-Classical behavioral theory, based on rational choice, where "rationality" was taken to mean following a preprogrammed optimizing algorithm, resulting in behavior that followed a strict stimulus-response pattern, representable by a single-valued function. But in the early phases it was still thought desirable to be able to obtain equilibrium results similar to those generated by Classical theory. Hence the neo-Classical theory of long-run equilibrium was developed, with its special assumptions designed to eliminate uncertainties, motivational conflicts, human failures, and subjective influences generally. So, as a result, stable neo-Classical equilibrium prices could be calculated with all the definiteness and certainty which attach to Classical "natural prices." But, as we have seen, the assumptions required for this distorted the neo-Classical concept of behavior.

(continued)

monopoly. It is a stage in the evolution of the market. Hence "competitive conditions" may prove to be temporary and unstable; they can be expected over time to give way to various forms of administered markets (Sylos-Labini, 1969; Eichner, 1976, 1986; Nell, 1992, ch. 17; see Chapter 10 below). In the competitive process many strategies will be employed, including innovation and cost-cutting.

The scarcity approach leads to the view that markets perform the social function of allocating scarce resources more or less optimally, according as they are more or less competitive. It also suggests that, whether they are competitive or not, markets are signaling devices; they coordinate activities. Prices are signals to which agents respond.[15] Supposedly, the price mechanism solves the problems of determining what? how much? and for whom? output will be produced, to quote the textbooks.

[14] (*Contd.*)

However, equilibrium, in the long-run or Classical sense, is not very important for a large class of neo-Classical analyses, which are concerned to predict the behavioral responses to market stimuli. It doesn't really matter whether the market signals, or the responses, are equilibrium ones or not. All that is necessary is that the signals – the stimuli – be clear and definite, and that the circumstances are propitious and the time sufficient, for the response to be forthcoming. Given the stimulus the response can be calculated; given a sequence of stimuli, a sequence of responses can be derived, with later market stimuli being the consequences of earlier responses. None of these need be equilibria, although they are optimally chosen. They might be market-clearing, or they might not be. So the shift to Temporary Equilibrium analysis, with its notion of intertemporal paths, and more recently the further shift to Disequilibrium Theory, can be seen as attempts to relax the assumptions, giving up the traditional notion of equilibrium in the process, so as to base the approach on a more plausible and more readily applicable concept of behavior. On the other hand, the Real Business Cycle approach has tried to adopt the idea of a sequence of optimizing responses to market signals, which are themselves the aggregation of an earlier set of such responses, while retaining the traditional concept of equilibrium. Yet in all these cases the earlier objections still apply: the notion of rationality is unacceptably burdened with tunnel vision and the behavior has not been properly related to its structural context.

[15] This view is technically defective. In a single market when demand slopes down and supply up, when the auctioneer calls the equilibrium price, both firms and households will respond with the correct quantity. But when constant or increasing returns prevail, the equilibrium price will *not* call forth the correct supply; firms will require a *quantity* signal to know what to supply, and even then the information is aggregate. It is still unclear what individual firms should do. The analogue in a general equilibrium setting comes when the production set is upper-semicontinuous, i.e., constant returns. Prices will not provide the information required to determine the equilibrium quantity, and even the aggregate quantity information will not tell individual firms how much to produce. In the general equilibrium case, however, the assumption of upper-semi-continuity is important mathematically (Takayama, 1985).

An obvious question arises: if the market is an efficient signaling mechanism, and allocates scarce resources to best advantage, as stimulus-response models indicate, why don't market relations hold between divisions of the firm, or between stages in a production process?[16]

The answer is that allocation is best carried out by an *organization* designed for that purpose. To put it another way: the stimulus-response approach is flawed, and the market is not an efficient signaling mechanism. To coordinate effectively, a bureaucracy is needed. A structural view, based on fieldwork, will suggest that the market, considered realistically, should never be thought to "allocate scarce resources efficiently." On the contrary, competition as a form of contained conflict is inherently wasteful. Each side seeks to *impose* costs on the other, at an acceptable cost to itself, to force the other side to cut a better deal. Such competitive conflict involves duplication of facilities, sabotage, industrial espionage, and waste in advertising, marketing, and packaging. It leads to destruction of productive assets when firms go under. Workplaces are left unsafe, and products may be dangerous, since firms will be motivated to cut corners in production whenever detection is difficult, or the dangers long-term and unlikely to be discovered before the firm can move on. Products and services are allocated to those who can afford them, not to those who need them. "Needs" will be created for products that will sell whether or not they do anything useful. Shortages will be met with price-gouging; overproduction in relation to effective demand may lead to bankruptcies. To call all these various forms of waste and inefficiency "transaction costs" would be absurd. They are generated by the market and are inherent in competition. It can plausibly be claimed that the market is wasteful, not efficient. But it is dynamic, and the same motivation that generates waste also drives innovation.

Dynamics can be complicated. Instabilities, cobweb cycles, and stagnation are also inherent in the market. The stimulus-response model conceals this by telescoping the actions undertaken in reaction to market signals, assuming that any action begun is successfully completed. As argued above, this is a serious mistake; any action can fail, and any decision can be revised – possibilities which give rise to uncertainty.

[16] This was the problem posed by Ronald Coase: why should administrative hierarchy replace the market in the internal activities of the firm, or in governmental bureaucracies, if the market is the best system for allocating resources? His well-known answer is that the market entails transaction costs, and these may be large enough to offset the market's greater efficiency (Coase, 1936; Williamson, 1980). "Transactions costs" are a weak reed on which to rest an important argument. The argument here is that bureaucracies are more efficient than the market at allocation, when both are understood realistically. But the market forces innovation, while bureaucracies stifle it.

Uncertainty

Uncertainty can arise for any agent in three kinds of relationships: to self, to others, and to the material world. The uncertainty stems from two necessary consequences of the "active mind," namely that intended actions can always fail, and new ideas, new ways of seeing the world, can always spring up. In regard to self, an agent must consider whether his plans will succeed, and whether he or she will or should change his or her mind. The same questions apply to others: who will succeed, in particular, who will prevail in conflicts, and who will change their minds? In market situations, this means picking winners and losers, on the one hand, and knowing who will keep their promises on the other. Both are essential information for investment or assessment of creditworthiness. Finally, in regard to the world, for the same reason that anything we do may fail, technologies may not work. Here, however, there is another source of uncertainty as well – nature may play tricks on us. Natural disasters – earthquakes, hurricanes, floods, meteors – may strike at any time; or we may be swamped by unexpected bonanzas – oil wells, gold strikes, bountiful harvests.

Three kinds of uncertainty typically predominate, arising from the possibilities that intended actions may either be changed or not be successful. First, concerning actions that affect the world, whether products and processes will work – "technological uncertainty." Second, whether goods will be sold, profits realized, or strategies succeed – this is "market uncertainty." And finally, whether economic promises will be kept, for example, payments made on time, bills honored – "financial uncertainty." The fact that these forms of uncertainty exist has led to institutional arrangements designed to cope with them. Money itself can be seen as a device to cope with the uncertainty attendant on promises to pay. Administrative hierarchy reduces uncertainty in regard to the execution of plans.

Many different exogenous "shocks" could cause any of these kinds of uncertainty to increase. Uncertainty will normally not affect plans – simply because it is *uncertainty*, and therefore cannot be represented by a probability distribution of outcomes. If it could, then we should call it "risk" – and an increase in risk changes a calculation in a determinate way. Risk therefore affects plans. But uncertainty does not, precisely because it does not lend itself to calculation. An increase in uncertainty will have an impact, however, but the impact will be on implementation, not on the plans. In general, a rise in uncertainty will lead to delays, in hopes of better information, or in some cases to quickly speeding up execution, before things get even worse, depending on which kind of uncertainty and the circumstances. Or it may lead to cancellation. Very often, variance will rise – different agents will react differently – so that, if the plans were coordinated to achieve an equilibrium, the effects of a rise in uncertainty would most likely be destabilizing.

If changes in uncertainty could be foreseen, they could also presumably be at least partly forestalled; so the circumstances would no longer be uncertain. Hence uncertainty is itself uncertain; it cannot be evaded, though its effects can be offset or mitigated, at least in part. The market, therefore, must be regarded as inherently liable to at least temporary destabilization from recurring fluctuations in the three kinds of uncertainty.

Administrative hierarchy eliminates many of these wasteful and destabilizing features, replacing them with planning and bureaucratic control (Robinson, 1965; Robinson and Eatwell, 1973). This significantly reduces the impact of all three kinds of uncertainty. Administrative planning can test processes and make sure they work; bureaucracy can ensure that plans are carried out and can prevent delays and alterations. Wasteful duplication can be prevented. Finance can be precommitted, and bureaucrats can enforce promises. Therefore, administrative hierarchy will normally bring about an allocation of resources superior to that generated by the market, provided that the bureaucracy is competently staffed, has been given clear directives, and functions smoothly. Comparing an "ideal" bureaucracy to the "ideal" competitive market (ideal, but realistically conceived, in both cases), and interpreting allocation strictly, the bureaucracy will always do the job of allocation better (Weber, 1920).[17] (Of course, a bureaucracy might be corrupted, but corruption is a market process!)

But recall what we have said about rationality. Knowledge is based on interpretation, guided by conceptual truths. The mind is inherently active and innovative. And the structure of a market system provides maximum scope – "the system of perfect liberty" – to choose a course of action. Further it is impossible not to act; competitive pressure will eliminate those who sit and wait. The rewards are great, the punishment severe; it is a system of carrots and sticks. The market will generate innovation, will force increases in productivity. By contrast, bureaucracies stifle creativity, and promote order and routine, the opposite of the competitive market. Schumpeter had it right; the market promotes innovation – it is wasteful and destructive, but it cuts costs and creates new products and new technologies. Administration may be more efficient at any *given* moment, but *over time* it will choke off innovation and lead to the growth of red

---

[17] This, of course, was the point the planners always had in mind, and was what Marx – and the Socialist tradition – meant when speaking of replacing the "anarchy of the marketplace" with conscious direction.

tape, so that costs will tend to drift upwards and productivity down. At any given time the market is wasteful, but over time it promotes innovation and raises productivity. Conflict encourages innovation, competition enforces it; a firm that does not innovate when its competitors do is likely to lose out (Schumpeter, 1942; Nell, 1991).

Markets encourage innovation in two ways, which will be explored later. First, as Adam Smith pointed out, competition straightforwardly develops pressures to innovate. The division of labor cuts costs; increasing returns both reward and encourage the attempt to capture larger market shares (Dosi, 1988; Freeman, 1992). Adam Smith's famous observation, "The division of labor is limited by the extent of the market," can be interpreted as a reason that markets will be systematically driven to expand. Secondly, in a Keynesian world, business in the aggregate always operates with a margin of excess capacity. Hence in general business faces a buyer's market; what is scarce is not resources, but demand. The system is demand-constrained in the aggregate. This creates pressure to cut costs and to innovate, in order to attract demand (Kaldor, 1980; Nell, 1991).

Besides creating pressure for innovation, the market acts as a selector. It picks winners and losers. Innovations that work will be rewarded; innovations that don't will fail. Profitable innovations will be financed; investment will flow to them, and they will spread. The rest will wither and die on the vine, or will carry on a furtive life in small niches. Capital invested in successful innovations will flourish and expand, that in failed efforts will shrivel or be lost altogether.

So the market system performs three evolutionary functions. First, as we saw earlier, the social system must reproduce itself. Production is consumed, people are born and die, new workers are trained as old workers retire, as inputs become outputs new inputs are needed, and so on. The market enables the cycle of production and consumption to operate more or less continuously. Second, the market generates pressure for variation, and third, it selects and rewards the successful variations, entering them into the cycle of reproduction, while excluding the failures. The system will therefore *evolve*, that is, it will tend to change systematically over time.

To understand the market's role in reproduction, we will have to examine production and exchange, together with distribution. To see how the market forces innovation, then, we will need to study competition more closely. And then we will have to consider how the pressures of demand work. To understand selection we will need to study profits, finance, and investment. These will all require a careful study of the monetary system.

Expectations

Returning to the allocation model, surely a defender might contend, forming expectations *rationally* would enable agents to cope with uncertainty? This argument misunderstands the role of expectations, on the one hand, and the way they interact with uncertainty, on the other.

Expectations feature strongly in prediction models. By contrast, structural analysis draws on what agents are supposed to do – their job descriptions – in their assigned roles. The object of the analysis is to determine how things work out when everyone does their appointed tasks. If a conflict is built into the structure, we analyze the possibilities. Programming analysis draws on the same materials to determine how to improve the way things work, or to delineate strategies in cases of conflict. Expectations about how others will react will be important in determining optimal strategies – but in a programming approach, it may make the most sense to determine their optimal actions. But prediction models try to tell us what *will* happen. For this it is not enough to know what agents *should* do; we need to know whether they will, in fact, do it. We also need to know if they will do it well or badly, fast or slow. In cases of conflict, we need to know who will win, or at least who is most likely to prevail.

It is not just the observer, however, who needs to know these things. A prediction model is based on what agents will do, and this often depends on their expectations of what other agents will do, or on their expectations of how things will work out. How much a consumer buys now, or a firm invests, depends on expectations of future prices and markets. But prices and markets in the future, in turn, depend on how much consumers buy and firms invest today. This can lead to cases of self-fulfilling expectations, but even apart from such problems, the interdependence of present and future variables makes it hard to obtain determinate solutions.

The favored approach is to assume agents form their expectations rationally. That is, they use the best available evidence, and base their expectations on the predictions of the best available theories. If a variable is the outcome of a well-understood process, then it would be rational to predict that the variable will have the value indicated by a model of the process. Normally, however, there will be random disturbances; processes will be *stochastic*. Hence predictions will tend to err – but because the errors are random, they will tend to average to zero over a number of trials, and they will exhibit no pattern. (Rational expectations are unbiased and orthogonal to the information set.) Over a period of time, predictions based on the model of the process will be the best that can be made. It is commonly assumed that agents have complete knowledge of the true deterministic patterns they face, and further that the stochastic components are represented by well-defined and known probability distributions. Under these conditions agents will form determinate and correct expectations – apart from random errors – and models built on the assumption that agents act according to such expectations will be soundly grounded.

Trouble begins when the outcomes of processes are uncertain. Outcomes may be determinate – but not unique. If more than one outcome is possible, agents may expect different outcomes, and thus act differently. They may also change their minds along the way. Even worse problems arise when the outcomes are uncertain in any of the senses explored above. These are senses which imply that we don't know something because it is not there to be known. A technique may not work for reasons of which we, as yet, have no inkling. It may not be possible to decide, given what we know, whether or not to expect a debt to be honored. Markets may break down or result in surprise outcomes because of unexpected shifts in parameters – or because of indeterminacy or multiple outcomes of processes. Uncertainty casts a cloud over the formation of expectations; even the best use of information may not be able to justify a firm commitment. Yet it may be necessary to act. In these circumstances, extraneous information – sunspots – may influence the final determination of expectations. Given the interdependence between current and future values, these expectations may influence the outcomes, so that, paradoxically, irrelevant noneconomic variables determine important economic processes.

No doubt this is possible, but the apparent likelihood of it being widespread is due to the absence of structure in the models commonly studied. Agents are not only rational; they hold positions and have commitments. These *bind* them in certain ways, which put limits on their possibilities of action and inaction. Taking structure into account can greatly reduce the apparent areas of indeterminacy.

On the other hand, the conventional approach normally ignores the problems of conflict and achievement. The outcome of a strike, or a price war, or a bidding competition cannot be predicted any more (or less) than a baseball game. Moreover, different agents will evaluate the same information differently, and normally issue different predictions. Achievement presents much the same problem. We may be confident that someone will do the job – design the product, make the sales pitch, construct the new building – but how well will it be done? Predicting success in the achievement of a high standard is inherently uncertain. Yet unless the job is done to the required standard, it may be useless. A miss is as good as a mile. The prediction may be wrong, even though there was every reason to think the job would be done properly.

Not only is prediction uncertain because others are fallible – predicting is, itself, a skilled activity, and is therefore uncertain on its own account. It can be done well or badly, some are better at it than others, and anyone can slip up. Even a good golfer can miss a short putt. Skilled activity is inherently fallible. Predictions can go wrong for no other reason than that the predictor missed a trick. (*Not* because the predictor behaved irrationally or failed to take into account relevant evidence, but because the predictor *weighed* the evidence imperceptively, and failed to see the patterns in it. It is a failure of insight, not of reasoning, that is at issue.)

All this provides good reason to reject the requirement that rationally based expectations ought to be consistent. On the contrary, rational agents will have good reasons to arrive at very different conclusions as to the likely outcomes of conflicts and levels of achievement to be expected.

Moreover they will often know that there are no good reasons to consider any particular outcome as more likely than any other. In such a case, forming an expectation on the basis of extraneous information is irrational. Instead, the best educated guess should be made – but it should be held with low confidence. When an expectation can be formed on solid grounds, it should be held with a high level of confidence. Sometimes it may be clear that there is only one reasonable outcome to be expected, but it may be quite unclear when it will take place. The expectation will then be definite, but the date at which the event is expected can only be held with a low level of confidence. We will see later how important it is to distinguish expectations and levels of confidence.

If uncertainty implies that different agents may reasonably arrive at different expectations, then behavioral analysis differs from structural in an important way. Conceptual analysis and fieldwork should be able to provide a single correct account of structure – at any given time the economic system has a definite structure,[18] which it is the job of fieldwork to understand and of a structural model to portray accurately. Such models can be better or worse, more or less detailed – but they cannot be *different*. By contrast, in a given context, similar economic agents, considering what to do and facing the same circumstances, and optimizing the same objectives, *can*, on reasonable grounds, reach different conclusions, and settle on different courses of action. Predictive model-building is an artform, building on a knowledge of the structure, including objectives and opportunities, but drawing, in part, on hunches and guesswork. It cannot be the case that, in general, all agents will always react the same

---

[18] Definite, but perhaps not wholly unambiguous. At any given time aspects of the structure may be changing under various kinds of pressures, including those generated by the working of the system itself.

way to the same economic stimuli – if it were, there could never be any innovation.

Against equilibrium

Economic theorists, including many attracted by the Classical vision, often tend to assume that innovations have no systematic effect of a kind that would change economic structure or the way that markets work.

Even when no such assumption is made consciously, theories still tend to run in terms of equilibrium and steady growth. Economic theories, especially those based on rational maximizing, are supposed to apply to all times and places – provided, of course, that the appropriate conditions hold. If there are "imperfections," then, of course, behavior will deviate from the optimal path, and in some periods and places imperfections will be more prevalent than in others. But this means that the model has simply to be adapted to reflect the imperfect conditions. Moreover, even those who are skeptical toward equilibrium and steady growth tend to replace them with an approach that runs in terms of regular cycles (Goodwin, 1992; Flaschel and Semmler, 1991).[19]

In general, then, economists tend to believe that economic principles are equally applicable to all periods. Some theories will work better than others, of course, in certain periods. But there are no fundamental problems with equilibrium (though some will add that dynamics is also necessary – and may be more important) and no reason not to approach all economic phenomena, whatever the period and place, in essentially the same way.

This view cannot be accepted. It is the result of a self-reinforcing dynamic. Economists tend to think in terms of equilibrium and steady growth because these follow from assuming "tunnel vision" rationality in individual choice. On the other hand, narrow rationality is assumed, together with extreme abstraction, because that will yield equilibrium conclusions. The historical reality, however, is change – persistent, directed and substantial change, sometimes accompanied by crises, brought about by innovative responses to market conditions. Moreover, economic changes are accompanied by institutional changes, which themselves result from

[19] Blaug, for example, certainly has no quarrel with the view that markets work the same way in all periods of history; he criticizes Hicks for insisting that causality in economics must be understood in the light of history and the consequent uncertainty, ignorance, and multiplicity of converging forces acting on any single event. Implicit in Hicks's view, of course, is the idea that causal configurations in economic matters may be signficantly different in different eras (Blaug, 1990, p. 110).

an evolutionary dynamic set in motion by the economic forces (Kelly, 1994). Steady growth may be a useful benchmark, but, as Hicks pointed out (1965, p.183), it is never found in reality.

So while equilibrium may be useful as a benchmark it cannot hold center stage in a realistic study. Models should always be solvable out of equilibrium.[20] Equilibrium has traditionally been considered a balancing of market pressures, of desire against scarcity, in the neo-Classical scheme, although it has also been interpreted to mean that all agents are in their chosen position – which implies very little about markets. Neo-Classical market equilibrium is the point to which adjustment tends, the point at which the forces of desire and scarcity are balanced.

A different notion appears in the Keynesian framework where expansionary injections are balanced against contractionary withdrawals. Neither desire nor scarcity is involved, at least in the neo-Classical senses of those terms. This, too, is a position of rest, but it need not result from maximizing nor be optimal in any sense.[21] It may be temporary and quickly reached; moreover, the dynamics may be realistic. If so, it may be helpful in constructing further dynamics. Understanding the conditions for such a balancing may help to show how markets work, providing the models are based on adequate fieldwork, so that they realistically portray the essential features of the institutions involved. But if the central feature of markets is contained conflict, in the course of which innovations are brought forth, too great an emphasis on equilibrium of any sort may direct attention to the wrong aspects.

Treating equilibrium as a position of rest has recently been called an "anachronism" by Lucas (1987). "Continuous equilibrium," a moving

---

[20] "Reproduction plus distribution according to a given rule" is *not* equilibrium in the usual sense of the term. No balance of forces is assumed; it is a calculation of what has to be achieved for the rules to be followed.

[21] Injections and withdrawals, e.g., investment and saving, *may* result from behavior decided upon by constrained optimizing, as the proponents of "microfoundations" would have it, but they could just as easily be the result of rule-following, where the rules were determined by political infighting and compromise. For example, investments could be decided by top management according to a formula worked out between the corporation's divisions; business saving by rules governing retained earnings, also reflecting a compromise between top management and shareholders. Household saving might reflect a deal between the interests of husband and wife, and so on. To say that those who follow rules that have not been derived from optimizing will lose out to those whose rules have is to make a dynamic claim. Such claims have not been demonstrated and, given the wide variety of dynamic possibilities, are certainly not plausible. In any case, the macroeconomic equilibrium requires only that injections be balanced against withdrawals; it is independent of the forces governing either.

path, should be seen as the result of optimizing. An equilibrium results from the optimizing behavior of economic agents, whether this is a position of rest or a path. It is an equilibrium because no one can become better off, without someone becoming worse off; Pareto optimality defines this approach to equilibrium.

In a world of conflict and innovation, such a view is strikingly irrelevant. In conflict one party becomes better off, sometimes much better off, at others' expense, and all strategies are directed either to that end, or to defending against such an outcome. Firms compete not only against other firms, but against their customers; employers compete against workers. To think in terms of Pareto optimality is to miss the point altogether. And to think merely in terms of optimizing, with given endowments, etc., in response to price signals, under conditions of competitive innovation is to miss what is happening.

Calculating optimal strategies should not be confused with equilibrium either; such calculations are ways of finding solutions to programming problems. It may be that agents will be assumed to adopt such solutions, so that the strategies will become actions attributed to suppliers or demanders in a market. But the agents must agree that the terms of the problem as posed in the rational choice model indeed describe their situation – and they must accept the solution. In that case optimizing strategies may become part of an equilibrium, but, as we have seen, it is equally likely that optimizing will lead to *changes*, so to deviations from the previous equilibrium, or normal position.

Moreover, a rational calculation is just that – it is a decision. It is not itself an equilibrium, even if it is reached by weighing additional rewards against extra effort. A decision is reached by a process of deliberation, a market position by a process of adjustment. Even if the strategy determined by the rational choice model is accepted and acted on, the market still has to proceed through a sequence of adjustments. Strategy and decision are not the same as dynamic interaction, which may converge, diverge, or cycle. Equilibrium has to be reached.

Given the kinds of uncertainty examined earlier, a *behavioral* equilibrium may be a notion of only limited usefulness. Even if reached there is no reason to suppose it will last; behavior may change whenever an agent has a new idea or reconsiders her preferences. By contrast, an equilibrium based on *structure* may be a useful analytical tool. Structural relationships will normally not be affected by changes in uncertainty. Structure changes slowly, and changes in structure must be brought about by formal actions which alter institutional practices. However, it may be possible to mix structural and behavioral notions in such a way that each complements the other. We will see this in the idea of the "closed and completed circuit," and again in the related concept of a "demand equilibrium." In each case the position of rest or balance is defined by structure, and achieved by behavior – but many different kinds of behavior, even changing behavior, will suffice.

Finally, even if equilibrium may be a useful assumption for limited purposes in short-run analysis, its only place in long-run thinking is as a reference. Far from being steady, as we have seen, growth is accompanied by systematic changes in the relative sizes of sectors, frequent changes in distribution (often systematic and taking place over long periods, but sometimes abrupt, as during a war), and constant but irregular changes in technology. To put it minimally, so far growth theory has not successfully addressed these questions. To meet this challenge, a new approach will have to recognize the role of markets and competition in driving innovation through conflict, and the price of this is dropping the idea that the function of markets is to allocate resources in an equilibrium manner. This will pave the way for the theory of Transformational Growth.

Transformational Growth

On the Classical view, then, the market engenders innovation; and in fact, the last century has seen innovations far beyond the dreams of earlier eras.[22] The market selects and spreads innovations through financing investment in successful projects. As innovations spread, the technology in use changes substantially, and this will affect the structure of costs, and therefore the way firms react to market signals. At a point when these changes have become sufficiently great, it appears that the market comes to work differently. As the working of the market changes, the institutions – the firms and households – will also change, to manage their market relationships better. This, in turn, can be expected to change their relationships to the other institutions of society.

This is an evolutionary process, and we can briefly sketch what that means.[23] In the abstract, such a process requires:

A population which reproduces and expands,
Pressing on a given environment which the population draws on to support itself,

---

[22] Related heterodox models (e.g., Kaldor, 1958; Kaldor-Mirrlees, 1960; Pasinetti, 1981; Harris, 1978; Boyer, 1989) have stressed the ubiquity of technological change and have explored its impact on the working of the economy.

[23] I thank Alice Schlegel for guiding me through the maze of writings on cultural and social evolution, and helping to clarify the analysis of evolutionary pressures on the economy. See Dennett (1995), Waldrop (1992), and Kauffman (1995), esp. chs. 9, 10, and 12.

A cause of variation in the population,
And a method of selection, picking out the variations which are successful and are likely to be preserved and passed on.

As a result, then, of the selection of successful variations, the characteristics of the population will change over time. The variations can be random or directed, and the changed characteristics which are passed on can be genetic or acquired. If the latter, the process is Lamarckian; if the former Darwinian. Very broadly, transformational growth is a non-Darwinian process of competitive selection, a Lamarckian process in which acquired traits can be inherited.

Firms and households grow, flourish, reproduce, and depart the scene (in many ways) – the institutions, not just the actors who fill the roles. Firms go bankrupt, or are weakened and fall prey to competitors who take them over. Surviving firms have to adapt to new conditions and new technologies, which means adapting their organization. Household property and position are passed along to the next generation according to kinship rules. In each case certain traits are identified and survive, so that they are incorporated into the successor firms and households. In this way the basic economic institutions change character over time. As they change, the way they interact also changes, which means that the character of the market develops.

In a sense what we see is the co-evolution of capital and labor, in a mutually dependent but competitive symbiosis. Competition leads to development and market selection, not to equilibrium. Co-evolution means that two different populations evolve together, in a mutually dependent way. (Classic examples range from prey and predator cases, to insects and flowers.) Each forms part of the environment for the other, and each acting on the other, helps to create the pressures which determine the new characteristics that will be selected. Clearly, learned or acquired improvement in both institutions and technology can be passed on. Equally, "mutation," that is, variations in characteristics or innovation, will itself be the result of a planning process, a process which, moreover, will be competitive and will be subject to market selection. (Better systems of R&D will be expanded, and worse ones shut down.)

Capital and labor, in turn, are tied together in a competitive symbiosis. Capital's gain tends to be labor's loss and vice versa – profits rise at the expense of wages. Real wages rise and labor gains when capital grows faster than labor and, conversely, capital gains and real wages fall, when labor grows faster – more clearly in the OBC, but there is still a modicum of truth in this in the NBC. Households benefit from business's success in developing better products and cheaper processes. Business benefits from household investment in improving the skills and education of the labor force.

Capital reproduces and expands, competing for markets by developing products and processes; labor reproduces and expands, with households competing for status (to rise in status). Capital needs profit to expand, labor needs pay to support status. Capital measures its success by the rate of return it earns on its advances; labor measures its success by the rate of net pay on its cost of living, its support. In a sense both measures are rates of return on advances. Moreover, both can be understood as rates of growth: capital grows by investment, and labor in terms of the increase in the amount of labor being supported, which may be the rate of investment in human (household) capital.

Capital and labor both grow, that is, expand, but they also develop.

Capital first circulates in the form of metallic money, then by means of paper, and finally as bank balances and electronic funds. At first capital is largely circulating capital, then it more and more becomes fixed, and finally it turns increasingly intangible, the result of research and development. In the early stages of free labor, most work is performed on short daily contracts, then contracts become longer and longer, with more fully defined rights. Labor becomes more specialized and some segments develop skills that embody higher and higher levels of general education.

Capital and labor both develop their institutional forms: capital begins embodied in family firms, then moves to limited liability companies, then to the modern corporation, and now, today, increasingly takes the form of networks of corporate allies. Labor is produced and reproduced by the family. Initially in some cases families may have taken the partially extended form; more commonly the form was the "embedded nuclear family" – that is, a system of nuclear families embedded in kinship networks that carry out domestic production through a division of labor, with exchange governed by norms of fairness and reciprocity. As domestic production is taken over by industry and domestic products become commodities, this system reduces to the isolated nuclear family, and now to the individual worker. Capital pursues its interests by developing cartels, monopolies, oligopolistic alliances, and political committees, white labor advances its cause by forming unions, and in some places, political parties. Households pursue community projects and engage in political activities to defend their interests.

Products and processes both evolve, and often evolve together, under pressure from both firms and households. Firms innovate, competing against other firms for the market, but households select, in competition with other households, for their use in reproducing and improving themselves. (This is the grain of truth in the doctrine of consumer sovereignty.) Firms develop new products, often in conjunction with the households that make up their market. Firms develop new processes to produce more cheaply, or more competitively. New products develop as firms seek niches in markets.

According to this idea, then, capitalism is a system in which value is reproduced and expands, and value competes for position in the environment, with successful strategies and innovations being selected by the market. (Perhaps this makes value analogous to life, in biology?) Value is embodied in two forms or species, business and households, or capital and labor, each of which reproduces and expands, and each of which depends on the other, while competing with that other in the sharing of the gains from their joint activity.

Although the co-evolution of capital and labor is the principal process in the development of capitalism, evolutionary pressures and processes are not confined to this case. Evolutionary processes could even be considered "fractal," that is, as occurring independently of scale, at all levels of the economic system. Firms evolve; but within a firm, management procedures evolve. Those who propose or carry out procedures that prove effective are promoted. Those who support ineffective or incorrect procedures – ones that don't work – tend to be dismissed. Firms that promote their own internal evolution will succeed, those that don't will lose out in the market. In a similar way, household forms evolve, and within each new form, activities and roles evolve. Successful assignments will be rewarded, unsuccessful punished. Products evolve, and so do subproducts; technologies evolve, as do production processes and subprocesses.

To analyze such transformational growth we have to be able to say that even though the system is changing, it remains the same system. This requires a criterion for "change" and "remaining the same" – when it is or is not the same system, though it has come to differ in important ways. To see how important this really is, consider that most of the products produced today were not produced a hundred years ago; whole sectors did not exist or were rudimentary then – aircraft, high tech, electronics, nuclear power, radio, and TV, even the automobile had barely begun. And prominent sectors of the last century have disappeared – steam engines, blacksmith shops, harness-making, currying and horse-care, gas lighting, and many more. But for all that, we want to say it is the same system. And the reason is that the core relationships, between wages, prices, and profits, have been reproduced and maintained through all the changes. These core relationships are clearly presented in the structural Classical model. The wage-profit trade-off or "factor-price frontier" gives the relation between wages and the rate of profit, whereas prices depend on capital-intensity, which itself changes as prices change. For this to be complete, however, prices must be expressed in money. Such relationships hold for all capitalist systems, however backward or advanced their technology.

That is a sketch, our vision of the economy as an evolving system. But the idea of evolution is little more than a metaphor – and one about which it is wise to be cautious![24] Transformational Growth is a way of thinking about economies and how they develop. It bears only a loose relationship to evolutionary thinking; its selection mechanism is the market. It is non-Darwinian; innovations are directed, and acquired characteristics can be passed on. Products and processes are "replicators"; households and firms are "interactors"; technologies, designs, and lifestyles are analogous to genes. Lineages can be traced in artifacts and in institutions. But this is all *analogy*. It helps to clarify a *basic vision* – of a system changing under the competitive pressures generated by the way it organizes the activities by which it maintains and reproduces itself. To spell out exactly what is implied in Transformational Growth requires developing theory and expressing it in specific models of markets and competition.

And this will be the task in what follows. But to relate markets and competition to innovation it will first be necessary to explore the role of money in markets.

---

[24] Khalil (1993) notes parallels between objectionable features of neo-Classical thinking and neo-Darwinism. Hodgson shows that his embrace of evolutionary ideas created inconsistencies in Hayek. Rosenberg (1994) objects to importing Darwinian evolutionary ideas into economics, on several grounds. First, he argues that "natural selection" is either a tautology or incapable of generating useful predictions, depending on how it is interpreted. Second, evolutionary equilibrium helps to explain gene frequencies – but evolution has a long time to work in, and adaptations *can* reasonably be taken as parametric. Populations can reasonably be assumed infinite, and genes "omniscient." By contrast, economics cannot be useful if it confines itself to the "long-run," has no stable set of characteristics comparable to gene frequencies to explain, and no grounds for assuming infinite numbers of omniscient agents. Evolutionary analogies just reinforce the worst tendencies in neo-Classical thinking. Transformational Growth, however, holds no brief for equilibrium, or for most of the neo-Classical framework. The tendencies in question are precisely the ones it avoids.

PART III

Money and the Golden Rule

Information from field work, assembled in the light of conceptual truths, will present a picture of how economic institutions work, and what their functions are. It will also enable us to understand how people occupying positions in those institutions are likely to behave, and why. A preliminary analysis led to a general conclusion, one that will be explored in detail in later chapters, that the function of the market is not to allocate scarce resources efficiently – markets are often wasteful – but is rather to force innovation through competition, and to select and support the winners. This, in turn, implies that the way economic institutions work will evolve over time.

Yet this conclusion was reached on methodological grounds only. Stimulus-response models, which support the allocation view of markets, were rejected for resting on a deficient concept of rationality, while ignoring structural relationships, and presenting an artificial and inadequate depiction of market behavior. Classical structural models were judged to provide an account of the way a surplus is generated, and of the framework within which competition operates to determine its division. But competition, from this point of view, can never lead to equilibrium in the long run. Equilibria may be reached in the short run, but any such positions must be considered provisional. For the human mind is inherently active; it never merely accepts what the world presents. And in the economic world it faces an ongoing conflict, in which those who stand still, tend to lose out. Hence at every point the active mind will be pressured to develop new strategies, new products, new ways to cut costs and produce more cheaply or turn out better products. Thus it was provisionally concluded that markets will tend to generate – and select – innovations.

To develop this perspective further will obviously require a full account of markets, of what they are and how they work, and how they change. An essential first step is to explore the idea of money, how it functions in markets, and how it furthers the development of competition, and of the market itself. To do this it is necessary to separate the idea of money into two parts – distinguishing the function it serves from the instrument that carries out that function. The function of money is to circulate goods according to a certain pattern, and at any time this function is carried out by a particular medium, for example, metallic coins, or paper or bank deposits, or, sometimes by a mix of these media. A monetary system should be understood as a medium, or several media, tracing out a well-defined and repeated pattern of circulation. When media emerge spontaneously, there will normally be many of them. To establish a workable currency there will have to be pressure to maintain stable relationship between them – pressure which will normally come from the state. Circulation both will be driven by competition and at the same time will enable competition to regulate values, while promoting innovation. The pattern of circulation, in conjunction with the cost of producing the medium, will determine how much is required for the circulation.

Innovation, in turn, will affect the monetary system. The pattern of circulation will become more complete, and will change from beginning with purchases by merchants, to starting from bank advances of working capital. But the money article will be even more dramatically affected, changing from metal to paper to bank deposits, that is, from a basis in intrinsic value to one depending on fiduciary relations, supplemented by government fiat, and finally to being grounded in double-entry book-keeping, together with liability management. Metallic money is expensive and inconvenient, but it has a "natural" value. However, its availability depends on mining. Paper money displaces it, and bank deposits supersede paper. At a certain point the medium will no longer have a "natural" value; the acceptability of money depends on the credit-worthiness of the issuer, and its value will be determined by the interaction of interest and prices. Credit proper also develops from commercial paper and bills of exchange to bonds to the modern stock and securities markets.

Patterns of circulation can be partial or complete. If partial, the system cannot be fully capitalist, and competition will be limited in its ability to carry out its functions. By the same token, there will be barriers to innovation. Complete circulation means that all goods and services produced can, in principle, exchange against money. Under these circumstances competition affects all economic activities, so that all are subject to pressures for competitive innovation.

The pattern of circulation calls for a specific quantity of money, equal in value to a bundle of goods for which the money exchanges. This bundle will not be the whole volume of goods produced; some sales are contingent on others, and the funds earned from the latter can be used to execute the former. The same quantity of money can be used to carry out a number of successive transactions, provided later are contingent on earlier. Such contingency, in turn, is explained by structural dependencies in production and distribution. In particular, the relation between the wage bill in the capital goods sector and the capital requirements in the production of consumer goods will prove to be the key to unlocking the mysteries of circulation, allowing us to trace the pattern, explain "velocity" and establish exactly what quantity of money is required.

The quantity of money in circulation will be adjusted by market forces. Metallic money will be provided by mining, up to the point where the value of money established in circulation just equals its cost of production. In different but analogous ways, market forces also regulate the quantity of paper and of bank deposits – "credit money." The key to this regulation will prove to be the "Golden Rule," that the rate of interest on capital must equal the rate of growth of output. So long as this holds, money in circulation will expand at a pace just sufficient to continue to circulate the growing output at a constant price level. Moreover, the Golden Rule implies and is implied by the wage-bill/capital requirements relationship, which underlies the process of circulation.

Besides the money in active circulation, that is, money used in circulating produced goods and services, money may be kept as a store of value, and/or used to effect exchanges of stores of value among wealth-holders. Initially known as "hoarding," this has developed into complex financial markets, which interact with the circulation for production. This interaction raises problems.

For all market-driven monetary systems suffer from an apparently inescapable flaw; they tend to be unstable. Such potential instability arises from the relation between the requirements of the pattern of circulation and the methods by which the medium can be provided – including interaction with financial markets. In metallic systems the instability may arise from a perverse response to prices – changes in the value of money. In paper systems it will be triggered by the relation between the rate of interest and the rate of profit (as Wicksell notes); while in modern economies it will arise from the interaction between the rate of interest and the rate of growth. To circumvent these problems governments have stepped in to manage the currency and control financial variables. This, in turn, has led to some of the classic problems of political economy.

CHAPTER 5

Circulation and production:
the need for money

Basic concepts

Perhaps market competition could take place without money, but it is hard to visualize. The market must bring about the uniformity of prices and wages – the "law of one price." With decentralized, multilateral exchanges it is difficult to see this happening under barter. Money provides the means by which competition exerts its influence. How in practice could people determine whether the same price was being charged for the same good in different transactions, if prices were expressed in different goods in different places or at different times? It would be necessary to have a complete list of barter ratios, and calculate the cross exchanges.[1] But the equilibrium barter ratios are supposed to be established by competitive pressures – supply and demand – which operate on *uniform* prices. This problem is nicely avoided by Walras's device of the "auctioneer," who calls out prices, expressed in a universal standard, and thus establishes that all exchanges will take place at the same price. If there is no auctioneer, however, then the law of one price must be established by competitive pressures. This is a problem in market dynamics, and it can be shown that under plausible assumptions, the

---

[1] In the Andes during the Inca Empire, exchange ratios were carved into the stone pillars in the center of the marketplaces. Accounts were kept by tying knots in pieces of rope. But these ratios were established by custom (based on labor times?), not by supply and demand (cf. Garcilaso de la Vega 1871, Part I).

processes need not converge to uniform prices.[2] In practice there are no auctioneers – but prices do tend toward uniformity, and this, it will be argued, is because money facilitates competition.

Of course, all prices could be calculated in the same standard, say, cowrie shells, and then goods could be traded, with records of net indebtedness kept, so that when all trades were complete the records could be taken to a clearing center, to settle the final accounts. If an equilibrium exists, then we know that the accounts will clear at the equilibrium prices. But using a universal standard of value and settling accounts in a central clearing system – Walras's "chamber of settlements" – comes very close to monetary exchange. Promises to pay – statements of debt – have been accepted in trade. This seems innocuous enough, until we ask, why should anyone believe a promise? What kind of information is available? In what ways can contracts be enforced – and how costly are the courts of law? Exchanges can be carried out by circulating promises and clearing them, it is true – but only if the promises to pay are regarded as reliable, which means they must be enforceable. Thus, for example, the issuer of debt must be known to possess assets that could be seized in the event of a default, and there must be effective procedures to seize them. In the conditions of early capitalism, as the monetary system developed, communications were slow, information difficult to obtain, courts were expensive and sluggish, so that promises were rightly regarded as risky. Money, on the other hand, was universally acceptable, since its value was secure. Money was needed precisely because promises to pay were not reliable (Burns, 1927; Hicks, 1989; Braudel, 1985).

Moreover, it might be argued that a clearing system could not in practice be established unless there were already specialized merchants accustomed to monetary exchange, and possessing information about reputations. As for rates of return, it is hard to see how rates of return in different sectors could be brought into line with one another, to establish a general rate of profit, unless exchange were carried out in monetary terms.

---

[2] Nell (1980, ch. 6, pp. 105–13) argues that neither Marshall, Wicksteed, nor Edgeworth had an adequate solution to the problem of competitive convergence to one price, and then shows that under standard assumptions, where a market-clearing equilibrium price exists, when buyers and sellers choose and revise their optimal marketing strategies in the light of changing market conditions, the frequency distribution of market prices will shift in a regular manner, with the modal price oscillating about the market-clearing level but displaying no tendency to converge on it. Nor will there be any tendency for the variance to shrink. The conventional account of competition cannot explain how a single market with a large number of buyers and sellers moves from a wide range of exchanges at a variety of prices to converge on a single price.

The cost of gathering the information about cross-exchange ratios, let alone of managing barter – bearing in mind that promises cannot be generally accepted – would be too great.[3]

The functions of money

The first question is to fit money into the structure of the economy. What is its role in the system of production and reproduction? This must be clarified before the effects of money on behavior can be studied.

The basic function of money is to serve as the medium of circulation, that is, as a medium enabling transactions to be carried out in monetary terms. Outputs can be sold for money, which will then be used to purchase inputs; wages and salaries will be paid in money, and household needs met by monetary purchases and payments. By calling it a medium of circulation it is implied that it is received in one transaction and passed along in others, successively making its way through the economy. This implies that the goods for which it exchanges can be – and are – valued in terms of money; hence being a medium of circulation implies money's function as a standard of value. Further, circulation requires that money should flow successively through sectors and social classes that are distant from one another, while retaining its value unchanged. Hence whatever circulates must also possess the ability to store value. Any goods possessing value can serve as a standard of value, and all durable goods can serve as stores of value. Neither a standard of value, nor a store of value, however, needs to circulate, and, indeed, most goods with these properties do not. Thus the function of serving as medium of circulation

---

[3] Neo-Classical theory assumes full market information, perfect foresight, easy mobility of factors, and then finds it has no role for money! If foresight concerning future payments were perfect, or even able to provide accurate probabilities, promises to pay – notes – would be fully acceptable and could circulate. (Where payment is merely probable, the notes would circulate at a discount.) But perfect foresight is impossible – it is inconsistent with the "active mind." If we could accurately foresee our future decisions we would already have made them, leaving no room for innovation, setting new courses, or failure! Money means an agent does not have to accept a promise of goods to be delivered later; nor must an agent accept goods he does not want, in the hope of later trading them for ones he does. Money is universally acceptable and cancels debt. It obviates the need for information about reliability and it reduces transactions costs. Metallic money is acceptable because it has "natural value," while fiat money is acceptable in payment of taxes and debts.

implies the other two textbook functions of money, while neither imply it.[4]

Of course, many different media of circulation could – and did – emerge spontaneously. Establishing and maintaining stable relationships between them – creating a uniform currency – has historically been the responsibility of the state.

Once a medium of circulation is established it permits the immediate and universal comparison of all kinds of economic activities. Previously, economic – and many other – activities were compared and evaluated, not on the market, but in the courts of law, where traditional obligations and privileges were weighed and adjudicated in the light of the doctrines of the "justum pretium" (Nell, 1992, chs. 12, 13). Monetary measures often figured in arguments and decisions, but in practice monetary circulation was limited, and awards, though often expressed in money, were normally executed in kind (labor services, assignments of rents, holdings of land or buildings; Coulton, 1945). Economic evaluation and comparability therefore depended on tradition and custom, as judged and enforced by courts.

Changes were therefore difficult to bring about, and challenges or interpretation of difficult cases were available only to those who could afford to go to court. The development of the market democratized this and made reliable monetary evaluations available to all, and equally if not more important, pried such evaluations free from the grip of custom and tradition. Changes in valuations came about "automatically" through the market, which reflected custom and tradition, but also made room for innovation.

The emergence of market values has had unsettling aspects. In the feudal world, as in the ancient, monetary calculations played a subordinate role. Universal monetary circulation, however, reduces the fabric of

---

[4] By contrast, Keynes argued that "Money of account, namely that in which debts and prices and general purchasing power are expressed, is the primary concept of a theory of money. ... Money itself, namely that by delivery of which debt contracts are discharged and in the shape of which a store of general purchasing power is held, derives its character from its relationship to the money of account, since the debts and prices must first have been expressed in terms of the latter" (CWJMK, V, p. 3) This is quoted favorably by Carvalho (1992), pp. 44–5, who considers the stability of money of account, and therefore of "money itself," one of the six basic postulates of a monetary economy. But money of account existed in manorial economies, as records from the feudal era amply show, yet little or no money circulated, and neither production, nor rent payments, nor most trade were monetized. At a later date, a limited form of circulation took place, cf. the discussion of the "exchange circuit" below, but this still did not represent fully monetized production.

life to the common denominator of money, with the reduction equally
evident to anyone taking part in the market. Activities, goods, services,
assets, jobs, works of art, love, fear, and the most intimate aspects of life –
and death – all could now be given a price and put on the market. As
Shakespeare says,

> Gold? yellow, glittering, precious gold? . . .
> Thus much of this will make black, white; foul, fair;
> Wrong, right; base noble; old, young; coward, valiant. . . .
> This yellow slave
> Will knit and break religions; bless the accurs'd;
> Make the hoar leprosy ador'd; place thieves,
> And give them title, knee and approbation
> With senators on the bench; this is it
> That makes the wappen'd widow wed again. . . .
> damned earth
> Thou common whore of mankind, . . .
> *Timon of Athens*

In the scramble for money, anything goes, black is white, foul, fair. The
cash nexus dominates – birth, rank, distinction, all melt before the power
of the purse. Any achievement can be compared to any other, in terms of
its monetary worth, and anyone can be motivated to do almost anything by
a sufficient offer of money.

In fact, the picture of money corrupting virtue is all too easily
overdrawn. Even in a fully developed monetary economy not everything
or everyone is for sale, and people have many other motives than making,
or keeping, money. Even in purely economic activities motives are often
mixed, valuations arbitrary, and commercial concerns are frequently pushed
aside. Consider someone deciding whether to start a business: family
responsibilities, the ideals of craftsmanship, aesthetic and moral considerations, all may enter into or even dominate the calculation, along with
the dollars and sense of it.

Nevertheless, it is true that economic considerations have a tendency to
spread into all fields of activity – think of sports, art, theater, or literature –
and once they do, competition brings pressures to rationalize, cut costs,
and develop effective marketing. Each of these activities requires inputs
of time, energy, and resources, which have to be paid for. Each therefore
must earn its way, covering these costs plus interest charges on initial
capital through sales or ticket admissions. Activities that don't pay their
way require fundraising and subsidies and will therefore always be under
pressure to keep costs down. Activities that more than pay their way
become targets for commercial investors, who may care nothing about the
intrinsic values at issue. Either way the pressures can lead to full-scale
commercialization, together with changes in the character of the activity

or product – we have the Super Bowl, the gallery scene in Manhattan, Broadway, the best-seller list. Art, theater, publishing and the like become businesses, in which the bottom line rules.

There are ways to prevent this. Philanthropy can raise enough capital to cover the costs; government subsidies can be provided. Non-profit organizations can manage an activity. Or it can find a special niche in the market, safe from competitive pressures – and not too profitable, so that commercial investors will not invade. But if motives other than economic are to predominate, the balance between an activity's costs and its returns must be covered somehow. Once monetary circulation is universal, the bottom line is always there.

But understanding the bottom line, and managing the problems it creates, requires a different set of skills than managing production or sales. To begin with, there is accounting, establishing the actual balance between the costs and revenues, and relating it correctly to time periods. Costs must be divided between current or running costs and overhead or capital; a separate but overlapping distinction must be drawn between fixed and variable costs. Revenues may be final sales or payments on account; receivables must be correctly dated and interest charges computed. When the balance and its course over time is correctly understood, the problem emerges of setting aside reserves and/or borrowing to tide over seasonal and other temporary shortfalls, that is, of managing the various gaps between revenue and outlay. Even when production and marketing are comparatively primitive these matters call for specialized skills and knowledge. Monetary circulation creates opportunities for specialization in monetary activities; it almost requires merchant specialists, and dealers in financial paper.

The irregularities and seasonal variations in supplies and demands open opportunities for speculation, by market specialists. Anticipating a shortfall in supply, say, a poor harvest, causing prices to rise, specialists – grain merchants – will hold back supplies, leading the price to rise earlier, and, in the best case, carrying supplies over from earlier periods to partially fill in the shortages in later. This will help to smooth out the fluctuations, and such activity by speculators can be considered stabilizing. This enables markets to deal with uncertainty and offers the opportunity for hedging a shortage by buying stocks of goods or claims to such stocks in advance.

However, a much more problematical situation can develop, where the *rising* price leads to withdrawal of supplies, in anticipation of further price increases; but the withdrawal of supplies itself causes price to rise further, leading, in turn, to further withdrawals. A cumulative process develops, transforming a moderate shortage into a famine. In this case the activity by speculators is radically destabilizing.

When there is a well-established normal price, to which temporary deviations will tend to return, speculation can be expected to take the first form, in which it is stabilizing and tends to dampen fluctuations. But when the normal level is uncertain, or itself liable to fluctuate, speculation will tend to take the second form, in which it exacerbates the swings of the market.

The effects of monetary circulation can be seen in both structure and behavior. In the case of structure, money furthers the division of labor and specialization of function, and by facilitating competition, enables it to establish the law of one price and thus to bring about a tendency toward a common rate of return, in ways that will be explored later. It also unites the entire system into a single economy, whose components all move together and respond jointly to outside influences. This unification reflects technological interdependence, but it is monetary circulation that establishes prices and channels of trade. That is, it defines the channels through which causality flows, and it marks the boundaries of the system, thereby indicating where "externalities" begin. Those agents who are not included in the monetary circuit are not part of the system; activities that are not "monetized" are not subject to the pressures of competition.[5]

To portray the economic system as one in which there are uniform prices and wages, and a general rate of profit, is to draw on the results of competitive pressures that are only possible in a monetary system. Barter – we shall argue in Chapter 8 – could never bring about such uniformities. General equilibrium models – neo-Ricardian as well as neo-Classical – assume the "law of one price" in a barter framework, applying it to goods, labor, and capital. Without an account of a payments system, the law of one price cannot be justified.

In the case of behavior it provides a simple, universal motivation. Money is thrown into circulation in order to make more money; activities are undertaken because they pay. Simple maximizing makes sense, because motivation is uncomplicated, and everything can be reduced to a common measure. If households have stable patterns of preferences, then what things cost can be compared to what households want. Households will wish to achieve their desires for the least money expenditure; firms will wish to make as high a return as they can. This is the germ of truth in

[5] But agents can be included on quite different terms. To be actively included means supplying a service or property – managing a factor – to obtain a flow of funds from market activity, and in turn, returning those funds to the market through expenditure or investment. Agents, however, can be passively included as recipients of grants over which they have no control, e.g., grants from the state, or philanthropy, or by virtue of legal entitlements. Competition will only operate on activities and goods that are monetized. "Public goods" are thus excluded.

the basic neo-Classical picture of the market, with its two sides interacting through the circular flow of money (Hollis and Nell, 1975; Nell, 1992, ch. 1). The picture is marred by restricting itself, as we shall see, to an "exchange circuit," but it captures an essential aspect of monetary circulation. Even in this regard, though, it misses the connection between money and innovation. Behavior is always open-ended. It *can* (sometimes) be reduced to a simple stimulus-response formula – but it can always change, too, in response to economic pressures. Money makes such changes readily possible.

Money and credit

To the extent that uncertainty over promises can be eliminated, credit can replace money. In ancient Rome credit relations prevailed in internal transactions, money being used for external ones; family ties, social pressures and severe penalties for infractions ensured payment (Mommsen, 1991). In the trading cities of the Renaissance, *cedules obligatoires*, circulating letters of credit, filled in when shortages of money developed – as in Antwerp in the 1500s (Braudel, 1985). Credit can enable transactions to be carried out, so that with suitable clearing arrangements, the actual barter swaps can be minimal. But such credit depends on trust. Trust, in turn, must be based on dependable institutions.

From this perspective money and credit can be seen to be the endpoints of a spectrum, in terms of what might be called the "perfection" of the settlement (Hicks, 1989). This is not quite the same as "liquidity." An instrument is liquid if generally acceptable. But paper may be acceptable, yet imperfect, in the sense at issue here. An instrument is imperfect to the extent that the value transferred is incomplete, uncertain or nonuniversal. Money is the standard for all of these; that is, payment in "money" *means* settling the transaction. The medium expressing the standard has usually been bullion, although bi-metallist regimes relied on both bullion and plate. Bullion acquires value from production and is naturally divisible, easily measurable, etc. When gold bullion of the requisite weight and fineness has been paid, the transaction has been settled "perfectly." Bullion is internationally acceptable, its purity is easily determined, and its value unquestionable. Payments in coin, though equal in value, may be a degree less perfect, since the value of the coin is given by its face. This bears a stamp, of a ruler or mint, declaring that in virtue of the stamp, the coin is to be accepted at face value, because no matter how worn, it will be exchanged for metal equal to the face value. When a coin is accepted the immediate transaction is finished, but the recipient of the coins must decide whether or not to convert them – if, indeed, the ruler or mint will honor the stamp. Coins may be declared acceptable in payment of taxes and in legal settlements of debt. Of course, such acceptability is circumscribed by national borders. Paper of equal value could also be accepted; such paper can be issued in place of metal, given legal standing, and made convertible. Inconvertible paper can be given legal status. Privately issued paper – pure credit – may be offered and accepted in transactions.

Thus we have, in descending order of perfection, from intrinsic value to fiduciary and fiat based money:

> Bullion
> Minted coins, legal tender
> Convertible paper, legal tender
> Inconvertible paper, legal tender
> Government credit
> Private credit

Whether imperfection matters or not, is another issue. If it does, then the instrument will be illiquid to that extent, and users will have to pay a premium. But if there is no uncertainty, such as to convertibility or acceptability in court, there may be no premium. Of course, in modern monetary systems – bank deposits, electronic funds – in which there is nothing into which legal tender could be converted, there is little difference between perfection and liquidity. Legal tender is more perfect than private credit, but that is all. This is "credit money," which has no "natural value," and its emergence is part of a general change in the character of the system.

Competition and innovation

When products are sold for money, and then the proceeds used to purchase inputs and hire factors, there is necessarily a point at which all or a part of the working capital of an enterprise temporarily takes a fluid form. This makes it possible for the capital, or a portion of it, to be readily shifted from one line of activity to another. Of course there will still be such questions as whether the owners possess the appropriate knowledge and skills to move into the new area, or whether the other available factors, not in fluid form, can also be adapted. But when working capital "turns over" and passes through monetary form, it is potentially *mobile* in a way that stocks of goods and labor are not. Purchasing power can be moved from one sector to another, or transferred from one class to another; it is liquid in a way that a truckload of ball bearings is not. It is also liquid in a way that a promise to deliver a truckload of ball bearings is not. Since no one can know for certain that the promise will be kept, it is not negotiable, or negotiable only at a discount, and only within certain markets.

The fact that capital can take liquid form makes it possible to buy and sell entire businesses, meaning not just the buildings and equipment, but the "going concerns" (Commons, 1968). It is sometimes forgotten that prior to the emergence of the modern monetized economy, land, for example, and many of the institutions that organized productive activities, such as guilds, could not be bought and sold. Nor could membership in them be purchased or sold – just as families and family membership cannot be bought or sold today. As monetary circulation expanded, of course, these barriers were first eroded, then swept away. Land could not be sold – but the rents could be; later the land could be leased, and the lease sold. Guilds and guild membership could not be sold – but the right to the use of a guild license could be marketed. But before the erosion of these rules, they governed the institutions of production. Family membership was closely tied to guild membership in towns or possession of a manor or a position in a manorial system in the countryside. Indeed, the household and its living space was not sharply distinguished from the place of work; they tended to be nearby or even the same – as can be seen even now on family farms.

The fact that revenues and costs can readily be measured, and each take monetary form, means that the earning power of a business can be calculated. The stream of prospective earnings can therefore be estimated and capitalized. This sum can then be compared with the sum invested in the project – a sum which would have to be borrowed and on which interest would presumably have to be paid. The price of a business, then, is the capitalized value of its stream of earnings, which should equal the value of the equipment, buildings, and working capital invested in it. The latter is its historical cost, the former its earning power.

Once businesses can be priced, they can be bought and sold, as going concerns. Hence new businesses, producing new products, or producing old ones in new ways, can also be priced, and compared to existing enterprises. If a new project appears superior to existing enterprises, then it would be worthwhile to raise the funds to establish it.

Capital can be assembled in fluid form, to be invested in the most advantageous line; nor need the time for assembling it be very long. Money can be raised by borrowing, and institutional lending permits flexibility; it is not necessary for *prior* savings to be assembled. Funds emerge at the "fluid" point in the circuit, and they can be shifted from maintaining and replacing present capital to developing an innovation.

Calculating in money makes it possible to cost out innovations, while raising capital in money form creates the practical possibility to introduce them. Competitive pressures encourage a race to introduce innovations, where success or failure in this race is measured in money.

Such success, however, comes in varied forms. Capital invested in a business can *grow* or it can yield a *profit*. The rate of profit is the return divided by the advance; the rate of growth is the expansion of the business divided by the investment. Either is a measure of the return on capital – and both are also comparable to the return in interest on a loan. "Capital arbitrage" is the process of shifting funds between these three fields of investment – placing capital funds for growth, for a share in profits, or for (presumably safer) interest, to achieve the highest return (given the level of acceptable risk). Other things being equal, the highest of these three rates of return, adjusted for risk, will tend to attract capital funds.

Patterns of circulation

Before money had developed fully, it was widely used in trade. In part, the precious metals were used to "make change." Merchants would travel to distant ports, bearing goods, which they would trade, bringing back other goods in exchange. But the two shipments would not normally be of equal value. Hence the difference would be made up by a compact, durable, easily divisible, reliable, and widely acceptable store of value – gold or silver. Early on, rulers realized they could standardize the means of payment and earn a profit by stamping coins of gold and silver, guaranteeing that they would accept the coin at full weight, in payment of taxes; alternatively, at the Mint, they would trade it for a new coin or an equivalent amount of uncoined metal. If the ruler were powerful the stamp provided a guarantee of value, for if the coin were not of the requisite fineness and weight, if it were worn or clipped, it didn't matter, since the stamp promised that it could be exchanged for an equivalent value, of silver or gold. Such a guarantee was worth something; hence the ruler charged a fee – seignorage – for stamping metal into coins (Burns, 1927; Hicks, 1969; Wray, 1990).

Once coins had become widespread, they began to circulate within localities as well. Towns grew at convenient meeting places for traders; then became concentrations of producers. A pattern of exchange began to develop, relying on money, which, in turn helped to promote the division of labor. But in the early stages, these patterns were incomplete and did not encompass the whole of production. These can be called "exchange circuits."

A "circulation" implies that funds start at some point and return there; the circuit closes. Circulations with this property can be designated "*closed.*" The circuit will be spelled out as a succession of activities – exchanges or payments – raising two further questions: First, do these activities each have a unique, determinate value? Funds move by way of exchange, so the value of the activities is the value of the funds needed to make the exchanges. The value of the circuit, then, is the value of funds needed to accomplish all the exchanges, which, in turn, will be the sum needed for the *largest single transaction* (or set of transactions) occurring at any one point on the circuit. If this is unique we shall say that the value of the circuit is unique. Second, is the order in which the transactions occur unique and determinate? If so, we say that the *path* is uniquely determined. Further, given a closed circuit with a determinate value and path, we can ask whether the value of the circuit is *stable*, meaning that forces exist to correct deviations and keep transactions at the proper values.

But circulations can be *partial*, encompassing only a portion of the goods currently produced, or *complete*, taking in all goods. But even a complete circulation covers only all produced goods; businesses still might not be marketable for money. For that there must be capital markets. A complete circulation implies that all produced goods exchange against money, but, as we shall see, capital and investment introduce the additional complications that the system expands and that money exchanges against claims and promises as well as goods.

To recapitulate: the theory of circulation studies circuits of money, to determine whether such circuits are or are not: closed, complete, determinate, unique in value and path, and stable in value. Defining a circuit spells out the need for a certain sum of money, to enable the activities on the circuit to be carried out monetarily. It can be thought of as a "demand for money," but it is not a behavioral function in the usual sense and rests on different foundations than conventional supply and demand theories of money. It is better to think of it as a "payments system" (Cartelier, 1996; Aglietta, 1996).

Pre-capitalist circulation

Partial circulation: the exchange circuit

Consider a medieval county fair.[6] The county may be assumed to consist of a set of feudal manors surrounding a town, populated by merchants and craftsmen. Individual choice is not at issue; people are born into definite roles and behave accordingly – yet they are nevertheless motivated by prospects of gain. On the manors peasants grow crops, owing rent for their land to the lords. From their crops they retain their subsistence and market a surplus equal in value to their rental obligations. (They may barter subsistence goods and services among themselves, with bailiffs keeping records and arranging clearings.) The surplus will be marketed to merchants whose working capital is a supply of silver coins. With the proceeds of their sale the peasants pay their rents.

---

[6] Fairs were held several times a year, notably once each season, as the manorial economy had different products to market each time.

The lords may then enter the market with the silver. Merchants, having traded their silver for goods, now keep a portion of the goods for themselves, and advance the rest to the craftsmen – or, better, obtain an advance of silver from the lords, and, in turn, advance this to the craftsmen, who then buy the materials and wage goods they prefer (gradually returning this portion of the silver to the merchants). The craftsmen next work up the materials into fine goods and luxuries, adding value to the goods equal to the support of the merchants. (The peasants produce a surplus over their subsistence to support the lords, the craftsmen produce a surplus over theirs to support the merchants.)[7]

At this stage the craftsmen repay the advances from the merchants by delivering the finished goods to them; the merchants then turn over these goods to the lords, receiving the rest of the silver, thus completing the transaction – and the circuit – as all the silver returns to the merchants. All the goods brought to market have exchanged for money, the peasants' rents have been paid in money, merchants have been supported, craftsmen have bought their materials and wage goods for money, and the lords have obtained a better selection of goods, and goods of finer workmanship than they could have had individually from the surplus of their own estates (Nell, 1992, chs. 12, 13, esp. pp. 253–4 and n. 58).

Without money, the peasants would have paid the lords in kind or in labor (the corvee); the lords, in turn, would have had to support craftsmen on the manor, working in the manorial shops. The use of money makes it possible for merchants to develop as a specialized activity. This, in turn, enables the craftsmen to concentrate in the towns, where they can exchange ideas, compete, and develop their skills. They benefit from economies of scale, and from the improvements in technique that come from interaction.

A simple diagram illustrates this circuit (Fig. 5.1). On the main circuit money flows from merchants to peasants to lords back to merchants. In a

---

[7] Three kinds of pressure will operate to bring the town and countryside rates of surplus value (surplus over subsistence) into equality. First, factor mobility: In spite of ties to the land, labor will be mobile enough to establish a rough equality in levels of subsistence. Thus craft workers and peasants will tend to have comparable standards of living, adjusted for investment in acquiring skills. The same applies to merchants and lords. If merchants earn high rewards, lords can set up merchant operations in towns; less readily but occasionally, merchants can become landed. Second, monetary mobility: if merchants are not earning enough, they can move to other towns, or take their funds to other fairs. The shortage of funds will lower the prices merchants have to pay. Third, the level of rents and feudal dues can be adjudicated in court; prices will be set by town aldermen and can also be appealed in court. The standard is the "just" price, which allows each station in life the right or normal standard of living. It can be shown that this implies that prices equal direct and indirect embodied labor, which in turn implies equal rates of surplus value in different sectors.

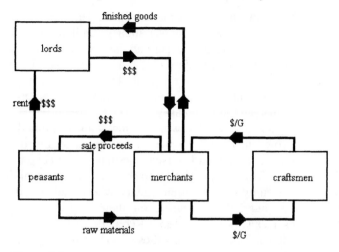

Figure 5.1

subsidiary circuit, money and goods circulate between merchants and craftsmen. Goods flow from peasants to merchants through craftsmen to lords, being partly consumed along the way by merchants and craftsmen, and worked on further by craftsmen. The diagram shows clearly that the circuit is both closed and incomplete.

Perhaps the most striking fact about this diagram is its great simplicity: like the famous textbook diagram, money flows one way, goods the opposite, and the two flows are equal – in the textbook case the whole amount of money swaps for the whole amount of goods. Here the value of silver = the value of the peasant surplus = the value of rents owed = the value of the lords' consumption = merchant income + craftsmen wages + raw materials. The supply of silver exchanges for each of these in successive stages of the circuit, clearly indicating that the value of the circuit is determinate and unique. The path, however, is not so clearly determined; more than one pattern of advances – from the lords to the merchants, merchants to craftsmen – is possible.

Goods are valued by their direct plus indirect labor content, and the same holds for silver. It is produced by a mining sector and may be coined or stamped by some authority, at the mine or elsewhere. But its value is measured by its weight and purity and depends on its labor content, not on the stamp of authority. Under normal conditions the labor value of silver will equal the labor value of the goods for which silver exchanges. When the goods brought to market exceed the normal level, however, prices will fall and the value of money in circulation will rise. Since more can be had for silver in circulation than it costs to produce, the output of silver will rise – or silver ornaments will be melted down. The increased silver in circulation will raise prices and lower money, until normal prices are reestablished. (As the mines get deeper and the veins exhausted, the labor content of silver at the margin may rise, raising silver permanently.) So the supply of silver will adjust through arbitrage to the demands of the circulation, and this pattern of adjustment is stable (though, as we shall see, this is an especially simple case).[8]

This is an exchange model which allows for a limited kind of production, devoted to improving the use-value of the already existing goods being marketed. The market exchanges have no effect on the initial supplies of goods. This is very similar to the Walrasian system, in which goods are marketed by self-subsistent agents, whose trade improves their utility positions, but has no bearing on their ability to support and maintain themselves (Koopmans, 1957; Rizvi, 1992).

So money exchanges only against those goods brought to market; the larger part of the goods produced remain in the countryside, where they are consumed directly by producers, or bartered, or circulated by makeshift coins of wood or leather – really tokens of indebtedness, markers or counters, rather than genuine coins. The system is closed, but it is not complete.

In such a system competition will encourage improvements and cost reduction in the towns but not in the countryside. Merchants and craftsmen compete with one another; a craftsman with improved wares will attract trade; one who can produce more cheaply may not be able to undersell (because of guild control over prices) but will be able to offer additional services or inducements. Towns will innovate, while the countryside stagnates.

Interdependence and production

The craftsmen of the towns are driven by money incentives and will look for ways to expand their markets. Living and working together they improve their skills. Consider what happens when they turn their attention to manufacturing, not luxuries, but, say, iron plows and leather

---

[8] Suppose some of the merchants at the fair are from abroad; they sell their wares and depart with silver, reducing the supply for the county. Consequently prices at the next fair must fall and the value of silver rise. It will therefore be worthwhile for merchants from abroad to travel there to make purchases – thus restoring the supply of silver. These are the circumstances that Hume had in mind, in arguing that "the same causes...must for ever, in all neighboring nations, preserve money nearly proportionable to the art and industry of each nation. All water, wherever it communicates, remains always at a level. ... [It] is impossible to heap up money, more than any fluid, beyond its proper level" (Hume, 1898, vol. I, pp. 330–41).

harness.[9] As a result of innovation and superior skills urban workmanship will be of higher quality than the products of the manors, with the result that when the urban products are introduced as means of production they will tend to raise the productivity of agriculture (Nell, 1992, ch. 13). As a result Agriculture comes to depend on Industry for tools, while Industry depends for wage goods and raw materials on Agriculture – and with the interdependence has come an increase in the surplus. For reasons similar to those mentioned earlier, pressures will develop to bring the rates of surplus value of the two sectors into equality.[10]

[9] Historically, the dependency of the countryside on the towns came about more from urban skills in making armor and weapons, than plows and harnesses. But this raises extraneous issues. More recently the question has been raised, e.g., by Persson (1989), whether the system of tied serfs working manorial lands was "efficient" compared to free labor working for money wages. This is not a well-formulated question. Lords did not "choose" between "alternative methods" of working their lands. (For one thing the lands weren't "theirs" in the modern sense of private property.) For another, part of the reason for farming the demesne was to ensure that the lords received the crops they needed to fulfill their military interests and obligations. And they needed peasant labor for other military supplies and purposes. The feudal system blended military obligation with economic claims; the two can't easily be separated. Furthermore, obligations, military and economic, were owed because of *status*, not because of contract. The peasant didn't *rent* his plot of land, in the sense that a modern farmer rents land. He owed "rent," feudal rent, because of birth, just as his lord owed obligations to his superior because of his status at birth. Such rents, due in virtue of status, were easily expressed, paid and collected in labor time, but proved to be difficult to calculate and collect in other terms. (If they are to be paid in grain, is it an absolute amount, or, as with sharecroppers, a percentage? How would the money equivalent to be calculated – on the basis of what prices?) It took the emergence of capitalism to make contractual rents and labor possible.

[10] Competitive pressures could not form a profit rate in these conditions, however, because the means of production were not available to be bought and sold. Land was still held in feudal tenure. It was not alienable, therefore no market for it could develop. Hence rents could not be capitalized. Similarly, trades and crafts were organized by guilds – but guilds could not be bought and sold. There was therefore no market for firms. Production was still organized largely through the household. (In fact ways were found around these obstacles, but they were costly.) It may be objected that these pressures are weak and imperfect. In that case rates of surplus value will only weakly and imperfectly tend towards uniformity. The labor theory of value and its corollary, the rate of surplus value, provide an organizing principle explaining relative prices and distribution in certain precapitalist conditions. Some writers (e.g., Samuelson, 1966, 1979; Morishima and Catephores, 1979) have contended that the labor theory of value should never be interpreted as a description of actual historical conditions. Then in Western Europe, prior to the emergence of markets for means of production, what was the organizing principle governing relative prices? If it is argued that there was no general principle, why does the price data exhibit regularity and order? Why did the commentators, from Aristotle to Petty, believe that prices reflected labor values?

Let us now consider a counterfactual supposition – that with growing interdependence lords seek to manage their holdings more efficiently, and decide to market the entire produce of the manor themselves, paying their peasants wages to work the demesne, instead of collecting rents. Lords never did this, at least not on any grand scale. But manorial relationships did become monetized, and many peasants did become agricultural wage-earners. By making this supposition we can examine the general circulation of money both against output and as wages. So we assume, likewise, that guild-masters market their output, and pay their craftsmen wages, keeping the money surplus to spend on themselves.

Although this system will now feature a complete circulation of goods against money, it will not be capitalist. For neither guilds nor manors are saleable – the enterprises are fixed in place; the lordship of the manor is inherited, and the manor entailed. Although peasants work the land for wages, they are tied to the manor of their birth and cannot move easily or freely from manor to manor, or from countryside to town. Masters of the guilds are elected, or appointed; membership in guilds, and sometimes masterships, will be inherited. In any event guilds and manors cannot be bought and sold. There is no fluid capital which can shift between one producer and another, or from Industry to Agriculture. But all goods will come to market, to exchange against money. Equally important, production will now be governed in the light of monetary incentives.

But with technological interdependence, and wage-labor in both sectors, it is possible to distinguish different categories of use for output. Some goods will be replacement means of production or subsistence, while others will be surplus goods. But production will not normally recommence until all output has been marketed. So the former will not be needed until the latter have been sold. Some transactions are now contingent on others.

There is another feature that has changed. In the exchange circuit the value of the goods in the circuit is always the same, and equal to the value of the money. But when all goods are brought to market this is no longer so. At the outset we have the means of production and subsistence, which presumably exchange for money. Production then takes place, resulting in gross output – replacement means of production and subsistence plus the surplus, a larger value of goods, which must now exchange for money. In Marx's notation (M for money, C for commodities, with primes indicating expanded quantities), the circuit is

$$M-C-P \rightarrow C'-M'$$

$C-P \rightarrow C'$ represents the production of the surplus. It is explained by the theory of surplus value, or in conventional economics by the theory of factor productivity. Whether either of these theories is adequate is not the issue; they nevertheless provide an account explaining how it is possible for C to become C'. But C' must exchange for a larger sum of money, M'. How does M become M' ? There is *no* theory to explain that – particularly if M is a sum of silver or gold (Nell, 1967, 1986). This is a problem the theory of the circuit must solve. (The Appendix discusses both Marx's and modern approaches.) As a first step, however, consider monetary exchange in an interdependent economy, without a surplus. This will enable us to see how the fact that some transactions are contingent on others makes it possible for a *given* amount of money to "circulate" first the inputs, then the larger amount of the gross product.

Elementary complete circulation: the two-sector "no-surplus" production economy

Consider a highly simplified economy consisting of two sectors. One sector produces a composite good which serves as means of production; the other produces a composite consumption good. This gives us two sources, two uses, and exchange. There are also two classes, workers and masters, who receive two different kinds of income. Both sectors are considered to be manufacturing; primary production is neglected here.

Assume that the means of production are fully used up each round of production, and that the consumption good supports labor for a definite period of time. The coefficients of production then are

$$a \quad b'$$
$$A \quad B'.$$

Only a single technique for each industry will be considered, namely the average technique for each industry currently in operation. Changing techniques requires scrapping and rebuilding plant and equipment, a process which cannot be studied until we understand circulation in given conditions.

The lowercase letters (a, b') stand, respectively, for the means of production and the number of labor units required to produce one unit of means of production. The uppercase letters (A, B') stand for the means of production and the number of workers required for the production of one unit of the composite consumer good. The column a, A shows the use of means of production, while the column b', B' shows the labor requirements in the two sectors. But the phrase "labor requirements" has to be understood carefully. What (b', B') show are the number of workers needed. So far, nothing has been said about how long or how hard they work. So we must define (b^*, B^*) as the number of workers, working for a unit period of time.[11]

Before defining this period, however, let us examine the question of the labor value of the two "commodities," or sectoral outputs. The labor value of an output is the sum of the labor directly and indirectly embodied in it, that is, the direct labor producing the output plus the direct labor in the inputs, and in the inputs to the inputs, etc. In a two-sector economy, however, it is simplified so that labor values are given:

$$\lambda_I = a\lambda_I + b^* \tag{1}$$

$$\lambda_C = A\lambda_I + B^* \tag{2}$$

where λ_I and λ_C are the labor values of the unit outputs of industrial and consumer goods, respectively. The first column of the righthand side then shows the indirect labor inputs and the second the direct labor inputs. According to the special labor theory of value, goods exchange in proportion to the direct and indirect labor time embodied in them. Hence

$$
\begin{aligned}
\lambda_I/\lambda_C = v &= \frac{b^*}{ab^* + B^*(1-a)} \\
&= \frac{\text{labor/capital goods}}{\text{labor/consumer goods}} \\
&= \frac{\text{consumer goods}}{\text{capital goods}}
\end{aligned}
\tag{3}
$$

The equations are the following.

Price equations:

$$P_1 = aP_1 + \omega b^* P_2 \tag{4a}$$

$$P_2 = AP_1 + \omega B^* P_2 \tag{4b}$$

where

$$
\begin{matrix}
a & b^* \\
A & B^*
\end{matrix}
$$

[11] Later we shall use b, B defined as equal to ωb^*, ωB^*, as the amounts of labor in the respective sectors, measured in terms of the consumption baskets required to support workers for the time required in production. The consumption baskets are assumed to be the same. It is less reasonable, however, to assume that the means of production in the two sectors are the same in all respects. The capital goods sector should therefore be considered partially "vertically integrated"; that is, goods which are used *only* as means of production in capital goods production will be expressed in terms of the means of production and labor used in common in the two sectors.

is the matrix of coefficients of machinery and labor inputs per unit output, with (b^*, B^*) representing the required number of workers, working for a *unit time period*; ω is the amount of the consumption good which supports one worker for the *unit time period*. Notice that this is not the wage rate as we shall define it later; furthermore we will see shortly that this can be set equal to unity. Continuing, P_1 is the price of machinery in accounting money and P_2 is the price of the consumer good in accounting money. Dividing both sides of both equations by P_2:

$$\frac{P_1}{P_2} = a\,\frac{P_1}{P_2} + \omega\, b^*$$

$$1 = A\,\frac{P_1}{P_2} + \omega\, B^*$$

Let $v = P_1/P_2$, and solve both equations for ω; then

$$v = \frac{P_1}{P_2} = \frac{b^*}{ab^* + B^*(1-a)} \tag{3'}$$

The units of accounting money cancel out.

Quantity equations

$$X_1 = aX_1 + AX_2 \tag{5a}$$

where X_1 is the amount of machinery produced;

$$X_2 = \omega\, b^* X_1 + \omega\, B^* X_2 \tag{5b}$$

where X_2 is the amount of the consumption good produced.
   Or rewriting,

$$X_1 = aX_1 + AX_2$$

$$\frac{X_2}{\omega} = b^* X_1 + B^* X_2$$

X_2/ω is the amount of the consumption good produced, measured as a multiple of the subsistence wage. It shows how many workers can be supported for the unit period. Dividing both sides of both by X_2/ω:

$$\frac{\omega X_1}{X_2} = \frac{a\,\omega X_1}{X_2} + \omega A$$

$$1 = b^*\,\frac{\omega X_1}{X_2} + \omega B^*$$

Let $q = \omega X_1/X_2$; then

$$q = aq + \omega A \qquad (5a')$$

$$1 = b^*q + \omega B^* \qquad (5b')$$

q is $X_1/(X_2/\omega) = \omega X_1/X_2$ which is the ratio of machinery to the consumption good, expressed as a multiple of the subsistence wage. Then, eliminating ω, we get

$$q = \frac{A}{Ab^* + B^*(1 - a)} \qquad (6)$$

giving us an expression for q which is symmetrical to that for v, the barter exchange ratio. Finally, eliminating v and q, respectively, we have from either set of equations:[12]

$$\omega = \frac{1 - a}{Ab^* + B^*(1 - a)} \qquad (7)$$

Next we put the quantities and prices together to form the complete system. Multiply the value equations by the quantities $(q,1)$ and the quantity equations by the values $(v,1)$:

$$qv = qav + \omega b^*q = vq = vaq + \omega Av \qquad (8)$$

$$1 = Av + \omega B^* = b^*q + \omega B^* \qquad (9)$$

It clearly follows that the condition for balance is that $av = b^*q$; the wage bill in the sector producing means of production (Department I) must equal the replacement cost of the means of production in the consumption good sector (Department II).[13]

We are here supposing that this economy produces just enough each round to replace the means of production used up in producing and to support the workers *during the time taken by production*. Remember we have redefined the coefficients, adding asterisks, to indicate that (b^*, B^*) now give the number of workers required *for the time it takes to produce* exact (direct and indirect) replacements, without surplus. At the end of the production round each sector has its own output but needs the output of the other. The consumption goods sector must trade its surplus beyond its own requirements for the means of production for the next round, while

---

[12] If ω is set equal to 1, this becomes $(1-B)(1-a) = Ab$, which implies, $(1/B - 1)(1/a - 1) = Ab/aB$, the expression we should have when the net wage and net rate of profit are zero, i.e., when the gross wage and profit rate are unity (Nell, 1973, 1992; Moss, 1980).

[13] Alternatively, from equations 3' and 7 and 6 and 7, $(w/v) = (1 - a)/b^*$, $w/q = (1 - a)/A$. Hence, dividing, we have $v/q = b^*/A$, so $Av = b^*q$.

the means of production sector must trade its surplus for the consumer goods it needs. Looking at the allocation of output another way, capital goods output goes to replace its own means of production and to replace the consumer goods means of production, while consumer goods output is divided between supporting its own workers, and those in the capital goods sector. Thus we have two sets of equations, expressed here as one for relative values and one for relative quantities. The value equations show the price ratio at which goods must be exchanged between the sectors in order for reproduction to take place; while the quantity equations show how the sizes of the sectors are scaled to each other, dividing the labor force between them. The quantity system defines the initial capacity of the consumer goods sector, and this will be an important basic unit, since its output defines a period of *time* during which the labor force can be supported. That is, this capacity, operated during a time, must produce an output which at the normal subsistence level will support the labor required both in it and in the other sector for that period. This defines the unit period, and the unit size.

The coefficients (a, A) and (b^*, B^*) are on a different footing. The first group (a, A) shows the means of production required to produce one unit of output, regardless of whether this unit of output is produced in a day, a week, or a year. The second shows the labor time reqired to produce exactly as much as is used up in production.

The first set is independent of time; the second set has two dimensions. On the one hand there is a *number* of workers needed to operate the process at unit level (think of the division of labor in an artisan shop); on the other hand, there is the time it takes to complete the process (how fast the workers work). These different relationships to time are the basis of Marx's distinction between "constant capital," which merely transfers its value to the product and "variable capital," which is capable of creating surplus value. The key to the creation of surplus value lies in the fact that while labor is supported for a given period of time by the subsistence wage, it can work more or less intensively (or for longer or shorter hours per day) during that time.

The circuit of money in the no-surplus economy
The key to the pattern of circulation will prove to be the condition for "balance" between the sectors, namely

$$Av = b^*q = \omega b^* \frac{X_1}{X_2} = b^* \frac{X_1}{X_2}$$

when ω is set equal to 1.

This states that the wage bill of the capital goods sector, which depends on the wage rate, the unit labor requirements in that sector, and the size of the sector relative to the consumer goods sector, equals the value of the replacement capital required in the consumer goods sector.

The actors in the drama of circulation are: merchants who bring money to the market and specialize in buying and selling; producers who sell their output for money, which they use to buy inputs and hire labor; and workers who receive money wages, which they spend on consumer goods while working. Producers are divided into capital goods and consumer goods firms. Circulation proceeds through the following steps:

1. At the outset, production is complete, and merchants have silver coins on hand equal in value, M, to the wage bill of the capital goods sector, b^*qX_2. Merchants buy the net output of the consumer sector, leaving that sector in possession of consumer goods equal in value to its own wage bill.

2. The consumer sector now buys its replacement means of production from the capital goods sector, with the funds which equal the value of that sector's wage bill. This is the crucial step, drawing on the balancing condition.

3. A secondary circulation now takes place within the capital goods sector. The capital goods sold to the consumer goods sector were produced by a subsector of the whole capital goods sector: call this subsector IB, and let the rest of the capital goods firms, who sell only to other capital goods firms, be subsector IA. The value of the replacement goods needed by IB, the subsector that sells to the consumer goods firms, will equal the wage bill of the firms that produce them in IA. Within subsector IA, however, the same subdivision can be made, between those firms who sell only to other firms in IA, and those who sell to IB; call these subsectors IAA and IAB, respectively. Then the wage bill of subsector IAA must equal the value of replacement goods produced by IAB. Repeating the procedure, we can see that the wage bill of subsector IAAA must equal the value of replacement goods produced in IAAB, and so on. As the wage bill successively circulates in the capital goods sector, it will enable each subsector to restore its wage fund and carry out its replacements – until we finally reach a set of producers who employ only their own goods as means of production.

The secondary circulation consists of an indefinitely long set of transactions, in which purchases of replacement capital goods provide revenue which subdivides into wages and gross profits.[14] Let W_I be the

---

[14] Neither the argument nor the formula depends in any way on the assumption that no surplus exists. The relationships hold generally.

initial wage bill, which, spent by the merchants, is next received by the capital goods sector in the form of replacement spending by the consumer goods sector. These receipts will be divided into wages and gross profit, the latter, $(1 - \omega\,b)W_I$, being spent on replacement. The wage bill of the first stage will be $W_{IC} = \omega\,b(1 - \omega\,b)W_I$, of the second stage, $W_{IIC} = \omega\,b[\omega\,b(1 - \omega\,b)W_I] = (\omega\,b)^2(1 - \omega\,b)W_I$, and so on. This converges to

$$\frac{1}{1 - \omega\,b}(1 - \omega\,b)W_I = W_I$$

Hence the proper wage funds are distributed to all producers in the sector; as they pay them out to workers, they will be spent on consumer goods, so the funds will return to the merchants. In the same way, the total gross revenue above wages of the sector can be calculated by summing the infinite series. The result is $[(1 - \omega\,b)/\omega\,b]W_I$ to which must be added the revenue above wages of the "final" subsector.

This "final subsector" produces its own means of production, as well as the means of production for the next-to-last subsector, the proceeds from the sale of which provide its wage fund. But the goods produced in the final subsector may nevertheless circulate against money. Consider a case where both products and processes are the same except for aesthetic differences, which appeal to different producers. Thus RED produces red machines using WHITE; WHITE in turn produces with BLUE, while BLUE will employ only RED equipment. All outputs sell for the same price. Either a promissory note or a sum of money starting with, say, RED, will first exchange for WHITE, then BLUE, and finally RED. Again, the actual transactions could be cleared, but in principle, the exchanges are against money.[15]

None of the transactions between the subsectors of the capital goods sector actually have to be carried out in sequence. Indeed, to do so might take an unconscionable amount of time. Instead, being desired and anticipated, they can be performed with promissory notes, which can be cleared in a central clearing house. All the firms are in the same sector and do business with one another, so they will know each other's reputations. Promissory notes can be passed along with endorsements. Since the balances cancel in the aggregate, all transactions can be settled, so that all firms receive their wage fund in money.

---

[15] This final subsector is more important than might appear at first sight. It produces its own means of production as well as the capital goods for other sectors; its productivity therefore sets an upper limit to the rate of growth. It is what Lowe termed the "machine tool" sector.

4. As the capital goods sector completes its "secondary" circuit, production will begin again, since output has been sold and the wage bill is on hand. As wages are paid, they will be spent, and the funds will return to merchants, as they dispose of their stocks of consumer goods. Upon the receipt of the first funds, some merchants, perhaps specialists taking advantage of the "short circuit" in the consumer sector, will buy additional consumer goods. These funds will provide the wages for consumer goods producers to pay their workers; as the workers receive their wages, they will spend them buying consumer goods, returning the funds to the merchants, who, in turn, will stock up with additional consumer goods, providing firms in the consumer sector another round of wage payments, and so on, until the revolving fund has completed the delivery of the stock of wage goods to the workers, over the period required for a full round of production. Thus a small fund, by revolving frequently, accomplishes a large set of transactions.[16] (Such revolving funds offer opportunities for financial innovations.)

At this point, when both sectors have paid out their wage funds, and the workers have spent their wages on consumer goods, all funds will have returned to the merchants, both sectors will have completed production, the consumer goods sector will need means of production, and both sectors will need to earn money to pay wages. The analysis has provided a constructive proof of the existence of a closed, complete circuit.[17]

[16] The number of times this wage fund turns over, in order to facilitate the sale of that portion of the output of the consumer goods sector destined for its own workers, will be its "velocity of circulation," using that term in the sense favored by both James and John Stuart Mill. "The essential point is not how often the same money changes hands in a given time, but how often it changes hands in order to perform a given amount of traffic." (Mill, 1848, book 3, ch. 8, sec. 3). Wicksell complains that this definition requires that we already know the value of money, so that velocity cannot become an independent factor in determining the value of money. But on the approach taken here, the value of (metallic) money *is* known, and the object of the analysis of circulation is to determine the required quantity – and to show that Mint regulations are sufficient to ensure that the monetary regime will be stable.

[17] But is the solution unique? The relation between the capital goods and consumer goods sectors is, but any number of solutions to the circulation of the consumer sector's wages may exist. The swap of the wage bill in capital goods for the capital replacements plus investment in consumer goods is the largest transaction that must be completed at one time. The capital goods must all be in place for a new round of production to take place; hence the purchase of the output of the capital goods sector must take place at one time. Hence the value of the capital sector's output represents the lower bound of the amount of money required for circulation. Mathematically, the problem is one in graph theory: to find the smallest amount of money that will exchange against all goods, which is to find the least cost Hamiltonian circuit for money (Christofides, 1975, ch. 10).

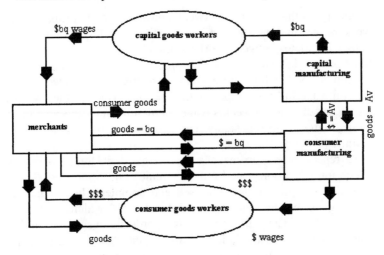

Figure 5.2

The circuit can be illustrated by a diagram (Fig. 5.2), in which the contrast to the exchange circuit is clear. It can also clearly be seen that there are three different patterns of exchange of money for good. Understanding these is the key to understanding velocity, and therefore the amount of money required in circulation.

There is first a straightforward swap, of a sum of money for a set of goods, when the merchants buy the consumer goods' surplus, and consumer goods in turn buys, all at once, its replacement capital goods. This exchange takes place at one moment; it cannot be spread out over time. The reason is that the consumer goods sector must have its entire means of production on hand in order to begin the next round of production. Production can't take place without the full complement of means of production being in place. This is an *assumption* which will be maintained throughout; but it is a reasonable assumption, designed to capture an important feature of reality, namely, that some goods must be on hand in full at the beginning of a round of production, while others can come in over time.

There then follows a sequential pattern of exchange, as the wage funds make their way through the capital goods sector, enabling each subsector to exchange its output, destined for use in that sector, for its wage bill, concluding with a sequential exchange of means of production in the "final subsector." Here again exchange must make it possible for the full complement of capital goods to be in place for the next round of production.

Finally, in the exchange between consumer goods firms and their own workers, we find a "rotation" in which a small sum of money continuously circulates between workers, merchants and firms, enabling wages to be paid and spent for goods, as production takes place. In this case the consumer goods which are exchanged against maney need not represent the full wage bill for the production period. Typically they will be much less, a day's wages, a week's wages, a month's salary. A small amount of money, circulating repeatedly, can bring about the exchange of a large volume of goods.

To repeat, money exchanges against goods in three distinct patterns, which must be coordinated for the circuit to be completed.[18]

Notice that the circuit is closed and complete. It also has a unique value, given by the requirement that $Av = qb^*$ which must be equal in value to M/X_2. The path is unique as regards the swap between the sectors and the secondary circulation. However, the "short circuit" depends on the length of the period for which wages are paid in the consumption goods sector, and this may be flexible. The question of stability will come up later.

These exchanges can all be made easier and speeded up through the use of bills of exchange and a clearing system. Less money will have to be moved around, and protected, and many transactions can be completed without waiting for shipments of money. Bills of exchange, in turn, will be more readily accepted if there is a market in which they can be sold (Wright, 1926, chs. 14, 15; Withers, 1937, ch. 4). The development of a commercial paper market will therefore speed up circulation, and thus, indirectly, increase productivity – since new production cannot be started, *ceteris paribus*, until the previous output has been marketed. (But once bills of exchange come into use, "bills of accommodation" will not be far behind.)

Fully developed complete circulation: the Mercantilist circuit

The "no-surplus" economy is interesting only as a hypothetical starting point that makes it easy to see how the circulation works. Because of its simplicity, the principles stand out. But it avoids the question raised earlier, namely, how can a given amount of money exchange first for a set

---

[18] In the literature examining monetary exchange these patterns are described, but usually as if all exchanges took place in the same manner. Thus the French authors suppose that all money swaps for all goods in one exchange; Wicksell describes sequential transactions, as does Marx; while Keynes refers to a "revolving fund" that ensures the circulation of goods. (Marx appears at different times to adopt all three, but never in combination.)

of inputs of a given size, and then later against a larger set of outputs? This must now be confronted.

The emergence of a surplus

The origins of the surplus will be examined more fully in Chapter 7. At this point a simple case will help explain the relation between productivity and velocity of circulation. Starting from the subsistence-level system above, consider a speed-up resulting from employers cajoling or coercing workers into working faster or harder. Inputs are processed into outputs in less time, but the amount of subsistence available to support workers during the production time remains the same. The actual conditions of work in the two sectors will very likely be quite different, but the speed-up will have to be the same in both.

Suppose Department I speeded up more than Department II. Then means of production would be completed before consumer goods, so the consumer sector would neither need them yet, nor have products to trade for replacements. The completed means of production will sit unsold until the products of the consumer goods sector are completed and ready to market. So the workers in Department I must either be laid off, or kept idle. If the latter, then the speed-up (beyond that of Department II) achieved nothing; but if they are laid off then worker resistance to speed-ups will be encouraged. Similarly, if Department I lagged behind Department II, when producing in the latter is finished, it cannot begin again, because means of production are not available, so workers will have to be kept idle or laid off. Furthermore, the faster producing sector will find it advantageous to encourage the fastest-producing firms in the slower sector by offering premium prices. Thus the faster sector will be slowed down while the slower sector will be encouraged to speed up. Hence the pace of work will tend to the same rate in both sectors, and will become institutionalized in formal and informal work norms. We will see later in Chapter 7 that if a sector cannot be speeded up, it can be made *larger*, thereby enabling it to continue supplying the needs of other sectors which are able to speed up.[19] (These conclusions generalize to any number of interdependent sectors, allowing for variations permitted by inventory policy. In general, the slowest rate of production sets the pace, a familiar result from critical path theory; Nell, 1992.)

Let t be the number of complete turnovers made within the period defined by the subsistence-level system. In the subsistence-level system

---

[19] Agriculture and mining cannot normally be speeded up. When manufacturing speeds up, demand for primary products will rise, driving up primary prices, creating profits which can then be plowed back to expand the size of primary activity to match the demand created by the speed-up.

itself, t = 1. Suppose there is a speed-up, so that t > 1. The production in both departments will be completed before the wage goods produced last period are exhausted. But the means of production for a complete production run are now available, and the newly produced consumer goods are sufficient to support a full period of production at t = 1. There is therefore a surplus of consumer goods – and *only* of consumer goods. Comparing this with the earlier equations for the no surplus economy, we see that the speed-up has produced a surplus consisting of the consumer good, while leaving the exchange ratio unaltered. Capital goods play a passive role. Both goods are available, so trade can take place and production can begin again.

The speed-up enables both sectors to complete a round of production in 1/t of the unit period. That is to say, Av is employed by (1/t)bq; exchanges between the sectors must take place t times before bq is expended. So tAv is needed for a wage fund of bq to be fully used. Each sector retains from its output such goods as it needs for its next round of production. They need 1/t of their wage funds. They can consequently devote the rest, 1−1/t, to paying surplus value. The rate of surplus value will therefore be:

$$s = \frac{(1 - 1/t)\text{wages}}{(1/t)\text{wages}} = (t - 1) \tag{15}$$

and this rate will be realized in the exchange which accomplishes the redeployment of commodities required to make the next round of production possible.

Notice the importance of exploitation in reaching this result. Because of pressure to produce a surplus, workers will be forced to work as hard and fast as they can, subject to the constraint that no sector can gain by working faster than the slowest basic workers in the slowest basic sector. Thus the pace of work is set by the *maximum* speed obtainable from the *slowest* workers, and competitive pressures force all to adjust to this speed. (Or, as we shall see later, adjust in size.) By contrast, suppose there were no exploitation and workers chose their own pace; suppose, in short, that the guilds were democratically controlled. There would then be no systematic pressure to maximize the pace of work, nor would there be pressures to adjust all production to a common work speed, for each group of workers would be free to pick its own position on the trade-off between the pace of work and the enjoyment of leisure.

What happens to the surplus? That, of course, depends on who appropriates it, and how. This requires some specification of the class system. Following our "stylized" history, consider a sketch of what might be called a "mercantilist economy." Each sector is operated by guilds, the

masters of which compete in the display of opulence and exploit their
journeymen and apprentices. These latter are forced to work both from
fear of the power of the masters and by lure of preferment – the possibility
of eventually rising to the status of masters themselves. The masters
control the sale of products and determine the disposition of revenues,
paying journeymen and apprentices only the socially defined subsistence
wage, appropriating the surplus for themselves. This can be thought of as
coming about through the degeneration of a system of once democratic
guilds, which have fallen under the control of the masters, who have
reduced wages to subsistence, while using their power to speed up
production. But the surplus is not used to expand or transform the production system; the *institutions* of production – guilds, manors – cannot
be bought and sold. The surplus is spent on various forms of consumption.
At first it may simply be used to support a nonworking population –
warriors, priests, and administrators, together with suitable retinues of
servants. But with development such classes will come to demand
specialized goods, and goods of higher quality. In general, then, the
surplus will be transformed into luxuries and items for display or osten-
tation. (As we shall see, this implies that the *composition* of the surplus
must be adjusted; it will have to consist of *both* goods, in the proportions
required to produce luxuries.) These, in turn, serve as a means to political
power and prestige, preferment at court, for example. Taxes and tithes will
also be levied per capita to support the State and the Church.

Under such conditions capital would not be fully developed. Workers
would be hired for pay, at a social subsistence level, and the surplus value,
in proportion to direct labor, would be appropriated by owners of the
means of production. But there would as yet be no investment nor
mobility of capital, which would be bound by mercantilist restrictions and
impediments, licensing, legal monopolies, and letters patent. Under these
conditions the surplus that emerges will stand in a uniform ratio to the
labor producing it. Moreover, institutional forces will tend to maintain this
ratio.[20] (Contrary to Marx, there is no need to assume uniform organic
compositions of capital, and in what follows I shall, in fact, assume that
the compositions in the two sectors are different.)

---

[20] Suppose in one guild – silversmiths – earnings per worker rose above those in
others. If this higher ratio were due to paying below normal wages, silversmiths
would come to face a shortage of apprentices, perhaps even lose journeymen. If
it were due to charging higher prices, bakers, butchers, carpenters and all the rest
would begin to raise prices to them, on the one hand, and on the other, petition
the municipal assembly and/or the courts to force them to reduce their prices.

Primitive duality: the late feudal system

So let us examine the system when a surplus emerges and its value is distributed in proportion to the direct labor employed. We begin with the quantities and coefficients of the no-surplus system; the speed-up implies that production will be completed before the wage fund, equal to the one round output of the consumer sector, is exhausted. When production is complete the sectors trade, means of production for means of subsistence. This trade is necessary since consumer goods cannot produce again without capital inputs. Capital goods cannot produce again without wage goods, which they obtain in this trade. The consumer sector holds back consumer goods equal to its original wage bill. Because of the speed-up, each sector will have on hand unexpended wage goods equal to its current obligation to pay surplus value, and will have after trade, their original wage bill.

So, dropping the asterisks and assuming $\omega = 1$, the equations for value and quantitites, respectively, will be

$$tv = tav + (1 + s)\,b \qquad (11)$$

$$t = tAv + (1 + s)\,B \qquad (12)$$

Clearly the solution for v will be unchanged. For quantities, initially, we have

$$q = aq + A, \text{ or in value terms } qv = vaq + Av \qquad (11^*)$$

$$t = (1 + c)\,(bq + B) \qquad (12^*)$$

From the way the surplus arises, it can be seen that $t = 1 + s = 1 + c$, so it follows from $tAv = (1 + c)\,bq$, that $Av = bq$. Since A, v, and b are unchanged, so is q. This is then the basis for the circuit.

Think of towns surrounded by small farmers. Towns produce tools for themselves and for the countryside; the countryside produces food for both. Calculating in per capita terms, merchants buy bq worth of consumer goods – agricultural products – for silver from the countryside, equal in value to M/X_2. Farmers then buy tools in the amount Av from the towns. This provides the towns with silver equal in value to bq; but only $(1/t)$bq is needed at this point. Hence silver equal to $(1-1/t)$bq will be available to pay as surplus value to the proprietors and masters in the towns. Production takes place again, and the same exchanges are repeated until the period is complete. That is, when t turnovers have taken place, $t(1/t)\,bq = bq$ will have been paid in wages, and passed through the secondary circulation, while $t(1 - 1/t)bq = (t - 1)\,bq = sbq$ will have been paid in surplus value. Farmers will then have bought tools equal to tAv. As merchants sold consumer goods and received sbq in silver from the towns' proprietors they will have bought cbq further from the countryside. With the funds from this sale, the countryside will both pay surplus value and engage its own labor – which it can do, since this is a "short circuit." That is, these funds will be paid out, spent on consumer goods, then used by the merchants at once to buy more consumer goods, enabling more funds to be paid out, which return to the merchants, etc., until the whole stock of consumer goods has been moved. The circuit follows the pattern of the circuit in the no-surplus case.

Developed duality: Mercantilism

But the elite can be expected to demand specialized luxury goods, produced by workers working with means of production. The surplus cannot therefore consist only of the basic consumer good; the relative size of the means of production sector will have to increase, to create a surplus in that good. How much of a surplus will be needed – and how much should the surplus of the consumer sector be reduced? What is required is the ability to produce higher quality consumer goods and luxuries. But these will be produced by the same methods and the same skills that are already in use. Thus the two surpluses should be proportional to the inputs of the consumer sector, so they can be combined in a "nonbasic" activity in that sector to produce luxuries.

To see what this implies we reexamine the equations for surplus value. First, multiply them by the quantities (q, 1). Then gross investment I can be written as the sum of the means of production produced for replacement use in its own sector and for the other sector, while total consumption C will be the amount of wages in consumer goods, plus the amount in capital goods, together with the surplus value paid to proprietors in each sector:

$$I = tqav + tAv$$
$$C = (1 + s)(bq) + (1 + s)\,B$$

When spending takes place on I and C, the capital goods sector will have to purchase consumer goods (pay wages to workers who will spend on consumer goods) and the consumer goods sector will have to acquire means of production. At the conclusion of a round of production, capital goods will need $(1/t)bq$, while consumer goods will need Av. So during a full period the terms tAv and bq will be traded (as opposed to tAv and $(1 + s)\,bq$). That is, the new q will be larger, indicating an increase in the relative size of the means of production sector to the consumer sector.

This gives

$$I = tqav + bq$$
$$C = (1 + c)\, tAv + (1 + c)\, B$$

where c, a quantity ratio, replaces s, the value ratio. These equations must also add up in quantity terms. Reading down the columns we find

$$tqv = tqav + (1 + c)\, tAv$$
$$t = bq + (1 + c)\, B$$

Canceling v and t in the first of these gives the *dual quantity system*:

$$q = aq + (1 + c)\, A \qquad (10')$$
$$t = bq + (1 + c)\, B \qquad (11')$$

Here (1 + s) represents the rate of surplus value, while (1 + c) is the ratio by which the output of each sector, minus the needs of the capital goods sector for that good, exceeds the amount needed in the production of consumer goods. (The luxury goods mentioned earlier will be produced – in unit amounts – by inputs of cA and cB/t, using the same workshops and tools, and the same workers, as in the consumer goods industry.)[21]

Solving both systems, we see that

$$(1 + s) = \frac{t\,(1 - a)}{Ab + B\,(1 - a)} \qquad (13)$$

and

$$(1 + c) = \frac{t\,(1 - a)}{Ab + B\,(1 - a)} \qquad (13')$$

---

[21] In a more general formulation, the luxury industry need not have exactly the same means of production to labor ratio as the basic consumer industry. Let the means of production per unit luxury output be m, and the labor per unit, measured in consumption good, be n (in italics to distinguish this case from that in the next chapter). So the luxury price equation will be

$$p_L = mv + (1 + s)n/t$$

and the dual will be

$$q = aq + A + m$$
$$t = bq + B + n$$

which becomes the same as that in the text when $m = cA$, $n = cB$.

so that, from equation 7, setting $\omega = 1$:

$$(1 + s) = (1 + c) = t \tag{14}$$

Further:

$$v = \frac{b}{Ab + B(1 - a)} \tag{15}$$

and

$$q = \frac{tA}{Ab + B(1 - a)} \tag{15'}$$

Further from equations 11 and 11', or from equations 15 and 15', we confirm that

$$tAv = bq$$

(As we will see in Chapter 7, these results generalize to many sectors; Nell, 1964; 1990.)

Moreover – and note the Marxian flavor – exchange at that ratio will bring about both the reallocation of goods necessary for reproduction and the distribution of the surplus, with surplus value being received in proportion to direct labor time. The surplus will consist of the quantities necessary to permit a nonbasic luxury activity in the consumer sector. Thus after a speed-up exchange takes place when a round of production is complete at the exchange ratio, v. The consumer goods sector will retain its full period wage fund, B, of which it only needs $1/t$, and swap $(1/t)bq$ for its required means of production, Av. But this provides the means of production sector with the wage fund it needs. That sector's period revenue, tqv, consists of tAv, the portion of its output swapped for wage goods, qtav, the amount of means of production held back for the next round, plus $(1 - 1/t)tAv$, its surplus value. The goods corresponding to surplus value, $(1 - 1/t)(tAv + B)$, thus make up a subsector of the consumer goods sector, capable of further refining consumer goods, converting them to luxuries. So this can be written $(1 - 1/t)(tAv + B) = c(Av + B/t)$. Hence the unit value of the luxury good would be

$$p_L = cAv + \frac{(1 + s)cB}{t}$$

The circulation of surplus value

Now consider circulation. Multiplying the value equations by the quantities, and adding the luxury good, we have

$$tqv = qtav + (1 + s)bq$$

$$t = tAv + (1 + s)B$$

$$tp_L = ctAv + (1 + s)cB$$

The luxury goods subsector will work at the same rate as the rest of the consumer goods sector; hence its equation must be multiplied by t, the turnover rate.

We assume that merchants have funds, equal or more than equal in value to the output of the luxury sector, but in any event no less than $(1/t)bqX_2$. Circulation then follows the sequence, calculating in per unit terms:

1. Merchants buy the output of the luxury good for silver coins.
2. The luxury sector distributes its revenue: it pays its proprietors surplus value, it purchases cAv from the capital goods sector, and it pays its workers wages equal to cB/t (or buys consumer goods cB/t, and advances them to its apprentices).
3. The capital goods sector pays surplus value equal to $s(1/t)bq$, and the consumer sector pays out sB/t.
4. Recipients of surplus value in all three sectors spend their incomes on luxury goods, returning the funds to the merchants.
5. Merchants now purchase consumer goods at least equal in value to $(1/t)bq$, the wage bill of the capital goods sector.
6. The consumer goods sector purchases replacement capital goods Av from the capital goods sector.
7. The funds from the sale of Av to the consumer goods sector now proceed through the secondary circulation in the capital goods sector. In this fashion they are paid as wages and are spent on consumer goods, while also circulating the capital goods for use in the capital goods sector. In the end the funds return to the merchants.
8. As the funds return to the merchants they buy further consumer goods.
9. As merchants buy consumer goods, the consumer goods sector pays wages, which are spent on the consumer goods, returning the funds to the merchants in a "short circuit."

This system is still mercantilist, not yet capitalist, since restrictions prevent the sale of enterprises, although labor is exploited and employers or masters are able to appropriate the surplus. These restrictions channel the forces of competition so that the appropriated surplus will be brought into proportion with direct labor, rather than total advances, that is, capital. A tendency to uniformity in the rate of surplus value, rather than a rate of profit, will be established.

With the speed-up of production, timing in the pace of consumption becomes crucial to the smooth working of the economy. A new set of consumers is brought into prominence – the recipients of surplus value, presumably the guild masters, the gentleman farmers, merchants, etc. If these new consumers spend at the same rate as workers, then if funds are paid out regularly, they will rotate so that the circulation will be completed at the same moment that production is finished. But if the recipients of surplus value spend more slowly, or if they save or hoard funds, the circuit will not be closed at the point when the round of production is finished. The merchants will not have their funds back; moreover, they will still have stocks of unsold goods, and a commercial crisis will ensue. As Mandeville and others saw, the obvious remedy is to encourage the regular and complete consumption of the surplus earnings; saving and hoarding, prudence and parsimony, should be discouraged. Profligacy may be a private vice, but it is a public virtue.[22]

Merchant earnings

But surely the merchants have to earn something? Merchants speed up the process of circulation; they help to make sales of output and purchases of inputs quicker and more convenient. Less time and effort is wasted in storage and bringing goods to market. This represents a saving on wage costs through a quickening of the pace of circulation, that is, a rise in t. Merchants, in the no-surplus economy, would be represented by a subsistence level of consumption. As turnover speeds up, therefore, they will obtain surplus value in proportion to their subsistence funds, exactly as in the producing sector.

We can write out the equations for this very easily. Merchants' labor time can be included in b and B, the coefficients giving the labor required for each sector. But merchants earn from turning over their funds. In the no-surplus case, merchants buy at the price representing the value, and sell to workers at a higher price, thus obtaining the whole fund back for less than the whole amount of consumer goods, keeping back enough for their own consumption. This consumption is equivalent to their labor. (If they kept back more, other guilds, prodded by disaffected workers,

---

[22] Prices are not very flexible in the guild system, so when demand for luxuries falls off, prices do not rapidly decline. Otherwise, worker consumption goods would fall, so that real wages and real consumption would rise in a compensating movement. Of course, as capitalism developed, saving for investment meant accumulation, in a system in which capital formed the basis of earnings, and competition engendered innovation. But this required a transformation of social life. In this case a fluctuation in investment relative to saving would lead to offsetting price changes, so that private virtues once again coincided with public – until the Keynesian era.

would help to set up rival merchants, undercutting them.) Thus, where μ represents the ratio of merchants' labor to total labor and λ_m is the support for merchants' labor measured in consumer goods,

$$\lambda_m = \mu\,(bq + B) \tag{16}$$

For this to work, merchants' markup must stand to price in the same ratio that merchant labor does to total labor:

$$\frac{p_S - p_B}{p_B} = \mu$$

With a speed-up, funds and goods will turn over faster, so that more sales will be made in the same time, and merchants will earn surplus value. In the no-surplus system merchants have funds equal to bq, the wage bill of capital goods, but they also turn over those funds separately in a "short circuit," exchanging a portion of them against B, the wage goods for the consumer sector. So their earnings arise from exchanges equal to bq+B, and when turnover rises, these funds must circulate the surplus value, and merchant earnings will increase in proportion. Therefore we have

$$(1 + s)\lambda_m = t\mu(bq + B) = t\,\frac{p_S - p_B}{p_B}\,(bq + B) \tag{17}$$

On the right-hand side, bq+B stands for the funds the merchants are turning over. When the rate of turnover rises, merchant earnings increase; t on the right-hand side is matched by 1+s on the left-hand side. Here p_S and p_B are the merchants' selling price and buying price, respectively, and must average to the value of consumer goods expressed in money. Under these conditions the rate of surplus value to merchants will be governed by t, the turnover speed, so will equal the rate elsewhere in the economy. A uniform rate of surplus value will prevail throughout the economy. In the next chapter we will see that the same rate of surplus value, governed by t, will also be established in the production of gold coins – mining and minting. Of course, as mercantilist barriers fall, and a rate of profit begins to form, the economy will cease to be characterized by a uniform rate of surplus value.

Capitalist circulation I: profits and growth

The capitalist production circuit
The regime of surplus value is an intermediate stage between feudalism and capitalism. As industry expands, the barriers preventing the

formation and movement of capital, and especially the buying and selling of enterprises, will be swept away, and in the new system the market will gravitate around prices corresponding to the rate of profits. Profits, in turn, will underwrite investment, as the system changes from stationary to expanding. The provision of the circulating medium will shift from merchants to banks. The evolutionary forces leading to these changes have been explored elsewhere (Nell, 1992, chs.12–15). For now the task is rethinking the pattern of circulation.

Equations

First, to show the characteristic trade-off in capitalism between wages and consumption on the one hand, and profits and growth on the other, the two-sector model must be adapted, showing changes in productivity, as before, but now exhibiting both a rate of profit and a rate of accumulation, while retaining the assumption of circulating capital. (Fixed capital will come later.) We begin with the same formulae as before, except we now write labor as b, B, measuring labor in the equivalent consumption goods; as before, t represents the number of turnovers in a unit period made possible by the speed-up; given a "no-surplus" system:

$$P_1 = aP_1 + bP_2 \qquad (18)$$

$$P_2 = AP_1 + BP_2 \qquad (19)$$

then let there be a speed-up, leading to a more rapid turnover of capital goods, producing a surplus of consumer goods. However, if the composition of this surplus is to include any of the capital good, the sizes of the sectors will have to be changed; this is done by multiplying by X_1 and X_2, keeping the total quantity of labor the same:

$$[tP_1 > taP_1 + bP_2] X_1$$

$$[tP_2 > tAP_1 + BP_2] X_2$$

Value relations do not depend on quantities, however. Hence, where $p = P_1/P_2$ and r the rate of profit,

$$tp = (1 + r) tap + wb \qquad (20)$$

$$t = (1 + r) tAp + wB \qquad (21)$$

Here w is the wage *rate*, meaning the competitively determined ratio of payments to labor to the cost of living, a process to be examined in Chapter 7. When this rate is unity, or 100 percent, labor's earnings just cover the cost of living; when it is, say, 110 percent labor earns 10 percent above the cost of living.[23] Then, solving for w in terms of $(1 + r)$, we have

$$w = \frac{t[1 - (1 + r)a]}{B + (1 + r)(Ab - aB)} = \frac{t[1 - (1 + r)a]}{D} \tag{22}$$

where $D = B + (1 + r)(Ab - aB) = B[1 - (1 + r)a)] + (1 + r)Ab > 0$, since the system's viability requires that $1 - (1+r) > 0$. From equation 22,

$$\frac{dw}{dr} = \frac{-tAb}{B + (1 + r)(Ab - aB)^2} = \frac{-tAb}{D^2} < 0 \tag{23}$$

which shows that profit and wage rates are inversely related, and

$$p = \frac{b}{B + (1 + r)(Ab - aB)} = \frac{b}{D} \tag{24}$$

so that $dp/dr = b(aB - Ab)/D^2 > $ or < 0 according to the sign of $aB - Ab$, that is, according to the relative capital/labor ratios of the sectors. (We will take it that the maximum rate of profit comes when $w = 1$, the basic cost of living. If $w < 1$, labor will not be able to function.)

The quantity relations start from the outputs in each sector in relation to their use as input. Initially we have

$$X_1 = aX_1 + AX_2 \tag{25}$$

$$tX_2 > bX_1 + BX_2 \tag{26}$$

By varying X_1 and X_2, a positive surplus of the capital good can be created. These two multiply the price equations, establishing the sector sizes. But profits will be invested, and wages consumed. Hence the uses of output must reflect this. Accordingly, the output of the capital good will be divided between replacement of its use in its own production, expansion in proportion to profits, and use in the production of consumption goods in proportion to consumer demand. Consumer goods output will be divided between replacement of basic wages in capital goods,

---

[23] To be analogous to other goods, labor must be represented by the means of subsistence that support it, and these, in turn, are produced by the consumer goods sector. Labor is shown as a column in the price equations, and the consumer goods sector will be the corresponding row. Since labor is represented by the goods that support it, the *basic* real wage will be unity. The *net* wage will then be a percentage of this, as market conditions drive the wage above or below unity.

expansion in proportion to the investment of profits, and providing wage good input in proportion to consumer demand in consumption goods:[24]

$$X_1 = (1 + g)aX_1 + cAX_2 \tag{27}$$

$$tX_2 = (1 + g)bX_1 + cBX_2 \tag{28}$$

so that letting $q = X_1/X_2$, and dividing by X_2,

$$q = (1 + g)\,aq + cA \tag{29}$$

$$t = (1 + g)\,bq + cB \tag{30}$$

from which we derive

$$c = \frac{t[1 - (1 + g)a]}{B + (1 + g)(Ab - aB)} = \frac{t[1 - (1 + g)a]}{D'} \tag{31}$$

where D' differs from D by having g in place of r.

Clearly g and c are inversely related, in the same relationship that holds between r and w. We also see that

$$q = \frac{tA}{B + (1 + g)\,(Ab - aB)} = \frac{tA}{D'}, \tag{32}$$

so that

$$\frac{\delta q}{\delta t} = \frac{A}{D'} > 0, \quad \text{and} \quad \frac{\delta q}{\delta g} = tA\,(Ab - aB)D'^2 > \text{or} < 0$$

according to the sign of Ab-aB, the ratio of the capital-labor ratios of the sectors.

When $g = 0$, c will be at its maximum level, and the whole net product will consist of the consumer good; that is the point from which we start when there is a speed-up. To form a net product which includes capital goods, there will have to be an adjustment of the relative sizes of the

---

[24] Notice how the dual is written here. To see what is at stake consider the famous Hicks-Spaventa model; the price equations are:

$p = rap + wb$ a, A are machines per unit machines, corn

$1 = rAp + wB$ b, B are labor per unit machines, corn

and p, r, and w and the price ratio, rate of profit, and wage rate. Then which way should the dual be written?

$q = gaq + Ac$ (Spaventa, 1970; Nell, 1970) or $q = g(aq + A)$ (Nell, 1986;
$1 = gbq + Bc$ $1 = c(bq + B)$ Park, 1994)

With slight adjustments in assumptions, each can be shown to work dimensionally. Each has a natural interpretation. The first shows the use of the relative amount of each good as divided between the two activities, investment and consumption. The use for replacement and/or expansion will be proportional

(*continued to next page*)

(Footnote 24 contd.)

to the activities, which depend respectively on the growth rate and consumption. The second shows the output of the capital good as devoted to growth by investing the capital good in each sector, while consumption is shown proportional to the labor employed in each sector. Solving each for g in terms of c gives the same relationship, parametrically identical to the w,r function in the price system. The first quantity system is fully equivalent to the price system. That is, the quantity variables behave in all respects the same way as the price variables. But why should this be taken as the correct way to write the quantity-growth relationships?

In the case of the "fully equivalent system" we find that Ap = bq. (By comparison the "corresponding system" above yields, wbq = gAp.) Let us take the normal condition to be described by the price equations and the fully equivalent dual, designating the quantities in that case by q*,1. Hence we have

$$q^* (p = rap + wb)$$
$$1 (1 = rAp + wB)$$

If all profits are invested and all wages consumed then

$$I = rapq^* + rAp = gapq^* + g(Ap)$$
$$C = wbq^* + wB = c(bq^*) + cB$$

If we were to cancel the prices (p, 1), we would have the "corresponding" form of the dual, with, however, the quantities of the equivalent system. But we have not taken into account that in carrying out expenditure plans, the consumer goods industry will have to purchase capital goods and the capital goods sector, or its workers, will have to acquire consumer goods. Hence Ap will exchange for bq* resulting in

$$I = gapq^* + gbq^*$$
$$C = cAp + cB$$

The first column shows the use of the capital good in the activities of investing and consuming; the second column shows the use of the consumer good in those activities:

$$pq^* = gapq^* + c(Ap)$$
$$1 = g(bq^*) + cB$$

The prices, (p,1), can be canceled out, giving the fully equivalent form of the dual. But when goods are used in the course of production, they give rise to costs and revenues. The revenues will have to be exchanged in order to cover the costs; hence Ap will swap once again with bq*, and g will be replaced by r (investment spending realizes profits), and c by w. This results in

$$pq^* = rapq^* + wbq^*$$
$$1 = rAp + wB$$

Here if we cancel out the quantities (q*,1), we have the original price equations.

The fully equivalent form not only allows duality to be properly expressed, but also permits the study of the spending of profits on investment and wages on consumption, the rationale usually given for writing the "corresponding" form. It is the correct form of the dual, and it subsumes the other.

sectors. The equations for the use of the outputs show a relationship between growth and consumption which is dual to that between profits and wages. The relative quantities of the sectoral outputs are related to the growth rate in the same way prices are related to the profit rate.

The Golden Rule

Perhaps we should not too readily assume that r = g. Even in a simple model why shouldn't there be saving out of wages and/or consumption out of profits? There is a clear answer: markets and competition. If workers are saving, their wages are above the cost of their socially defined lifestyle and can be forced down (by encouraging immigration, to build up a "reserve army," or by labor-saving innovation). If capitalists are consuming out of profits, there is an opportunity for some among them to invest instead, grow faster, reap economies of scale and take over the markets of the spendthrifts. This will keep consumption out of profits down. Many expenditures which appear to be such consumption, however, should more properly be considered competitively necessary expenses of maintaining "intangible" capital, such as information networks, goodwill, brand loyalty, and the like. The issue will be considered further in Chapter 7.

With this in mind, note that since

$$p = \frac{b}{D} \quad \text{and} \quad q = \frac{tA}{D'}$$

if $D = D'$, that is, if r = g,

$$pq = \text{value of capital goods per worker} = \frac{tAb}{D^2}$$

But $dw/dr = -tAb/D^2$. Hence,

$$\frac{dw}{dr} = -pq \tag{33}$$

where the right-hand side could also be written $-K/N$. That is, the slope of the wage-profit trade-off equals the value of capital per worker when the Golden Rule holds. This means that Ndw, the increase in payments to labor, exactly equals $-$ Kdr, the reduction in payments to capital, or vice versa. The implication for circulation is that a change in distribution causes no net change in financing requirements; any change can be managed by shifting funds from financing capital payments to financing payments to labor or the reverse. Banks can shift the composition of their loan portfolios, between financing investment and financing wage advances; merchants can switch between purchases of capital goods and purchases of consumer goods.

It also follows that the elasticity of a point on the wage-profit trade-off equals the ratio of the share of capital to the share of labor (Nell, 1970).

Look again at the equations for prices and quantities, above. When $r = g$, $D = D'$; hence, dividing prices by quantities,

$$\frac{p}{q} = \frac{b}{tA}$$

the *ratio* of prices to quantities is invariant to the distribution. This implies that $dp/p = dq/q$, or that $(p/q)dq/dp = 1$. The relative quantity of capital goods to consumer goods moves uniformly with the relative price. Cross-multiplying, we obtain

$$tAp = bq$$

the cross-sectoral balancing condition! This holds for all distributions between wages and profits, so long as the Golden Rule is satisfied.

This conclusion can be reached by another route. Multiply the price equations by the quantities and the quantity equation by the prices, and cancel. The result is

$$(1 + r)\,tAp = (1 + g)\,bq, \text{ or since we shall assume } r = g,$$

$$tAp = bq \tag{34}$$

Given the speed-up, then as long as all and only profits are invested, whatever the distribution, prices and quantities will stand in a fixed, that is, invariant, relationship. To put it another way, the value of the relative output of capital goods, evaluated at the relative price of consumer goods, is a constant. When the real wage (or rate of profit) changes, the central relationship in monetary circulation is not affected.

The circuit of profits

First, continuing with the assumption that money is based on a gold-silver-copper combination, we shall explore how it may be earned by an export industry, a subsector of consumer goods, which produces a good for sale abroad. The size of this sector is determined by the amount of additional money needed each period to finance accumulation and profits. Since the basic funds for circulation are M/X_2, which must be equal in value to bq, the additional money will be $gM = \$$ equal in value to $rbqX_2 = [(1 + r)tmp + wn]X_2$, where m is the coefficient of capital goods in the export sector and n is the labor coefficient. Since the size of the sector is determined by the monetary requirement, we will fix it arbitrarily as unity, so that the amount that has value $\$$ is equal to 1, the per capita unit quantity of gold. (Later we will consider paper money, issued by banks in proportion to their equity.)

So we must rewrite our equations, adding the export subsector, and setting everything out in monetary terms. Labor is expressed in terms of the consumer goods that support it, and the wage is a pure number representing a percentage of the basic wage:

$$q\,[tP_1 = (1+r)\,taP_1 + wbP_2] \qquad P_1\,[q = (1+g)\,aq + c(A+m)]$$
$$1\,[tP_2 = (1+r)\,tAP_1 + wBP_2] \qquad P_2t = P_2\,[(1+g)\,bq + c(B+n)]$$
$$\$ = (1+r)\,tmP_1 + wnP_2 \tag{35}$$

where we assume that the export good is exchanged for gold of equal value. Thus exporting is simply another way to produce gold. (It is convenient to aggregate the export subsector into the consumer sector proper. The exported good is simply a modified version of the consumer good, and m and n stand for the inputs of means of production and labor that make the modifications and carry out the exporting activity.)

Here the price equations are multiplied by the quantities and the quantity equations by the prices. When both sides of the right equation on the top row are multiplied by t, both equations on the top row equal tP_1q; and when $r = g$, the first terms cancel. Since $w = c$, we obtain

$$t(A+m)\,p = bq, \quad \text{where } p = \frac{P_1}{P_2} \tag{36}$$

So, from the assumptions above, we have

$$rbqX_2 = \$\Pi \tag{36'}$$

where $\Pi = 1$ in equilibrium, that is, when the natural value of money is equal to the value in circulation.

The stock of money will serve as working capital, that is, to purchase capital goods and pay wages. The amount required will be equal to the wage bill of the capital goods sector. The previous period's output of both goods is available. The consumer sector buys its *new* equipment, providing the capital goods sector with the wage bill for its *new* investment. Funds thus circulate, as each sector spends on investment. So production can commence, businesses will pay workers, who, in turn, will buy consumer goods, returning the funds to the businesses in the consumer goods sector. As these funds return to it during production, the consumer goods sector will order its replacements from the capital goods sector, in which the returning funds will circulate. Both sectors will earn profits, which in turn will support the payments for the net product. Part of this will be in the export sector, which will earn the funds needed to augment the funds for working capital. For in a growing system, the money supply must be expanded each period. Let's consider the sequence step by step.

The system is expanding at a regular pace. Capital and labor are growing in tandem, and the balancing condition holds since each side grows at the same rate. At any point in the process, the amount of gold is fixed from the previous period, but the increment required for the next period will have to be earned during the current period. Circulation proceeds in a variant of the manner already described for the mercantilist case. Each sector "invests," that is, purchases investment goods; each sector adds to its next period's operating capital, so that it will set additional means of production to work with additional workers. *This period* the capital goods sector's per unit wage bill was bq, but the consumer sector produced consumption goods for that sector equal to $(1+r)bq$; *this period* the consumer sector employed $t(A+m)p$, per unit, but the capital goods sector produced capital goods for delivery to that sector of $(1+g)t(A+m)p$.

Circulation proceeds in two stages, in which the first circulates the net product, the second the replacement goods and basic wages. Merchants begin the process by purchasing $rbqX_2$ worth of consumer goods, the new output destined to exchange against the new wage bill in the expanded part of the capital goods sector. With the funds from the sale of these goods, the consumer sector buys its *new* means of production from the capital goods sector. (They may also buy part of their replacement goods.)

This sets off a secondary circulation in the capital goods sector, in which the *new* capital goods for that sector are circulated, and during which the wage funds return to the merchants in exchange for the new wage goods. So rbq has exchanged against $gt(A+m)p$. At this point, the merchants recycle the funds that have returned to them and purchase further consumer goods so that they have on hand goods corresponding to the full wage bill of the present capital goods sector, bqX_2. At this point the consumer sector, in turn, can complete its acquisition of replacement capital goods from the capital goods sector, $t(A+m)pX_2$, providing that sector with its wage fund.

A secondary circulation ensues, during which the replacement capital goods in the capital goods sector are exchanged, and the wage funds are spent on consumer goods, returning the money to the merchants. So, in this second round, in per unit terms, bq will have exchanged against $t(A+m)p$. Finally, the merchants buy consumer goods for sale to consumer goods workers, who receive wages and spend them on the goods, returning the funds to merchants in a "short circuit." In effect the same exchange process repeats itself, first on a small scale to circulate the new capital goods and new wage goods, then on a larger scale to circulate the replacement of established capital goods and wages.

Notice that in the process of circulation it is not necessary to distinguish between the net wage and the basic wage. For in the quantity and price equations, q is defined to cover c, net consumption, and p has to cover w, the net wage. Therefore rbq and bq represent amounts of consumption goods that include net consumption $c = w$, and $gt(A+m)p$ and $t(A+m)p$ represent values of capital that cover net as well as basic wages. When r changes, in such a way that g changes with it, then w and c will change in an offsetting manner, and we have seen that $\Delta rK = -\Delta wN$. Funds which merchants had previously intended to devote to the purchase of capital goods will now be redirected to the purchase of consumer goods, or vice versa. Alternatively, funds which banks had intended to loan for investment, can now be redirected to providing advances for wages. But the change in distribution does not change the total funds required.

This accounts for the circulation of all currently produced goods. But the investment means that output next period will be greater, as will the wage bill and the replacement funds. Hence the stock of gold must also expand. This takes place through a subsector of consumer goods. In addition to the purchase of its replacements for consumer goods worth bq, the consumer goods sector received the capital for the export subsector, tmp (and in the trade for investment, it also received the increment of this period over the last in its means of production). The export sector produces at the same time as the rest of the consumer sector; its wage bill is managed as part of the short circuit, and investment takes place through the same exchange that circulates the investment goods of the two principal sectors. At the end of the period the export sector sells its output, equal in value to rbq, abroad for gold. The stock of gold therefore expands in the ratio $(rbq)/bq = r$, which in turn equals g, the rate of capital accumulation.

Variations: banking and services

Alternatively, instead of merchants, the circulation could begin with banks advancing investment funds to the consumer sector equal to their new capital. The firms in the consumer sector purchase these capital goods, providing the capital goods sector with the wage funds needed for new operations. These pass through the secondary circulation, enabling the exchanges of capital goods in the capital goods sector, and returning to the consumer sector in exchange for consumer goods. Consumer goods can now repay the banks (or roll over the funds for the next round of exchanges). For the banks now advance (additional) funds for the purchase of replacement capital goods; this provides the capital goods sector with its full wage funds, which enters into the secondary circulation, as before. As the funds return to the consumer sector, they repay the banks, or roll over the funds again, as the banks finance the short circuit of wages in the consumer sector.

Suppose there is a special service or luxury subsector, and that the size of the consumer goods sector has been adjusted to enable them to produce the goods needed to support such a subsector. As funds return, as part of the short circuit, the merchants or banks advance a loan for its wage bill to this sector. Let the wage bill be W_s, where the loan is arranged so that the amount to be repaid is L_s, but the funds actually advanced are equal to $W_s = (L_s)/(1 + i)$, where $i = r$. The profit from this industry will go to those making the advances. The spending of the loan on service wages will ensure that the corresponding consumer goods, in turn, are sold; while the spending of the profits by the merchants and/or bankers will put funds in circulation, which added to the receipts of the consumer sector, will sum to the value of the services. The expenditure of those funds on services will provide the service sector the receipts that enable them to repay their loan, returning the funds to the merchants/bankers. This subcircuit will expand as the consumer sector expands; since the earning of funds abroad is adjusted to the size and growth of the consumer sector as a whole, the new funds for the service sector will be earned along with the rest – assuming that the service sector grows pari passu.

This circuit is closed and complete, and it has a unique value determined by the wage bill/capital requirements equation. But there are a number of ways in which the path could vary. The short circuits could take various forms. The turnover of funds for replacements could proceed more or less rapidly. Many different arrangements can be imagined between consumer sector businesses and the export industry. In each case market pressures will tend to engender innovations.

Notice that while the new money may be deposited with the banks, initially it comes into the hands of the export industry. This is an expensive way to expand the money supply. In the next chapter we will consider how a banking system could supply an equally reliable paper currency more cheaply, and with further advances in communications, it could provide for the needs of circulation through bank deposits. Market pressures will tend to support such innovations. The Golden Rule will prove basic to each of these.

Capitalist circulation II: the modern economy

So far the argument has been developed in the context of circulating capital. What we have examined may be called the *current circuit* or the *income circuit*. But Mass Production requires fixed capital, which brings with it a new set of problems for the analysis of circulation. Fixed capital turns over slowly – it lasts longer than the normal period of production; its production and sale must therefore be financed. This finance, however, should be considered a normal – inescapable – part of the circulation. Here again the key to successful circulation will prove to be the Golden Rule.

The circulation of fixed capital

The chief peculiarity of fixed capital goods is that they outlast the period of circulation, with the result that their producers cannot sell replacements to their customers at normal marketing times. A further but related problem arises because of their size, which means that new investments will not take place regularly each period. New investment – increasing size and capacity – may take place only at the time of replacement. Yet producers of these goods must somehow continue to produce if they are to stay in business and meet the demands that will eventually come due, and for this they need circulating capital – to cover wages and current inputs. This means that the producers of circulating goods, who use fixed capital, must lend to producers of fixed capital goods, who require circulating capital to continue operations, and the effects of this interchange, including the interest payments, must be accounted for.

Certain simplifying assumptions will be useful. First, circulating goods can all be assumed to move in one common period of time. Second, let us suppose that the industries that produce fixed capital goods employ circulating capital, while circulating capital industries use fixed capital; the use of fixed capital in fixed capital can be ignored for the moment. Third, all fixed capital goods of a given kind will be traded at the same time; a common period of depreciation will be assumed.

Prices, wages and profits will be taken as given. During the course of time the durable goods are gradually used up, so that at the close of a period of circulation an industry employing fixed capital must set aside a certain part of its earnings to cover depreciation. Producers of fixed goods must acquire circulating capital in order to continue production, but they cannot sell their output as yet. Hence they must borrow the depreciation funds to cover their circulating capital needs.

The amount of borrowing a fixed capital producer needs to undertake depends on the ratio of the write-off period to the period of circulation (i.e., the period in which fixed capital goods are produced). Bearing in mind that the circulating capital for the first period after sales will be earned by the sale, the borrowing required, β, in a given industry, i, will be

$$\beta_i = \frac{wp - cp}{cp} \, c_i$$

where wp is the write-off period, cp is the circulation period, and c is the circulating capital. In the same way, the amount that producers of circulating goods in industry j, will be willing and able to lend, L, will depend on the same period, and on the level of depreciation per period, d:

$$L_j = \frac{wp - cp}{cp} \, d_j$$

To show that borrowing will equal lending in an economy in balance, consider a system with fixed capital but no surplus (or one in which the surplus is devoted to paying wages). The circulating capital industries use fixed capital, while the fixed capital industries use circulating capital. If there is no surplus, then the circulating industry's demand for fixed capital must just equal the fixed industry's demand for circulating capital. All that is necessary is to adjust to a common write-off period, say, n periods of circulation. Then every nth period, the fixed capital producers will deliver the durable goods, F, and each of the following n periods, circulating capital producers will pay off 1/n of the price. Thus c = F/n.

Now suppose that the system is expanding each period through the reinvestment of profits, so that r = g. Then the output of both fixed and circulating goods must grow each period at 1+g; which implies that the circulating goods consumed by producers of fixed goods will expand at the rate $(1 + g)\mathbf{q'C_f p}$, while the fixed goods consumed by the producers of circulating goods must also expand at that same rate, to maintain balanced growth. But the transaction in which this balance is realized can only take place at the end of the write-off period, that is, after n periods of production/circulation. Hence compound growth must be figured in each case: we require that

$$(1 + g)^n \mathbf{q'C_f p} = (1 + g)^n \mathbf{q'F_c p}$$

This approach will be developed in Chapter 7. Here $\mathbf{C_f}$ is the matrix of circulating inputs in the fixed capital industries, and $\mathbf{F_c}$ is that of fixed inputs in the circulating capital industries; $\mathbf{q'}$ is the vector of industry sizes in balanced growth proportions, and $\mathbf{p}$ is the price vector corresponding to a uniform rate of profit. The output of fixed capital goods for eventual sale to users, who produce circulating capital goods, must grow over the write-off period at the same compound rate as that at which circulating capital goods output – and rest of the economy – is expanding.

How can this be arranged? Fixed capital producers must borrow circulating capital at a growing rate in order to reproduce an expanding output. Fixed capital users must accumulate funds at an expanding rate in order to have the wherewithal on hand to make the eventual purchases of appropriately expanded capacity. Thus each period fixed capital users will invest their (growing) depreciation allowances in a sinking fund, which will earn compound interest. This sinking fund, in turn, provides loans of both current allowances and its own earnings, to producers of fixed capital goods, enabling them to continue production on an expanding scale. So long as this interest compounds at the rate $i = g$, and provided the depreciation allowances each period are correctly calculated, and grow from period to period at rate g, the sinking fund will be just the amount required to enable the users of fixed capital to purchase the expanded output of durable goods, thus providing them with appropriately expanded productive capacity.

Note that users of fixed capital are assumed to be able to expand their output before the fixed capital is augmented. When installed the capacity of fixed capital will be assumed to rise, first as it is broken in, then as the labor force learns to use it, and finally as usage suggests various improvements. Hence its capacity level will increase over time, and this will be figured into the planning of investment. To the extent that this is not so, in a growing economy, producers will have to overbuild at the outset. Excess capacity will have to be built into the system.

Money, capital, and the sinking fund

When producers of circulating goods complete a period of production they may be assumed to sell their output to traders for money. An appropriate portion of this money is put into a sinking fund to cover eventual replacement costs; the rest goes to pay income and to cover the new period's circulating charges. The sinking fund is then loaned (by a bank or broker) to fixed capital producers, who use the proceeds of the loan to purchase circulating goods, thus returning the funds to the traders. This puts them in position again to purchase the next period's production of circulating goods. Each period the size of the circulating industries expands at the rate $1+g$; each period a given fraction of the value of the growing fixed capital is set aside for replacement and, together with the accumulated compound interest, is available for lending. The fixed capital called for at the time of replacement will not be the original amount, but an amount reflecting growth. During the period of its use, fixed capital *expands in capacity*, as a result of learning by doing and/or planned improvements that take place as it is broken in. The output produced by fixed capital grows, as do circulating inputs. Hence the amounts set aside each period will also grow, as they are a fixed proportion of the productive capacity of the fixed capital. Each such portion will grow at compound interest, which is the rate at which circulating capital grows, compounded over the write-off period.

Depreciation allowances should be calculated so as to keep capital intact while growing at the prescribed rate. Thus if fixed capital is F and the allowance is A, where g is the rate of growth and i the rate of interest, capital at the end of the first period will be $(F - A) + (1 + g) C + (1 + g) A$, where C is circulating capital (wages, materials, energy, etc.). That is, $(1 + g)A$, rather than A must be set aside, because capital is growing. At the end of the second period capital will be $(F - 2A) + (1+g)^2 C + (1+g)A + (1+g)(1+i)A$; at the end of the third period, $(F - 3A) + (1+g)^3 C + (1+g)A + (1+g)(1+i)^2 A$; and so on. As the existing stock of fixed capital is worn down, the sinking fund is built up, and compounds at a rate of interest which in equilibrium will be equal to the rate of growth.

In this way fixed capital can expand in pace with circulating capital. The sinking fund will be loaned to producers of fixed capital, providing them with the circulating finds they need to expand their operations so as to produce and warehouse the increased set of fixed goods that will be needed at the replacement date. Each period they will borrow the new allowances plus the earnings from the sinking fund. Their accumulated borrowings will equal the value of the output of fixed capital goods; which is to say that the sinking funds will be just the sum required to purchase the expanded output required for the growth-augmented replacement.

(The loan in period 3, say, was that period's depreciation allowance, so equal to $(1+g)^3$ times the initial allowance, A. By the end of the write-off period, this will have compounded to

$$(1 + i)^{n-3} (1 + g)^3 A = (1 + i)^n A, \quad \text{since } i = g$$

So at the end of the write-off period all loans of depreciation allowances will equal the current period's depreciation charge, since the amounts borrowed have grown at rate g, while the amounts loaned have compounded at rate i.)

At the end of the write-off period circulating producers sell their current output, and, instead of putting the last period's depreciation allowances into the fund, these allowances will be employed to make the first payment toward the replacements. At this point the loans will be called in; as we saw the funds will be exactly the sums needed. As loans are called in, payments are made; the funds will be deposited, and the increased deposits will justify banks in expanding their note issues. Hence there will be no shortage of funds. The repayments of loans will recirculate as payments toward the purchase of capital goods, until all loans are settled and all durable goods paid off.[25]

The condition $i = g$ here enables the output of fixed capital to expand in pace with the growth of circulating capital (wage goods), by providing a way to finance continuous and expanding production of durable goods, even though these cannot be sold except at widely spaced intervals.

Continuous production and circulation

Underlying the discussion so far has been the assumption that production processes take the "batch" form: they begin and end at definite moments of calendar time. It has been tacitly assumed that these moments are coordinated, so that all producers in an industry operate more or less together – as for example, farmers do. Further, the producers of inputs and purchasers of outputs of any given industry coordinate their activities with the timing of that industry. In the history of the advanced world these assumptions would not be problematic prior to World War I, and perhaps even up to World War II. In the postwar era, however, batch processes have been replaced or are being phased out. Mass production replaces the batch system with continuous output.

Batch production means that a plant begins processing a certain amount of raw material, running it through a number of successive stages. Each stage basically occupies the attention of the entire plant while it is running. At the end, the batch is complete, ready for marketing; then, and not before, another batch can begin. (Think of farming.) Continuous production means that as soon as one batch of raw material has begun to be processed, another can be started up. Input goes in continuously – or at regular but frequent intervals – and output emerges steadily. It is not necessary to wait for a process to be complete to begin work on another set of inputs. (Think of an assembly line.)

Continuous production implies continuous circulation; there will be no clear beginning and ending points on which all processes converge. It might seem, therefore, that a different approach is required. If production is continuous, for example, it will not be the case that producers in the

<hr/>

[25] A similar condition can be shown to hold for the residential housing market (and for commercial construction, also). The average length of mortgages should be keyed to the turnover rate of the housing market. Then the real interest rate on mortgages should be set equal to the real growth rate of housing – which will depend on population growth, and the rise in per capita income, on the assumption that higher per capita incomes will imply a demand for better and more expensive housing. If the earning of Savings and Loans are proportional to the real rate of interest, then when $i = g$ the S&L bank capital will grow at the same rate that deposits and loans expand (since these expand at the rate of real growth, g).

capital goods sector all finish at the same time, ready to sell replacements to the consumer sector, and to one another in the secondary circulation. Instead of exchanges of given amounts at a moment of time there will be a flow of exchanges over a period of time.

The fact that processes run continuously, however, need not imply that the pattern of circulation will work differently. Processes take time; production starts at the point where the raw material is acquired and set up for processing, and finishes when the good is turned over to be marketed. There is still a definite *sequence* to the operations, and this sequence defines the relation of the process to other processes. Moreover, completing these sequences takes a definite amount of time, during which labor must be paid. Thus, for example, each day work on a new set of auto bodies will begin, requiring as input a certain amount of appropriately alloyed steel. The output of this steel from the capital goods sector must be adjusted to this rate of input, with appropriate precautionary inventories. Steel inputs of a different kind of steel for gear shifts may be required only weekly, and steel for engine blocks in larger amounts, monthly, perhaps, but only in the six months prior to the assembly of the new models. These input rates must be coordinated, "backward" with steel producers, and "forward" to adjust the timing of the production of components, so as to assemble the whole car without delays caused by waiting for some process to be completed. Once the processes have been coordinated, it will be possible to designate the point at which inputs enter, and the later time when the finished product is complete. The wage bill will then be the flow of wages of the number of workers involved for that period of time. Summed over all capital goods industries, the wage bill so calculated for the capital goods sector will have to equal the flow of capital inputs into consumer goods industries, for a similar period.

In short, the processes producing capital goods and those producing consumer goods must still be coordinated. Even if there are no common starting up and completing points, outputs and inputs must be delivered in the right amounts, and in the required sequences, so that they are available at the points when they are needed in the continuous process. Producing too much will require storage; producing too little will cause costly delays.

The essential coordination is adjusting the flow of capital replacements and investment for the consumer sector, so that it will be available as the consumer sector produces a flow of output to meet the spending of the wages of the capital goods workers. The relationships between the different subsectors within capital goods must be similarly coordinated, so that the capital sector's wage fund can circulate without a hitch. Such coordination is a principal function of monetary circulation.

The case of fixed capital showed how mutually dependent processes with different turnover periods are adjusted to one another through credit. This can be applied to circulating processes with different turnover periods as well.

"Continuous output" should not be overstressed. Even under Mass Production the seasons, traditional holidays and social customs provide a framework that sets definite marketing dates toward which manufacturers aim. Think of summer vacation time, Christmas shopping, the New Year sales, spring fashions, the fall Back-to-School sales.[26] So, while under continuous production there need be no common starting and finishing points, these will often exist, nevertheless. In any case, there must still be coordination of inputs and output, and this requires that different processes be adapted to one another. Outputs designed to serve as inputs in other processes must emerge in a flow that provides them in the required amounts at the necessary times. Hence the timing and size of processes must be adjusted.

Conclusions

Circulation analysis shows in detail how money flows through the economy, and exchanges against produced goods. A closed complete circuit means that the economy is fully monetized. Understanding the circuit makes it possible to calculate exactly how much is needed to exchange against the total output. This has been shown in two circuits, the current or income circuit, and the fixed capital circuit.

As we shall see further, it also helps us to see how money adapts to the needs of trade. In terms of modern debates, all money is "endogenous," in one way or another. Money is supplied by commercial enterprises for commercial motives – mines, merchants, even mints, which charge seigniorage. Later as technology develops, bringing changes in markets, the medium of circulation will be provided by banks and financial institutions, again with profit in mind. But the fact that the medium of circulation is supplied in response to economic motivation does not mean automatic adjustment. On the contrary, as we shall see, the economic motivation can destabilize the system. To prevent this, the process has to be regulated by the State. These regulations, however, must be designed to work with, and not against, the market forces. The next chapter will deal with the issue of monetary stability.

---

[26] As late as 1926 seasonal considerations were so significant that open-market commercial paper in New York also had a seasonal character (Wright, 1926, pp. 514–15).

Appendix: Recent theories of circulation

According to the modern, largely Franco-Italian, account of monetary circulation, in industrial economies in which the surplus, or part of it, is invested banks provide the funds for circulation, yet the basic pattern can be considered a swap of money for goods. The idea is simple, and in its contemporary form appears to owe its origin to Marx. Money circulates against goods in the circuit,

$$M - C - P \rightarrow C' - M'$$

where M is the funds capitalists advance, C represents the goods and labor capitalists purchase, P is the production process resulting in C', where $C'C$ (the difference being surplus value), and M' is the money which capitalists receive back from the sale of C'.

In Marx the theory of surplus value explains how the total *value* of goods and services produced can be greater than the total *value* of the inputs entering into production. This accounts for $C - P \rightarrow C'$. (In conventional economics this is the job of the theory of factor productivity – although this latter is often little more than an assumption.) In linear production theory the relationship between relative industry sizes, consumption, and growth can be used to show how the *quantities* produced can be greater than the input *quantities* – a result which is mathematically dual to the value relationships.[1] But there is no theory which tells us how the initial sum of *money* which circulates the inputs can expand into the larger sum which circulates the outputs. How does M become M'? (Luxemburg, 1951; Bukharin, 1972; de Brunhoff, 1973, pp. 58–72). Marx wrestles with this, but arguably loses the match when he settles for the position that the actual sum in circulation is M'.[2]

---

[1] Strictly speaking, most linear production theory starts by *assuming* that the matrix of coefficients is "productive." By contrast, Marx provided a theory explaining this: workers were paid for an amount of time in the form of command over goods that would support them for that time. Capitalists then put them to work, and cajoled or coerced them into producing more goods in that time than were required to support them and replace the means of production. In other words, capitalists *speeded up* production, i.e., did just what Taylor's management science said they should do. However, Marx did not foresee the enormous changes that Mass Production would bring.

[2] Marx posed the problem, "How the capitalist manages always to withdraw more money from circulation than he throws into it. ... The question at issue ... is not the formation of surplus value. ... [A] sum of values employed would not be capital if it did not enrich itself by [creating] surplus value. ... The question ... is not where the surplus-value comes from but whence the money comes into which it is turned" (Marx 1967, vol. 2, p. 330). Capitalists as a class advance the

Critique of the "single swap" pattern

On this view, *theoretically*, it is correct to speak of M becoming M′, but *in practice* there is no initial sum of money, M, followed later by a larger sum, M′; there is *only* M′. The gross product swaps in a single exchange for an equivalent sum of money – just as the agricultural surplus swaps for silver in the markets of the Middle Ages (Nell, 1992, ch. 12). There is an annual market; all goods are brought to market at one time, and exchanged against an equal value of money. Merchants now have goods, producers money. Incomes are paid and spent, and the goods are bought back again. The surplus is consumed, means of production and subsistence are employed productively, and a new gross output is produced, ready to be exchanged for money again.[3]

(contd.)

total money capital; it is then converted into productive capital, "which then transforms itself within the process of production into commodities worth [more than the original capital] . . . [so that] . . . there are in circulation not only commodities . . . equal to the money-value originally advanced, but also a newly produced surplus-value. . . . This . . . surplus-value . . . is thrown into circulation in the form of commodities. . . . But such an operation does not by any means furnish the additional money for the circulation of this additional commodity-value" (p. 331). He offered several solutions to the problem. The gold-mining sector could produce the additional funds; but then the entire surplus would end up in the hands of gold producers. He abandons this idea and toys with the notion that "the problem itself does not exist. All other conditions being given, such as velocity . . . a definite sum of money is required in order to circulate commodities worth [a definite sum] . . . independently of how much . . . of this value falls to the share of the direct producers" (i.e., of how much is surplus value). In other words, M′ exchanges for C′, the proposal discussed in the text. Later Marx notes that money is thrown into circulation by capitalists, who advance only their money capital, equal to the means of production and wages, therefore to M, not M′. At this point he tries to set out a sequence of intersectoral transfers which will circulate surplus value. Methodologically, this is the correct approach, although his account is defective (Nell, 1986).

[3] A peculiar exchange circuit appears in Blanchard and Fisher (1989), pp. 178–9. The government issues money and provides it to "individuals" as a transfer payment! Individuals deposit their endowments of goods with firms in exchange for bonds (the firms are supposed to operate a "storage technology," in which goods deposited with them increase over time); individuals buy goods from firms with money. Firms deposit the money with banks, who use it to redeem the bonds. This is not a plausible account of circulation. The government's behavior is unexplained and unbelievable. The banks do not seem to be necessary – or even convenient. Without employing labor the firms magically produce goods which individuals want; and finally, taking the system on its own logic, neither money nor bonds appear to be necessary, since the firms could issue claim checks to the individuals (as interest on their bonds) which could be used to redeem their own goods or buy those of others.

This approach faces a number of problems. First, as Marx himself noted, the money-capital advanced is the working capital; the only money "thrown into the system," as he puts it, is that advanced by capitalists. This will always be less than the gross output, at least under conditions in which the net product is entirely consumed. Capitalists seek to minimize their costs; they will therefore *advance less money* than they draw out from the sale of output. This is what created the problem in the first place.

The weakness of this position is evident in a second difficulty. This framework has trouble dealing with productivity changes. When productivity rises, the quantity of goods will be larger, and so it seems that to keep the price level constant the quantity of money must increase proportionally, since the money must exchange with the goods at the marketing date. But when productivity rises in an interdependent production system, the quantity of goods exchanging against money need not increase. Productivity may increase because goods are being processed *faster*, as in the case labeled the "late feudal system." In this case the same quantity of goods may be taken to market as before; the only difference is that they will be taken more frequently, but there will be no increase in the quantity being marketed at any given time. (To say that *over* a period of time, the quantity of goods on the market has risen just obscures the fact that at any *given* occasion of marketing the same quantity comes to market). This means that the quantity of money must likewise turn over faster. But wages are paid, and households spend, at the same rate as before; the difference is that workers are working faster. Of course, in other cases the amount of goods exchanged against money does change – but the circulation is also speeded up. The "single-swap" approach cannot distinguish these cases, or analyze the speeding up of turnover.

Third, to base circulation on M' is expensive. The total quantity of money needed must equal the value of the gross output. In the case of circulation by a metallic currency, that is a large sum to keep tied up unproductively. In the case where circulation is funded by bank loans, it will require substantial interest charges, which will eat into profits (and which must be accounted for in the circuit). If a less expensive system could be devised, there would be opportunities to profit by introducing it.

Fourth, in an interdependent production system, not all goods can be marketed at the same time. No one will be in the market for means of production until net output has been sold. Production takes time, and during this time, wages must be paid, and household consumption goods purchased. There is a *sequence* of transactions, which follows the sequence of activities in production and distribution.[4] This means that total output cannot swap at one time for the total money supply.

The circulation approach

The approach favored by the French Circulation writers described by Deleplace, and represented in Deleplace and Nell (1996), does not rest on merchants or on token money, metallic or paper. Instead, money consists of bank deposits, in a system in which banks make loans to enable production and exchange to be carried out. Banks create deposits by making loans. That is, a loan is an asset, a deposit a liability, hance when a loan creates a deposit, the books stay in balance. But the liability must be managed; the bank must be able to meet all demands on it for liquidity, which means that it must be part of a *system* which makes liquidity available upon demand to members. Although it is bank-based, there is one striking similarity between the French approach and Marx – in the French account total bank advances appear to equal total GNP[5] (see Poulon, 1982, ch. 2; also Parguez, 1995). The circuit works as follows.

Firms anticipate the demand they will face, and plan employment and production accordingly. They turn to the banks for funds. Banks advance working capital to firms (to those they consider creditworthy), who pay wages to households, who, in turn spend the wages for consumption, returning the funds to firms, who repay the banks. Banks make short-term

---

[4] This did not appear in the case of the medieval county fair, because manorial production was not governed by monetary transactions (Nell, 1992, chs. 12, 13; 1986).

[5] The Franco-Italian circulationists are by no means alone in making this assumption. In setting out a transactions-costs model introducing money into GE, Ostroy and Starr assert (1990, p. 34) that "agents must have, at each trading instant, sufficient money to finance their current purchases." This is then embodied in a definition: "The economy is said to be monetary if there is a good, 0, so that for all households,

$$p_o b_o^h > \text{ or } = p\,x^h$$

(where p represents price, b is the endowment vector, and x the vector of net planned purchases). A monetary economy is hence described ... by the property that there is a zeroth good universally held in a quantity sufficient to finance all purchases" (p. 35). Apart from the meaning of "net" in the above, this appears to be the same idea as in the circulationist scheme: the supply of money covers all planned purchases. However, for Ostroy and Starr money is among the fixed endowments.

loans to finance purchases of equipment to firms, who buy investment goods, returning the funds to firms, who repay the banks (Graziani, 1985, p. 161).[6] (The funds so advanced are *initial* investment funds, making the production and sale of capital goods possible. *Final* investment funds, fixing the long-term debt-equity structure of firms, are raised separately through the capital market.) Output equals C + I; banks make loans equal to C + I, and these funds are spent buying the output. At the end of the period banks are supposed to be repaid; if there is a failure to complete the circuit, as will often happen, financial instruments will be issued, enabling the circuit to begin again. Over time financial obligations can be expected to build up, until they become sufficiently burdensome to cause a crisis, and are wiped away in a restructuring.[7]

This can be illustrated with a simple diagram (Fig. 5.3). Banks advance working capital and investment purchasing funds to entrepreneurs in each sector. Wages are paid to households, who spend them on consumer goods, returning the entire wage bill to the firms of the consumer sector. Entrepreneurs in both the consumer goods and the capital goods sectors spend on investment, buying goods from the capital goods sector, returning the entire advances for investment to the firms of the capital

---

[6] Funds must be advanced for production in both sectors: Keynes wrote, "The production of consumption goods requires the prior provision of funds just as much as does the production of capital goods" (Keynes, 1973, xiv, 282).

[7] A modern variant of the "single-swap" circuit sets the rate of growth of the money stock equal to the rate of growth of output (Foley, 1984, 1986; Semmler and Franke, 1988). In Foley's version lags in the processes of production, sale, and "recommittal" (respending net revenues) will make it impossible to sustain growth unless there is net new credit creation. Only in the implausible case of a zero recommittal lag – said to correspond to an infinite velocity of money – would net new credit not be required. This would imply that capital goods replacement and new investment were undertaken instantly in each industry, as soon as its round of production was complete. Supply would therefore confront an instantaneous demand, and there would be no need for credit – presumably because producers could finance their purchases with receipts from their own sales. But this could only happen if the stock of money were sufficient to enable all exchanges to take place at once, which would require that it be equal to the gross product. Foley has not answered Marx's question – how can gross output be purchased for money when capitalists only put total costs into circulation? Foley argues that because credit must be advanced to fill the recommittal lag, the money supply will grow at the same rate as output. But the growth of the money in circulation from period to period is a *different* problem, to which there are several answers, e.g., earning money from sales abroad, new output from mines, new issues of paper justified by growth of bank assets, etc.

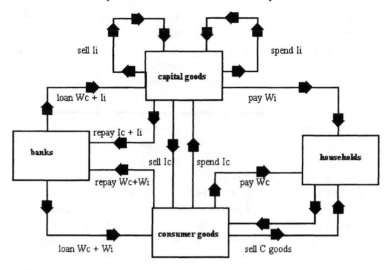

Figure 5.3

goods sector. The funds thus return to the firms and are supposed to be repaid to the banks.

Critique of the Franco-Italian circulation approach

Criticism of this scheme should not obscure the merits of the French circulation approach. For example, it develops the theory of money on the basis of circulation – money as medium of exchange – rather than portfolio behavior – money as an asset. This perspective has been largely overlooked in mainstream writings. Taking this approach leads one to draw a sharp line between *banking activities* which provide the medium in response to market signals, and *capital market activities* which are concerned with asset management. (This suggests that it is important to distinguish between short-term and long-term interest rates, and that the influences on the two may not be the same.) Given that production and exchange depend on the willingness and ability of banking to supply the medium, money can hardly be "neutral." Moreover, money adapts; the approach explains just how and why "loans create deposits," which helps to clarify the relationship between money and credit, and it also explains the way "money adapts to the needs of trade." Finally, the circuit itself provides a useful framework for periodizing time in macro-economic analysis, replacing the cumbersome and ambiguous "short-period/long-period" distinction.

However, the working of the circuit requires that funds be advanced and then return to their starting point. All goods and all labor must exchange against money. Further, all movements of funds must be consistent with economic motivation. No unnecessary expenses can be incurred. The usual account of the circuit in the Franco-Italian literature faces difficulties.

For when part or all of the net product is invested, and banks advance gross investment funds, which together with working capital add up to gross output, even if the funds are promptly spent, it does not follow that the banks can be repaid. While, in the aggregate, the funds return to the aggregate borrowers, namely the firms, these are divided into two kinds, operating in different sectors. Firms in *both* sectors will have borrowed investment funds, but only those in capital goods earn them back in sales, and *both* sectors will have borrowed wage funds, but only consumer goods firms get them back in sales. Will the borrowing and earning from sales balance overall? This will depend on an equilibrium condition, which itself reflects the interdependence between the sectors.

The equilibrium condition, of course, is that the wage bill of the capital goods sector should equal the investment expenditure of the consumer goods sector. This can easily be seen from the social accounts: the capital goods sector spends on wages and capital goods, but earns from the sale of capital goods; the consumer goods sector spends on wages and capital goods, but earns from the sale of consumer goods for wages. The net sales of the capital goods sector are its sales to the consumer goods sector for investment. The net sales of the consumer goods sector are the goods it sells to the workers of the capital goods sector. For equilibrium these two must be equal.[8] But the French authors do not explain how this comes about.

Indeed, the strength of the French approach is precisely that it recognizes that the sectors are interdependent. Yet it retains the idea that the total money supply equals the value of total output. This results in a pattern of circulation that appears at times to run counter to economic common sense. If the total wage bill and funds for total investment are both advanced, then the capital goods sector will find itself awash in unnecessary cash. For when the capital goods sector sells investment goods to the consumer sector, it will receive funds equal to its wage bill.

---

[8] The French account of the circuit depends on this condition being fulfilled, through some form of adjustment, yet many authors take the very strong position that markets provide wholly inadequate coordination (Cartelier, 1995; Deleplace, 1996). How then is this condition established? It has been argued in the text that a correct account of the circuit explains this.

Yet it has just borrowed its wage bill! Why does it borrow this money (incurring expenses and interest) when it can earn it?

The problems can be seen by reexamining the diagram. Unless it is expected that its sales to the consumer goods sector will cover its wage bill, the capital goods sector would not borrow working capital from the banks, for they would not be able to repay it. But if such sales in fact will cover the sector's wage expenses, why should its firms borrow? Would they not be better off asking for advance payments on orders? Or working out a mutually acceptable clearing date? (Nothing is specified in the diagram about timing or priorities, e.g., that wages have to be paid before the goods are sold.) To borrow working capital in these circumstances means to incur unnecessary expenses.

Following the account in the diagram, when the capital goods sector sells to consumer goods, why do they repay the loan of their wage capital at once? If they kept even a fraction of the proceeds, they could use such funds to circulate the remaining investment goods among themselves, working out a clearing system for transactions within the sector. Brokers and dealers would see the opportunity to take business away from the banks by providing a clearing service that would be cheaper than taking out loans. They would then not need to borrow investment funds from the banks, saving themselves expenses and interest.

(If the capital goods sector did not borrow its working capital from the banks, but instead earned it from sales to the consumer goods sector, a "secondary" circulation within the capital goods sector would ensure that all firms obtain the working capital they need – as we see in Chapters 5 and 6. As the wage bill successively circulates in the capital goods sector, it will, in the limit, enable the unexchanged amount of produced goods to be reduced below any arbitrary level – until we finally reach a set of producers who employ only their own goods as means of production. This "final subsector" produces its own means of production, as well as the means of production for the next-to-last subsector, the proceeds from the sale of which provide its wage fund. But the goods produced in the final subsector may nevertheless circulate against money, in the manner indicated in Chapter 5. None of the transactions between the subsectors of the capital goods sector actually have to be carried out in sequence – being desired and anticipated, they can be performed with promissory notes. Since the balances cancel in the aggregate, all transactions can be settled, so that all firms receive their wage fund in money.)

In a similar fashion the consumer goods sector also incurs unnecessary expenses, in the Franco-Italian account of the circuit. It is assumed to borrow its entire wage bill from the banks. Yet the sector sells its own products to its own workers. Would it not be advantageous to find a way to reduce the borrowing costs, by advancing claim checks to workers and clearing them among themselves? A small amount of money would suffice for this, an amount which could continuously circulate.

Even if no clearing system were devised, the whole wage bill would not have to be borrowed. A fraction of the whole wage bill would do, since it would be advanced, then spent, returning to the firms, who would pay the funds out again. Suppose the production period is two months; firms might borrow one week's wages for that period, paying it out eight times, receiving it back each week as workers spend it on their household consumption. They certainly need not borrow the full eight-week wage bill for eight weeks.

A further problem concerns repayment: when all wages have been paid and spent, and investment purchases have been made, the funds return to firms, who must now repay their loans. But surely they must now pay interest? Where do the funds for *this* come from? The only money in the system is that previously advanced by banks. This money – and no more – is paid out and spent, and so this amount is earned back from the sale of goods. How do the firms come up with the money with which to pay interest?[9]

This blends into a more general difficulty. Firms are assumed to borrow their entire working capital plus their prospective investment, and to repay both at the conclusion of the circuit. Ignoring interest for the moment, this implies that they borrow the entire gross product, and repay it.[10] How do they make a profit? Under these conditions, only the banks could earn anything – namely, interest – and that will only be possible given an answer to the previous objection.

Monetary circulation in a surplus-producing system should not be treated as a "single-swap" circuit. Interdependence implies sequencing in exchange, and competitive pressures will lead to economizing on the use of money.

The strengths of the French approach could be preserved by adopting the account of circulation presented earlier. This can be illustrated on a two-sector diagram (Nell, 1992, chapter 20):

[9] The payment of interest also presents a problem in Wicksell's theory of circulation (Wicksell, 1985, ch. 9). He proposes that interest expands the money supply in the same proportion that productivity expands the supply of goods – thus providing M' to purchase C', in Marx's notation. But he requires banks to accept costly deposits for which, in the context of his scheme, they have no possible use, and he fails to explain how they find the money with which to pay the interest (Nell, 1967).

[10] The aggregate Quantity Equation for this system would have to read $M = \pi Y$, with $V = 1$. That is not consistent with any of the estimates for any industrial economy.

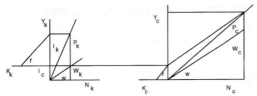

The diagram on the left shows output of capital goods on the vertical axis, Y_k, and employment in the capital goods sector, N_k, on the horizontal. The steep line is the output function, the shallower shows the wage bill, assumed equal to consumption. Investment demand is marked off on the vertical axis; this determines output and employment in the sector, and so its wage bill. This wage bill is then mapped onto the diagram on the right, where the steeper line shows the output of consumer goods as a function of employment in the consumer sector, while the shallower shows the consumer sector wage bill. It is clear that the wage bill of the capital goods sector equals the gross profits of the consumer goods sector, $W_k = P_c$. Total output is $Y = Y_k + Y_c = W_k + W_c + P_k + P_c$, where $P = I$.

Inventories of consumer goods are assumed on hand; production times in the two sectors are coordinated, and all firms act in concert. Circulation starts with banks advancing funds as needed, which will eventually equal W_k. Capital goods firms begin production, paying wages to households, who begin purchasing consumer goods, running down inventory, but building up the bank balances of consumer sector firms, so that their capital remains intact. As soon as these balances begin to accumulate, consumer goods firms can draw on them to pay wages, which promptly return to them as consumer goods workers buy consumer goods. When production is complete in both sectors, the consumer sector will have the entire wage fund, and its inventories will be wholly depleted. It will then spend its profits, P_c, on its desired gross investment, I_c returning the wage fund to the capital goods sector.

At this stage W_k, W_c, and P_c have been spent on C_k, C_c and I_c, all monetized by the circulation of bank advances equal to W_k. To complete the circulation I_k must be sold for money and the advances repaid. The capital goods sector may be divided into two subsectors, one of which sells to the consumer goods sector, the other being the rest of the firms. Then the first subsector receives the whole revenue: it repays its wage advances and spends its gross profit purchasing its desired investment goods. The second subsector receives its revenue, repays its wage advances, and spends its gross profits on its desired investment goods. This process will be repeated until a subsector is reached the 'machine tool' sector – which makes its own capital goods. At this point the sale of I_k will be complete and all advances repaid. Bank advances equal to W_k will have circulated the entire output Y.

CHAPTER 6

Circulation and instability: the supply of money

PAN: *The country banks are breaking*
The London banks are shaking
Suspicion is awaking:
— —— —— —— ——
Experience seems to settle
That paper is not metal
And promises of payment
Are neither food nor raiment
Thomas Love Peacock, 1825

To maintain the established price level at any given time, the circulation will need a certain amount of the medium (which may or may not possess an intrinsic value). In a stationary metallic monetary system lost, damaged and worn-out currency must be replaced; in a growing system the currency must also expand, but this will be difficult for a metal currency. Paper credit will first supplement, then replace metallic money. The monetary medium evolves through metal, and paper, to the bank deposit system, and now to electronic funds transfer. Credit goes through a similar development, from commercial paper and bills of exchange to loans and bonds, to stocks and modern securities, including derivatives. But the methods of supplying the medium of circulation must be carefully regulated in each period. In the Craft economy if too much or too little is supplied, the price level may change; under Mass Production, the interest rate will be affected. The problem arises that these changes may prove self-reinforcing, especially as they interact with money held as an asset for various (including speculative) reasons. Such money functions as part of the "financial circuit," which becomes more complex as fixed capital increases. The relationship between the two circuits – roughly, the ratio of "active" to "idle" funds, in an older terminology – depends on the difference between the real rate of interest and the rate of growth. In general, however, this relationship tends to have unstable features. To prevent such instability from damaging the economy, the State tends to establish institutions to stabilize the

216

currency – the Mint in metallic systems, the Central Bank in paper systems, and monetary policy in the modern economy.[1]

The quantity and value of money

The problem is both simple and general. It can be seen most easily in Craft conditions, but it is present in more complicated form in modern economies. The money needed for circulation depends on the characteristics of the circuit. It will be supplied either from the money-supply industry – mining and minting, and/or banking – or from hoards and/or the financial markets. In each case the supply responds to market incentives, but in each case unstable responses are likely in plausible circumstances. To explore this it will be necessary first to establish the value of money and relate it to the quantity in circulation.

Pre-capitalist conditions

In metallic regimes (and to some extent in paper) the value of the *medium* (as opposed to the value of the monetary metal) tends to be inverse to its quantity in circulation, while spending and hoarding are competitive options for possessors of money. Earlier we saw that a difference between the value of money in circulation, and its natural value, resulting from production, could lead to adjustment. For example, when the former exceeded the latter, production of money would increase, and additional money in circulation would drive up prices, reducing the circulation value of money.

But this is not necessarily how the adjustment will work. A deviation from the "correct" amount in circulation could lead to an inverse movement in value, bringing an adjustment of stocks of hoarded money that worsens the initial deviation. Thus, an excess amount in circulation drives up prices, reducing the value of money. The fall in the value of money leads to reductions in hoards, which further increases the amount in circulation, adding to the upward pressure on prices. Conversely, a deficiency in the circulation will raise the value of money, which could lead to further hoarding, that is, to withdrawals from circulation, bringing lower prices and a still higher value of money.

These responses are by no means certain, but they are possible. The art in designing monetary systems, and in the practice of money management, lies in finding ways to prevent these reactions and to create stabilizing responses instead. To study this, first consider the supply of the

---

[1] Hayek (1937, 1976) distinguishes between *holding* money and *using* money, and he proposes that there be two forms of currency, one adapted to each of these functions. Money for use in circulation should be government-issued money, while money designed to serve as a store of value should be privately issued.

metallic medium, and its value. We can derive an expression for the supply price of the monetary metal, which at the same time shows how expensive such a system is.

Cost of a metallic monetary system in primitive conditions

Gold (standing for monetary metals in general) is supplied by gold mining, which uses labor and means of production and produces enough gold each period to replace the coins worn out or lost during the year. Recall the labor value equations of the previous chapter. Let m be the means of production required to mine and mint gold, and n be the labor required. The labor value of gold (at the present depth of the mines), $\lambda_\$$, will be

$$\lambda_\$ = m\lambda_I + n = \frac{mb^*}{1-a} + n \tag{1}$$

Hence

$$\frac{\lambda_\$}{\lambda_C} = \frac{mb^* + n(1-a)}{Ab^* + B^*(1-a)} \tag{2}$$

which is to say, the labor value of money expressed in terms of consumer goods, in other words, labor time (the reciprocal of this will be the gold price of consumer goods):

$$\frac{\lambda_C}{\lambda_\$} = \frac{\text{labor/consumer goods}}{\text{labor/money}} = \frac{\text{money}}{\text{consumer goods}} \tag{3}$$

$$= \frac{Ab^* + B^*(1-a)}{mb^* + n(1-a)} \tag{4}$$

The labor value–money exchange equations multiplied by the quantities are

$$tqv = tavq + (1+s)\,\omega b^* q \tag{5}$$

where from equation 7 of the previous chapter

$$(1+s) = \frac{t(1-a)}{Ab^* + B^*(1-a)\,\omega} = t \tag{6}$$

As before, we can normalize by setting $\omega = 1$, allowing us to drop the asterisks:

$$t = tav + (1+s)\,B \tag{7}$$

$$tP_\$ = tmv + (1+s)\,n \tag{8}$$

where $P_\$$ is the consumer goods price of gold, and

$$v = \frac{b}{Ab + B(1 - a)} \qquad \text{(cf. Equation 3' in Chapter 5)}$$

Clearly the q cancels out in the first equation. The unit size of the gold equation is taken as the amount of gold needed each period to replace worn-out and damaged coins in the no-surplus system. With the speed-up of production generating a surplus, circulation also speeds up, and coins are worn out, damaged, and lost relatively more frequently. Workers in mining and minting are a fixed labor force, required to man the mines and the mint, whether much or little output is required. Hence the output of the gold mining and minting sector depends on the speed of production in the other sectors, or equivalently, on the *size* of the surplus. The output of gold adapts to the need for it in circulation. As a direct consequence the rate of surplus value will be the same in gold production as in the other spheres, since it is brought into being by the same speed-up of activity.

Solving for $P_\$$, the consumer goods price of gold, by substituting and rearranging:

$$P_\$ = \frac{mb + n(1 - a)}{Ab + B(1 - a)} = \text{the labor value of money} \qquad (9)$$

We can assume that gold mining and minting is controlled by the crown, and that the surplus value in mining and minting is appropriated by the crown as seigniorage – the charge for minting new coins or for exchanging old coins for new. These funds will then be spent on consumer goods, supporting the crown's retainer, judges, public officials, etc.

Introducing new coins into circulation takes place during the normal course of exchange. Coins will be lost or damaged at some average rate which will be proportional to the circulation, and production of new ones will be adjusted to this. Using stock on hand the Mint will buy (mv) worth of means of production t times during the period, paying wages to n laborers to use these means to mine, refine, and mint coins.

Late feudalism: the primitive dual
    The capacity output of the consumer goods sector must cover the needs of the workers in the mines and mint:

$$1 = qb^* + (B^* + n)\omega \qquad (10)$$

The quantity equations now read

$$q = qa + (A + m)\omega \quad \text{(from equation 5a' in Chapter 5)} \qquad (11)$$

$$t = (1 + c)[qb^* + (B^* + n)\omega] \quad \text{(from equation 5b' in Chapter 5)} \qquad (12)$$

So,

$$
(1+c) = \frac{t(1-a)}{\omega[(A+m)\,b^* + (1-a)\,(B^*+n)]}
$$

$$
= \frac{t\,(1-a)}{[Ab^* + B^*\,(1-a)]\,\omega + [mb^* + n\,(1-a)]\,\omega} \tag{13}
$$

Substituting for ω using equation 7 in Chapter 5,

$$
(1+c) = \frac{t\,[Ab^* + B^*(1-a)]}{[Ab^* + B^*(1-a)] + [mb^* + n(1-a)]}
$$

Now divide by $Ab^* + B^*\,(1-a)$, and substitute $(1+c) = t/(1+P_\$)$ where $t = (1+s)$ from the previous chapter. So

$$
(1+P_\$)\,(1+c) = (1+s), \quad \text{and}
$$

$$
c = \frac{s - P_\$}{1 + P_\$} \quad \text{or} \tag{14}
$$

$$
P_\$ = \frac{s - c}{1 + c}
$$

Now multiply the quantity equations by prices, and then look back to the price equations multiplied by quantities. From the first equation of each set we have an expression for (qv); hence, subtracting we obtain

$$
(A+m)\,v = b^*q \quad \text{(new balancing condition)} \tag{15}
$$

A word on dimensions. How can $P_\$$ be subtracted from the pure number s? Surely $P_\$$ seems to have the dimensions of labor time per unit output of gold, whereas s is a pure ratio, so dimensionless. But the discrepancy is only apparent. Labor time translates directly into consumer goods, because we have assumed that consumer goods support labor for a definite time, which defines the period of circulation. Since $Ab^* + B^*(1-a)$ is labor time per unit of consumer goods the dimensions cancel. The same is true for $mb^* + n\,(1-a)$, which has the dimensions labor time per unit of gold. But equation 1 provides the labor value, the direct plus indirect embodied labor, and we have seen previously that the initial (no-surplus) net output of consumer goods defines the unit amount of labor time. Therefore, the unit of gold and the unit of labor are interchangeable measures, as are labor and consumer goods, so the dimensions cancel, and $P_\$$, like s, is a pure number. It shows by how much the real costs of keeping the circulating medium intact reduce the productivity of labor below the value ratio.

Hence, this also provides us a measure of the saving that would be had by shifting from a metallic currency, with "real" value, to a paper one with only a nominal value. If the nominal value of the paper equals the real value of the gold that it replaces, then prices will be unchanged in nominal terms, but c, consumption per head, will rise in proportion to the difference in the real resources required to produce the paper and those required to produce the gold.

Mercantilism: the developed dual

As the system develops demand will form for the specialized and luxury products, which will be produced by a subsector of the consumer sector. The surplus will therefore have to consist of both goods, in proportion to their use as inputs into the consumer sector and the money industry. The equations, using m and n as defined in Chapter 5, are

$$q = aq + (1 + c)(A + m) \tag{16}$$

$$t = bq + (1 + c)(B + n) \tag{17}$$

which implies

$$(1 + c) = \frac{t(1 - a)}{b(A + m) + (1 - a)(B + n)} \tag{18}$$

and this, as we saw earlier, reduces to

$$(1 + c) = t$$

Hence we reach the same results as in the case of the primitive form, where the surplus consists only of the consumer good.

The evolution of capitalist money

With free labor in conjunction with the formation of a rate of profit, so that enterprises can be bought and sold, capitalism comes into its own. But profits will be competitively reinvested. Capitalism expands exponentially, and this creates pressures for the monetary system to become more flexible. A metallic monetary system depends on mining – a form of primary production – for its basic supply. Far from growing exponentially, mines eventually become exhausted. Moreover, a metallic money supply is expensive and, for large transactions, inconvenient. And finally neither mining output nor supplies of metal can be adjusted quickly, to match rapidly changing requirements of circulation, as required, for example, in conditions of Mass Production.

Currencies are national; they have to be, because their stability depends, as we shall see, on national management. Nations whose currencies are widely accepted gain a competitive advantage in trade over those whose currencies are less readily accepted. Hence there exist competitive pressures to improve the acceptability and reduce the costs of using a currency. It is not our purpose to explore the evolutionary dynamics of money here. Rather it is to show, first, that a common pattern of circulation holds for all three main forms of money; second, that the determination of the value of money, and therefore of the real wage, is very different for fiat money and money with a natural value; and finally, that all three face similar problems of instability in the relation between the circulation and the holding of money as an asset, although these problems will take different forms in the different monetary systems.

Money

Money is not gold; rather, gold and other metals are suitable, because of their value, durability and malleability, to serve as money. Money is whatever serves the monetary functions uniformly within a region.

The primary monetary function is that of providing a medium of circulation, since it implies the others. Because production takes time and circulation mirrors production, a medium of circulation must also be a convenient store of value, and since circulation requires calculation and the ability to make change, the circulating medium must also be divisible into convenient accounting units.

A medium of circulation, then, is whatever enables circulation to be accomplished; it must be the means of settlement, and the pattern of circulation must complete all transactions so that a new round of production can begin with the same price level and the same relative burden of debt. Credit may be issued during the circuit but must be extinguished by the end. The distinguishing mark of credit is that it earns net interest during circulation. Credit-money – credit functioning as the medium of circulation – on the other hand, does not. Any interest earned will be offset by interest paid over the course of the circuit.

But the medium of circulation need not consist of a metal. The advantage of a produced commodity is that the value of money is set by the cost of production; it does not have to be determined by a convention enforced by regulation. The disadvantage is that a commodity money system is both expensive and hostage to the fortunes of mining and exploration. Primary production cannot be expected to grow exponentially over an indefinite future; but capitalism grows.

Manufactured commodities can grow, and such commodities would have values determined by their cost of production. They have been used as money, though never widely, for two reasons. First, it may be too easy to invest in their production and they may tend to be overproduced. Second, such goods have a practical or normal use, with the result that, at times, the medium of circulation might become scarce, because it has been used up. Commodity money must be tied to primary production.

Instead of either primary or produced commodities, however, the medium of circulation could be a token, produced at minimal cost and otherwise largely useless, endowed with value by convention, limited in issue by some form of backing or collateral, and designated as an acceptable means of settlement by government fiat. (Contracts will be considered legally settled when such tokens are presented in payment.) Paper money is an example; but the tokens need not be material – bank deposits, for instance. What makes such tokens money, is not merely the government's fiat, but the fact that by means of them the transactions of a circuit can be completed and the circuit closed.

But token money does significantly enlarge the role of government. Under a precious metals regime the government sets and enforces standards, operates the Mint, and regulates the currency. It is also, of course, responsible for enforcing contracts and preserving public order in the conduct of trade. But with token money it must also regulate the issue, in order to preserve the value of the currency, since it is purely conventional. This, of course, opens the door to temptation.

To see the gain from introducing paper we must examine how a metallic system works when a rate of profit is earned and the economy is growing. Under these conditions the currency must expand in pace with accumulation.

Metallic money for the capitalist circuit

Precious metals are produced by mining and prepared for circulation by minting. Mining and minting must be arranged so as to regularly replenish the money in circulation, since a certain percentage of coins will normally be lost and misplaced each year and others will become too worn to be acceptable. (Still others will be withdrawn into hoards.) Moreover, each period additional money will be required, since the system is expanding. Since precious metal is produced under Mercantilist arrangements it will have a value determined by the labor needed, directly and indirectly, in its production. Under Capitalism its value will be determined by a price-profit equation. When the wage is set in terms of such money, the real wage is fixed, and prices are expressed in money terms.

Paper money

Instead of earning the required increment of funds by selling abroad, a financial services sector could be developed to generate the supply endogenously. The advantages would be threefold: first, there should be a saving of resources, in the form of lower labor and capital inputs into financial activities than in exports. Secondly, if the new system were properly designed, the ability to adjust funds to the requirements of circulation would be greater. Thirdly, investment transactions are typically large-scale and often require amassing funds one place and spending them in another. For such purposes, coins made of precious metal are inconvenient, compared, for example, to paper notes or other written promises to pay.

Let m' and n' be the coefficients for the banking sector, where the unit size is the scale that provides the services and funds necessary for a closed, complete circuit. If, as seems likely, both are less than m and n, the respective export coefficients, then banking is unambiguously a less expensive method of obtaining the circulating medium than exporting. But if one coefficient were less and the other greater, then the decision would depend on the level of the rate of profit. At a certain level there will be a "switch" – which implies the possibility of "reswitching," as well.

The circulation of profit under a banking/paper money system (an inconvertible system) will follow the pattern already examined, bearing in mind, however, that a medium of paper combined with a deposit/checking system has no value arising from production. Hence there is no process of market adjustment to align the value in circulation with "natural" value. The important point in this context is that aligning the value in circulation with the value arising from production constrains the *quantity* of money in circulation. In the absence of such a process, automatic constraints no longer exist.

This does not imply either for individual banks, or for the banking system as a whole, that they can accept any amount of deposits and make any volume of loans. They are constrained by two kinds of risk. There is first the risk of an unexpected run on deposits – withdrawals at levels too great to be met by orderly liquidation of assets. Against this contingency banks hold reserves; what is necessary is that the *system* have adequate reserves, and quick methods of mobilizing them to assist banks facing unexpected withdrawals. This is the responsibility of a Central Bank.

The second risk, however, is the decisive one. It is an essential and widely misunderstood point that an individual bank's (and the system's) long-term ability to issue depends on its *own independent capital*. Banks accept deposits, which are liabilities, but cost little or no interest, and make loans or investments, which are its assets. A portion of the deposits must be held in reserve accounts, against unexpected runs, and so will not earn. The bank's earnings arise from the difference in the rates earned on the assets and the rates, if any, paid on the liabilities. But there is the long-term risk that some investments will go sour, and this risk is what bank independent capital must cover. If a bank accepts deposits and makes corresponding loans in excess of what its capital can cover, it puts the funds of all its depositors at risk. On the other hand banks will be tempted to underestimate risk, especially in good times, while depositors will normally not have access to bank books and records, and so will not be in a position to judge whether banks are undercapitalized. The issue calls for independent regulators with appropriate expertise.

The total deposits that a bank can accept can easily be calculated. Let D be bank deposits, A assets, and R reserves, with ρ indicating risk and r being the reserve ratio, so that $rD = R$ (the "deposit multiplier" would be $D = (\frac{1}{r})R$); then $D = A + R$, and bank capital $K_B = \rho A$. Bank capital must cover the risk in the investments. From these it follows that

$$D = K_B/\rho(1 - r).$$

Given risk and the reserve ratio deposits are constrained by bank capital. This expression can be applied to the banking system as a whole. If deposits are greater than the above, there are uncovered risks; if less, additional lending could be encouraged to create more deposits.

Bank capital will increase as bank profits are earned and reinvested. Bank profits derive from interest earnings; in the simplest case, if deposits pay no interest bank profits will grow at compound interest. Deposits on the other hand grow at the same rate as income. If the rate of interest equals the rate of growth, bank capital will expand *pari passu* with deposits, and risks will remain covered. Thus, instead of earning the increment of new money from exports, the financial sector will earn profits that will increase its capital assets enough to justify an increment in the note issue of just the amount needed to service the increase in capital and employment.

In a "convertible" paper money system, paper is issued to supplant or augment the precious metals in circulation, a stock of these metals being held, in the form of bullion or plate, as a reserve against the paper. If paper is overissued, its price in circulation would fall against gold and silver, so that the paper would quickly be presented at the banks for conversion into metals. Arbitragers will be able to make a profit borrowing paper and presenting it for conversion. (Even if convertibility is suspended, banks will have to accept their own notes back in repayment of debts; so gradually excess notes will be withdrawn.) So long as the gold and silver exchanged for the paper does not circulate, but is held as an asset, such as bullion, conversion removes money from circulation, reducing inflationary pressure, and raising the value of money. If paper were underissued, compared to the needs of circulation, prices would fall and paper rise, signaling banks that it would be profitable to issue additional notes. The precious metals act as the standard for money, and as an asset, but the reserve metals do not and cannot circulate. (For if they did, an over-issue could not be corrected by the reflux of paper, since the amount of money in circulation would not change, the metals simply replacing the paper.)

This corresponds to the traditional account of the "law of reflux" (Mehrling, 1996); but there is no reason to require that the reserves be precious metal. They could be securities, titles to land, or claims of any kind; all that is necessary is that when the paper issue exceeds or falls short of the needs of circulation, it will fall or rise, and consequently swap against whatever assets are held as reserve.

Banks will advance paper notes to business firms for the purpose of paying wages. So the capital goods sector, anticipating orders, will begin hiring and producing, and with their wages workers will purchase consumer goods. So the consumer goods sector will also hire and start production, activating the "short circuit." Beginning a new round of production, however, requires the consumer goods sector to purchase its full complement of replacement goods plus its net investment, since these must be on hand from the beginning. They will do so with their sales receipts plus, if necessary, advances from the banks, in anticipation of further receipts. The purchase by the consumer goods sector sets off the secondary circulation within the capital goods sector, enabling it to realize profits and make its own investment purchases. Thus, as before, all goods will be circulated by money, where the amount required will equal $bq, the wage bill of the capital goods sector.

However, this amount was borrowed from the banking system. (Even if the capital goods sector could borrow less for wages, the consumer sector would have to borrow the remainder, in order to pay for its replacement plus investment purchases.) Hence interest will have to be paid when the loan is settled. These payments must be deducted from the respective sector's earnings. One implication, of minor concern here, is that to maintain a uniform profit rate, prices will have to be adjusted to reflect each sector's borrowing requirements. But the major point is that as a result of these earnings the capital assets of the banking system increase. If the interest rate earned equals the growth rate of capital, then given fixed coefficients, employment, and output, bank assets will expand *pari passu* with the system. If the banking system's assets initially supported an issue of notes sufficient for the circulation, they will now support an issue just sufficient for the new level of activity, following the expansion due to investment.[2]

Notice that the expansion could take place in a way analogous to the way the export system worked. Banks, finding their assets and actual or expected deposits (and therefore reserves) increased, will need to expand their services. To do this they issue additional notes – justified by their

---

[2] The earned interest will have to be invested in safe securities, such as government bonds, in order to be considered part of the capital – nonborrowed reserves – of the banking system. There will therefore have to be a regular flow of newly supplied such securities.

earnings – to pay for their replacement and capital expansion, and for their wage advances. So far so good; then in analogy with the export earning model, they would pay new funds equal to $rbq = G(tm' p + wn')$ to the consumer goods sector. When the consumer sector concluded its transactions with the capital goods sector, these funds would reappear as deposits, as anticipated. Thus the "probleme du profit" is solved, and

$$\frac{dM}{M} = r = g \tag{19}$$

But there is no reason to suppose that the banking sector is anywhere near this expensive. Indeed, if it were, it would hardly be worthwhile to shift from metal to banking. To earn gold, exports had to be worth what they sold for; the export sector therefore absorbed real resources. Banking and financial services will absorb resources, too; but the difference is that these services do not exchange for the new funds. The new money is created. This creation is justified by the earnings of the sector, which are added to bank capital, but there is no reason that the real resources absorbed by the sector have to equal the value of the new money. Paper money, being cheaper, can therefore be expected to displace a metallic currency.

Bank deposits

Mass Production cannot function effectively with only paper money; fixed capital requires bank finance, and so do purchases of consumer durables. Carrying about large amounts of paper money is dangerous and inconvenient. But it is easy to shift from paper to bank deposits and the writing of checks, although this does call for careful monitoring to prevent fraud. It also requires a central clearing system (Wright, 1926, ch. 12). Consumer purchases can be made easier with credit and debit cards. The circuit will work in the same way – banks create deposits by making loans, instead of issuing notes.

Banks create deposits by making loans, and they are able to do this by virtue of double-entry bookkeeping. Loans are assets, deposits liabilities; when loans create deposits, the books balance, and the issue becomes liability management. The bank must meet demands on it for liquidity; hence banks, taken together, must support each other by providing liquidity wherever it is needed.

Banks issue lines of credit to firms, to cover their wage bills. These will be drawn upon as firms adjust their output to demand. As credit is drawn down, the deposits of firms rise; as wages are paid, business deposits fall, and those of households rise. But the level of deposits – money in circulation – adjusts in pace with the level of employment, which in turn is governed by the level of demand. Bank deposit money is "endogenous."

There is an important difference, however, between this and the earlier circuit. Commodity money enters circulation with a value derived from production. Paper notes acquire a derivative value from commodity money by being convertible. We shall see that when the wage is set in terms of either of these, competition will ensure that prices adjust to fix the real wage. But bank deposits are purely nominal; they have no "anchor" in a value derived from a production process. Such money therefore, has no "natural" or normal value. A level of the money wage, in a pure bank deposit system, does not necessarily imply a level of the real wage.[3]

Indeed, if it is anchored to anything, bank deposit money appears to be tied to the money wage,[4] rather than the other way around. Nor is this a particularly good anchor. Money wages are or can be market-driven, so there is nothing to prevent the market from reacting to *movements* in the value of money, as opposed to deviations from the normal level. As we shall see, if the market reacts to movements, rather than levels, instability will tend to result.

A corollary is that changes in the Quantity of Money in circulation no longer directly affect its value. The price level will now be determined by different forces, and since the Quantity of Money in circulation is set through bank lending, changes in that quantity come through changes in bank activity, and therefore affect the rate of interest.

The pattern of circulation

In the no-surplus case, the circuit began with merchant purchases; in the Mercantilist and early Capitalist economies, the circuit began with the issuance of wages. In each of these cases the output of both sectors was assumed to be used up within the period. But modern capitalism is driven by investment in durable capital goods. Here the French theorists (see Deleplace and Nell, 1995) surely have a point; the circuit cannot get under way unless there is financial backing for the proposed investment spending. Current production – whether for stock or to order – can be financed in the usual way, by drawing on lines of credit for working capital. These lines of credit will be repaid as firms receive (or borrow)

[3] In neo-Classical theory it is difficult to demonstrate that when the value of money is determined by supply and demand, nominal money will have a positive value (Bennetti, 1995; Deleplace and Nell, 1995, "Afterword"). The problem arises because supply-and-demand equilibrium has been defined and shown to be optimal in the absence of money, leaving it no role. It is not obvious that money is the most desirable store of value – it can easily be outcompeted by safe bonds – and developing its role as a medium of exchange would require specifying the pattern of bilateral exchanges, an aspect of exchange normally suppressed in neo-Classical thinking.

[4] Keynes, 1936, pp. 236–9.

depreciation funds. For the sale of large-scale, long-lived capital goods requires long-term finance as well, which means that the capital goods will be paid off over time. As we have seen, financing the sale of currently produced long-term investment goods calls for mobilizing depreciation funds.

New investment demand in the consumer goods sector provides the starting point, purchasing the output of the capital goods sector. This output will have to expand at the going rate in order to keep pace with the growth of the consumer sector. The capital goods sector will therefore produce output for investment both in itself and in consumer goods. Since these are long-lived goods, their sale will have to be financed by drawing on the accumulated depreciation funds, which have been growing at the rate of interest. As the investment sector produces in response to the investment demand, it will pay out its wages, which, when spent, then provide the profits of the consumer goods sector, enabling the consumer goods sector to service its debt. The consumer goods sector issues its own credit to pay its wages, on the "short circuit," as before. In practice, in any modern economy, the actual circuit is likely to be a mix of advances from lines of credit for working capital and advances for investment spending.

The "real" money wage and the fiat money wage

When money circulates, and wages are paid in money, prices will also be realized in money terms. There are two very different cases, however. When money is itself produced – a metallic currency, gold – or convertible into such a currency, then the money wage in the mining and minting industry becomes a *real wage*. For the (fixed) normal output will be gold, input prices will be expressed in gold, and wages will be gold. The output price will be unity. The rate of profits in the gold industry will therefore depend on the prices of its *inputs*, and competition requires that these prices adjust so that all industries have the same rate of profit. But when the money wage in gold rises, the output of gold net of wages – the surplus plus the amount to be traded for inputs – necessarily falls. Hence a rise in money wages in gold requires a general fall in the rate of profits.

In terms of the model of the previous chapter:

$$P_1 = (1 + r)\,taP_1 + wbP_2 \tag{20}$$

$$P_2 = (1 + r)\,tAP_1 + wBP_2 \tag{21}$$

$$\$ = (1 + r)\,tmP_1 + wnP_2 \tag{22}$$

Here $\$$ is the unit output of gold (previously expressed as unity, but given a symbol now to distinguish it from the other unit outputs). The prices of the capital and consumer goods are now money prices, to be expressed in gold. There are three equations and four unknowns: P_1, P_2, r, and w. The wage is now fixed in terms of gold:

$$w = \mu\$, \quad 0 < \mu < 1 \tag{23}$$

(the wage is a certain fraction of unit gold output)

There are now three equations and three unknowns. From the equations for the prices we find

$$\frac{P_1}{P_2} = \frac{\mu\$b}{1 - (1+r)\,ta} = \frac{1 - \mu\$B}{1 - (1+r)tA}$$

Solving for r, we have

$$r = \frac{\mu\$(b+B) - 1}{\mu\$t\,(aB + Ab) - a}$$

From the equation for $,

$$P_2 = \frac{\$ - (1+r)\,tmP_1}{\mu\$n}$$

and substituting into the expressions for P_1/P_2,

$$P_1 = \frac{\$}{1 + (1+r)\,t\,(m-a)} \tag{24}$$

$$P_2 = \frac{\$\,(1 - (1+r)\,tA}{[1 - (1+r)\,t\,(m-a)]\,(1 - \mu\$B)} \tag{25}$$

It is apparent that when the wage is set in terms of gold, which is taken as the standard (so valued in terms of itself, thus dimensionless), the dimensions of the inputs into gold production cancel out in both terms, so the rate of profit is immediately fixed in the gold industry, thereby establishing a competitive yardstick against which to measure other industries.

It should be clear from this that in metallic money systems the Homogeneity Postulate does not hold. If money in circulation increases or decrease, so that Π, the value of money, is no longer equal to $P_\$$, the resulting adjustments will have real effects, for resources will move into or out of the mining and minting sector.

When money is not produced, or is no longer tied to a produced metal, no such convenient relationship exists. The money wage is set in terms of *fiat* money, that is, inconvertible paper, or bank deposits. No industry or activity exists in modern economies with an output that is identical to money;[5] hence money prices will have to be set in every industry. As a consequence the money wage is not automatically translated by competitive pressures into the real wage. In the case of a "real" money wage, setting the money wage in the labor market determines the real wage; but when the wage is paid in fiat money, the real wage cannot be determined without considering the markets for goods as well. Changes in the money wage no longer imply equivalent changes in the real wage – or even changes in the same direction. Money wages and money prices must *both* be determined. When the money wage is expressed in terms of fiat money, a rise in the money wage could be matched by a rise in prices in fiat money, holding the real wage down, providing the banking system is willing to increase advances. In the case of gold, output is fixed exogenously (by assumption). In the case of fiat money, however, the availability of loans is constrained by banker's prudence, which may not be up to the strain of refusing a risky opportunity to turn a profit.

When the value of money is no longer anchored by attachment to metal, it is liable to fluctuate sharply, in particular, to fall. A popular response has been to try to limit or at least to control the quantity of money. As we shall see, this does not and cannot work. But there appear to be several potentially practical ways to prevent a fall in the value of money. One, of course, is simply to maintain whatever wage and price levels have been established, by allowing, or even encouraging, slack in the economy; weakness in demand will keep wage demands down and prevent prices from rising, or from rising too fast. This works, after a fashion, but it is costly in terms of lost output and forgone opportunities. It sacrifices economic growth on the altar of price stability. Another way might be to rely on the rate of interest: given a money wage, established for example by bargaining, the Monetary Authority could fix and maintain a rate of interest, which, if established for a long enough period, would engender capital movements that would bring the rate of profit into line. This would fix prices and the real wage. But as we shall see, the Monetary Authorities may not have the requisite degree of control, and the capital movements need not be stabilizing. A more workable variant of this would be to use government market power – both purchasing power and regulatory power – in labor markets and goods markets to set "yardstick" prices and wage norms. However, to prevent an upward drift of the price level, this would likely need to be supplemented by an incomes policy. Together these

[5] Even financial services earn money from handling money – they don't produce it. And the amount they handle is determined by the market. The money wage in gold became the real wage because the output of the mining and minting industry was fixed.

could fix the value of money in terms of the real wage. The rate of interest could then be set with an eye to encouraging full employment.

When the currency is anchored to metal, it is convertible, so the issue of notes and loans cannot exceed the limits imposed by the available reserves of precious metals. The reserves, in turn, will be adjusted, so that the value of metal in circulation together with the value of the currency circulating as proxy for metal will equal the "natural" value of the precious metal, as established in mining and minting. The adjustment of values will ensure an adjustment of the volume of notes and loans. But when the value of the currency is no longer anchored to the value of metal the issue of notes and loans by the banking system is no longer constrained by the quantity of metal. Reserves of precious metal do not constrain note issue, nor do they limit the making of loans. "High-powered"money does not *determine* the money supply. To be sure, the public must retain confidence in the banks, and must believe that risks are covered. But that simply means, first, that adequate reserves must be set aside as activity proceeds, and second, that bank capital must be adequate to cover the risks arising from bank portfolios. If banks exceed the limits imposed by their capital, the system becomes riskier. But there is no automatic pressure to adjust. The habit of thinking that reserves determine deposits is a carryover from a currency anchored to metal.

The required quantity of money

The total quantity required, M, can therefore be calculated from the equation

$$M \Pi = t (A + m) pX_2 = bqX_2 \qquad (26)$$

where we define $\Pi = 1$, when

$$\frac{bqX_2}{M} = P_\$ \qquad (27)$$

where $P_\$$ is the value of money, as determined by the equation for mining and minting. The annual output, then, must be such as to keep this at the correct level (assuming a stationary economy).[6]

This is not a supply and demand equation in the usual sense. It is not derived from summing over the individual equations of all those in the

---

[6] This is not the Quantity Equation. For that we would need the Velocity of Circulation, to tell how often this sum must turn over to circulate the entire output. M exchanges once for Av, then again for bq, in a succession of transactions, and finally a small portion of money, revolving repeatedly, turns over the wages of the consumer sector. The average velocity will be total output divided by bq.

market, nor does it depend on the solution to any specific maximizing problems – although it is assumed that agents will seize on profitable opportunities for arbitrage.

In the history of monetary thought, the Cost of Production and the Quantity theories have been seen as alternative explanations of the value of money (Wicksell, 1895, chs. 4, 5; 1967, vol. 2, pp. 141–53). This formulation reconciles them. The quantity of money in circulation determines its value in the short run; the difference between this value and the "cost of production" value, will, in the long run, lead to arbitrage which will adjust the quantity so that the two values are equal – provided that adjustments of asset holdings of money or bullion and plate do not lead to instability.

This is a theory of the adjustment of the quantity of money; it is *not* a theory of inflation, where that refers to a general rise of prices over a long period of time. Such sustained inflation requires an explanation of the persistent pressures driving up prices. In particular, it cannot be inferred from this approach that general price increases are normally caused by a rise in the quantity of money in circulation. This may sometimes be the case, but the mere availability of money, for example, does not explain the motivation to use it in ways that will drive up prices of goods. Other options exist: it could be hoarded, loaned, or expended in the asset market. In metallic currency systems, hoarded money is always available in large amounts. So the cause of an inflation will not be such money; it will be whatever draws such money – or paper issued against it – into circulation.

The value of money

In the Mercantilist economy, and in early Capitalism, the commodity which serves as money, and/or as the standard for money, enters circulation through the Mint with a value acquired during the process of production. Adjustments will take place when the demands of circulation vary; money will be brought into or removed from the circulation according to the needs of trade.

Rules have to be designed to ensure that market pressures will adjust the amount of money minted and put into circulation in the right way, for destabilizing reactions are always possible. The Mint price, the conditions of convertibility at the banks, and laws governing melting coin, clipping, and other practices, and the rates at which different coins are acceptable in settlement of debts and for payment of taxes, all can be adjusted to help ensure that the money supply is kept adequate and that reactions to deviations tend to stabilize it.

In order to design and adjust these regulations and laws, it will be useful to know the cause when an apparent change in the value of money in circulation takes place (see Appendix to Chapter 7). Some causes call for adjusting the quantity of money in circulation to restore the original price level, but others do not. First note that the causes of apparent changes in the value of money may arise on the goods side, or on the money side. For example, the value of money might change because of a change in the quantity of goods on offer, prices constant, due perhaps to productivity advances, or to accidents of the harvest. A different quantity of goods exchanges for the presently circulating quantity of money. Or distributional changes may affect the value of goods on offer, quantities constant, so that the present quantity of money will trade for a different value of goods. But the change might originate on the monetary side – as a result of a rise in hoarding, or a fluctuation in the production of precious metal, so that a different quantity of money exchanges for an unchanged sum of goods. Or, finally, variations in productivity or distribution could alter the value of precious metal, so that a given sum of money commands a different value of goods.

When the quantity of goods changes, and relative prices remain constant, money prices will change, that is, the value of money changes, so the amount of money in circulation must adapt to restore the price level. Similarly, if accidental forces cause the amount of money in circulation to rise or fall, goods and relative prices constant, its current value will fall or rise, and to correct this the quantity must be brought back to the proper level. In both these cases the original price level should be restored, and the appropriate aim of policy will be to see that market forces maintain a constant price level in the long run.

By contrast, when the quantities are unchanged, but relative prices change, there will almost certainly be a change in the price level. And this may or may not require a change in the supply of money in circulation, depending on the effect of the price changes on the cost of production of the money metal. But even if the quantity of money must change, if relative prices have changed, the objective of monetary policy will *not* be to restore the previous price level. For the value of money, as measured by its cost of production, will now be different (leaving aside special cases).

To see what adjustment will be required it is necessary to know whether relative prices have changed, money the same, or whether the value of the money commodity has changed, other prices the same. In the first case (the case Ricardo hoped would be the norm), the quantity will change, but the price level will not; in the second, both the price level and the quantity will change. Of course, all relative prices – goods and the money commodity – might change. It was in order to analyze and distinguish these cases properly that the search for an "invarible measure of value" was undertaken.

When distribution changes, relative prices change. In the next chapter we will see that, when r = g, the wage-profit rate trade-off is linear. This is equivalent to a Standard Commodity; hence it is possible to tell which prices are rising and which falling (see Appendix to Chapter 7). Changes in the value of precious metals relative to other goods can therefore be determined and isolated. Let us assume either that such changes are insignificant[7] or that they are *absorbed by the Mint*. Then if merchants can adjust the funds they bring to market, so as to maintain $M\Pi = bqX_2 = X_2t\,(A + m)\,p$, as q and p change, these variations may be accommodated without requiring a change in the price level – even though both the prices and the quantities of goods have changed.

Banking and changes in distribution

Merchants may find it difficult to adjust the amount of their working capital. Banks, however, can rather easily adjust the level of their advances, to keep them equal to the wage bill of the capital goods sector. A circulation based on banking and finance is better placed to adapt when income distribution changes, compared to a mercantile system. Suppose between one period and the next, the wage rate rises, so that the rate of profit falls. Suppose further that r = g continues to hold before and after this change; the relative sizes of industries are adjusted to the change in distribution. When r = g, the wage bill of the capital goods sector will equal the replacement demand of the consumer goods sector (plus the requirements of the banking sector). It was further shown in the previous chapter that this implies that the slope of the wage-rate/rate of profit trade-off equation equals the (negative of the) labor to capital ratio. That is,

$$-Kdr = Ndw$$

Suppose the circuit is partly financed by lines of credit for working capital and partly by advances for the purchase of capital goods. The wage rate then rises. The funds needed to pay the increased wage bill are exactly those released by the fall in profits due to the change in the profit rate, given the adjustments in industry sizes. Finance can thus be shifted from capital spending to underwriting wage payments. There will be changes in relative prices, but there need be no change in the price level, nor, in pure banking systems, in the quantity of money. For if the available supply of money, circulating in the normal pattern (at the normal velocity), drawing on the available sources of credit, sufficed to circulate all goods at the

---

[7] As Ricardo apparently believed, implausibly arguing that the gold mining industry had an "average" capital/labor ratio, so that gold would not change in value with changes in distribution.

initial rate of profit, then, when (for example) the rate of profit falls, the decline in the profit flow is exactly matched by the rise in the wage bill. The funding and credits have simply to be transferred, for example, from financing investment to financing wage payments and consumer spending. Of course, there may be practical difficulties, but in principle, the new circulation can be accomplished with the monetary resources that managed the old.

This shows that the financial system can readily adapt to changes in distribution – provided that there are market forces that tend to bring r and g into line with one another. This will be explored later.

Additional remarks

Bills of exchange will speed up transactions and reduce costs; they will therefore come into general use. But once they do, bills of accommodation follow – credit drawn on the general reputation of a house, rather than against specific goods. These latter will develop because they are useful – and people will pay for them. As long as they are not issued in excess, no harm is done, and commercial activity will be further assisted. However, only the prospect of deterioration against metal stands in the way of an excessive issue – a weak defense at best when the pressures to expand are strong. Even worse, when paper falls against metal, Gresham's Law comes into play – metal will be hoarded, rather than brought back into circulation. Hence a fall of paper will encourage faster circulation of paper, intensifying the inflation. Further withdrawal of coin and bullion into hoards could then cause a shortage among banks and goldsmiths, leading to a suspension of convertibility, and the collapse of paper.

In practice a metallic currency has always consisted of a number of metals, even if only one served as the official standard. Most frequently, the combination included gold for large transactions and taxes, silver for commerce, and copper for small change and daily wages.[8] (Copper will often be valued, not by weight, but as a token.) This means that after

---

[8] Maintaining an appropriate balance between the amounts of the various metals and the various denominations was not simple. Berkeley and others devoted a good deal of effort to devising solutions for the shortage of silver in Ireland in the early and mid–eighteenth century (Johnston, J., ed., *Berkeley's Querist*, Dundalk: Dundalgan Press (1970), esp. queries 460–482). Pamphleteers in Swift's group proposed a mint, Berkeley a bank. The Irish were reacting to what was widely seen as a deliberate campaign by the Master of the Mint, Sir Isaac Newton, to drain Ireland of silver, to make up for the loss of silver by London, consequent upon a deliberate effort to attract gold. (In 1717 Newton slightly overvalued gold, in what may have been a crucially important strategic decision for the development of London as a money market.)

producers of large or expensive goods sell their output they will have to convert at least part of their receipts into copper. Merchants selling consumer goods will receive copper which they will have to convert to silver. Silver receipts, in turn, will have to be converted to gold for taxes and large transactions. First gold or silversmiths and then banks will find it profitable to provide this service, and whoever does so will have to have on hand stocks of the various metals and/or appropriate coins. To save on such services, shopkeepers, artisans, and households will also keeps stores of coins of all denominations. Such holdings do not circulate themselves, and they are held as an asset, since earnings arise from the service of making change. (When change is made, the holding remains the same in total value, apart from the service charge.) By the same token, businesses and households who hold money in order to avoid paying the service charges for conversion are earning an implicit return. Such holdings can be correctly described as "transactions demands" for money as an asset, and they are clearly separate from money which is temporarily on hand, pending expenditure.

In the same way paper notes of various denominations might be held for transactions purposes. But when money consists of bank deposits and credit cards there can be no "transactions demand" in this original sense. There is no need to make change, or convert coins of one size into another. In a paper money regime, funds may still be needed to bridge the gap between normal receipts and planned outlays. In a bank deposit system, however, this will not be necessary, since a line of credit can be negotiated to cover such exigencies; there will therefore be no need for firms – or given credit cards, households – to carry "transactions balances." (Electronic funds transfers with a "just-in-time" payments system can reduce the need to draw on lines of credit.) The usual "transactions demand" is, in fact, "money on the wing," in Robertson's phrase. It is money in circulation, being used to execute expenditure plans, rather than money being held idle as an asset. The "demand for money," which adds together money in circulation and money being held as an asset, is misconceived.

(Sometimes transactions holdings are identified as the *average* balances over a period of time in accounts into which wages or salaries are paid and from which household expenditures are made. Thus an account into which a monthly salary is paid, and which has been run down to zero at the end of the month, will have an average balance of one-half the monthly salary, and this is treated as the transactions demand. But in fact, no money is being *held* at all; money is being *used*. The salary is a flow, as is the household expenditure. Money held as an asset is a stock. It is true that the more the time shapes of the flows of income and expenditure differ, the more money will be required to facilitate the circulation (Neisser, 1928; Ellis, 1937). But this just means that more money will be tied up at various points in the circulation, e.g., in the short circuit. Of course, if calculated properly, the money held to bridge income-expenditure gaps could turn out to be the very quantity required to accomplish the circulation – since the sum the consumer sector needs to purchase capital replacements will have to be "held" at an appropriate point in the circuit. But it will be "held" for spending purposes which can only be fully understood by studying the entire circuit, rather than focusing on individual motives and behavior. In any case the fact that this money is held temporarily does not justify adding the money flowing in circulation to the stock of money held as an asset.)

Digression: distribution and the quantity theory

The "Quantity Equation" expresses the usual view of the aggregate relationship between money income and the circulating medium. As we just saw, a variant of that equation can be justified – the funds that purchase the output of the consumer goods sector circulate in such a way as to monetize the wages in the capital goods sector. These funds also enter into a "short circuit" to circulate the wages of the consumer sector. But a long tradition has worked with the aggregate equations

$$PY = MV, \text{ and also, } PK = MV_k$$

$$(\text{or, sometimes, } rPK = MV_k; \quad \text{e.g., Meltzer, 1995}) \quad (28)$$

These aggregate equations are expressed in money terms. To achieve this the wage must be paid in money, and then prices must be set at money levels so that the money wage divided by prices will equal the given real wage:

$$PY = (1 + r) PK + wN \tag{29}$$

where w = money per unit labor time, and w = w/P.

The underlying concept of the wage here, w, is that of a pure *ratio*: money per unit labor divided by money per unit goods, where labor and the goods that support it are interchangeable. The money wage is a derivative notion, which can be interpreted as the money payment per unit labor time, where labor time is measured in the commodities that support labor for that time. The money prices, then, are the prices of those goods. If labor is paid money wages, then as the wage bill circulates, all trades will be monetized and prices will be expressed in money.

To return to the Quantity Equations above: as written they could be considered merely truisms, with no causal significance. (Money prices, *P*, and money wages, w, will be written in italics here.) The first merely states that the money value of aggregate output (income) is equal to the money in circulation times its "income velocity," whatever that is; the second that the money value of capital, or of profits, equals the money in circulation times "capital velocity."

However, the Monetarist tradition interprets the Quantity Equation to make a causal claim, namely that, given velocity, V, the supply of money, M, determines the level of nominal income.[9] According to proponents of this approach, real output, Y, will be determined in equilibrium by real forces alone (absolute quantities are determined together with relative prices, which also permits the outputs of different sectors to be aggregated), so the money supply fixes the price level, P, making allowance for temporary exogenous influences.[10]

There is another aggregate equation which highlights the relationship between prices and wages:

$$P = kwN/Y, \quad \text{where k is the mark-up of prices over money wages} \tag{30}$$

The Markup Equation (Weintraub, 1978) states that, given labor requirements per unit output, N/Y, the price level, P, stands in a fixed ratio, k, to

---

[9] Of course, causality is a matter of interpretation. Either equation could be true, and yet the causality could run the opposite direction. For example, the banking system could adapt the money supply to its expectations of the prices leading firms wish to set, so that P or changes in P would determine M or changes in M. Or, in the other case, w, the money wage, might be adjusted to keep up with the expected rises in prices, so the direction of causality would run from P to w. But these are not the ways these equations have normally been understood.

[10] he approach has always had its detractors. Joan Robinson, for example, never thought highly of the Quantity Theory of Money: "The archetypal quantity theory formula MV = PY, like any identity, has to have its terms defined in such a way as to make it hold" (Robinson, 1971). Once we have managed to set it up, however, we are still only halfway to a useful tool of analysis; the next problem is whether causality runs from left to right, as monetarists maintain, or from right to left: "Suppose that between one year and the next PY rises; either activity has increased – employment and output are higher this year than last, or the general price level has risen because of a rise in costs in money terms; then if the quantity of money has not increased, the velocity of circulation must have risen." This, Robinson says, makes sense. But to read it the other way, to invert the causality, is to take flight from reality: "There is an unearthly, mystical element in Friedman's thought. The mere existence of a stock of money somehow promotes expenditure."

the money wage rate. Hence whatever determines money wages, like collective bargaining, fixes the price level.[11]

Both equations suffer from faults resulting from aggregation. Neither treats changes in the composition of output or of labor, nor is it always clear what should be included in some of the aggregates, for example, in the money supply, or in the total amount of labor. It is not easy to make unambiguous adjustments to either equation for changes in the degree of utilization. Nevertheless both equations have been widely

---

[11] Both equations purport to explain inflation. Differentiating the first gives

$$V \frac{dM}{dt} = Y \frac{dP}{dt} + P \frac{dY}{dt}$$

and dividing by equation 28 yields

$$\frac{dP}{P} = \frac{dM}{M} - \frac{dY}{Y}$$

Differentiating equation 30 and dividing yields

$$\frac{dP}{P} = \frac{dw}{w} - \frac{d(Y/N)}{Y/N}$$

Ignoring many obvious qualifications, in the first case inflation is determined by the excess of the rate of growth of the money supply over the rate of growth of output – the Central Bank is to blame. The policy implication is therefore to control the rate of growth of the money supply, an austerity approach. In the second case, inflation is caused by the pressures that push up money wage rates faster than labor productivity is growing – the trade unions are to blame. The policy called for is one that ties growth in money wages to growth in productivity, an incomes policy. But the markup equation used here *presupposes* both the distribution of income and the pattern of consumption and growth. Policy questions concerning changes in either of these cannot be examined on the basis of a given markup, and hence cannot assume a fixed relationship between the real wage and the productivity of labor. As a policy tool, therefore, markup theory must be used with caution.

used to analyze questions concerning the causes of, and policies toward, inflation.[12]

An implication of the theory of monetary circulation

The true nature of the Quantity Equation can be seen by allowing for the fact that the money in circulation adapts to the need for it. To see this let us take the special but plausible case (Nell, 1983) where all money is in the form of bank deposits and banks have agreed to provide lines of credit for business sufficient to cover their working capital needs, up to the full capacity level. (To build a plant a business must negotiate a line of credit to cover working capital needs. Providing its own would mean holding unnecessarily large liquid reserves much of the time, but not to have immediate access to working capital could mean costly delays.) Working capital covers variable costs, the main item of which is the wage bill. So as a first approximation, we can say that the banking system has agreed to advance automatically whatever funds business needs to cover its labor costs.

In fact, the basic funds needed to enable circulation, as we have seen, equal the wage bill of the capital goods sector. The advance of this money creates the initial deposits, and activity then creates further deposits. But as this money returns to the banks, they will readvance it for the "short circuit." Such deposits (together with cash, which we are ignoring) can be treated as the conventional "money supply," M_1, and will vary proportionally with activity.

As employment expands and contracts lines of credit will be activated or reduced, and so bank deposits will fluctuate accordingly. The resulting simple equation is

$$M = wN \tag{31}$$

---

[12] Weintraub (1959, 1978) has argued for many years that k is a reliable constant over the short and medium term, with only a slight trend in the long run. By contrast, he contended with Joan Robinson that V has shown significant variability. Friedman, and Monetarists generally, have held that V is more stable than any competing explanatory parameter, and that, therefore, the quantity equation should be preferred. Kaldor (1970) has argued, however, that V is stable only when and because the money supply itself is adjusting. Friedman (1970) has accepted Kaldor's contention that the money supply is (at least partly) endogenous, but holds that it does not affect the validity of the quantity theory. Recently, Basil Moore (1988) has argued that contemporary economies operate a system of credit money the quantity of which is determined by borrower demand.

But when we combine this with the aggregate Quantity Equation, and compare the results to the aggregate Markup Equation, the consequences are striking:

$$MV = PY \qquad\qquad P = k\frac{wN}{Y}$$

$$wNV = PY \qquad\qquad P = k\frac{M}{Y}$$

$$P = V\frac{wN}{Y} \qquad\qquad Mk = PY$$

Hence it is clear that $V = k$, and that, starting from either equation, the other can be derived.[13]

(This result does not depend on the money supply adjusting to demand for working capital, as for example, when Mass Production firms adjust employment to anticipated demand for goods. It fits as well into the case of the Craft economy, for it would follow equally in a situation where merchants or holders of wealth sat on a fixed sum, M, with which they purchased the goods serving as the "wages fund." There would still be a question as to the direction of causality, and $V = k$ would still hold.)

When the Quantity Equation is written in the "Cambridge" or "cash-balances" form,

$$M = \frac{1}{V}\, PY$$

it is clear that $(1/V)$, the proportion of money income the public wishes to hold in active money balances, is equal to the share of wages in money income, wN/PY. This follows directly from the fact that business money balances will equal working capital, that is, the wage bill, and household balances will only rise when business balances fall, that is, when and as wages are paid. When household balances fall as consumer purchases are made, then business balances rise once again, in exactly offsetting measure.

These results do not depend upon an exact equality between the money supply and the wage bill. If, for example, the money supply were *proportional*, rather than equal, to the wage bill, as would be the case if we treated bank advances as the money supply (equal to the wage bill of the capital goods sector),

$$M = xwN$$

---

[13] In the case of capital velocity, we have $V_k = rPK/M = rPK/wN$. The reciprocal of velocity is the coefficient for the "demand for money."

where $0 < x < 1$ is the ratio of the wage bill of the capital goods sector to the total wage bill, then we would derive

$$P = xV\frac{wN}{Y} \quad \text{and} \quad \frac{k(M/x)}{Y} = P$$

Clearly $xV = k$. That is, when the money supply is proportional to the wage bill, velocity will be proportional to the mark-up, by the same factor. So the same conclusions follow.

It should be clear that k, and its equivalent V, depend on the rate of profits and relative prices. Further, however, k and so V depend on the quantities of the system.[14] The markup depends not only on distribution and prices, but also on the relative sizes of industries and sectors. But money prices do not directly affect k or V if the real wage (the rate of profit) is given, for the money dimension clearly cancels out. But endogenous money, provided by a banking system in return for interest, raises k and V through the effects of interest.[15] Thus, for example, a change in the equilibrium rate of growth would be likely to change the velocity of circulation, since it would normally affect quantities.[16]

---

[14] Given $PY = (1 + r)\ [PK + wN]$, as the money form of $\mathbf{q'p} = (+r)\mathbf{q'Ap} + w\mathbf{q'L}$ (assuming circulating capital), we can see exactly what k, and therefore V, means. Rearranging, we find

$$P = \frac{Y}{Y - (1 + r)\,K} \frac{wN}{Y}$$

from which it follows that

$$k = \frac{Y}{Y - (1 + r)K} = \frac{\mathbf{q'p}}{\mathbf{q'p} - (1 + r)\,\mathbf{q'Ap}}$$

and when interest is earned upon the wage bill, this becomes

$$k = \frac{(1 + r)Y}{Y - (1 + r)K} = \frac{(1 + r)\mathbf{q'p}}{\mathbf{q'p} - (1 + r)\mathbf{q'Ap}}$$

See Nell (1990) and a more extended treatment, Pasinetti (1976), appendix to ch. 5, esp. pp. 130–32.

[15] Interest must be paid upon lines of credit used to provide working capital. This requires rewriting the matrix equation for prices as a function of the real wage and the rate of profit, to show earnings of profit, equal to interest, upon the wage bill, just as some of the classical economists used to argue. The difference when interest is figured upon the wage bill comes in the last term, which must then be $(1 + r)wN$.

[16] Notice that equation 6 and the earlier equations from which it is derived, imply that endogenous money, supplied as lines of credit for working capital, not only raises the markup as compared to a barter world, but also affects relative prices. Such money is not neutral. Even more strikingly, if the average rate of profit and the long-term rate of interest are correlated, equation 6 implies that velocity rises with the long-term rate of interest, quite independently of any "economizing" on money balances.

The "velocity of circulation" has usually been taken to mean the number of times the stock of money turned over in the course of the transactions making up the period's income, while its reciprocal, the coefficient of the demand for money, has been understood as the average interval of rest between transactions. The length of this interval was held to depend partly on institutional arrangements and partly on decisions determined, in turn, by other economic variables.

The income velocity equation apparently tells us that the "demand for money" is inversely related to the markup, the capital velocity equation that it is inversely related to the ratio of wages to profits. Truisms are true, of course, but we need to ask whether these equations can reasonably be interpreted as showing a "*demand for money.*" In the case of coins, a genuine "transactions demand" for money as an asset, to be used in transactions, can be indentified. This does not seem to be possible in the case of bank deposit money, but even if it were, it could not be identified with the truism expressed by the Quantity Equation.

How, then, should the truism be interpreted? The connection of velocity with the markup factor is not readily apparent. The average markup simply shows the average profit being earned over variable costs, and reflects market pressures interacting with the firms' decisions about their cash flow requirements. But whatever the traditional concept of velocity was supposed to mean, it has always been measured by the ratio of money income to the stock of money. This provides the clue to understanding the connection here. As Joan Robinson repeatedly pointed out, the traditional concept of velocity is little more than a metaphor; no precision has ever been given to it, for no definition has ever been provided explaining just what was to be counted as one round of circulation, that is, when a sum of money should be counted as having returned to its starting point (Nell, 1968, pp. 157–62).

The previous chapter, however, provided an exact concept. The pattern of circulation was identified precisely and the quantity of money required was determined. But the Quantity of Money involved was different from the traditional quantity, and there were three different patterns of exchange, defining three different kinds of velocity – for circulating funds. In addition, a separate pattern of turnover was identified for fixed capital. The pattern of circulation is seen to reflect the interdependence of production. "Velocity" indicates the volume of goods which a certain sum of money can "turn over" by passing from one industry to another, as in the "secondary circulation" or in the "short circuit." When productivity increases, a given wage bill generates more output, so the volume of goods circulated by the money representing the wage bill increases.

According to this perspective, there is a direct connection between productivity and velocity.

In short, in regard to the Quantity Equation, what has been thought of as a turnover relationship is actually not one at all: it is simply a proportionality between money income and a sum of money. What it means in the traditional form is undetermined. But we have given it a meaning, which identifies the sum of money circulating goods with the sum going to wages. This sum is therefore equal to wage costs, so its ratio to income is the markup.

Moreover, if this is correct, any evidence apparently supporting the Quantity Equation is also ipso facto evidence supporting the markup equation.[17] Further, the reason that "velocity" is relatively constant is simply that it *is* the markup factor, appearing in a different context. Velocity rises with the long-term rate of return because the markup does, not as a result of economizing on money holdings. Interpreting the Quantity Equation as showing the demand for cash balances, we see that household and business balances rise and fall in an exactly offsetting pattern, reflecting the circuit of expenditure. Also, the markup and therefore velocity reflect not only given distribution, but also the given relative sizes of the different sectors. Hence neither explanations nor policy conclusions can be derived from the quantity theory, for the Quantity Equation is simply another way of writing the Markup Equation. Insofar as demand-for-money equations are derived from the quantity theory, they

---

[17] But as noted earlier, while the markup equation used here is valid, though oversimplified, care must be taken in drawing policy conclusions from it. For it *presupposes* both the distribution of income and the pattern of consumption and growth. Policy questions concerning changes in either of these cannot be examined on the basis of a given markup, and hence cannot assume a fixed relationship between the real wage and the productivity of labor. As a policy tool, therefore, markup theory is more limited than has sometimes been supposed. An example: the PE curve in Marglin and Bhaduri (Marglin and Schor, 1992; Nell and Semmler, 1991) is supposed to show aggregate markup – i.e., profit share – as a function of capacity utilization, on the basis of firms' pricing policies in relation to money wages. In the analysis this curve remains fixed, while movement takes place along an IS curve, or while the IS curve shifts position. But the aggregation to form PE requires an assumption about relative quantities, and relative quantities will normally change when the IS shifts, and may change even as a consequence of movement along IS. Hence the curves are not independent of one another in the way needed for the analysis.

are likewise suspect – even if they are more complex in form, for they may simply be mirroring a more complicated Markup Equation.[18]

The reason, then, that the Quantity Equation cannot be treated as a serious explanatory tool of analysis is that when money is endogenously determined (or functions as a wages fund), the Quantity Equation is not an independent relationship at all. It is simply a disguised form of the Markup Equation, and so is a reflection of whatever forces have determined the markup, which itself depends both on distribution and on the relative sizes of the sectors, that is, on growth.

Clearly neither the traditional Quantity Equation nor its cash balance form can be supported. As Joan Robinson pointed out, though for somewhat different reasons, when the definitions are properly chosen, "it is obvious that the formula is correct... but then it has no causal significance whatever." The traditional "transactions demand for money," and, in fact, the whole edifice of demand and supply in monetary theory, call out to be reformulated, with the theory of circulation as the new foundation.

The potential for instability

When the value of money in circulation deviates from its production value, an adjustment in the quantity of money is required.

---

[18] To take an interesting example, Goldfeld's pre-1974 "demand for money equations" overpredicts money demand for 1974 and later; after the oil shock "money demand," in real terms, is lower than would be expected from the estimations based on earlier years. But the oil shock set off an inflationary wage-price spiral in which money prices rose faster than money wages, leading to a rise in the average aggregate mark-up. Thus $1/V$ fell; hence a transactions term derived from $M/P = (1/V)Y$, showing the demand for money based on velocity and the level of current real income, would be reduced compared to earlier periods. A simple regression analysis suggests that this might account for as much as half of the "missing money"; moreover the equation based on the mark-up tracks the turning points quite well (Nell and Delamonica, 1995). It opens an avenue of investigation wholly unexplored to now, but which may be particularly interesting in view of the fact that the troubles arose in the demand deposit equation, rather than the currency equation. It is demand deposits that we would expect to be endogenous (cf. Goldfeld, 1976). To carry this into the 1980s and 1990s requires shifting from M_1 to M_2 or even to a composite measure, in view of the changes in market practices in regard to the payments system.

Market forces may bring this about – but they could also take the system in the wrong direction.[19]

Hoarding and instability on the exchange circuit

Suppose, for example, in our first case, the exchange circuit, that in addition to serving as medium of circulation, silver were hoarded as a store of value; that is, suppose a portion of household, guild or manorial wealth to be held in the form of precious metal. Hoards, however, will not necessarily be held in currency. Coins are easy to lose. Moreover, being small, portable and readily concealed, they can easily be stolen. Hence it is dangerous to store wealth in the form of coins. But precious metal can be molded into candalabra and other decorative and/or useful artifacts, neither portable nor easily concealed. When needed, these can easily be melted down and minted into coins.

When more than the normal quantity of goods is brought to market, prices will fall, and the value of money will rise. But the rise in the value of money will raise the value of hoarded silver, also. Here we must consider two effects: first, the effect on circulation of a higher value of silver/money, and second, the effect of rising value.

At first glance, it might seem that the higher value of hoarded silver would lead to an increase in current spending by the wealthy – a real balance effect. This would be a mistake. As Hicks pointed out in reviewing Patinkin (Hicks, 1967), a higher value of hoarded money will first lead to a portfolio adjustment, a rearrangement of asset holdings, which will affect asset prices, but not put any additional money into circulation. Thus holders of silver might spend on (or trade plate or bullion for) paintings and objets d'art, or they might purchase additional land and buildings, or further rights to monopolies or positions in the

---

[19] The efficient markets hypothesis would have it that speculation must always be stabilizing, for otherwise speculators who bet on the system moving away from equilibrium would lose their shirts when the system returned to its fundamental position. Hence speculators will learn that the best strategy is to anticipate equilibrium, and they will therefore tend to drive the market there. There are two difficulties here. First, distribution depends in part on a *contest*, a clash of forces, the outcome of which can be predicted no better or worse than that of a sporting event. We simply can't say, for sure, who will win. It also depends on questions of technology, and we likewise cannot say, for sure, what will work. So there is an inherent element of indeterminacy in the fundamental position. The knowledge is just not there to be had. Second, even in periods when distribution is or seems to be clearly settled, and no major technological changes are on the horizon, so that the fundamental position could be determined, the time that will be required to reach it cannot be known. Moreover, this time may be long enough that some of the fundamentals could change. There is plenty of room for destabilizing speculation.

market. None of these are currently produced; such spending therefore does not put money in circulation; it simply redistributes existing assets, changing asset values in the process.[20]

However, this is not the end of the matter. There is also ongoing expenditure on long-term projects. These have been planned and are under construction, or are being improved, like merchants building counting houses. As a result of the higher value of money, the same effects can be achieved for less money. Hence money allocated to these projects can be added to hoards, or could be spent otherwise. In particular, the projects could be expanded. Will the elasticity of expenditure on continuing projects with respect to the value of money be greater or less than unity? If it is unity, the project will be expanded to absorb the originally allocated funds; if it is less, then funds will be saved; only if greater than unity will additional funds be thrown into circulation.

Thus the effect of a *higher* value of money on current spending is ambiguous, but it is not likely to lead to an increase. Indeed the more likely effect, particularly in the short run, is that the lower price level will lead to less money being put into circulation than before. But this will be destabilizing since it will put further downward pressure on prices.

The effect of the *rising* value of money is not ambiguous at all, however. As prices fall and money rises, the prudent course of action is to withhold money, since over time its value is increasing.[21] Such withholding, practiced by everyone, reduces the quantity of money in circulation, so that prices fall further, and the value of money continues to rise. The expectations are self-fulfilling, and the effect is destabilizing.

[20] Hicks argued that changes in the value of money would not be symmetrical in their impact on current spending. A rise in the value of money, as seen, will chiefly affect portfolio composition, with little or no impact on current spending. But a fall in the value of money (a general price rise) may require additional saving, to build money assets to the level required for transactions and precaution (Hicks, 1967). It should be remembered, however, that the burden of debt is similarly affected, but provides an opposite influence to that of the value of real balances, so even these admittedly tentative conclusions need to be considered cautiously.

[21] Expectations of rising or falling prices are not always destabilizing, as is sometimes believed (Lowe, 1965; Hicks, 1939). In Craft conditions, working-class households might expect prices to fall, but they would not be able to hold back spending, since they have no accumulated stores of goods to support them. They must spend their income as it comes; nor can they move their spending forward when they expect prices to rise. They have no assets, so cannot borrow. Only *wealth-holders* can speculate – merchants and bankers, not to mention grain-dealers, whose speculative activity figures among the chief causes of famine (Sen, 1981).

A fall in the value of money, rising prices, will similarly tend to be destabilizing, although the impact of lower silver values may lead to saving to build up precautionary hoards.

Taken together, then, when a significant stock of hoarded silver exists, changes in the value of money are no longer stabilizing. To put the point sharply, when money serves both as a medium of circulation and as a store of value, changes in value tend to be destabilizing.

A very simple measure deals with this, however. By introducing the Mint, and the concept of money as legal tender, acceptable in taxes and settlement of debts, the link between the value of money in circulation and the value of hoarded gold or silver can be broken, thereby eliminating the connection essential to the instability. This will be explored in detail below.

Hoarding and regime instability

Once a surplus exists, the formation of hoards can be expected.[22] We have already noted the transactions demand for holdings of money. Part of wealth will be held in the form of money, for precautionary and other reasons, and money held as an asset will be kept for safety in the form of bullion or plate, or objets d'art. (In an economy with a metallic currency, the latter are doubly significant as stores of wealth: the workmanship will increase in value with inflation, partly because embodied labor will rise to keep pace with current labor, and partly because the scarcity value of the artistic work will rise to reflect the general price level. But the value of the precious metal will serve as a hedge against deflation.) The chief motive for hoarding, however, must be speculative. Bullion will be hoarded in anticipation of a rise in its value – falling prices of goods and of other assets. It will be dishoarded when prices are believed to be rising.

As we saw earlier, such hoarding tends to create instability. A rise in the value of money, if reflected in the value of hoards, will tend to divert new gold and silver into hoards and will likewise encourage the melting down and withdrawal of currency. Both effects will further depress prices, adding to the rise in the value of money. A fall in the value of money will encourage dishoarding and further fuel the inflation.

The instability in question does not arise from particular patterns of expectation, nor does it depend on any special assumptions about risk aversion or any other patterns of preferences. It does not depend on the kind of maximizing practiced, or on the degree of perfection of markets. It

---

[22] Galiani believed that in the mid-eighteenth century, in Naples, hoards amounted to four times the amount of money in circulation. Others have argued for even higher ratios (Braudel, 1979, vol. 1, pp. 463–464).

is not tied to any particular model and does not depend on whether solutions are unique or multiple.[23] It is a kind of meta-instability which will be present in the institutional arrangements governing monetary circulation, so long as there is a motive for gain, *whatever* the other expectations and preferences, or the exact form of maximizing objectives or degree of market perfection. No general problems of rational choice are involved. It arises from no more than the simple pursuit of self-interest in given circumstances, market arbitrage.

This form of instability can be called "regime instability," to distinguish it from the more commonly discussed market instability. Market instability, by contrast, depends on the precise assumptions made about the individual agents operating in the market – their sources of information, their degree of mobility, their exact motivation (the form of the function they maximize), and the position from which the market began. From these data, first, supply and demand functions are derived – usually for "representative agents" – to determine the equilibrium, and then reaction functions are defined to describe the response to deviations from the equilibrium. It is then shown that for some or all values of parameters the reactions move the market away from equilibrium, or cause it to cycle around equilibrium.

None of this is relevant to regime instability, which instead, refers to the rules of the system, in this case the monetary regime. Here the question is simply one of arbitrage. If the value of money in circulation rises or falls, opportunities for arbitrage will be created – and anyone might take advantage of them at any stage of the circuit. No assumptions need be made about "representative agents," nor does a full maximizing model have to be defined. Nor is equilibrium significant. All that has to be assumed is that agents, of whatever class or occupation, will take advantage of simple opportunities for gain. The question is, if such opportunities arise, will the arbitrage tend to drive the system back to the proper position, that is, the position considered correct by the authorities who make the rules.

Formally, the issue is simple. Given the circulation equation

$$M \Pi = X_2 (A + m) v = \kappa \tag{32}$$

we have

$$M \Delta \Pi + \Pi \Delta M = 0, \quad \text{which implies}$$

$$\frac{\Delta M}{M} = -\frac{\Delta \Pi}{\Pi} \tag{33}$$

---

[23] Blanchard and Fischer (1989), ch. 5, survey the assumptions that give rise to instability in contemporary models.

Thus an increase in the money in circulation will lead to a fall in the value of money (a rise in the price level), a fall to a rise. But will, say, a rise in prices, that is, a fall in the value of money, lead then to stabilizing behavior that reduces M, that is, to actions that withdraw M from circulation, increasing hoards, or will it lead to more money being thrown into circulation, reducing hoards, and driving the value of money down further? Will we have, as the behavioral response,

$$\Delta M = \varepsilon\,(\Delta \Pi), \quad \varepsilon' > 0, \text{ or} \tag{34}$$

$$\Delta M = \phi\,(\Delta \Pi), \quad \phi' < 0? \tag{35}$$

If the behavior is given by ε then a rise in prices (fall in Π) leads to a withdrawal of money, and prices will come back down. But if it is given by ϕ, then the pattern of behavior will be such that a rise in prices will lead to an increase in money in circulation and further rises in prices.[24]

Circulation involves a sequence of transactions in a number of different markets; some may, others may not, be in equilibrium. Whether or not they are in equilibrium, the question for a monetary system is, will its normal working create incentives to provide the correct amount of money and adjust an incorrect amount? Market stability or instability does not imply a corresponding condition for the regime. Regime instability does not imply any particular market instability, although clearly there must be some kinds of disorder in markets.

Such instability is not to be confused with the commercial problems that may result from an "insufficient rapidity of circulation," to use the old-fashioned phrase. This last is a manifestation of an effective demand failure, and so is on a different level. Effective demand does depend on expectations, etc.; it is the outcome of a particular pattern of behavior, whereas the instability under discussion exists for all patterns of behavior – though it may well come to the fore more readily in some. Nevertheless it could easily interact with effective demand problems, and the consequences could be indistinguishable in practice. A failure to spend rapidly enough, on the part of the recipients of surplus value – Mandeville's problem – will lead to a shortage of demand and a fall in prices; therefore,

---

[24] Nothing has been said about the form of either function. Given suitable parameters, even in the stable case, overshooting could take place, leading to cycles. Persistent destabilizing movements in either direction could peter out. Behavior of either kind could change after a point, in a "regime switch." These matters depend on the exact nature of the objectives and how they are being pursued. The issue in the text is simpler and more general – will the market adjust the Quantity of Money in the right direction?

to a rise in the value of money, stimulating a tendency to increase speculative hoarding, which further reduces the circulation – and so on.

The role of the Mint

To prevent such instability it will be useful to break the link between the value of money in circulation and the value of bullion or plate as an asset. If bullion or plate could simply be melted down and stamped by private individuals, and if coins can be melted into bullion and plate, then the value of hoards will reflect that of circulating money. So much metal will yield so many coins. But if melting and minting is made the prerogative of the state, a partial barrier is interposed: conditions must be accepted (e.g., prohibitions or limitations on melting), seigniorage must be paid, and an official price for precious metal has to be accepted. Delays will occur. A certain weight of metal will no longer translate directly into command over goods. Moreover, coins with the official stamp will be accepted at face value, in settlement of taxes and debts, irrespective of weight and fineness. Of course, the Mint cannot ignore the values established in commerce. On the contrary it will have to impose itself on the market; it can do this partly by size and partly by virtue of its official position, but it will not succeed unless it uses both position and size skillfully.[25]

Once the Mint has been established, the value of money in circulation will no longer be the value of the metal in the market, considered as an asset. As a result when prices fall and money rises, bullion and plate will not necessarily rise, or will not rise by as much. This will be especially true for short-run variations. Hence it reduces the temptation to withdraw money and melt it down, that is, to hoard for speculative reasons. On the other hand, it also makes it more difficult to take advantage of lower prices, since it will take longer to convert hoards into coin. Hence, as well as blocking instability, whatever stabilizing influences may exist may be reduced. When additional money is needed for circulation, it will no longer reliably be provided from hoards, nor from the mines. It will have to come from somewhere else – from the creation of paper capable of circulating.

[25] A well-managed Mint also establishes a uniform coinage. Latouche (1967), chs. II.5, III.1, gives an example of unsuccessful management of minting. The Merovingian kings lost control of minting in their realm to private coiners, who overissued and debased the coinage, bringing a deterioration in the coin and a loss of its acceptability abroad. Coins of the same denomination by different private minters had different and fluctuating values. Coin became scarce, as did metal, and as the value of money fell, good coins were spent abroad. It became difficult to collect taxes. Charlemagne had to struggle to reestablish control over the coinage, which was lost again under Charles the Bald, with similar results.

The fixed Mint price of bullion and plate stabilizes the nominal values. When prices rise, so that money falls, the nominal value of silver and gold will not fall, or fall much. The Mint price acts as an anchor; the market will not react to the *rise* in the price. Hence the value of plate and bullion held as an asset will not tend to decline relatively to fixed nominal debts. Were it to do so, it would pay to coin at once and pay off debts – thereby increasing the quantity in circulation, further contributing to the rise in prices. Similarly when prices fall and money rises, if the nominal value of plate and bullion increases, thus raising plate and bullion relative to fixed debts, it would pay to roll over debts rather than discharging them, thus further reducing the circulation and worsening the deflation. But a stable Mint price tends to prevent this.

However, the establishment of a Mint presents a new set of problems. Faced with a shortage of revenue in relation to need, a government may resort to debasement of the currency (alternatively: overissue of inconvertible paper money) as a short-term measure to cover its deficit. This will set off a monetary inflation. But such an inflation is a self-stimulating cumulative process. For the rise in prices – fall in the value of money – will lead holders of money balances, which are being accumulated with an eye to specific future expenditures, to spend these balances sooner, that is, to move their purchases ahead in time, so as to avoid the loss due to the falling value of money. But this puts additional upward pressure on prices.

Thus a given debasement, setting off an inflation, will cause a further loss in the value of money. Hence whatever the initial shortfall, it will end up being larger. To avoid (or at least control) these problems rules are required prohibiting, or limiting, debasement and overissue of paper.

Metallic currency: export adjustment

In Chapter 5 it was shown that when the rate of profit exactly finances the rate of growth, so that $r = g$, an equilibrium circuit exists – provided the export industry can earn the required increment of gold. An informal analysis shows that this is by no means guaranteed, although there are forces working toward such an end.

The export sector sells its product abroad for gold or precious metals, or most likely for coins made of such metals. If the export sector earns more than $rbq each period, then the domestic supply of money will increase; under certain circumstances this will tend to cause prices to rise, thereby creating pressures for money wages to keep pace. If either p or w rise, and all the more if both rise, then the price of the export on world markets will rise. Initially, this may increase earning still more (and destabilizing activity by hoarding speculators may also exacerbate the inflation), but eventually the rise in price will ruin the market, especially since the absorption of gold may lower prices elsewhere. Earnings will then fall.

If they fall below $rbq, then the domestic economy will not have enough gold, and prices will tend to fall, exacerbated by the actions of speculators. This will lower p and w, cheapening the export good and improving its competitiveness.

This specie-flow mechanism, however, is suspiciously oversimplified. If the earnings imbalance is large enough and continues long enough, no doubt prices will adjust; but there are certainly other possibilities along the way. Suppose, for example, that the earnings are only a little greater. Instead of – or along with – a permanent rise of prices, activity levels might increase, perhaps from a temporary price surge, leading to higher employment. The lower real wages could be sufficiently offset by higher employment to result in higher earnings for labor, thus increasing the requirements of circulation. This cannot be a likely effect in Craft conditions (for reasons we will see later), but in some circumstances it might be significant.

As another possibility, the excess money may not, at first, enter circulation at all; it might go into hoards, raising bank reserves, and leading to lower lending rates of interest. These lower rates, in turn, could well facilitate both accumulation and luxury consumption, leading to higher growth rates and greater levels of activity. Again, the needs of circulation for money will be raised, eventually absorbing the excess.

On the other hand, while there will normally be some room for expansion in activity levels, in a Craft-based economy there cannot be much. But the possibility does help to validate the beliefs of the Mercantilists that earning precious metals abroad promoted domestic prosperity. A shortage of funds would depress prices, and very likely depress them at least initially, relative to money wages, squeezing profits. It would also tend to raise interest rates and reduce activity levels, perhaps leading to adjustment, but at the price of prosperity.

Suppose that activity has expanded as far as possible, interest rates have fallen, and now the excess funds do enter circulation, raising prices. But now suppose that there is an excess supply of labor, perhaps because of labor-saving investment in the countryside, so that money wages do not keep pace with the increases in prices. Profits will therefore increase, and so will accumulation. Insofar as this results simply from increased activity and employment, the requirements of circulation will be raised – but let us ignore this effect, since we have just considered it. A higher rate of accumulation due to a higher rate of profit and a lower real wage, implies a change in q. If $aB > Ab$, that is, if consumer goods is labor intensive, then q will rise with G, and the higher rate of accumulation will imply a greater need for circulating funds; but if the consumer goods sector is capital intensive, q will move inversely with G, and the rise in accumulation – apart from its effects in increasing the level of activity – will reduce the needs of circulation, making the situation worse.

Metallic currency: specie flow and balance of payments

If there is an insufficient influx of gold the currency will not expand to keep pace with the increase of production, and prices will tend to fall. If it is excessive, they will tend to rise. In each case the value of money will change inversely to prices. When prices are low, according to the traditional argument, money will be pulled in either because the low prices make goods attractive, or because the higher value of money makes it an attractive market for gold merchants to sell in. In other words gold may flow in (or out) either on current or on capital account.

Suppose gold flows in on current account, in excess of the amount required. Let us consider the effects, and the relationships to a trading partner. Given capitalistic production, the rise in commodity prices would translate into a rise in profit (fall in the real wage). (In a relatively early stage of capitalism, the techniques of production may be considered fixed and difficult to change. Substitution effects would be unlikely.) An inflow of gold leads to a "profit inflation." But if profits are higher, then ownership of enterprises will be more desirable than ownership of fixed-interest financial assets. Hence bonds will be sold and enterprises bought. So bond prices will *fall*, not rise, and interest rates will rise, *pari passu*, with prices – as in fact they did during most of the formative years of capitalism, giving rise to what Keynes called "Gibson's Paradox."

This also suggests that a gold inflow could be destabilizing. By driving up prices and profits, it leads to a shift in favor of real capital (or shares, claims to real capital) thereby raising interest rates and attracting a further inflow. The process could come to an end as labor resists the erosion of its real wage and demands a comparable rise in money wages, converting the profit inflation into a wage-price spiral. But labor may not be in a position to make such demands; for example, there may be surplus labor. The high prices, however, are likely to cause problems for exports, and perhaps lead to a shift of spending in favor of imports. So the price rise will eliminate the initial inflow. (The proportional loss of exports plus loss of sales due to substituting imports for domestic goods must be less than the proportional increase in spending brought about by the influx of gold. Otherwise domestic prices would fall back to their original level.) Hence the domestic country could move to where it would simultaneously *lose* gold on current account, and *attract* gold on capital account. Prices will stop rising when the loss on current account just balances the gain on capital account.

For example, consider two trading partners, one an advanced CBF economy, the other a largely artisan primary producer. Call them "Britain" and "Argentina." Suppose that the initial inflow of gold to "Britain" had come from the trading partner, "Argentina." Britain exports manufactures/capital goods, and imports consumer/primary products. Argentina has the opposite trading pattern. The inflow of gold drives up prices in Britain, the loss of circulating gold lowers prices in Argentina. Britain will experience a profit inflation, which will drag up interest rates, and while losing exports, it will begin to attract funds on capital account. Argentina, with lower prices, will export more and will therefore begin to earn on current account, but as its profits have also fallen, its interest rates will decline, and it will begin to experience an outflow on capital account. Its prices will stop falling when its earnings on current account just balance its losses on capital. Britain's prices will stop rising when its losses on current account are just balanced by its gains on capital. If there are only two countries they will arrive at this "balanced disequilibrium" together.

British workers will find the higher prices partly or wholly offset by the cheap imports; their standard of living need not be undermined. Moreover, the higher profits and capital inflow will lead to higher growth. The lower prices in Argentina and capital outflow, however, will lead to slower growth. Hence the gap between the two will widen over time.

This suggests, then, that adjustment through current account, and through capital account need not be complementary, but, as in this example, can fight against each other. Of course, it would require a more complex model to work out what finally will happen.

A similar conclusion might be reached by a different route. We need not accept the claim that an influx of gold *directly* bids up prices. (But since output and employment are fixed – at least in the short run – not money wages, or money wages only with a lag.) Suppose, instead, that we took the position that the influx of gold expands bank money. This would initially lower interest rates, and expand investment. If the influx came on current account the lower interest rates might lead to a corresponding outflow on capital. (If the influx were on capital account, the decline in interest rates would presumably bring it to a halt.) Again the two could conceivably reach a balance, supporting not higher prices but an expanded level of activity.

Suppose, however, that the output of investment goods could not be expanded, because of capacity constraints. Prices would be bid up relative to money wages, and profits would increase. So interest rates would have declined and prices risen – this is the case closest to the traditional discussion. The problem is that the decline in interest rates can only be temporary. If profits do increase, and growth takes place at a higher rate, interest rates will be pulled up as investors abandon fixed-interest bonds for higher earning ownership shares. The higher profits and growth rates will attract overseas investors and funds will flow in on capital account, while the higher prices will damage exports and encourage imports, leading to an outflow on current. We are back to the earlier case, with a temporary detour through lower interest rates.

Adjustment, then, is possible, but by no means certain. The Mercantilists were surely correct to emphasize the importance of export earnings in maintaining high activity and prosperity. But it is an expensive system; a significant level of real resources must be devoted to earning the funds necessary to underwrite the circulation of investment.

Paper instability

As the development of capitalism leads to more sophisticated financial arrangements, various kinds of promissory notes begin to displace metallic currency, eventually leading to a system of circulation in which notes and bank deposits are the chief media. With the passing of metallic currencies, the Mint no longer plays a significant role. But systems of paper money, and the similar checkable deposit systems, face their own forms of instability.[26]

First consider a government paper system. The danger is simple; once confidence in the currency fades, for whatever reason, good or bad, there will be a "flight to quality." The paper – "greenbacks" – will be exchanged as fast as possible for anything else, bidding up the prices of goods, services and securities. Bidding up security prices will lower interest rates, adding fuel to the pressure driving up prices. Since taxes are normally collected only at stated times, while government spending is continuous, the rising prices will put pressure on the government to issue more paper notes, in order to maintain its spending commitments.

To understand the principles of capitalism it is important to consider how it works with the role of government reduced to the minimum. So let us consider the case where paper notes are issued by private banks. A pure

---

[26] Friedman (1959) argued that private paper money would tend to be overissued; indeed, he contended that the incentive would be to overissue until marginal revenue equalled marginal cost – the latter being merely the cost of paper and printing! The result would be a flight from paper, resulting in the collapse of such currencies. As many critics have noted, this surely overstates the case (and is certainly not compatible with rational expectations), but Friedman's historical point is sound, " this (overissue) is what happened under so-called 'free banking' in the United States and under similar conditions in other countries." (1959, p. 6)

paper money system is one in which paper is not convertible into metal. The banknotes issued by each individual private bank are backed by the holding of securities, which, in the face of demand for redemption could be sold for more acceptable notes. Such a system, however, is subject to its own particular form of instability. Competition forces banks into a race for customers; hence there is pressure to overissue. For any one bank at any given time, this will be profitable; moreover, for any individual bank, a temporarily excessive issue of banknotes need not create any special problems. As loans are repaid, notes can be retired and reserves built up; curtailing new loans will reduce the loan portfolio to the right level. This will bring things back to the correct relationship without any fanfare, provided, of course, that no run develops.

But matters are different for the system as a whole. Since the monetary circulation has to grow to keep pace with the expansion of output that results from accumulation, banks will tend to hold reserves in the form of earning assets, rather than metal. Competitive pressures will virtually require this – but it has important consequences for stability. If competition leads all banks to overissue, then paper notes will depreciate relative to metal reserves and hoards. Holders of paper will tend to redeem it for metal, reducing the paper in circulation; this tends to stabilize the issue. But when securities are held as reserves, the process is different.

When competition leads to an overissue, prices will tend to rise. In an early capitalist system, output and employment will be inflexible. Initially, therefore, the overissue will lead to expansion and a profit inflation; paper depreciates, but profits are up, so interest will be pulled up, and bank earnings will rise. Hence with securities earning more, a higher issue will appear to be justified in terms of the value of reserves. Banks will tend to raise lending rates to keep pace, and the competitive pressure will continue to stimulate still further overissues. This can lead to difficulties, which can only be sketched here.

Profit inflation implies higher capital goods prices. Competition and higher costs will lead consumer prices to rise in pace. The real wage therefore falls. Inflation means money is worth less, so money will be traded for securities and other assets. But the immediate effect of a tendency to shift from money into securities will be to support or even drive up the price of securities, so that bank assets will appear to be high, and may even seem to justify continuing to issue notes.

With lower real wages, after a point consumer demand will be constricted, and once this is appreciated, the consumer sector's demand for new capital goods will rise more slowly – or even fall off. (Moreover, money wages may begin to rise, perhaps even catching up, in any event making the inflation general. This would support consumer demand, but would tend to reduce profits back to the normal level.) With the weakening of consumer demand and the consequent slump in the consumer sector's demand for capital goods, profits will no longer increase. The effect of a weakening in the consumer sector's demand for capital goods will be to reduce the capital goods sector's incentive to invest. The all-around weakening of investment will lead to prices and profits slumping.

The inflation, meanwhile, will have eroded the earning power of banks; their old loans are now being repaid in depreciated currency. Banks will feel the pressure of lower earnings; moreover, with lower profits as growth falters and/or wages rise, the securities they hold will be worth less in fundamental terms, even as the shift away from money is supporting their inflated prices. Having overissued, and finding securities shaky, banks will now find themselves without adequate backing, at a time when the public is inclined to exchange notes for reserves; banks may be compelled to sell securities, such as for bullion, to redeem their notes, a move which will tend to drive down security prices. But with a drain of securities their earning position will be further undermined; moreover, as securities begin to decline, they will no longer have adequate resources to redeem their notes, which may well provoke a run. A likely result is that general overissue, followed by loss of reserves, will continue until a collapse takes place. (This will be examined further in Chapter 9.) To prevent this regulations restricting note issue have been widely adopted.

Banking and the Central Bank

Moreover, just as competition drove banks to overissue notes, so it will tend to drive banks to overextend loans, creating a new form of instability. As in the case of paper, the instability arises because, for each individual bank, overlending tends to be profitable in the short run; moreover, if a given bank restrains its lending, while its competitors overlend, it will lose market share. But for the system as a whole, overlending leads to expansion, driving up prices.[27] In a Craft economy this lowers the real wage, raising profits, which, in turn, drives up the prices of securities, and thus encourages further overlending, together

---

[27] Does overissue of paper, or overlending (leading to inflation) drive down interest rates? Real and/or nominal? If the interest rate were equal to the profit rate, would this prevent overissue and/or overlending? These questions have been extensively debated, but the discussion cannot be surveyed here. However, we shall see that the key lies in the relation of the rate of interest to the rate of growth. The Golden Rule requires that these be equal; then the growth of money in circulation will just keep pace with the growth of output. And the growth of capital funds will just equal the growth of fixed capital requirements. But this equality is easily destabilized, although there are also market forces tending to establish it.

with arbitrage against reserves. The result is a steady pressure weakening the system.

The corrective in this case is the imposition of reserve requirements, together with a system of bank inspection to ensure compliance.

More is needed, however. Even well-managed banks can run short in a crisis, and *any* bank can be ruined in a panic. The reason bank liabilities can serve as currency is that they are perfectly liquid; they are available on demand. But bank assets have to be marketed, so are not perfectly liquid, especially in a time of financial disorder. A lender of last resort is needed, to protect well-run establishments in crises, by providing them with the liquidity to meet their liabilities. Such an authority should also possess the powers to preside over liquidations, to prevent a disorderly collapse whenever overissuing gets out of hand, and to prevent closures from injuring innocent parties. This is necessary to prevent any general tendencies to withdraw funds permanently, for this would endanger the whole banking system.

The danger is that such a lender will encourage banks to make excessive and/or risky loans, knowing they will be bailed out. To prevent this, the Central Bank must design policies that will protect a bank's customers in the event of a failure of its assets, but will punish the stockholders and management (Wray, 1990, ch. 10).

Just as the Mint establishes a price at which it will buy and sell metal, so as to tend to stabilize the currency, and also to redeem underweight coins, thus validating the coinage, so the Central Bank establishes a rate of interest that tends to stabilize the system of notes and bank deposits, while at the same time making loans and accepting deposits at that rate in its capacity as lender of last resort, "validating" deposits. And just as the Mint must recognize that, although larger and carrying the stamp of authority, it is still one player among many in the market, so the Central Bank must face the limits to its powers. By virtue of size and authority, it can – within limits – fix the nominal rate of interest. It is the "price leader." But market forces adjust the quantity of circulating medium to "the needs of trade," to use the old-fashioned phrase, and the Bank cannot affect this. Of course, by restricting credit and raising interest rates, it can alter the level of investment spending, thereby influencing the composition (Craft economies) or the level (Mass Production economies) of aggregate activity. Since the quantity of money in circulation adapts to activity, this *indirectly* modifies the money supply. But the Central Bank cannot directly set the quantity of money.

In particular it is a mistake to think that the Central Bank can manipulate the "money supply" by adjusting reserves – "high-powered money" – which will then be converted into increases or decreases in deposits by the money multiplier. On the contrary, reserves adjust to the level of deposits, which, in turn, reflects the level of activity. When private demand rises, for example, lines of credit are activated, and as the borrowed funds are spent deposits of business receipts rise, out of which appropriate reserves will be set aside. Thus as deposits rise, reserves rise; similarly when demand falls and deposits of receipts decline, reserves will be drawn down. (In severe declines, excess reserves are likely to appear, as in the early 1930s (Bradford, 1932, chs. 8, 9; Currie, 1968)).

Note that private investment demand creates offsetting profits which are then available to finance the final payments for the capital goods, making it possible to repay advances all through the production process. Public deficit spending, however, sets in motion the same process, and likewise creates private profits equal to the spending, but the government pays in government-created money. The profits so created will be in excess of investment needs; when firms place them in securities they will drive up security prices and lower interest rates. Alternatively, when deposited in banks, they will appear as excess reserves tending to drive down interest rates (Mosler, 1994, 1997) – unless absorbed by new government bonds or sterilized by open-market operations,

The Monetary Authority cannot reliably affect deposits simply by adjusting reserves. Suppose it created additional reserves. This will tend to drive interest rates down (assuming a pool of credit-worthy borrowers.) Investment spending *might* be increased, in which case deposits would rise. But they would rise because activity has increased, not because of the additional reserves. If activity rose autonomously, lending would rise, deposits would increase, and, as banks set funds aside, reserves would increase in pace.

As a matter of policy, to limit inflation, and to control the volume of speculative funds, the Central Bank's interest rate, of course, should lie close to (slightly below) the "Golden Rule" rate, that is, the rate at which the proportional increase in money in circulation would equal the proportional growth of capital. This point will be developed in a moment.

Digression: a New Monetary Economics?

Recently, a "New Monetary Economics" has emerged, arguing that a competitive payments system could dispense with managed money altogether, making payments in the form of electronically transferred claims that will be settled in securities (Greenfield and Yeager, 1983; Hall, 1982; Black, 1970; Fama, 1980). The idea is that market forces would supply the circulation with a stable quantity of privately issued assets capable of supporting transactions at a stable price level. These financial assets would be tied to real assets and would therefore be of varying quality, term, and so on. No asset would have a privileged place in the payments scheme; all would presumably be acceptable in settlement of debts – or if they were not, they could be traded for others that were.

This cannot be taken altogether seriously, but investigating such systems illuminates the way money works. The proposal assumes fully developed markets of all kinds, continuous equilibrium, and no uncertainty. Under these conditions all assets are equally liquid. Money, however, is necessary precisely because no one can know for certain that a debt will be paid, or that an asset will yield the promised sum. To assume that all agents are reliable, that all assets will yield the expected return, and that technology works as expected is to remove the need for cash payments. If a promise is as good as gold, then there is no need for gold! Gresham's Law tells us that everyone will pay with promises. So if the economy were assumed to be the least bit realistic, agents and assets would be very different in quality and reliability, future assets values and payments would be uncertain, and Gresham's Law would ensure that only the *least* reliable assets circulate. Some of these will default, upsetting circulation. That is why money is needed.

Worse, it is arguable that securities will be systematically overissued.[28] Consider an agent A and its nearest competitors. Each has two strategies – LI, limited issue, OI, overissue. LI returns normal profits, expressed as rate r; if one party overissues and the other does not, the overissuing party will expect to make a gain, G, expressed as a gain per unit capital, at the other's expense. A larger portfolio will provide economies of scale, making larger issues at lower unit costs, plus giving possibilities of diversification. With such economies better terms can be offered, and additional business attracted. But if both sides choose OI, then both will suffer a small loss, l, expressed as a loss per unit capital, which arises from the fact that they have undercut their earnings and taken on additional risk. The strategic situation looks like this in the diagram.

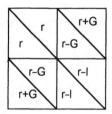

28 The other side of overissuing is overlending, in relation to the financial institution's own capital. The portfolio becomes too large for the capital to cover the risks, which means that the quality of the institution's paper declines.

Clearly $r - 1 < r$, so for both to choose OI is to ensure that the final outcome will be worse for both. But this is what they will choose, since

$r + G > r$

$r - 1 > r - G$, provided $G > 1$

That is, assuming the financial market to be sensitive to small differences in the terms of offerings, each reasonably believes that a substantial gain can be made by attracting customers from the rest, which can be done by issuing on a larger scale. Hence all do, and no one gains. But all securities are overissued, and hence will tend to depreciate. No one will want to hold them, so velocity will rise, and with it, the price level.[29]

Financial markets and the financial circuit

Money in circulation turns over against goods; it is not *held*. Goods are sold for money revenue, and then this revenue is exchanged again for inputs and labor. The circulating capital does not consist of goods plus the medium of circulation; rather, the goods turn over against the medium, so that the form of the capital changes while the amount remains the same. This applies as well to the funds that circulate depreciation and fixed capital. By contrast, money held as an asset does not circulate in the same way, exchanging against goods and services; it is a part of financial capital.

Financial markets

The funds that circulate fixed capital are not held as assets. Moreover, the loans of depreciation funds from producers of circulating capital to fixed capital producers could, in principle, be made directly. However, this would require firms to develop a different kind of expertise, that of dealing in loans. Instead, specialists will emerge, to intermediate, and they will find it convenient to trade with one another, borrowing and lending, or issuing securities. Intermediation is essentially a banking affair, but it does also create securities for financial markets.

The chief source of such securities, however, will be the shares of limited liability corporations, and the bonds issued to finance long-term investments. To these must be added securities issued by financial companies, and government bonds. Financial markets exist to float new issues

---

[29] It could be argued that repeated games would lead to learning. No doubt, but there would then be a free rider problem. That, in turn, could be controlled by an institution representing the collective interest of banks – which leads straight to the Central Bank!

of stocks and bonds, to rearrange existing portfolios, and to cash in earnings from the appreciation in the prices of existing securities

The financial circuit

The financial circuit, then, is the circulation of money held as an asset, among the holders of financial assets, allowing them to reorganize or restructure their portfolios, and, of course, allowing them to speculate. Such money may turn over quickly or slowly; it passes from hand to hand among asset-holders. In the circulation of produced goods and services there will be a point at which firms are fully liquid, and another at which they are completely illiquid. But in the case of the financial circuit, there are no systematic points of liquidity and illiquidity. Some money and near-money will normally be held in portfolios. Debts are repaid, and the funds reloaned. Each year a certain fraction of the total of bonds outstanding falls due and must be repaid. These loans must then be renewed. To put it another way, as time passes the average maturity of bond-holdings will fall, so portfolios must be lengthened, to maintain the desired time structure. Thus a sum of money equal to the reciprocal of the average maturity of the total value of bonds must circulate, first as repayments, then as new loans. A similar turnover will take place among stocks. The securities of stagnant and declining sectors will be sold, those of expanding ones bought, as portfolios are shifted to reflect the growth pattern of the economy. As the prices of the former fall, more and more shares will exchange for the same money, which will then buy progressively fewer of the latter.

The money needed to update loan portfolios and shift growth stocks will be the minimum that must circulate in the financial sphere. But this money need not be held while waiting for an "updating transaction." It can be used to shift portfolios speculatively – some asset-holders are bullish about some stocks, bearish about others. Other financial funds hold different opinions. So the money can circulate to change the composition of portfolios. Every day the news will affect some stocks and bonds in some small measure – sometimes in large measure – offering opportunities for gain and risk of loss. Anticipating the news – and since the future cannot be accurately foreseen, opinions will differ – financial capitalists will try to take advantage of opportunities for gain and hedge against loss.

Speculation and growth stocks

Unfortunately, this is not all. The financial side gives rise to a more significant instability. Once family firms have developed into modern corporations, financial markets become more complex. Besides commercial paper, credit for working capital and finance for fixed capital, there are stocks, in particular, growth stocks. And here the relationships are problematical. For if a company has a certain number of shares outstanding, and retains its profits to reinvest them, the price of its shares must rise, since they are now claims to a larger enterprise. (Of course, this presumes what the market must judge, that the new investment is well considered and likely to succeed.) Since such companies retain their earnings, they will not distribute more than nominal dividends; hence their stocks and shares will be held for appreciation, reflecting the market's judgment of the success of the investments of such companies. If $ is the price of a share, then d$/$ will be the rate of share price appreciation, and if investments are well planned, so that with a given technology and market conditions, new investments earn the same rate of profit as old, d$/$ will equal g, the rate of growth. Arbitrage should then ensure that, allowing for risk, d$/$ = i, the rate of interest.

But at this point speculative activity enters the picture, creating a problem analogous to that which arose with hoarding in the case of a metallic currency. Suppose a flood of opportunities opens, raising g above i; this will lead the stock market at the margin to consider d$/$ > i. The initial quotes will reflect a small number of purchases based on expectations and the knowledge of well-placed traders. Reservation prices will rise. As information about the successful investments spreads through the market long-term bonds will be sold, and the funds will be shifted into growth stocks and shares. This will tend to depress bond prices, raising i, a move in the correct direction. But the shift of funds into stocks and shares will bid up prices, tending to *raise* d$/$ further, maintaining the inequality. (Whether the difference widens or narrows depends on the relative volume of bonds and growth stocks; but even if it narrows, it will never disappear.)[30]

At this point the effect of arbitrage with the bond market will have been to raise d$/$ above the rate of growth of the real capital stock. Hence companies considering investment must pause to reflect – stocks are risky, but they are rising. It might be worth delaying projects to take advantage of the rising stock market. As in the case of hoarding, this is a self-justifying move. If money is diverted from real investment, and from bonds, to investment in stocks and shares, d$/$ will continue to rise. The

---

[30] As interest rates and stock prices rise, the value of bank portfolios will rise. It is true that high interest rates imply lower values for bonds, but once the rise in the stock market begins to attract investment funds from companies the increase in the value of stock can easily outweigh the loss in bond values. In any case earnings from interest will be high, and demand for credit – for speculation on the market – will be strong. Hence banks will expand their offerings of credit.

effect, of course, will also be to pull up interest rates. Both of these moves tend to justify additional issues of notes or credits by the banking system, adding fuel to the speculative fire. But as funds are diverted from real investment, g will fall. The problem began with g > i; market pressures will drive i up, and the diversion of funds will bring g down. But there is nothing to make the process stop once g and i are in balance! For the arbitrage is not between g and i directly; it is between i and d$/$, and the process has raised d$/$ above g. As long as funds are diverted into the stock market, d$/$ will continue to rise.

Moreover, rapid price appreciation means that d$/$ is out of line with current real growth. At some point firms will not wish to further curtail their real investment, and portfolio managers will begin to pick up bond bargains, looking for a capital gain. At this point funds will begin to flow into the stock market at a reduced rate, and d$/$ will slow down. Considering that the stocks are overvalued and that bonds are a bargain, portfolio managers will begin to sell, and stock prices will come crashing down. But once again there is no obvious stopping point. Selling sets off a self-justifying downward spiral.

Stabilizing a modern monetary system, then, requires stabilizing stock market booms and busts, or at least insulating the real economy from them.

Funds for production, funds for finance

The money in active circulation is money devoted to production and distribution. It circulates as rapidly as production and consumption are carried out; it is "money on the wing" in Dennis Robertson's phrase (Robertson, 1926) and has also been termed "active balances" (e.g., by A. J. Brown, 1955, pp. 81–8, passim). In the modern world these funds can be more or less identified with M_1 – the qualification arising from the fact that checkable savings accounts, NOW deposits and some time deposits (usually included in M_2), which earn interest, are increasingly being used for transactions. But the money in the "financial circuit" may be idle a good part of the time; it is money held as an asset or used to shift assets among portfolios. Robertson called it "money in the vault," although, of course, it would not actually be kept in a vault in a modern economy. Rather it would be an entry in a computer or record book respresenting a claim in a time deposit, a savings account or a mutual fund or other money market account. A reasonable contemporary measure of such "idle balances" might be $M_3 - M_1$.

Contemporary monetary analysis tends to lump together all balances; the argument here suggests that a strong and useful distinction can be drawn between active and idle balances, that is, between the funds in the circuit of production, and those in financial markets. If this is correct we should expect to find that active and idle balances depend on different variables.

Capital arbitrage and the Extended Golden Rule

"Capital arbitrage" designates the shifting of capital funds in search of the highest return. But the familiar tendency of capital to shift from industry to industry, ideally culminating (if certain conditions hold) in a uniform rate of profit, is only one form of capital arbitrage. In addition, *funds* move about seeking the highest return in interest; moreover, funds can be shifted from financial assets to real assets, if the rate of profit in some line, adjusted for risk, exceeds the available interest rates. Finally, interest compounded is financial growth; earnings and capital gains ploughed back result in real growth. Funds can be shifted between financial assets and real holdings in pursuit of *growth*.[31] In short, capital will shift between types of assets, pursuing the highest yield or increase in value, whether that is expressed as a rate of profit, a rate of interest or a rate of growth. Financial-real interaction takes place in the arbitrage between r and i.

In the Craft economy, firms were family businesses and the market for corporate control as we know it today did not exist. It was more difficult to "buy into" growth; one had to approach a firm and arrange to become a "silent partner," or persuade them to make shares available. Arbitrage therefore chiefly took place between r and i, at the margin, that is, in connection with investment and replacement. By contrast, in Mass Production the market for corporate shares is fully developed, and arbitrage between i and r need not be confined to the margin. Moreover, arbitrage between i and g becomes possible, adding another strand to financial-real interaction.

Capital arbitrage does not necessarily lead to the formation of a uniform rate of profits, still less to a uniform growth rate. Interest rates, adjusted for risk, will tend to be pulled together, allowing for imperfections. In Craft conditions, Chapter 9 will show, in the long run, growth determines the real wage, and therefore the profit rate, which in turn determines the long-run interest rate. But given the uniformity of interest rates, arbitrage

---

[31] As used here, "interest" designates a contractual payment of funds, "profit" the realized money earnings of an enterprise, and "growth" the increase in the capital value of assets or holdings. Each of these can also be considered as an expected magnitude.

between bonds and capital holdings, and the need to borrow to invest, will tend to establish a degree of uniformity in profit rates. Technological interdependence will tend to pull growth rates together, but innovation and the displacement of older products by new will create wide disparities. The interest rate will be the most uniform, then the profit rate, and the growth rate least.

Under Mass Production this order of uniformity will be preserved, but both growth and profit rates will be even more disparate. Innovation will take place at a higher rate, with more sweeping waves of "creative destruction," and with corporations investing retained earnings, the influence of the rate of interest on profit rates will be much weaker.

However, in both Craft and Mass Production economies there will be market forces set in motion when the *average* levels of these rates are unequal. In some cases, these forces will tend to bring the average levels of these rates together, that is, to establish what we shall term the "Extended Golden Rule," the equality: $i = r = g$, on average, net of rents, and adjusted for risk and taxes (and in the case of profits, for what we shall call "maintaining social capital"). However, in other cases, the inequality of the rates may be preserved or widened.

Most clearly, in both cases, there will be a tendency for the average, realized rates of g and r to come together, and to move together. In the Craft economy, in the long run, as we shall see, the real wage will adjust to provide the profits to finance the growth of capital that will employ the growing labor force. Since laboring households save little or nothing (other than saving for consumer durables), profits (net of management salaries) must be very close to investment. In the short run price changes, relative to money wages, will adjust profits to investment. Hence profits adjust to investment in both the long and the short runs, and realized profit rates will tend to equal or lie close to actual growth rates. Under Mass Production (as we shall see in Chapter 10), corporate pricing plans will establish a markup over variable costs designed to generate profits sufficient to finance the investment required to meet the expected growth of demand. (The expected growth of demand, g_d, must be distinguished from other notions of growth. The term designates the *growth of markets*; cf. Chapter 10.) The expected growth of demand determines planned growth of capacity, and these plans also set the benchmark prices that will establish the planned rate of profit. In the short run, investment spending will generate offsetting profits. So again, in both the long run and the short, the rate of profits will adjust to the rate of growth.

But how does the rate of interest fit into this picture? The answer is not simple.

Stability in monetary growth

It has traditionally been held that competitive pressures will tend, in the long run, to bring the rate of profit and the rate of interest into line with one another. Consider holding assets for income: a given sum invested can bring a (risky and somewhat uncertain) return in profit, when placed in a venture, or it can bring a (safer) return in interest, when placed in bonds of suitable rating. In addition, in a growing economy, in which there is a market for shares in enterprises, investors can choose between growth through investment in real assets, on the one hand, and growth through compound interest on bonds, on the other. In these circumstances arbitrage should tend to bring the real rate of interest into line with the rate of growth (Nell, 1992). That is, investors can choose to hold shares in partnerships or to own companies, which will plow back their earnings and grow; or they can hold long-term bonds, reinvesting the interest. A well-managed portfolio will have a selection of safe and risky income-yielding assets, and a similar potpourri of growth assets, the mix of income and growth itself being determined by the needs of the asset-holder. So it appears that competitive pressures exist, tending, in theory, at least, to bring the rate of profit, rate of interest and rate of growth into line with one another.

When the rate of interest equals the rate of growth, the earnings of the banking system enable it to expand its monetary issue – or its credit lines – at the same rate as the economy is growing, without increasing the banking system's degree of risk. That is, the banking system's earnings will be proportional to the interest rate; hence bank capital will accumulate at the rate of interest, while deposits and loans will grow at the rate output is growing. When the two are equal the expansion of the money supply, to keep pace with the needs of trade, will take place at a constant ratio of bank capital to liabilities. But does this arrangement provide a stable monetary regime?

First, consider deposits and lines of credit for working capital. When i = g credit for working capital will expand at the same rate as the demand. When i < g, *justified* credits will expand by less than the demand; alternatively, granting the credits needed by business will be riskier, since the capital to back them is lacking. Hence there will be a tendency for i to rise. The higher level of i together with the shortage of credits will tend to reduce activity and investment; hence there will be a tendency for g to fall. Conversely, if i > g then the ability to offer justified credits will grow faster than the demand for them. Competition will tend to bring interest rates down, while the lower rates and the easy availability of credit will tend to encourage expansion. The relationship therefore might appear to be stable.

On the other hand when i < g, g will tend to rise. This is, after all, the traditional relationship – when credit is cheaper, investment will expand. If there are diminishing returns in the capital goods sector, then g will not be able to expand indefinitely. If i rises and g cannot, after a point, then the relationship would seem to be stable.[32] Diminishing returns suggest a Craft economy; adapting the discussion to a Mass Production economy, i < g would imply an expansion of output, up to full employment. Hence g, the growth of output would tend to rise as capital shifted. At full capacity, g would no longer rise, but with inflation the real rate of interest would fall. Suppose i > g; this would lead to contraction rather than deflation, but g would fall. In short, the relationship may be unstable – with the proviso, however, that the impact of interest rates on investment may not be very strong.

Next, consider the financing of fixed capital. As we have seen when i = g sinking funds expand at the rate necessary to purchase and install the expanded replacement capital. But when i < g, the sinking funds will not be growing at the required rate; there will be shortages of credit for working capital to finance the current operations of the fixed capital producers, and the funds will not be sufficient to purchase a full complement of replacements expanded at the rate g. Demand for credit to expand operations at the rate g will tend to drive up interest rates, while the shortages will curtail operations, tending to reduce the growth rate. Similarly, when i > g, the sinking funds will be growing faster than demand for working capital, tending to bring interest rates down, while the easy availability of credit will tend to encourage additional expansion. Again the relationship appears to be stable.

But once again, other things being equal, a low rate of interest will encourage expansion, and a high rate discourage it. A discrepancy between i and g will tend to perpetuate itself through the stimulus provided to investment. This needs to be explored, both for Craft economies and for Mass Production.

Rates of profit, growth, and interest in Craft economies

In a Craft economy growth is "extensive." That is, profits are saved and (deposited in banks and) loaned to new entrepreneurs, who, drawing on established technology, set up new firms of optimal or normal size, which serve new markets. Growth comes about by mobilizing savings – chiefly profits distributed to owners of financial assets – in order

---

[32] Wicksell (1898, 1967) argued that when the money rate of interest lay below the real rate of return, a *cumulative* inflation would take place. And when i > r, according to Wicksell, a cumulative deflation would take place.

to set up new enterprises. (Think of the westward expansion of the United States in the nineteenth century, or of the expansion of industrial cities.) The rate of profits and the rate of growth of capital will tend to be close in such an economy. If investment falls short of profits, prices will fall, and real wages rise, bringing profits down. (The lower prices may encourage investment.) If investment tends to exceed profits, prices will be driven up, lowering the real wage and raising profits. (The shortages and higher prices may also discourage investment.) When there is a discrepancy, investment and profits move to close the gap. There will therefore be a tendency for the actual rate of profit to be brought into equality with the rate of accumulation of capital.

But the situation is different with respect to the rate of interest. Instead of overcoming a difference between the (average, current) rate of profit and the real rate of interest, by moving them towards one another, market forces will tend to move them both in the same direction. A discrepancy will tend to be preserved. The rate of interest and the rate of profits will tend to track one another.

Suppose $i < r$, where we assume that r is in line with normal investment, so that $r = g$, allowing for discrepancies due to extraneous factors. The difference, we assume, may be the result of monetary policies, or it may be due to accidental factors. Because $i < r$, entrepreneurs will eagerly borrow all existing savings, equal to profits, and invest them. But because borrowing is cheap, they will seek to borrow more, and banks, eager for business, will lend. The heavy demand will tend to drive up interest rates. But as they invest their loans, the entrepreneurs will drive up prices, and the higher prices, in turn, will lower the real wage and raise profits. Thus r and i will rise together. Since the discrepancy is preserved the stimulus will remain, and the two will continue to rise, along with prices, in a process not dissimilar to that suggested by Wicksell.

(While prices rise, money wages do not. This is therefore a profit inflation, in Keynes's terminology, not a general inflation. So it is not appropriate to adjust the current nominal rate of interest by subtracting the rate of price increase. But if one did make some sort of adjustment, the effect would be to reduce the rate of interest, thus increasing the discrepancy.)

But this cannot go on indefinitely, or even for very long (contrary to Wicksell, 1985, 1962, ch. 7; 1967, pp. 193–208). For investment will be rising while consumption is being constricted. Yet a substantial fraction of investment will be in industries supplying consumer goods, and a further fraction will be in industries supplying means of production for the consumer goods sector. Investment takes place to meet expected new demand; but with real wages declining, the growth of consumer demand will be less than anticipated at the previous level of the real wage. When this becomes apparent, investment will be cut back – and as we shall see later, this may be part of the explanation of the characteristic pattern of expansion and crisis in Craft economies.

Suppose, on the other hand, that i > r. It will be too expensive to borrow, except for the most profitable projects. Hence investment will decline, and with this decline, prices will fall, raising real wages and lowering profits. The slack demand for borrowing will lead interest rates to sag, so that the rate of profits and the rate of interest will decline together, in periods of falling prices. But once again, this cannot continue indefinitely. For real wages are rising, which will lead to an expansion of consumption, eventually creating additional investment opportunities.

Rates of growth and rates of interest in Mass Production

We will first examine the implications of a divergence between the rate of interest and the rate of growth, and then consider whether there are pressures tending to bring them together with one another and with the rate of profit. We shall see that market forces do not tend to eliminate divergences, but they do lead the rates to track one another.

For the sake of argument, let us assume that the Monetary Authority has the power to peg "the" (long-term) *nominal* rate of interest. (In effect, we are assuming that the term structure is orderly.) We consider a Mass Production economy, in which large corporations invest by plowing back profits. The rate of profits therefore is the net earnings of corporations divided by the value of the productive plant, equipment, etc., operated by the corporations. This implies that in making *portfolio* decisions asset-holders do not react directly to the rate of profits; instead they concentrate on the relation between the nominal rates of interest on bonds and the nominal yield from holding stocks. This last is made up of dividends and share price appreciation. Here we meet an important difference between the Mass Production and the Craft-based economies.

In the latter profits are earned by the family firm, and are either saved or used to pay interest. Interest in turn supports rentiers and is itself largely saved. Savings, in turn, are loaned for investment by new firms. A representative asset-holder has a choice between owning a business of a *given* size, earning the going rate of profit, and owning a bond yielding the going rate of interest. Asset-holders arbitrage between the rate of profit and the rate of interest.

By contrast, in Mass Production economies, asset-holders have a choice between stocks or shares in a corporation, and bonds. Profits are ploughed back, so the value of the share reflects not only, and perhaps not chiefly, its current and near term future earnings, but its future *growth* prospects. In other words, asset-holders arbitrage between the rate of interest and the rate of growth. Let's consider this more closely.

Asset-holders – households, trusts, nonprofits, institutions, pension funds, etc. – will be supposed to have portfolios composed of two kinds of holdings: financial assets, such as bonds, which are commitments to pay principal and interest on certain dates, *independently* of the economic success of the payer, and "real" assets (in inverted commas, to indicate that these are *claims*), such as growth stocks, holdings of venture capital, and participation in direct ownership, where payments, and market value, are conditional on economic performance.[33]

We assume that there are good reasons to expect the economy to realize a certain rate of growth (allowing for a normal degree of variation, comparable to the variations one might expect in the rate of profits), and further that this rate of growth is expected to hold for the indefinite future. Under these circumstances the value of an asset that constitutes a claim to real resources (or is held in terms that vary strictly with real growth, such as growth stocks) will compound at the rate of growth. (Assuming that stock markets have the appropriate information, the price of the asset – the claim on real invested resources – will appreciate at a rate equal to the rate of growth.) A similar position held in financial assets would compound at the rate of interest. A representative asset-holder must build a portfolio by comparing the rate of growth to the rate of interest.

We assume that portfolio holders manage their assets to best advantage, but we need *not* settle on any particular formulation. They may maximize the present value of the portfolio or the rate of growth of asset values, they may or may not be subject to any of various plausible constraints, they may or may not have a finite time horizon, and the markets may or may not be "imperfect." But in *any* of the usual formulations, when the rate of growth and the rate of interest are equal, after making appropriate allowances for risk and liquidity, etc., portfolio holders must be indifferent at the margin between what we have just termed "real" and "financial" assets. And when i is *not* equal to g, one will compound faster than the other, which creates an incentive to shift capital until increasing risk or some other factor offsets the advantage provided by the difference between i and g.

At this point we can ask, what will be the effects of portfolio managers selling one kind of asset to buy into the other?

---

[33] Even "blue chips" depend on growth. They pay a "normal" dividend, but their total return is dividend plus appreciation, and the stock price will vary with Wall Street's assessment of company performance, especially judgments as to the wisdom of their investments. Hence even on blue chips the return reflects expected growth.

First suppose market pressures and the Monetary Authority have combined to establish a position in which nominal i < real g. "Real" assets compound faster than financial; hence asset-owners will tend to shift into real holdings, buying into partnerships, growth stocks, and takeovers, and putting up funds for new investment projects.[34] As a result,

Bond prices will tend to drift down, raising i.

Investment in real terms will tend to rise, raising g.[35]

When i < g debt will be falling as a share of income; hence the rise in nominal i will have little or no effect on g. But it will mean higher interest costs. Since demand is strong firms will tend to pass these along in higher prices, setting off a wage-price spiral. The resulting inflation may offset the tendency for nominal i to rise, keeping *real* interest rates low.

To maintain the level of nominal i initially established, the Monetary Authority would have to infuse funds into the bond market at a rate equal to that at which funds are shifting into "real" assets.

Since both nominal i and g tend to rise, as a result of a certain amount of capital shifting from financial to real, the difference, g − i, will tend to be maintained. They need not rise in exactly the same proportion, but there seem to be no grounds for expecting i to rise faster. Hence the tendency for capital to shift will persist.

Arguably g cannot rise indefinitely. But this will *not* lead to i catching up to g. At a certain point g will hit a capacity-constrained maximum, where one or more basic industries cannot expand production to keep pace with the proposed level of investment. Such bottlenecks will lead to rising prices among basic goods, which will then be generally diffused as a cost inflation. (Moreover a rise in growth might cause prices to rise in primary sectors). But inflation *reduces* the *real* rate of interest. Hence at the point where g can no longer rise, inflation emerges to prevent the *real* rate of interest from achieving parity with the growth rate. The difference still persists, and the tendency to shift capital remains.

Next suppose that market pressures and the Monetary Authority have established a position in which nominal i > g. "Real" projects will

[34] For precautionary and insurance reasons, asset-holders will wish to maintain both kinds of assets in their portfolios. Given a difference between earnings on real and financial assets, only a fraction of funds will be considered shiftable. This fraction, however, will be an increasing function of the difference, i − g.

[35] This is surely plausible intuitively; in Chapter 10 the case will be developed in terms of a model of corporate pricing and investment (see also Nell, 1992, ch. 17; 1993; also cf. Wood, 1978 and Eichner, 1976, 1986).

compound more slowly than financial assets. Funds will therefore be shifted out of "real" into financial; new savings will be channeled into financial markets rather than put into new construction. Intermediation will increase. As a result,

Bond prices will tend to rise, lowering i.

Investment in real terms will tend to fall, lowering g.

When i > g, debt is rising, leading firms to cut back and trim costs. As a result the rate of inflation will tend to fall, offsetting the tendency of nominal rates to decline.

To maintain the level of nominal i initially established the Monetary Authority will have to sell securities, withdrawing funds from the financial markets, at the rate at which they are being injected by the transfer from "real" holdings.

Since both i and g tend to fall, as the result of a certain amount of capital shifting from real holdings to financial ones, the difference, i − g, will tend to be maintained, as the two move in the same direction. Hence the tendency for capital to shift will persist. Further, as growth declines, so will the rate of inflation, which will tend to *raise* the real rate of interest, offsetting any tendencies that might bring them together.

There are symmetries and asymmetries here. Just as there is a maximum to the growth rate, so there appears to be a minimum. Growth will not fall below the replacement level. But although inflation will slow, prices will not necessarily ever decline. A fall in investment and growth, in the modern economy, will lead to a clear-cut and determinate decline in capacity utilization, but prices will tend to remain steady or decline only slightly and irregularly. Once inflation has fallen to zero no rate of price decline will develop to raise the real rate of interest.

However, other factors may limit the decline in i. When bond prices are high above normal, there will be speculative fears of a collapse, inhibiting further bidding up of bonds. When interest rates are too low, the earnings of financial institutions will be low. Hence risk will rise in financial markets, and this will inhibit further shifts of funds. Thus the difference between i and g will not be eliminated, yet the two will tend to move together.

The claim is not that a persistent gap between i and g will last forever; the point is rather that it will not close until there is a change in basic parameters. Hence the argument here is relevant to the analysis of "long swings" in economic behavior. Thus i < g is consistent with the long postwar boom, from the end of the war to the first oil shock, and i > g with the growing stagnation since then (see figures at the end of this chapter).

The gap in each case will tend to close with the changes that bring about the turning points – but rather than an equilibrium, a new regime of persistent divergence is likely to be established (Nell, 1992, ch. 21; 1997, ch. 13).

Unbalanced growth

Next consider the case where there is no clear-cut expectation of a single, economy-wide rate of growth. Instead, different sectors will be expected to grow at different rates. *These* expectations are definite, but there will be no overall uniform, balanced rate of growth for the economy as a whole.[36]

A distribution of growth rates can be drawn, with g_i on the horizontal axis and K_i, the amount of capital in the i-th sector, on the horizontal.

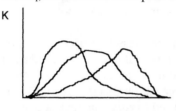

Three possible distributions are illustrated.[37] The areas under the curves, total K, should be equal (the drawings may not be accurate). Consider the middle distribution, approximately normal. Holders of "real" assets in low growth sectors will prefer a safe financial asset if the

[36] Unbalanced growth must be consistent with the simultaneous expansion of the basic sectors. Final output as a whole, basic or nonbasic, cannot expand over time at a rate faster than the rate of expansion of the slowest growing basic sector. A *component* of final output, produced using the slowest growing basic product, could expand faster than that slowest basic if another component, relying on that same basic, were growing more slowly, or shrinking, so that the first could expand at the expense of the second. But once the second becomes a small proportion of final output, this will not matter. In general, the interdependence of basics tends to pull growth rates together.

[37] Many others are possible. A two-humped distribution might be particularly interesting. Negative growth rates could be included. But too much cannot be made of this; whether a sector will attract funds depends on many other characteristics besides its size and expected growth rates, not least the judgments as to its riskiness, the likelihood of its technology developing further, etc. The point, however, is that when (always, in practice) growth rates are divergent, the rate of interest that balances the flow *into* financial markets from low-growth sectors with the *outflow* from financial markets to high-growth sectors will be the rate that is the analogue to the rate i = g, when g is uniform.

interest rate is above their growth rate. Holders of financial assets, on the other hand, might well be willing to shift to a "real" asset growing more rapidly.[38] If it is set equal to the *mean* rate of growth g, then, *ceteris paribus*, the flow of funds from low growth sectors to financial markets will just offset the flow from financial markets to high growth sectors. Growth rates and sectoral sizes are symmetrically distributed above and below the mean – creating symmetrical incentives to shift capital. There will therefore be no pressure on the rate of interest to change, and no need for the Monetary Authorities to continually intervene.[39]

But this rule is not without problems. Consider the other distributions. In the first, a few large sectors have low growth rates, which, however, will be close to the average, while a large number of small sectors have high growth rates, many well above average. The large sectors with low g's close to the mean are unlikely to shift much capital into financial assets; but this may not be a problem since the small high-growth sectors may not be able to absorb that much. So this distribution may stimulate a lower degree of capital-shifting than would take place under a normal distribution. The third distribution, on the other hand, with many small low growth sectors, and a few large high-growth ones is likely to face a difficulty. Since the low-growth sectors are small, not much capital will shift out of them, but the large high-growth sectors are likely to strongly attract funds from the financial markets. The rule, setting i = mean g, may have to be modified here.

Nevertheless, the principle is clear. When there are different growth rates in different sectors, the mean of the growth rates is the rate to which i should be pegged (allowing for minor adjustments). If i lies below or above that rate persistent movements of funds will take place, just as if i were below or above a uniform balanced growth rate.

[38] Why don't holders of low-growth assets shift directly into high-growth assets? Because this is not a barter model. Holders of low-growth assets must *sell* the assets first, and, in general large real assets are sold, not for cash, but for *financial* assets, usually at the liquid end of the spectrum. Then, of course, these same asset-holders might wish to buy into a high-growth industry, but the point about purchasing real assets is that direct knowledge of the business is required. Acquiring this, and making informed judgments will take time.

[39] If capital shifts *out* of low-growth sectors, and *into* high-growth areas, then the low end of the distribution is shrinking and the high end expanding. The mean will drift upwards. From period to period the rate of interest would have to be raised. This would be true if the shifts of capital had no effect. But as capital moves into high growth areas, the markets will tend to become saturated, and growth will slow down, while, very possibly, opening up elsewhere. The shape of the distribution will change, but perhaps not very rapidly.

Two conclusions may be drawn from this. First, the difference between g and i is not eliminated by the shifting of capital, and as long as it is not, the incentive to shift will remain. This can create difficulties for the Monetary Authorities, if they wish to try to peg the interest rate at a level significantly different from the growth rate. Second, the growth rate and the interest rate will tend to *move together*, that is, in the same direction, and even to the same degree. They will tend to "track" one another, and this may be the most important effect of "capital arbitrage." Since prices and inflation will tend to move in the same direction as the growth rate, prices, inflation, and the interest rate will also tend to move together.

A table of capital arbitrage relationships

At the danger of oversimplifying, these various aspects of capital arbitrage can be summed up in a table. Note that financial arbitrage is listed both as short run and as long run under Mass Production. When current profitability rises because of greater demand it is easy in modern financial markets to switch from bonds to stocks in the short run. But successful long run growth will also encourage arbitrage. It should be borne in mind that many of these relationships have only been described; the mechanisms will be analyzed more carefully in later chapters.[40] Nevertheless, it may be useful to group them all together at this point

Variables	Causality	Relationship	Mechanism	Adjustment
Craft Economy, Long Run				
g & r	g → r	positive (strong)	growth/real wage	stable
r & i	r → i	positive (weak)	arbitrage	stable
g & i	no relationship			

---

[40] Some recent writings have argued that the rate of interest, set by the Central Bank, can determine the long-period, normal rate of profit (Pivetti, 1991; Panico, 1988; Bharadwaj and Schefold, 1990.) Changes in the rate of interest change financial costs, and require businesses to adjust their markups accordingly. Unless debt and equity are perfect substitutes, a difference between equity/labor ratios and debt/labor ratios would imply that the prices so established would not be long-period normal prices. But there is no reason to suppose that all industries have the same debt/equity ratio. Debt and equity cannot be perfect substitutes, however, since their values are determined by different forces. In addition, in Joan Robinson's words, it is "excessively fanciful" to suppose that the Central Bank could set a *long-period* interest rate (or that there is such a thing!) and could do it independently of other forces in the economy. However, a cautionary note: interest costs *are* important and certainly do influence prices. When real i > g, debt increases more rapidly than output, and over the *long run* we can expect pressures to develop to increase markups.

Craft Economy, Short Run

g & r	g → r	positive (strong)	effective demand/ price	possibly unstable
r & i	i → mgl r → g	inverse (strong)	MEC	stable
g & i	no relationship			

Mass Production, Long Run

g & r	g → r	positive (strong)	corporate pricing	stable
r & i	r → i	positive	financial arbitrage	possibly unstable
g & i	g → i	positive	growth stocks	unstable

Mass Production, Short Run

g & r	g → r	positive (strong)	effective demand/ quantities	possibly unstable
r & i	i → mgl r → g	inverse (weak)	MEC	stable
g & i	g → r, i	positive	financial market	unstable

Rejected

r & i	i → r	positive	prices based on interest costs

*Implications of the Extended Golden Rule
for the monetary system*

On the one hand, then, the working of the monetary system depends on the Golden Rule, and, on the other, the working of the system also tends to enforce that rule, provided that government keeps the system's tendencies to instability under control. That is, changes from one point on the wage-profit trade-off to another tend to be matched by correlative movements between equivalent points on the consumption-growth trade-off.

This matching is brought about by two kinds of market forces. On the one hand there are expenditure pressures. Changes in the normal wage tend to bring equivalent changes in consumer spending. Changes in normal investment spending tend to generate equivalent changes in normal profits. On the other hand, if the actual, average rates of g and r, or either and i, are out of alignment, arbitrage will tend to pull them back together – barring instabilities and outbreaks of speculation. On this interpretation the Golden Rule is not an equilibrium condition; it is an extension of the "law of one price" to the earnings from capital.[41]

[41] If i = r, the "valuation ratio," v, must be unity – and, if both also = g, following Kahn's formula literally, it will be undetermined (Kahn, 1972, ch. 10). But there is a conceptual difficulty in Kahn's approach, which illuminates our argument. Let N = number of shares, s = the retention ratio (corporate saving), r = profit rate, p = price of shares in money, i = rate of interest, g = growth rate of capital,

(*continued to next page*)

The system depends on the rule in four ways. First, when r = g, the "balancing condition" holds in a simple form. This is the foundation for the circuit. Second, if i = g, then changes in the wage bill, due to a rise or fall in the real wage, will exactly equal the corresponding changes in

> (*contd.*)
>
> K = capital stock expressed in standard value (or labor value, anyway, in real terms). Kahn proceeds from v = pN/K, to dv + vdK/K = v(dN/N + dp/p), after manipulating. In steady growth, dv = 0, so vg = v(dN/N+dp/p), i.e. g = dK/K = dN/N + dp/p. Kahn then sets forth two definitional equations:
>
> Investment is financed partly from retained earnings and partly from selling shares
>
> $$g = sr + \frac{pdN}{K} = sr + \frac{(pN/K)dN}{N} = sr + \frac{vdN}{N}, \text{ implying } \frac{dN}{N} = \frac{g - sr}{v}$$
>
> The return on shares, i, equals the dividend yield plus appreciation
>
> $$i = \frac{(1-s)\,rK}{pN} + \frac{dp}{p} = \frac{1}{v(1-s)r} + \frac{dp}{p}$$
>
> Combining these,
>
> $$i = \frac{i}{v}(1-s)\,r + g - \frac{dN}{N} = \frac{(1-s)r}{v} + g - \frac{g - sr}{v}, \text{ so that}$$
>
> $$i - g = \frac{r - g}{v}, \quad \text{and} \quad v = \frac{r - g}{i - g}$$
>
> Hence if i = r, v = 1, and if i = r = g, v is undetermined.
>
> But neither of the definitional equations makes sense dimensionally. Both i and g are pure numbers, as are dp/p and sr. But pdN/K has the dimensions "money/Standard Commodity" (or other measure of real value.) These are also the dimensions of v, which therefore cannot equal an expression made up of dimensionless variables. Similarly the first term of the second equation, [(1 − s)rK]/pN, has the dimensions "Standard Commodity/money," so cannot be added to dp/p, which is dimensionless. Here dN/N is likewise dimensionless, so it cannot be equated to (g − sr)/v, which is not.
>
> Two changes must be made to correct these problems. The second term of the first definitional equation must be multiplied by 1/v, to express the amount of real capital equipment the money raised by the new shares, pdN, will purchase. It then follows that g = sr + dN/N. From the basic identity, Kv = pN, we can derive dK/K = pdN/Kv + dp/p = dN/N + dp/p. Hence dp/p = sr. Correcting the second definitional equation requires rewriting its first term, to show the amount of share value the commodity dividend represents
>
> $$i = \frac{(1-s)rKv}{pN} + \frac{dp}{p}$$
>
> Then canceling and substituting we have
>
> $$= (1-s)\,r + sr = r$$
>
> The Kahn formula cannot be used to object to our approach.

payments to capital, due to the correlative change in the profit rate. So when distribution changes, funds that went to underwriting investment can be switched to financing payments to labor, or vice versa. Third, when i = g, bank capital will expand at g, enabling the growth of justified bank liabilities to keep pace with the growth of the economy. That is, bank capital will grow at the same rate as deposits and the demand for loans, maintaining a constant coverage ratio. Fourth, fixed capital producers can borrow from sinking funds to spend on circulating capital while continuing to produce and expand to keep pace with the rest of the economy.

The Golden Rule is a condition relating *flows* of profit, interest, and investment. It is not to be understood here as an optimality condition, nor as relating to "best-practice" techniques. The Golden Rule is the condition for the circuit to work, and it tends to be established and maintained by competitive pressures, provided that the instabilities of the system are controlled.

The monetary system tends to pull i and g together through arbitrage. As we just saw if the two are not equal, funds will shift; in some cases, particularly in a Craft economy, this will pull them closer, but in Mass Production the discrepancy may remain, and it may even set off destabilizing speculation.

Finally, it is worth remarking that there seems to be no tendency for i = g to prevail in the international economy, and there appear to be few curbs on destabilizing speculation, either.

Empirical issues and policy

Of course, monetary policy could keep i and g unequal, even though it will require the authorities to work against the pressures of the market. But the effect of such inequality over the long run, would be for bank capital to grow faster than the needs of circulation, when real i > g, and more slowly when i < g. In the first case, funds will tend to shift into financial markets, and the banking system will be able to make additional loans; in the second, funds will tend to shift into real assets, and speculation in financial markets will tend to be restrained. In the era of the Craft economy market forces will tend to pull i and g together, but in the modern world these pressures will not be present.

In the modern world, M_1 (plus, today, a number of components of M_2) represents the funds in active circulation, which will always reflect the "needs of trade," so will grow with output, while M_2 and M_3 add to this measure various "idle" funds, to use an older terminology. Idle funds are those held as assets and/or used to adjust portfolios in the capital market, plus other funds held for financial speculative or precautionary purposes.

Hence $M_3 - M_1$ (adjusted M_1) would be a good measure of speculative holdings and funds available for managing capital transfers, while the ratio $(M_3 - M_1)/M_1$ shows the relationship between financial and circulating balances.

Contrary to the conventional approach these two sets of funds should not be added together; they have different turnover rates, and they respond to different economic forces. When real i < g there will tend to be a shift of holdings into real assets; funds will tend to be withdrawn from, or not to enter, financial markets, while the opportunity cost of holding funds (rather than financial assets) will be low. Hence, due to this shifting of funds into liquid holdings, $M_3 - M_1$ will tend to rise, relative to M_1, but because bank capital (depending on i) is growing more slowly than activity (depending on g), and therefore than bank lending, the financial system will become weaker and more dependent on the lender of last resort. (This ratio did indeed grow significantly during early postwar periods of low nominal rates, when the real interest rate fell short of the growth rate of output.) By contrast when i > g, the opportunity cost of idle holdings will be high, and speculative opportunities will be available; funds will therefore shift into financial markets, and the ratio $(M_3 - M_1)/M_1$ will tend to fall. Bank earnings will be high and financial institutions will feel free to take risks. (Again this accords with the U.S. picture of markets in the 1980s. Figures 6.1–6.4 below. Cf. Bernanke and Blinder, 1988, repr. in Mankiw and Romer, 1993, vol. 2, esp. pp. 330–2. For a more comprehensive picture, see Nell, 1996.)[42]

The correct policy, from this perspective, might seem to be to maintain the long-term nominal rate of interest at a level equal to the normal growth rate – or even a little above to allow for the upward price drift corresponding to improvements in quality. The stock of money would then grow *pari passu* with the demands of circulation at a constant price level. Inflation would therefore have to be financed through monetary innovations, and the consequent costs and difficulties should slow it down. There would appear to be no incentive to shift funds either into or out of financial markets.

However, this would be to underestimate the inventiveness of modern finance. The great danger is speculation, which leads to costly crashes, and during a boom will attract funds away from productive investment. To prevent this, financial innovation has to be restrained, which can be done by keeping the growth of bank earnings low. Moreover, as far as monetary policy can assist, funds should be channeled into productive investments.

---

[42] There is a political significance to the fact that when i > g, funds flow into financial markets. It helps to explain the near-universal support in the financial community for austerity policies, which raise i and hold g down.

Keeping i well below g, which also tends to strengthen growth, will clearly be required.

To further discourage speculation a turnover tax might be imposed on holdings of financial assets. Assets liquidated after being held for a full cycle would be exempt from this tax; assets held for, say, half a cycle would pay a reduced tax. All other asset transactions would be liable for the full tax. Asset-holders will have to think carefully about speculative transactions – the tax will be certain, the gain risky. The prospect of gain will have to be large enough and likely enough to warrant paying the tax. This will encourage long-term asset holding. Such a tax will also tend to prevent deposit insurance from encouraging speculation; such insurance and other "lender of last resort" measures are needed to prevent disasters from spreading. But for that reason they reduce the costs, and therefore the risks, of such failures, thus encouraging risk-takers. A tax on speculation will tend to counter this.

Speculation in foreign exchange has become an impediment to effective national monetary policy. To control this, a similar tax could be imposed, with a portion of the tax going to an international agency to be used to develop a stabilization fund. Foreign exchange purchased for the purposes of trade would be exempt. On purchasing foreign exchange the tax could be placed in escrow, to be released when shown that the funds had been used in commerce.

The details of such schemes are not at issue here. The point is that holdings of money as an asset tend to move into speculation, which, in turn, can easily become destabilizing. This must be controlled if monetary circulation is to work smoothly.[43]

Conclusion: the significance of circulation analysis

The shortcomings of the supply and demand approach, treating money as an asset, have long been evident (Hicks, 1988; Moore, 1988). Until recently, there has been nothing to put in its place. Circulation analysis offers an alternative, which emphasizes that aspect of money most conspicuously neglected in the normal approach, namely, its role as a medium of circulation. It, too, relies on "supply and demand," that is to

---

[43] A counterpart to a turnover tax on speculative holdings would be "stamp duty" on idle funds and the unused balances of lines of credit. Such a duty would be levied after funds had lain idle for, say, two quarters; its purpose would be to encourage circulation and economic activity. "Social credit" experiments with such a tax in Austria and in Canada in the 1930s have generally been deemed modestly successful. Silvio Gesell made similar proposals, which were regarded sympathetically by Keynes.

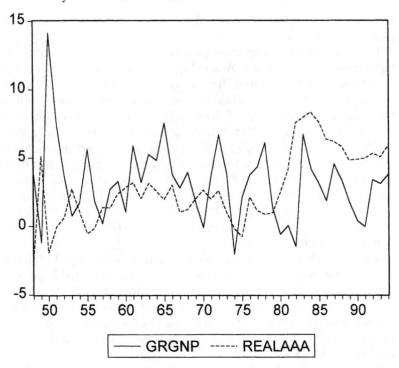

Figure 6.1 The rate of growth and the real rate of return on AAA bonds for the postwar United States. Source: Citibank Macroeconomic Database 1995.

say, on market forces, but in a context in which the institutional arrangements are paramount.

The study of circulation reveals how expenditure takes place. The circulation approach relates the pattern of monetary expenditure to production and distribution. It shows how sectors are related to the categories of income distribution. It makes it possible to study who is included and who excluded, and on what terms.

Competition and interdependence determine the form of the circuit. This form can be seen most clearly when production is carried out by means of "batch" processes: there will be a definite beginning and ending, and a clearly defined sequence of activities. This provides a new way of treating time. Instead of the "long run" and the "short run," defined by whether or not the stock of capital can be varied, time is classified by the relation of activities to the circuit. There are first, those that take place entirely within one circuit, next there are those variables

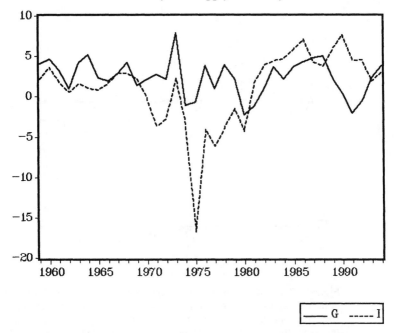

Figure 6.2 U.K. rate of growth and the real rate of return on Treasury bills, 1959–1994.

that can be altered from one circuit to the next, and finally, there is the write-off period for fixed capital. Activities within a circuit, for example, consist of current spending, and current lending; whereas prices and money-wages (in the modern economy) will only change from one circuit to the next. Embodied technology and the capital stock change with the write-off period.

Drawing on this makes it possible to describe the exact way a sum of money effectuates the exchange of goods, and the analysis defines the time required for circulation. There are three different patterns of exchange between money and goods in the circuit – a straightforward swap, a sequential process, and a rotation. Completion of the circuit requires that these be coordinated.

Once the circuit is properly described, it is possible to explain the way the supply of money for circulation adapts to the need for it, both in the current circuit, and in the circuit for fixed capital, without reference to the specific hypotheses concerning expectations, preferences or market behavior which appear to be required by the more conventional approaches. All that is necessary is that agents should be competitive in

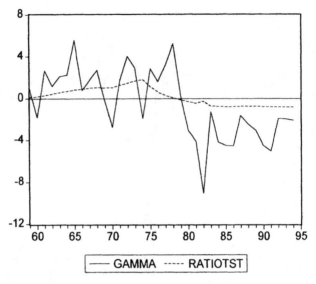

Figure 6.3 The difference between the rate of growth and the real AAA bond rate and the ratio of idle to active funds normalized around zero. Note that the normalization around zero was achieved by subtracting one from the ratio. If idle balances are exactly equal to active balances, then the nonnormalized ratio will take on a value of one. We have adjusted this so that when idle balances are exactly equal to active balances, the ratio is equal to zero. Thus, zero becomes our reference point for the two regimes for both series.
Sourse: Stephanie Thamas, 1997, dissertation proposal, New School for Social Research.

some broad sense, motivated by the desire for gain in some form, and generally aware of trends in the market. The argument depends only on the most general features of monetary institutions.

These features develop over time however, and as they change monetary needs change. The ratio of money required for the fixed capital circuit to that required for the current circuit will tend to rise with the advance of industry. In the era of Mass Production the ratio of these two kinds of money will itself become a variable dependent on interest rates in relation to growth rates.

Spelling out the working of different monetary systems reveals a new set of problems: these systems tend to generate instability. Metallic currencies run risks from hoarding; paper money systems are endangered by the tendency to overissue. Bank deposit systems tend to overlend. Maintaining monetary growth depends on an unstable stock market. In each case the instability can be seen to arise from a common source – the

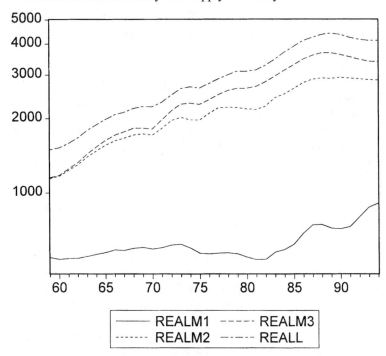

Figure 6.4 Movements of monetary aggregates, 1959 to 1994. Source: Citibank Macroeconomic Database 1995.

complex relationship between money's role as medium of circulation, and its ability to function as an asset.

As the currency system evolves from metal to paper to bank deposits, it loses its tie to a value grounded in production. But a currency no longer anchored to metal is also not constrained by reserves. It adapts to the needs of circulation, and the system's reserves also adapt. Such a system can be expected to be prone to inflation.

In each case, as money develops and new problems of instability emerge, a new set of questions for practical policy analysis can be defined. To control or eliminate such instabilities there will have to be appropriate rules and restrictions. These will have to be set and enforced by the state; monetary analysis becomes political economy.

In addition, it is possible to develop expressions for the costs of the various monetary regimes. A metallic regime depends on the costs of equipment and labor in the mines; a banking system depends on the labor and resources required to provide financial services at the appropriate level.

The study of monetary circulation shows the need to control regime instability, and to cover the costs of the monetary system. In addition, it can be misleading to examine the economy as if all behavior were settled simultaneously, as tends to be implied by supply and demand functions. In circulation, activities take place in sequence. The economy is a "monetary production system," in which monetary circulation drives innovation and development, but carries a potential for instability.

The two-sector treatment of monetary issues must now be generalized. This means developing the Classical Equations – which describe the basic structure of capitalism – in monetary form, so they can serve as the basis for the study of effective demand. This will then make it possible to examine the changing patterns of market adjustment during the process of Transformational Growth. As we follow these changes we will also keep note of the evolution of the monetary and financial system.

A note on the charts: The first two charts show the relationship between the rate of growth and the real rate of interest for the postwar United States and Great Britain, respectively. Note that the real rate of interest frequently moves opposite to the growth rate, but the overall pattern of the real rate does tend to track the growth rate over the long run. But prior to 1980, the real rate of interest always lies below the growth rate, while after 1980 it lies above the growth rate. The third chart relates the ratio of $(M3-M1)/M1$ to the difference between the rate of growth and the real rate of interest. Notice that the turning point is just before 1980. The fourth chart plots the logarithm of the chief monetary aggregates and once again it is notable that the behavior of M1 changes shortly after 1980 and the behavior of the other aggregates also changes markedly during the 1980s. (Thanks to Stephanie Thomas for preparing the charts.)

A note on bimetallism: Bimetallism tends to promote stability, for, in general, when the amount of metallic money in circulation is too great or too small, the deviation from the correct level will not be the same for both metals. Suppose there is an excess of metal, so that prices are rising (the value of money is falling). But let us assume that the excess of silver is greater than that of gold, so prices are rising faster in silver. Hence the value of silver will tend to fall relative to that of gold. This will tend to divert speculators, who instead of moving against metals in general, will move against silver; they will buy gold for silver and hoard gold. This will pull gold out of circulation, while increasing silver. Thus the divergence between gold and silver will tend to widen, but the speculation in favor of gold will tend to constrain and offset the price inflation as a shortage of gold develops. Thus a bimetallist system will normally tend to generate a stabilizing movement in one of the metals.

PART IV

The wage-profit trade-off

For more than a decade a summer school met in Trieste to try to broker a marriage between Keynes and Sraffa, that is, to base the theory of effective demand on the Classical equations of production, as developed by Sraffa. This objective also provided a basic theme for the 1985 Sraffa Conference in Florence, and the idea runs through *The New Palgrave*, in the entries of Eatwell and Milgate, among others. By now it has inspired dozens of books and articles. Yet in spite of its obvious appeal, developing a Sraffa-based Classical foundation for Keynesian theory remains a minority project, and resistance is strong, particularly among macro-economists of all schools.

Part of the problem stems from a methodological difference: the Keynesian approach stresses uncertainty and its effect on investment, money and finance, which, in the hands of post-Keynesians, has led to dynamic analysis and the study of fluctuations, whereas neo-Ricardians, following the Classicals, have stressed the tendencies of the economy to gravitate toward stable "long-period positions." If the economy is sufficiently orderly that it tends to gravitate toward, or around, such positions, how can there be Keynesian uncertainty? Alternatively, if the latter is prominent, how could the economy center on long-period positions? The questions concern market adjustment. We will argue that patterns of adjustment depend on characteristics of the technology; since these tend to change over time, the way markets adjust will also tend to change. These issues will be examined in Chapters 8 and 9, and again in Part 6.

This project cannot even get off the ground, however, until the Classical equations have been expressed in monetary terms. This we have accomplished for a two-sector approach in the preceding part, showing in detail how the circulation of money takes place. But circulation, like productivity, depends on the time required for production in relation to the time supported by the wage. Yet we explained the origin of the surplus – changes in productivity – very briefly, and we introduced our concept of the wage, as a ratio between the goods that support the basic standard of living and the total pay, in a very limited model, and without showing how it might be determined by competition. Chapter 7 will begin with a closer look at the tactic of representing labor by the wage goods that support workers for a period of time. This approach solves the problem of heterogeneous labor, allows for an explanation of the origin of the surplus, and requires defining the net wage as a pure number, analogous to the rate of profit.

It also has implications for the representation of duality, since the column of labor inputs (wage goods) in the price equation will interchange, in forming the quantity dual, with the row of inputs into the production of wage goods. Writing out both price and quantity equations, then, starting from a self-reproducing no-surplus economy, a speed-up can be shown to result in a surplus which defines a unique maximum wage, Wm, with positive relative prices. Corresponding to this – representing the distribution of this surplus as profits – will be a unique maximum rate of profit, R, also with positive prices. It will equal a unique maximum rate of growth, G, associated with positive relative quantities.

This framework can then be used to show, first, that when $G = R$, the wage bill of the capital goods sector equals the capital requirements of the consumer goods sector. Extending this result shows that the wage bill/capital requirements balance will hold for any r, $0 \leq r \leq R$, so long as $r = g$, $0 \leq g \leq G$, where $r = g$ implies that the net wage, in value terms, equals the net rate of consumption in quantity terms, the Golden Rule.

Drawing on both the Golden Rule and the wage bill/capital requirements equality makes it possible to construct a *linear* wage-profit trade-off. This can then be extended to the case of fixed capital – an extension that provides insight both into the function of capital markets, and into the need for and role of underutilized capacity.

The wage bill/capital requirements condition has been shown to be the law of monetary circulation. Drawing on this result and the linear trade-off, the Classical Equations can be converted to the usual variables and relationships of macroeconomics. This convertibility is conditional on the Golden Rule, but it is argued that plausible adjustments to profits and investment will bring them into close alignment, and further that deviations of capital and revenue from their Golden Rule values will tend to be small in a moderm industrial economy.

In Chapter 8 the convertibility of the Classical equations into macroeconomics is viewed from the opposite perspective. Since monetary circulation takes place, and macroeconomic relationships hold, the Classical Equations must hold. In the Craft economies it may also be true that the Classical Equations describe a center of gravitation, toward which the system tends to move. But gravitation requires flexible prices and regular investment, neither of which are to be found in Mass Production. The Classical Equations describe the structural core of the economy, in abstract and simplified form, but it is misleading to term them a "long-period position." Rather, the Classical Equations show the slowly evolving framework in which behavior takes place.

Chapter 9 then draws on this framework to examine the pressures leading to the drift of labor from the countryside to cities, the mechanization of work, and the development of greater flexibility in craft production. This last, in turn, leads to a pattern of market adjustment very close to that described by conventional supply and demand – Marshallian adjustment. But such adjustment, in which something akin to "marginal productivity" relationships play a crucial role, even though the system is demand-driven, creates problems that lead to innovations that bring business into the era of Mass Production.

CHAPTER 7

The Classical system: the Golden Rule, labor, and the wage-profit trade-off

In Chapters 5 and 6 we considered a simple two-sector model and developed an account of circulation based on the specializations of the sectors, and their interdependence. This must now be generalized. The model concerns manufacturing; we take it for granted that a primary sector supplies raw materials – and earns differential rents. A tertiary sector supplies services to both households and buinesses; its circulation was examined in Chapter 5. Neither will be considered in this chapter. First, we need to explain the conception of the wage and of labor here, then we must develop a general account of the relationships between the Golden Rule, the wage-bill/capital requirements condition, and the wage-profit trade-off, and finally we must extend these results to fixed capital and relate the analysis to macroeconomics.

This will not be an easy project, and, paradoxically, familiarity with the subject may make it even more difficult. The problem is not analytical – neither the model nor its mathematics is especially complex. The issues are conceptual. In particular, while the analysis put forward here *looks* very similar to many recent Classical studies, it differs from most in at least the following respects:

> Two distinct sets of industries are defined, capital goods and consumer (wage) goods[1]

[1] No good serves both purposes. There are simple paradigms: machine tools are capital goods, t-shirts consumer goods. But there are complexities; these can be handled by careful classification. Clothing worn on the job, and meals provided in company canteens can be treated as part of the wage, along with company cars. Education is "human capital formation"; see Chapter 10. Home heating oil, electricity for domestic use, and home telephone service, for example, should be considered *separate subindustries*, with petroleum, electric power, and telephone lines and switching equipment supplied as *inputs* from the corresponding parent industry in the capital goods sector. As long as *costs are incurred* in preparing a capital good for consumers, or delivering it to them, the resulting product is a distinct consumer good and that portion of the industry can be assigned to the consumer sector – or, *mutatis mutandis*, to the capital goods sector. (Many apparently problematical cases – taxicabs, fax and xerox machines, advertising, etc. – are properly part of the *service* sector, and also have significant externalities. They enter into the tertiary circulation and are not at issue here.)

The amount of labor is measured in basic wage goods
The wage is divided into basic and net portions
The net wage rate is defined as a ratio of the net to the basic, and
   is a pure number
The analysis begins from a hypothetical no-surplus system,
   which serves as a point of reference
The surplus emerges as a result of speeding up production
The production equation for wage-goods interchanges with labor
   (measured by wage-goods) in the dual

These points should be familiar from the two-sector model, and will be examined further, but it is worth stressing, nevertheless, that if they are not borne in mind, it will be easy to misinterpret the argument.

Wages and labor

According to the account of circulation presented above, when a sum of money is paid to workers for their labor-time, and is spent on consumption goods, then prices will be realized in money terms, and the pattern of respending will ensure that all goods will circulate against money. Paying workers a sum of money for their labor-time which they then spend on the consumption goods which support them (and their families) establishes a *ratio* betwen money wages and consumption goods. It is usual to divide the money wage by the prices of the consumption goods, to form the "real wage." This approach will be followed here, in a sense, but it will be interpreted somewhat differently, making it possible to define a wage *rate*, analogous to the profit rate.

Alternative concepts of the wage

The modern revival of the Classical approach is based on Sraffa. He defines the wage as the wage per unit of labor, "which like prices will be expressed in terms of the chosen standard" (p. 11). He then provisionally sets "the national income of a system in a self-replacing state" equal to unity, thereby expressing prices and wages in terms of this sum. The whole of the wage is taken as variable, ranging from 0 to 1, though he notes that this "relegat[es] the necessaries of consumption to the limbo of non-basic products" (p. 10). He suggests that "it would be [more] appropriate ... to separate the two component parts of the wage and regard only the 'surplus' part as variable; whereas the goods necessary for the subsistence of the workers would continue to appear, with the fuel, etc., among the means of production" (pp. 9–10). Thus he notes the difference between basic and net wages, but he considers the former to be on a footing with means of production. In Part III, however, it was noted that the subsistence of workers and the means of production differ in regard to time in a most important way – a difference that underlies the ability to produce a surplus. It is to capitalize on this difference that we have defined the real wage as a ratio.

By contrast, Sraffa's definition treats the wage as a fraction of national income; it is labor's share, rather than a wage rate. Later, he suggests that instead of actual national income (net product or surplus) the wage should be taken as a fraction, ranging from 0 to 1, of *Standard* national income. This has the advantage of establishing a simple linear relation between the wage so defined and the rate of profits, but it implies that the wage as a share of actual national income varies in relation to the Standard wage as the latter changes.

Such wage concepts were designed to clarify the relationships between prices and the rate of profits, but they are not useful in an analysis of the ways competition impinges on the labor market. The wage share – let alone the Standard wage share – is not a concern either of laborers or of employers; the issue for the former is the ratio of pay to the cost of living, for the latter, pay in relation to productivity and product price (with cost of living as a long run consideration.) Competition among workers will imply that they will move to the employers paying the most in relation to cost of living, driving down such pay, while competition among employers will imply that those paying the least in relation to normal productivity and price will earn the highest profit. To keep pace, others will have to cut their wage offers. However, those who pay the least will tend to get the poorest workers, those who pay the most, the best. Those swamped with workers will be able to cut wage offers, those facing shortages of labor will be obliged to raise them. Hence there will be a tendency for competition to force a convergence to a common wage rate, defined in terms of the ratio of money payments to cost of living for workers, and money outlay to the excess of productivity over the funds required needed to meet fixed costs for firms. More precisely, firms will need to meet interest costs on capital, management salaries, other contractual obligations and depreciation. These must be subtracted from output; their desired wage payments will then be:

$$\frac{\text{money}}{\text{labor} - \text{time}} \times \frac{\text{labor} - \text{time}}{\text{output} - \text{fixed charges}}$$

while labor's desired wage receipts will be

$$\frac{\text{money}}{\text{labor} - \text{time}} \times \frac{\text{labor} - \text{time}}{\text{cost of living}}.$$

(So long as the money wage is "money/labor-time" prices will be expressed in terms of money.) In terms of units, these come to the same thing. The money wage per unit time divided by the cost of living for a unit time has the dimensions money per goods. The money wage per unit time divided by productivity minus fixed charges, running costs per unit labor time, has the same dimensions. Competition affects both these ratios. But different pressures lie behind each, and for these ratios to be the same – for an acceptable level of payments by business to equal an acceptable level of receipts for labor – there will have to be a balance of forces. (As we shall see, in Craft economies, where setting the money wage fixes the real wage, a shortage of labor in relation to capital tends to lead to bidding up the wage, a surplus to driving it down.)

In Part III, these ratios were taken as equal by assumption, and the wage was presumed to be either unity (net wage of zero) or greater than unity, a positive increment over the basic cost of living.[2] In the no-surplus economy labor's net productivity was zero. The speed-up created a positive rate of productivity, while competition ensured that the wage would be a uniform fraction of the surplus, which consisted of the same goods as basic consumption. Any nonuniformity in payments to labor could be seen directly; competition could therefore be presumed to iron it out.

Heterogeneous labor
But matters are this simple only because we have assumed that all labor is the same. Imagine a Labor Day parade. First, perhaps, will come the older craft workers – carpenters, plumbers, electricians, masons, bricklayers. Then more modern skilled workers, from the auto and steel and chemical industries. Oil workers and atomic workers, machine tool workers. Computer programmers, software engineers. Many of us don't even know the precise names of their jobs and skills. But even in

---

[2] This could also be thought of as a return to the "investment" in the household. The "cost of living" can be considered analogous to circulating capital; education, training, etc., would then be long-term intangible investment, analogous to research and development expenses for firms. Building personal connections can be considered analogous to developing brand-name recognition and "goodwill" on the part of business. Households will plan their investments in education, training, etc., in order to maximize their rate of earnings – but this should not be thought of as competitive with the return on *capital*. To invest capital, to buy a bond, for example, it is necessary to put money down; household investment typically involves more *time* than money – the time can be used to acquire an education but could not be used to buy a bond.

categories we know there are many complexities. Consider drivers – they range from racing car drivers to less heroic but still skilled and licensed workers – everything from chauffeurs, to forklift operators to truckers to taxicab drivers. Then there are the women assembly line workers, clerical workers, secretaries, teachers. And men, too, in these lines. And professionals and semiprofessionals – computer salesmen, TV producers, cameramen, photographers, X-ray technicians, janitors, hospital administrators, nurses, teachers of science, of social sciences, athletic coaches and trainers, doctors, lawyers, dentists, subway conductors, medical secretaries. The list goes on and on, covering traditional jobs as well as new ones – and virutally every innovation now leads to new jobs, requiring new skills or new combinations of skills.

Once we allow for the actual heterogeneity of labor, competitive pressures in the labor market will become more complicated. Both basic consumption and the composition of the surplus goods on which net wages are spent can be different in different sectors, raising the question whether the wage rates for different kinds of labor are comparable. Lumberjacks in Maine do not need swimsuits, water-ski instructors in Florida do not need snowshoes. Different lines of work often require very different levels and kinds of training or education and also often either require or characteristically lead to very different patterns of consumption, even to different lifestyles. And, of course, the various kinds of work are paid very differently, and while wage and salary differentials are usually stable in the short and medium term, they do shift over time, and they sometimes change very suddenly and sharply.[3]

Yet, in spite of this all-too-obvious variety, virtually all schools of economic theory insist on treating the total labor of society as a homogeneous magnitude, measured by the total number of workers, multiplied by the average workweek, with the differences in skills and intensity and level of work, somehow reduced to simple common average labor. Or in the case of Marxian theory, to some notion of "abstract" labor.

This would not matter much if such a treatment of labor were a mere simplification, which could be dispensed with in more advanced work. Or if there were, in practice a workable and widely accepted method of expressing skilled labor in terms of simple labor (which kind of simple

---

[3] Just to get an idea of the extent of the differences, the U.S. Dictionary of Occupational Titles currently lists over 80,000 entries (twenty years ago it listed 60,000 titles).

labor?). But homogeneous labor is not a mere simplifying assumption, and there is no such method.[4]

Attempts to define the "Quantity of Labor"

To date there have been four chief approaches to the problem of expressing heterogeneous labor as a homogeneous magnitude. Taking them in arbitrary order: first, more complex or more advanced labor may be reckoned a multiple of simple common labor in proportion to the cost of education and training required for it. Units of various kinds of labor and other inputs combine to produce a certain kind of skill as output – as in an education or training program. A matrix of these programs can be written and provided we can measure the inputs in simple labor, we can express the outputs as multiples of such labor. Alternatively, we can treat each kind of labor as a separate output, produced by other kinds of labor and commodities; then the system can be solved for the relative prices of the different labors, along with other prices. But, of course, the rate of profit must be known in order to solve the system – and to determine the rate of profit we must know the wage rate and the labor coefficients.

---

[4] Consider the problems. Standard neo-Classical theory needs to know the capital-labor ratios for production functions. These enter into the determination of prices, so they must be known independently of them. Even at the micro-economic level setting out a production function requires aggregating all the different types of labor employed in a firm or in a process. Neo-Classical growth theory must be able to write capital-labor ratios for aggregate or for sectoral production functions. Conventional development theory must be able to compare the relative labor intensities of different projects and development plans. Keynesians and their Monetarist opponents need measures of total, marginal and average productivity and how these are changing over time both for sectors and for the economy as a whole. Post-Keynesians need a measure of total labor, in order to determine income distribution in a theory in which changes in wage rates relative to average productivity determine the price level. Neo-Ricardians and most general equilibrium theorists use the total labor of society to fix the scale of the system in studying growth, changes in proportions, and choice of technique. This is particularly important when there may be variations in returns to scale, and it is especially problematic when changes in proportions or technique are at issue, since then the proportions of different kinds of jobs and of skills required in the economy will normally change, too. Finally, of course, Marxists and all who draw on the Marxian tradition need a measure of the total labor of the economy and of its distribution among industries in production, both in order to define labor values, and in order to determine the deviations of prices of production from values, which depend on the ratios of means of production to labor, in the different industries. Further, the total surplus (expressed in labor value) divided by the total labor is the expression for the rate of surplus value, or rate of exploitation, which according to what has come to be called the Funda-mental Marxian theorem, determines the rate of profit.

Moreover, this approach assumes that degree of skill is the only problem in regard to heterogeneity. It does not deal with how to aggregate two different types of simple labor, where the workers both perform wholly different tasks using different talents, and consume different vectors of subsistence goods. So it begins by taking for granted that we already have a *general* measure of "simple" labor. The second approach, taken by Sraffa, based on Adam Smith, and endorsed by Keynes, is to aggregate on the basis of wage differentials – or differentials in consumption patterns. Keynes, for example, writes

> in so far as different grades and kinds of labor and salaried assistance enjoy a more or less fixed relative remuneration, the quantity of employment can be sufficiently defined for our purpose by taking an hour's employment of ordinary labor as our unit and weighting an hour's employment of special labour in proportion to its remuneration; i.e. an hour of special labour remunerated at double ordinary rates will count as two units. . . . This assumption of homogeneity in the supply of labour is not upset by the obvious fact of great differences in the specialised skill of individual workers and in their suitability for different occupations. For, if the remuneration of the workers is proportional to their efficiency, the differences are dealt with by our having regarded individuals as contributing to the supply of labour in proportion to their remuneration. (Keynes, 1936, pp. 41–2)

But this calls for a prior theory which explains the wage differentials. If they depend on relative efficiency then we require an account of "efficiency" which is independent of prices and productivity – since these last cannot be determined without quantities of labor. How is the ratio of one kind of output to one kind of labor to be compared to the ratio of *another* kind of output to *another* kind of labor? If the differentials depend on differences in the cost of living, it means already knowing the values of the consumption goods. Moreover, this approach assumes that two workers who are paid the same wage per hour do the same amount of labor, even though they may do completely different tasks, calling for different skills, and consume different bundles of goods. So appealing to wage differentials also begs the question.

Moreover, if "wage differentials" are assumed they must express a difference in value terms, since the bundle of goods which supports a day of one kind of work may be quite different from the bundle that supports a day of another type. Yet this creates a further problem. If real consumption is given, different consumption bundles in different industries, then as price changes with changes in distribution, the value ratio of these bundles will change. Hence the differentials would apparently change even though real consumption relativities were constant.

The third way, suggested by Marx, contends that different labors are "set equal by the market." One kind of labor can be taken as the standard; it will then be equated through exchange to various amounts of other kinds of labor. It is possible to determine a set of "reduction coefficients" which express the various kinds of labor in terms of an abstract standard (Krause, 1981), but to do this, it is necessary to assume that the market establishes a uniform rate of exploitation in terms of abstract labor, that is, actual labor multiplied by the reduction coefficients. There seems to be no reason whatever to suppose that the market would do this.

Finally, the whole idea of homogeneous labor could be rejected, and, instead, a number of different kinds of labor could be assumed. At first, this looks promising, and some writers have been led to the interesting and perhaps important problem of the potential conflicts in distribution between different groups of workers (Bowles and Gintis, 1977). But, this is no answer to the problem of aggregating or measuring labor. It will still be necessary to aggregate the working time of various workers who do not do exactly the same thing in the same way, or consume exactly the same goods, yet their working times must be counted (as the same kind of labor). (Moreover, in many respects, even very different kinds of labor stand in the same relation to capital, and it should be possible to analyze this in theory.) Consider a farmer. He does different work summer and winter, and he consumes different baskets of goods. How can his different labors in different seasons be aggregated? So even if we identified as many different kinds of labor as there are workers, we would not have enough heterogeneity.

Indeed, in its simplest form the difficulty is similar to that for capital: we cannot use prices to aggregate labor, because we need aggregated labor to determine prices.[5] But the approach developed here does not need to aggregate different kinds of labor in order to determine the productivity of labor in general – and once this is determined, the maximum rate of profit follows. Further, it will be argued that competitive pressures determine the wage rate as a *ratio* of net pay to base pay, *regardless* of the specific sector (waterskiing, lumberjacking) and the corresponding composition of base consumption. So the problems posed by heterogeneous labor can be handled by the same analysis determining the level and

---

[5] Unlike the case of capital, to some extent, at least, the aggregation of labor has been justified, but the justification has a price, one with implications for theory. "Scarcity" cannot be expressed by the ratio of capital to labor, not only because both depend on prices and distribution, but also because the latter depends on the speed of turnover, nor can the choice of technique be legitimately determined by the rate of profits, since the latter cannot be known until a technique has been put into operation (see below).

changes of productivity which already proved central to understanding monetary circulation.

Defining the wage as a ratio

Now let us develop the wage-labor concept to be used here. First consider the ratio between money wages and spending on worker consumption. If all and only money wages are spent on consumer goods, then the wage ratio is unity. If a portion of the money wage is saved, the ratio is greater than unity. If worker consumption has to be supplemented from previous saving or from outside sources, the ratio is less than unity.

A further refinement can be introduced: consumption can be divided into basic household support and luxury consumption. Basic household support is what must be convered in order for the workers to be able to perform their functions; it is the minimum which they must be paid, in the long run. The ratio of the money wage to basic consumption can now be examined; if there is luxury consumption and/or saving, the wage ratio will be greater than unity. The "net" wage will consist of luxury consumption plus saving, and the net wage rate will be the ratio of luxury consumption plus saving to basic consumption.

It may be objected that the wage should not be defined in terms of ratios of bundles of goods, since if prices or preferences change, households will change the composition of the bundle they consume, as can be shown by conventional demand theory. This is not the way we propose to analyze the problem. Conventional demand theory is not suitable for a theory of reproduction.

Instead let us consider an approach that derives a theory of expenditure – and later, *growth* of expenditure – from the social characteristics and positions of the various categories of households. Households are reproduced, with the appropriate characteristics. Households are divided into social classes, and further divided by sectoral occupation. Each class, and within each class, each occupation, is characterized by a "lifestyle" that has evolved historically, and rests on values, preferences, skills and information handed down from one generation to the next. Lifestyles are selected competitively, in the sense that the pattern of consumption they represent permits the corresponding households to work and reproduce effectively.

A lifestyle can be expressed in a vector of standards, where the elements of the vector designate the level and style that must be achieved in terms of various "characteristics" – nutrition in diet, elegance in clothing, comfort in furnishings, and so on. The lifestyle is specified by a vector of *levels of characteristics*, not by quantities of goods, nor by the "utility" or level of indifference resulting from quantities of goods (Lancaster, 1966,

1979; Ironmonger, 1972; Nell, 1992, pp. 393–7). Trade-offs are possible to some extent at the level of the characteristic – taste vs. nutrition in diet, comfort vs. elegance in clothing – but make little or no sense as between characteristics. Items of food cannot be substituted for items of clothing, or shelter for transport. The proportions between the elements in the vector of standards are fixed, because they represent the *inputs* required to produce the ability to work and to reproduce the next generation of similarly skilled workers. (The levels, of course, are not absolute; there will be a normal range of variation, just as in producing steel, there will be a range of variation in the amount of coke required. But given such ranges, the proportions are fixed.)

Households face a minimizing problem, then, which is to achieve the standard at minimum cost (thus releasing income for luxuries, for saving, or for investment in self-improvement). The actual combinations of specific foods, items of clothing, etc., will be chosen to achieve the standard at the least cost, and when prices change, these combinations of specific goods will change.[6] But the relations between the elements in the standard will remain unaltered.

Over time lifestyles will develop and change, and so the standards will be respecified. Competition enforces changes in lifestyles; those who do not keep up with the Jones's lose out on the best opportunities. The chief causes of such changes will be developments in technology that lead to

---

[6] The optimizing problem faced by working-class households is to minimize the cost of consumption while at least achieving the standard of basic consumption. The vector of standards is $\mathbf{s} = s_1, s_2, \ldots, s_m$; prices are $\mathbf{p} = p_1, \ldots, p_n$; the choice variables are $\mathbf{x} = x_1, \ldots, x_n$; and the coefficient matrix $\mathbf{A} = a_{ij}$ indicates the contribution of each of the n goods toward attaining the required levels of each of the m standards. The problem can then be expressed in a linear programming format as

Min : $C = \mathbf{px}$
S.T. : $\mathbf{Ax} \geq \mathbf{s}$
        $\mathbf{x} \geq 0$

Over a range price changes may not affect the optimum purchasing plan; then, e.g., if the slope of a price ratio comes to coincide with the slope of a constraint, a small change could lead to a radical change in the budget – an item could be dropped entirely, for example. However, families may be assumed to have small differences, so that the aggregate will be less volatile than the individual family. The introduction of new goods can easily be handled in this framework. Later chapters will develop this approach.

redefining jobs and the skills required for them, on the one hand, and changes in the geographical patterns of living, on the other. New jobs call for new skills, which means new forms of education and training. Urbanization and suburbanization lead to new patterns of living, new transportation and communication problems, and different family relationships. Moreover, the choice of goods that can be used to meet the standards will also evolve. New goods will be introduced; new elements will be added to the vector of standards. But these changes will be gradual; at any given time, the vector of standards may be taken as fixed.

Returning to the analysis of the wage: form a price index of the elements of the vector of standards, taking each element at its "normal" level of attainment. This last will be the unit basic level of real consumption, and multiplying each element by its price index (an average of the cost of the different ways of satisfying it), will give the cost of the basic consumption. Then the real wage rate will be the ratio of the money wage to this basic level of consumption. This ratio can be unity or greater than unity, where this indicates that workers are receiving a portion of the surplus. It can also be less than unity, at least for a time, provided worker consumption can be supplemented from other sources.

If the wage rate is unity, or some other ratio, prices must be such that the portion of output in each sector going to pay wages (sold for money used to pay wages) must stand in that ratio to the wage-earners' consumption bundle.

Competition and net pay

Competition in the labor market will lead workers to move around, change jobs, retrain, etc. They will be motivated in part, of course, by "job satisfaction," by the desirability of location, status, image, etc. These we must take as given. But pecuniary reward also matters, and that we can examine. We have argued that what counts here is not the absolute level of per capita pay, but the level of such pay over and above what is needed to support a worker and family in a particular line of occupation. Suppose that a worker is equally qualified for two different occupations, between which he or she is otherwise indifferent, both in regard to on-the-job activities, and in respect of job-related lifestyle (consumption patterns, social activities, location, etc.). The choice between them will depend not on the absolute level of pay, but on the extent of the surplus pay, over and above the necessities. A job with higher real pay than another, but an even higher level of necessary consumption, will be less attractive (less "profitable" to the household) than one with lower pay, but proportionally even lower necessary consumption. The latter will afford a larger "surplus" consumption.

This "surplus" household consumption, in turn, should be measured, not in absolute terms, but as a ratio to necessary consumption, that is, to the consumption required to maintain the family as a unit capable of delivering a well-fed and well-rested worker to the labor force every day, equipped with the necessary skills – and also producing that worker's eventual replacement.

But we must be careful. The claim implies that a household would prefer a lower payment in real purchasing power, compared to a higher one, if the job required a proportionally even higher level of necessary – job-related – expenditure. It further implies that if job-related necessary spending is very large, and the rate of surplus low, households will prefer jobs, not only with a lower level of pay, but also with a lower absolute magnitude of surplus pay! Thus suppose that necessities for Job A come to $100 per week, and the rate of surplus is 15 percent. For Job B the necessities come to only $70, but the rate of surplus is 20 percent. The two jobs are otherwise equally attractive. The claim is that Job B will be rationally and competitively the correct choice, even though $15 > $14.

Why should this be the case? At first glance it seems strange, for surely more money is better than less? But remember that the worker is indifferent between the $70 lifestyle and the $100 lifestyle. Household consumption patterns even today, and far more so in the past, have been strongly tied to occupation and status. The necessities of Job A come to $100; but the first thing an A household will do with extra money is buy somewhat more of these very necessities. The first luxury is to have a little more than you need of what you absolutely do need. Then consumption can begin to expand – but the next thing will be to get what is needed in a higher grade – to obtain better quality versions of the same things. Only then can we expect to find branching out into different goods, new entertainments and enjoyments. But even these are likely to remain closely tied to the social life centering on the occupation – a point both Raymond Williams and Edward Thompson have made. Hence the first expenditures from the surplus will tend to follow the same pattern as the spending on necessaries. Thus the real value of the surplus in Job B is higher – at least so long as we are comparing comparatively low rates of surplus.

Thus the wage should be thought of as pure number, 90 or 110 percent of the basic standard of living – the pure maintenance-and-replacement level – associated with a job. Wages conceived in this way can be rendered uniform by competition among workers, analogous to the working of competition among businesses, even though jobs require heterogeneous skills and patterns of basic consumption differ.

Duality, the Golden Rule, and the wage-profit trade-off

Once the equations for wages, prices and profits have been written down, usually in "normal" form, it is common to write down the "corresponding" system of consumption, quantities and growth, where these are held to reflect the same forces and institutions underlying the price system. The two systems are said to correspond to one another because they reflect the same economic pressures.

But exactly what set of quantity relationships "corresponds" to a given price system may not be easy to determine. As a consequence dual quantity systems have been written in a variety of ways. In particular, the treatment of wages and consumption in relation to labor has varied. This will prove to be significant because it bears directly on our understanding of the Golden Rule.

The mathematical dual is an *equivalent* system, based on the same technology, subject to the same pressures, but examined from the perspective of uses and quantities, rather than costs and prices. It is equivalent in the sense that it solves for the same "surplus" variable – whether this is the rate of surplus value, the rate of profit, or whatever. The study of duality is the *comparison* of such equivalent systems, to determine the structure which they have in common, that is, the properties which are invariant to a change from one to the other. Because the systems are equivalent, they can each be multiplied by the variables of the other, and the results equated, making it possible to extract the common structure.

The key to properly writing the dual is to treat labor like other goods. It is produced and used. But we must be careful to observe the way we have defined labor and the wage. Labor is *measured* by the consumer goods that support workers for a unit period of time, and the real wage rate is a pure ratio. In other words, labor is *equivalent* to the consumption of consumer goods; such consumption *supports* labor as it works. The consumption of specific consumer goods by labor working in a certain industry, *is* the labor input into the production process in that industry. Hence, *labor is "produced" by the consumer goods sector.* In writing the dual, then, we must interchange the column which serves as labor coefficients, namely, consumer goods per unit of the various outputs, with the row of coefficients showing the inputs of various goods into the production of consumer goods.

A common practice is to write the labor-time input coefficients into the various processes, forming a column. Then a further process is written (one which is not capitalistic, and so does not earn a rate of profit) consisting of a row of consumption coefficients, as if consumption were an industry that produced labor time (Kurz and Salvadori, 1995, pp. 101–2). A unit of labor is supported by the consumption of a certain set of goods. Many writers form the dual by interchanging this row and column.

This procedure is not correct. An industry – a row – shows the inputs used up *now*, to produce a good which *then* has to be allocated/distributed through exchange: "each commodity ... is found at the end ... to be entirely concentrated in the hands of its producer" (Sraffa, 1960, p. 3). But consumption supports labor *on the job*, i.e. now there is no process of production resulting in the concentration of a product in the hands of the producer, which *then* must be exchanged for the inputs to enable a further round of production. Labor, already in place and working, receives its wages, which it then exchanges for the consumption which supports its *current* ability to work. So labor and the consumption that supports it are *equivalent*. Hence labor can be measured by the consumer goods that support workers for a period of time. This means that inputs of the consumer goods sector produce the output – the consumer good – which is allocated through exchange and is the measure of labor.

Writing labor as measured by the consumption that supports it (the column of labor inputs), and treating the consumer goods sector as the corresponding industry (row), not only allows duality to be properly expressed, but also permits the study of the spending of profits on investment and wages on consumption, the rationale usually given for writing the other forms. It is the mathematically correct form of the dual, and it subsumes the others.[7]

When the growth-consumption-quantity dual to the price equations is written correctly, the Golden Rule can be shown to be equivalent to two historically important propositions – the linear wage-profit trade-off, and the Marxian relationship between the wage bill of the capital goods sector and the capital requirements of the consumer goods sector.

The system

First let us define the relevant matrices:

A will be chosen to be square, n × n, semipositive and indecomposable. The argument will be developed for the 3 × 3 case; but a more general case is implied – there can be as many capital goods industries (goods 1 and 2), and as many consumer goods industries (goods) as one likes, so long as the postulated relationships between the two sectors hold.

Writing out a 3 × 3 version of the input matrix we have

$$\mathbf{A} = \begin{matrix} a_{11} & a_{12} & a_{1s} \\ a_{21} & a_{22} & a_{2s} \\ a_{s1} & a_{s2} & a_{ss} \end{matrix} \tag{1}$$

[7] Recall the footnote on duality in Chapter 5.

from which three other matrices can be defined

$$
\mathbf{K} = \begin{array}{ccc} a_{11} & a_{12} & 0 \\ a_{21} & a_{22} & 0 \\ 0 & 0 & 0 \end{array} \qquad \mathbf{K_s} = \begin{array}{ccc} 0 & 0 & 0 \\ 0 & 0 & 0 \\ a_{s1} & a_{s2} & a_{ss} \end{array} \tag{2}
$$

and

$$
\mathbf{S} = \begin{array}{ccc} 0 & 0 & a_{1s} \\ 0 & 0 & a_{2s} \\ 0 & 0 & a_{ss} \end{array} \quad \text{with } \mathbf{ss}^* = \begin{array}{ccc} 0 & 0 & 0 \\ 0 & 0 & 0 \\ 0 & 0 & a_{ss} \end{array} \tag{3}
$$

so that

$$
\mathbf{A} = \mathbf{K} + \mathbf{K_s} + \mathbf{S} - \mathbf{ss}^* \tag{4}
$$

$\mathbf{A}$ is defined as follows: $\mathbf{K}$ is productive, and $\mathbf{K_s}$ and $\mathbf{S}$ are chosen so that the workers required and the cost of producing the goods that support them make $\mathbf{A}$ a basic no-surplus matrix, that is, semipositive and inde-composable, but only just able to reproduce. The consumer good is basic, and enters into the production of every good, including itself. Each capital good enters directly or indirectly into the production of the consumer good. As a result, a unique set of strictly positive prices will ensure that for each sector revenue just equals cost. There will be a unique set of strictly positive quantities such that the columns of $\mathbf{A}$ multiplied by such quantities all sum to the amounts produced, implying that a system in which it appears as an input matrix producing unit outputs will exactly reproduce itself with no surplus. Let $\mathbf{p}$ be a column vector of prices, and $\mathbf{q}$ a column vector of quantities. Then we have,

$$
\mathbf{Ap} = \mathbf{p}, \text{ implying } (\mathbf{A} - \mathbf{I})\mathbf{p} = \mathbf{O}, \mathbf{p} > 0 \tag{5}
$$

and

$$
\mathbf{A'q} = \mathbf{q}, \text{ implying } (\mathbf{A'} - \mathbf{I})\mathbf{q} = \mathbf{O}, \mathbf{q} > 0 \tag{6}
$$

Writing out one line of each, to check the dimensions, we have

$$
a_{11}p_1 + a_{12}p_2 + a_{1s}p_s = p_1
$$

The coefficients give the material inputs needed to produce one unit of output of good 1, plus the subsistence needed for workers for the time required to produce one unit. Each coefficient is multiplied by the price of the input, or of the subsistence good; hence the equation shows the *costs* of producing equal to the *value* of the output. The dimensions of the first term are those of p_1, that is, price numeraire per unit good 1. The second term is good 2 per unit good 1 times price numeraire per good 2 = price numeraire per good 1, and likewise for the third term. For quantities,

$$a_{11}q_1 + a_{21}q_2 + a_{s1}q_s = q_1$$

The coefficients show the amount of good 1 *used* in the production of each of the goods, multiplied by the relative size of each industry. The equation shows that the uses, added up, are equal to the amount produced, relative to the numeraire. The dimensions of the first term are those of q_1, that is, good 1 per unit quantity numeraire. The second term is good 1 per unit good 2 times good 2 per quantity numeraire = good 1 per unit quantity numeraire, and likewise for the third term. So the dimensions check.

Hence we can premultiply 1 by $\mathbf{q}'$ and 2 by $\mathbf{p}'$:

$$\mathbf{q}'\mathbf{A}\mathbf{p} = \mathbf{q}'\mathbf{p} \tag{7}$$

and

$$\mathbf{p}'\mathbf{A}'\mathbf{q} = \mathbf{p}'\mathbf{q} \tag{8}$$

Hence

$$\mathbf{q}'\mathbf{A}\mathbf{p} = \mathbf{p}'\mathbf{A}'\mathbf{q} \tag{9}$$

The wage bill in capital goods, in relation to capital requirements in consumer goods

Sectors 1 and 2 (more generally, 1, ... k) produce means of production, sector s (more generally, k + 1, ... n) produces the basic consumption good – the good which is necessary for subsistence. (When a surplus is produced, consumption out of the surplus can consist of many different combinations of goods, provided the quantities are adjusted appropriately.) In the no-surplus system, the aggregate wage bill of the sectors producing means of production is equal in value to the capital goods requirements of the consumption goods sector. This can be seen by considering equations 7 and 8 written out in full:

$$q_1[a_{11}p_1 + a_{12}p_2 + a_{1s}p_s] = q_1p_1$$
$$q_2[a_{21}p_1 + a_{22}p_2 + a_{2s}p_s] = q_2p_2 \tag{7a}$$
$$q_s[a_{s1}p_1 + a_{s2}p_2 + a_{ss}p_s] = q_sp_s$$

$$p_1[a_{11}q_1 + a_{21}q_2 + a_{s1}q_s] = p_1q_1$$
$$p_2[a_{12}q_1 + a_{22}q_2 + a_{s2}q_s] = p_2q_2 \tag{8a}$$
$$p_s[a_{1s}q_1 + a_{2s}q_2 + a_{ss}q_s] = p_sq_s$$

(When the same equation is written in a different format, we will attach a letter to its number.)

First take the two equations for the consumption good, s. Both cost of production and the value consumed equal $p_s q_s$. Moreover both equations contain a common term, $p_s a_{ss} q_s$, which can be canceled. Hence,

$$q_s a_{s1} p_1 + q_s a_{s2} p_2 = p_s a_{1s} q_1 + p_s a_{2s} q_2 \tag{10}$$

The left-hand side is the capital requirements of the consumer goods sector; the right-hand side is the aggregate wage bill of the capital goods producers.

This conclusion can be reached by another route. Take the cost-revenue equations for goods 1 and 2 and add them; do the same with the aggregate value of the uses of the goods. Both are equal to $p_1 q_1 + p_2 q_2$. We have,

$$q_1 [a_{11} p_1 + a_{12} p_2 + a_{1s} p_s] + q_2 [a_{21} p_1 + a_{22} p_2 + a_{2s} p_s]$$
$$= p_1 [a_{11} q_1 + a_{21} q_2 + a_{s1} q_s] + p_2 [a_{12} q_1 + a_{22} q_2 + a_{s2} q_s] \tag{11}$$

It is easily seen that the a_{11} and a_{22} terms cancel out. Closer inspection shows that the a_{12} and a_{21} terms also cancel. This leaves

$$q_1 a_{1s} p_s + q_2 a_{2s} p_s = p_1 a_{s1} q_s + p_2 a_{s2} q_s$$

exactly the expression, equation 10, found earlier.

In terms of the matrices defined earlier, we can write this

$$\mathbf{q'Sp = p'K'_s q} \tag{12}$$

Technical progress

A surplus can be created by a speed-up of turnover, so that more work is done in the time during which labor is supported by its consumption. Such a speed-up affects unit costs in relation to revenues, on the one hand, or it may have an equivalent effect on unit replacement requirements in relation to quantities, on the other. We must compare its effects on prices and quantities.

There are two different kinds of coefficients in our model, which may correspond to Marx's distinction between "constant" and "variable" capital. "Constant capital" coefficients are those which show the inputs required per unit of output, regardless of the time required for production. So much wood per table is needed, whether it takes a day or a week to make the table. "Variable capital" coefficients are those which vary with respect to output, according to the speed of production. If workers work faster, then fewer units of subsistence goods will be required per unit of output, since workers will need to be supported for less time. (Alternatively, we could think of it this way – if a surplus exists it could be eliminated, by slowing down the speed of production sufficiently.)

This, then, is where the surplus comes from. Workers must be induced, cajoled or coerced to work faster, to convert inputs into outputs in less time.[8] This implies an institutional conflict of interest between workers and the capitalist firms that employ them. Thus the speed of production is not determined by "choices"; it is settled through day-to-day conflict, over the rules and practices and expectations governing everyday life on the job.[9] A resolution of this conflict is embodied in work norms, but disagreements may continue over how to interpret the norms.

So a surplus emerges because employers make workers do more work in a given time than is needed to reproduce the goods they consume during that time. A set of consumer goods will support workers for a given time; if they work faster or harder, then, in that period of time, they can convert more inputs into output than is needed to replace the consumer goods.

This must now be shown in the equations of our simple model. We introduce a variable, t, to indicate the speed of production, the number of times input is converted into output during the period for which wages

[8] Essentially, this is Taylorism, increasing output per worker *within* the context of a given production process. Time and motion study speeds up work in a given context; scientific management will enable tasks to be carried out in the most efficient order. Tasks will be broken down and labor divided in the most effective way (Taylor, 1911).

[9] Methodologically, *any* situation can be interpreted as a choice in neo-Classical terms, given a little imagination and Humpty Dumpty's approach to words. For *any action whatever* will be carried to a certain point and stopped (necessarily, since we don't do one and only one thing all day, forever), so can be interpreted as the result of the interaction of forces favorable to continuing the action but declining in intensity with those unfavorable to it but constant or gaining in strength. This is all the easier, since most neo-Classicals accept a methodological position which does not require explicit empirical identification of such forces. It is enough to show that outcomes are "as if" such forces were operating. The question is not, therefore, can such an interpretation be given, but does it make sense?

are paid.[10] Since each sector uses the other's output as means of production, the two sectors must coordinate the timing of production. If one sector finished before the other, it would not be able to sell its product, since the other would not yet be in a position to buy. If the faster producing sector nevertheless kept its workers on, at make-work jobs, it would lose the advantages of speeding up. But if it laid its workers off while waiting for the other to finish, it would confirm workers in the belief that working harder or faster just leads to unemployment. This is likely to create resistance. Thus the faster sector faces incentives to slow down.

On the other hand, the faster sector provides incentives to the slower to speed up, since it is clear that gains are possible and a surplus can be produced. Furthermore, the fastest firms in the slower sector will obtain a competitive advantage – they will be able to trade first, and begin the next round ahead of the rest. Hence the faster sector will tend to pull the slower along, while the slower will act as a drag on the faster. Thus the two will tend to settle on a common speed of production.

Two related qualifications deserve mention. First a slower producer could be a proportionally larger size, meeting demand from inventory. But increasing size requires investment. For example, following an unequal speed-up, the slower producer's price might be driven up by the shortage, creating profits, which invested, would increase the relative size enough to allow prices to fall back to normal. This will be examined later in the chapter. It is a complexity that is not relevant here. Second, more generally, carrying inventories will allow industries to operate at different speeds, but at a cost which must be compared to the costs of finishing up and waiting. *Given* an inventory policy, however, the slowest producer sets the pace; no one can produce at a faster clip than that, apart from the inventory leeway.

---

[10] But what determines t? This is not the subject at issue here; market forces, bargaining power, and sociopolitical factors are all surely a part of the story. A neo-Classical might argue that t will reflect the increasing disutility of additional effort, balanced against the diminishing returns from such extra effort. Then if pay is proportional to the marginal intensity of work, i.e., compensating marginal disutility, neo-Classical forces would determine t/w, the ratio of intensity to pay. Yet, by the same token we should have to accept a time preference determination of the rate of profit. The decision how much work to offer will be made by workers in each sector, on utility grounds; and the decisions as to how much capital to offer will be made by capitalists on time-preference grounds. That means adding two equations, one for t/w, and one for r, to a system which has only one degree of freedom. The theory ends in inconsistency. (Unless the two kinds of decisions are both made by the same agents, and so are coordinated. But in that case we would have only one social class, capitalists and workers would be the same people. So neo-Classicism faces a dilemma: either it is inconsistent, or it cannot describe class society.)

(The relationships here seem to depend on workers doing work, in the physicists' sense. If machinery and energy do the work, and workers merely supervise repair and assist equipment – if the tools use the worker instead of the worker using the tools – then this approach may not apply, and it might not be appropriate to describe the system as one in which exploited labor is the source of the surplus and profit.)

Let **T** be a diagonal n x n matrix, with zeros off-diagonal, and t's on the diagonal for all those elements corresponding to rows or columns of **K**, after which the diagonal entries, corresponding to either $\mathbf{K_s}$ or **S**, will be unit entries. Note that t represents *turnover* per unit period, so that $1 \leq t \leq \alpha$. If t = 1 the system reduces to the no-surplus case. When t > 1 there will be a surplus of the subsistence good or goods. **T** is an elementary operator and does not affect rank. So a surplus arising through a speed-up of production is shown by post- or pre-multiplying **A** and **A'** by **T** and **I** by the scalar t.

Starting from a no-surplus system,[11] a uniform decline in production times can be represented by the equation

$$(\mathbf{AT} - t\mathbf{I})\mathbf{p} + (t - 1)\,\mathbf{Sp} = \mathbf{O} \tag{13}$$

where **A** is the n × n matrix of inputs, rows showing the inputs into an industry, columns showing the use of an output, **T** has zeros off-diagonal, and t's on, except in the positions corresponding to subsistence goods, where it has unit entries, **I** is the unit output matrix, **p** is the price (column) vector, and **S** is an n × n matrix consisting of zeros and the vector of subsistence goods. Each output will be increased by t, while inputs will be multiplied by **T**. The difference will be a surplus, (t − 1)**S**, consisting entirely of the subsistence good, since output is increased by t, while input remains the same, whereas for every capital good output and input, initially equal, both increase by t.

Evidently this will hold even if there are *many* subsistence goods, and *different consumption patterns* in different industries. The makeup of the elements of **S** – the support for labor in different industries – need not consist of the same goods in the same proportions. However many subsistence goods and whatever the composition of subsistence in different

---

[11] The system $(\mathbf{A} - \mathbf{I})\mathbf{p} = \mathbf{0}$, will have a unique, all-positive solution, **p**, since all goods are basics, which satisfies the Hawkins-Simon conditon. This can easily be adapted to joint production; a no-surplus system will have a unique positive solution. A uniform speed-up will then yield the same results; nonuniform speedups can also be studied. It is possible to permit negative joint outputs – what Hobson called "illth," pollutants, etc. – and positive inputs, e.g., improved health from exercise on the job. Care must be taken to insure that prices remain positive, however.

industries, starting from a "no-surplus" system, a uniform speed-up in all industries will lead to a surplus vector that consists of the same set of goods, in the same proportions, as the aggregate vector of subsistence. Labor can be as heterogeneous as capital; what is important is that production times be coordinated. If the speed-up is such that all finish together, then the aggregate surplus will consist of the same "composite" commodity as the basic support for labor. This means that each industry would be able to appropriate a surplus, at the common rate, consisting of the same goods in the same proportions, as those which support its labor.

If this surplus is then paid to labor, or distributed to guild masters in proportion to the labor they employ, $t-1$ will equal the maximum wage, w_m, or the rate of net productivity of labor:[12]

$$ATp + (t - 1) Sp = tp \qquad (14)$$

and

$$TA'q + (t - 1)TK_s'q = tq \qquad (15)$$

In equation 14 the speed-up is reflected in the different movement of costs in relation to revenue. Material costs and revenue move together, maintaining a constant relationship; labor costs, however, *fall* relative to revenue – there are differential rates of *cost* turnover, so that the revenue/labor ratio rises.[13] (This can be considered an interpretation of Marx's "constant and variable capital.") The economy experiences technical progress; hence it is intuitively reasonable that both the price and quantity systems should show the *same* form of technical progress. (Given a way of making the systems interact, such as a Theory of Monetary Circulation would provide, this would necessarily be the result.) Accordingly we

---

[12] The private and the social rate of surplus can be distinguished. The private rate of surplus arises from a speed-up in a no-surplus economy in which only private inputs and outputs are shown. But if externalities can be measured, they can be included as joint products or joint inputs. The social rate of surplus will come from solving the system when these are included.

[13] We can also write

$$A'Tq + (t - 1)K_s'q = tq$$

In this equation the speed-up increases output and replacement needs in proportion, but leaves consumption unchanged – more input is converted into output, for the same level of necessary consumption – there are differential rates of *use* turnover so that the means of production/consumption ratio rises. We will not consider the other two possibilities. They are not as appealing intuitively, and the argument can easily be adapted to them. Writing the matrices TA and $A'T$, both price and quantity systems experience technical progress in the form of differential use turnover. In the case of TA and TA', the two systems have different forms of technical progress, but opposite to the case examined.

focus on the case in which equivalent price and quantity systems both undergo technical progress in the form of differential rates of cost turnover.[14] Equations 14 and 15 are both equivalent systems in the sense defined earlier: in equation 14 we solve for the prices consistent with a ratio of surplus to labor of $t - 1$, in equation 15 we solve for the quantities consistent with that same surplus ratio.

To see these equations, including the one in the footnote, we write them out in full:

$$\mathbf{ATp} = [ta_{11}p_1 + ta_{12}p_2 + a_{1s}p_s \qquad (t-1)\mathbf{Sp} = (t-1)[0 + 0 + a_{1s}p_s$$
$$ta_{21}p_1 + ta_{22}p_2 + a_{2s}p_s \qquad\qquad 0 + 0 + a_{2s}p_s$$
$$ta_{s1}p_1 + ta_{s2}p_2 + a_{ss}p_s]; \qquad\qquad 0 + 0 + a_{ss}p_s]$$

$$(14a)$$

$$\mathbf{TA'q} = [ta_{11}q_1 + ta_{21}q_2 + ta_{s1}q_s \qquad (t-1)\mathbf{TK'_s q} = (t-1)[0 + 0 + ta_{s1}q_s$$
$$ta_{12}q_1 + ta_{22}q_2 + ta_{s2}q_s \qquad\qquad 0 + 0 + ta_{s2}q_s$$
$$a_{1s}q_1 + a_{2s}q_2 + a_{ss}q_s]; \qquad\qquad 0 + 0 + a_{ss}q_s]$$

$$(15a)$$

When written out, equation 15a simplifies to

$$a_{11}q_1 + a_{21}q_2 + ta_{s1}q_s = q_1$$
$$a_{12}q_1 + a_{22}q_2 + ta_{s2}q_s = q_2$$
$$\frac{1}{t}\,[a_{1s}q_1 + a_{2s}q_2] + a_{ss}q_s = q_s$$

We also have

$$\mathbf{A'Tq} = [ta_{11}q_1 + ta_{21}q_2 + a_{s1}q_s \qquad (t-1)\mathbf{K'_s q} = (t-1)[0 + 0 + a_{s1}q_s$$
$$ta_{12}q_1 + ta_{22}q_2 + a_{s2}q_s \qquad\qquad 0 + 0 + a_{s2}q_s$$
$$ta_{1s}q_1 + ta_{2s}q_2 + a_{ss}q_s]; \qquad\qquad 0 + 0 + a_{ss}q_s]$$

$$(15a^*)$$

As can be seen, $-1\mathbf{Sp}$ and $-1\mathbf{K'_s}$ have the effect of removing the respective last columns from the expressions on the left, and $+t\mathbf{Sp}$ and $+t\mathbf{K'_s}$ replace them, respectively, with the vectors of wages and of the inputs into

---

[14] We will not consider the other two possibilities. They are not as appealing intuitively, and the argument can easily be adapted to them. Writing the matrices **TA** and **A′ T**, both price and quantity systems experience technical progress in the form of differential use turnover. In the case of **TA** and **TA′**, the two systems have different forms of technical progress, but opposite to the case examined.

the production of the consumption good, each multiplied by t. Hence we can write

$$\mathbf{ATp} + (t-1)\mathbf{Sp} = t\mathbf{Ap} = t\mathbf{p} \Leftrightarrow \mathbf{Ap} = \mathbf{p}, \quad \text{and}$$
$$\mathbf{A'Tq} + (t-1)\mathbf{K'_sq} = t\mathbf{A'q} = t\mathbf{q} \Leftrightarrow \mathbf{A'q} = \mathbf{q}$$

The systems following the speed-up have the same solutions as the no-surplus system; that is, the relative prices and relative quantities are the same. Hence it will continue to be true that the aggregate wage bill in capital goods equals the capital requirements in the consumer good sector.

Premultiplying equation 14 by $\mathbf{q}'$ and equation 15 and 15a^* by $\mathbf{p}'$, we obtain:

$$\mathbf{q'ATp} + (t-1)\mathbf{q'Sp} = t\mathbf{q'p}, \quad \text{and} \tag{16}$$

$$\mathbf{p'TA'q} + (t-1)\mathbf{p'TK'_sq} = t\mathbf{p'q} \tag{17}$$

$$\mathbf{p'A'Tq} + (t-1)\mathbf{p'K'_sq} = t\mathbf{p'q} \tag{17*}$$

Writing the matrices out we have,

$$\mathbf{q'ATp} = q_1[ta_{11}p_1 + ta_{12}p_2 + a_{1s}p_s] + q_2[ta_{21}p_1 + ta_{22}p_2$$
$$+ a_{2s}p_s] + q_s[ta_{s1}p_1 + ta_{s2}p_2 + a_{ss}p_s];$$
$$(t-1)\mathbf{q'Sp} = (t-1)[0 + 0 + q_1a_{1s}p_s + 0 + 0 + q_2a_{2s}p_s$$
$$+ 0 + 0 + q_sa_{ss}p_s] \tag{16a}$$

$$\mathbf{p'TA'q} = p_1[ta_{11}q_1 + ta_{21}q_2 + ta_{s1}q_s] + p_2[ta_{12}q_1 + ta_{22}q_2$$
$$+ ta_{s2}q_s] + p_s[a_{1s}q_1 + a_{2s}q_2 + a_{ss}q_s];$$
$$(t-1)\mathbf{p'TK'_sq} = (t-1)[0 + 0 + tp_1a_{s1}q_s + 0 + 0 + tp_2a_{s2}q_s$$
$$+ 0 + 0 + p_sa_{ss}q_s] \tag{17a}$$

$$\mathbf{p'A'Tq} = p_1[ta_{11}q_1 + ta_{21}q_2 + a_{s1}q_s] + p_2[ta_{12}q_1$$
$$+ ta_{22}q_2 \, a_{s2}q_s] + p_s[ta_{1s}q_1 + ta_{2s}q_2 + a_{ss}q_s];$$
$$(t-1)\mathbf{p'K'_sq} = (t-1)[0 + 0 + p_1a_{s1}q_s + 0 + 0 + p_2a_{s2}q_s$$
$$+ 0 + 0 + p_sa_{ss}q_s] \tag{17a*}$$

Now if we set equations 16 and 17 equal and cancel we will get

$$t[p_1a_{s1}q_s + p_2a_{s2}q_s] = q_1a_{1s}p_s + q_2a_{2s}p_s$$

(as can be seen from multiplying equation 15a by prices and comparing to equation 7a, which is equivalent to equation 14). Here the rate of turnover times the capital requirements in the production of consumer goods equals the aggregate wage bill in the capital goods sector.[15] Notice that this result is the same as the no-surplus case, for in that case $t = 1$.

The wage rate and the profit rate

Notice that $w_m = t - 1$; for each level of t there will be a unique maximum level of the wage. Below we shall prove that corresponding to each level of t there exists a unique maximum rate of profit, with positive prices, which is equal to the maximum rate of growth, associated with positive quantities. The equations for profits and prices, growth and quantities, are

$$(1 + R)[ta_{11}p_1 + ta_{12}p_2 + a_{1s}p_s] = tp_1$$
$$(1 + R)[ta_{21}p_1 + ta_{22}p_2 + a_{2s}p_s] = tp_2 \qquad (18)$$
$$(1 + R)[ta_{s1}p_1 + ta_{s2}p_2 + a_{ss}p_s] = tp_s$$

$$(1 + G)[ta_{11}q_1 + ta_{21}q_2 + ta_{s1}q_s] = tq_1$$
$$(1 + G)[ta_{12}q_1 + ta_{22}q_2 + ta_{s2}q_s] = tq_2 \qquad (19)$$
$$(1 + G)[a_{1s}q_1 + a_{2s}q_2 + a_{ss}q_s] = tq_s$$

Taking the first equation in the R-set, and dividing numerator and denominator by $1/t$;

$$R = \frac{p_1}{[a_{11}p_1 + a_{12}p_2 + 1/t(a_{1s}p_s)]} - 1. \qquad (20)$$

Clearly R will increase or decrease as t increases or decreases. Moreover, a change of t will affect R more in those industries that have a large ratio of wage-goods to means of production. To restore the equality of the rates of profit, prices will have to change. A similar result holds for G.

Next we consider the distribution of the surplus between wages and profits. The wage rate will be a fraction applied to the value of labor,[16] and

---

[15] A similar relation holds for equations 16 and 17*. When $\mathbf{q'ATp}$ and $(t - 1)\mathbf{q'Sp}$ are added, the column of wage payments is removed and then replaced by the same column with each element multiplied by t. Exactly the same happens in equation 9a – the effect of the addition is to replace the column by one with the same elements, each multiplied by t. Consequently in equations 16 and 17* t appears as a scalar on both sides of each equation and therefore cancels out. The equations 16 and 17* reduce to equations 7 and 8, respectively. Hence the condition holds that the aggregate wage bill in the capital goods sector will equal the capital requirements in consumer goods production.

[16] Remember that labor – measured in worker-hours – is represented here by the consumption goods required to support the necessary workers for that length of time, where "support" implies the normal, socially expected standard of living for workers. Hence the "value of labor" is the value of the means of subsistence supporting that amount of labor.

the profit rate a fraction applied to the value of capital. The surplus then will be distributed by providing labor with a percentage increase above subsistence, for example 110 percent of subsistence. Profits will similarly be distributed as a percentage rate of the value of capital. When the rate of profit is zero, the wage will reach its maximum – the whole surplus will go to labor. Hence the wage will equal $t-1$, the productivity of labor. But we are interested in the *net wage*, that is, the wage above subsistence, and in the net rate of profits. So it will be useful to rescale the wage, taking the subsistence level as a zero net wage, and the maximum wage as the full net wage, 100 percent of the net wage, or 1. When the wage, so defined, is zero, the rate of profit will reach its maximum, R, which, for a given t, follows from the characteristic root of the matrix. So we have

$$0 \leq r \leq R, \qquad 0 \leq w \leq 1$$

In the same way we divide the uses of the surplus into augmenting the means of production, that is, investing for growth, or increasing consumption. We can therefore define a rate of growth, comparable to the rate of profit, and a rate of consumption, analogous to the wage rate. The ranges will be

$$0 \leq g \leq G, \qquad 0 \leq c \leq 1$$

where G is derived from the opposite characteristic root of the matrix. Without drawing on the notion of a characteristic root, we shall see that $G = R$.

Hence we write,

$$\mathbf{tp} = (1 + r)[\mathbf{ATp}] + w(t - 1)\mathbf{Sp}, \quad \text{and} \tag{18a}$$

$$\mathbf{tq} = (1 + g)[\mathbf{TA'q}] + c(t - 1)\mathbf{TK'_s q} \tag{19a}$$

The first says that revenue, relative price per unit, equals total relative unit costs plus profit figured on all costs advanced, both material and labor, plus the net wage paid to labor. The second[17] tells us that relative sectoral

---

[17] The alternative dual equation would be

$$\mathbf{tq} = (1 + g)[\mathbf{A'Tq}] + c(t - 1)\mathbf{K'_s q} \tag{23*}$$

which when multiplied by prices and written out is

$$(1 + G)p_1[ta_{11}q_1 + ta_{21}q_2 + a_{s1}q_s] = tp_1 q_1$$
$$(1 + G)p_2[ta_{12}q_1 + ta_{22}q_2 + a_{s2}q_s] = tp_2 q_2 \tag{23*}$$
$$(1 + G)p_s[ta_{1s}q_1 + ta_{2s}q_2 + a_{ss}q_s] = tp_s q_s$$

It is interesting to note that when this equation – showing use-reducing technical progress – is combined with the primal equation reflecting cost-reducing technical progress, the Golden Rule relationships still hold! This will be shown in footnotes below.

sizes are the sum of each good's various uses as means of production, expanded by investment, plus its uses in the consumption sector, expanded in proportion to consumption demand.[18]

When $w = c = 1$, $r = g = 0$, and equations 18 and 19 reduce to equations 14 and 15. The maximum wage means that labor is paid the full surplus, so that the ratio of payments to labor to its basic cost of living is equal to labor's productivity, $t-1$.

The maximum rate of profit and the corresponding rate of growth

Next consider $w = c = 0$; we will show that when this is the case $R = G$, and that these values are unique.

As a preliminary we must show that when a speed-up has taken place, resulting in a maximum wage, with prices equal to those of the no-surplus system (unique and strictly positive), corresponding to this w_m, there exists a unique maximum rate of profit, R, which also has strictly positive prices. That is, instead of the surpus being distributed in proportion to labor, it will now be distributed in proportion to capital.

Assume that, initially, the prices are those corresponding to w_m, indicated by a subscript w, but that the surplus in each industry is appropriated as profit, so that each industry may have a different rate of

---

[18] Duality is defined as discussed, following Spaventa (1970) and Nell (1970). The price system has three kinds of costs, input costs, profits, and wages, in different categories of industries; so the dual must have three kinds of uses, replacements, investment, and consumption, for the outputs corresponding to those industries. By contrast, Pasinetti, for example, develops a "dual growth-consumption system" (1976, pp. 190–9) which is not set up this way. He defines a row vector of direct labor inputs, a_n (p. 72) and a column vector of per capita consumption, c, (p. 194) and then compares the "dual" equations (pp.199–200),

$$p = a_n [I - (1 + r)A]^{-1}, \quad \text{and} \quad q = [I - (1 + g)A]^{-1}c$$

where p and q are vectors of relative prices and relative quantities, r and g are rates of profit and growth, respectively, I is outputs, and A inputs. These are said "to express the price system and the physical quantity system as perfectly dual to each other ... independently of time." But these are not properly dual to each other. There *has* to be a relationship between the amount of labor and the consumption that supports it. So there is *no difference* between a_n and c; the latter simply shows the consumer goods that support labor, element by element. A similar objection can be raised against the discussion in Kurz and Salvadori (1995), pp. 101–2 and 113–14. What is needed for duality is the vector of inputs into the production of the consumer good that supports labor.

profit, and consider the equations:

$$(1 + R_1)[ta_{11}p_{w1} + ta_{12}p_{w2} + a_{1s}p_{ws}] = tp_{w1}$$
$$(1 + R_2)[ta_{21}p_{w1} + ta_{22}p_{w2} + a_{2s}p_{ws}] = tp_{w2} \qquad (21)$$
$$(1 + R_s)[ta_{s1}p_{w1} + ta_{s2}p_{w2} + a_{ss}p_{ws}] = tp_{ws}$$

Without loss of generality we can assume $R_1 > R_2 > R_s$. (The example will stick with three industries, but any number can be imagined.) To find the uniform rate of profit corresponding to this speed-up, we shall reduce the price of the product with the largest R, holding input prices constant, until it falls to the level of the next largest, and similarly raise the price of that good with the smallest R, holding input prices constant, until it rises to the level of the next highest R. Then we allow input prices to vary, recalculate, and repeat the procedure. So p_{w1} falls to p'_{w1} and p_{ws} rises to p'_{ws}. We have

$$1 + R'_1 = \frac{tp'_{w1}}{ta_{11}p_{w1} + ta_{12}p_{w2} + a_{1s}p_{ws}}$$

$$1 + R'_2 = \frac{tp'_{w2}}{ta_{21}p_{w1} + ta_{22}p_{w2} + a_{2s}p_{ws}} \qquad (22)$$

$$1 + R'_s = \frac{tp'_{ws}}{ta_{s1}p_{w1} + ta_{s2}p_{w2} + a_{ss}p_{ws}}$$

The principle can be established by considering the price reduction. By reducing p_{w1} to p'_{w1}, R_1 falls to $R'_1 = R_2$. Then input prices are allowed to adjust, and we recalculate. Let us suppose that good 2 is the industry which uses the *largest* amount of good 1 per unit output, and consider an extreme case: good 1 uses good 2 and labor as inputs but good 2 uses *only* good 1. Then

$$1 + R'_1 = \frac{tp'_{w1}}{ta_{12}p_{w2} + a_{ss}p_{ws}} = 1 + R_2 = \frac{p_{w2}}{a_{21}p_{w1}}$$

In this case, when input prices are adjusted, to give R'_2, we multiply the above by the ratio p_{w1}/p'_{w1}, and obtain

$$1 + R'_2 = \frac{p_{w2}}{a_{21}p'_{w1}} = 1 + R_1 = \frac{tp_{w1}}{ta_{12}p_{w2} + a_{ss}p_{ws}}$$

The reduced price of the first round, when entered as an input price, has produced a largest rate of profit in the second round that is no lower than that of the first. This is because the second good used *only* the first as input. Once the second good uses other inputs, for example labor, R_2' will always lie below R_1. Let

$$1 + R_2 = \frac{tp_{w2}}{ta_{21}p_{w1} + a_{2s}p_s}, \quad \text{and} \quad 1 + R_2' = \frac{tp_{w2}}{ta_{21}p_{w1}' + a_{2s}p_s}$$

So we have

$$1 + R_1' = \frac{tp_{w1}'}{ta_{12}p_{w2} + a_{ss}p_{ws}} = 1 + R_2 = \frac{tp_{w2}}{ta_{21}p_{w1} + a_{2s}p_s}$$

Now when we multiply both sides of this equation by the ratio of initial to adjusted prices, p_{w1}/p_{w1}', we obtain R_1 as before, but instead of R_2' we have

$$1 + R_2^* = \frac{tp_{w2}}{ta_{21}p_{w1}' + (a_{2s}p_s)(p_{w1}'/p_{w1})} \tag{23}$$

Since $p_{w1}'/p_{w1} < 1$, $R_2^* = R_1 > R_2'$. This will hold in every round of price adjustments, so long as the good which uses the largest amount of the good with the largest profit rate (the latter being the good whose price is reduced) also uses other inputs. Since we have assumed that every industry uses labor (wage goods), the largest rate of profit in each round will be less than the largest rate of the previous round.

A similar argument will show that raising the lowest price will lead to a higher minimum rate of profit each round. Since the largest rate of the rates of profit of the various industries falls from period to period, and the lowest rises, the difference between them can be made as small as we like. In short, they converge to a uniform rate.

The prices resulting from this process will still be strictly positive. When the wage is at its maximum, the price vector will be the same as in the no-surplus case. When the net wage falls to zero and the rate of profit rises to its maximum, for any price to turn zero or negative, some element in its equation must become negative. But all coefficients and variables are positive; the only variable that could become negative is a price. Since no price can turn negative first, all must remain positive (Sraffa, 1960, pp. 27–8, sec. 39). (In the same way, adjusting quantities to growth, no quantity could become negative, unless some other had first.)

When the net wage is zero a maximum rate of profit must exist, with positive prices. Now we must find its relation to the maximum rate of growth. Next write out the equations in full, with the price equations multiplied by quantities, and the quantity equations multiplied by prices.

$$(1 + R)q_1[ta_{11}p_1 + ta_{12}p_2 + a_{1s}p_s] = tq_1p_1$$
$$(1 + R)q_2[ta_{21}p_1 + ta_{22}p_2 + a_{2s}p_s] = tq_2p_2 \qquad (18b)$$
$$(1 + R)q_s[ta_{s1}p_1 + ta_{s2}p_2 + a_{ss}p_s] = tq_sp_s$$
$$(1 + G)p_1[ta_{11}q_1 + ta_{21}q_2 + ta_{s1}q_s] = tp_1q_1$$
$$(1 + G)p_2[ta_{12}q_1 + ta_{22}q_2 + ta_{s2}q_s] = tp_2q_2 \qquad (19b)$$
$$(1 + G)p_s[a_{1s}q_1 + a_{2s}q_2 + a_{ss}q_s] = tp_sq_s$$

To simplify exposition, let us define a number of shorthand expressions. Capital letters will be used for the various terms in what follows; these definitions are temporary:

$$
\begin{aligned}
A &= q_1a_{11}p_1 & M &= q_sa_{s1}p_1 \\
B &= p_2a_{22}q_2 & N &= p_sa_{1s}q_1 \\
C &= q_1a_{12}p_2 & X &= q_sa_{s2}p_2 \\
D &= q_2a_{21}p_1 & Y &= p_sa_{2s}q_2 \\
& & Z &= q_sa_{ss}p_s
\end{aligned}
$$

All these terms are non-negative.

In addition, let $Q = A+B+C+D$, and let $H = M+X$, and $J = N+Y$.

Using this shorthand notation we can rewrite the equations 18a and 19a:

$$(1 + R)[t(A + C) + N] = tp_1q_1$$
$$(1 + R)[t(D + B) + Y] = tp_2q_2$$
$$(1 + R)[t(M + X) + Z] = tp_sq_s$$
$$(1 + G)[t(A + D + M)] = tp_1q1$$
$$(1 + G)[t(C + B + X)] = tp_2q_2$$
$$(1 + G)[N + Y + Z] = tp_sq_s$$

Clearly the sum of the R-equations is equal to the sum of the G-equations. Hence, aggregating and substituting

$$(1 + R)[t(Q + H) + J + Z] = (1 + G)[t(Q + H) + J + Z], \text{ and}$$
$$R = G \qquad (24)$$

The maximum rate of profit exists, prices are positive, and the maximum rate of profit equals the maximum growth rate. To show that R is unique, we must show that no other rate could exist, satisfying these equations with prices and quantities positive. Suppose we interpret R as one such value, and G as another. In equation 18a, R and the prices can be taken as one solution. But instead of the G and quantities corresponding to that solution, in equation 19a, let G and the quantities represent another, *different* solution, in which G is not equal to R. We multiply each set of equations by the solutions of the other. We have shown that the prices are positive. Then for equation 24 to hold when R is not equal to G, both sides would have to be zero, which would imply that some of the quantities would have to be negative (Sraffa, 1963, pp. 28–9, sec. 41). A second solution with positive quantities therefore cannot exist. (Assume the quantities to be positive, and the argument can be reworked to show that some of the prices in a second solution would have to be negative.)

Now take the top two equations from each set; this becomes

$$(1 + R)[tQ + N + Y] = (1 + G)[tQ + t(M + X)], \text{ or}$$

$$(1 + R)[tQ + J] = (1 + G)[tQ + tH]$$

which gives us

$$\frac{1 + G}{1 + R} = \frac{t(A + C + D + B) + N + Y}{t(A + D + C + B) + tM + tX} = \frac{tQ + J}{tQ + tH} \tag{25a}$$

and from the production and use of the consumer good we have

$$\frac{1 + G}{1 + R} = \frac{t(M + X) + Z}{N + Y + Z} = \frac{tH + Z}{J + Z} \tag{26a}$$

Setting these equal and canceling gives

$$t^2QH + t^2H^2 + tZH = tQJ + J^2 + ZJ, \quad \text{which simplifies to}$$

$$tH(tQ + tH + Z) = J(tQ + J + Z) \tag{27}$$

By inspection, if tH > or < J this cannot hold. Hence tH = J.[19]

---

[19] A similar procedure can be followed for the alternative dual. The equations in 18a and 19a* for the production and use of the consumption good both equal tp_sq_s. Setting them equal, we obtain

$$(1 + R)[t(q_sa_{s1}p_1 + q_sa_{s2}p_2) + q_sa_{ss}p_s] = (1 + G)[t(p_sa_{1s}q_1 + p_sa_{2s}q_2) + p_sa_{ss}q_s]$$

The equations in 18a and 19a* for the production and use of means of production, when added, are both equal to $t(p_1q_1 + p_2q_2)$. Setting them equal we have,

$$(1 + R)[t(q_1a_{11}p_1 + q_1a_{12}p_2 + q_2a_{21}p_1 + q_2a_{22}p_2) + q_1a_{1s}p_s + q_2a_{2s}p_s]$$
$$= (1 + G)[t(p_1a_{11}q_1 + p_1a_{21}q_2 + p_2a_{12}q_1 + p_2a_{22}q_2) + p_1a_{s1}q_s + p_2a_{s2}q_s]$$

From above, using these shorthand expressions,

$$\frac{1 + G}{1 + R} = \frac{t(M + X) + Z}{t(N + Y) + Z}$$

(*continued to next page*)

When r = g, w = c

The preceding proof can be adapted to the case where $0 < r < R$, but $r = g$. When this is the case, we shall now show that $w = c$. This is the Golden Rule. The equations are[20]

$$tq'p = (1 + r)[q'ATp] + w(t - 1)q'Sp, \quad \text{and} \tag{28}$$

$$tp'q = (1 + g)[p'TA'q] + c(t - 1)p'TK_s'q \tag{29}$$

$$(1 + r)q_1[ta_{11}p_1 + ta_{12}p_2 + a_{1s}p_s] + w(t - 1)q_1a_{1s}p_s = tq_1p_1$$

$$(1 + r)q_2[ta_{21}p_1 + ta_{22}p_2 + a_{2s}p_s] + w(t - 1)q_2a_{2s}p_s = tq_2p_2$$

$$(1 + r)q_s[ta_{s1}p_1 + ta_{s2}p_2 + a_{ss}p_s] + w(t - 1)q_sa_{ss}p_s = tq_sp_s$$

$$\tag{28a}$$

$$(1 + g)p_1[ta_{11}q_1 + ta_{21}q_2 + ta_{s1}q_s] + c(t - 1)tp_1a_{s1}q_s = tp_1q_1$$

$$(1 + g)p_2[ta_{12}q_1 + ta_{22}q_2 + ta_{s2}q_s] + c(t - 1)tp_2a_{s2}q_s = tp_2q_2$$

$$(1 + g)p_s[a_{1s}q_1 + a_{2s}q_2 + a_{ss}q_s] + c(t - 1)p_sa_{ss}q_s = tp_sq_s$$

$$\tag{29a}$$

[19] *(Contd.)*

And also, using the same shorthand notation,

$$\frac{1 + G}{1 + R} = \frac{t(A + C + D + B) + N + Y}{t(A + D + C + B) + M + X}$$

Let $Q = A+B+C+D$. Then setting these equal, substituting and cross-multiplying,

$$[t(M + X) + Z](tQ + M + X) = [t(N + Y) + Z](tQ + N + Y)$$

Let $H = M+X$, $J = N+Y$; then multiplying out and canceling tZQ,

$$t^2QH + tH^2 + ZH = t^2QJ + tJ^2 + ZJ, \quad \text{or} \quad H(t^2Q + H + Z) = J(t^2Q + J + Z)$$

Suppose, $H > J$, or $H < J$; by inspection, term by term, the equation could not hold. Hence, $H = J$, and therefore, from the initial equations above, $R = G$. Remember that $M + X = N + Y$ states that the wage bill of the capital goods sector equals the capital requirements of the consumer sector, as in equation 10.

[20] The alternative dual is

$$tp'q = (1 + g)[p'A'Tq] + c(t - 1)p'K_s'q, \text{ which written out is}$$
$$(1 + g)p_1[ta_{11}q_1 + ta_{21}q_2 + a_{s1}q_s] + c(t - 1)p_1a_{s1}q_s = tp_1q_1$$
$$\tag{29*}$$

$$(1 + g)p_2[ta_{12}q_1 + ta_{22}q_2 + a_{s2}q_s] + c(t - 1)p_2a_{s2}q_s = tp_2q_2$$
$$(1 + g)p_s[ta_{1s}q_1 + ta_{2s}q_2 + a_{ss}q_s] + c(t - 1)p_sa_{ss}q_s = tp_sq_s$$
$$\tag{29a*}$$

We proceed as in the previous case.[21] Rewriting the equations for the production and use of the consumer good

$$
\begin{aligned}
(1+r)[t(M+X)+Z] &+ w(t-1)Z \\
&= (1+g)[(N+Y)+Z] + c(t-1)Z
\end{aligned}
\tag{30}
$$

$$
(1+r)(tH-J) = (c-w)(t-1)Z
\tag{30a}
$$

From the equations in 28a and 29a for the production and use of goods 1 and 2, we have

$$
\begin{aligned}
(1+r)[tQ+N+Y] &+ w(t-1)(N+Y) \\
&= (1+g)[tQ+t(M+X)] + c(t-1)t(M+X)
\end{aligned}
\tag{31}
$$

Using H and J, $r = g$, and rearranging, this becomes

$$
(1+r)[tH-J] = (c-w)(t-1)(J-tH)
\tag{31a}
$$

Hence from equations 30a and 31a we have

$$
(c-w)Z = (c-w)(J-tH)
\tag{32}
$$

Clearly $Z > 0$. But $J-tH$ is zero. This equation can only be true if both sides are zero, for any level of the rate of profit. Hence $r = g$ implies $w = c$.[22]

---

[21] At this point it might be useful to restate the point about duality made earlier. From equations 28a and 29a, assuming all profits invested and all wages consumed, using the short-hand notation above, we have

$$
I = (1+r)[tQ+J] + (1+r)[tH+Z] = (1+g)[tQ+\{J\}] + (1+g)[\{tH\} + Z]
$$
$$
C = w(t-1)J + w(t-1)Z = c(t-1)\{J\} + c(t-1)Z
$$

Applying $tH = J$, and substituting the bracketed terms on the right-hand side, this becomes

$$
I = (1+g)[tQ+tH] + (1+g)[J+Z]
$$
$$
C = c(t-1)tH + c(t-1)Z
$$

The first column on the right-hand side shows the use of the capital goods; it corresponds to the first two rows of equation 28a. The second column shows the use of the consumer good and is the same as the third row of equation 29a.

[22] This logic can also be followed in the case of the alternative dual. From the equations 28a and 29a* for the production and use of the consumer good, we have

$$
(1+r)[t(M+X)+Z] + w(t-1)Z = (1+g)[t(N+Y)+Z] + c(t-1)Z
\tag{30*}
$$

(*continued to next page*)

Now look again at equation 30. When $r = g$ and $w = c$, the Z terms cancel,

$$(1 + r)t(M + X) + \cancel{(1 + r)Z + w(t - 1)Z}$$
$$= (1 + g)(N + Y) + \cancel{(1 + g)Z + c(t - 1)Z} \tag{30}$$

r and g divide out, and we are left with

$$t(M + X) = N + Y, \quad \text{that is,} \quad tH = J$$

which, written out in full, is

$$t(q_s a_{s1} p_1 + q_s a_{s2} p_2) = p_s a_{1s} q_1 + p_s a_{2s} q_2 {}^{23} \tag{33}$$

or in matrix terms

$$\mathbf{q'Sp = p'TK'_s q} \tag{33a}$$

---

[22] (contd.)

Assuming $r = g$, and remembering H and J, we have

$$(1 + r)(H - J) = \frac{1}{t}(c - w)(t - 1)Z \tag{30a*}$$

From the equations in 28a and 29a* for the production and use of goods 1 and 2, we have

$$(1 + r)[tQ + N + Y] + w(t - 1)(N + Y)$$
$$= (1 + g)[tQ + M + X] + c(t - 1)(M + X) \tag{31*}$$

Using H and J, $r = g$, and rearranging, this becomes,

$$(1 + r)[H - J] = (c - w)(t - 1)(J - H) \tag{31a*}$$

Hence from equations 30a* and 31a* we have

$$\frac{(c - w)Z}{t} = (c - w)(J - H) \tag{32*}$$

It is clear that $Z/t > 0$, but we know that $J - H = 0$. Hence this expression can only be true in general if $c = w$, so that both sides are zero. Hence $r = g$ implies $c = w$.

[23] In the case of the alternative dual following the same procedure yields $H = J$, or writing it out in full,

$p_s a_{1s} q_1 + p_s a_{2s} q_2 = q_s a_{s1} p_1 + q_s a_{s2} p_2$, which in matrix terms is

$$\mathbf{q'Sp = p'K'_s q}$$

This shows that equations 7a and 8a hold in general, that is, for all levels of the rate of profit, so long as $r = g$. It is clear that the above equation together with equations 28 and 29 also implies that

$$\mathbf{q'ATp = p'A'Tq}$$

It is clear that equation 33a together with equations 28 and 29 also implies that

$$q'ATp = p'TA'q \qquad (33b)$$

Finally, it is clear from equations 24a, 25a, and 30, that if r is *not* equal to g, then it cannot be shown that tH = J, that is, equations 33 and 33a cannot be derived. The Golden Rule and the condition that the wage bill in capital goods equals the capital requirements in consumer goods mutually imply one another.[24]

Constructing the trade-off

With given methods of production, **A**, and a given speed of production, t, consider a change in r, such that r = g holds, implying a corresponding change in w, where w = c. That is, r and g vary together,

---

[24] This is demonstrated for a Hicks-Spaventa two-sector model in Nell (1995). An earlier version has circulated since 1988. Some historical notes may be appropriate. The property of the Golden Rule, that the slope of the wage-profit trade-off is equal to the capital-labor ratio, was first demonstrated in Nell (1970) (von Weizacker, 1971, proved a similar theorem, but employed a different approach and did not appreciate the proposition's significance outside of the narrow question of measuring capital.) The analytical point seems to have been first shown geometrically by Nell in a seminar following a presentation by Spaventa in 1968. (Draw the dual w, r and g, c, curves for the two-sector model. When r is not equal to g, the value of capital per worker can be shown to be equal to (c−w)/(r−g), which can be represented on each curve by the slope of a chord connecting the corresponding r, w and g, c, points on the curves. Then as r approaches g, the tangents to the wage-profit and consumption-growth trade-offs are reached the same point, and are the limits approached by the chords. Cf. the diagram and discussion in Nell, 1992, p. 118.) The argument was then analyzed algebraically by Bailey and Schweinberger and finally written up by Nell. The relationship between the wage bill of the capital goods sector and the capital requirements of consumer goods production was, of course, developed by Marx in *Capital*, vol. II, in connection with the circulation of money. Rosa Luxemburg carried the argument forward, but after her little advance was made until Lowe (1954, 1965, 1976) extended the analysis to three sectors in the context of growth – but without considering money. Nell (1976) and Hagemann (1992) have studied Lowe's model mathematically. Erlich (1960) suggested that the wage bill/capital requirements condition provided a basis for studying "traverses" between one growth path and another, an argument developed in detail by Lowe (1976) and studied further by Hagemann. Coontz (1950) suggested that the relationship between the wage bill of capital goods and capital plus profits in consumer goods could be the basis for effective demand. Nell (1975, 1992) developed an account of the multiplier based on that relationship and then extended the argument to monetary circulation (1978, 1986, 1995), returning the discussion to Marx's original concern, and in the last of these studies, making the connection with the Golden Rule. Juillard (1992) has carried out an empirical study of the relationship for the U.S. economy.

and so do w and c. This should not be construed as *movement*. Rather it is a comparison of differences – two different but equivalent systems are being compared for different rates of profit.

The initial situation will be indicated by a time subscript 0, the new by a time subscript 1. These subscripts apply to quantities, prices and the rates r, g, w, and c. Both patterns of technical progress will be considered, though the argument will be developed for the important case, with the second simply indicated in a footnote. The equations are

$$t\mathbf{q}_0'\mathbf{p}_0 = (1 + r_0)[\mathbf{q}_0'\mathbf{ATp}_0] + w_0(t - 1)\mathbf{q}_0'\mathbf{Sp}_0 \tag{34}$$

$$t\mathbf{p}_1'\mathbf{q}_1 = (1 + g_1)[\mathbf{p}_1'\mathbf{TA}'\mathbf{q}_1] + c_1(t - 1)\mathbf{p}_1'\mathbf{TK}_s'\mathbf{q}_1 \tag{35}$$

In both periods both equations hold, but in drawing the comparison from the initial period to the later, we contrast the price-profit equation, multiplied by quantities, with the growth-consumption equation, multiplied by prices.

Since the inverse exists (by the assumptions on **A**), these can be rewritten

$$t\mathbf{q}_0'\mathbf{p}_0[\mathbf{q}_0'\mathbf{ATp}_0]^{-1} = (1 + r_0) + w_0(t - 1)\mathbf{q}_0'\mathbf{Sp}_0[\mathbf{q}_0'\mathbf{ATp}_0]^{-1}$$

$$\tag{34a}$$

$$t\mathbf{p}_1'\mathbf{q}_1[\mathbf{p}_1'\mathbf{TA}'\mathbf{q}_1]^{-1} = (1 + g_1) + c_1(t - 1)\mathbf{p}_1'\mathbf{TK}_s'\mathbf{q}_1[\mathbf{p}_1'\mathbf{TA}'\mathbf{q}_1]^{-1}$$

$$\tag{35a}$$

At this point we are entitled to choose normalization rules for prices and quantities, respectively. Here we draw on equations 18b and 19b, which imply that

$$(1 + R) = \mathbf{q}'t\mathbf{p}[\mathbf{q}'\mathbf{ATp}]^{-1} = (1 + G) = \mathbf{p}'t\mathbf{q}[\mathbf{p}'\mathbf{TA}'\mathbf{p}]^{-1}$$

We normalize prices and quantities respectively by setting the following expressions equal to unity:

$$\mathbf{q}'\mathbf{ATp} = \mathbf{p}'\mathbf{TA}'\mathbf{q} = 1$$

Hence

For the price equation we have: $\mathbf{q}'(t\mathbf{p} - \mathbf{ATp})[\mathbf{q}'\mathbf{ATp}]^{-1} = R$ (36)

For the quantity equation we have: $\mathbf{p}'(t\mathbf{q} - \mathbf{TA}'\mathbf{q})[\mathbf{p}'\mathbf{TA}'\mathbf{q}]^{-1} = G$

$$\tag{37}$$

We have already shown that R = G.

We can therefore rewrite equations 34a and 35a as

$$1 + R = (1 + r_0) + w_0(t - 1)\mathbf{q}_0'\mathbf{Sp}_0[\mathbf{q}_0'\mathbf{ATp}_0]^{-1} \quad \text{and} \tag{38}$$

$$1 + G = (1 + g_1) + c_1(t - 1)\mathbf{p}_1'\mathbf{TK}_s'\mathbf{q}_1[\mathbf{p}_1'\mathbf{TA}'\mathbf{q}_1]^{-1} \tag{39}$$

Subtracting equation 39 from equation 38, we have

$$0 = (r_0 - g_1) + w_0(t - 1)\mathbf{q}_0'\mathbf{Sp}_0[\mathbf{q}_0'\mathbf{ATp}_0]^{-1}$$
$$-c_1(t - 1)\mathbf{p}_1'\mathbf{TK}_s'\mathbf{q}_1[\mathbf{p}_1'\mathbf{TA}'\mathbf{q}_1]^{-1} \tag{40}$$

When $r = 0$ and $w = 1$, we know from equation 34 that

$$t\mathbf{q}_0'\mathbf{p}_0 - [\mathbf{q}_0'\mathbf{ATp}_0] = (t - 1)\mathbf{q}_0'\mathbf{Sp}_0,$$

and multiplying by the inverse,

$$\{t\mathbf{q}_0'\mathbf{p}_0 - [\mathbf{q}_0'\mathbf{ATp}_0]\}[\mathbf{q}_0'\mathbf{ATp}_0]^{-1} = (t - 1)\mathbf{q}_0'\mathbf{Sp}_0[\mathbf{q}_0'\mathbf{ATp}_0]^{-1}$$
$$\tag{41}$$

But the left-hand side of equation 41 is equal to R by equation 36. So equation 40 becomes

$$0 = \Delta r + w_0 R - c_1(t - 1)\mathbf{p}_1'\mathbf{TK}_s'\mathbf{q}_1[\mathbf{p}_1'\mathbf{TA}'\mathbf{q}_1]^{-1} \tag{42}$$

Drawing on equations 33a, 33b, and 36, we can rewrite the last term as

$$(t - 1)\mathbf{p}_1'\mathbf{TK}_s'\mathbf{q}_1[\mathbf{p}_1'\mathbf{TA}'\mathbf{q}_1]^{-1} = (t - 1)\mathbf{q}_1'\mathbf{Sp}_1[\mathbf{q}_1'\mathbf{ATp}_1]^{-1} = R$$
$$\tag{43}$$

Hence,

$$0 = \Delta r + (w_0 - c_1)R = \Delta r + \Delta wR, \quad \text{or}$$
$$\frac{\Delta r}{\Delta w} = -R \tag{44}$$

Integrating equation 44, we have

$$r = \int_0^R \Delta r = -R \int_0^1 \Delta w = C - Rw \tag{45}$$

where C is the constant of integration. But C must equal R, since $r = R$

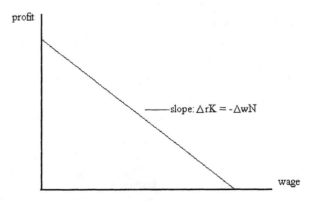

profit

slope: $\Delta rK = -\Delta wN$

wage

when w = 0. Hence,

$$r = R(1 - w) \qquad (46)$$

where it is understood that r = g, R = G, and w = c. R will rise or fall with increases or decreases in labor productivity, that is, with changes in t, but the equation will otherwise be the same.[25]

Equation 46 is a construction based on comparisons of differences between equivalent (dual) systems, for different profit rates.[26] We have not been considering movement from one such system to another.

---

[25] A linear wage-profit trade-off has usually been associated with Sraffa's Standard Commodity. This, in turn, has been judged to be an interesting, and perhaps revealing, but essentially auxiliary construction. It renders the relationship between wages and profits "more transparent," but in itself, it is wholly arbitrary. There is no reason why anything should be valued in standard terms (Dobb, 1973; Eatwell, 1975; Roncaglia, 1978; Schefold, 1986; Kurz and Salvadori, 1993). The Standard Commodity has often been associated with Ricardo's search for an "invariable measure of value" – but while it has some of the properties he sought, it is widely agreed that it is not the measure Ricardo wanted, and further that the search for such a measure is a chase after a will o' the wisp. (Harcourt, 1983; Kurz and Salvadori, 1993). But the linear relationship derived here is based on the Golden Rule, in a system in which the real wage is a pure ratio. Unlike Sraffa's, it is not an auxiliary construction – the Golden Rule is based on market pressures and reflects, in idealized form, the actual relationships of the system.

[26] Here are the equations for constructing the linear wage-profit trade-off using the alternative dual equation:

$$t\mathbf{p}_1'\mathbf{q}_1 = (1 + g_1)[\mathbf{p}_1'\mathbf{A}'\mathbf{T}\mathbf{q}_1] + c_1(t - 1)\mathbf{p}_1'\mathbf{K}_s'\mathbf{q}_1 \qquad (35^*)$$

$$t\mathbf{p}_1'\mathbf{q}_1[\mathbf{p}_1'\mathbf{A}'\mathbf{T}\mathbf{q}_1]^{-1} = (1 + g_1) + c_1(t - 1)\mathbf{p}_1'\mathbf{K}_s'\mathbf{q}_1[\mathbf{p}_1'\mathbf{A}'\mathbf{T}\mathbf{q}_1]^{-1} \qquad (35a^*)$$

$$\text{(normalization of quantities)} \quad \mathbf{p}'(t\mathbf{q} - \mathbf{A}'\mathbf{T}\mathbf{q})[\mathbf{p}'\mathbf{A}'\mathbf{T}\mathbf{q}]^{-1} = G \qquad (37^*)$$

$$1 + G = (1 + g_1) + c_1(t - 1)\mathbf{p}_1'\mathbf{K}_s'\mathbf{q}_1[\mathbf{p}_1'\mathbf{A}'\mathbf{T}\mathbf{q}_1]^{-1} \qquad (39^*)$$

$$0 = (r_0 - g_1) + w_0(t - 1)\mathbf{q}_0'\mathbf{S}\mathbf{p}_0[\mathbf{q}_0'\mathbf{A}\mathbf{T}\mathbf{p}_0]^{-1}$$
$$\qquad - c_1(t - 1)\mathbf{p}_1'\mathbf{K}_s'\mathbf{q}_1[\mathbf{p}_1'\mathbf{A}'\mathbf{T}\mathbf{q}_1]^{-1} \qquad (40^*)$$

$$\{t\mathbf{q}_0'\mathbf{p}_0 - [\mathbf{q}_0'\mathbf{A}\mathbf{T}\mathbf{p}_0]\}[\mathbf{q}_0'\mathbf{A}\mathbf{T}\mathbf{p}_0]^{-1} = (t - 1)\mathbf{q}_0'\mathbf{S}\mathbf{p}_0[\mathbf{q}_0'\mathbf{A}\mathbf{T}\mathbf{p}_0]^{-1} \qquad (41^*)$$

$$0 = \Delta r + w_0 R - c_1(t - 1)\mathbf{p}_1'\mathbf{K}_s'\mathbf{q}_1[\mathbf{p}_1'\mathbf{A}'\mathbf{T}\mathbf{q}_1]^{-1} \qquad (42^*)$$

$$(t - 1)\mathbf{p}_1'\mathbf{K}_s'\mathbf{q}_1[\mathbf{p}_1'\mathbf{A}'\mathbf{T}\mathbf{q}_1]^{-1} = (t - 1)\mathbf{q}_1'\mathbf{S}\mathbf{p}_1[\mathbf{q}_1'\mathbf{A}\mathbf{T}\mathbf{p}_1]^{-1} = R \qquad (43^*)$$

Hence,

$$0 = \Delta r + (w_0 - c_1)R = \Delta r + \Delta wR, \text{ or}$$

$$\frac{\Delta r}{\Delta w} = -R \qquad (44)$$

Integrating this gives us the linear relationship.

Some implications for capital theory might be noted. The neo-Classical production function is constructed on stationary assumptions; the corresponding wage-profit trade-offs will therefore normally be nonlinear, creating the well-known possibilities of reswitching and capital-reversing. This critique is, of course, valid, and the construction of a linear wage-profit trade-off, on the basis of the Golden Rule, in no way affects it. However, the conventional view that there might be positive profits in a stationary equilibrium is not tenable. Capitalism grows; capital, in value terms at any point in time, is a *price times quantity* magnitude, and both the price and the quantity components must be determined at their equilibrium values. The result, above, is a linear wage-profit trade-off, and it is therefore conceivable, if wage pressures were appropriate, that a *dynamic process* might trace out, over time, a pattern in which the value of capital per worker rose as the rate of profit tended to fall (Nell, 1977; Nell, 1992; Laibman, 1992, chs. 5–8; Dickinson, 1956–7; Foley and Marquetti, 1997).

Extension to fixed capital

So far the analysis has assumed circulating capital. It must now be extended to fixed capital. Since it should be clear that the argument is easily adapted to all the forms of technical progress, we will only consider the normal case, in which the form of technical progress is equivalent in the price and quantity systems.

The problem posed for the present argument by fixed capital is that it is durable, in the sense that the period over which it is consumed is typically several times longer than the period of time required for its production. Durable equipment will be used over several periods, but must be paid for on acquisition. It therefore tends to be an exceptionally large purchase for users, while posing the problem for producers that replacement sales will be delayed; they may have to produce over several periods without a sale.[27]

This means that users will typically set aside funds each period, in order to accumulate the sums needed for replacement purchases, while

---

[27] Sraffa chiefly addressed a different issue. When the rate of profit is given, the price of an industry's output and the prices of its inputs will be constant; the input coefficients for circulating inputs are given. The industry uses fixed capital, and the *amount* is given. But its *efficiency* may vary from period to period, e.g., it is likely to decline with age. How then can the price and the rate of profit be the same, in different periods? Sraffa's answer was to treat the durable good itself as a joint output of the industry each period, along with the normal product. The partly depreciated durable good can then be assigned a price – a book value – which will reflect its changing efficiency in such a way as to maintain the uniformity of the rate of profit and the constancy of output price. Such prices can be determined by writing out the price equations for each period in the life of the durable good.

producers will be obliged to borrow or negotiate advance payments in order to be able to pay for circulating inputs during the periods between sales. Users are therefore in a position to finance producers.

Notice, however, that *growth* is taking place continuously. Fixed capital will be operated with a growing complement of circulating capital, and when it is replaced, the replacement amount must be larger by the compound growth rate. This means that the capacities of the producers of fixed capital must be growing at that same rate. We shall see that this problem is solved by the appropriate investment of the depreciation funds – which brings us to the rate of interest.

We begin from the simplest case. Consider a no-surplus matrix of the kind studied earlier. But now we add a column showing fixed capital usage and a row giving the coefficients of fixed capital production.

$$\mathbf{A_f} = \begin{matrix} d\,f_1 & a_{11} & a_{12} & a_{1s} \\ d\,f_2 & a_{21} & a_{22} & a_{2s} \\ d\,f_s & a_{s1} & a_{s2} & a_{ss} \\ d\,f_f & a_{f1} & a_{f2} & a_{fs} \end{matrix}$$

Here d represents the rate at which the durable good is depreciated in each round of production, assumed for the moment to be the same in all sectors. The coefficients f_1, f_2, f_s, and f_f represent the amount of the durable good which is acquired at the outset of production, including the use of the durable good in its own production. The coefficients a_{f1}, a_{f2}, and a_{fs} represent the inputs into the production of the durable good.

Here fixed capital is represented by the *flow* of depreciation per period. Hence we can write the equations for prices and quantities, for the no-surplus economy, multiplying the price equations by the quantities, and the quantity equations by the prices:

$$q_1[df_1p_f + a_{11}p_1 + a_{12}p_2 + a_{1s}p_s] = q_1p_1$$
$$q_2[df_2p_f + a_{21}p_1 + a_{22}p_2 + a_{2s}p_s] = q_2p_2$$
$$q_s[df_sp_f + a_{s1}p_1 + a_{s2}p_2 + a_{ss}p_s] = q_sp_s$$
$$q_f[df_fp_f + a_{f1}p_1 + a_{f2}p_2 + a_{fs}p_s] = q_fp_f$$

(47)

$$p_1[a_{11}q_1 + a_{21}q_2 + a_{s1}q_s + a_{f1}q_f] = p_1q_1$$
$$p_2[a_{12}q_1 + a_{22}q_2 + a_{s2}q_s + a_{f2}q_f] = p_2q_2$$
$$p_s[a_{1s}q_1 + a_{2s}q_2 + a_{ss}q_s + a_{fs}q_f] = p_sq_s$$
$$p_f[d(f_1q_1 + f_2q_2 + f_sq_s + f_fq_f)] = p_fq_f$$

(48)

For the sectors producing goods 1, 2, and s, exchange works as before. But for fixed capital the value of the depreciation incurred by its use must be transferred to the producers of fixed capital, to enable them to purchase circulating inputs. Such transfers can be treated either as down payments on replacement goods, or as paying off earlier purchases.

The rate of depreciation, d, is taken as uniform. That means that the output of fixed capital will be "sold" – delivered for replacement – every $1/d$ periods. But d need not be uniform. The calculation of prices or quantities would proceed in the same manner, and the properties of the system would be unchanged if d were replaced by d_1, d_2, d_s and d_f. But the sale of fixed equipment would then take place at different times for different sectors.[28]

Next consider the effect of fixed capital on the relationship between the wage bill of the capital goods sector and the capital requirements in the consumer goods sector. The use and production of the consumer good are both equal to $p_s q_s$. So

$$q_s[d\,f_s p_f + a_{s1}p_1 + a_{s2}p_2 + a_{ss}p_s] = p_s[a_{1s}q_1 + a_{2s}q_2 + a_{ss}q_s + a_{fs}q_f]$$

and canceling the common term, we have

$$q_s a_{s1}p_1 + q_s a_{s2}p_2 + q_s d\,f_s p_f = p_s a_{1s}q_1 + p_s a_{2s}q_2 + p_s a_{fs}q_f \qquad (49)$$

The first two terms on each side are the same as before. The third term on the left-hand side is the capital depreciation in the production of consumer goods; on the right-hand side it is the basic wage bill in the production of fixed capital equipment. The expanded expression shows the current capital requirements, including fixed capital, in consumer goods production to be equal to the wage bill of the capital goods sector, including the production of fixed capital.

Now consider the equations for the production and use of durable equipment, both equal to $q_f\,p_f$. Canceling the common term, $q_f\,d\,f_f\,p_f$, we have

$$q_f[a_{f1}p_1 + a_{f2}p_2 + a_{fs}p_s] = p_f[d(f_1 q_1 + f_2 q_2 + f_s q_s)] \qquad (50)$$

which says that the value of the circulating goods required in the production of fixed equipment is just equal to the depreciation of fixed equipment used in the production of circulating goods. In matrix terms we can write

$$q'C_f p = p'F_c q \qquad (50a)$$

---

[28] To draw on Sraffa's approach here would require writing out the *dual* to his fixed capital price equations. Such a dual would show how quantities would have to vary during the period over which the durable instruments are used, to compensate for the varying efficiency, while still maintaining a uniform balanced growth rate and constant outputs of the principal product. This would be the equivalent problem.

where C_f and F_c, the circulating goods used in production of fixed equipment, and fixed equipment used in circulating goods production, are defined in analogy to S and K_s.

Subtracting equation 50 from equation 49, we obtain

$$q_f[a_{f1}p_1 + a_{f2}p_2] - p_s[a_{1s}q_1 + a_{2s}q_2]$$
$$= p_f[d(f_1q_1 + f_2q_2)] - q_s[a_{s1}p_1 + a_{s2}p_2], \quad \text{or} \tag{51}$$

$$q_f[a_{f1}p_1 + a_{f2}p_2] + q_s[a_{s1}p_1 + a_{s2}p_2]$$
$$= p_f[d(f_1q_1 + f_2q_2)] + p_s[a_{1s}q_1 + a_{2s}q_2] \tag{51a}$$

This states that the sum of the circulating inputs into the production of fixed equipment and the consumption good equals the depreciation of fixed equipment plus the wage bill in the capital goods sector.

The effects of a uniform speed-up of production are easily seen. Both output and use – depreciation – are increased in the same proportion. In this respect it is no different than circulating capital, and the results will be the same. However, since it is being used up faster, fixed capital sales will take place sooner.

The real complexities of fixed capital come with the rate of profit. For then the transfer of depreciation funds from the use of durable goods to the producers of such goods must be treated as a loan, on which interest must be paid. Such interest will compound, so that the sum available at the time when the durable good is used up will be greater than the replacement value. However, in a growing economy, this is exactly appropriate. For the circulating part of the capital will have been growing each period; which means, as well, that the intensity with which fixed capital equipment is used must rise at the rate of growth. Output, that is, must grow steadily, which means that the ability to use fixed capital must increase each period. From the time of its installation until its replacement, users of fixed capital must be able to expand their output, while working with their given stock of fixed capital. By the end of its life the fixed capital initially installed must be capable of producing the level of output that will be achieved as a result of the economy's growth over its lifetime. This suggests that fixed capital must be *overbuilt*, which means that economies of scale will gradually be realized as it is used more fully over its lifetime. It also builds in excess capacity and is an important structural reason for the flexibility of Mass Production.

However, there is another aspect to this. The *capacity* of given plant and equipment will rise over time, first as fixed equipment is broken in, then as the labor force learns to use it, and finally as usage suggests various improvements. Hence the capacity level will increase over time, and this will be figured into the planning of investment. Thus as growth takes place and circulating capital increases, the intensity of use, but not the amount, of the fixed equipment will rise in proportion. The amount of fixed plant and equipment, in physical terms, will be the same but its *productivity* will increase at the rate of growth.

A change in distribution will normally change the intensities with which processes are operated – quantities – which means that fixed capital in different sectors will be used up more or less rapidly. The *rate* of depreciation in sector i, d_i, will be unaffected, but the *amount*, $q_i d_i f_i$, will change as q varies.

To maintain the proper relationship, then, the replacements must be increased at the same rate. But in that event, the producers of durable equipment must *also* expand; hence each period the circulating capital of the producers of fixed equipment will have to be increased.

The solution draws on the condition that the depreciation from the use of fixed capital in the production of circulating goods equals the current circulating requirements in the production of fixed equipment. We assume the Golden Rule and write out the equations. Note, however,

$$(1+r)q_1[ta_{11}p_1 + ta_{12}p_2 + dtf_1p_f + a_{1s}p_s] + w(t-1)q_1a_{1s}p_s = tq_1p_1$$
$$(1+r)q_2[ta_{21}p_1 + ta_{22}p_2 + dtf_2p_f + a_{2s}p_s] + w(t-1)q_2a_{2s}p_s = tq_2p_2$$
$$(1+r)q_f[ta_{f1}p_1 + ta_{f2}p_2 + dtf_fp_f + a_{fs}p_s] + w(t-1)q_fa_{fs}p_s = tq_fp_f$$
$$(1+r)q_s[ta_{s1}p_1 + ta_{s2}p_2 + dtf_sp_f + a_{ss}p_s] + w(t-1)q_sa_{ss}p_s = tq_sp_s$$
$$(52a)$$

$$(1+g)p_1[ta_{11}q_1 + ta_{21}q_2 + ta_{f1}q_f + ta_{s1}q_s] + c(t-1)p_1ta_{s1}q_s = tp_1q_1$$
$$(1+g)p_2[ta_{12}q_1 + ta_{22}q_2 + ta_{f2}q_f + ta_{s2}q_s] + c(t-1)p_2ta_{s2}q_s = tp_2q_2$$
$$(1+g)p_f[tdf_1q_1 + tdf_2q_2 + tdf_fq_f + tdf_sq_s] + c(t-1)p_ftdf_sq_s = tp_fq_f$$
$$(1+g)p_s[a_{1s}q_1 + a_{2s}q_2 + a_{fs}q_f + a_{ss}q_s] + c(t-1)p_sa_{ss}q_s = tp_sq_s$$
$$(52b)$$

Fixed capital is treated here as circulating capital. The rates of profit and growth, r and g, are therefore rates calculated on *circulating* expenses. To get the corresponding rates on total invested capital, the fixed capital inputs in the r-equations, 52a, and the usage of fixed capital in the g-equations, 52b, must be multiplied by 1/d.

Since fixed capital is treated as circulating, the analysis performed earlier can be carried through in the same way. First consider the case where w = c = 0, so that R and G are at their maximum. To simplify the discussion, we introduce the following temporary notation:

$$F1 = q1\,df_1p_f \quad 1F = p_1a_{f1}q_f \quad FF = q_fdf_fp_f$$
$$F2 = q_2\,df_2p_f \quad 2F = p_2a_{f2}q_f \quad V = F1 + F2 + FF + 1F + 2F$$
$$FS = q_s\,df_sp_f \quad SF = p_sa_{fs}q_f$$

The proof that $R = G$ follows that previously given. Apart from the third row and column the equations are the same. In the R-equations the third row gives the inputs producing the fixed capital good, while the third column shows the uses of that good. The G-equations, being transposed, show the use of the fixed good in a row, with the corresponding column showing the inputs into it. $V + FS + SF$ have been added to both sets of equations. Since $R=G$ held before, it will also be true of the augmented equations.

The next step is a little more complicated. Set consumer goods production equal to consumer goods use, both $= tp_sq_s$, and capital goods production equal to capital goods use, both equal to $t(p_1q_1+p_2q_2+p_f\,q_f)$.

Then from the equations for consumer goods production and use, we form

$$\frac{1+G}{1+R} = \frac{t(FS + M + X) + Z}{(SF + N + Y) + Z} = \frac{tU1 + Z}{U2 + Z}, \tag{53}$$
$$U1 = FS + M + X, U2 = SF + N + Y$$

and from the equations for production and use of capital goods,

$$\frac{1+G}{1+R}$$
$$= \frac{t(F1 + F2 + FF + 1F + 2F) + t(A + C + D + B) + N + Y + SF}{t(1F + 2F + FF + F1 + F2) + t(A + D + C + B) + t(M + X + FS)} \tag{54}$$

Substituting V, Q, U1, and U2, and setting the two expressions equal, we have

$$\frac{t(V + Q) + U2}{t(V + Q) + tU1} = \frac{tU1 + Z}{U2 + Z} \tag{55}$$

Cross-multiplying and canceling yields

$$t^2(V + Q)U1 + (tU1)^2 + tU1Z = t(V + Q)U2 + (U2)^2 + U2Z$$
$$= tU1[t(V + Q) + tU1 + Z] = U2[t(V + Q) + U2 + Z] \tag{56}$$

This can hold if and only if $tU1 = U2$, that is,

$$t(M + X + FS) = N + Y + SF \tag{57}$$

When $r = g < R$, consumption and the wage will be positive. The proof that $w = c$ follows in the same way as before. From the equations for the production and use of the consumer good, we have

$$
\begin{aligned}
(1 + r)[t(FS + M + X) + Z] + w(t - 1)Z \\
= (1 + g)[(SF + N + Y) + Z] + c(t - 1)Z
\end{aligned}
\tag{58}
$$

that is,

$$(1 + r)(tU1 + Z) + w(t - 1)Z = (1 + g)(U2 + Z) + c(t - 1)Z \tag{58a}$$

Since $r = g$ we can write

$$(1 + r)(tU1 - U2) = \frac{(c - w)Z(t - 1)}{t} \tag{59}$$

From the capital goods sector, we have

$$
\begin{aligned}
(1 + r)[t(V + Q) + U2] + w(t - 1)U2 \\
= (1 + g)[t(V + Q) + tU1] + c(t - 1)U1
\end{aligned}
\tag{60}
$$

Since $r = g$ this can be rewritten (changing the signs),

$$(1 + r)(tU1 - U2) = (c - w)(t - 1)(U2 - tU1) \tag{61}$$

Hence, canceling $t-1$,

$$\frac{(c - w)Z}{t} = (c - w)(U2 - tU1) \tag{62}$$

As before, $Z/t > 0$, but we have already seen that $U2-tU1$ equals zero. For the equation to be true, therefore, $c = w$ must hold.

Going back to equations 52a and 52b, we see that in general

$$t(F1 + F2 + FS) = t(1F + 2F) + SF \tag{63}$$

The fixed capital used in the production of circulating goods is equal to the circulating capital required in the production of fixed equipment. Given this and equation 52, it would seem that the wage-profit rate trade-off could be constructed as before.

However, fixed capital equipment raises new issues: when an industry's fixed capital comes to be replaced, will enough have been produced to provide replacement compatible with growth at the balanced rate? Will sufficient funds have been accumulated to pay for such replacements? We must know that the exchanges called for can actually take place within the framework of the Golden Rule, before we can construct the trade-off. That the exchanges are all compatible can be shown as follows.

Each period users of fixed equipment invest their depreciation allowances in sinking funds, which earn a rate of interest equal to the rate of profit.[29] As a result, the capital of users of fixed equipment is maintained intact: the equipment depreciates a certain amount and so is worth less than in the previous period – but exactly the amount of such depreciation is deducted from earnings and invested. The devalued equipment earns the rate of profit, the depreciation allowance is invested at the rate of interest – equal to the rate of profit. Hence earnings on capital are unchanged, as, period by period, the capital changes in composition, from durable equipment to sinking fund. Each period the profit on circulating capital will be invested in expanding the circulating capital.[30] Each period the profit on fixed capital, along with the depreciation funds and the

[29] In Section 82, p. 69, Sraffa (1960) writes "as soon as the rate of profits rises above zero, equal depreciation quotas [in successive years on machines of constant efficiency] would entail different charges (the charge consisting of depreciation plus profit) on machines of different ages, since at any given rate of profits less would be payable for profits on the older and partly written-down machines; and therefore equal depreciation would be inconsistent with equal prices for all units of the product." Sraffa's solution is to increase the annual depreciation quotas on older machines relative to newer ones (bearing in mind that the sum of all depreciation quotas over the lifetime of the machine must equal its original price), thereby restoring the equality of the charge at different ages. However, this point (and a fortiori the discussion in Section 83) does not appear to be correct. The depreciation quotas are invested and earn interest equal to the rate of profit. Hence the total earning of the industry's capital – machines and circulating capital *plus* financial investment – is always the same as the profit on the original capital, consisting only of machines and circulating capital. As the machines age the capital shifts from machines to the investment fund, but the amount remains the same. Of course, the investment fund expands, because profits and interest are ploughed back, so that when the machines are scrapped the fund will be sufficient to purchase the larger complement of durable goods needed to keep pace with the general rate of growth.

[30] Remember, users of fixed capital are assumed to be able to expand their output before the fixed capital is augmented. In practice, the capacity of fixed capital will to rise, first as it is broken in, then as the labor force learns to use it, and finally as usage suggests various improvements. Capacity will increase over time, and this will be figured into the planning of investment; coefficients would then be averages over the lifetime of equipment. But the argument takes the coefficients as given; capacity is fixed. To the extent that capacity does not rise with time, in a growing economy, producers will have to overbuild at the outset. Excess capacity is therefore a normal feature of an economy using fixed capital. A change in distribution will normally change the intensities with which processes are operated – quantities – which means that fixed capital in different sectors will be used up more or less rapidly. The *rate* of depreciation in sector i, d_i, will be unaffected, but the *amount*, $q_i d_i f_i$, will change as q varies.

interest earned on such funds will be invested in the sinking fund, to accumulate the sum needed for the replacement of fixed equipment at the appropriately expanded size. And each period the *existing* fixed capital will increase its productive capacity through learning by doing and reaping economies of scale, enabling it to be worked by a growing volume of circulating capital. (This can be thought of as an increase, each period, by a factor of $1+g$ in the book value of the depreciated fixed plant.)

The demand for circulating funds by fixed capital producers increases with the rate of growth. Hence if durable equipment lasts for n periods, this will be $(1 + g)^n \mathbf{q'} \mathbf{C_f} \mathbf{T} \mathbf{p}$. The supply of such funds will be the depreciation funds accumulated at compound interest, that is, $(1 + r)^n \mathbf{p'} \mathbf{F_c} \mathbf{T} \mathbf{q}$. If $r = g$, then given equation 58, these expressions will be equal,

$$(1 + g)^n \mathbf{q'} \mathbf{C_f} \mathbf{T} \mathbf{p} = (1 + r)^n \mathbf{p'} \mathbf{F_c} \mathbf{T} \mathbf{q} \tag{64}$$

This implies that the output of fixed capital equipment will have been produced and stockpiled at a rate reflecting the rate of growth, so that when existing equipment is worn out and ready for replacement, the new equipment will be larger in the same proportion that circulating capital has grown.

There is therefore no problem in constructing the Golden Rule trade-off in systems using fixed capital.[31]

Issues and implications

These are surprising results. They link the Golden Rule, a product of modern growth theory, with the capital-goods wage-bill/consumer-sector capital requirements condition, which goes back to Marx and the Classics, and then tie both to a Sraffa-like construction that recalls concerns of Ricardo. And we have shown in earlier chapters that all three play significant roles in the Theory of Circulation. Before we can put this analysis to work, however, we have to show that it is not a special case, and we need to explain just how it links up with macroeconomics and monetary circulation. To this end we will first examine nonuniform technical progress, that is, nonuniform "speed-ups" of work, and show that they can be incorporated in the argument. "Choice of technique" – a favorite topic in mainstream analysis – need *not* be considered, however, for the reason that techniques develop in practice and are never chosen in

---

[31] Using the fixed capital matrix $\mathbf{A_f}$ in place of matrix $\mathbf{A}$, we rewrite equations 34 and 35 and carry through the derivation in exactly the same way to equation 42. Then, to reach the equation corresponding to equation 43 we must substitute drawing on equations 57, 63, and 64, after which the construction proceeds as before.

the abstract. The empirical objection that profit rates and growth rates are far apart, the former normally lying well above the latter, will be shown to rest on two misconceptions about the social accounts. When these are corrected, profits and investment appear to be quite close. This permits an easy translation of the Classical system into macroeconomics, and that in turn allows us to explore the implications for circulation.

The "uniform" system when production speeds change nonuniformly

A uniform speed-up in production times may not be possible, so we must consider what happens when the "speed-up" is not uniform in all sectors. An alternative to speeding up is to increase the capacity size of the sector; if the sector cannot produce faster, turning more input into output in a given time, then it can produce larger batches at a slower pace in that time. In this way the system can adjust to increases in productivity when production times are raised at different rates in different sectors. An auxiliary "uniform system" can be constructed for any set of nonuniform reductions in production time by (hypothetically) reallocating labor; in this auxiliary system the rate of labor productivity will appear as a pure ratio of surplus consumption goods to the wage bundle.

To analyze this, first define a vector $\mathbf{x}$ such that $\mathbf{s}'\mathbf{x}\mathbf{S} = \mathbf{s}'\mathbf{S}$, where $\mathbf{s}'$ is the row sum vector, $\mathbf{S}$ a matrix consisting of the subsistence vector and zeroes in all other entries. The x-multipliers will be expressed as fractions of the labor supply. It follows from elementary algebra that

$$\mathbf{x}(\mathbf{AT} - t\mathbf{I})\mathbf{p} + w_m\mathbf{x}\mathbf{S}\mathbf{p} = \mathbf{O} \tag{65}$$

has the same solutions as the original equation. (Each element of each row of the determinant will be multiplied by the same member of the set of x-multipliers. Hence each of the x's can be factored out of the final result.) But w_m will no longer be a pure quantity ratio, independent of prices. It will now be

$$w_m = \frac{\mathbf{p}'[\mathbf{I}t - \mathbf{T}'\mathbf{A}']\mathbf{x}}{\mathbf{p}'\mathbf{S}'\mathbf{x}} \tag{66}$$

A nonuniform speed-up will replace t and $\mathbf{T}$ with a matrix $\mathbf{Y}$ consisting of off-diagonal zeros and the different speed-up times for each sector along the diagonal. Then we separate $\mathbf{K}$, the matrix of capital inputs, from $\mathbf{S}$, that of subsistence. The equation for the nonuniform speed-up will be

$$(\mathbf{YK} + \mathbf{S} - \mathbf{YI})\,\mathbf{p} + w_m\mathbf{S}\mathbf{p} = \mathbf{O} \tag{67}$$

and to reduce this to a corresponding uniform system, in which the surplus will stand in a purely quantitative ratio to the subsistence, we need to find a set of x-multipliers that will adjust sectoral size inversely to the speed-ups. Call this X. Disregarding **p**, since it will not be affected by elementary operations, the problem becomes determining w_m from the equation

$$\mathbf{X}(\mathbf{YK} + \mathbf{S} - \mathbf{YI}) + w_m \, \mathbf{S} = \mathbf{O}$$

The problem is to find the multipliers that will solve the determinantal equation

$$|\mathbf{YK} + \mathbf{S} - \mathbf{YI} + w_m \mathbf{S}| = 0 \tag{68}$$

Written out in full, simplified to three sectors, 1, s, and n, where 1 and n are capital goods, and s, the subsistence good, this will be

$$\begin{vmatrix} t_1(a_{11} - 1) & (1 + w_m)a_{1s} & t_1 a_{1n} \\ t_s a_{s1} & (1 + w_m)a_{ss} - t_s & t_s a_{sn} \\ t_n a_{n1} & (1 + w_m)a_{ns} & t_n(a_{nn} - 1) \end{vmatrix} = 0 \tag{69}$$

We now multiply

$$\text{row 1 by } 1/t_1$$
$$\text{row s by } 1/t_s$$
$$\text{row n by } 1/t_n$$

and then add the first and third columns. They sum to zero, since the original system was a no-surplus economy. The second column can therefore be set equal to zero. Rearranging,

$$w_m = \frac{1}{a_{1s}/t_1 + a_{ss}/t_s + a_{ns}/t_n} - 1 \tag{70}$$

This is the new wage, or productivity of labor, resulting from the nonuniform speed-up. The multipliers can be interpreted as reallocating labor so that the surplus consists only of the subsistence good or basket of goods, creating an auxiliary or imaginary uniform system with the same maximum wage and prices. Prices in both the nonuniform system and in the corresponding auxiliary uniform system will be the same as in the original no-surplus economy, except for the price of the subsistence good.

Relation to the market

The auxiliary system is not simply an analytical device. Under appropriate conditions, market prices may shift resources in its direction. Industries that speed up will find themselves ready to sell before customers are ready to buy. In conditions of AS/TA and CBF, their prices will fall. Customers for slower industries will face a shortage, and prices will rise. Hence the faster sectors will contract and the slower expand. The market will tend to adjust sizes inversely to speeds. We will leave the details for the next chapter, but the idea may be sketched now.

Assume that the rise or fall in price is proportional to the deviation of the speed-up from the weighted average value, $t - 1 = w_m$. If the speed-up in sector i is greater than average, $t_i > 1 + w_m$, output exceeds demand and market prices fall, if it is less, $t_i < 1 + w_m$, market prices rise. We further assume that industries will distribute the surplus in the ratio w_m; if earnings are greater they will go into expansion, if less, the size of the industry will be contracted, but the "correct" surplus will be distributed on the existing size. (The distributed claims are assumed promptly spent on the subsistence goods.) So the ratio between the price charged at a given time and the price of the previous period, $p_{\tau+1}/p_\tau$, is governed by the ratio between the current demand and the normal output, that is, cost of production plus surplus. Then (denoting time period by subscripts τ and $\tau + 1$) we can write

$$\frac{p_{j\tau+1}}{p_{j\tau}} = \frac{q_{j\tau}}{\Sigma_i a_{ij} q_{i\tau} + w_m \Sigma_i c k_{sij} q_{it}}, \text{ so}$$

$$p_{j\tau+1} = \frac{p_{j\tau} q_{j\tau}}{\Sigma_i a_{ij} q_{i\tau} + w_m \Sigma_i c k_{sij} q_{i\tau}} \tag{71}$$

But low speed-up industries have low quantities and so high prices, and vice versa. That is, prices and output are constrained by

$$p_{j\tau} = p_{jn} q_{jn}, \text{ hence}$$

$$\frac{p_{j\tau+1}}{p_{jn}} = \frac{q_{jn}}{\Sigma_i a_{ij} q_{i\tau} + w_m \Sigma_i c k_{sij} q_{i\tau}}, \text{ so}$$

$$\frac{p_{jn}}{p_{j\tau=1}} = \frac{\Sigma_i a_{ij} q_{i\tau} + w_m \Sigma_i c k_{sij} q_{i\tau}}{a q_{jn}}, \text{ hence}$$

$$\frac{p_{jn} q_{jn}}{p_{j\tau+1}} = q_{j\tau+1} = \Sigma_i a_{ij} q_{i\tau} + w_m \Sigma_i c k_{sij} q_i \tau$$

In vector terms, this becomes,

$$\mathbf{q}_{\tau+1} = (\mathbf{A'}) \mathbf{q}_\tau + w_m \mathbf{K}'_s \mathbf{q}_\tau \tag{72}$$

and it can be shown that this converges on the set of q's in which the surplus consists of the same goods as the inputs into the subsistence sector.[32]

---

[32] The proof is a simple adaptation of that given in the next chapter for the rate of profit. We have shown that the maximal characteristic root is uniquely and positively related to w_m. We therefore substitute $x(w_m)$ for the characteristic root in the Lyapunov function, where $x(..)$ is the function relating w_m and $1/(1+R)$.

An important corollary follows: the adjustment of sizes during a nonuniform speed-up implies that output per head and circulating capital per head increase in the same proportion, in the nonuniform case as well as in the uniform case. Hence if wages increase in the same proportion, the rate of profit will be unchanged. This sets the stage for productivity bargaining and will be examined further later.

Productivity versus change of technique

The preceding has defined an increase in the productivity of labor as a uniform or nonuniform reduction in the labor-time coefficients. This increases both the maximum rate of profits and the maximum wage, but lowers the labor-capital ratio. We can analyze an increase in the productivity of nonlabor inputs in the same way. Suppose an innovation uniformly cuts the materials and energy required to produce unit outputs in all industries by a certain fraction – and therefore also reduces wear and tear on tools in the same proportion. The same labor turning over the same volume of inputs will therefore produce a larger output. The analysis will be the same except that nonlabor inputs will take the place of labor. Both the maximum wage and the maximum rate of profit will be raised, but the labor-capital ratio will increase. The chief difference is that a generalized speed-up of work has some intuitive appeal, whereas a generalized reduction of material and energy inputs does not seem plausible. *Particular* improvements in materials use and energy conservtion, however, are certainly common, but these will change relative prices.

Productivity advances take place when coefficients are reduced. When there is a uniform reduction of coefficients, then two outcomes are possible: either output and nonlabor (labor) can be increased, keeping labor (nonlabor) the same, or output and nonlabor (labor) can be the same, with labor (nonlabor) the same. If the reductions are general, but nonuniform, then the same outcomes are possible, if supplemented by appropriate changes in the relative sizes of industries. (For some productivity improvements to take place, it may be necessary to reach a certain minimum scale of operation.)

As we have seen, a linear wage-profit trade-off can be constructed for this system. Hence since both kinds of productivity improvement raise both intercepts, the new trade-offs lie above the old at all points, that is, for any wage rate, the rate of profit will be higher after the productivity increase.

Suppose some coefficients are reduced, but, in order to make this possible, some others are increased. In this case we are no longer considering productivity changes: we now face a *change of technique*. (Not a "choice of technique; cf. Pasinetti, 1981, pp. 188–9). It is no longer certain that both intercepts will be increased; the new trade-off may cut the old at some level of the wage. Changes of technique will be considered in Chapter 9.

Wages and consumption, profits and growth

Earlier in connection with the two-sector case, it was argued that competition and market pressures would tend to eliminate worker saving (holding it to the level needed for consumer durables), and would also forestall capitalist consumption. Further we suggested that what looks like capitalist consumption is often expenditure needed to maintain intangible capital. This suggested that the Golden Rule should not be considered implausible. Yet, in practice, rates of profit are much higher than rates of growth. This does not call into question our treatment of capital arbitrage, because the obtainable rates are close enough: *dividend* rates, when adjusted for risk, are not so far from interest rates, and while price appreciation rates on growth stocks are all over the place, so is risk, not to mention speculation. But to develop the Classical Equations on the basis of the Golden Rule, as a foundation for the Theories of Money and Effective Demand, it is necessary to show that the rate of profit *in industry* is reasonably comparable to the rate of growth of output.

First consider consumption and wages. For all but the highest levels of salary earners, saving out of disposable wages and salaries is approximately equal to the servicing requirements of debt on consumer durables. This suggests that, in fact, wages tend to be wholly consumed. High salary earners must be considered in a different category; they tend to have substantial investments in education, training and on-the-job skills, including informational assets. Part of their salary should be treated as a return on this capital. But we expect capital income to be saved. Hence high savings by the higher categories of salary income cannot be taken as a counterexample to the claim that all earned income is spent.

Saving for retirement – pensions and social security – by current workers is approximately offset by the current spending of the retired (see below).

The question remains, whether there is consumption out of profits. This would appear to be the case, since from an empirical perspective there is a wide gap between the rate of profit and the rate of growth. As calculated by most investigators, the rate of profits is more than two times the rate of growth (Dumenil and Levy, 1992; Michl, 1992; Shaikh, Ochoa,1984). Normal calculations deduct taxes and depreciation. That still leaves rents and business expenses for intangible capital. But these items are unlikely to account for the difference. On the other hand, Asimakopulos (1992) finds that while adjusted U.S. gross after-tax profits exceed gross investment expenditures, gross retained earnings are normally less.[33] This suggests that to bring these measures together, we should reconsider what is done with dividend and interest income. Two important recipients of dividend and interest income are nonprofit institutions and pension funds. Both of these spend their portfolio income, and together they account for a large percentage of consumption out of profits.

But is this properly considered consumption? Nonprofits – foundations, universities, hospitals, churches – spend on activities like education, research, culture and religion, medical care, and similar matters, all of which can be considered *maintaining the social fabric*. None of these are productive in the sense that the outcome of the activity directly enters into the production of any goods and services. On the other hand, these activities support the stage on which the economic drama in enacted, they knit together the social cloth. No single economic process would be directly injured if less were done by universities or foundations or churches. But in time *all* economic processes would be undermined or damaged. To the extent that this is so, such expenditures should be thought of as generalized maintenance of the social fabric, and deducted from profits as capital expenses.

As for pensions, they are clearly spent on consumption. The question is whether they are really part of profits. For they represent not so much worker saving, as the transference "through time" of worker consumption. The vehicle for such a transfer is capital, but what it is carrying is postponed consumption. The wages of today's workers are lowered (as premiums are funneled into the pension fund) and the difference between what today's workers earn and what they get is paid to yesterdays' workers (as they draw their pension checks from the fund). In return, today's workers are guaranteed that tomorrow's workers will do the same for them. The capital market ensures this guarantee, by investing it in bonds and securities. The securities accumulated during the wage-earning life of a worker are translated into an annuity, which provides income that will be consumed by workers in retirement. Accumulated pension rights are normally proportional to current wages, and pensioners are part of worker households. So the annuities represent a proportional addition to household income, as if the household wage had been increased.

[33] Gross after-tax profits is reached by adding interest and dividends to gross retained earnings. During the period 1950–88, gross investment always lay above gross retained earnings (approximately, on average, $I = 1.33 GRE$) and below gross after-tax profits (approximately, on average, $I = .75 GAP$) (Asimakopulos, 1992, pp. 30–4).

But there is a secondary element in the transaction: the profit paid as dividends on pension fund securities is added to the eventual retirement package, part of the inducement to today's workers to turn over a portion of their wages to yesterday's. Pensions are compensation spent on consumption and therefore represent net earnings redirected from growth to consumption. But – in line with the assumptions above – these earnings are paid to working households in proportion to current wages, in the form of annuities purchased with the retirement accumulation. Therefore when annuities are bought, the capital of pension funds representing dividends and capital gain earnings must be subtracted from *growth* and added to *consumption*: $-\Delta gK = \Delta cN$. The purchase of annuities redirects funds from investment to income designed to support consumption. But the rate of profits will *appear* to be unchanged, although a portion of the stream of profits has been designated to support consumption. Because this stream is proportional to current labor and has been directed to consumption, its actual effects are exactly as if the wage had been increased: $-\Delta rK = \Delta wN$.

Taking the first of these two items into account will tend to bring profits closer to investment, bringing the rate of profit nearer to the rate of accumulation. Taking the second into account explains how the apparent rate of profit can exceed the rate of growth, and consumption the wage, consistently with the Golden Rule – a portion of profits is set aside regularly, and distributed to workers in proportion to their wages, to be consumed.

Nevertheless, it will normally be the case that r > g. How is this likely to affect the wage-profit tradeoff? In fact, it has been shown that, typically, in a modern industrial economy the effect will be very slight (Bienenfeld, 1988). This depends on there being similar proportions of inputs in different industries; however, the similarity need not lie between ratios of *direct* inputs, but rather in the ratios of the inputs needed in the aggregate to produce the direct inputs, and the inputs needed in the aggregate to produce those (first-round) inputs and also later-round inputs. That is, ratios of *vertically integrated* capital and labor inputs, and vertically integrated inputs to produce those inputs, and so on, are to be compared.[34]

---

[34] Bienenfeld examines the deviation of prices of production from labor values, using the Standard Commodity as numeraire (equivalent to setting g = G.) He then lets r vary from 0 to R, and shows that the price-rate of profit curve can be reliably approximated by a parabola – the curve has no inflection, implying a similarly simple wage-profit curve. Our concern is with a smaller variation: when g is given, what happens when r not equal g? Bienenfeld shows that even the largest deviation, r = 0, g = G, has little effect on the shape of the curve when the input proportions are similar in the deeper layers.

If these turn out to be similar, then the effect of a deviation from the Golden Rule will be small. This will tend to be the case, if – as in Mass Production – the technology is such that certain commodities are widely used as inputs, and production techniques are shared among industries, for then there will tend to be similarities in the ratios of inputs in the various "layers." Indeed the ratios will tend to become more similar in deeper layers.

Empirical work can therefore be based on the Classical Equations. But they do not have to be written out in full detail. Instead we can work with a macroeconomic system, for when the Golden Rule holds, there is a very close relationship.

Relation to macroeconomics
The macroeconomic equation for income is

$$PY = wN + rKP \qquad (73)$$

where P is the price level, w the money wage, and Y, N, and K are aggregate net income, labor, and capital respectively. When the wage is paid in money, prices and therefore incomes and capital will be expressed in money. The spending of wages will ensure that revenue is realized in money.

The macroeconomic equation can be rewritten

$$r = \frac{Y - wN}{K} = \frac{Y}{K} - \frac{wN}{K}, \quad \text{where } w = \frac{w}{P} \qquad (74)$$

Y/K is the inverse of the capital-output ratio, and N/K that of the capital-labor ratio. This can be rewritten

$$r = \frac{Y}{K} - w \frac{N}{K} \frac{Y}{N} \frac{N}{Y} = \frac{Y}{K(1 - wN/Y)} \qquad (75)$$

Next suppose that Y/K and N/K are related by

$$\frac{Y}{K} = \frac{aN}{K} \quad \text{(as in equation 41), so that } Y = aN, \text{or } a = \frac{Y}{N}$$

We can then write

$$r = \frac{Y}{K}\left(1 - w\frac{N}{Y}\right) = \frac{N}{K(a - w)} = \frac{Y}{K}\left(1 - \frac{w}{a}\right) \qquad (76)$$

If macroeconomic variations were restricted to shifts between equivalent systems, the kinds of change in r and w considered above, it would be reasonable to take Y/K = aN/K = R, for all distributions. So long as the Golden Rule held – bearing in mind the adjustments just discussed –

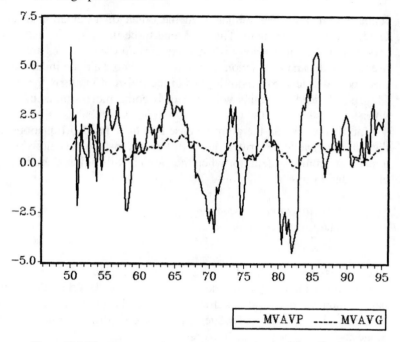

Figure 7.1 12-quarter moving average: growth rate of profits and growth rate of GNP. The volatility of the growths rate of profits is much greater, but the two normally move in the same direction; the tracking is very close.

changes in prices and quantities would have offsetting effects on the numerator and denominator. Then w could be redefined as a fraction of the inverse of labor productivity, ranging from 0 to 1, and we would have equation 41.

Under these conditions, with r = g, and t given, letting P = (Y −wN), $\Delta P = r\Delta K$. But $\Delta K = gK = I$. Hence $\Delta P/P = r(I/P) = r$. So the *growth rate* of profits is a good proxy for the rate of profits (a moving average to avoid short-run fluctuations) obviating all the difficulties of valuing capital stocks. Furthermore, the growth rate of capital will equal the growth rate of output and the growth rate of investment.

It has to be admitted, however, that macroeconomic price and quantity variations can take all kinds of forms not considered here, especially in the short run. Moreover, actual price and quantity systems differ from those in equivalent form.

Nevertheless, such systems may often be partially equivalent, and can therefore be compared with this idealized construction, which can be considered as a map of the possibilities. But there is a further reason to believe this construction may be relevant to actual economies – actual economic systems operate through the circulation of money, and the case can be made that circulation depends on the same relationships (Nell, 1996).

Relation to the Theory of Circulation

Ricardo originally raised the question of "the invariable measure of value" in relation to changes in the value of money in circulation.[35] The significance of this may not have been fully appreciated by modern commentators. Both neo-Ricardians, following Sraffa, and mainstream commentators have worked almost exclusively with equations defined in "real" terms, without relating them to circulating money. In any case most contemporary monetary thinking approaches money in terms of its function as a store of value, rather than as the medium of circulation. (Deleplace and Nell, 1995, introduction and afterword). Current thinking is not attuned to the problems on which Ricardo and his contemporaries were focused.

Ricardo and Malthus sought to establish a measure of value in order to distinguish three causes of widespread price changes:[36] those that arose from the (mis)management of the currency, those stemming from changes in the value of precious metals, and those arising from changes in distribution or productivity (real causes). A complication arose from the fact that the third set of causes could also affect precious metals. Never-

---

[35] As did Malthus, whose 1823 *Measure of Value Stated and Illustrated* was subtitled "With An Application of It to the Alterations in the Value of the English Currency since 1790." On p. 62 we find "Among the questions for the determination of which a standard measure of value is most particularly required, are those which relate to alterations in the value of the currency." There follows a discussion of the value relations of the precious metals to other commodities based on his earlier analysis of the effects on values of different ratios of labor, means of production, and time required in production.

[36] Ricardo first raised the question of an "invariable measure of value" in 1810, in connection with the Bullion Controversy. He opposed the popular view that the value of the currency should be measured by its purchasing power over the mass of commodities, arguing instead that it must be measured by its command over the commodity which served as standard. As a consequence, he contended that the management of the monetary system would have to keep the purchasing power of money over the standard constant. Then, if the standard were, as it should be, "an invariable measure," changes in the monetary prices of commodities could be unambiguously traced back to real causes; changes in the value of money, on the other hand, would show up in money's relation to the standard (Kurz and Salvadori, 1993; Marcuzzo and Rosselli, 1986; Ricardo, *Works*, vol. 3, esp. p. 64).

theless, by comparing items whose value had changed with an invariable measure, it would be possible to trace the origins of the changes.

Their picture rested on an incomplete, ambiguous and inadequate Theory of Circulation, namely, the Quantity Theory of Money. As a consequence, even with an "invariable measure" it would be impossible to tell whether price changes had arisen from monetary causes or not, since the causal lines were not clearly defined.

The preceding chapters have provided a more adequate Theory of Circulation (Nell, 1986, 1995; Deleplace and Nell, 1995). Once the causal lines are clear the issues concerning the measure of value no longer seem compelling. By contrast, the *real* relationships that seem to be involved in circulation stand out. First, the flow of funds through the capital goods industries, and the realization of profit, appear to depend on the balance between capital requirements in the production of consumer goods and the wage bill of the capital goods sector. Second, balance between the growth of the money supply (bank's ability to lend) and the growth of output require equality of the rate of interest, the rate of growth and the rate of profit, an equality that "capital arbitrage" will tend to bring about. This is the "Golden Rule," and an important result follows: notice that equation 35, together with equations 32 and 27, implies

$$\Delta rK = -\Delta wN$$

that is, that the change in the flow of capital payments is equal and opposite to the change in the flow of wage payments. As noted earlier, banks which service both kinds of payments can simply switch their activities when distribution changes.

Consider a banking system that funds wage payments, on the one hand, and capital goods purchases on the other. Suppose at a certain rate of profit, it is able to provide the circulating medium required for the economy (required, that is, to enable both wage payments/consumer spending, and investment spending/profit realization), and that when it does so, it is making as many loans as its capital will justify, given the prevailing views about risk. Suppose now that the rate of profit changes, for example, rises.[37] The above condition tells us that such a system can

[37] If r increases, then in the long run, i will also rise, and the banking system will accumulate capital faster. But g will rise, and therefore the demand for loans facing the banking system will also increase. This concerns the balance between bank capital and banking activity over time and is irrelevant to the question facing a banking system, *now*, before either its earnings or its additional business have developed. At the point at which distribution changes, with r rising (falling), and w falling (rising), the monetary/financial system must switch its provision of funds. If the changes in wages and in profits are exactly offsetting, then it can do so without disturbing its capital-to-activity ratio.

switch loans from one sphere, say, wage payments, to the other, underwriting capital goods purchases, and remain fully loaned up. Conversely, if the above condition is not met, then when distribution changes, a banking system will either have to make more loans than its capital justifies, given the risks, or it will make fewer loans, implying that it is overcapitalized (Nell, 1995).

Chapter 5 argued that these two relationships – cross-sectoral wage-bill/capital and the Golden Rule – are basic to the Theory of Circulation. Now we have shown that together they imply the existence of a linear wage-profit trade-off. Given the Theory of Circulation, however, this linear trade-off will not be an "auxiliary" construction, based on an arbitrary choice of "numeraire." Since it rests on capital arbitrage and monetary circulation, it is the competitively determined equilibrium relationship between wages and profits.

Significance

The value relationships here turn on the proposition that the wage bill of the capital goods sector equals the capital requirements in the consumer sector,

$$q'Sp = p'TK_s q \qquad (77)$$

This states that the value of labor in use is equal to its cost of production, that is, to the cost of producing the consumer goods that support labor. This relationship is invariant to changes in productivity brought about by speeding up the processes of production, and has been shown to imply and to be implied by the Golden Rule. On this basis a linear wage-profit trade-off can be constructed, showing all the relationships between prices at various levels of the rate of profit. When the rate of profit is zero the prices will be those of the no-surplus system, which reflect "labor values"; at other rates of profit prices diverge from these, but the core value relationship between the wage bill and capital requirements remains unaffected. Moreover, the construction of a linear wage-profit trade-off provides a definitive solution of the "transformation problem." What remains invariant when a rate of profit is formed from a rate of surplus value, is the ratio of the net product to the means of production.

The wage-profit trade-off shows the set of possibilities inherent in the technology and economic institutions currently in place. That is, the *current* "normal" level of the wage shows the corresponding current "normal" rate of profit and prices, and given capital arbitrage and the Golden Rule, these will correspond to consumption and growth rates. Theoretical discussion works with an idealized form of these relationships – and a linear wage-profit rate relationship can be used to examine the connections between growth rates and real wages. In practice, of course, the wages, rates of return and prices brought about by the market – especially the growth rates – will be nonuniform and liable to erratic fluctuations. But under certain conditions the theoretical constructions will be guidelines or "centers of gravity" for the market. Competition and capital arbitrage will tend to maintain orderly relationships, and the establishment of "benchmark prices" provides markets with guidelines. It is possible to argue that market practices will tend to keep actual prices close to those guidelines.

Alternative levels of the wage will show the corresponding alternative levels of the profit rate and prices (growth and quantities), on the assumption, of course, that the same work norms hold. This should provide a note of caution: it is not always legitimate to suppose that the economy can easily move from one level of the wage to another. The adjustments required may be far-reaching, and may not converge. Even when they do, however, the path is likely not to be simple.

This is therefore a Theory of Value. But, as we have seen in the previous chapters, the central relationships here – between the wage-bill and capital goods, and the Golden Rule – are the foundation for a Theory of Monetary Circulation. The former defines the Quantity of Money required for a closed, complete circuit, and also identifies a key transaction which proves to be crucial in defining the sequence of transactions. The latter is the condition for the balanced growth of the money supply, and the slope of the trade-off, implying $\Delta rK = -\Delta wN$, tells us that a banking system that could finance the payments required for any one distribution of profits and wages, can also finance any other. If this is correct, then the wage-bill/capital requirements condition will be established and maintained by market forces. In that event the linear wage-profit trade-off will not be an "auxiliary construction"; it will be the equilibrium relationship between wages and profits. Moreover, the invariance implied, which resolves the issues known as the "transformation problem," is required by the system of monetary circulation, and so is not arbitrary. We may have here a basis for the long-sought integration of the Theory of Value with Monetary Theory.

The system we have just explored expresses the structural core of capitalism, what remains "the same" during the processes of transformational growth. Whether capitalism is small-scale or large, Craft-based or Mass Production, whether markets are flex-price or fix-price, wages, prices, and profits, in relation to money and productivity are central. Prices in real terms are exchange ratios that enable reproduction together with the distribution of the surplus. Labor must be treated on a par with other goods. The Classical equations show that the real wage and the rate of profits are inversely related, that the surplus increases with a speed-up, and capital arbitrage tends to bring about equality between the rate of profit and the rate of growth, which, in turn implies that the wage bill of the capital goods sector equals the capital requirements of the consumer sector, or, in fixed capital systems, the circulating goods of the fixed capital sector equals the fixed capital requirements of the circulating goods sector.

Appendix: A numerical example

A surplus value regime

Imagine a very simple society consisting of only three industries, producing wood, iron, and turkeys, indicated by a, b, and s respectively, the last being the consumption good, which supports a unit of labor for the time initially required for production (Nell, 1981, 1986). Suppose that originally in this time only as much is produced as is consumed or used up during production. The methods of production are given, and the formulae, showing inputs combining to produce outputs, are

$$a \quad \& \quad b \quad \& \quad 2s \rightarrow 8a$$

$$3a \quad \& \quad 0 \quad \& \quad 2s \rightarrow 5b$$

$$\frac{4a}{8a} \quad \& \quad \frac{4b}{5b} \quad \& \quad \frac{4s}{8s} \rightarrow 8s$$

Prices are found from

$$ap_a + bp_b + 2sp_s = 8ap_a$$

$$3ap_a + 0 \quad + 2sp_s = 5bp_b$$

$$4ap_a + 4bp_b + 4sp_s = 8sp_s$$

Solving, and writing out the prices taking each commodity successively as numeraire, we have

$$p_a = 1 \quad \frac{3}{5} \quad \frac{3}{8}$$

$$p_b = \frac{5}{3} \quad 1 \quad \frac{5}{8}$$

$$p_s = \frac{8}{3} \quad \frac{8}{5} \quad 1$$

In other words, prices stand in the proportions $3:5:8$. Notice that the wage bill of the capital goods sector is column [2s, 2s], while the capital requirements of the consumer sector are row [4a, 4b]. Taking turkeys as the standard, the former is $2 + 2 = 4$, the latter, $(4 \times 3/8) + (4 \times 5/8) = 32/8 = 4$.

A surplus can be generated by running the processes faster, so that production is completed in less time than the available consumption goods can support (Nell 1990). Inputs of wood and iron per unit output will be unchanged, but let us suppose that the period of production in each industry is cut by one-half. Since only half as much consumption will be required, the resulting system will be

$$
\begin{array}{ccc}
1 & 1 & 1 \to 8 \\
3 & 0 & 1 \to 5 \\
\dfrac{4}{8} & \dfrac{4}{5} & \dfrac{2}{4} \to 8
\end{array}
$$

The new level of consumption will be 4s, and there will now be a surplus of 4s. But the same amount of labor could be employed (working at twice the speed for half the initial time), and gross inputs and outputs would be doubled, producing a surplus of 8s. In either case the rate of surplus will be 100 percent, and will be a pure ratio, in which both numerator and denominator consist only of the consumption good.

By reallocating labor between the industries, the composition of the surplus may be varied, but its value-ratio to basic consumption will be unchanged. For example, the labor employed in the production of the consumption good might be cut in half, and the labor so released allocated to the production of wood. Premultiply the matrix of inputs minus outputs by the row vector [2, 1, 1/2]; the result is a system with a surplus of nine units of wood and one of iron – [9a, 1b, 0s]. But the ratio of this surplus to the consumption supporting labor is

$$
\frac{9 \times 3/8 + 1 \times 5/8}{4 \times 1} = \frac{32/8}{4} = 1
$$

To solve for prices the surplus must be distributed. Initially let us assume that it goes to labor, in proportion to work performed. On the assumption that all units of labor are of equal efficiency, and equally supported by the consumption good, the surplus will be distributed when a net wage of 100 percent is paid, equivalent to the ratio of the old period of production to the new, minus one:

$$
\begin{array}{ccc}
1 & 1 & 1+1 \to 8 \\
3 & 0 & 1+1 \to 5 \\
\dfrac{4}{8} & \dfrac{4}{5} & \dfrac{2+2}{8} \to 8
\end{array}
$$

Here the maximum net wage, $w_m = 1$, and the same prices will clearly rule, since the arithmetic of the original system has been re-created.

The sequence of transactions

Initially, merchants have silver on hand, worth $4. They buy bq for $4, putting the money in the hand of the consumer goods sector. Consumer goods then buys $Av = 4ap_a + 4bp_b$, where $p_a = 3/8$, $p_b = 5/8$. $12/8 + 20/8 = 4$, so the $4 is now transferred to the capital goods sector, in return for the capital goods required in the consumer goods sector.

The capital goods sector now has $4, its wage bill, on hand, as follows:

$$\text{The producers of a have } \$12/8 = \$1\,\frac{1}{2}$$

$$\text{The producers of b have } \$20/8 = \$2\,\frac{1}{2}$$

$$\text{The producers of a need } 1b = \frac{\$5}{8}$$

$$\text{The producers of b need } 3a = \frac{\$9}{8}$$

The internal transaction in the capital goods sector is

$$\text{Producers of a keep } \frac{\$7}{8} \text{ and buy } 1b \text{ for } \frac{\$5}{8}$$

$$\text{Producers of b keep } \frac{\$11}{8} \text{ and buy } 3a \text{ for } \frac{\$9}{8}$$

(The price of b in terms of a is $\frac{\$5}{3}$; see table above.)

Each subsector therefore has on hand the money it has retained from its sale to the consumer goods sector plus the money it has earned from the sale of its product to the other capital goods subsector:

$$\text{Producers of a have retained } \frac{\$7}{8} \text{ and have earned } \frac{\$9}{8},$$

$$\text{making } \frac{\$16}{8} = \$2$$

$$\text{Producers of b have retained } \frac{\$11}{8} \text{ and have earned } \frac{\$5}{8},$$

$$\text{making } \frac{\$16}{8} = \$2$$

Both therefore have the funds they need to pay wages and surplus value, in accordance with the scheme of "primitive duality," corresponding to the late feudal system, in Chapter 5. As wages and surplus value are paid and spent, the funds flow back to the merchants, who buy the remaining consumer goods. As such funds are received by consumer goods producers, they likewise pay wages and surplus value, the spending of which returns the funds to the merchants.

Prices and the rate of profit

Now let us calculate the maximum rate of profit and corresponding prices for our example, in which we have assumed given, single-product industries, and where $w_m = 1$. Taking $w = 0$, that is, setting the net wage at zero, so that workers live at the socially defined subsistence level, we have

$$\begin{vmatrix} .125 - x & .125 & .125 \\ .600 & -x & .200 \\ .500 & .500 & .250 - x \end{vmatrix} = 0$$

Expansion by cofactors, after regrouping, gives

$$x^3 - .375x^2 - .20625x - .01875 = 0$$

The maximum (and only relevant) root of this polynomial (Wood, 1978, ch. 2) turns out to be .705. Substituting for a check, we find

$$(.705)^3 - .375(.705)^2 - .20625(.705) - .01875 = .000137$$

which is as close to zero as we can reasonably expect, given the need to round decimals.

The rate of profit, R (the maximum value of r, but the lowest of the possible rates calculated from the polynomial), is then,

$$R = \frac{1}{.705} - 1 = .418, \text{i.e.}, R = 41.8\%$$

Substituting this into the price equations,

$$[p_a + p_b + p_s][1.418] = 8p_a$$
$$[3p_a + 0 + p_s][1.418] = 5p_b$$
$$[4p_a + 4p_b + 2p_s][1.418] = 8p_s$$

we find (setting the price of good b equal to unity):

$$p_a = .592$$
$$p_b = 1.000$$
$$p_s = 1.749$$

The problem of expressing changes in value

Returning to the original example, remember that when $r = 0$, and $w = w_m$, prices are

$$p_a = 1 \qquad .60 \qquad .375$$
$$p_b = 1.666 \quad 1 \qquad .625$$
$$p_s = 2.666 \quad 1.60 \quad 1$$

When the rate of profit rises to its maximum, reducing the net wage to zero, prices will be

$$p_a = 1 \qquad .590 \quad .339$$
$$p_b = 1.69 \quad 1 \qquad .574$$
$$p_s = 2.95 \quad 1.742 \quad 1$$

Now we can plainly see the problem in choosing an arbitrary numeraire. When wood is taken as numeraire, so that $p_a = 1$, the price of iron, p_b, *rises* from 1.66 to 1.69 when the rate of profit rises from zero to its maximum level of 41.8 percent. But when turkeys, the subsistence good, are chosen as numeraire, so that $p_s = 1$, the price of iron, p_b, *falls* from .625 to .574, when the rate of profit rises.

This is not a contradiction, since different numeraires are chosen, but it is certainly an ambiguity. The long-run prices in the core have changed, but how, and by how much? Are the economic agents facing a price rise, or a price decline? Consider an example which illustrates an issue debated in the era of metallic currencies: the medium of circulation is fixed in terms of the good that has arbitrarily been taken as standard (say, as an accident of history). Agents know only how the prices move in terms of the medium, which, of course, expresses values as they appear in terms of the standard. In two adjacent countries, therefore, one of which has taken wood as the standard and the other turkeys, but which operate the same technology, when both experience the same change in the real wage, we will find opposite patterns of price movement for iron and consequently opposite beliefs by agents about changes in the value of their currencies.

This illustrates Sraffa's remark at the beginning of the chapter on the Standard Commodity (1960), p. 18: "The necessity of having to express the price of one commodity in terms of another which is arbitrarily chosen as standard, complicates the study of the price-movements which accompany a change in distribution. It is impossible to tell of any particular price-fluctuation whether it arises from the peculiarities of the commodity which is being measured or from those of the measuring standard." This need pose no difficulties for agents, who can calculate prices in terms of their wage goods, and decide their actions accordingly, but it can be a serious problem when managing the currency is at issue. A falling value of money suggests that policies should encourage money to be withdrawn from circulation, a rising value, that policies should stimulate the injection of funds. In this example the value of money may not have changed at all, but appeared to change, because tied to a variable standard. If money prices are to express changes in relative prices correctly, then the value of money must be established in a standard which does not change with distribution.

The Standard System

Sraffa constructs the Standard System by peeling away portions of the actual system in such a way that what remains is an economy in which the commodities stand in the same ratios to each other as inputs as they do as outputs. This ratio has been identified as equivalent to the "critical proportion" of labor to means of production, such that the revenue required to pay profits at a higher rate will be exactly supplied by the corresponding decline in the wage, with no necessity for prices to change (Sraffa, 1960, ch. 3). When the industries are arranged so that this ratio prevails in each, the composite commodity produced will not change in value.

To discover how the system should be pared down, we need to solve the following system of equations, to find the appropriate multipliers:

$$(1q_a + 3q_b + 4q_s)[1.418] = 8q_a$$
$$(1q_a + 0q_b + 4q_s)[1.418] = 5q_b$$
$$(1q_a + 1q_b + 2q_s)[1.418] = 8q_s$$

Since the system is linear and homogeneous, we can expect to find only relative values for the q's. Arbitrarily setting q_s equal to unity, we find that the solution to the system is

$$q_a = 1.95$$
$$q_b = 1.69$$
$$q_s = 1$$

These values for q_a, q_b, and q_s indicate how the scale of each industry should be adjusted to construct the Standard System. Taking turkeys as the numeraire, wood must be expanded by 95 percent and iron by 69 percent. (Had we taken the iron industry – industry b – as numeraire, that is, left it unchanged, the vector would have been [1.157, 1.00, 0.592], so that industry a – wood – would be expanded by 15.7 percent, while industry s – turkeys – would be cut to 59.2 percent of its original scale.)

We can now write the original system in the form of a table, listing the three industries, and showing both net and gross outputs and the surplus rates:

Industries					Net output	Gross output	Surplus rate
	Wood	Iron	Turkeys	$Q_i - Y_i$	Y_i	Q_i	R_i
a	1 +	3 +	4 =	8	0	8	0
b	1 +	0 +	4 =	5	0	5	0
s	1 +	1 +	2 =	4	4	8	100%

outputs

    8   5   8

Now using the same format we can write the Standard System:

	a	b	s				
a	1.95 +	5.07 +	4	11.02	4.58	15.60	41.8%
b	1.95 +	0.00 +	4	5.95	2.50	8.45	41.8%
s	1.95 +	1.69 +	2	5.64	2.36	8.00	41.8%

Gross outputs are

    15.60a   8.45b   8.00s,   or normalizing on s,
     1.95   1.056   1.00

The Standard Net Product is therefore:

    4.58a,   2.50b,   2.36s,   or normalizing on s,
    1.94,   1.059,   1.00

And the Standard Capital is

    11.02a,   5.95b,   5.64s,   or normalizing on s,
    1.95,   1.055,   1.00

Allowing for rounding errors, these are evidently all in the same proportions.

Removing the anomaly
Since the Standard Commodity does not vary in value when the rate of profit changes, a solution can be provided by expressing the prices of each commodity in terms of the Standard Net Product. Precisely – and only – because this composite commodity does not vary in value, it can legitimately be taken as *constant* – unchanged in value in the two cases, $r = 0$, $r = 41.8$ percent. The constant, in turn, can be arbitrarily set at unity. In this sense only is the numeraire arbitrary.

So for the case where r = 0, we have

$$1.95p_a + 1.056p_b + 1.00p_s = 1$$

which we now add to the price equations. Noting that $p_a = 3/8p_s$ and $p_b = 5/8p_s$, we can rewrite this in terms of p_s,

$$(1.95)(.375)p_s + (1.056)(.625)p_s + 1.00p_s = 1$$

which becomes $2.391p_s = 1$, i.e., $p_s = .418$.

Hence, when expressed in the Standard Commodity the prices corresponding to r = 0 are:

$$
\begin{aligned}
p_a &= .1567 \\
p_b &= .2614 \\
p_s &= .418
\end{aligned}
$$

(These can be checked by substituting them back in the original equations for prices when r = 0; e.g., for "wood":

$$.1567 + .2614 + 2(.418) = 8(.1567)$$
$$1.254 = 1.2536$$

accurate to three decimals; similar results hold for the other two.)

When r reaches its maximum, r = .418, then $p_a = .592p_b$, and $p_s = 1.749p_b$. So, rewriting this time in terms of p_b:

$$(1.95)(.592)p_b + 1.056p_b + (1.00)(1.749)p_b = 1$$

which reduces to $3.959p_b = 1$, or $p_b = .2526$. Expressed in standard terms, then, the prices are

$$
\begin{aligned}
p_a &= .1495 \\
p_b &= .2526 \\
p_s &= .4417
\end{aligned}
$$

Returning to the original question, then, we see that when r rises from 0 to its maximum level:

p_a falls from .1567 to .1495

p_b falls from .2614 to .2526

p_s rises from .418 to .4417

There is now no ambiguity. Two components of the price vector, both of them the industrial goods, fall, and one component, the consumption good, rises. All the price movements are rather small. (Notice, finally, that the initial productivity increase, because uniform, did not affect relative prices; productivity changes can therefore be analyzed by the device of constructing a weighted average of changing speeds of turnover, thereby separating the real change from the effects of varying prices.)

The wage-bill capital requirements balance

When the Golden Rule is observed, the wage bill of the capital goods sector will equal the capital requirements of the consumer goods sector. This can be seen by looking at the initial system, in which the rate of profit is zero, and there is no growth, and then at the system in Standard proportions, which corresponds to a growth rate equal to the maximum rate of profit.

Wage Bill of K-goods	Capital Inputs in C-goods
When $r = 0 : 2s, 2s$	When $r = 0 : 4a, 4b$
$4 \times .418 = 1.672$	$4 \times .1567 + 4 \times .2614 = 1.6724$
When $r = .418 : 1.95s, 1.69s$	When $r = .418 : 4a, 4b$
$(1.95 + 1.69)(.4417) = 1.607$	$4(.1495 + .2526) = 1.608$

In each case, the rate of profit equals the rate of growth, and the prices are those corresponding to the rate of profit, while the quantities are those which correspond to the rate of growth. In each case, therefore, the wage bill of the capital goods sector equals the capital requirements of the consumer goods sector. At any other prices or quantities this equality will not hold.

Distribution and circulation

The condition derived earlier, that $Kdr = -Ndw$, can now be illustrated. The initial position is the no-surplus economy. A uniform increase in production speed takes place, creating a surplus of 100 percent, or a new wage $= 1$. When this falls to 0, the net rate of profit rises from 0 to .418. Hence, $\Delta w = -1$, and $\Delta r = .418$. N is the support for the labor force, used in producing the standard net output. K is the standard capital, at which $r=g=.418$. Capital will be evaluated at the prices in the table above at which $r = g = .418$. The support for labor will be evaluated at the initial price, in standard terms, when $r = 0$, namely $p_s = .418$. Hence we have

$$K\Delta r = [11.02, 5.95, 5.64] \times [.1495, .2526, .4417]' \times \{.418\}$$
$$= [1.647 + 1.503 + 2.49] \times .418 = 2.35$$
$$-N\Delta w = -5.64 \times .418 \times -1 = 2.35$$

These are exactly equal.

CHAPTER 8

The Classical system:
gravitation and market adjustment

None of the elements in the Classical equations represents a directly observable entity. The coefficients, as usually interpreted, represent physical inputs for "best-practice" techniques (which at most only a few firms yet use). Data are actually collected as *costs*, that is, in monetary terms, not in physical terms. The products are likewise idealized and defined in physical terms, whereas output data are also mostly collected as revenue.[1] Neither firms nor households, nor any other decision-makers appear. Prices are uniform – the "law of one price" is assumed – and prices are expressed in barter terms; rates of profit and wages are similarly smoothed out and hold in barter value. No trace of the unruly market or the hurly-burly of actual competition can be found. Nor could there be, for the Classical equations do not represent behavior at all. They portray, in simplified and idealized form, the essential structure of the capitalist system, the core relationships between wages, prices and profits, on the one hand, and consumption, quantities, and growth on the other.

Yet this poses an *epistemological* problem. If all our empirical data come from the market, while the Classical equations are independent of the market, how can we verify that those equations correctly describe the core relationships of the economy? An answer was supplied by Smith and Ricardo: market prices and other market variables "gravitate" toward, or around, the corresponding "natural" prices or other variables, where "natural" values are those determined by the Classical structural equations. The Classical equations are justified because they are the centers of

[1] Collecting data in physical terms presents serious problems. Sales frequently involve a service component which is hard to disentangle from the good – even when the production process is single-product – but which may vary with different customers, and with different levels of output. Costs pose even worse problems; physical depreciation is a matter of educated guesswork, as must be any estimates of the physical counterpart to maintenance charges. The apparent constancy of running costs over a wide range of output levels does not necessarily mean fixed coefficients, let alone that these coefficients could be identified and measured. It could just as plausibly mean that small and difficult to measure declines in some running costs were approximately offset by small increases in others.

gravitation, that is, the points toward which market processes converge, or around which they oscillate. This has sometimes been presented as showing that the natural solutions are the fixed points of market dynamics. In the modern revival of Classical theory, demonstrating "gravitation" has developed into a substantial exercise in dynamic theory, almost a cottage industry.

But the gravitation approach suggests that the validity of the Classical equations rests on the way they underlie market processes. They are verified because they are necessary to understand and interpret market behavior. By the same token, then, if market processes prove to be unstable, as they apparently are in some periods, the Classical equations would not be verified. This does not seem plausible. That market behavior is or has become unstable need not imply a different underlying structure – a small change in behavior may be sufficient to create instability. There must be independent ways to verify the Classical equations.

Moreover, it will be shown here that gravitation only holds for Craft-based economies. Demonstrations of gravitation all appear to require, among other conditions, that market prices be flexible – in particular, that they vary inversely with outputs – that sectoral investments and outputs vary directly with sectoral differences in profit rates, and that there be a given level and composition of normal effective demand. Such circumstances are not to be found in the era of Mass Production, in which market prices tend to be fixed over large ranges of output, and multiplier-accelerator type processes lead to unstable movements in output, while investments are planned on the basis of projected future growth, and a large proportion of total demand depends on government.

But if gravitation only holds for certain periods, the Classical equations cannot be justified, in general, on the grounds that they describe the position towards which the economy is tending. Instead we shall argue that they are true because they correctly, although abstractly, describe the core structure of the system. To understand how this can be empirically verified we must reexamine the traditional distinction between "market" and "normal" prices.

Normal prices

The idea of gravitation

Gravitation is thought to resolve issues on two levels. It answers the epistemological question as to what *justifies* the claim that the Classical equations provide a reasonable and useful account of capitalism – when it is all too obvious, for example, that the rate of profit is not uniform and growth is anything but steady. The equations describe the position toward which the system is *tending*, even though it may never arrive there. Secondly "gravitation" is proposed as a solution for two distinct problems in market adjustment, which should be kept separate, although they are often treated as if they were the same. The first is the establishment by market pressures of the uniformity of prices, wages, and especially (both because it is the most problematic and because it is the key to the others) the rate of profit. The second is the shift from one composition of demand to another, through changes in relative market prices and profitability.[2] (A third variant can be stated also, but it is really a combination of the other two: the convergence to a new level of the rate of profit, after a change in distribution.) In all of these the rate of profit is the connecting link. When demand changes, or when there is a temporary deviation of market from natural prices, relative profitability will change, meaning that the rate of profit will become nonuniform. Market reactions will then tend to drive it back to uniformity, bringing outputs back into line with normal demands.

But the uniformity of the rate of profit is not the same as the uniformity of prices and wage rates. These latter do not necessarily imply capitalism. We have already examined the competitive pressures in the labor market. Earlier we mentioned that market pressures would operate, albeit weakly, in the manorial/guild economy, tending to equalize earnings per capita, adjusted for training, etc. If demand rose in one area relative to others, such as because of sales abroad, this would drive up prices and per capita earnings, attracting apprentices and also journeymen from other fields whose skills were related or transferable. Activity would expand until prices fell to a normal level of per capita earnings. To these market pressures there would be added political pressures for regulation. Earnings would not be figured on "capital," because business establishments could not be bought and sold. Neither guilds nor manors were marketable, nor was there a regular pattern of investment creating new guilds and manors. Feudal society was static.[3]

---

[2] What counts is the relationship between composition of demand and the production system. *Given* a composition of demand, when productivity – the speed of production – changes in some sectors relative to others, sectoral sizes will have to be adjusted. The process will be the same as when the composition of demand changes, given productivity and capacity sizes.

[3] Static, that is, in the sense that it did not grow regularly through a market-driven process of investment. For what it is worth, this accords well with economic theory; if there is no rate of profit, prices will be as if the wage were at its maximum, and so will equal direct plus indirect embodied labor. Under these conditions it is a condition for the optimal "choice of technique" – implied by the Golden Rule – that the rate of growth also be zero or nonexistent. Of course, in actual fact, feudal society was quite innovative, as we would expect on the basis of our account of the role of the market.

The formation of a rate of profit, therefore, must be seen as, initially, the development of capitalism itself. Once established, however, the pressures that maintain the rate of profit are the same as the pressures that also support the mobility of capital. These, we shall argue, function through monetary institutions and will be found in the pattern of circulation. The pressures that establish and maintain the rate of profit cannot be adequately understood in a barter context.

Lifting the veil of barter

Gravitation processes are dynamic patterns of market movement. The argument here will be that market dynamics cannot be adequately modeled in barter terms; models that seem to be barter actually implicitly incorporate some aspects of a monetary system, while ignoring others. To explore this consider different conceptions of barter:

Current barter – goods traded for goods, in spot transactions
Extended barter – goods traded for goods, in spot and future transactions

In both these cases, settlement takes place only upon joint delivery of goods to both sides.

Negotiable barter – goods for goods, spot and future, but *claims* for spot or future delivery are transferable to third parties. Delivery may be demanded for settlement.
Generalized negotiable barter – claims or goods for goods, spot or future, generalized transferability of claims. Claims ranked by quality; high-quality claims will settle debts.

In the first two cases shifting capital in response to price changes may very well not be feasible. The price changes would be figured in barter ratios, and to know precisely how much a price had changed it would be necessary to compute all the cross-exchanges. Then capital would have to shift in the form of inventories of goods. It would be necessary either to carry unwanted inventories, in the hope of swapping them for the desired ones, or to assemble the desired combination at the outset. In either case a shift would require seeking out new channels of trade. The transactions costs would be substantial and could easily swamp potential gains.

The third form is a way station en route to the fourth. If claims are negotiable, then a market must develop to make that negotiability general. Such a market requires specialists in the vetting, ranking and insuring/ endorsing of promissory notes. It also requires developments in merchant law and in the practice of courts. But the negotiability of claims will lead to *discounting*, selling longer-term claims for shorter ones. This implies a rate of interest, expressed in generalized purchasing power. Once discounting is established, generalized lending will follow.

In other words, as we move from limited negotiability of promises to deliver goods to generalized negotiability, we move toward a system with loans of generalized purchasing power, on which interest will be paid, in that same purchasing power. It is a short step from this to a monetary system.

A monetary system permits the full development of competitive pressures. Moreover, the money rate of interest acts as a universal norm for profitability. The uniformities do not have to be explained by gravitation; they are established in the course of monetary circulation. And we shall see that it is monetary circulation which provides the justification for the use of the Classical equations, answering the epistemological question.

Observing normal prices

First, consider how – and whether – we might observe natural or normal prices. According to the Classics they depend on distribution, and tend to be fixed for long periods of time, whereas market prices, determined by the fluctuating forces of supply and demand, reflecting the current pressures of competition, can be observed in the events of the marketplace. This suggests (Garegnani, 1992, pp. 345–7) that the relation between market and natural prices is the same as the relation between actual, observed magnitudes and theoretical ones. But this does not seem correct. "Market price" is itself a theoretical construct: according to the Classics, market prices are determined by the ratio of current demand and supply. Market price equals "natural price" when the current ratio of demand to supply is unity. This is a theoretical claim, which can be tested against observation. (In the early Classics, market prices were not well defined, but even there the rule relating them to natural prices was clear; cf. Roncaglia, 1990.)

Natural prices, on the other hand, are independent of current demand and supply. The relation between the two concepts is not that of particular to general, but rather of a current or "market" theory to a "natural" or capital theory, in which specific, theoretically defined current variables determine the divergence of market values from natural. Both "market price" and "natural price" must be understood as theoretical concepts, and both must be related to empirical counterparts.

Marshall blurred this sharp distinction, by making it a matter of degree – the longer the period of time under consideration, the greater would be the influence of the long-term forces of distribution and productivity, the shorter the period, the more pronounced the influence of supply and demand in the market. For Marshall, by implication, all prices are market prices, but some are current, and reflect short-term forces and constraints, while others are long-term averages.

This is a coherent position at least; long-run prices can be constructed as averages of current prices. But neither the Classics nor the modern revivalists would accept Marshall's position. For the Classics, natural prices are independent of the current market; the Classical equations represent a different set of forces, not just the same forces with some constraints relaxed.[4] Market prices are determined by current supply and demand, natural prices by distribution and technology.

But market prices are visible in the market; they could be seen, measured and recorded; but how could natural prices be observed, if they were *not* present in the market? Are they logical constructs, figments of the theoretical imagination, which will show up, if at all, as special kinds of equilibrium positions in the market? How could any prices be said to exist, if they are not realized in the current transactions of the market?

A "price," however, is a more complex idea than it might seem at first glance. An actual transaction takes place at a particular time and place between a certain buyer and a certain seller, for a specific set of goods. This transaction will take place at an agreed price; on a different day, or a different hour, the same buyer and seller may transact at a different price. Even on the same day, or the same hour, for a different set of goods, there may be a different price – even in the same set of goods, some may be paid for in cash, some on 30-day credit, some on 90, or 120, at different interest rates, resulting in different effective prices. Transactions can be decomposed into virtually any degree of detail. We choose the level of observation, guided by conceptual truths. But when economists, or business planners, reporting observations, speak of "*the* price," what is meant is some *norm* or *average* that has tended to prevail in standard transactions between buyers and sellers of good standing (Hollis and Nell, 1975). Market pressures then shift this norm or average. So even market prices are a kind of construct, an average or a tendency precipitated in the analysis of raw observations.

Clearly natural prices or normal prices must be such a construct, but they need not necessarily rest on observations of current or recent

---

[4] Garegnani (1992, *New Palgrave*) appears to imply that market prices are empirical instances of natural prices, differing from them only by virtue of accidental random effects, which cancel out in the long run. The relation suggested is that between concept and object. "Natural price" is the theoretical concept, "market price" the empirical counterpart. But "markat price" is itself a theoretical construct – market price reflects the balance of supply and demand.

purchases and sales. Prices will normally be quoted or estimated in *contracts raising capital*. Such prices will be needed to calculate the normal earnings of a period, in order to compute the discounted value of a stream of returns. To determine the price of a company, for example, it will be necessary to estimate the normal prices of its products, and its normal costs. Similarly, to determine the value of an asset pledged as collateral, it will be necessary to estimate its normal earnings – and to show that these earnings can be expected to continue. In each case the prices quoted will be the norms expected to prevail over long periods, prices, therefore, that reflect the best available estimates of the underlying and permanent cost and profitability conditions. If change is expected, however, the norms of the past will be useful only insofar as they help to calculate the new estimates of costs and earnings.

Market prices are established in the goods markets; natural prices are quoted and used in transactions in the capital market. From such quotations, embodied in long-term contracts, and therefore "observable," it will be possible to distill a set of normal prices, prices which govern the behavior of agents and which are wholly independent of current or recent transactions in the market. But this depends on the institution of long-term contracts quoted in monetary terms, and such contracts – in which the rate of interest plays a central role – are drawn up on the basis of long-term expectations.

Price formation

How, then, are these prices determined? By general consensus, through the establishment of a uniform rate of profit, the level of which, of course, depends on distribution, that is, the real wage, or its reciprocal, the average markup. But that just poses the question, how is the uniformity of the rate of profit established and maintained, if not by the incentives of supply and demand in the market? (And how can distribution be determined if not by supply and demand in the labor market? We return to this in Chapters 9 and 10.) For if the uniformity (and/or the level) of the rate of profit *depends* on the market, then natural prices, after all, turn out to be nothing but market prices in a special kind of equilibrium.

In particular, many contemporary theorists regard natural prices as those which result from the interaction between investment in pursuit of profit and market prices governed by supply and demand. Price changes, resulting from changes in supply and demand, affect profitability; capital then shifts in response to the altered pattern of profits, changing supplies, which, in turn, creates a new pattern of market prices – and so a new pattern of profitability. When this process converges – if it does – to a uniform profit rate, there will no longer be any market incentive for capital to shift from one industry to another. And all markets will clear, because if some did not, prices would change, altering profitability again (Goodwin, 1953, 1980; Semmler, 1984).

Yet we shall see shortly that this by now well-studied process faces difficult questions. Is it rational to make investment decisions on the basis of price changes that merely reflect the current balance of supply and demand? Investment depends on long-term expectations, not on the current state of the market.[5] Perhaps the two processes that make up a cross-dual dynamic pattern (see below) should not be combined in general; for one is short-term, the other long.

If there were a *permanent* change in demand, investment undertaken to adapt supply accordingly would be appropriate. The Classics argued this; capital would be shifted when normal demand changed, and the change was known to be permanent. Smith, for example, is quite definite: "The quantity of every commodity brought to market naturally suits itself to the effectual demand [the demand at the natural price]. It is the interest of all those who employ their land, labor, or stock in bringing any commodity to market, that the quantity never should exceed the effectual demand; and it is the interest of all other people that it never should fall short of that demand" (Smith, 1961, ch. 7, p. 57). Such a change in demand will normally result in price and profitability changes – as we shall show below. It will not be random. However, the price and profitability changes are not necessarily *signals* to shift investment. The changes in prices and profits may be part of the *process* by which capital moves.

But if the formation of natural prices comes about through processes in which capital is shifted in response to changes in profitability brought about by fluctuations in current supply and demand (leaving open how it is known that the changes are permanent), there would be no separate "natural processes" which establish natural prices, as market processes bring about market prices. Natural prices, in this case, simply *are* market prices that have reached a certain kind of equilibrium. Alternatively, they could be considered the potential resting places, or attractors, for market prices. In either case, natural prices would have no *independent* existence; they appear in the market or not at all.

---

[5] If long-term expectations are adaptive, current changes will have an effect, but a small one; if they are rational, the current state of the market, if considered accidental or transient, may have no effect at all. How do agents know that *current* prices correctly presage *future* profitability? If they do not know this, they should not invest on the basis of current prices; but if they do know the pattern of future profitability, they need not wait for current market developments.

For Marshall, this surely makes sense, since the distinction between two kinds of prices has been converted into a difference of degree in the determining forces. But it is less clear for Smith and Ricardo, both of whom describe natural prices not only as determined by different forces from those which fix market prices, but also as prices that can be, and are, widely known independently of the current state of the market. This suggests that such prices actually exist, in some observable sense, even though they are not recorded in the supply-and-demand transactions of the marketplace. We have suggested that they form the basis for long-term monetary contracts, and thus exist, independently of current market conditions, in the provisions of such contracts. Let us consider how capital decisions, made in monetary terms, might establish normal prices.

Investment, the rate of interest, and the rate of profit

Profit is earned by capital, and capital is the relationship in which labor and means of production circulate – turn over against money – to produce and reproduce output. The question, then, is how are prices established and maintained so that the rate of profit in different industries tends to be the same? Why are these prices considered "natural"? Are they independent of the market, that is, of current supply and demand? Consider these questions, on the one hand, in the context of an economy operating a Fixed Employment technology, and on the other in a Mass Production economy.

Output is circulated by money, and capital moves through borrowing and lending at a market-determined (and therefore uniform) rate of interest. At a certain point in the circulation, capital is maximally liquid; at this point the anticipated profit from operations can be compared to the rate of interest obtainable on bonds. This clearly sets a minimum level to the rate of profit. Of course firms should not make rash decisions on the basis, say, of a short-term weakness in sales, but that is precisely the advantage of liquidity: the firm can *pause*, put its working capital in the money market, and wait for better times. On the other hand, if the *long-term* expected earnings are below those obtainable from long-term bonds, capital should be withdrawn permanently. But if the earnings are above those from bonds, will the threat of competition compel firms to keep planned prices down to the level where the profit on capital exceeds the rate of interest by only the margin appropriate for risk and inconvenience? Perhaps, but not necessarily. Entry is difficult and expensive; moreover, it depends crucially on entrepreneurial and technical skills – qualities that may or may not be transferable from one line of business to another. Consider an important case: $r > g_d = i$, where r and g_d refer to the profit and growth (of demand) rates for that industry, and i is the general rate of interest. The low interest rate will discourage buying bonds, but reinvestment of industry profits will lead to expansion of capacity in excess of the growth of demand. So prices will tend to fall, bringing profits down toward g. (On the other hand, if $r = g_d > i$ the investment of profits will just match industry growth.) In other words, the uniformity of the rate of profit tends to reflect the fact that the earning of invested capital anywhere must be compared with the interest which could be obtained on the same funds.[6]

First consider additional investment, using existing techniques, in goods already being produced. The new facilities will have the same costs, and will produce the same products, for the same prices, as the old. If the costs were greater, the new firms would lose out competitively; if the costs were less, they would displace the old – but if the technique is the same, the normal costs have to be the same, and competition would obviously be able to undercut any higher price, while any lower would fail to cover interest costs. The output of the new investments will therefore have to be priced so that they earn the same rate of return as the old capital. The capacity added must match the growth of demand. In a Craft economy profits will be saved and invested. A rise in the expected growth of markets will bring a rise in investment, bidding up prices and raising profits; a fall will in the same way lower profits. Hence r and g will tend to be close and they will move together. As we have seen, the rate of interest will both tend to bring uniformity to the rate of profit, and will interact with r, so the two will track one another. But in Mass Production the rate of interest will interact with the rate of growth.

Next, consider investment in goods already produced, but where the new facilities will employ a new and superior technique. The new technique will produce at lower costs; the new producers will therefore capture the new markets at a lower price. If they were to continue to sell at the established price, it might appear that they could earn supernormal profits. But it would not be reasonable to place these profits in the bond market, since the rate of interest is less than the rate earned on the new

---

[6] Money is fluid, mobile, and universally acceptable; money can therefore be easily shifted in pursuit of the highest returns. It will be assumed here that the rate of interest on money, adjusted for risk and liquidity, will be uniform. How the rate of interest is determined is not at issue here; cf. Chapters 12 and 13. The rate of interest establishes and enforces the *uniformity* of the rate of profits, but it does not set its *level*. Later it will be argued that market pressures cause the rate of interest to move in line with the rate of growth.

technique. Nor would firms be well advised to try to enter other lines of business, in which they lack the relevant expertise. Hence they will either invest, expanding capacity ahead of demand, or they will distribute the profits, raising the price of shares, and lowering the cost of capital for the firm. Either way, the result will tend to lead to expansion ahead of demand, driving prices down. Such prices, however, will undercut existing firms, thus enabling the new firms to utilize their capacity fully. This will gradually force the older firms either to modernize or to exit – in the end reestablishing normal profitability with the new technique. Any prices quoted in contracts would have to be based on this judgment.

Finally, a new product. Whenever investment in a new line of activity is undertaken, the investors providing the capital must be assured that the venture will earn enough to pay its way. To establish this, the prices of products and costs will have to be known; prices will therefore be set at the time of investment. They cannot be set at a level below that which would earn the prevailing rate of interest, or general rate of return elsewhere. If, however, they are set so high as to return earnings greater than the normal rate (allowing for risk, for recoupment of development costs, etc.) there will be room for entry undercutting this price – or for an existing firm, facing a temporary or permanent loss of customers, to recoup by undercutting. A new firm would therefore find it extremely risky to try to get away with a high price; it would not be prudent, or acceptable, to quote such a price in long-term contract. Moreover, were such a high price to hold for any time, entry of capital and reinvestment of profits, leading to expansion at a higher than normal rate, would take place, changing any estimates of potential competition in supply. Moreover, precisely because of the high price, the development of demand would be slower than otherwise; in the long run, therefore, a high price, expected to yield higher than normal returns, would be driven down, and the industry will be forced to carry excess capacity. It is thus likely that excessive investment, and undercutting by existing firms, plus the threat and occasionally the reality of entry from "outside," would combine to drive prices down to the level consistent with normal profits, as expressed in the rate of interest.

So investment in existing products, using the current process, will earn the going rate of return; new and better processes will drive out old, and settle down to earnings at the going rate, while the pricing of new products will tend to settle at a level that will return profits at the going rate. Think of prospective returns – outputs minus wage costs, summed over the future. Assuming an infinite time horizon, discounted by the interest rate, this stream of earning should equal the value of capital –

"capital asset pricing." In matrix form:

$$\frac{p(I - A - wS)}{i} = Ap$$

Obviously if i = r, this is the Classical price equation, and if r = g then setting prices to earn r will generate the profits that, invested, will provide the capacity to expand output at a rate equal to g, where g is the expected growth of demand. In any event the rate of interest on money sets a floor, while the process of circulating new investment goods tends to drive the price down to that floor, making allowance for risks and the special difficulties of various trades.

Natural prices then are determined along with investment, as part of the decision to commit capital to a project. (This will be explored in detail for Mass Production in Chapter 10.) These prices will then be the guidelines in production and marketing; as the basic list prices, they will be the target that sale prices will be expected to meet, on average. They are the results of the pressures determining decisions in the capital market, and they are established long in advance of any production. Market prices are determined by the interaction of current demand and supply; natural prices are determined before any supply is produced. Anticipated market reactions play a role in establishing natural prices, since such anticipations enter into the investment decision. The long-term configuration of technology, distribution, and the expected pattern of growth are the basis for natural prices. Of course, this configuration also provides the background for market prices. But these factors are considered directly in the planning of natural prices, while actual market pressures, determining market prices, do not materialize until much later, and include other factors, such as capacity constraints and cyclical forces, which, though short-term, may also be systematic. Natural prices therefore exist and are determined, independently of supply and demand in the market.

Gravitation and price flexibility

What then is the status of gravitation? Before we can examine this, however, we need to know what kind of economy is under consideration. Capitalism can exist in relatively primitive technological circumstances, or it can operate modern, sophisticated equipment. Or, of course, somewhere in between. Market prices, and therefore, gravitation, depend on whether technology requires employment to remain fixed, or permits it to vary with demand. Gravitation is the outcome of behavior, we shall argue, and it will hold in Craft-based economies – although not always – but not in conditions of Mass Production. But the Classical equations, which describe the core relations of capitalism, hold in both periods.

Convergence

Since Adam Smith it has been claimed by the Classical tradition that market prices gravitate toward natural prices.[7] Market prices reflect the ratio of demand in money terms to current supply on the market and are subject to temporary influences and particular pressures. Natural prices reflect the permanent forces, the deep structure, underlying the market. As Ricardo argued, and the Sraffian equations of today claim to show, natural prices are independent of supply and demand and depend only on the selection of the best-practice coefficients of production and the rate of profits (or the real wage) for a given set of outputs. They are determined by the pressures of competition and are unaffected by the temporary influences impacting the market and so provide a "center" around which market prices gravitate. The idea is that "there would be underlying tendencies for the system to move towards the dominant methods of production and towards uniformity of the rate of profit" (Bharadwaj, 1989, p. 235). This seems to imply that market prices will gravitate toward, that is, tend to converge on, natural prices, and in the process, bring together divergent rates of return, although under some conditions market prices could oscillate around natural prices, without any tendency to converge on them, no matter how long the period considered, so long as natural prices emerged as the average.[8]

This has an important implication: if market prices tend to converge on natural prices, and if the agents of the economy know, even roughly, what the natural prices are – that is, if they know the normal rate of profit – then natural prices will be the ones that agents should rationally expect to hold. Moreover, it will be worth their while to move to the long-period position

---

[7] Ricardo, however, invariably linked profits and interest in his analysis (cf. ch. 21, "Effects of Accumulation on Profits and Interest"). In his view, the rate of interest simply expressed the rate of profit on capital in its money form. When capital shifted from one industry to another it would most readily move, not through "a manufacturer ... absolutely changing his employment," but as a reduction of loans to one industry coupled with expansion in another (Ricardo, ch. 4, pp. 48–9). Indeed, for Ricardo, natural prices themselves had to be expressed in money, which in turn is fixed in terms of the commodity which serves as standard (Kurz and Salvadori, 1993; Marcuzzo and Rosselli, 1986).

[8] "If what is asserted in the theory ... is to be valid, there must ultimately exist some forces that bring the actual magnitudes towards the levels determined in the theory." (Garegnani, 1992, p. 347)

(LPP). The long-period method (LPM) thus appears to imply, or at least suggest, rational expectations.[9]

"Gravitation," then, is the relationship between market and natural positions. Market variables gravitate toward their natural levels as the market adjusts to new natural positions, to a new long-term composition of demand, for example, or to a changed level of productivity in a basic industry, or an altered price of a basic import. The market reacts to temporary shocks and exogenously caused displacements, for example, to the volatility of investment, by gravitating around its normal position.[10] In any of these cases we can be understood to be referring either to prices and quantities in specific industries, or to prices and the composition and/ or level of output in the economy as a whole. Let us explore the issue by examining a "Fixed Employment" or Craft system, which we will later contrast with a Mass Production economy.[11] In the former, employment is

---

[9] Garegnani (1992, *New Palgrave*) appears to imply that market prices are empirical instances of natural prices, differing from them only by virtue of accidental random effects, which cancel out in the long run. Random effects would have a mean of zero. Natural prices would therefore be the outcome of a deterministic process, with a zero-mean error term, and so would be the rational expectation (see also Parrinello, 1990). Note that this apparently implies that market prices and natural price will be determined by the same systematic forces.

[10] "The short-period outputs and employment are not only amenable to analysis, but Keynes's analysis of the degree of utilization of capacity is obviously of the greatest importance. The question is whether we need nothing else. And once the terms of the question are correctly understood, everybody will agree that we do need something else" (Garegnani, 1992, p. 352). The LPM does not reject short-period analysis, but, on the contrary, claims that such analysis *requires* an LPP to provide the center of gravity for the short-run behavior. An alternative view would claim that short-period analysis needs not a center of gravity but to be set in a fixed or slowly changing context. What is needed is an explanation of *economic structure*.

[11] Parrinello (1990), arguing against the view that gravitation might hold in an earlier period but not in the present, writes, "According to one view the validity of the [LPM] should be assessed relatively to the historical circumstances of the economy ... the method is apt [when] the technical conditions of production are relatively stationary or slowly changing, but [not] in an historical era, like the present one, in which innovative activity is so pervasive. ... [But] the flow of innovations which affect the adjustment of economic variables (in particular the innovations in information technology and in plant flexibility) can also be faster" (p. 123). Innovations change the LPPs, but they also increase the ability to adjust rapidly; gravitation has to take place more frequently, but it can also move faster. There does not seem to be any evidence for this, but in any case it is irrelevant to the points argued here, which concern the changing character of the market adjustment process, from a (weakly stable) price mechanism, to a (locally unstable) quantity adjustment system.

fixed by the establishment of firms, which face two choices: to operate using all hands, or, if prices have fallen below variable costs, to shut down. This represents the process of market adjustment in the conditions of early capitalism, operating Craft technologies. In the latter, employment is easily varied, while keeping running costs constant, in response to fluctuations in current demand.

An intuitive account of convergence

Now let's consider the Craft economy case, in which demand changes for goods that enter into the production of other goods. This means that the new market prices will affect costs – and therefore profits. Market prices will differ from natural, rising above or falling below, according as demand has increased or decreased. But the changes in profits, and therefore the capital gains or losses, which lead to the adjustments in outputs, can no longer be directly calculated. Since the goods enter into the production of each other, the new market prices imply new patterns of costs. A price could be bid up by increased demand, but the prices of the good's means of production might have been bid up even more; hence instead of profit increasing, implying a capital gain and expansion, profit could be squeezed, leading to contraction! The initial adjustment could run in the wrong direction.

But the good whose demand, and therefore market price, has increased the most, relative to its natural price, will necessarily show a capital gain, and will expand. Since its market price has risen the most, relative to its natural price, none of its means of production could increase as much; therefore its revenues must increase more than its costs, and its profit rate must rise. Consequently, when the capital gain is reinvested, this industry will expand. By the same token, the good the ratio of whose market price to natural price is lowest will necessarily experience a capital loss, since none of its means of production could fall as much. So, according to our assumptions, it will contract.

In each case the change – expansion or contraction – will tend to move market prices back towards the natural levels, and will therefore tend to remove the largest causes of changes in costs. In the next period, the same will be true – whatever the variations in costs and therefore in profits among the various goods, the one with the largest and the one with the smallest ratio of market to natural price will adjust in the "correct" direction.

Moreover, we can see that these adjustments will converge. For in the second round, it can be shown that the highest (lowest) ratio of market price to natural price will be less (greater) than the maximum such ratio in the first round – and this will be true in every subsequent round.[12] (This is similar to the convergence argument in Chapter 7.) But the details of this adjustment are important, since we are going to argue that they are not compatible with Mass Production. We had better examine this formally.

A formal model of gravitation

Let us make three assumptions: Short-run supply will be perfectly inelastic – vertical. Long-run supply will be perfectly elastic – horizontal. Average revenue curves will have unitary elasticity. Then we can posit that the ratio of the market price, p_τ to the natural price, p_n, is equal to the ratio between normal demand, q_n, and current supply, q_τ, multiplied by an adjustment factor, k, which allows for some laxity in the assumptions. So:

$$\frac{p_{j\tau}}{p_{jn}} = k \, \frac{q_{jn}}{q_{j\tau}}, \quad \text{where j indicates the industry} \tag{1}$$

This assumption facilitates the algebra enormously.[13] It states that actual

---

[12] We assume (1) the pattern of demand is given; (2) a later market price can only rise from its earlier level, as a result of supply contracting; (3) supply will only contract if there is a *capital loss*, i.e., if the market prices of means of production (in relation to natural) have risen more than the market prices of the product. Finally, all goods use labor (the consumption good).

Consider a good, its market price unaffected in the first round, whose means of production consist only of the "highest" priced good. If its production employed no labor at all, so that all costs rose in the second round to the level of the "highest" first round good, then the capital loss would be proportional to that level, and the reduction of supply (with demand of unitary elasticity) would raise the price of the good being considered to the level of the highest first round price. But all production processes employ labor; hence the capital loss cannot be equiproportional to the increase of the first round highest price. So long as there are positive labor costs, so long as the ratio of means of production to output is less than unity, the cost increase can never create a capital loss sufficient to raise market price to the level of the price of the means of production causing the capital loss. Therefore the highest ratio of market to natural price in the second round must be less than the highest ratio in the first – and similarly for every subsequent round. Successive such adjustments will eventually bring outputs into line with the new composition of demand (see Chapter 7).

[13] Thanks to Enrique Delamonica for suggesting this approach.

market demand equals normal expenditure, adjusted by k.[14] That is, if the three assumptions hold, the system would permit no more than the normal level of expenditure, and market incentives would see to it that actual expenditure will be that amount. But k permits actual and normal expenditure to diverge. This reflects the conditions of early capitalism, operating Craft methods, in which employment and output tend to be fixed; it takes this feature to the extreme in the vertical short-run supply curve. The horizontal long-run supply curve reflects the fact that expansion of the industry occurs simply by replication of optimal-sized firms. It is usual to assume that when there is excess demand, that is, when $q_n/q_\tau > 1$, the market price will rise above the natural price, and vice versa; we are simply requiring that the changes be in the same proportion. Note that if supply and demand are equal, the ratio is unity, and market and natural prices coincide.

We also assume that the ratio between the quantity supplied at a given time and the quantity supplied in the previous period, $q_{\tau+1}/q_\tau$ is governed by the ratio between the market price and the cost of production, including the normal profit. Profits above normal are reinvested, so output expands. Profits below normal require proportional contraction. (Here again deviations from the assumptions are captured by k.) This can be written as

$$\frac{kq_{j\tau+1}}{q_{j\tau}} = \frac{p_{j\tau}}{(1+r)\Sigma_i a_{ji}p_{i\tau} + \Sigma_i w s_{ji}p_{i\tau}} \tag{2}$$

(Here the coefficients, a_{ji} and s_{ji}, and later, k_{sij}, differ from those in Chapter 7 in this way only: to simplify the equations by dropping the t's, they are defined as the coefficients *after* a speed-up has taken place.)

[14] Garegnani, for example, builds his argument on the "classical postulate of given effectual demands" (1990, p. 332), which, he notes, implies that "*aggregate economic activity* ... can be taken as given in analyzing market prices. A first view ... is that deviations of the actual outputs from the respective effectual demands ... will in general broadly compensate each other with respect to their effect on aggregate demand and its determinants (the saving propensities and the level of gross investment.)" Such "compensation" can legitimately be assumed in conditions in which investment and consumption move inversely to one another – early Craft-based capitalism – but not in economies where they characteristically move together, as in Mass Production.

Then from equation (1),

$$
kq_{j\tau+1} = \frac{q_{j\tau}p_{j\tau}}{(1+r)\Sigma_i a_{ji}p_{i\tau} + \Sigma_i w s_{ji}p_{i\tau}}
$$

$$
= \frac{(q_{j\tau}p_{jn}q_{jn})/q_{j\tau}}{(1+r)\Sigma_i a_{ji}p_{i\tau} + \Sigma_i w s_{ji}p_{i\tau}}, \text{ which implies} \tag{3}
$$

$$
\frac{kq_{j\tau+1}}{q_{jn}} = \frac{p_{jn}}{(1+r)\Sigma_i a_{ji}p_{i\tau} + \Sigma_i w s_{ji}p_{i\tau}}
$$

Inverting this expression and using equation (1) again,

$$
\frac{q_{jn}}{kq_{j\tau+1}} = \frac{(1+r)\Sigma_i a_{ji}p_{i\tau} + \Sigma_i w s_{ji}p_{i\tau}}{p_{jn}}, \text{ which implies}
$$

$$
\frac{p_{jn}q_{jn}}{kq_{j\tau+1}} = p_{j\tau+1} = (1+r)\Sigma_i a_{ji}p_{i\tau} + \Sigma_i w s_{ji}p_{i\tau} \tag{4}
$$

Divide through all prices by w, the real wage; then let $p_{ij} = p_{ij}/w$, and $A = (A + S)$. In matrix form this then becomes

$$
\mathbf{p}_{\tau+1} = (1+r)\mathbf{A}\mathbf{p}_\tau \tag{5}
$$

This could be interpreted by saying that capitalists when facing excess demand (supply) adjust quantities in such a way that the prices in the following period cover the cost of this period. This, of course, does not end the matter, because quantities may be changing in all the sectors at the same time. Thus, there are going to be excess demands and supplies in the following period, which will prompt new adjustments.[15]

However, since the market prices depend on the relation between demand and supply the stability of the system may be studied using equation 5. If it converges it means that following a disturbance, caused, for example, by a change in the normal levels of demand, prices have once again reached their natural levels, which means that supply and (the new) demand are equal. This can be seen in equation 1, and also is evident in the way market prices are written in equation 4. There it can be seen that adjustments of the market prices are actually adjustments of supply to demand, because the p_{jn}'s, the natural prices, are constant.

If all the prices are strictly positive, equation 5 is a discrete-time positive linear system. Its stability properties may be analyzed with the help of the following Lyapunov function, where f_m represents the

[15] Observe that this is essentially the same process we discussed in connection with the adjustment of industry sizes in the case of nonuniform productivity increases.

strictly positive left eigenvector of $\mathbf{A}$, corresponding to the maximum eigenvalue of $\mathbf{A}$, namely, λ_m. Since the economy is assumed to be viable, the Perron-Frobenius Theorem implies that λ_m exists, is positive and is less than one:

$$V(\mathbf{p}_\tau) = f_m|\mathbf{p}_\tau - \mathbf{p}_n| \tag{6}$$

Evidently, $V(\ )$ is continuous, positive and has a minimum when the market prices equal the natural prices. Drawing on equation 5 we can write

$$V(\mathbf{p}_{\tau+1}) = f_m|(1+r)\mathbf{A}\mathbf{p}_\tau - (1+r)\mathbf{A}\mathbf{p}_n| \tag{7}$$

Substituting $\lambda_m f_m = f_m \mathbf{A}$:

$$
\begin{aligned}
V(\mathbf{p}_{\tau+1}) &= (1+r)f_m\mathbf{A}|\mathbf{p}_\tau - \mathbf{p}_n| = (1+r)\lambda_m f_m|\mathbf{p}_\tau - \mathbf{p}_n| \\
&= (1+r)\lambda_m V(\mathbf{p}_\tau)
\end{aligned}
\tag{8}
$$

The difference between $p_{\tau+1}$ and p_τ can be made as small as we like by repeated iteration. This expression shows that the system will be stable as long as $(1+r)\lambda_m < 1$. Notice that $1+r$ could be $1/\lambda_m$. In this case the system has a closed orbit. However, if workers receive a wage above the subsistence level, $1+r$ will be less than $1/\lambda_m$ and the system will be stable. That is, after a shift in demand, market prices will converge to natural prices, in the process adjusting the quantities to the new pattern of demand through capital gains and losses.[16]

The same analysis can be used to examine the adjustment of quantities. Drawing on the equations above, we assume that the ratio between the price charged at a given time and the price of the previous period, $p_{\tau+1}/p_\tau$, is governed by the ratio between the current demand and the normal output, including normal growth, adjusted as before by k. Then we can write (using the coefficients defined in Chapter 7)

$$\frac{1}{k}\frac{p_{j\tau+1}}{p_{j\tau}} = \frac{q_{j\tau}}{(1+g)\Sigma_i\, a_{ij}\, q_{i\tau} + \Sigma_i\, ck_{sij}\, q_{i\tau}}, \text{ so} \tag{9}$$

$$\frac{1}{k}\, p_{j\tau+1} = \frac{p_{j\tau}q_{j\tau}}{(1+g)\Sigma_i\, a_{ij}\, q_{i\tau} + \Sigma_i\, ck_{sij}\, q_{i\tau}} \tag{10}$$

---

[16] The shifting of capital implies shifting of labor. But ratios of labor to means of production will differ from industry to industry; for the process to work without a hitch there must be a pool of reserve labor which can be drawn down or increased as capital shifts. Full employment cannot be part of the picture.

But we know that $p_{j\tau}q_{j\tau} = p_{jn}q_{jn}$, which is the monetary constraint above. Hence

$$\frac{1}{k}\frac{p_{j\tau+1}}{p_{jn}} = \frac{q_{jn}}{(1+g)\Sigma_i\, a_{ij}\, q_{i\tau} + \Sigma_i\, ck_{sij}\, q_{i\tau}},\ so \tag{11}$$

$$\frac{p_{jn}}{(1/k)p_{j\tau+1}} = \frac{(1+g)\Sigma_i\, a_{ij}\, q_{i\tau} + \Sigma_i\, ck_{sij}\, q_{i\tau}}{q_{jn}},\ hence \tag{12}$$

$$\frac{p_{jn}q_{jn}}{(1/k)p_{j\tau+1}} = q_{j\tau+1} = (1+g)\Sigma_i\, a_{ij}\, q_{i\tau} + \Sigma_i\, ck_{sij}\, q_{i\tau} \tag{13}$$

Again we can divide through by c, expressing the relative quantities, q_{ij}, in relation to c, while adding the matrices $\mathbf{A}'$ and $\mathbf{K}'_s$ to make $\mathbf{A}'$.

In vector terms, this becomes

$$\mathbf{q}_{\tau+1} = (1+g)\,\mathbf{A}'\mathbf{q}_\tau \tag{14}$$

With a Lyapunov function, and drawing on the Perron-Frobenius Theorem, but this time using the *right* eigenvector, this can be shown to converge.

Now premultiply the price vector adjustment equation by the quantity vector, and the quantity adjustment equation by the price vector:

$$\mathbf{q}_{\tau+1}\mathbf{p}_{\tau+1} = (1+r)\mathbf{q}_{\tau+1}\mathbf{A}\mathbf{p}_\tau$$

$$\mathbf{p}_{\tau+1}\mathbf{q}_{\tau+1} = (1+g)\mathbf{p}_{\tau+1}\mathbf{A}'\mathbf{q}_\tau$$

So long as $r = g$, so that the wage bill of the capital goods sector equals the capital requirements of consumer goods, as $\mathbf{p}_{\tau+1}$ tends to $\mathbf{p}_\tau$ and $\mathbf{q}_{\tau+1}$ tends to $\mathbf{q}_\tau$ these will converge on $\mathbf{qAp} + \mathbf{qSp} = \mathbf{pA'q} + \mathbf{qK'_sp}$.

This simple model illustrates a number of important propositions. First, a clear distinction is drawn between market prices, which depend on supply and demand, and natural prices, which depend on the coefficients and the rate of profits. Second, convergence depends on movements of *both* prices and quantities. Third, *both* prices and quantities are stable; there is no "dual instability" problem, contrary to, for example, Nikaido (1985). Fourth, the model illustrates in strikingly simple form both the "Law of Excess Demand" and the "Law of Excess Profitability" (in the terminology of Flaschel, Franke, and Semmler, 1996). The former states that prices rise whenever demand exceeds normal supply, and vice versa; the latter that quantities rise whenever profits are above normal, and vice versa. Finally, the market mechanism itself *automatically* shifts capital from low to high profit sectors. Firms do not have to make decisions to "enter" or "exit."

Extending the approach

Instead of a change in demand, consider a change in supply conditions, bringing cost changes, disrupting the uniformity of the rate of profit, affecting its level, and causing firms to make adjustments in prices. If the change is permanent, however, a sequence of adjustments, analogous to that for changes in demand, will take place, converging to the new natural prices.

Suppose now that a permanent change takes place in the cost of production of a basic resource, raising its price, say, because a basic import in its production rises. The price increase, of course, restores the rate of profit in the production of that basic, but now costs will rise to all users of the good. Hence users will have to raise their prices, in order to restore their rate of profit, and this, in turn, will create cost increases for other users down the line. In each round the price increases restore the initial rate of profit; in each round, also, the ratio of the new price to the initial price of the product, of the goods with the largest and least such ratio, is less (greater) than the corresponding ratio in the previous round (see Chapter 7). And the reason is the same as in the case of an adjustment to a new pattern of demand: the increase of price in a later round is due to a rise in the price of means of production in an earlier round. But the percentage rise in product price caused by the largest price rise among means of production must be less than this increase of the means of production – since every production process also employs labor. Hence in each round the maximum increase will be less than that in the previous round, and since each round restores the initial rate of profit, the process moves towards reestablishing the initial rate with a new set of prices – and can be brought arbitrarily near this limit by proceeding long enough. (An important application of this extension: Chapter 10 shows how firms in an industry form price policies and investment plans. But one industry's prices will often be another industry's costs. As this shows, the adjustment will converge – unless other effects have been triggered.)

Unfortunately, in Mass Production this will often be the case. There has been both a change and a general rise in money prices; the consequence of the latter is a decline in the real wage. This decline is what has made it possible to maintain the initial rate of profit in the face of a rise in real costs, and it is entirely possible that labor will insist on an increase in money wages – to restore the initial real wage. This, of course, will set off another round of adjustments, as business tries to pass these higher wages along in higher prices. In other words, there is a possibility, perhaps a likelihood, of pressures tending to create a wage-price spiral, which, in turn, may have an impact on aggregate demand. But these issues will have to wait.

Gravitation and historical conditions

Gravitation and Classical price mechanisms

The preceding is designed to capture the essentials of the gravitation process as simply as possible. A number of models have been advanced, attempting to demonstrate that a *general tendency* to converge to long-period positions exists in capitalism, without regard to particular historical conditions. (Asymptotic stability can be shown, but global stability remains elusive.) All appear to be similar to the above model, in that they require that prices be flexible, and that investment adjusts to relative profitability, conditions peculiar to the Craft economy.[17] For example, the models of Garegnani (1990), Kubin (1990), and Dumenil and Levy (1990), all assume that price changes lead to changes in relative sectoral profitability, and that when sectoral profit rates fall below normal, or when one sector's rate is below that of another, capital will be cut back in the low-profit sector, and expanded in the high.

---

[17] Two examples from recent literature: Dumenil and Levy (1987, 1990) provide a model which converges to a uniform profit rate with homothetic growth, provided the starting set of prices is "not too far" away from the final. Although the model is one of long-run dynamics, it rules out technical progress, increasing returns and variability in demand, all of which certainly occur in Mass Production in the long run. There is no accelerator; investment responds only to profit rate differentials – otherwise, it is tacitly assumed that all savings are invested – i.e., anything produced will find a market at some positive price, thereby ruling out Keynesian problems. Prices are adjusted in the light of inventory movements, raised when inventories fall, and lowered when they rise. Prices are therefore demand-sensitive. Yet there is no demonstration that this is either a profit-maximizing strategy, or required by competition. (They address some of these issues in later works.) A more plausible investment function would upset the convergence, as would a different pricing strategy. (Concerned over the exclusion of Keynesian issues, Dumenil and Levy have proposed a revised model, in which questions of "proportion" are distinguished from questions of "dimension," the latter referring to aggregate activity levels. However, their revised model still has prices depending on supply and demand, and technical progress is still excluded. And from a Keynesian point of view it rests on unacceptably restrictive assumptions about investment – the aggregate level is fixed and independent of aggregate demand, capitalists are mainly concerned with allocation of capital between industries, etc. (see Dumenil and Levy, 1990; Cartelier, 1990).

As a second example consider a similar but more elegant model, developed by Kubin. It likewise rests strongly on assumptions that are unacceptable in Mass Production: e.g., a given expected long-term growth rate is assumed; investment responds to current differences in sectoral profit rates that, in turn, reflect current market prices, which are assumed sensitive to output changes.

Perhaps the most carefully developed argument is that of Garegnani. He has presented a detailed examination of the process by which an economy producing the "wrong" outputs, so that market prices deviate from natural, and rates of profit are nonuniform, will adjust back to its natural position. The crucial mechanisms are that lower than normal profit rates lead to an outflow of capital, reducing output, and that a fall in output raises prices, and so raises the sectoral profit rate. Interdependence creates difficulties (Steedman, 1984) which can be avoided by focusing on the industries that respectively have the lowest ratio of market to natural prices, and the highest. These will have below-normal and above-normal profit rates, respectively. The one will therefore contract, and its price rise; the other will expand, and its price fall. Repeated movements will gradually bring the profit rates to converge on the normal rate.

Implications

An important feature of the process modeled here is that the changes in market prices actually bring about the shifts of capital, which in turn adjust output appropriately. The market price changes are not signals or stimuli, to which agents react by optimizing, for example, or exiting and entering industries; they are the adjustment process itself. The market mechanism shifts capital *automatically*, through capital losses and gains. Capital gains are ploughed back, and output rises; capital losses reduce activity and output shrinks. Hence investment adjusts, in good Keynesian fashion, to demand – through flexible prices. But the adjustment process is "regular"; there is neither a multiplier nor an accelerator.

In Craft conditions this makes sense; capital is largely borrowed, so that fixed interest costs must be paid, absorbing normal profits. Depreciation allowances will have to be used to maintain capital, or will be invested in sinking funds. Thus when receipts fall, fixed charges will have to be met, leaving a depleted fund of working capital, which cannot be augmented by borrowing, given that the market outlook is poor. Hence financial constraints will dictate cutbacks of invested capital. Conversely, firms with above average receipts will have profits in excess of their financial charges. They can invest them anywhere, but the business they know best is their own, which is expanding. If there are any economies of scale, they would be foolish to overlook them.

But under modern conditions of Mass Production, this no longer holds. A large part of capital is equity, so that firms face no binding financial constraint. They do not have to cut back; they could use the opportunity of a slack market to retool and reorganize their sales approach. The Craft firm had to adapt to the pattern of demand; the Mass Production corporation can try to influence demand. A fall in current demand need not bring down prices; sales will drop off, but profits will be protected somewhat by the fact that labor can be laid off. Nor need the change necessarily indicate a long-term decline; under conditions of Mass Production, a significant portion of demand is discretionary. Even if it does, Mass Production firms might react by redesigning their product, and/or retargeting their sales. Rather than contract, they might innovate. In short, an imbalance between current demand and supply in a Mass Production industry need not result in changes in prices, and even if indicative of a long-run shift, need not necessarily – in itself – lead to a shift in investment. Gravitation will not be the norm.

The cross-dual model and Craft conditions

Consider what is probably the most widely developed approach – the "cross-dual" model, first introduced by Goodwin (1966). The underlying idea is that prices provide profits and are assumed to depend on quantities, through supply and demand, while quantities are assumed to depend on prices, through the reinvestment of profits.[18]

There are three problems here. First there is a general question concerning the linkages in the causal sequence: prices change because of excess demand, a current market relationship. This changes current profitability, creating profit rate differentials, which in turn leads to changes

---

[18] The basic "cross-dual" model gives rise to limit cycles. That is to say, it shows market prices, wages and profits *orbiting around* the LPM, *not* moving toward it. Such orbits can be quite wide. The system can be made to converge by assuming "demand substitution," but this is unrealistic when techniques are embodied in fixed capital. Further, the assumptions required to assure convergence to a long-period position appear to rule out variations in effective demand, while requiring prices to be demand-sensitive rather than cost-determined (Special Issue, *Political Economy*, 1990). Flaschel, and Flaschel and Semmler (1986 and 1987), following Mas-Collel, have shown that when the reaction functions are augmented by responses to rates of change of key variables, models of cross-dual dynamics (Goodwin, 1967) will exhibit a very general stability, meaning that market prices will both revolve around, and return to, the long-run "natural" prices, and the uniformity of the rate of profit will be restored if upset. The basic "cross-dual" idea is that the change in prices in any industry depends on excess demand for that industry's products, while the change in outputs in any industry depends on that industry's level of profits (usually modeled as the difference between the actual and the normal rates of profit, Semmler, 1990). The augmentation terms then add that prices change more, the greater is the change in excess demand, and outputs change more the greater is the change in profits.

in investment, as capital shifts to the most profitable sectors. But investment is a long-term commitment. Is it reasonable to change long-term commitments, on the basis of current market relationships? Surely not; investment depends not on current, but on *expected future* profitability, to which the current state of the market is a weak and unreliable guide. This objection applies to all cross-dual models.

Second, cross-dual models assume that in the process of adjustment prices are demand-sensitive, although governed Classically in the long run by technology and the rate of profit. This raises a problem of interpretation. If there are constant returns in the long run, why not in the short? But if returns (costs) were constant in the short run, would prices respond to demand changes? Very likely not, as most studies have shown.[19] Transformational growth suggests an answer here, but it limits gravitation to a particular period or set of circumstances.

To stabilize the cross-dual model plausibly, the reaction functions must be "augmented," but there are problems interpreting the system even with the augmented relationships. The basic model treats investment as independent of demand, current or future. Hence it would be possible to have a situation in which excess supply existed, but because profits were positive, investment would still take place. Excess supply would drive down price, but in versions based on neo-Classical relationships, the expansion of output would lower the marginal product of labor, hence lead to a lower real wage. Thus profits would not necessarily contract, and investment would continue! In a Goodwin/Classical model a similar phenomenon can be seen: investment is presumed to expand, because profits have expanded, even though output is expanding faster than consumption. Then when real wages rise, and consumption expands faster than output, investment falls! Under the augmented system, these relationships become even less intuitive – investment expands further the more slowly consumption grows relative to output, and investment falls more, the more rapidly consumption grows relative to output. These relationships, in the augmented form or not, make sense only on the assumption that savings determines investment.

Cross-dual models may be the preferred representation of what has come to be called "the Classical adjustment mechanism." But more

---

[19] Post-Keynesians, for example, argue that prices tend to remain steady, while outputs adjust to demand – and while this gives rise to interesting patterns of dynamic adjustment, such models take the set of prices/markups/profit rates as given at the outset.

generally, Classical adjustment is based on patterns of behavior that require two general and two specific conditions to be met.[20]

The general conditions are

That the level of aggregate demand should be stable, and
That investment should be "regular," as defined below, and should not introduce technical progress.

The specific conditions are

That changes in output in relation to demand should lead to inverse changes in prices, and
That changes in current profits should lead to changes in investment and in output in the same direction.

For example, consider a Craft economy in which prices are responsive to changes in demand, technical progress is sluggish and occasional, and investment, averaged over several years, is governed by changes in the labor force and in per capita income, where the labor force (adjusted for productivity), in turn, also grows at a stable exogenous rate. We can define such investment as "regular"; note that "regular investment" does not necessarily imply full employment, even in the long run. The fact that a moving average of investment is a stable function of growth in the labor force and per capita income merely means that the capital stock expands pari passu; it does not imply that the expansion is large enough to employ all additional workers. A "reserve army" could exist and grow at the same rate. But regular investment does imply a determinate trend, and it excludes multiplier-accelerator effects.

But these conditions are not rationally justifiable in advanced capitalist economies, where we characteristically find short-run constant costs (prices independent of demand), competitively driven technical progress,

---

[20] Kurz and Salvadori (1995), p. 19, write, in regard to "gravitation," that "The discussion ... is based on two main propositions. ... First, the market price depends on the difference between current supply and 'effectual demand,' where the latter is defined as 'the demand of those who are willing to pay the natural price of the commodity' (Smith, 1961, 1:8). If the difference is positive, negative, or zero, then the market price is taken to be lower, higher, or equal to the natural price. Second, the difference between market and natural price triggers movements of capital (and labor) and, as a consequence, adjustments in the composition of production: the output of a commodity increases, (decreases) if the market price is above (below) the natural price. In this view the constellation in which actual outputs equal 'effectual demands' and actual prices are at their natural levels is a center of gravitation. The notion of gravitation and the related concept of the uniform rate of profit were basically adopted by all economists up to the 1920s and were abandoned only later."

market power and Keynesian investment, governed by the multiplier-accelerator. In addition a large part of demand depends on government, rather than the market.[21] As a consequence,

The level and composition of aggregate demand will not be stable from one period to the next

Investment will be irregular and will always both introduce and depend on technical progress

Changes in current output will be independent of changes in current or long-term prices and vice versa

Changes in investment and output will be largely independent of (or depend in a complex and unstable way on) changes in current profits.

But patterns of behavior other than the Classical adjustment mechanism tend to produce unstable or nonconverging results (Caminati, 1990; Boggio, 1990). These may be interesting in themselves, but they cannot justify the idea that the Classical equations represent the position to which a Mass Production economy tends.

This is a strong claim. Suppose that government neutralized the multiplier-accelerator and that relative price changes appeared as relative inflation rates. Then could we find a place for the gravitation process in a modern economy? First, even successful countercyclical policies would not make investment "regular" in the above sense. The economy would be stabilized, but investment behavior, in relation to demand, would not. Second, in the modern world, inflation is normally cost-driven, not demand-driven, since excess capacity is the rule. Neither theory nor evidence suggests that inflation plays any kind of stabilizing role. Finally, under Mass Production, investment and pricing *plans* are determined together, in

[21] A further difference, to be explored later, is that, under modern conditions, best-practice techniques do *not* by themselves govern prices; that is, prices are not forced by competition to the levels implied by lowest-cost methods. Since each firm operates all vintages, prices will tend to the level that will provide the required profit on the *average* level of costs, that is, on the average level of productivity of all vintages still operating, weighted by the normal percentage of output each vintage produces. No firm can force the price down since all continue to operate old equipment, as well as new. (If a newcomer tries to enter, employing only new equipment, it would have to attract customers from existing firms, since they build to match demand growth, and there would be difficulties in doing so even at lower prices, since existing firms have invested in customer relations – and moreover, will adopt strategies to block entry.) Benchmark prices will not be set on the basis of best-practice coefficients. They will earn the profit required by the company's financial and investment plans, on the costs of all vintages kept in operation. (Best-practice productivity, along with wages, will be important in determining the scrapping margin.)

the light of the expected impact of prices on the *growth* of the market. Corporations do not adjust investment or set prices according to the levels of sales; instead they chiefly react to rates of growth; moreover, they also seek to affect the development of demand; they do not necessarily take it, or changes in it, as among the givens (Nell, 1994, ch. 10). In short, successful stabilization would not restore the conditions for gravitation.

It is surely no accident that the LPM began to be abandoned in the 1920s. As noted, the breakdown of the International Gold Standard posed new problems of a dynamic nature, as did the Great Depression. But in fact deeper changes were taking place: the economy was changing its characteristic patterns of adjustment[22] (Sylos-Labini, 1993). As we saw in Chapter 2, in early capitalism, roughly from the Napoleonic Wars to World War I, capitalism adjusted through a *price mechanism*. Prices were flexible in both directions, more flexible than money wages and more flexible than output, which in turn tended to be more flexible than employment. In other words short-term productivity varied procyclically. With wages less flexible than prices and with prices exhibiting a downward trend, long-term productivity gains are distributed via lower prices. Following World War II a different pattern is evident. Prices move in one direction only – upwards; but money wages also only move up – but the changes are larger! Wages rise faster than prices (at least up to the 1980s). Long-term productivity increases are transmitted via higher wages. Short-term productivity variations are much less marked; short-term employment and output are quite procyclical.

Prior to 1914 advanced capitalist economies adjusted by means of a price mechanism. Demand above (below) normal (reflecting fluctuations in Investment or Foreign Trade) drove up (down) prices relative to money wages; the real wage moved inversely to demand, hence to output and employment. But employment was relatively sticky; hence changes in the real wage implied changes in the same direction in consumption. In short consumption and investment tended to move inversely, which kept the overall level of aggregate demand more or less stable (Nell, 1992, ch. 16; Nell and Phillips, 1995; Nell, 1997).

---

[22] Studies have been carried out for the United States, Great Britain, Canada, Germany, Japan, and Argentina, comparing the adjustment patterns of the economy 1870–1914 to the patterns of 1950–90. (Nell, ed. 1997) These appear to confirm that a weakly stabilizing "price mechanism" existed in the earlier period, whereas adjustment tended to follow a multiplier-accelerator pattern in the later. (The "stylized facts" outlined in the text are discussed in detail in Nell and Phillips, 1995, and Nell et al., 1996. The explanation for the working of each of the patterns is given in Nell, 1992, ch. 16, and below in Chapters 9 and 11.)

Under these conditions, gravitation makes good sense. The overall level of demand tends to be stable, increases in one sector being offset by decreases in others. Prices are flexible, employment and output less so. Firms tend to settle at an optimum size and remain there. Growth takes place through the creation of new firms. Earnings are distributed; savings tend to come out of capital income. Windfall profits will tend to be saved and invested. Otherwise investment will tend to be regular.

By contrast, under Mass Production there is no "price mechanism" of the earlier kind, and the economy's adjustment tends to take place through the adaptation of quantities. The level of aggregate demand (capacity utilization) is not stable and changes in one sector, far from being offset in other sectors, tend to generate further changes in the same direction in other parts of the economy. Prices tend only to rise, as do money wages, which rise faster. Output adapts to demand through the multiplier, but output changes in turn influence investment spending through the Capital Stock Adjustment Principle. *The system is inherently unstable.*[23] In particular, this instability is the result of the investment-savings interaction. It is stabilized by government policies and by the large slow-moving government sector. (In the late nineteenth century in advanced economies, government activities, including transfers, accounted for barely 5–10 percent of GNP; in the late twentieth century that has risen to 35–50 percent and higher. Modern governments have developed "built-in stabilizers" and regularly, if often unsuccessfully, undertake counter-cyclical policies.) It should be clear why the LPM is likely to be inappropriate: LPPs are stable, and are reached through a dynamics of price movements interacting with regular investment. Convergent "gravitation" models exhibit the *wrong* dynamics for contemporary conditions.

[23] Minsky advanced this claim in his contribution to the Sraffa Conference, to which Garegnani replied, "Minsky's 'destabilizing' relations cannot but generate fluctuations, and therefore fluctuations on either side of a *trend*, that is, generate those very oscillations whose possibility he seems to be denying" (Garegnani, 1992, p. 348; Bharadwaj and Schefold, 1992). Instability would oscillate around a stable trend, which, however, need not be one of full employment, thus bringing Keynes and the Sraffian interpretation of the Classics together. Minsky replies, "Within the general pattern of expansion implied by accumulation and innovations that lead to technical dynamism, the result of endogenous instability and a structure of constraints and interventions is a path without an equilibrium or a centre of gravity" (1992, pp. 370–71). That, ex post, a time series can be decomposed into a trend and oscillations around the trend is dismissed as "arithmetic not economics." It might be added that we now know that such decompositions are generally not unique (Canova, 1992), so that deciding which is the correct trend and pattern of oscillations calls for assistance from theory. Appealing to the existence of a trend can therefore provide no support for one type of theory – long-period – as opposed to others, e.g., dynamics.

To state the point in an oversimplified, schematic form: the LPM and neo-Ricardian theory, considered as a theory of long-period positions, applies to the Craft economy, and to the early period of capitalism. But later, mass-production-based capitalism is unstable and requires a dynamic analysis. The LPM therefore may or may not be appropriate; the system does not necessarily converge on, or oscillate about, any definite position. There may be a theoretically determinate trend – but, equally, there may not.[24] Keynesian, post-Keynesian and capital-stock adjustment theories – which tend to be fixed-price approaches – are likely to be useful. This should not be taken to rule out many other forms of dynamics, linear and nonlinear – including some in which prices, or some prices, are flexible under various circumstances. The point is simply that the economy is institutionally and technologically different in different eras, and these differences find expression in changes in the patterns of market adjustment. Under Mass Production markets may follow many different kinds of dynamic paths.

Convergence in early capitalist economies

Nevertheless, "convergence to LPPs" is undoubtedly part of the adjustment process in some economies. These adjustment processes will be explored in more detail in the later chapters, but some of the conclusions may be previewed now. Consider a Craft economy in which prices are responsive to changes in demand, technical progress is sluggish and occasional, and investment, averaged over several years, is governed by changes in the labor force and in per capita income, where the labor force (adjusted for productivity), in turn, also grows at a stable exogenous rate. We have defined such investment as "regular"; note that regular investment does not necessarily imply full employment, even in the long run. The fact that a moving average of investment is a stable function of growth in the labor force and per capita income merely means that the capital stock expands pari passu; it does not imply that the expansion is large enough to employ all additional workers. Regular investment implies a determinate trend, and it excludes multiplier-accelerator effects.

But these cannot be excluded simply by fiat. As Garegnani was careful to note, the level of aggregate activity must be taken as given, in analyzing gravitation, since some prices will be rising and others falling. He suggests (as a first approximation) "that deviations of the actual outputs from their respective effectual demands ... will in general broadly

---

[24] What, then, keeps the system from flying off in all directions? Previously fixed commitments and obligations tie down the bulk of economic activity at any given time.

compensate each other with respect to their effect on aggregate demand." It seems that this can only happen, and investment can only be regular, if there are no multiplier effects. For if there were multiplier effects, then each discrepancy between actual output and normal would produce such an effect. If investment depends on income growth, as it surely does, then a multiplier change in income will be likely to affect investment. There is little justification for assuming that such chains of effects will be broadly compensating.[25] (As we shall see in Chapter 11, a condition can be found, which, based on technology and cost behavior, divides economies into those that do and those that do not, have multiplier relationships in aggregate expenditure. This condition also implies that in economies without a multiplier prices will be responsive to demand.)

This suggests that Classical Gravitation is not the only form of adjustment in Craft economies. As technology developed it became more flexible; in the factory system employment could be varied, though still not easily. Output would then become even more flexible. But both variations would continue to depend on prices. The next chapter will study the question by reexamining a version of adjustment by way of prices, which, with poetic license, we will call a "Marshallian" system.

Arguably, modern technologies are vastly more flexible, in that they allow for changing levels of output and employment while keeping productivity and costs constant. So the more advanced the technology, the more likely that short run costs will be constant; hence prices will be insensitive to demand, and a multiplier relationship will exist, so that investment cannot be assumed regular. Convergence is therefore out of the question. But when the technology is inflexible, as in underdeveloped regions or in earlier eras, cost curves will be U-shaped, prices will depend on demand, and fluctuations in autonomous components of aggregate demand will tend to generate offsetting movements in consumption. Investment therefore could be regular (this would not be a post-Keynesian world). Under these conditions (always assuming limited technical progress) convergence might well be plausible – although the inadequacy of the investment theory implied in cross-dual analysis remains a defect in the case.

---

[25] Even if two deviations in opposite directions had exactly the same multiplier effect if one took place before the other, its impact would change income and act on investment first, changing the circumstances in which the other will take place. For example, an expansion might raise income to where it presses on full capacity, so that the full impact cannot be realized. Or a decline might precipitate a slump, and the opposite deviation beginning later while the economy was on the downswing, might be swamped.

The Classical equations and Mass Production

Keynesian investment and "long-period positions"
The chief factors undermining the "long period," considered
as a position towards which the economy is moving, in fact arise from
investment. The movement of the economy toward the LPP cannot be
separated from its movement in the process of accumulation. (Caminati,
1990). This gives rise to a dilemma: if investment were the movement
from the actual, current capital stock to that desired in the stationary state
(or along a steady growth path with only disembodied technical progress),
then it would make sense to consider such investment as part of the
movement towards the LPP, and it would have no disruptive influence.
But such an assumption would also preclude a union of Classical (Sraffian)
theory with Keynes. On the other hand, if investment is conceived in the
Keynesian manner as an ongoing, perpetual activity, driven by competition and grounded in uncertainty, and therefore highly volatile, then its
effects are likely both to be disruptive and to change the desired LPP.
Such investment will be disruptive because it is not tending to a definite
endpoint, and so, from period to period, the direction as well as the
amount of investment may fluctuate.[26] Sectors will tend to expand dis-
proportionately. Moreover, since competitive pressures operate on technical progress, each period's new investment may embody superior
techniques or produce superior products, so that from period to period, the
best-practice position will shift – without, however, moving in a determinate direction. (It cannot be assumed, for example, that costs will
always fall – they could rise, while quality rose more. Yet quality might
well be hard to measure.) It is illegitimate to "abstract" from technical
progress, since it is as much a product of competition – and a part of the
process – as any other economic variable. Once a Keynesian conception
of investment has been adopted (as it will be in Mass Production), it is no

---

[26] Even if a long-period position could be defined, towards which the system could
be shown to move, changes along the way would alter the initial data, so that the
target would move as well. The LPP could turn out to be unattainable in
principle. Robertson foresaw, and attempted to forestall, this objection, writing,
"It may be that the long-run equilibrium is *never* attained. It is the state of affairs
which would be attained if all the forces at work had time to work themselves
out" (1957, pp. 92–3). But this permits nonattainment only as the result of
insufficient time. It cannot be the case that "the forces at work" include forces
which cause the LPP to shift, or to change character. Nor can they include forces
whose working would tend to cause patterns of movement driving the system
away from the LPP.

longer possible to claim that the economy is always moving toward a determinate destination, and the long-period method becomes problematical.[27]

Consider a standard "marginal efficiency" calculation, writing out, now, in year 1, the yields – pq-wn – for each of the years in Project 1's expected lifetime, and discounting the sum of them to equal the supply price. Let one year pass. Now – a year later – the firm will undertake another project, Project 2; but because of technical progress Project 2 will very likely improve in certain ways on the previous year. These improvements may render part or all of Project 1 obsolescent; or they may complement it and improve its efficiency, and/or the quality of its product. Consequently Project 1's yields after the first year must be rewritten, and its MEC recalculated. In the following year the firm will again invest, and Project 3 may compete with or complement either or both of the earlier ones, again requiring rewriting their yields and recalculating their MECs, and so on. It is no longer possible to point to what Garegnani has called the pattern of persistent forces, moving through accidental disruptions toward a definite long-period position, or in the terminology of rational expectations, to deterministic processes with random disturbances. Once firms engage in technical research and invest in expanding and improving their own activities, long-term expectations will shift continually, but in ways that cannot be anticipated reliably.

Indeed Garegnani tries to tame the unruly behavior of investment by arguing that, whatever the pattern of fluctuations, the course of investment "inevitably describes a trend" (1992, p. 347), which it is the aim of theory to explain. This cannot be accepted; a trend can always be constructed ex post, but for the reasons given, in an advanced economy no trend can be established ex ante. Nevertheless his argument provides an important clue, when he refers to observed short-term fluctuations as standing to long-term trends in the same relationship the Classical economists described as holding between market prices and natural prices. His claim is

---

[27] Roncaglia (1990) appears to agree with this when he writes, "whenever the current and normal degree of capacity utilization differ, ... realized profits will be affected, and this will affect financing conditions, which in turn may affect investment expenditure, ... as well as technology both through embodied technical progress and through cumulative 'learning' processes; on the other hand, the current level of investment expenditure will affect aggregate demand, and hence ... current capacity utilization." Roncaglia's concern is to criticize the notion that Sraffa's equations define a "stable or persistent" position representing "an average of booms and slumps," about which the economy in some sense gravitates. This appears consistent with the argument in the text, which claims that Keynesian investment precludes an interpretation of the Classical equations as representing the position toward which the system tends.

questionable,[28] but the natural-market relationship may offer solid ground for theory in the sea of Keynesian uncertainty. Natural prices are explained by technology and distribution; market prices are based on natural prices, but deviate from them in accordance with current conditions. This suggests that there should similarly be two appropriately related accounts of investment, dealing separately with long-term factors, and with the modifications introduced each period by current conditions. The first would explain the investment plan, the second the actual implementation. The account of the plan would lay out the influences determining the construction of capacity, finance, and the pricing policies, in the light of technological innovation and the anticipated growth of the market, while the second would take into account the influence of current sales and

---

[28] At a certain point (1992, pp. 345–7; 1989, pp. 355–7; *New Palgrave*, pp. 23–5), Garegnani appears to advance a different argument, one drawing on a long Classical tradition: that the relation between actual, observed magnitudes and theoretical ones is the same as the relation between market and natural prices, as these were understood by Adam Smith and the Classical economists. This must be considered problematical. Actual, observed magnitudes are empirical reports, based on the rules of observation and evidence. These rules provide the ways of relating the observations to the appropriate theoretical constructions. The first question to be answered is always, are the observations relevant, that is, are they instances of the theoretical concepts? If not, then, of course, they cannot confirm or disconfirm any theoretical propositions. Actual magnitudes are therefore related to theoretical concepts as particular to general – they are instances, usually very imperfect ones, and the rules of evidence are designed to tell us when these imperfections are so great that the observations no longer count. Both the criteria of relevance and the applicable rules of evidence must be determined by methodological reflection, guided by conceptual truths – in this instance, about market behavior and strategy, on the one hand, and the maintenance of structure, on the other, since market prices chiefly reflect behavior, and natural prices structure. "Market price" is itself a theoretical construct, related to "natural price" by a relationship which states that the market price diverges from natural price according to the current ratio of demand to supply. In later treatments, demand and supply are each the subject of theories. The claim that market price diverges from natural as a function of specific factors is a theoretical claim, which can be tested against observation. (In the early Classics, market prices were not so very clearly defined, but even there the rule relating them to natural prices was clear; cf. Roncaglia, 1990.) Natural prices are determined by distribution independently of current demand and supply. Nor need distribution be determined by supply and demand. Thus the relation between the two concepts is not that of particular to general, but rather of a current or "market" theory to a "natural" or capital theory, in which both sets of variables are determined by distinct theories. The degree of divergence of market values from natural will then itself be a theoretical issue. Specific, theoretically defined current variables will determine the divergence of market values from natural. Both "market price" and "natural price" are theoretical concepts, and both must be related to empirical counterparts.

earnings, and current financial conditions on the rate at which investment plans are carried out (Nell, 1992).

This is exactly the kind of theory to be developed in Part V. Investment plans and benchmark prices will be determined together; arbitrage in financial markets will generate a persistent pressure pulling rates of return together; but current investment expenditure will deviate from the plan in the light of current conditions, being influenced in particular, by current interest rates and the level of aggregate demand. While investment plans will be subject to frequent revision, and will not tend to move toward a determinate position, investment spending may be quite determinate in a given short period, although sometimes fluctuating sharply from period to period. Market prices, however, will not deviate that much from benchmark levels, because modern technology produces with approximately constant variable costs. But while at any time there is a long-run plan, there is no long-run position toward which the system is moving; and the plan itself is continually changing, so there is no established trend, either. There is a relationship between current decisions and capital, or long-run decisions; but there is no long-run position to gravitate toward.

Benchmark prices

But none of the arguments rejecting gravitation implies a similar rejection of benchmark prices, corresponding to investment calculations by firms. Benchmark prices, of course, are calculated by firms, but they will hold industry-wide where there is price leadership by a dominant firm, and where there is strong competition, and all firms are similar.

Competition tends to establish benchmark prices as industry prices, for the prices planned by an individual firm must be consistent with its competitive position in the industry. A caveat, however: while benchmark prices are designed to generate the profits needed to build the capacity for expected growth, it may not always be possible to establish these prices in real terms. If they have done their marketing homework business may be able to set the desired money prices in the market for goods, but they may not be able to control the level of money wages. The attempt to establish benchmark prices may end up in a wage-price spiral.

Nevertheless, they are very close to being behavioral counterparts to Classical prices. The rates of return relevant to the firm are those expected, respectively, from new investments and from existing operations. The prices are those not only expected, but planned for – they are the prices the firms plan to charge, in the light of their competitive circumstances, and have good reason to think will lead the market to grow at a rate that will justify the construction of the capacity they plan to build.

Moreover, these prices are based on the average productivity of plant and equipment, which changes slowly as new investment introduces higher productivity equipment. Similarly, invested capacity will shift from line to line only slowly, as depreciation funds are accumulated and reinvested. Hence benchmark prices, though subject to change, will change slowly, and so provide a reliable set of valuations for macroeconomic purposes.

Rather than centers of gravitation, we should consider benchmark prices as guideposts, determined as part of the investment plan, and serving as the standard or target to be reached in marketing. They are "long-run" because they result from capital decisions; market prices are short-run, because they reflect current decisions, which modify the long-run plan in the light of current conditions. But it can be shown that even when demand and other conditions vary considerably, it will still be optimal for firms to keep their prices at the benchmark levels.

Benchmark prices are designed to earn the required rate of return on the average value of the firm's invested capital, the rate required by the firms' financial position and investment targets. (They also reflect the discounted value of the industry's expected normal earnings, summed over the relevant future). This rate states what firms plan to achieve, as they begin operations. They begin from the benchmark, and so long as conditions remain within a certain range, the optimal strategy will be to stick with them. Thus benchmark prices will be the relevant prices for the analysis of the short run and will tend to remain steady as other variables change.

Centers of gravitation are the endpoints of a dynamic process; by contrast, averages of benchmark prices, weighted by vintages, provide a starting point. But as the process unfolds, from short period to short period, the plans and the benchmarks themselves will be modified, often in unpredictable ways, although the changes (based on averages) will normally be small. At any time, however, there will be a plan providing guideposts and setting standards for current decisions. Our argument now will be to show that this plan, and the benchmark prices, will embody the Classical system. First, however, we must clear up issues relating to technical change and choice of technique.

Technical change and the Classical equations
Technical change is far more common and far more regular in the era of Mass Production than in earlier periods. According to proponents of the LPM this should have no effect on how we understand the Classical equations. Technical progress will reduce one or more coefficients, while possibly increasing others. If the technical innovations prove to work as expected in practice the result will be a new dominant technique; that is,

at the prevailing wage (rate of profit) technology embodying the innovations will offer the least-cost methods. Accordingly the LPP will be defined as the prices and rate of profit corresponding to such technology, given the prevailing wage.

This is not unreasonable in an era in which technical changes are comparatively rare. But when large corporations and governments underwrite continual, competitively driven research the picture is not so easy to accept. Technical progress takes place *continuously*. That is, at every moment, labor and other costs are being lowered. That is what we mean by defining a rate of technical progress, and by saying that real wages rise steadily with productivity over each short run. Of course, for some purposes we can neglect this and take productivity and/or real wages as fixed in the short period – the error will only be a couple of percentage points.

But this is *not* allowable in relation to the LPP. Each innovation leads to a new LPP; since innovations are continuous, prices and the rate of profits in the LPP will be continuously changing. Moreover, even though the changes in costs may be small, of the order of 1 or 2 percent during any single short period, the impact on *prices* may be very great. For instance, if any two rows or columns are very close, small changes in coefficients could lead to very large price changes (Semmler, 1984, provides numerical examples). In other cases small changes in coefficients which lead to changes in the ratios of means-of-production to labor ratios, can be sufficient to change the direction in which prices will move with changes in the rate of profit or wages. A small change in productivity could imply that a previously rejected type of fixed capital equipment now dominates that currently in use. Current equipment should therefore be replaced by the new type; the current type should only be kept if it can earn quasi rents. The required revaluations, and consequent price changes, could be dramatic. (Yet the next period might see improvements in the previously dominant technique, restoring the original position.) So ordinary technical progress of the sort which takes place during any short run could lead to large swings in long-period prices from one short period to the next and could also, at the same time, change the pattern of price movements when the wage changes.

Short-period Keynesian and dynamic analyses need to draw on prices, or "aggregators" – value indexes – that remain relatively unchanged from period to period, in order to make comparisons of output and its components. Long-period prices will not be suitable for this if they can fluctuate sharply from one short period to the next.

Not only will long-period prices be useless as a basis for short-period analysis; they will also be unreliable guides to *actual* prices and *actual* profit rates. For actual profits will depend on all the "vintages" of capital in use, embodying technological "vintages." Hence best-practice techniques will overestimate profit. In conditions of Mass Production each firm invests its profits and grows; hence each firm will operate many vintages. No firm will be able to market exclusively on the basis of the current best-practice technology. Hence prices will not be forced to the level of the current best-practice technique; they will be set to ensure that the least efficient vintage will still cover its costs and so will reflect a weighted average of the techniques operating.[29]

Choice of technique

The analysis of choice of technique grew out of the effort to generalize the principle of diminishing marginal returns. Its origin lay in the extension of cultivation to marginal lands, in which a rise in the land-labor ratio led to decline in the returns to land. Marginalist theory sought to extend this principle to the case of changes in the capital-labor ratio, and analogous inverse changes in the rate of return to capital. An increase in the land being cultivated by a given labor force was seen as no different in principle from a rise in the amount of equipment being operated by a given labor force; the same economic relationships were therefore expected to hold. This proved not to be the case; capital is more heterogeneous than land and the measure of its quantity depends on prices – a problem that did not arise with land, although it turns out that the order of fertility cannot be defined independently of prices. But while the inverse relation does not hold, even the critics appear to accept the analogy between changing the land-labor ratio and the capital-labor ratio in other respects.

[29] This is a very different perspective on technologically obsolete equipment than that of Sraffa (1960), p. 78, who holds that "Machines of an obsolete type which are still in use are similar to land in so far as they are employed as means of production although not currently produced. The quasi-rent . . . received for those fixed capital items which, having been in active use in the past, have now been superseded but are worth employing for what they can get, is determined precisely in the same way as the rent of land." Like land they are nonbasics. This picture is justified when firms invest just enough to reach their optimal size, and replace old machines with best practice ones, retaining outmoded equipment only so long as it earns a positive quasi rent. Under these conditions new firms, and firms that have just completed replacement, will operate *only* best-practice equipment and hence will set a competitive standard. Price will *tend* to be driven to the level justified by best-practice costs. But under Mass Production there is no optimal size for firms; all firms reinvest and grow, and all firms operate all viable vintages. Hence no firms price on the basis of the current best-practice technique.

Yet there is a significant difference. When a given labor force is required to work additional land, or is applied to less land (alternatively, more labor is applied to given land), the work being done is still recognizably farming – cultivating the land. The list of *activities* to be performed by labor, together with the list of required skills, remains the same: laborers will till, plough, rake, hoe, weed, fertilize, tend animals, milk cows, slaughter pigs, shear sheep, cut hay, thresh oats, etc. Some activities may be cut back, others intensified, but no new activities will be added or skills required.

In the era of early Craft-based capitalism it may well have been true that changing the labor force in an artisan shop led to a reorganization of work, but did not change the list of activities or the skills called for. Additional workers in Adam Smith's pin factory would have permitted a finer division of labor, and allowed the reduction of set-up time, for example, but would have required no additional skills. After a point, however, further subdivision of tasks would very likely prove inefficient, and create awkwardness. A reduction of the labor force, similarly, would call for a single worker to perform several tasks, increasing set-up time, and perhaps decreasing efficiency. There would likely be a middle point where average efficiency was greatest. But in each case no new activities would be called for by the changes.

The same could not be said for a shift from artisan shops to water- or steam-based factories – or for a shift from water to steam power in an already existing factory. In the first case, a group of artisans, previously working in small shops, or at home under the putting-out system are gathered together into large factories, running on a central power source. There will be first of all a change in mentality; previously workers set their own pace. Now even master craftsmen must work under the supervision of a foreman. The way work norms are set and enforced will be different. Second, some workers must operate the central source of power and manage its distribution to workstations; this will entail new activities, requiring new technical knowledge and skills. In the second example, shifting from water to steam power, operating and maintaining a water wheel requires different technical expertise than running a steam engine, although the systems of distributing power to workstations may not be that different. In each case, activities will be different, different technical prowess will be called on, and work norms cannot be known for certain in advance.

Yet perhaps a simple emendation of the argument might make it possible to use the traditional approach: assign probabilities to the various possible levels of unit labor cost in the new technologies, and then choose the cost-minimizing technique as before. Of course, these probabilities will be purely subjective at first, but as the new systems are tried out, for example, by those with a preference for risk-taking, the numbers can be expected to improve. So it could be argued that there would be nothing different in principle. Techniques would be chosen on the basis of cost-minimizing, and the study of the relations between wages, prices and profits would at the same time be the study of the choice of technique, with different techniques proving competitive at different levels of the wage rate – but not tracing out a well-behaved neo-Classical function.

Unfortunately the labor requirements per unit time – and, indeed, the skills required of labor – cannot be reliably known in advance. They depend on work norms that must be established in practice. The reason is that the way to operate a technology can only be *learned* in practice. It can be planned in advance, but it will be improved and developed only when put into operation. The probability approach is appropriate when the issue is *risk*, but it cannot be used when the difficulty arises because the knowledge is simply not there. It is not that the labor cost might be this or that; until the technique is developed in practice, no one even knows exactly what the various workers will do.

Moreover, the technology will continue to develop; improvements will come with use in many unexpected ways. A technology that has never been put into practice is not on the same footing with one that has – it is a plan, a set of guidelines, but it has no content. It is embryonic, a direction, but not yet a destination. It is not merely that the coefficients are not known, nor that the probabilities will be nothing more than educated guesses. It is that an untried technology will flower in unexpected ways – and also break down at unanticipated junctures. What will happen when it is put into practice can be predicted in outline but cannot be predicted in any detail.

The reasons are easily seen. For any new technology or new product there are possibilities of *internal* and *external* increasing returns as the project is put into operation. These will be difficult to foresee, as a quick survey will indicate. Internal returns – to specific firms – might arise from traditional returns to scale, or from returns to learning that occur naturally over time, or they might come from improved organization of work. External returns – to the industry as a whole – could arise from co-operation among producers, from more appropriate schooling/training of labor and/or management, from wider dissemination of knowledge about the product or technique among the general public, from better transportation and communication services applying to the industry, from the development of support and repair services for the industry or its customers, or finally, from lower prices of inputs, due to economies of scale among suppliers.

Precisely what will happen cannot be foreseen. But it is virtually certain that when a new technology is put into practice *some* pattern of increasing returns will develop. The exact path of development, however, will depend on where, when and how it was put into practice, and later developments along the path will generally depend on earlier – the development of a technology is path-dependent. The choice of technique, then, is not a choice between two well-understood, fully mature processes, analogous to a choice between one kind of hammer and another for a carpentry job; it is a choice between a well-understood existing process, and a set of possibilities which will have to be developed. It is a choice between the known and the unknown, between a job and an adventure, between home and a journey, security and challenge. A new process may very well be more expensive, initially, but hold prospects for becoming vastly cheaper, and perhaps also for operating on a wholly different scale.[30]

Indeed, the typical change of technology leads to many more changes than the theory allows for. According to the "choice of technique" framework, a cost minimizing decision leads to a change in methods of production, in an economy with a fixed list of goods and qualities of labor, with given wage differentials or a uniform wage. The new methods supposedly produce the same goods as before, but more cheaply, at the "switching" prices. This is very implausible; a change in methods may well affect the quality of the product. If fiberglass is substituted for steel, cars will be lighter. Different machinery will have to be used to shape the body parts; workers will have to learn to operate and maintain different kinds of machinery. If steam is substituted for water power, a new power plant will have to be operated and maintained, although it may run the same equipment; when electricity is introduced, not only will there be a new source of power, there will be also be a new kind of transmission and it may be desirable to develop new kinds of equipment. So new industries and products may be added. New jobs calling for new

---

[30] Cigarette rolling machines, when first introduced in the 1870s, were initially more expensive than hand rolling – and less reliable, but within a few years had reduced the unit cost to less than one-hundredth of its earlier level. Moreover a few machines in a few weeks could supply the normal demand of the whole country for a year. As the song makes clear, John Henry could outdo the early coal-mining equipment – at the cost of his life! But even he could not have kept pace with later innovations.

skills will be defined. The choice of technique framework is seriously misleading.

Moreover, a technology in use has many variants, and over time technical progress and learning will reduce the costs in these variants at different rates. Firms will continue to use a variant, even when another is more cost-effective, if they have reason to think that the presently more expensive method will become cheaper in the future – as often happens. Hence the currently least-cost method may not be the true least-cost method. Nor will it always be the case that one single method turns out to be dominant "in the final analysis." One method may be dominant for a period, and then another at a later time; they may alternate until both are displaced, or the product is superseded. Even looking back, it may prove to be difficult and/or arbitrary to calculate which method was "actually" the cheaper.[31]

At any given time the technologies known to an economy can be classified in four categories according to their stage of development. First, there are those that can be considered obsolescent. Better methods exist, either cheaper, or capable of producing higher quality products, or both. But, because they are embodied in plant and equipment, and in the skills of present-day labor, they continue to be used, while being phased out gradually. Then there are the presently mature technologies, working in their prime. Plant and equipment has been fully developed, all "bugs" have been worked out, labor is fully trained. Work norms are clear and products are well defined. These are the current "best-practice" techniques. Third, there are the newly emerging technologies; these are just being built. Neither workers nor engineers fully understand them; innovations are still taking place, bugs are still being worked out, and products and product quality is still developing. These are the technologies of tomorrow, and will displace the mature technologies of today. Finally, on the drawing boards – in the computers – we can find the technologies of the day after tomorrow, the technologies that have yet to be built. Only prototypes exist and the properties are not fully understood. The possibilities can be seen, but even these are changing. Some will not work out at all; others will develop, but in unexpected ways.

---

[31] The calculation will depend on how much the methods differ and over what period each was superior. But later "gains" should be discounted, if we are looking at the decision at the outset. Or should we view it historically? And what rate of discount? Should actual prices be used, or should normal prices? And so on.

Only in the case of the first two could a "choice" of the sort analyzed in theory be made. Only yesterday's and today's technology, technologies actually embodied in equipment and operated by labor under concrete work norms, can be represented by precisely defined coefficients (whether fixed or variable). But the only choice that appears to fit the framework of theory is to scrap the old and build more of the new – but even this requires special analysis, since it could lead to a change in the list of goods (Flaschel and Semmler, 1992). Worse, the firms operating yesterday's and today's technology have committed themselves to *specific paths*. Such commitments may override the abstract choice in terms of least cost.

In short, the choice of technique model presents an important comparison, but its practical relevance is limited by a dilemma: on the one hand, the data required are only available for techniques that have actually been put into operation, but on the other hand, where techniques are already operating, the firms that have adopted them have committed themselves to a path of development – so that decisions are likely to reflect "path-dependency." The ranking of techniques studied by the theory concerns a central question – cost! – but in practice cannot describe behavior. The analysis of the choice of technique is *prescriptive* economics and is of little or no use in descriptive modeling.

Understanding the Classical equations in a modern economy
In a Craft economy, "gravitation" might very well hold, and the Long Period Method would be justified, but short-run average cost curves would be U-shaped, prices flexible, and capacity utilization could not easily be varied. Under Mass Production, capacity utilization is easily varied, and average costs are flat, so prices don't change readily, but "gravitation" is not justified – market adjustment through the multiplier-accelerator is unstable – and the LPM must be reconsidered.

The LPM approach to the Classical equations is sometimes presented (Eatwell, 1983, 1997; Garegnani, 1987, 1992) as a way of studying prices on the basis of three sets of givens – the real wage, the composition of output, and technological possibilities – in contrast to mainstream theory, which takes as its three sets of givens, preferences, endowments and technological possibilities, but studies prices and outputs together. From our perspective each of the mainstream givens is defective: "individual preferences" are the wrong starting point for understanding consumption. "Lifestyles" would be better, as we shall show in Chapter 10. Any "endowments" that importantly constrain behavior will be good candidates for innovation; and we have just examined the problems in the "choice of technique," problems which are common to both approaches. More generally, any theory of prices and outputs based on given circumstances, whether short-run or long, must provide an appropriate foundation for *evolutionary* thinking about the path of development that arises from variation and market selection. This requires a *reproduction* perspective, which mainstream theory lacks but the neo-Ricardian has. Unfortunately the three sets of givens render that approach problematical for Mass Production: the "real wage" is not a given. At best, it emerges as the outcome of the simultaneous setting of prices and wages in labor and product markets; at worst, it is settled – if at all – only as the outcome of a wage-price spiral. The "composition of output" reflects the distribution of income and – special cases apart – the two interact. (This is a different form of interaction than that modeled by neo-Classical "supply and demand," but in general it is not possible to separate the determination of prices from that of output.) Finally, techniques evolve.

In short the LPM may be a reasonable approach for the era of Craft technology, but it is not appropriate for Mass Production. But this leaves one facing the epistemological problem: since market adjustment is unstable in Mass Production, the Classical equations cannot be considered as defining a center of gravitation. If the system does not tend to converge upon the position defined by the Classical equations, how can those equations be justified? The Classical equations are basic to the study of distribution and accumulation; they show us how prices, wages, and profits are related, and the dual equations relate growth, consumption, and relative sectoral sizes. Transformational growth – *any* serious approach to growth – needs these equations.

We have presented a different interpretation. First, the Classical equations are justified by monetary circulation. Circulation – on the theory presented here – rests on the wage bill of the capital goods sector equaling the capital requirements of consumer goods, which in turn is based on the Golden Rule. Further, as we saw, the rate of interest generates pressures that tend to pull profit rates together. This supports the equations developed in Chapter 7. But on *any* theory of circulation, investment spending will generate corresponding realized profits, while wages will support consumption spending. Then prices and relative industry sizes must be compatible with this. The fact that circulation takes place tells us that *some* form of the Classical equations must hold.

Moreover, the Classical equations must be understood as *long-run* (not long-period). That is, they represent consequences of capital decisions. An important post-Keynesian insight is that pricing and investment plans are made together (Wood, 1976; Harcourt, 1982; Eichner, 1976; Nell, 1992; Nell, 1993a). The prices so determined are *benchmark* prices, and they are determined together with the investment needed to keep pace with market growth – and the determining factor is that the markup must be sufficient to provide the profit required to finance that investment.[32] If markets are generally growing at a common rate, on average, which is confidently expected to continue, the rate of profit, on average, will reflect this.[33]

The Classical equations then show the profit rate and prices, in real terms, which correspond to long-run benchmark pricing. In particular, they permit the analysis of the implications of technological interdependence, so that the relationships between real wage costs, prices, and profits can be calculated. Such calculations can be used for comparative statics, for example, to study the implications of wage changes or even of innovations. They can also be used as the starting point for dynamic analyses, and as describing the framework in which short-term effective demand problems are to be studied. But for these purposes the equations

[32] Post-Keynesians typically argue that a long-term rise in the expected growth of markets will lead to a rise in money prices relative to money wages, increasing the rate of profit – the growth rate influences the profit rate. (Also, typically, they fail to examine the adjustments required by the interdependence of industries, cf. Steedman, 1994.) Neo-Ricardians typically argue that a long-term rise in the expected rate of interest will lead to a rise in money prices relative to money wages, increasing the rate of profits – the interest rate influences the profit rate. (Likewise these arguments typically fail to examine interdependence.) Stated in long-run terms, the two lines of argument are closely parallel, but it can be argued that the former is substantially more plausible (Nell, 1988; 1992, ch. 22; 1995). Joan Robinson commented, "Whichever way we [understand the rate of profit], the suggestion that it could be determined by monetary policy seems to be excessively fanciful." Neo-Ricardians complain that it is difficult to define a "long-term expected rate of growth of demand" (but see Chapter 10), yet it is equally or more difficult to make sense of a "long-term expected rate of interest," determined by the Monetary Authorities. By contrast, it is straightforward to argue that the mark-up must be adapted to the former; it is much less clear that it must accommodate the latter (Nell, 1992, ch. 22). On the other hand, given a rate of interest and an active capital market, arbitrage and the shifting of capital through the capital market (*not* entry and exit of firms) will provide pressures tending to pull rates of profit of different companies and conglomerates together. (This does not imply that the rate of interest governs the rate of profits – a divergence between the two will put pressure on the Monetary Authorities which they may not be able to resist.)

[33] Moreover, when demand is strong technological interdependence will tend to pull the growth rates of basic sectors (including necessary wage goods) together, since the growth rate of the system as a whole cannot exceed the rate of the slowest-growing basic (Pasinetti, 1977, pp. 211–12; Nell, 1994). Of course, there may be basic sectors being phased out and replaced by new products, and there may be entirely new sectors emerging, difficult at first to classify as "basic" or "non-basic," e.g., high-tech in its early days, biotech today.

must be interpreted as showing the *average* rate of profits and the average input coefficients, averaged that is, over all the vintages currently in operation, weighted by their normal contributions to output.

This need not necessarily be considered such a radical departure from Sraffian normal practice. Garegnani, for example, suggests that the normal rate of profit can be considered an *average*; if so, then the technical coefficients should also be thought of as averages. What has to be modified is the idea of a *single* dominant technique; instead the rate of profit must be calculated on the basis of the inputs and outputs of *all* equipment and/or techniques still considered worthy of being kept in operation.

Such averages will change slowly, no faster than the rate of growth plus the rate of depreciation, in fact. Such slowly changing averages will be suitable to serve as the coefficients in equations determining normal prices (e.g., for the purposes of short run analyses. That is, such Classical equations can be thought of as the disaggregated barter equivalents to markup equations.)[34] Actual rates of profit, of course, are not uniform, but a uniform rate can be thought of as a simplified ideal form. Such a rate of profit is indeed a "Platonic Idea," but it does not float above historical time. It is firmly anchored in the present by the fact that it rests on technical coefficients defined as averages of the vintages currently in operation.

It may be objected that Classical prices are exchanges designed to make reproduction (and growth) possible. Why would businesses undertake exchanges to carry out replacements for suboptimal techniques? Surely all current exchanges would reflect only the *best* techniqe currently known? Such an objection misunderstands the nature of fixed capital (including "capital" fixed in the long-term training of the labor force). As long as it is fixed, equipment must be operated in the way it was designed, allowing for improvement. Exchanges necessary to replace operating capital – materials, energy, repairs, replacement of parts, etc. – must be carried out on a regular basis, as dictated by the design of the plant, until the time comes to *scrap* that plant and replace it with a newer

---

[34] Given the wage, if labor costs in different industries are marked up in proportion to their respective capital/labor ratios, the resulting profit to capital ratios will be uniform. In other words, given the capital/labor ratios, and the real wage, a rate of profit can be converted into a set of markups for the different industries and vice versa.

vintage (and differently trained labor). The fixed capital itself will be carried at a value which reflects its current productivity (which will normally increase – for a time – through learning by doing). So normal exchanges for replacement *do* reflect the vintages currently in operation.[35]

Finally, the Classical equations are sometimes interpreted in line with a very traditional concept of the long period, namely as the hypothetical fully adjusted position of the economy, "as if" the system were stable and everything had worked out. This might also be considered the "abstract, normal position" of the economy, abstracting from all temporary or accidental influences, and allowing for competitive pressures to fully

---

[35] As noted in Chapter 7 Sraffa's treatment of fixed capital chiefly addresses the issue of normal prices over the lifetime of fixed equipment, and does not consider changing utilization or learning by doing. When the rate of profit is given, the price of an industry's output and the prices of its inputs should be constant. The input coefficients for circulating inputs are given, but the industry uses fixed capital, and while the *amount* is given, its *efficiency* may vary from period to period, e.g., it is likely to decline with age. How then can the price and the rate of profit be the same, in different periods? As we saw Sraffa's answer was to treat the durable good itself as a joint output of the industry each period, along with the normal product, so that the partly depreciated durable good could be assigned a book value reflecting its changing efficiency in such a way as to maintain the uniformity of the rate of profit and the constancy of output price. But this raised a new problem: "equal depreciation quotas [in successive years on machines of constant efficiency] would entail different charges (the charge consisting of depreciation plus profit) on machines of different ages, ⋯ equal depreciation would be inconsistent with equal prices for all units of the product." Sraffa's solution is to increase the annual depreciation quotas on older machines relative to newer ones (bearing in mind that the sum of all depreciation quotas over the lifetime of the machine must equal its original price), thereby restoring the equality of the charge at different ages. However, the depreciation quotas are invested and earn interest equal to the rate of profit. Hence the total earning of the industry's capital – machines and circulating capital *plus* financial investment – is always the same as the profit on the original capital, consisting only of machines and circulating capital. As the machines age the capital shifts from machines to the investment fund (cf. Clark, 1895; Marx, 1967, vol. 2), but the *amount* remains the same. Of course, the investment fund expands, because profits and interest are ploughed back. But this is exactly as it should be: when the machines are scrapped the fund will be sufficient to purchase the larger complement of durable goods needed to keep pace with the general rate of growth.

work themselves out.[36] But for a Mass Production economy this position *cannot* take the methods of production as given, once and for all, such that they can be compared and a set of dominant techniques defined. Technical progress is driven by the same pressures of competition that tend to shift capital about in pursuit of the highest return. The same pressures that tend to establish uniformity in prices, and wage and profit rates, tend also to generate regular technical progress, potentially upsetting any ordering of

[36] This appears to be close to the most recent position taken by Eatwell (1997): "The long-period position is simply the position which the forces of competition will, in any given set of circumstances, tend to establish. ... The simplest version...would result in a tendency to a uniform rate of profit and the associated prices. There is nothing in this conception to lead us to suppose that this long-period position is ever attained. ... There is nothing to make us suppose that it will not itself continuously change as circumstances change. ... There is nothing that decrees that the long-period position should not be in part determined by the process by which it is approached (that is, ... be path-dependent). And there is nothing which requires that the economy should have been in that long-period position ... 'as far back as Adam'" (p. 389).

If behavior depends on expectations (as Joan Robinson assumed) the last two sentences are in conflict. For if the system is not, and has not been, in equilibrium, expectations will tend to form on the basis of its actual, disequilibrium positions, and these will govern its motion. Then if it is path-dependent, the system may head toward another, or a changing long-period position (or perhaps not toward any definite position at all.) Only if it is *already* in the long-period position can we be sure that expectations will be consistent with this position and be such as to return it to equilibrium in the event of a small displacement. This difficulty can be overcome, however, if we understand Eatwell to be saying simply that the LPP is nothing more than the solution for a set of equations describing the economy in the case of a uniform rate of profits. To be sure, this verges on stipulative definition – if a model, any model, models, then we will call whatever it portrays, a "long-period position." He writes, (p. 392), "If we are to be able to talk about any phenomenon whatsoever, abstractions will need to be made and solutions worked out which, if the model is indeed well constructed, will be centres of gravitation. And the only way to assess the impact of changes in any parameters of the model will be to compare alternative solutions." But a closer look reveals that his argument is neither so innocent, nor so tautological: "The competitive tendency to establish normal prices implies that any deviation of the rate of profit from uniformity will set up competitive forces which will tend to re-establish that uniformity" (p. 391). So a solution must *exist*, and the phrase "the position which the forces of competition ... tend to establish" suggests that it must be *unique*. In any case, it definitely has to be *stable*. But stability cannot be presumed in Mass Production. In any case the argument here is that the Classical equations describe neither behavior nor its outcome; they describe how the *structure* must be aligned in order for circulation to ensure successful (expanded) reproduction. In performing that task the Classical equations do indeed focus on "normal" conditions, which they then further abstract and simplify.

techniques. No "separation ... justified by the necessity of ... orderly scientific procedure" can get around this or its damaging consequences. But a set of *slowly changing* coefficients *can* be defined, on the basis of which the Classical equations can be written. With some qualifications these can then be used in traditional ways – comparative statics – but, more importantly, can also be drawn on to define the framework for short-period analyses of effective demand, and to provide a starting point for dynamic analyses.

Method

This, then, suggests an alternative to the LPM. The first step is to establish a structural model, relating prices, wages and the rate of profit, on the one hand, and quantities, consumption and growth on the other. In actual fact, neither profit rates nor growth rates will be uniform. Profit rates will normally lie closer together than growth rates, however, since even industries that are declining – negative growth – maintain a semblance of profitability, while the profit on new products may be kept down to promote their growth. However, this picture can be "stylized" – fading industries can be dropped, and emerging ones entered at what is expected to be their mature size; prices can be adapted to remove the influence of artificial barriers and imperfections. Such a "stylized" structural model can then be used to conduct theoretical investigations of the implications of interdependence for prices and profits, growth and quantities, and can serve as the basis for theoretical studies of dynamics, including gravitation.

Once a structural model is in place many possible patterns of behavior can be considered (Nell and Semmler, 1991). Economic agents – households and firms – can be modeled on the basis of (a stylized representation of) their presently given circumstances and commitments. Short-run Keynesian adjustments might be examined, as might questions of the business cycle and longer-run issues concerning growth dynamics and the processes of technological change. The appropriate "stylization" of the structural model may be different for different questions, such as regards the degree of uniformity assumed in profit and growth rates, or how fading and emerging industries are treated. Moreover, certain features of the model can be expected to be different in different eras, for example, the relative flexibility of prices, and of output and employment.

The resulting behavior in each case will tend to give rise to path-dependent dynamics. An important issue will be arbitrage between different possible sources of earnings for capital, arbitrage not only between different industries, but also between financial and real holdings, and within the latter, between opportunities for profit and opportunities for growth. But it cannot be presumed that such dynamics will be stable, nor in the case of capital shifting, that the movement will tend to establish uniform rates.

Such modeling of behavior within a structural framework can be considered *descriptive*. An alternative approach, useful for different purposes, would be to consider the agents in a fully abstract manner, as if they had no partiular characteristics or commitments, so that choices would in effect be made a completely impersonal and timeless perspective. Such outcomes would be *prescriptive*, that is, they would represent what an ideally rational agent would do, on various assumptions about information, mobility, etc. (Hollis and Nell, 1975). Such outcomes can provide a critical standard by which to judge the path-dependent working of descriptive dynamics, but they cannot illuminate the actual working of the economy. (This is *not* a question of "perfect" versus "imperfect" markets. The contrast is between dynamics in which agents start from given circumstances, and *timeless* constructions in which agents are only studied in equilibrium.)

Thus an alternative method may be proposed, in which Structural Analyses are first conducted, and then Behavioural Options are considered (SA/BO). The LPM with gravitational analysis is one possibility within this framework, although it interprets the Classical equations in too restrictive a manner. Much recent work in dynamic economics, including Keynesian and Kaleckian dynamics, fits squarely into this approach, as does Transformational Growth.

CHAPTER 9

Cycles and growth:
market adjustment in Craft conditions

In both the Artisan era and in the time of Craft-based factories prices tended to be more flexible than output or employment. The cycle showed itself in prices and real wages, which tended to vary inversely to employment. Cycles in the early period of the Craft economy tend to converge toward a normal position, which is periodically disrupted by financial crises, resulting in a Slutsky-Frisch dynamic pattern. But price fluctuations put great pressure on business. By developing greater flexibility in production, output and employment – hence running costs – could be varied in response to sales, making it possible to keep prices stable, thus protecting profits, but leading to a different pattern of market adjustment. But the necessary technological changes – mechanization – lead to changes in the proportions of the sectors, in particular to an outflow of labor from agriculture.

Initially, the ability to vary employment and output was limited, and still costly, requiring price changes to make it worthwhile. However, this ability, even if limited, was in the interests of firms individually. But developing it brought changes in the way markets adjusted. Limited flexibility in adjusting output and employment characterizes the Marshallian adjustment process. Deviations of employment from the "normal" position will increase unit costs.

As we shall see, further development, resulting in Mass Production, greatly increased the ability to vary output and employment, without affecting unit costs, but it made the system as a whole more unstable. The Keynesian system can be seen as the theory of the intermediate formation between Marshallian adjustment and that of Mass Production. Post-Keynesian theory, on the other hand, has addressed the modern Mass Production economy.

Trends and transitions

In each of the periods, AS/TA, CBF/MA, and MP/CA, definite trends can be found in the relationships between sectors. In all three, for example, employment in agriculture declines as a proportion of the labor force; agricultural output/GNP also declines, though not as markedly. In

410

the first two employment in manufacturing rises, but it is steady or declining in the third, and so on. Both declining and rising trends can be described by logistic curves – the process begins slowly, picks up speed, then winds down and draws to a close. In each stage there is a characteristic pattern of such logistic trends, and as the stage develops, as the product or process cycles unfold, risks and rewards will shift, until the risk-reward calculation warrants a transitional investment.

These characteristic patterns are causally linked. In AS/TA, for example, the rise in manufacturing leads to greater concentrations of craft workers, which promotes competitive interaction between them. This leads to better work and to innovation. The increase in manufacturing also allows economies of scale to be realized. Both effects promote increased productivity and superior products. These improved products, in turn, are employed in agriculture, where they raise productivity, which leads to a further outflow of labor. These workers drift to the cities, where, seeking employment, they tend to depress wages. The lower wages encourage firms to risk operating on a larger scale, thereby further enhancing productivity (Nell, 1992, ch. 15). So the two trends, declining employment in agriculture and rising employment in manufacturing, are mutually dependent.

The process begins slowly; workers displaced from agriculture make their way to towns and cities, where they can live more cheaply, by banding together. Economies of scale exist in living arrangements. But the greater concentration of people and industries drives up land values, creating economies of scale in industry. This in turn leads to greater productivity, and gradually to better products, which eventually increase the outflow of labor from agriculture. (We will explore the fluctuations in wages which drive this process later.) As productivity rises in the towns the process will speed up. Thus we have two related logistic curves:

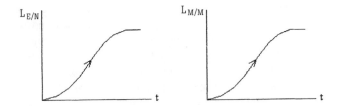

On the left we have the curve showing the outflow from agriculture. L_E is the total exodus from agriculture at any time, N is the total employment in agriculture, and α is the reaction coefficient. On the right we have the rise in employment in manufacturing. L_M is the total labor that has flowed in to manufacturing, M is maximum level that can be supported (e.g., given the requirements of the service sector), and μ is the reaction coefficient. (The notation here is specific to this model.) The equations are

$$\Delta L_E = \alpha_{(\tau)} L_E (N - L_{E(t-1)}), \text{ and}$$

$$\Delta L_M = \mu_{(\tau)} L_M (M - L_{M(t-1)})$$

The time subscript on the final term builds in a delay factor. The subscript τ on the reaction coefficient indicates that the coefficient is time dependent; it changes over time, but not according to the same periodicity as the labor-flow variables. It is used here to indicate a moving average of $\alpha_{(t)}$ over, say, three periods, so the effects will be dampened. The curves are interdependent, according to the following:

$$\alpha_{(t)} = \alpha(L_{M(t-1)}), \alpha' > 0; \quad \mu_{(t)} = \mu(L_{E(t-1)}), \mu' > 0$$

The reaction coefficient in agriculture, indicating the speed at which the outflow of labor rises, will itself rise in proportion to the labor employed in manufacturing, though the effects will be felt only as the moving average rises. Similarly the reaction coefficient in manufacturing will rise in proportion to the outflow of labor from agriculture, though again, the effects will come only through the moving average.

The result is a system of positive feedback. As labor flows out of agriculture and finds employment in manufacturing, the outflow is speeded up; but the faster labor leaves agriculture, the more the expansion of industry is encouraged.

Once the process begins in earnest the need for a central source of energy in manufacturing will be evident. Steam and water power fill this, but water, like wind, depends on location, and also on the weather and the seasons. Steam can be used anytime, anywhere, but adpating it to various conditions requires skilled engineers. With it, however, the factory system can be developed, separating workspace from domestic space, and moving production to a much larger scale. But the risks are commensurably great. As the size of the manufacturing sector rises, however, the risks of sticking with the small-scale system of artisan shops, r_A, will tend to rise, while r_C, the risks of investing in the new large-scale factories, fall, as the technologies improve, and more and more skilled craftsmen become familiar with them. Moreover, the rewards, R_C, of the craft-based factory system will tend to rise as economies of scale, both internal and external, are reaped, while the rewards of the artisan system, R_A, will tend to decline under the pressure of competition from the factories. In sector after sector, when it becomes apparent that $(r_C - r_A) < (R_C - R_A)$, the transition to CBF/MA will begin.

The decline of employment in agriculture and the commensurate rise in manufacturing continue in this period; but within manufacturing a new trend emerges. In manufacturing output and employment both, the share of capital goods rises relative to consumer goods. Machine tools and precision metalworking develop; the increase in the scale of capital goods brings improved alloys, methods for producing interchangeable parts, improved methods for transmission of power (cams, rods, belting, gearing, etc.), larger and more efficient waterworks and steam engines, among other advances. And, although the share of services in total employment remains fairly steady, within services, a related new trend develops: the share of professional services rises relatively to domestic. Professional schools emerge, training managers, engineers, and scientists. Formal education spreads to the whole population, requiring expanded schools and teacher training programs. With the increase of literacy, publishing expands, and with publishing, advertising.

These trends are likewise causally connected. The rise in capital goods, by providing the means to raise the productivity of processes, and to substitute machinery for labor, contributes to increasing the size of firms. Larger firms, in turn, require trained and professional management. Further, to develop capital goods requires workers skilled in draughtsmanship and engineering, again calling for training. Engineering and professional schools develop and begin to do basic scientific research. Technology ceases to rest on the traditional lore of the crafts, and becomes science-based. As it does, the capital goods sector can expand; as capital goods expands, it calls for more professional services. The relationships can be modeled in the same way as before:

$$\Delta Y_{mk} = \kappa_{(\tau)} Y_{mk}(Y_m - Y_{mk(t-1)}), \quad \text{and}$$

$$\Delta L_P = \Pi_{(\tau)} L_P(S_P - L_{P(t-1)})$$

Here Y_{mk} is the fraction of capital goods in manufacturing output, Y_m is the total size of manufacturing output,[1] and κ is the reaction coefficient. L_P is the number of professional service workers, S_P is the total number of service workers, and Π is the reaction coefficient. As before, each reaction coefficient depends positively on the lagged value of the other equation's main variable, and τ indicates that at any period the operative value of the coefficient will be a moving average of its value in the current period and

---

[1] This will, of course, not be constant; but it depends on a variety of other factors, hence is taken as a parameter in this context. The same applies to S_P, which depends on the labor requirements of manufacturing and agriculture, and therefore also on the forces that determine the size of manufacturing.

the previous two. Then,

$$\kappa_{(t)} = \kappa\big(L_{P(t-1)}\big), \ \kappa' > 0; \quad \Pi_{(t)} = \Pi\big(Y_{mk(t-1)}\big), \ \Pi' > 0$$

Again we have a positive feedback system. As the output of capital goods rises in manufacturing, the growth of professional services will accelerate; the larger the size of professional services in relation to total services, the faster will be the growth of capital goods.

Innovation: Craft to Mass Production

An industrial factory, even a craft-based one, may produce a sizeable fraction of what the total market can absorb and may put at risk a considerable fraction of the total capital in the trade. A later innovation could render the factory obsolescent before the investment has been re-couped. The new technology on which the new plant is based may contain "bugs," which will cause costly delays and may alienate customers. Later firms may then benefit from the bug extermination efforts of early innovators. The market may not develop as rapidly or in the directions initially forecast.

On the other hand the rewards for well-chosen innovation are also great. If the plant is successful the firm will not only earn superprofits, but will also develop a dominant market position, from which it can continue to expand on the basis of low and falling unit costs.

The decision to innovate, moving from one kind of technology to another, thus depends on a balancing of such risks and potential rewards. In terms of cost alone, the new technology may be clearly superior to existing techniques; once established, unit production costs will be lower, perhaps even dramatically lower. But because of the scale on which it must be introduced, and in view of the possible initial difficulties, business may be loathe to take the step. Indeed, there may be a barrier to anyone taking the first step. Consider two firms, A and B, contemplating an investment in a new technology:

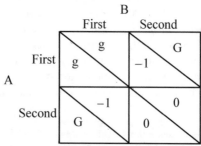

If both go first, assuming the market is large enough for both, both will make a reasonable gain. But if A goes first, and B waits and builds later, then B makes a large gain (learning from A's initial troubles), while A makes a loss, losing out to B's superior plant. Similarly, if B builds First, A will reap all the benefits and B stands to lose. But if both wait, neither will gain anything. Yet, clearly, to wait and invest Second is the dominant strategy.

But with a manifestly superior technology waiting in the wings something has to give way. One answer is to form cartels; everyone will go in together. Another is for government to assume the risks and guarantee profits for those who take the plunge.

More generally, a firm considering whether to invest in a new technology or to expand further using the old, must balance the risks of the new compared to those of the old against the rewards from the new compared to the earnings of the old. Let subscripts C and M indicate the old and the new, r will designate risks, and R rewards. Since the old method is tried and true, r_C simply depends on the likelihood of investment in the new technique. Similarly R_C will be the normal rate of return in the industry in question. For the new technology, however, r_M will be greater the newer and more untried the technology, the larger the scale required, and the larger the number of competitors adopting it, while it will be less the higher the anticipated rate of growth of the market. The potential rewards, R_M, will be greater the lower the unit costs with the new technology, the larger the market share, and the faster the growth of the market, while they will be lower the greater the competition.

As the process takes off, the risks associated with Mass Production, r_M, will fall, while those of Craft-based factories, r_C, will rise; similarly the rewards from investing in Mass Production, R_M, will rise, while those resulting from Craft production, R_C, will tend to decline. Taking the risks first, as science and the professional services advance, both technology and management will become more reliable, reducing the risks associated with innovation. By the same token, and, in addition, with the output of capital goods increasing (and because of economies of scale, improving), the danger of continuing to rely on the older methods – the danger that competitors will adopt the new – will increase. As for the rewards, as science, technology, and management improve, the new methods will become more and more cost-reducing. Hence, when $(r_M - r_C) < (R_M - R_C)$ it will pay to begin to invest in Mass Production, and the transition will begin.

Each of these transitions leads to changes in the pattern of market adjustment. In AS/TA market adjustment is based on "gravitation," in CBF/MA, it is based on Marshallian marginal productivity, which, at the point of transition to Mass Production, exhibits Keynesian features.

Finally, in MP/CA we have a fully operative cycle in employment and output, driven by fluctuations in investment spending.

The early cycle in the Craft economy

We have seen that when aggregate demand varies, because of volatile investment spending, prices will fluctuate, leading consumption to vary inversely with investment. In the extreme case, as well shall see in Chapter 11, the elasticity of consumption with respect to variations in investment will be -1, which implies that

$$\left[\frac{dC}{C}\right]_{(t)} = -\left[\frac{dI}{I}\right]_{(t)}$$

This can be termed the "elasticity condition," and it is achieved within the same period.

Investment, however, must keep pace with changes in consumption. The capital stock must be adjusted to the requirements for output, which in turn depends on demand. If consumption demand rises, for example (driving up prices and raising profits), capital in the consumer goods sector must be increased, which requires that capacity in the capital goods sector in turn be expanded (raising prices and profits there, too). If this growth is expected to be permanent, then the capital stock will have to expand continually, hence investment will have to grow at this rate. Investment will therefore grow in pace with the growth of consumption:

$$\left[\frac{dI}{I}\right]_{(t+1)} = \left[\frac{dC}{C}\right]_{(t)}$$

This can be termed the "profit-investment gravitation condition," and, like the elasticity condition, it is obviously a great oversimplification. The implication that the rate of growth of *investment* will adjust *pari passu* is clearly extreme. The equality of dI/I and dK/K is a condition for balanced growth, which may seem out of place in a cycle analysis. But in a Craft economy capacity will tend to be fully utilized; hence investment, as the output of the capital goods sector, will tend to vary with the capacity of that sector, which in turn will vary with the growth of the capital stock. The capital stock of the two sectors will tend to expand in tandem, governed by the normal relationship between them. Thus dI/dK = I/K. But the new output of investment goods must follow the previous adjustment of the capital stock; hence the one-period lag.

These two equations give rise to a very simple cycle. Plot dC/C on the vertical axis, positive and negative, and dI/I on the horizontal. The origin will mark, not zero, but the normal rate of growth, the rate at which prices are at normal levels. Thus a negative rate of growth of consumption or investment signifies a deviation below the normal level, a positive a deviation above. Now draw in the two equations. The profit-investment condition rises from left to right at a 45-degree angle, splitting both the negative and positive quadrants. The elasticity condition similarly bisects the other two quadrants, falling from left to right.

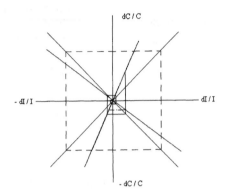

Start from an autonomously given positive (that is, above normal) rate of growth, dI/I. This will drive up prices, reducing the real wage, and lowering consumption, leading to a negative (below normal) dC/C, in the lower right quadrant. But a negative dC/C, expected now to be permanent, calls for a reduction in the growth of capital, and hence in investment. (This is *very* implausible, but we will see the point in a moment.) The new, negative rate of investment growth then leads to a fall in prices, raising real wages, and so bringing a rise in the growth of consumption. This in turn calls forth a rise in investment, restoring the original level of dI/I. The cycle swings around indefinitely, so that prices rise and fall (real wages fall and rise), causing investment and consumption to fluctuate in an offsetting pattern.

Both equations represent extreme cases, but the profit-investment gravitation is particularly unrealistic. When dC/C turns negative, the prices of investment goods will decline, and profits will collapse. This will pull investment down; the rate of growth of *capital* will fall, but it clearly need not fall to the exact level of the rate of growth of consumption. Suppose, instead, that the equation read

$$\left(\frac{dI}{I}\right)_{(t+1)} = a\left(\frac{dC}{C}\right)_{(t)}, 0 < a < 1$$

Then, as can be seen from the dotted line for the profit-investment condition in the diagram (drawn so that $a = 1/2$), the process will tend to converge to the origin. If, further, the elasticity condition were also more realistic, so that

$$\left(\frac{dC}{C}\right)_{(t)} = -\ell\left(\frac{dI}{I}\right)_{(t)}, 0 < \ell < 1$$

the process would converge even faster, as can be seen from the dotted elasticity line, in conjunction with the revised gravitation condition. (The coefficients are in a different script, to distinguish them from other a's and b's.) Analytically, in the first case,

$$\left(\frac{dI}{I}\right)_{(t+1)} = -\left(\frac{dI}{I}\right)_{(t)}, \text{which oscillates, while in the second case}$$

$$\left(\frac{dI}{I}\right)_{(t+1)} = -a\ell\left(\frac{dI}{I}\right)_{(t)}, \text{where } 0 < a\ell < 1, \text{which converges}$$

In short, revising the equations to make them more realistic shows that the process tends to gravitate towards the normal position. This is therefore a "short" cycle, relevant to gravitation. The Craft economy tends to be stabilized by its price mechanism, so long as its financial system does not upset the applecart. But the interaction of the price system with an imperfect banking sector creates a mechanism for cycles: optimism leading to overissue followed by credit collapse provides "exogenous shocks" that set off convergent fluctuations. As the system settles into equilibrium, optimism in the banking sector begins to rise again. The Craft economy, therefore, is a stable "real" system, destabilized by its financial sector, resulting in Slutsky-Frisch dynamics.

Growth and the labor market in early Craft conditions

Having related the Classical equations to money, and in the process, established the wage-profit trade-off, we will now draw on the latter to examine the determination of money wages, through wage-bidding in a growing economy. In a system with metallic money, or convertible paper, this tends to determine the real wage. That is the implication of our discussion of the Golden Rule.[2] Wage-bidding, in connection with

[2] To repeat (it's important!): When the money wage is fixed in a metal standard that is itself produced (or in paper tied to such a standard) a rise in the money wage is equivalent to a rise in the *real* wage in the metal-producing industry. (The output is metal, the inputs are *valued* in metal, and the wage is fixed in metal.) Competition requires that all industries conform to one another; hence wages and profits in other industries must adjust to those in the industry producing the standard. The general rise in the wage bill implies an exactly offsetting decline in the payments of profit, permitting circulation to carry on as before.

growth, will be shown to determine the real wage, and therefore the rate of profits, and hence normal prices. This is the long-run theory of the Craft economy – and we will see that it has as its correlate, a long-term cycle.

In conditions of labor shortage, when capital is growing faster than the labor force, money wages will be bid up. The wage-profit trade-off provides us a way of determining how much wages will increase; similarly in conditions of labor surplus, when the labor force grows faster than capital, we can determine how much wages will be forced down. But this analysis will only hold for Craft-based production. In fact, it leads to systematic pressures to mechanize, which, in turn, tend to generate a cyclical outflow of labor from agriculture. As this takes place, small-scale craft production is replaced by large-scale factories, which, however, still employ technologies based on the traditional skills and methods. But markets work differently in a number of ways, explored here as "Marshallian adjustment."

Suppose, however, that we consider the presently installed and operating technology, with a given real wage and rate of profit, leading to a corresponding pattern of consumption and investment, and consider what happens in a Craft economy when market pressures lead to a change in the money wage rate. Further, let us suppose that these pressures lead to such a change over a considerable period of time, during which prices may vary over a normal range. We shall assume at first that the technology is Craft-based, so that prices are more flexible than money wages. So we may take the average of their movement over the normal range as the normal price, and, in effect, consider prices to be fixed. Hence a change in the money wage will change the real wage. This will lead to a stable or cyclical adjustment. In later chapters we will consider price changes, and we will see that labor markets in Mass Production do not adjust in a stabilizing manner.

Market pressures on the money wage arise, in Craft economies, from the ratio of the labor force to the number of places offered, including self-employment. The number of places depends on the number of establishments, and their average size. Investment will create new establishments, which we can assume, like existing ones, to be of optimal size, where this depends on the technology and the market they serve. Starting from a point of initial adjustment, investment must take place at a rate sufficient to create new positions at a rate equal to the rate of growth of the labor force, where this depends on population growth and immigration/emigration. The technology will be that already in place, assumed to be given – new investment will embody the same technology being operated by existing firms. Scrapping of obsolete technology will be ignored for the moment. Any changes in technology will be determined in the course of the analysis.

The argument will proceed in several stages. First, if the growth rate of capital exceeds the growth rate of the labor force, money wages will be bid up; if it sinks below that of labor, money wages will be driven down. If the two are equal, money wages will remain stable. Drawing on the wage-profit trade-off, we will be able to analyze the extent of the effect on money wages of a shortage or surplus of labor. Second, the change in money wages has effects on both the growth of capital and the growth of the labor supply.

Take the growth of capital first: A higher or lower money wage, with the price level given, will lower or raise the profit rate, and therefore the growth rate – provided that the effects on investment are symmetrical with those on saving! The qualification that the price level is "given" will also need to be examined. But the argument will be that the movement is in the correct direction for adjustment.

When employment is strictly fixed, even though output may be variable, when capital moves from one sector to another, there will be an impact on the size of the reserve of unemployed labor, owing to the fact that capital/labor ratios vary. When the wage is rising there will be a tendency for demand to shift to consumption goods, which, at least in the early stages of capitalism, might be expected to have relatively low capital/labor ratios. Higher wages will therefore tend to lead to higher employment, reducing the reserves of unemployed workers, and thus leading in the direction of increasing the pressure to push up wages still further. Falling wages and rising profits, on the other hand, will tend to shift demand to capital intensive goods, reducing employment, raising the unemployed reserves, and intensifying the downward pressure on wages.

As for the labor supply, the Classics argued that a rise or fall of the real wage would lead to a fall or rise in the marriage age, resulting in an increase or decrease in births, and so to a rise or fall in the net reproduction rate. A decline in real wages probably does lead to a rise in the marriage age and a decline in births – the Irish example is a good one – but a rise in real wages does not lead to a rise in births. Historically, the Iron Law of Wages has proved wrong; rising real wages has led to a decline in population growth. However, this does not mean that the labor supply has not risen; the fewer new workers produced by smaller but more prosperous families have been healthier and better educated, so may well count as an increase in labor supply, measured in efficiency units. There may also be effects on willingness to work. But higher wages may call forth additional workers, or since incomes are higher, fewer. Lower wage rates may lead to a rise or a fall in labor force participation. It depends on other circumstances – income distribution, the kinds of jobs available, social factors. Finally, a change in real wage rates may influence decisions to move to other geographical areas – but levels of wealth and poverty, income distribution, and unemployment may matter as much or more, and here again, cultural and social factors will be highly significant (Hobsbawm, 1995; Julca, 1997). In short, changes in real wage rates may lead to changes in the number or quality of workers households produce, and may lead to changes in the amount of labor households will offer. They probably do encourage immigration and emigration. But so many other factors, both economic and noneconomic, are involved that it is unlikely that any general relationships can be established.

A special case: r = w

Nevertheless it may be interesting to explore a special case. The set of subsistence goods has been defined as that set of goods which will support the labor force, permitting it to exactly reproduce itself, for the unit period of production. A positive net wage rate means that the set of wage goods is, say, 110 percent of that needed to support an exactly reproducing labor force. It could therefore be interpreted as a wage permitting the labor force to *expand* by 10 percent at the "normal standard of living." Alternatively, it could mean that the normal labor force could be made 10 percent better educated, producing a labor force capable of correspondingly greater productivity. Neither of these interpretations may be very plausible, but both have roots in the history of economic thought, the first in the Classical, the second in the "human capital" literature (and hints of both may be found in Walras, 1954, Lessons 17 and 18).

So let us provisionally assume that the real net wage rate can be interpreted as the growth rate of the labor force. In that case, since we shall assume that r = g, equilibrium between the growth of the capital stock and the expansion of the labor force, will require that r = w. This can be shown on a diagram, marking off the rate of profit on the vertical axis and the wage on the horizontal, with the respective maxima, R and w_m as indicated. Here it is clear that

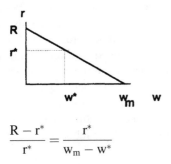

$$\frac{R - r^*}{r^*} = \frac{r^*}{w_m - w^*}$$

which implies that

$$r^* = w^* = \frac{Rw_m}{R + w_m}$$

This says that the rate of expansion at which the growth of the labor force and the growth of the capital stock will just balance is equal to the product of the maximum rate of profit and the maximum wage, divided by their sum. This formula, of course, could easily be adjusted for the case where the rate of growth of the labor force was a function of the net wage rate, rather than equal to it.

In actual fact, however, the net wage bears neither a simple nor a stable relationship to the growth of the labor force. Even more important, labor and capital need not, and normally do not, grow in tandem. Far from it; the increased labor force needed for industrial expansion has been largely drawn from agriculture, and this does depend in part on the net wage.

When the real wage rises, demand for consumer goods increases. This tends to raise agricultural prices and profits, encouraging mechanization. The mechanization of agriculture displaces rural labor. (This could be reinforced by a rising demand for textiles in urban areas, resulting from higher wages. An increased demand for woolens would encourage enclosures and sheep farming, displacing workers.) Thus a rise in money wages, price level constant, would tend to increase the labor supply for industry. A decline, however, would not lead to a shift back to agriculture.

Balanced growth is neither plausible nor interesting. But a divergence between the growth of labor supply and growth of capital may lead to changes in the money wage, with the price level constant, where these changes in turn, may tend to adjust both the growth of capital and the growth of the labor force in a manner that will lead to further expansion.

Market adjustments of the "real" money wage

In important respects the effects of labor shortage and labor surplus may be symmetrical in a Craft-based economy. In each case competition will lead to changes in the money payments to labor, what we have called "real" money wages. When there is a surplus of labor, for example, employers will tend to, or threaten to, replace presently employed workers with presently unemployed workers, who will be willing to work for lower wages rather than remaining unemployed. When there is a labor shortage, workers, particularly skilled workers, will move or threaten to move to firms willing to pay higher wages rather than carry underutilized capacity. (In each case, the first action of the market will tend to be on skilled workers, with the rest of the labor market tending to follow along, maintaining normal differentials.)

Consider a labor surplus first. The total labor force is given, and can be taken as unity. Then the percentage that can be employed by the capital stock – jobholders – will be j, and the unemployed percentage, u, where $j + u = 1, 0 < u < j < 1$, and $1/j > u/j > u$. In these circumstances, consider, in the abstract, the calculation whether a worker should accept a wage cut or not. Initially, we may suppose that all jobs are equally open. In fact, of course, some workers will have close ties to some employers, who will not wish to offend them or risk losing them. This will be considered in a moment. But if all jobs are equally open, then the probability for any worker of obtaining employment is j. Let w be the initial money wage, and $-\Delta w$ be the proposed cut in the wage. (Because this money wage is also a real wage, it is not written in italics.) A worker must decide whether to accept or reject the proposed wage cut; each worker's decision can be considered successively in relation to those of others.

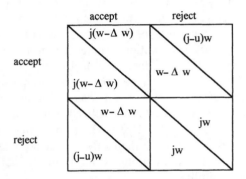

If Others Reject, and the given worker, Worker A, also Rejects, each will have a probability of earning equal to jw. But if Others Reject and Worker A Accepts, then Worker A's probability of employment will increase to unity, but the earnings will be reduced to $w - \Delta w$. Others will hold out for w; their chance of obtaining this, at best, will be j. However, the percentage of workers who do not receive job offers at w, u, will certainly accept a cut. Hence this number must be subtracted from j, so that the chance of obtaining w will be $j - u$. If Others Accept, and given workers, taken in succession, choose Reject, holding out for w, the workers' chances will be j reduced by u. But if given Workers Reject, Others, choosing to Accept, will appear individually to be certain of employment, so will expect to earn $w - \Delta w$. If both Accept, there are no holdouts, so the probability of employment is the same for both, j, and the earning will be the reduced wage, $w - \Delta w$.

Provided two conditions on the reduction of the wage are met, it will be advantageous for both sides to choose to Accept, in spite of the fact that both would be better off if they chose to Reject. The conditions are

$$j(w - \Delta w) > (j - u)w, \text{ which holds iff } \frac{u}{j} > \frac{\Delta w}{w}$$

$$w - \Delta w > jw, \text{ which holds iff } \frac{w - \Delta w}{w} > j, \text{ i.e } \frac{\Delta w}{w} < 1 - j = u$$

But we know that $u/j > u$, so that if the second is met, the first will also be satisfied. The percentage wage cut must be less than the percentage of the labor force that is unemployed. If this is the case, it will be abstractly rational for workers seeking the largest expected earnings to accept the cut.

Before turning to an analysis of ways to resist or minimize wage cuts, we can develop an analogous account of firms facing demand for wage increases in conditions of labor shortage. Once again we will consider a given firm confronting its rivals, in conditions where each firm must make its calculations individually. Total productive capacity can be taken as unity; if the entire labor force works, percentage J of capacity, and therefore of potential output, and with a given wage, of profits, will be realized. Percentage U of capacity will be unutilized, where, as before, $J + U = 1, 0 > U > J > 1$, and $1/J > U/J > U$. Firms must decide whether to Bid or Stick, that is to raise their offer of money wages, or hold the line. Raising wages lowers profits from R to $R - \Delta R$. The strategy matrix looks very similar.

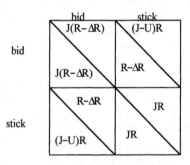

It would be best for both sides if both chose to Stick. But choosing strategies individually, each will find it optimal to choose to Bid for labor, at the expense of reducing profits, provided the wage-bidding reduces profits subject to the conditions:

$$J(R - \Delta R) > (J - U)R, \text{which will hold iff } \frac{\Delta R}{R} < \frac{U}{J}$$

$$(R - \Delta R) > JR, \text{which will hold iff } \frac{\Delta R}{R} < U$$

As before $U/J > U$ so the first condition will be met if the second is. So the proportional decline in profits due to bidding up wages must be less than the ratio of unutilized capacity to total capacity. If this is the case, then it will be optimal for firms to bid against each other to attract or keep labor.

But the impact of both labor surplus and labor shortage may be prevented from reaching its maximum level. Labor can resist the effects of labor surplus, and firms can act to offset the pressures of labor shortage. In each case the analysis draws on the wage-profit trade-off.

Take labor surplus first. Workers can resist wage cuts by slowing down production, "working to rule," and indulging in minor acts of sabotage.[3] Such behavior will shift the wage-profit trade-off inward, by raising labor time per unit of output; that is, it will reduce productivity and raise the labor-capital ratio, thereby reducing the rate of profit for any level of the wage. (In the same way, firms paying higher wages, under conditions of labor shortage, can intensify work norms, and increase pressure for greater productivity, shifting the trade-off outward, in the same way, reducing the labor-capital ratio.) At the wage w^* an inward shift of the wage-profit trade-off reduces the rate of profit by Δr.

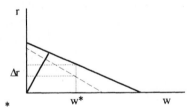

Under these conditions we can calculate the wage cut that business will try to enforce. A wage cut raises the rate of profit by an amount given by the slope of the wage-profit trade-off. But faced with a wage cut, workers will resist and perhaps retaliate by reducing productivity. For small wage cuts this might be almost insignificant, but as the wage cut becomes larger, worker resistance can be expected to intensify. This will be particularly the case when the cuts begin to interfere with normal living. This can affect workers' health and state of mind, leading to inattention

---

[3] The word comes from "sabot," a wooden shoe, which workers allowed to fall into the machinery, causing damage.

and lower morale. Hence with large cuts, productivity can be expected to fall off significantly.

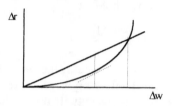

Here the curved line shows *decline* in the rate of profit due to productivity reducing resistance on the part of labor, shifting the wage-profit line inward, as a function of the cut in the wage, while the straight line indicates the *rise* in the rate of profit resulting from a wage cut, moving along the wage-profit trade-off. At the point where the rising curve cuts the line the gain in the profit rate due to lower wages is wholly offset by the loss due to lower productivity. At the lower level of the wage cut, where the tangent to the productivity-reduction curve is parallel to the wage-profit line, the increase in profitability is at a maximum. At this point the marginal loss in profit due to lower productivity just equals the marginal gain due to a lower money wage.

Labor shortage is different in the important aspect that firms bidding against one another may not in fact always be able to increase the number of workers. Higher wages may encourage immigration; it may lead to an increase in participation, and most commonly it can lead to an outlflow from agriculture. But all of these require institutional changes, and are therefore time-consuming. Hence, while firms will bid to try to get more workers, they may in fact find themselves frustrated for long periods. For any given firm, however, it will believe that it can attract workers by increasing wages, that is, reducing profits. Initially, a small wage bid may be expected to attract numerous, loosely attached or "floating" workers; but then it will take progressively larger wage bids to attract further workers, these being bound to their jobs by ties of loyalty and habit. Thus at first, the increase in utilization from the expected extra workers will raise profits more than the higher wages will reduce them. But at a certain point, the additional workers do not add enough to offset the additional wage cost.

In the same way, in conditions of labor shortage firms will try to benefit from the fact that they must pay higher wages. They can use this as an occasion to intensify work norms, bringing pressure to work harder or faster, in ways that labor will be less inclined to resist precisely because it is being paid more. Initially, as wages are raised, it will be comparatively easy to increase productivity, but as the wage increase becomes larger, it will become progressively harder to raise productivity further. This can be seen on a diagram resembling the one examined earlier.

At the point where the curved line cuts the rising straight line, the fall in profits due to the rise in the wage has fully offset the increase in profits due to the effect of higher wages in encouraging higher productivity. At the point where the tangent to the productivity curve is just parallel to the wage-profit line, the increase in the rate of profit is at a maximum. At this point, the additional increase in profits due to the higher productivity brought about by a further rise in the wage just equals the marginal decrease in profits due to that further addition to wages.

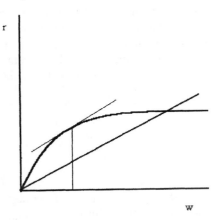

The price level

In Craft-based economies, prices are more flexible than money wages, because employment is fixed. But the fluctuations in prices revolve around a central value, which in each market depends on the value of the good in relation to the value of money. The general level of prices is the inverse of the value of money, itself a produced output, whose quantity is adjusted in circulation to maintain a constant value. Hence in a Craft economy with commodity money, changes in money wages imply corresponding changes in real wages in the long run, that is, after adjustments have settled down.

Machinery and the Golden Rule

When a speed-up of production takes place the wage-profit trade-off flattens, indicating that the labor-capital ratio has decreased. Hence at every level of the wage, the rate of profit rises. As can be seen by inspection, the higher the wage, the larger the proportional change in the rate of profit that will be caused by a given increase in the speed of production. When the wage is zero $\Delta r/r$ is minimal and rises only slightly at low levels of the wage, but as the wage increases, it begins to rise sharply, finally accelerating to approach infinity as the wage approaches its initial maximum. The higher the wage the more worthwhile it is to increase productivity.

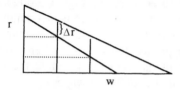

Now consider an innovation in which the use of capital goods, in standard proportions, is increased, reducing the maximum rate of profit, in order to enable output to be produced more swiftly, raising the maximum wage. The new wage-profit trade-off will cross the old at a certain level of the wage, as indicated. At all levels of the wage above that point, the new method will yield a higher rate of profit.

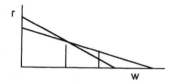

Of course, such a comparison of all possible levels of the wage depends on the system adhering to the Golden Rule. But if r = g, and, if, as the wage moves, r and g adjust to maintain their equality, then at any particular level of the wage a similar calculation can be made. If, by using more capital, production can be speeded up enough to maintain the level of the rate of profit with that higher capital, then at any *higher* wage, the rate of profit will be increased. Hence if wages are expected to *rise*, if, for example, a labor shortage exists, ways to speed up production by using more capital should be explored. When r = g, and the two move together, the expectation of upward pressure on wages creates an unambiguous incentive for capital-using, labor-saving innovations, substituting machinery for labor.

This is a particularly important type of technical change, one which has often been treated as the prototype "choice of technique." Labor productivity – the maximum wage intercept – rises, while capital productivity – the maximum rate of profit intercept – falls. (Foley and Marquetti (1997) call this "Marx-bias" technical change, capital-using and labor-saving.) For reasons given earlier thinking of such a change in terms of choice of technique is misleading. But an actual or expected rise in the wage generates an incentive to develop and purchase machinery that will perform the operations of some set of the workers, at a cost that will be covered by the wages of the displaced workers, which will constitute the annual payment on that machinery. Such mechanization was particularly important in the nineteenth century, but appears to have given way in the 1930s and 1940s to Hicks-neutral technical change, in which the wage-profit trade-off shifts outward, indicating increased productivity of both capital and labor.

Such substitution will trace out a path on which as the wage rate rises and the rate of profits falls, capital-intensity will rise. This would appear to be indistinguishable from moving along a "surrogate production function." But there is an important difference; this is movement during a process of growth, rather than choice from among given alternatives. What is traced out is a potential path of movement, showing that as the wage is driven up, bringing the rate of profit down, capital-intensity will rise. Mechanization is the appropriate response to labor shortage.[4] And pressure for mechanization in AS/TA will tend to push the system towards CBF/MA.

The "General Law of Capitalist Accumulation" – revised

From the analysis so far we can derive a picture that is reminiscent of Marx, or perhaps, "Marx after Goodwin." Growth determines the money, and therefore, in a metal-convertible paper system, the real wage. This, in turn, fixes the rate of profits and normal prices. But this process is not likely to proceed smoothly.

The wage in relation to labor shortage and surplus can be drawn as shown. A labor shortage, measured to the left, will drive up the wage, but there will be a maximum height to which it can rise. A labor surplus will drive it down, but not below a certain minimum. This diagram can be combined with a labor-output function for Agriculture, and a wage-profit trade-off for industry, to present a simple "long" cycle (Nell, 1992, ch. 15).

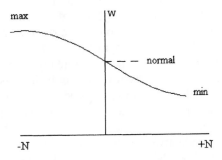

---

[4] It is sometimes argued that mechanization, raising capital-intensity, will *drive down* the rate of profits. Such an interpretation is not justified. There can be no incentive to move to a less profitable method of production; but faced with rising wages, mechanizing to reduce labor costs makes sense.

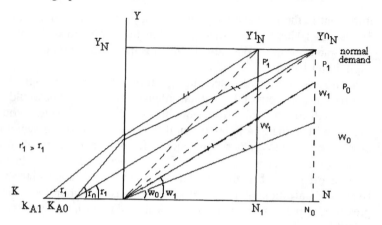

Figure 9.1

A labor shortage develops as industry grows rapidly, absorbing all the available (healthy, adequately skilled) labor in the towns. The real wage begins to be driven up, threatening profits, and slowing growth. This then leads to the development of machinery, substituting for labor, relieving the shortage, and enabling accumulation to continue at a subdued pace, as explained in the previous section. The machinery, however, leads to superior products and higher productivity, and this will be of great interest in agriculture. For the labor shortage and higher wages in industry will have had an effect on wages in the countryside. This can be seen in the diagram, measuring agricultural labor to the right, agricultural capital to the left, and output – at a fixed level representing normal demand – on the vertical axis. The labor-output function rises from left to right intersecting demand and fixing the normal level of employment.

The rise of wages from w_0 to w_1 brings the profit rate down from r_0 to r_1. But new machinery is available, and by investing in it, increasing its capital from K_{A0} to K_{A1}, agriculture can raise its productivity, swinging the labor-output line to the new position indicated. This increases profits, raising the rate of profit from r_1 to the higher level r_1'. (If the new machinery were sufficiently productive this might be as high as r_0.) The effect of the substitution is to release labor $N_0 - N_1$, which migrates to the cities, forming a pool of surplus labor and driving down the wage again. This raises the rate of profit, leading to higher growth, which absorbs the labor, leading to upward pressure on the wage, starting the cycle once more. Over time mechanization will rise, and labor will move to the cities. This process has been observed everywhere in capitalist development.[5]

Marshallian adjustment

We noted earlier that a system of Craft-based factories is not so inflexible that no changes whatever take place in employment when demand changes. Given a sufficient period of time, existing firms can vary the pattern of work, extra hands can be taken on, extra workstations can be set up, using spare parts or reserve equipment, or fallow fields can be brought into cultivation. Work crews can be augmented or cut back, work norms can be adjusted, and all such changes will see corresponding changes in output. The different ways of organizing work can be ranked, and if tradition is to be believed the ranking will exhibit diminishing returns in output to additional employment. This employment-output function is not a conventional production function; the technical coefficients are given and relative prices are fixed. Only the labor coefficients vary, and the ranking associates increased output with increased labor employed (but with progressively smaller increases in putput).

The advantages of flexibility

Let us consider a market for a manufactured product facing a demand curve of at least unitary elasticity. There is a normal level of output and a normal level of variable costs. Fixed costs can be shown spread over output by a total average cost curve. Normal profits, then, are given by price minus variable cost times normal output. Now suppose the average revenue curve shifts in and down, from AR_1 to AR_2. If output cannot be varied, then price will be driven down to variable cost and profit will be completely eliminated.

But suppose that output could be increased, though only by rearranging work in a less efficient manner, so that variable costs at the margin rose. Then instead of a vertical line rising at the "normal" level, somewhere near this point variable costs will rise as shown by the steep line, to meet

---

[5] Accompanying mechanization we find the rise of *joint production*, as primary products are developed into appropriate inputs for households and industry. For example, assembly-line Chicago slaughterhouses produce beef and hides (and many other products) jointly. The joint prices must lie below the *separate* production prices of beef, discarding hides, and hides, discarding beef. The lower prices will be used to introduce the products into the lifestyles of new sets of consumers – which will require developing the products to suit their tastes and circumstances. Finding markets for joint products – or new uses for by-products and waste products – leads to further development of the production processes. Joint production leads to innovation.

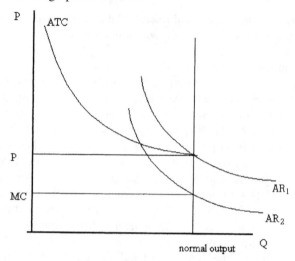

the falling curve of average total costs. The point where these two meet (beyond which marginal costs will be greater than average, pulling the latter up) will be the new normal level. Now consider a fall in average revenue from AR_1 to AR_2. The lower demand will force price down, but only to the new level of variable costs, since output can be reduced, reducing the pressure driving price down. Instead of being wholly eliminated, profit will be reduced to the shaded area in the diagram. (In each case the ratio of profit at normal output to capital is assumed to equal the normal rate.)

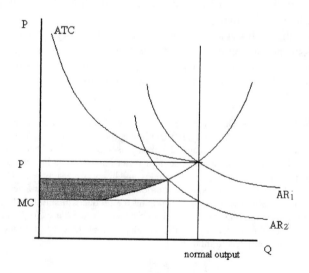

Adjustments will center on the normal position, which will reflect the least-cost work norms and level of employment. The average cost curve will be U-shaped, and the adjustments will resemble a Marshallian market. Further development will lead to Keynesian forms of adjustment. The output and cost functions will be examined further in later chapters; market adjustment is the issue here.

Gravitation focuses on adjustments to permanent changes in demand. But variations in demand pressure can occur for many reasons, some permanent, some transitory – and some which might be either, depending on later developments. In the latter two cases, business may be assumed to continue to expect markets, on average and in the long run, to grow at the natural rate. But their confidence in such expectations will change. In particular, while they may believe that their market will *eventually* expand as indicated by the natural rate, they may lose faith that it will do so in the near term. When confidence falls, they will be inclined to delay planned investments; when it rises they may speed up their projects.

These fluctuations in demand pressure lead to variations in prices, outputs, real wages, and employment that trace out very familiar patterns – the traditional relationships of Marshallian "partial equilibrium" analysis. But we shall now show that these relationships can be derived, not from choice according to given preferences in conditions of scarcity, but from the optimal reactions, by firms and households, to fluctuations in demand pressure on a system of given plant and equipment. The normal operating levels of this given capital stock are the minimum points of average cost curves, with normal prices, yielding normal profits, equal to such costs. We will construct supply and demand curves and show that this system adjusts according to what look very much like the familiar "marginal" conditions. This is not Marshall, but it has a Marshallian flavor.

The optimal size of the firm

In a Fixed Employment economy firms will seek their optimal size and continue to operate at that size indefinitely (Robinson, 1931). When technical progress is irregular and economies of scale rare, existing firms will normally not risk overextending themselves by making new investments, especially when markets are local and transport and storage costs are high. By contrast under conditions of Mass Production, firms will invest and grow, keeping pace with the growth of the market. Growth under conditions of Fixed Employment therefore implies investment in the formation of new firms, rather than investment in the expansion and renovation of existing firms, "extensive" rather than "intensive" expansion.

The reasons can be seen quite easily. Plot plant size on the horizontal axis against costs and gains on the vertical. The larger the size the larger the earnings, but economies of scale are relatively few; assume for the sake of the argument that there are also relatively few diseconomies.[6] We assume constant returns to the basic technology – for lack of better general alternative. Then there will be gains to having a larger plant size (or a cluster of plants), in the form of reduced unit management costs, economies of scale in handling, benefits from bulk buying, and a monopsony position with local suppliers. The relation between size and earning can therefore be approximated by a rising straight line, implying constant marginal returns to size; plant and equipment costs and extra earnings due to size rise together. Now consider the costs incident upon size. At low sizes the costs and risk due to size will be nil; at some point, however, a larger size will begin to imply significant transport and storage costs – the plant will be larger than the immediately accessible market. As the area served increases, these will rise rapidly, under conditions of Craft technology.

Moreover, with costly transport and slow communications it is risky for firms to try to serve distant markets. Local conditions in distant markets may change, new competitors may emerge, or tastes may change. The costs of a loss, due to market change or obsolesence, is also greater. A larger plant represents a larger commitment of the family's resources to a single project – there is less diversification in the family portfolio.

The cost-size relationship will begin from a positive point on the horizontal axis, rise slowly at first, and then turn up steeply, eventually cutting the size-earning line. The optimal size of the plant will be found at the point where the marginal cost just equals the marginal gain due to extra plant size.

The optimal number of plants which a given management coordinates depends on the organizational arrangements and on the effectiveness of communication. It may be wiser for a family to diversify its portfolio, rather than invest it all in one line of business.

Each firm in each sector will be assumed to operate the same technology, choosing the minimum point of average variable costs as the normal operating level.

---

[6] This is to avoid the traditional argument that there are first economies of scale, then diseconomies, yielding a U-shaped average cost curve, the optimal size of the firm being that at which average costs are a minimum. Sraffa (1926) has argued that the ideas of increasing and diminishing returns cannot be combined in this way. But when demand varies, output and work intensity and/or output and employment can vary, and it will be argued that such variation normally leads to lower average productivity both below and above the point of normal operation.

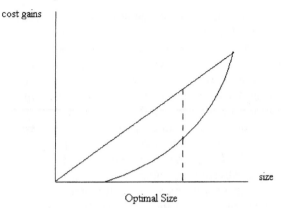

Optimal Size

Given that growth takes place by founding new firms of optimal size, the savings-investment nexus must now be explored.

Finance and savings in the Craft-factory economy

The financial market mobilizes savings, which will come chiefly from capitalist families and channels them into new investments, normally made by new firms. So firms will pay out all their profits, since they have no need of retained earnings. The market therefore collects household savings and makes them available to entrepreneurs founding new enterprises.[7]

Savings out of profit income, and therefore loanable funds, will obviously be higher, the higher the rate of return.[8] Two observations are relevant at this point. First, the cross-sectional pattern of saving is easily explained by this approach; high-income households normally have a large component of property income, most of which they save. Low-

---

[7] The loan market will be assumed to be fully competitive, in the sense that new firms will be able to borrow all the funds that they can effectively manage at the going rate of interest. The security for the loan is the project in which the proceeds are invested; such projects must therefore be judged riskworthy in the eyes of the bankers.

[8] Utility theorists normally introduce "time preference" – Irving Fisher called it "impatience" – to explain the relation between interest and savings. Consumption now is preferred to consumption later; interest is needed to compensate for the delay. The higher the rate of interest, the more consumption will be postponed, etc. Of course, some kinds of delays are extremely annoying, but to generalize this to all consumption of all kinds is absurd. Many consumer items, especially consumer durables, stand in natural sequences; Act II cannot be appreciated before Act I, dessert follows dinner, marriage follows graduation, buying the house waits on the promotion, and the furniture and appliances on the house. When we set out a plan for the household's lifecycle this kind of natural order becomes apparent. There is no role for generalized impatience.

income and most middle-income households have only wage income, and save little or nothing. The majority of households own no income-yielding property and depend upon the breadwinner's wages to support the family. The normal level of real wages will normally just cover the long run socially necessary expenses of family life – allowing for normal variation, etc. Normal saving out of wage income, therefore, can only be what is required to provide precautionary funds, or to finance expenditures later in life. It is therefore a reasonable approximation to assume that all wage income is spent on consumption. But a minority of households receive profit incomes. The profits of business include normal remuneration for management by the capitalist owners; such remuneration will likewise normally just cover the expenses of living in the style appropriate for proprietors. So savings out of these "wages of superintendence" – as Mill termed them – can only cover precautionary funds and funds set aside for later consumption. When such remuneration is subtracted from general business earnings the result will be the true net earnings. The savings out of these earnings – profit income – will then be available for lending.

Second, in the period of Craft-based factories, we would not expect saving to be tightly related to current income – since consumption and investment are inversely related. A family's status in the community will depend on the size of its wealth-holding relative to those of others; hence there will be strong pressures to save at the same (or a higher) rate as other families. Capital-owning families will compete for status and prestige in the community, and will save in order to accumulate wealth. Wealth will enhance their access to investment opportunities, and improve the ability of family members to command high salaries. Competition to accumulate wealth will thus tend to establish a common or general rate of financial accumulation. Moreover, it will require that the rate of interest at which savings are supplied to the market as loans to be the same as the risk-adjusted rate of net earnings on marginal projects.

The marginal efficiency of capital schedule

Using the prices determined in the analysis of the long-run setting the various possible investment projects can be evaluated and ranked.[9] In

[9] The marginal efficiency schedule has been severely criticized in recent years, and correctly – for an industrial, corporate economy. But the criticisms do not apply in the present context. For example, "reswitching" objections do not arise, since the "normal" MEC schedule is defined only for the prices, etc., derived from the long-run setting, and shifts in the MEC schedule are understood to be deviations from the normal position. The problem that changes in interest rates require the

(continued)

a Craft economy the number of possible projects can be considered given at any time; innovation is irregular and infrequent. Each project creates a new firm, and each such firm will be of "normal" size – an optimum given by the number of employees a family firm can effectively manage, balancing the limited economies of scale against the increasing risks of size.[10] Each project will have a cost – the price of plant and equipment – and each will generate a stream of returns – gross revenue minus wage and other current costs – over the expected lifetime of the project. The marginal efficiency is the discount rate that equates the discounted sum of this stream to the project's cost. Projects can be ranked by their marginal efficiencies, and the amount of investment in successive projects can be measured on one axis with the progressively lower marginal efficiency on the other. If investment projects now are as numerous and as profitable as expected when the capacity in the capital goods sector was installed, the amount of investment demanded at the marginal efficiency just equal to the current rate of interest will equal the normal capacity of that sector, and this will also just equal the saving out of normal profits (and wages). This is the normal position.

Under normal conditions new investment projects will produce the same goods by the same methods as existing firms. The new projects will

[9] (contd.)
readjustment of prices to reestablish equilibrium (and thus the reranking of projects, so one can never "move along" a schedule) does not arise since the interest rate only changes when the MEC schedule is out of its normal position – the argument only applies to adjustments around equilibrium. When a firm regularly invests, as in conditions of Mass Production, the MEC calculation cannot govern its choices, because the firm's future investments will have "external" effects on future earnings from the present investment; to know the returns of future periods correctly these effects must be estimated – but the future investments might not be made unless the present one is successful. Thus the MEC calculation is caught in a dilemma – it cannot determine its present investments until it decides its future ones, but it cannot determine those until it decides on its present plans. This will be a problem for industrial corporations engaged in regular investment (who will make their decisions on other grounds than the MEC), but it poses no difficulties when each new project is carried out by a separate new firm.

[10] Growth has a different character in the two systems. In the Craft economy it will be extensive, spreading out to new locales, reaching to new populations, but new investments will simply replicate old. There will be little innovation, and new capital will not necessarily compete with existing. In the Mass Production economy, however, each firm will grow indefinitely through internal investment, because new capital is technically superior to old; hence new firms would have a competitive advantage over old. Firms are thus compelled to grow by investing, and growth will be intensive, involving increases in capital per head and output per head, as well as innovations in processes and products.

differ in location and in the experience and skill of the workers and entrepreneurs. But in a period of innovation many investment projects may be designed to produce new products or to employ new techniques.

Now consider the effects of changes in the state of confidence, due to anything which affects a business's estimate of future sales. If the future looks bleak, prices will be expected to fall; if it looks rosy, they will rise, relative to normal prices. Consider the evaluation of an n-period project:

$$p_1q_1 = \frac{p_2q_2 - wn}{1 + \rho} + \frac{p_2q_2 - wn}{(1 + \rho)^2} + \cdot \frac{p_2q_2 - wn}{(1 + \rho)^n}$$

If p_1, p_2 and w all rose or fell in the same proportion, the marginal efficiency, ρ, would be unaffected; but if p_1 and p_2 change in the same proportion, but w remains fixed or changes by less, then the earnings stream changes relative to the cost, and ρ will change. Changes in prices relative to money wages will therefore shift the marginal efficiency curve; a rise will shift it outwards, a decline inwards.

Such changes will also affect the rankings of projects; when prices rise relative to their normal levels, the real wage declines, and the more labor-intensive a project is, the more profitable it becomes. Thus the largest increases in profitability will take place in projects requiring the least investment; projects previously far down the line may be moved up to the forefront. Hence the shift will be irregular and the shape of the curve may change. The intercept on the horizontal axis shifts out because the cost of each project increases as the prices of capital goods rise. In the new ranking the most profitable projects will be grouped near the origin on the horizontal axis; thus at these levels the curve will shift up the most.

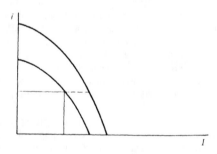

Shifting the marginal efficiency curve in the short run

Start from the marginal efficiency curve in its normal position, and consider an improvement in the state of confidence; the curve will alter position and shift up and out, as shown in the figure. (Employment in existing firms is taken as fixed during the period under analysis; variations in employment due to reorganization will be examined later.) At the initial level of the interest and profit rates, savings – loanable funds – equaled investment. Investment demand is now higher, but the anticipated profit rate on existing operations will also be higher, since the real wage is lower. (The increase in profits on current operations will be more pronounced in labor-intensive industries.)[11] Hence, funds will not be loaned at the old rate of interest; new loans will be made at a higher rate, reflecting the higher rate of profit. In turn, the demand for loans will be higher, but at the higher interest rate investment demand will, of course, be less than at the initial rate. Hence both the interest rate and the level of savings rise, the first reflecting the increase in the realized rate of profit, the second reflecting the level of profit generated by the larger investment demand, resulting in the higher level of prices. Neglecting savings out of wages, the relationship will be

$$i = \frac{s_p P}{K}$$

where s_p is the saving propensity out of profits, and K is the capital stock evaluated at the long-period prices. An exactly parallel analysis holds for a decline in the state of confidence, leading to lower prices relative to money wages, resulting in a downward shift of the MEC schedule, bringing lower profit and interest rates. These in turn bring lower levels of saving and investment, but because of the lower interest rate, the decline in investment will be less than indicated by considering the initial shift in the marginal efficiency curve at the normal level of interest. The relationship between interest and savings $(= s_p P)$ thus is a line from the origin with a slope of 1/K. The diagram of the two schedules thus presents a falling marginal efficiency curve coupled with a rising interest-rate/savings curve, which looks for all the world exactly like the traditional picture. Savings and investment therefore adjust in a stable manner to changes in the state of confidence, and so to changing actual rates of growth.

[11] The fact that the profit rate has temporarily become nonuniform, because of the unequal capital-labor ratios of the sectors will have little or no influence. Higher profit rates are an inducement to movements of capital; in an artisan economy this would suggest that firms should leave the low-profit sector and enter the high. But once fixed capital is used firms would be ill-advised to act in response to *current market* prices and profits; major investment decisions should only be made on the basis of long-term judgments about the relative future courses of prices and profits.

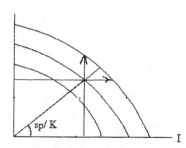

An implication of this analysis is that when the marginal efficiency schedule shifts, prices and interest rates will move together; as prices rise and fall relative to the money wage, they generate movements in the same direction in realized profits, which in turn (no doubt with some lag) bring a corresponding movement in the rate of interest. Since these are movements around the "normal" level, we should see deviations in both directions, corresponding to periods of boom and slump. This corresponds to the pattern discussed by Keynes under the heading "Gibson's Paradox" (cf. Keynes, 1933, 2: 198–208).

Construction of demand and supply curves

An increase in investment demand will not only bid up the price of capital goods; it will also increase their output, as workers in the capital goods sector intensify their efforts – and thereby raise costs. In the same way, a decline in investment demand, due to a collapse of confidence, shifting the marginal efficiency curve in, will lead to a fall in both price and output. Let us explore this in a diagram.

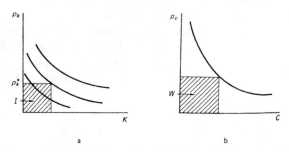

Plot price on the vertical axis and the output of capital goods on the horizontal. Then since the marginal efficiency-saving-interest market fixes investment demand in nominal terms as an amount, I, the demand curve for capital goods will be a curve of unit elasticity, the area under which will be I. An increase in I means a shift outwards of the curve.

When p_k is at the normal level, p_k^*, I will purchase the number of capital goods needed for expansion to supply the expected growth in the market.

This indicates the normal level of operation of the investment goods sector. When investment demand rises, p_k will be driven up. So when $p_k > p_k^*$, too few goods will be purchased, resulting in a growing shortage of output in consumer goods, leading prices there eventually to rise, also. Similarly when investment demand falls, p_k will fall as output tends to remain near the normal level. But when $p_k < p_k^*$, initially firms may buy the desired (smaller) complement of capital goods more cheaply; but competition will tend to force all prices down, leading to expansion of consumer demand, and hence to increased derived demand for capital goods, tending to restore the initial level of demand, now at a lower price. Variations in investment spending will lead to inversely related price and output movements – demand curves.

Household spending

Like firms, households will be seeking to make the best adaptations they can to changes. A version of marginal utility analysis can help to explain the way wage-earning consumers adjust their spending when conditions change. A normal pattern of consumption must be taken as given at the outset – explained by household technology and social conventions. Income will be divided into categories of expenditure: food, clothing, shelter, entertainment, transportation, etc. Each will be further subdivided into subcategories of specific goods. In normal circumstances fixed ratios will hold both between the goods within a category, and between the categories. Then two kinds of variation can take place: money incomes can change, with prices fixed, or prices can change, with money incomes fixed. On the assumption that marginal utilities are in equilibrium in normal conditions, changes in expenditure patterns can be explained by the *relative* changes in marginal utilities. But the changes will affect households differently. With prices given, when money income changes, expenditures on *categories* A and B (e.g., food and entertainment) will change according to the relationship between MU_{A1}/MU_{A2} and MU_{B1}/MU_{B2}, where 1 indicates the initial position and 2 the position after the change.

This is the Engel Curve relationship; as incomes rise, the proportion of income spent on necessities falls, while that on conveniences and luxuries rises. Similarly when prices change, incomes given, the change in spending will depend on a comparison of the changing marginal utility of specific good α, as quantity varies, with the changing marginal utility of good β. It will never be necessary to compare the marginal utility of α directly to that of β, nor of category A to category B. That is, the utility from another armchair per unit money need not be measured directly against that from another rug (let alone additional hours of heating), nor need the utility from further spending on food be compared to that from entertainment per unit money. Only the rates of change have to be compared – the rate of armchair utility is falling faster than the rate of rug utility, furniture utility is falling faster than entertainment utility. Colloquially; we need another armchair less than we need another rug, we need a longer vacation more than we need more furniture. So long as marginal utility is always positive but diminishing, the desired *changes* in expenditure on the different categories of consumption will be determinate, as will the division within each among specific goods. The normal pattern of consump-tion is assumed given – what is explained are the *changes* in consumption due to changes in income or prices. Marginal analysis explains marginal changes, but it must take the starting point as given.

A shift in saving

Traditional theory has been thought to imply that a rise in the propensity to save would lead to a fall in the rate of interest, bringing a concomitant rise in investment. Keynes pointed out that a rise in the propensity to save was equivalent to a fall in the propensity to consume, the multiplier effects of which would lower income. Hence actual saving would not rise, so interest rates would not fall, and investment would be unchanged. Here, however, there is no multiplier. The economy, although industrial, still operates in the Craft mode. Nevertheless, the traditional story cannot be justified. A rise in the propensity to save, implying a fall in the propensity to consume, leads to a fall in consumer prices. This implies a rise in the real wage, which will tend to prevent a decline in the volume of consumer demand. The lower prices will tend to bring lower profits, and interest rates will therefore also tend to decline. But the lower profits will dampen the state of confidence, so the lower interest rates will provide no stimulus. A decline in consumption is *necessary* to raise investment, but it is not *sufficient*.

Responses of production

Next consider the ability of production to respond to an increase in demand to a point above the long-term normal level. For small increases in demand the intensification of effort will yield the required output; but as the increases become larger, the proportion which can be supplied falls off.

The positive vertical axis measures output supplied, while the negative section shows output diminished; the horizontal shows demand above and below normal. The 45-degree line indicates that demand and supply are equal. In the positive quadrant the curve falls away from this fully adjusted line, indicating that higher levels of demand are progressively harder to supply through intensification of work effort, working longer

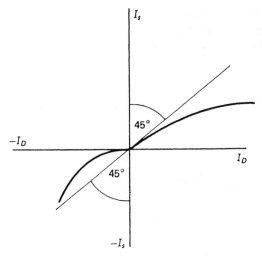

and harder, or adding more laborers.[12] When demand falls below normal capacity there will be a tendency to maintain production – adjust very little – and only when the shortfall is large will supply drop off significantly, reducing labor. This is shown by a curve which runs along the horizontal axis to the left, then turning in toward the 45-degree line.

Now consider an outward shift of the unit-elasticity curve, representing a horizontal shift in investment demand. The production-response function determines the proportion of this increase in demand which can be supplied. Mark this off, and then read upward to the new unit-elasticity curve to find the new price. Consider a different, larger or smaller, shift in demand, and repeat the process. Joining the resulting points traces out a "rising supply curve." The same procedure, using the negative section of the production-response function, can be followed to trace out declining price and output. This curve, rising from left to right in price-quantity

---

[12] Even where it is possible, working at greater intensity than normal can be presumed to be harder, more dangerous, or less healthy. Surely there were good reasons for establishing the normal level in the first place. Hence such work can only be temporary; in traditional terms, it involves an increase in the disutility of labor, at the same time that the real wage is diminished. Working harder for less pay requires that the conditions of work will have to be changed. This means investment, the construction of additional capacity, so that the pace of work can be returned to its normal level. The role, then, of the "marginal disutility of labor" is to require that the adjustment be merely temporary, and that the additional output brought about by more intensive use of existing facilities be replaced with additional output from new capacity. The converse case will also have to be short run; when output is reduced through working at less than the normally intensive pace and this is coupled with higher wages, employers will eventually want to dispose of workers who are both expensive and underutilized.

space, shows the supply offered in response to changes in demand, where the variations in demand lead to changes in price – and the changing supply, in turn, leads to changes in costs.

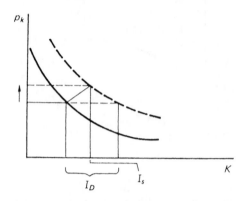

The curve traced out in this way looks exactly like a traditional supply curve, but it does not reflect scarcity of endowments in relation to preferences, and in particular it does not reflect marginal costs. It does, however, reflect the difficulty of adjusting output through intensification of work using given produced means of production – something like Marshall's "real costs."

Variations in demand may require the use or release of "marginal" factors, marginal in the sense not only that they are dispensable, but also that they are not of the same quality as the regular labor force or equipment. Thus expansion will be accompanied by diminished productivity at the margin, so by rising costs. Less well-trained or educated workers, less fertile land, older or previously retired equipment may be brought into service, at higher costs, which can be met out of the increased prices. These increased costs are *permitted* by the higher prices; they do not *explain* them. In the traditional theory the higher costs, resulting from intensive and extensive diminishing returns, are the cause of the higher supply prices – the supply curve of the individual firm is the rising section of its marginal cost curve, and the supply curve for the industry is the horizontal sum of the individual supply curves of the firms. Because these rising costs both governed prices and reflected diminishing returns to the employment of scarce factors, a price could be said to measure opportunity costs, and an equilibrium to imply efficient allocation. In spite of the apparent similarities in the shapes of the functions, however, nothing of the sort can be inferred here.

A shift in production at the margin

Demand pressure raising the price of capital goods will signify higher costs of production in consumer goods, where prices will now have to be raised. Since money wages and, initially, employment are fixed, the higher price of consumer goods means a decline in real wages and so a proportional decline in demand. Pressure will therefore develop for work to be done less intensively in consumer goods, more intensively in capital goods. It is plausible, therefore, to consider a shift, in the long run, of some marginal workers from the consumer goods sector to the capital goods sector. Such a shift, it emerges, will tend to restore the initial equilibrium in the same way that gravitation did.

The "response of production" diagram just considered implies that average productivity is at its peak at the origin. Above this point more is demanded than can be supplied at costs covered by the revenue obtainable at the initial price. Hence price is driven up. Below it a drop in demand will not be matched by reduction in supply, because costs cannot easily be adjusted downwards. Prices will therefore be driven down by "over-supply." However, as demand weakens, average productivity will decline as the pace of work becomes less intense.

Assuming firms to be of approximately the same size and efficiency, we can aggregate them into production curves, for each sector – consumer goods and capital goods. Each curve will center on the most efficient point, to either side of which average productivity will be lower (average cost higher). Above the most efficient point, additional workers or additional worktime add progressively less to output while below it, removing workers and reducing worktime progressively reduces output by more; the marginal product diminishes as employment increases, and increases as employment declines. The curves are defined only for a limited range.

Consider equilibrium in the consumer goods sector. The vertical axis shows output of consumer goods, Y_c, the horizontal, employment in consumer goods, N_c. The capital stock, K_c, will be plotted on the horizontal axis to the left of the origin. The normal rate of profit, r, will be shown by the angle formed by a line from K_c to the vertical axis. The segment from the origin to that intersection measures the share of profits, P_c. Normal output, Y_c^*, and normal employment, N_c^*, are determined by a ray from this intersection which is just tangent to the production curve. Such a ray measures average variable productivity, which will be at a maximum (average variable cost a minimum) at the point of tangency. The vertical segment $W_c = Y_c^* - P_c$ represents the share of wages, and the angle $w = W_c/N_c$ measures the wage rate.

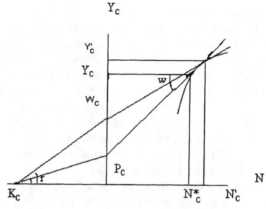

If demand should rise above the normal level to Y'_c, the price will be driven up, and the real wage will fall and the profit rate rise, as indicated by the dotted lines showing the new angles. The new tangency line will touch the production curve at Y'_c, N'_c, so employment will rise to produce the new output at the lowest cost. The share of wages will fall, while that of profit rises.

The equilibrium of the economy as a whole can be shown by exhibiting both sectors on a single diagram. The consumer sector may be presumed to be much the larger of the two. The capital stocks of each are measured to the left. The rate of profit is expressed by the angle $r = P_c/pK_c = P_I/K_I$; the two angles must be equal. The resulting parallel lines mark off the share of profits in each sector on the vertical axis. From each intersection point parallel lines run to the points of tangency on the production curves, forming the wage angles, $w = W_c/N_c = W_I/N_I$. The normal *relative* price of capital goods in terms of consumer goods, p, must be set so that the height of the production curve of capital will be just the level required to ensure that the profit and wage angles of the two sectors are equal, that is, that the respective lines are parallel.

Now suppose there is a permanent decline in investment demand; assume that the natural rate has fallen, so the economy is going to grow more slowly. The market price of capital goods falls, and the new optimal position in capital goods calls for reduced employment. The decline will tend to pull down the price of consumer goods, but more slowly. The rise in the real wage ensures that consumer demand will remain strong. With profit higher in consumer goods, marginal firms, whose workers and owners have relevant skills, will exit capital goods and enter consumer goods. Thus K_I will decline and K_c will rise. Since best-practice capital-labor ratios are fixed, employment will tend to vary in proportion, and therefore so will output. The capital goods sector's production curve will

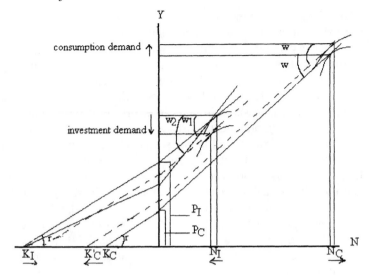

Figure 9.2

shift down, while that of the consumer sector will tend to rise, both in proportion to the shifts in the K's. P_I has fallen, and pK_I moves in, tending to restore r, as the production curve moves down, restoring employment and lowering the wage. K_c rises, and Y_c rises in proportion. The movement will continue until the wage and profit lines are parallel again, with no change in the normal relative price, p. Entry and exit thus tend to adjust the system back to the original equilibrium.

But since pK_c/N_c will normally not equal pK_I/N_I, it does not follow that all the labor that seeks to shift will find work, or that all the equipment can be transferred or converted. For the adjustment to work reliably there must be reserves of unemployed labor which can be drawn down or increased as firms move from sector to sector. If transferring the required amount of capital released too little labor, the adjustment would not be possible, unless there were unemployed reserves to draw on. If it released too much labor, the excess must be added to the ranks of the unemployed.[13]

The adjustment process works, then, so that a rise (fall) in investment demand leads to an increased (diminished) output of capital goods together with a diminished (increased) output of consumer goods: investment, profits, and savings rise (fall), consumption and real wages decline

[13] The adjustment will apparently work whichever sector is the more capital-intensive, unlike the neo-Classical two-sector growth model, provided that unemployed labor is available.

(rise). Bearing in mind that all the changes are fluctuations around a normal position, these adjustments are stabilizing.

Some very traditional relationships can be seen here in several different forms. The increase in demand drives up prices relative to money wages; therefore the real wage declines, but at the same time, work effort and output increases. Hence we have a declining real wage accompanied by increased work, measured in units of effort, or, if the pressure is persistent, increased employment. Also, as the real wage declines, marginal workers will be transferred to the sector where output is increasing, and price has risen. As output increases in response to demand, with a given labor force, work effort becomes progressively less effective; when output and the labor force are given, and demand declines, price falls; in either case, the "marginal value product" declines. When the demand change is persistent, employment will be adjusted. In all of these we have an echo of the traditional theory, but nowhere have we assumed scarcity conditions in regard to employment, nor does the relationship say anything about opportunity costs.

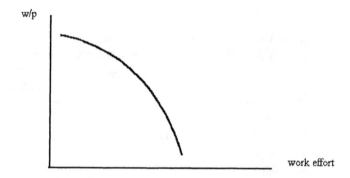

This process of stabilization by entry and exit differs from Classical gravitation in one important respect, however. In gravitation, the *process itself* carries out the shift of capital. Price adjustments imply capital gains and losses, and these combined with strict financial constraints force firms to lower or raise their working capital, thus changing output levels. If the price changes are permanent the shifts will take place automatically. But in the Marshallian system firms can vary their levels of employment and output and are not subject to the same types of constraints. They must *judge* whether to move or not, and such judgments must be based on their expectations of prices and patterns of demand. Shifting means disinvesting in one sector and reinvesting in another, a costly and risky affair. Even if it is called for, it might not take place.

So it seems that there are circumstances in which convergence to a LPP will be plausible. Prices must be flexible when the ratio of demand to supply changes. Price changes must change profitability, and changes in profitability must lead to changes in investment. Investment, however, must otherwise be stable, and the initial change in demand must not set up a sequence of cumulative additional changes. These conditions are not unreasonable in economies operating craft-based technology, in which there is little or no regular technical change, and in which the development of new markets can be clearly foreseen, perhaps on the basis of population growth. On these assumptions new investments set up new firms that replicate existing ones. Firms are established at, or quickly grow to, their optimal size. The trend of investment will be clearly defined, and current investment spending can be expected to gravitate around it. Since technical change is occasional, irregular and unforeseen, and since each firm operates plant and equipment of only one "vintage," competition will tend to drive prices to those corresponding to the best-practice technique. It is plausible that investment will iron out differences in profitability between sectors. No investments will be made that do not pay at least the rate of interest; areas of investment that return more will attract capital, and competition will drive prices and earnings down. Over time accidental variations can be expected to average out so that the system will more and more closely approximate the long-term position. The crucial assumptions are no technical change, "regular" investment, and prices determined by supply and demand (Nell, 1992, ch. 16).

Aggregate employment and marginal productivity
The argument so far has focused on small changes and shifts of factors between sectors. But as the Craft economy develops employment becomes more variable, and larger adjustments can be considered – in particular, the relationship between changes in employment and changes in output. We can develop a Marshallian macroeconomics!

The Marshallian production function
First, we must look more closely at production. Keynes accepted what he called the "first postulate of Classical economics . . . this vital fact which the classical economists have (rightly) asserted as indefeasible. In a given state of organization, equipment and technique, the real wage earned by a unit of labor has a unique (inverse) correlation with the volume of employment. . . . If employment increases, then, in the short period, the reward per unit of labor in terms of wage-goods must, in general, decline and profits increase. This is simply the obverse of the familiar proposition that industry is normally working subect to decreasing returns in the short

period during which equipment etc. is assumed to be constant. . . . But when we have thrown over the second postulate [that the real wage equals the marginal disutility of labor], a decline in employment, although necessarily associated with labor's *receiving* a (larger real) wage . . . is not necessarily due to labor's *demanding* a larger quantity of wage-goods" (1935, pp. 17–18).

In other words, if the money wage is fixed in the short run, any change in demand will change employment in the same direction, according to a function which exhibits diminishing returns, but it will do so by changing prices, which at the same time changes the real wage in the opposite direction. Employment changes as a combined consequence of demand pressure and a fall in the real wage, itself induced by that pressure.

A. C. Pigou (1940) described the functions relating output of consumer goods and investment equipment to labor, in conditions of given plant and equipment as follows: "Thus when industry is in a state of moderate depression with a fair amount of idle equipment, marginal prime cost may be approximately constant over a considerable range. Obviously this will not be so when the industry is working at or near full capacity. Then marginal cost must be rising" (pp. 51–2; also see p. 9, note).

Marshall himself held that "For short periods [we may] take the stock of appliances for production as practically fixed; and . . . are governed by expectations of demand in considering how actively . . . to work those appliances. . . ." (1965, p. 374). Viner defined the short-run "to be a period which is long enough to permit of any desired change of output technologically possible without altering the scale of plant, but which is not long enough to permit of any adjustment of scale of plant" (Viner, 1952, p. 202). That is, output is a function of given plant and equipment, which remains unchanged in quantity and form, to which is applied a variable amount of labor.

J. M. Clark (1923), commented on the growing importance of large-scale fixed plant and equipment in creating overhead costs, which might apply to a variety of outputs. Not only could the level of employment and output be varied, the composition – the proportions of different products or services and thus of different kinds of labor – could also vary. Clark considered this a new development, dating from the end of the nineteenth century.

Hicks (1966), refers to the derivative of such a function as a "short-period marginal product," defined as "the additional production due to a small increase in the quantity of labour, when not only the quantity but also the form, of the co-operating capital is supposed unchanged" (pp. 20–1). He doubts whether this conception can be given any "precise meaning" or "useful application," for a worker added to unchanged plant will produce less additional output than the "true" marginal product, which would result from optimizing the form, leaving the amount unchanged; while a worker subtracted will reduce output by more. The clear implication, nevertheless, is of diminishing returns.

Moreover, it seems that Marshall's famous example of the shepherd was of precisely this sort, as Dennis Robertson admits (1931, p. 226). Robertson supports Clark's view that the "Principle of Variation" is the essence of marginal productivity; the marginal product is only properly defined when the "co-operating capital" changes in form, to achieve an optimal configuration, while remaining fixed in amount. But he admits that this renders the conception practically useless (p. 228). By contrast, the Marshallian approach was eminently practical; the marginal principles were developed not with an eye to theoretical consistency and elegance, but as guides to understanding and policy. On the continent it was quite otherwise; and marginal productivity there was always understood to require variations in methods of production – as Hicks and Robertson also insisted.[14]

The Marshall-Pigou conception is *not* the modern neo-Classical production function; it represents the employment of labor to operate given plant and equipment – to operate the production system – at various levels of intensity. It is what Joan Robinson later dubbed a "utilization" function, and it has no implication that technique is different at different points. Instead, different points represent lesser or greater degrees of utilization of the given plant and equipment. But this brings up the question, Why do returns diminish?

Craft-like technology

Diminishing returns – increasing cost – at the margin when more workers operate given equipment arguably is a feature of craft-based technology. The middle and late nineteenth century may be seen as a period of transition; the factory system had developed, but production was still largely organized around craft methods. A review will be useful.

Firms are still small and family owned, but no longer family operated. They are companies, and separate from the domestic economy. Production takes place in factories, but skilled work crews operate equipment; all members of the crew must be present for any level of production. The layoff system has not yet developed; layoffs and short time are difficult to organize. Employment can be varied, but only slowly and with difficulty, since it requires reorganizing the work crews, and redividing the work.

---

[14] The Capital Controversy has shown that the well-behaved neo-Classical production function, in which the form of capital changes as techniques vary, is unacceptable. Of course, this critique does not affect the Marshallian function.

Adding a worker reduces intensity, but raises output; subtracting one raises intensity, reducing output. Labor will generally resist pressures to increase work intensity; and employers will be reluctant to allow intensity to decline, for reasons of "spoiling the market." In each case, therefore, the changes must be compensated by a changing wage.

These changes rest on characteristics of the technology. In a craft-based factory or shop, a comparatively inflexible central power source (steam, water, wind) runs a number of workstations, with power transmission conveyed by a mechanical system – rotating shafts, wires, belts, pulleys, etc. The layout of workstations is severely constrained by this transmission system (and in the case of water by the geography of the river). Fixed costs are substantial, and plant and equipment can be utilized more or less intensively, but start-up and shut-down costs are large. "Batch" production prevails, requiring a full detail of workers, whether the batch is large or small. Average costs therefore fall as output increases, up to a "normal" operating position. Operating beyond the normal range may lead to breakdowns in equipment, or may require activating equipment that is substandard. Or it may simply be that additional workstations will have to be positioned so far from the central power system that transmission will be costly and inefficient. Plant and equipment unit costs therefore fall as output rises to the normal range, and then rise thereafter.

Running costs per unit, by contrast, rise steadily. These are chiefly labor costs. From somewhere below the normal point they begin to rise, for two reasons. Core skilled workers must increase their intensity of effort, and while this will pay off in extra output, it will do so at a diminishing rate, simply from the effects of tiredness and inattention, as has been well documented. Secondly, additional workers will have to be hired to assist the core workers, to overcome bottlenecks, and help with various time-consuming but less essential tasks, such as set-up, materials and product handling, and cleanup. The workers hired for such tasks will have to know the process; they cannot be wholly unskilled, so they cannot be paid at a much lower rate. But the gains from such additional employment will be limited. (There may also be one-time costs due to reorganizing work.) Hence unit labor costs will tend to rise. At some point rising unit running costs will offset the falling fixed costs – particularly if average equipment costs also begin to rise, because of breakdowns or the need to draw on inferior or outmoded machinery.

Consider an increase in demand pressure, starting from the normal position. Since output and employment are inflexible, prices will rise relative to money wages. The real wage will decline, stiffening resistance by labor to intensifying work, but making it possible for employers to try to add workers. Additional workers will require a reorganization of the work-force, which will result in higher output, but at a diminishing rate. A fall in demand will lead to a fall in prices, leading employers eventually to reorganize, dividing the reduced amount of work among fewer workers, each of whom, however, will have to do more, since (choosing the normal level of output and employment both as unity) because of the need for core workers to be in place in batch production, the proportional reduction of employment will be less than the proportional reduction of output.

Unemployment can also arise from shutting a plant down completely; when times are poor in an industry, and prices fall relative to money wages, low-cost plants and superior equipment will stay in operation, while high-cost plants will be temporarily closed. As demand expands, prices will rise, and progressively higher-cost plants and equipment can be brought back into service. Thus we have an aggregate relation between output and employment that shows diminishing output at the margin; how fast it diminishes will depend on technology.

There are therefore good reasons to expect the function relating output to employment with given plant and equipment, in conditions prior to Mass Production, to exhibit diminishing returns to labor at the margin. Expanding operations above or reducing them below "normal" will require reorganizing the work; unit labor costs tend to rise with output, average fixed costs to fall. The normal operating level will be set where these meet. Beyond this point, equipment costs are likely to rise also.[15] In addition, for the system as a whole competition should ensure that the relatively lower cost, more efficient operations produce a higher proportion of output when demand is at lower levels, while high-cost producers are pulled in or increase their relative share at higher levels.

Cost curves

These utilization functions imply a set of corresponding cost curves. For if the real wage equals the marginal product of labor, then price equals marginal cost; that is,

$$p = w \frac{\Delta N}{\Delta Y}$$

which is the change in cost over the change in output, assuming that labor is the only variable cost.

---

[15] Keynes mentions another reason – that as output and employment expand, less suitable labor will be brought in, working with lower efficiency and less well adapted to the equipment.

Each sector will construct plant and equipment of the size and type required to meet normal demand at the lowest unit cost. The (vertically integrated) capital goods sector will have built plant to meet the expected normal demand (from the consumer goods sector and from itself) for replacement plus new growth (anticipated, perhaps, on the basis of population increase). The consumer goods sector (not vertically integrated) will have constructed capacity to supply the consumer demand of workers in the capital goods sector, as well as its own workers, at the normal real wage.

Each firm will be assumed to construct and equip plant of optimal size. Firms at any one stage of production can be assumed to sell competitively to firms at the next stage. When demand for the final product of the sector varies, it can be assumed that the same proportional variation is transmitted to all supplying firms at earlier stages. This requires, however, that cost changes at the final stage incorporate the increased costs of earlier stages. To avoid making adjustments for this and to keep our focus on the effects of price and real wage changes on the employment of labor, in light of changes in labor's productivity, we can provisionally assume that each firm in each sector produces the final output of that sector directly from its equipment and labor. (Alternatively, the outputs and prices for the sectors will have to be considered composites.)

The normal output of the consumer goods sector must meet the normal consumption demand from the workers of the capital goods sector, considering both in real terms. This demand corresponds to the normal profit of the consumer goods sector, and if we assume that all and only profits are invested, it also corresponds to normal growth. Similarly the profits of the capital goods sector will equal that sector's investment.

Arbitrage by potential investors, in turn, will ensure that the rate of interest and the rate of growth are equal, allowing for risk, since investors have the choice of lending or hiring managers and setting up new firms. Subtracting the earnings of management, then, the normal profit on business operations, in real terms, will just cover the costs of capital.

Corresponding to normal profits, equal to normal capital costs, there must be an equal level of net or surplus output. The consumer goods sector must produce the output required by the capital goods sector, and the capital goods sector must produce net capital goods to provide for the consumer goods sector's normal growth as well as its own. Plant and equipment will be designed to generate this normal level of net output, together with output that will cover the current costs of production.

On this basis the utilization functions for both sectors can be drawn (see left-hand figure below). Measure employment along the horizontal axis and output on the vertical. Mark off the level of net output, corresponding

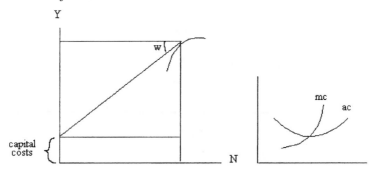

to capital costs. A plant can be built to produce this, plus the output that will cover the costs of production. There will be several levels of employment at which such a plant can be operated, as indicated by the curved line. At low levels of employment, adding workers can increase output quite a lot; after a point adding workers will not increase output much. A ray drawn to the utilization function from the point on the vertical axis corresponding to normal capital costs measures average variable product; the point where this is tangent to the function will be the optimal level of operation, and will correspond to the desired level of employment. The angle of this ray represents the real wage, and this will equal the marginal product when the ray is tangent to the curve.

This converts neatly into average and marginal cost diagrams (see the right-hand figure). Multiplying employment by the real wage gives average variable cost, while the cost of additional output is the additional labor cost. The average cost curve is therefore U-shaped, and cut at its lowest point by the rising marginal cost curve.

If we now suppose that investment stands at its normal level, and that all and only wages are consumed, the ray from the point measuring capital costs, equal to investment spending, can be interpreted as aggregate demand. If the real wage is at its normal level, output and employment will take place at minimum average cost, and the real wage will equal the marginal product of labor.

The outputs of the two sectors can be aggregated in terms of the prices embedded in the system. Normal output and normal employment are also known; but there will be a range of possible variation above and below normal levels, provided a longer period of time is considered, during which work can be reorganized. Even below and certainly above the normal level, additional employment or increased work intensity will yield diminishing increments of output. The normal level of activity is the most efficient. Plot aggregate output on the vertical axis, and total employment on the horizontal. The origin indicates the lower limit of the range of variation. This function shows the variation in output with changes in employment or intensity of work *with given equipment and technique.*

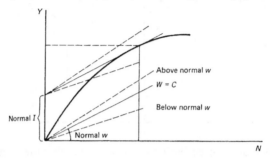

Next, a straight line rising from the origin with the slope w/p will indicate the wage bill corresponding to each level of employment, which, by assumption, will also equal worker consumption. (State transfer payments are negligible in early capitalism.) If we neglect capitalist consumption, this will also be total consumption. (The argument is unaffected by a constant level of capitalist consumption.) Then normal investment can be designated by a point on the vertical axis, and total demand, C+I, will be given, for each level of employment by a line starting from the Investment point, rising parallel to the Wage-Consumption line. Suppose this line lies wholly above the output curve; demand would exceed supply at every level of employment. Prices would be bid up, so the real wage would fall, until the demand line swung down so as to just touch the output curve at the level of normal employment (since we began from normal investment). If the demand curve intersected the output line below normal employment, profit could be increased by cutting prices, raising real wages and employment, and so increasing output to the point at which the demand line is tangent to the output curve. Hence at the level of normal activity, the real wage will equal the short-run marginal product of employment, in the given technical conditions.

Now consider fluctuations in the level of investment. If investment falls below its normal level, total demand will intersect the output curve below the level of normal employment; it will be possible then to raise profits by cutting prices and raising real wages and employment. Output will rise until the demand line is tangent, which will be at a point near to but below the normal level. When investment is above its normal level, total demand will be excessive, and prices will be bid up, until the real wage falls, bringing the demand line down, until it is tangent again, at a point near to but above the normal position. Changes in prices adjust the real wage, and employment as well as output, so that movements in consumption offset variations in investment.

Such a system is not without potential problems. Suppose that after a point the marginal product diminishes sharply, so that the average cost curve is markedly U-shaped. In these circumstances consider a substantial rise in investment spending, well above normal. To justify the required level of employment, the marginal product will have to fall to a negligible level; in other words, prices will rise sharply, reflecting the steep increase in average and marginal costs. With a given money wage, a sufficiently steep price increase will push the real wage below what Robinson called the "inflation barrier," a conventionally defined minimum standard of living. The result will be a push for higher money wages, and a wage-price spiral.

Another possible adjustment must be considered. When the aggregate demand line lies wholly above the normal output-employment curve, pressures will build to increase the intensity of work at every level of employment, so that the normal output curve will tend to shift up – and conversely when aggregate demand lies below the normal output curve at the normal level of employment. These shifts due to changing work intensity may not be very large, but they will increase the stability of normal employment, and reduce the size of the price fluctuations required to restore equilibrium. With variable prices, then, the real wage will adjust to equal the marginal product of employment, and aggregate profit will be maximized, at a rather stable level of employment.

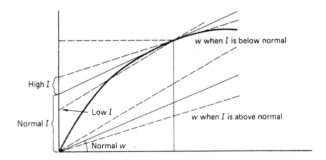

When money prices change relative to the money wage, changing the real wage, the most labor intensive industries will be affected the most. For example, when the real wage falls, money prices have been bid up, and output increases, but labor productivity falls when reorganization takes place and production rises above the normal level. Similarly, when the real wage rises, output will fall below the previous norm, and productivity will increase when employment is cut back. The more labor-intensive the industry, the greater the proportional change. This would seem to call for appropriate changes in relative prices.

Productivity

Starting from such a position of equilibrium we can consider another adjustment, which will help to explain the generally downward pattern of price movements in the nineteenth century (Sylos-Labini, 1989, 1992). Let there be an increase in labor productivity, either due to improved organization, or better work methods. This will shift the utilization function up over a substantial range. As a result, at the initial level of employment there will be excess output, driving down the level of prices, until the real wage line – and so the aggregate demand function – swings up enough to restore the tangency at a higher output and somewhat reduced level of employment. Since employment changes only slightly, and only after prices, the effect on money wages will be negligible. The benefits of increased productivity are passed on to consumers in the form of lower prices – but, it should be noted, increased productivity does have a small tendency to reduce employment, as representatives of the working classes have always argued.

Notes on the cycle

These relationships suggest a mechanism that would generate "short," crisis-ridden cycles. Suppose there is an unexpected fluctuation in demand – perhaps a sudden surge in exports. Prices will be driven up relative to the more stable money wage. Profits will boom, and the real wage will fall. The lower real wage, however, will only sluggishly encourage a rise in employment – and the rise will be proportionally smaller than the decline in the wage. Hence aggregate consumption will fall. But over the long term investment must be proportional to consumption. The decline in consumption will thus lead in the next period to a decline in investment.

This would appear to be a stabilizing pattern of adjustment – although cycles could result if the relationships were lagged. However, the working of the monetary and financial system is likely to lead to a pattern of systematic "overshooting," pushing the system into instability.

When profits rise, so that $r > i$, investment will be stimulated. Capital arbitrage will pull up interest rates. The higher rates, in turn, will encourage an increased note issue, more lending, and a greater acceptance of commercial paper. The additional money and credit will tend to push up prices even further, further increasing r, and maintaining the difference. But in the process the banks and discount houses will become over-extended – they are basing their issue and acceptances on a rate of interest which has risen because of a *temporarily* higher rate of profit. When aggregate consumption declines and begins to pull investment down again, the level of activity will fall back toward normal, but the excess funds will keep prices up for a time. While prices are still high, but real activity does not warrant investment on the scale of earnings, funds will shift into financial markets, which will continue to boom. The lower level of investment will lead to lower prices and profits. The mismatch between paper and real positions will soon show, and the failures to meet commercial paper will reveal the weakened position of the banks – leading to a run. At this point there will be a collapse, and the resulting shortage of liquid funds will lead to a collapse of business activity, driving it far below normal levels. Once the failures and reorganization have taken place, however, recovery will tend to be rapid, for the low prices will support strong consumer demand, and the failures will open space for new firms, stimulating investment.

In short, the cycle exhibits what has been called the *"saw-tooth"* pattern: a sharp, rapid upturn, followed by a crisis and a crash, then a recovery leading to normal operations, followed again by a boom. The monetary and financial system will keep prices too high for too long, and then will lead to a crash bringing them down too far. The monetary system destabilizes otherwise stable commodity and labor markets. This will appear to be in sharp contrast with a Mass Production economy.

Conclusions

Relative prices depend on the normal position and the normal rate of profit, where "normal" indicates something we have a right to expect. What is normal is not supposed to be realized at any particular moment of time. It is an average over good times and bad. Such good and bad times result from variations in investment; these fluctuations affect the realization of profits. If the fluctuations in investment average out to the normal level, then gains from above normal earnings and losses from below normal in the various sectors will average out also, regardless of differences in labor intensities. Relative prices will therefore remain fixed, while employment and the real wage will adjust in such a way as to maximize profit. The real wage will equal the marginal product of aggregate employment. But this is not an equilibrium relationship. It is the condition determining the pattern of adjustment around the normal long-term position of the system. So investment spending in the Craft economy can be expected to fluctuate around a normal level. But this creates incentives to render employment and output more flexible in relation to changes in demand.

Conclusion: the pressures for change

Transformational growth arises from the fact that the normal working of the economy sets up pressures for innovation. The inflexibility of output and employment creates problems which help to explain the

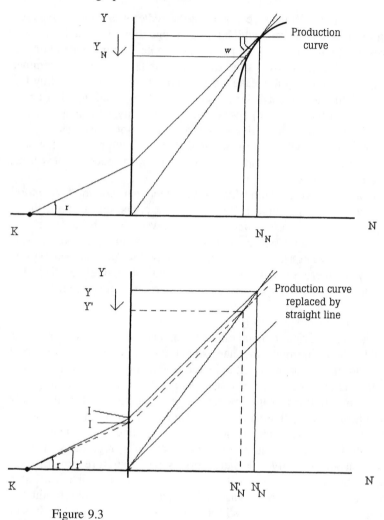

Figure 9.3

changes in the Craft economy which have brought about the system of Mass Production.

A Craft economy has a pattern of stabilizing adjustments, which depend on market-clearing prices. From the point of view of the system this may be good, but from the point of view of individual producers these adjustments have some undesirable properties. For example, when demand falls off, production will be run more slowly, but, exceptions aside, a nearly full labor force will still have to be on hand. From the individual owner's point of view, this is an unfortunate expense, which necessitates injurious price competition. Conversely, when demand is strong, the rate at which production can take place will depend on the morale of the labor force, and its willingness to put in extra effort. It will be expensive to increase output; additional employees will add diminishing increments of output. The capitalist does not have full or satisfactory control over the pace of work, or the level of costs.

What the capitalist needs, first, is greater control over the process of production, especially over the productivity of labor and the pace of work. Motion and time studies were developed to provide just this (Barnes, 1956). Second, firms need to be able to vary costs when sales are fluctuating, which requires being able to lay off and rehire labor easily, which, in turn, depends on being able to vary the scheduling of production. To do this start-up and shut-down costs must be minimal. Third, firms need to be able to store output without spoilage. If when sales fall, output can be cut back, and along with output, costs can also be cut, and at the same time, unsold inventory can be stored without significant loss or other costs, a great deal of pressure for potentially ruinous price-cutting will be lifted. Part of the problem thus reduces to a technical question, which is, How can production be run at less than full blast, without all workers having to be present? Further, how can production be started up and shut down, easily and costlessly, so that a drop in demand can be met by running short-time?

The normal working of the system throws these questions up; the answers will help businesses to compete more effectively. Inflexible output and costs, resulting in overproduction and cutthroat price competition, is potentially ruinous. The problem can be illustrated by drawing on the diagrams developed earlier in this chapter.

Measure Y on the vertical axis, N on the horizontal to the left of the origin and K to the right, as before. Mark the point of "normal" output and employment, and draw a production curve through it. Then when demand falls below this level, prices will fall, the real wage will rise, employment will decline somewhat and both the share and rate of profits will fall. Mark off a level of lower-than-normal demand such that prices fall, the real wage rises, and employment falls just enough to reduce profits to zero.

Now consider on a second diagram what happens when the production curve flattens out. Prices fall less, and employment more, so the real wage rises less, and profits do not decline as much. When the curve becomes a straight line, prices do not fall at all, and the full burden of the adjustment

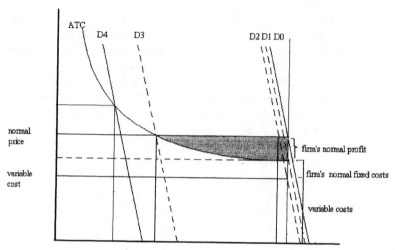

D0: normal price, normal output, normal profit
D1: flex price - firm's profit falls to zero   fix price - very small drop in firm's profit
D2: flex price - shut down point   fix price - about 1/4 of firm's profit lost
D3: flex price - long since bankrupt   fix price - no profit, only fixed costs
D4: flex price - long since bankrupt   fix price - shutdown point

Figure 9.4

falls on employment. Profits still decline, of course, but the decline is minimized.[16]

Now consider a price diagram, in which we contrast a perfectly inflexible, early Craft system with a perfectly flexible Mass Production pattern of output adjustment. We assume a steep and inelastic steep

[16] The role of demand here plays to a concern of Patinkin's, who has recently written, "For a long period before the writing of the first edition I had been puzzled by the apparent contradiction between the intuitive feeling, on the one hand, that there was a connection between a firm's product-output and its labor-input, and the traditional demand curve for labor, on the other hand, that did not depend explicitly on output and whose sole independent variable was the real wage" (p. xvi). Patinkin's answer, of course, was to develop quantity-constrained demand curves (although he admitted that these were problematical) to overcome the conventional labor market theory's independence of aggregate demand. But the above marginal productivity analysis faces no such problems; for it is precisely variations in aggregate demand that give rise to the changes in the real wage that make (the relatively difficult) changes in employment and output profitable.

demand curve, on the grounds that empirical studies have shown this to be common for manufactured goods. Further, the theory of demand based on lifestyles, developed in the next chapter, will show this to be a plausible case. In any event the object of the exercise is to present a vivid, even dramatic, contrast between flexible and fixed prices, and such a demand curve serves admirably.

The diagram is similar to those used earlier, but here the ATC only includes contractual costs. The normal price, however, is set at the level which will provide normal profits. Variable costs are constant. Then in the Craft economy, with normal output fixed, when demand falls, price will fall; but in Mass Production, with output and employment flexible, output will adjust and price will remain fixed. The effects on profit are indeed dramatically different:

When demand successively shifts down and in

D_0: we have normal price, normal output and normal profit

D_1: Flexprice – Profit falls to zero
Fixprice: Small decline in profit

D_2: Flexprice – Shutdown point
Fixprice: Less than a third of profit lost

D_3: Flexprice – Long since bankrupt
Fixprice: Zero profit; fixed costs covered

D_4: Fixprice: Shutdown point

This should clearly demonstrate the benefits of developing technologies that will make it possible to respond flexibly to changes in demand.

The implications of this change need to be explored. The market creates the incentives which lead to the technological developments that make labor a cost that varies with output, which in turn varies with demand. The result is an improvement in the flexibility of the firm's response to changing market conditions. But because prices no longer adjust, when one sector experiences a change there will be no compensating variation in the other sector. Nor will there be shifts of resources between sectors. By successfully stabilizing prices, business has eliminated the mechanisms that stabilized output.

But, in fact, business will not be satisfied by stabilizing prices. It will attempt systematically to gain greater control over the production process, substituting mechanical power for labor power, and mechanical or electronic control for human skill, as far as possible. In general the larger the scale on which operations take place, the better the prospects for doing this – an aspect of economies of scale – "the division of labor is limited by the extent of the market."

These pressures also tend to change the nature of competition; previously it centered on prices in a comparatively simple way. But now the chief focus will become technological development, especially in relation to market share – since an increase in share can permit a larger size, which in turn will make it possible to extend the division of labor, reaping economies of scale that in turn will permit the consolidation of a lower price, and so on. Once an advantage is achieved, a "virtuous" cycle develops, enabling the successful firm to establish a leading or even dominant market position.

With competition centering on a race for improvements in technology, the strategic situation of firms changes. Firms will no longer seek to establish their optimum size and remain there, and it will no longer be possible to permit new firms to supply the growth of the market. For new firms will be able to build plants and buy equipment embodying the latest technology; they will therefore be able to establish a cost advantage and invade the markets of established firms. Hence existing firms must invest regularly, incorporating new technology into their plants, and growing enough, so that taken together, they will supply the expansion of the market. Prices will therefore have to be set with an eye to providing the profits that will finance growth. To facilitate such growth, firms will withhold profits, investing them directly, rather than distributing them to shareholders. Both the pattern of competition and the working of financial markets is altered by the move to Mass Production technology and the shift from extensive to intensive growth.

Consequently, when the new technology becomes widespread the system as a whole will begin to work differently. First, when demand for a particular good falls, say, due to a general decline in investment spending, workers will be laid off, and output reduced. Prices will tend to fall comparatively little – if at all. Thus instead of consumption varying inversely with investment, as it did when a decline in demand led to a greater proportional fall in prices than in money wages, consumption will now also decline, since the laid-off workers and their families will now have to curtail their consumption spending. But this reduction in consumer spending will itself lead to further layoffs. The elasticity of consumption with respect to investment will now be positive.

PART V

Investment and Mass Production

In Craft conditions growth is driven from the supply side. The growth of the labor force sets the pace of capital accumulation, through its influence on real wages. Capital accumulates as savings are assembled by the financial system and made available to new firms, of the same size and character as the old. All this changes with the advent of Mass Production.

In the era MP/CA growth is demand-driven. Investment plans depend on the expected growth of the market; this is essentially a generalization of the "accelerator" principle. So we need an understanding of how and why markets grow, to provide the basis on which business can develop firm expectations of market expansion. Market growth depends on the expansion of incomes, on the one hand, and on the willingness of households or firms to extend their use of a product or of new groups to adopt it, on the other. A concept of normal market growth can be defined but can also be seen to depend somewhat precariously on particular conditions.

These investment plans cannot be developed without a consideration of pricing policies. When demand increases, capacity must rise in pace. Yet the growth of demand is not itself independent of pricing – lower prices will encourage market expansion, higher will discourage it. On the other hand prices provide the profits on which to base the financing of investment – higher prices enable a larger expansion of capacity; lower, a smaller. These are not actual prices, of course, but planned prices, the benchmarks that will be used to evaluate projects and arrange financing. Two cases can be distinguished, where the market growth rate is independent of the choices of firms, and where firms can affect the growth rate. Markets tend to evolve from the former in the latter, through investments designed to attract customers and develop brand loyalty. Yet it can be shown that actual prices tend to stay close to the benchmarks. Prices and investment plans will be the subject of Chapter 10.

Investment plans must be put into practice (see Chapter 11). Plans can be speeded up, or slowed down, depending on current conditions. These effects can be neatly contrasted with the adjustment process of Craft economies in an elasticity formula. Price changes in Craft economies involve changes in spending due to changes in real income – except when firms shut down. This last is a change of status, a discontinuous change, from solvency to insolvency, from employed to out-of-work, and signifies a breakdown. Under Mass Production, however, such changes are institutionalized, for example, in the "layoff" system, where workers are laid off in an orderly fashion when demand slows down, and hired back when it picks up, in the meantime supported by unemployment insurance. A simple theory of consumption can be formulated on this basis and can be developed into a multiplier-based account of the labor market, showing a positive, rather than inverse, relation between real wages and employment. Moreover, labor market adjustments will tend to be ineffective or destabilizing.

A major feature of the Mass Production economy is a built-in propensity for inflation, a base rate deriving, surprisingly, from productivity growth. Inflation is a process which reflects conflict: it is defined here as a wage-price spiral, which will normally tend to benefit capital at higher levels of unemployment, but may squeeze profits at low levels.

Effective demand is indeed the key to understanding the behavior of the economy as a whole. Capitalism is "demand-constrained." To understand what this means calls for contrasting a demand-constrained system with a resource-constrained one – such as socialism. In doing so, we also discover that the Harrod-Domar approach actually is no help in analyzing growth. Growth must first of all be approached as the *growth of demand*, a point we meet in Chapter 10.

CHAPTER 10

Demand growth, pricing, and investment plans

In Craft-based economies investment will tend to be "regular," that is, to be governed by the growth of the labor force, which itself is the result of the structural tendency of labor to be released from agriculture and flow from the countryside to urban areas (Nell, 1992). Growth takes place through the establishment of new firms on borrowed or entrepreneurial capital, so the rate of interest imposes a tendency to uniformity of the profit rate. Thus we saw that investment in Craft economies could be broadly described as supply-driven, that is, governed by the natural rate. But in Mass Production economies investment will be governed by technical progress and the expected growth of demand; investment will not be "regular," and, moreover, the growth of demand will be influenced by the pricing policies adopted in the course of investment planning. Financial markets will encourage seeking out the highest profit rate, but any tendency to uniformity will be weak, because growth in different markets will be different.

Prices and investment plans will therefore be determined together, in the light of expected growth. These will be "benchmark" prices and will serve as guidelines to current behavior; however, there are good reasons for market prices to stay fairly close to the benchmarks.

Market adjustment under Mass Production will take a different course than in Craft economies. Adjustment in those economies depended on prices; but in the era of Mass Production prices tend to be stable, and quantities adjust, even in the face of large swings in demand. To understand this it is necessary to examine the way prices are set and adjusted. That will be the chief purpose of this chapter.

There are two important qualifications to the claim that prices are insensitive to demand in Mass Production, however, and each will play a role in later discussions. The primary sector and within manufacturing, machine tools both tend to retain certain Craft-like characteristics, as a result of which prices retain a certain degree of flexibility with respect to variations in demand. As the argument of this chapter will show, this is not the case for Mass Production manufacturing in general.

Even when carried out with technology adapted from Mass Production, primary production tends to retain Craft characteristics. Once a level of activity has been determined – fields planted, oil wells drilled, mines opened – varying the level of output tends to be difficult and costly. Moreover, even if output can be varied, current running costs are a low proportion of total operating costs, so the firm's cost position will not be much affected. To add to the difficulties of adjusting output to demand, storage costs tend to be high for primary products. Either they tend to spoil, as with agricultural goods, or they are bulky and costly to handle, as in the case of oil. Carrying excess output over from one period to another may not be worth it. Hence primary products tend to be price-flexible even in the Mass Production era, and the analysis developed below does not apply to them.

Besides the primary sector, one section of manufacturing itself, and paradoxically, the most advanced, also retains certain Craft-like characteristics. *Machine tools*, the very heart of Mass Production, seldom if ever produces for the market. Its orders are specific, and specialized, and it tends to develop working relationships with customers. Workers are highly skilled, with a good knowledge of practical engineering, and take as much as four years to train. Work teams and morale are important. Firms are small or medium-sized, run by hands-on managements. Because of the strong connections between customers and firms, and because orders are design-specific, direct price competition is not frequent, and bidding wars do not occur in times of slow growth and weak demand. Moreover, given the engineering versatility of this sector, in slack times, firms are able to shift part of their capacity into other lines of activity – producing consumer goods, or engaging in government-sponsored R&D. Weak demand does not lead to falling prices. By contrast, however, in times of rapid growth, the machine tool sector constitutes the ultimate bottleneck; long-run growth cannot exceed the capacity of the machine tool sector. And if growth is to move to a higher rate than present capacity will allow, to expand that capacity will require *reducing deliveries to other sectors*, for a period of time (Lowe, 1976). High growth and strong demand pressure can therefore be expected to drive up prices in the machine tool sector.

Growth and demand

Demand and prices

To plan their market strategies, firms must know how the market will react to changes in prices. This depends on the composition of the prospective customers in terms of social class and income level, on the one hand, and on the way they have planned their household budgets, on the other, bearing in mind that this will take different forms in different classes. To keep matters simple we will distinguish only two classes here – the professional and managerial, and the working class. The professional/managerial class is hierarchically organized; managers and professionals exercise authority and receive pay in proportion to their rank, which in turn must be achieved on a foundation of skills and education, which are then manifested in job performance, earning promotion. The working class does not compete for advancement; promotions are based on seniority. Within each class there may be other hierarchical relationships also; some groups will tend to set the fashions for the rest, and in some but not all fields, the highest level of the professional class will set the standards for the whole society. (A class is a class of families each earning similar incomes by performing jobs of similar status, exercising the same kinds of skills, and producing a new generation capable of occupying the same social positions, while also reproducing themselves.) We shall assume that these are the only two classes.[1]

Significant differences in household economic behavior exist between these classes – they do not just characteristically consume in different patterns. Technological innovation brings new categories of employment. In particular with the shift to MP/CA (but already evident in CBF/MA), the development of the modern corporation opened a whole new class of managerial and white-collar jobs. Formerly working-class families could see opportunities; traditional middle-class families could see the need for reskilling, redirecting or upgrading their educational attainments. "Self-betterment," rising in the world, could become a realistic ambition for a large fraction of the population. So the aspiring professional and managerial classes will seek advancement along a career path; they hope and expect to rise in the social order, to find regular promotion to positions of higher status and pay – and they expect and hope that their children will also advance. For these families, the consumption problem breaks into two parts: First, there is the choice of the cheapest means to achieve the standards required by their lifestyle, and then, secondly, there is a further choice of the best strategy of *investing* in education and career. They therefore plan their normal household budget with an eye to setting funds

---

[1] In early or Fixed Employment capitalism an appropriate simplification would be to consider the two classes of families to be those of the owners of firms, and the families of workers. But in advanced or Flexible Employment capitalism, capital is no longer lodged with families; it is held in professionally managed portfolios and/or by financial institutions. The former owners are now beneficiaries or largely passive recipients of interest and dividends; their interests in life center around their careers (cf. Chapter 13).

aside, after meeting the requirements of the household, to invest in furthering their or their childrens' careers.

By contrast, members of the working class do not enter on careers, cannot normally expect much in the way of promotion (as opposed to raises) and do not expect to rise in station in life. Their extra time and income will be spent on hobbies, vacations, and leisure-time activities.

Families in each class must therefore spend their incomes in ways designed to meet the expectations of the world. Social life is governed by norms regulating the foods we eat, the clothes we wear, the kind of houses we live in, and so forth. These norms will differ for the two classes, but families in each must maintain the ability of their breadwinners to work adequately and attentively – at a minimum they must be well-fed and rested, dressed properly, with suitable transport to work from homes comfortable enough for them to bring up the next generation. The consumption of each class will be guided by a conception of its appropriate lifestyle, given its place in the social pyramid. (Such styles are displayed and debated in the media of the day, whether radio, TV, and movies – or in earlier times, newspapers, novels, and traveling players, or even sumptuary laws and the pulpits of the churches.) Failure to "keep up with the Joneses" will result in loss of status, hence of opportunities, preferential treatment, contacts, and so on. Social competition enforces adherence to the norms.

A lifestyle will be expressed as a vector of standards, $s_1, \ldots, s_m$, where the elements express the level and style that must be achieved in each area, such as nutrition in diet, elegance in clothing, size of housing, and so on. Observe that a lifestyle is specified by levels of characteristics (Lancaster, 1975), not by quantities of goods.[2] The consumption of goods contributes in various ways to achieving levels of these characteristics, posing the problem of choosing the right mix of goods.

Given the basic requirements of life and the appropriate designations of style (including normal saving for retirement), each class will face a similar budget problem: to minimize the expenditures needed to cover these basics in adequate style, in order to release resources for investment

---

[2] A fuller and more sophisticated treatment would consider the possibilities of trade-offs between the characteristics composing the lifestyles. This must be done carefully: nutrition and taste may trade off in diet, and elegance and comfort may trade off in clothing, or shelter. The possibilities of substitution are much less, however, when it comes to the broader categories of diet, clothing, and shelter. A smaller house for better clothing? Eat cheaply to dress smartly? On occasion, no doubt; but as a general rule a certain standard will have to be achieved in each category, but the significant trade-offs will be in the way it can be reached. For our purposes, however, these questions can be set aside.

in advancement, or for hobbies and leisuretime activities, respectively. Thus both classes face a straightforward optimizing problem:

$$\text{Min}: \; C = p_1 q_1, \ldots, p_n q_n$$

$$h_{11} q_1 + \cdots + h_{1n} q_n \geq s_1$$

$$h_{21} q_1 + \cdots + h_{2n} q_n \geq s_2$$

$$\overline{\phantom{h_{11} q_1 + \cdots + h_{1n} q_n \geq s_1}}$$

$$h_{m1} q_1 + \cdots + h_{mn} q_n \geq s_m$$

$$q_1, \ldots, q_n \geq 0$$

Here C is the cost of living, $p_1, \ldots, p_n$ are the prices of consumer goods, $q_1, \ldots, q_n$ are the choice variables, namely the quantities of the various goods that will be purchased by the household budget, $s_1, \ldots, s_m$ are the lifestyle parameters, representing the level and quality required of the various characteristics making up the standard of living, and the coefficients h_{ij} indicate the contribution of each of the n goods consumed by households toward attaining the required levels of each of the m standards. Given the prices, C will be minimized by the appropriate choice of goods, subject to meeting the standards making up the lifestyle.

This problem has a dual, namely to maximize the value of the lifestyle, by assigning imputed values to each of the standards. This maximization will be subject to a set of constraints. These will show that the sum of those valuations, when each valuation is multiplied by the coefficients giving the contribution of each good to the respective standard, must be less than or equal to the price of each good:

$$\text{Max} \; L = w_1 s_1 + w_2 s_2 + \cdots + w_m s_m$$

subject to:

$$w_1 h_{11} + w_2 h_{21} + \cdots + w_m h_{m1} \leq p_1$$

$$w_1 h_{12} + w_2 h_{22} + \cdots + w_m h_{m2} \leq p_2$$

$$\text{---}$$

$$w_1 h_{1n} + w_2 h_{2n} + \cdots + w_m h_{mn} \leq p_n$$

where

$$w_1, w_2, \ldots, w_m \geq 0$$

The value of the lifestyle is maximized by imputing values to its component standards. For each good, these values times the contributions of the good to each standard, cannot add up to more than the price of the good. Since the optimal solution to the primal is also the optimum of the dual, choosing the right combination of goods maximizes the value of the lifestyle.

This is not "simple maximizing." It is one thing to assume that agents buy cheap and sell dear, or pick the best product or the most desirable job from among those readily available. It is quite another to suppose that they solve a complex programming problem, or act "as if" they did. A justification is needed. First, the motivation is clear; the cost of meeting the standard of living must be kept down, if there are to be funds to support the drive to rise in the world, or to permit leisure time activities, depending on class. Second, the calculation does not require any esoteric or unavailable information. Third, those who get the solution wrong will find themselves short of funds for advancement and so will tend to lose out in the race to rise (or will have less to spend on leisure – and will be subject to pressure by family and friends). Finally, what households consume is open to everyone to see; so a better solution, a cheaper set of goods that does the job, should quickly become evident, and will be widely imitated. Thus, while *calculating* the solution cannot be assumed, *trial and error* can be expected to find it and competition will lead to its adoption.

Once it is determined how to achieve the standard for minimum cost, the second problem emerges. For the working class (and those in the professional class whose careers have peaked), the question is how to spend the discretionary component of their income most agreeably. Hobbies, vacations, sporting events, entertainment, and do-it-yourself activities, all will figure here. But precisely because these are luxuries, spending on them may be subject to relatively sharp fluctuations resulting from changes either in fashion, or in views of the future.

Professional households, on the other hand, must examine their opportunities for advancement and decide, given the uncertainties, what their best investments will be. Each opportunity may be considered to have a present cost (perhaps in foregone earnings as well as actual payments) and to yield an (uncertain) stream of future benefits, in the form of higher status and salary-plus-benefits. When present values are calculated, the projects can be ranked in terms of internal rates of return; investment will be limited by the excess of income over the cost of meeting the expected standard of living, plus whatever can be borrowed. (A more sophisticated treatment might analyze possible trade-offs between investment in career advancement and meeting the normal standard of living. Here, however, we assume the decision ordering to be lexicographic.)

A simple example will illustrate the principle. The cost of an education is the foregone earnings during the time spent learning, wN_{hi}, where w is the normal wage and N_{hi} is the labor-time of the i-th household. This sum must be borrowed at the going rate of interest, or assembled from savings. But in the latter case, it could be used to buy a bond, so the interest rate will be its opportunity cost. Each period after the acquisition of learning there will be a gain to the household, resulting from the higher wage now obtainable by the household's labor; this will be $(w^* - w)N_{hi}$. The one-period rate of return on an investment in education will therefore be $(w^* - w)/w$. Now consider the long-term: both w^* and w will increase at rate g, reflecting the growth of productivity and the advance of capital. So the stream of returns will be

$$(w^* - w)N + (1 + g)(w^* - w)N + (1 + g)^2 (w^* - w)N$$
$$+ (1 + g)^3 (w^* - w) N + \cdots$$
$$= (w^* - w)N [1 + (1 + g) + (1 + g)^2 + (1 + g)^3 + \cdots]$$

But this stream of returns must be discounted each period by the rate of interest, appropriately compounded, forming a series of discounting terms:

$$1 + (1 + i) + (1 + i)^2 + (1 + i)^3 + \cdots$$

Thus if the rate of interest equals the rate of growth, the discounting and growth effects will cancel, and the long-term rate of return will also be $(w^* - w)/w$. This can therefore be directly compared to the rate of interest, and it will be worthwhile for households to invest in projects of learning and the acquisition of skills so long as their anticipated rates of return are greater than or equal to the rate of interest.

Such calculations ensure that spending for advancement will not be squandered. But the objective of such spending is not simply to increase earnings; it is to rise in *status*. As firms compete with one another for market shares, households compete for status.

Now consider the equilibrium of a representative family, say, of the professional class, assuming two goods and three constraints (see diagram).

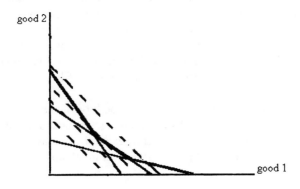

So long as the price ratio lies between the slopes of the first two constraints, price variations will not affect the optimal purchasing plan. But if the ratio comes to coincide with the slope of the first constraint, good 2 will drop out of the budget altogether, and there will be a large increase in the purchase of good 1 – which could be the result of a rise only in the price of good 2; the price of good 1 itself could be unchanged. The new corner optimum will be at a higher cost level, leaving less for investment in career advancement, which will therefore have to be cut back or postponed. So a small rise in good 2's price could lead to its complete elimination, together with a large rise in expenditure on good 1, and a cutback in spending on education. (Notice that this could also happen with given prices, as a result of a change in the required standards.)

The moral is that price changes may have little or no effect over a significant range, and then suddenly have large effects, sometimes in unexpected directions. Of course, households will not all be the same, so in the neighborhood of the point at which a representative household would change its consumption pattern, we can expect to find a distribution of actual households, some changing at prices above, some at prices below the representative point. Changes in social standards and expectations may have similar effects. New goods and services can easily be introduced if they can be shown to achieve the same results at lower cost. (As the price of a new good falls, and it is increasingly adopted, it displaces one or more older goods; the demand curve for the new good has an echo in the displacement curve of the old.) Emulation effects normally follow the social hierarchy; the consumption styles of the rich and famous set standards to which the rest aspire (or sometimes, against which they react). When prices can be reduced or money incomes raised sufficiently, elements of more prosperous lifestyles will be incorporated into those of lower levels, usually in modified form. (The stereo systems of the rich become the boomboxes of the poor.)

Demand, then, is based on class structure, reflecting money incomes, and on price. There are two kinds of effects, which we may call "composition" and "incorporation" effects. Price changes lead, as a rule discontinuously, to changes in the composition of household budgets. Even substantial price changes may have only small effects over some ranges; then small changes may have large effects. In general, however, quantities purchased will move inversely to prices, as a result of household budget planning.

The other effect depends on the cost of a good, or of the meeting of a standard, in relation to income. Substantial changes in prices will alter the range of possibility for income groups. (For example, a substantial cut in the price of television sets, in the mid-1950s, dramatically increased the market.) Lower prices (or cheaper versions) bring goods previously available only to the well-to-do into the purview of lower-income groups, allowing them to incorporate those goods into their budgets, creating larger markets. Alternatively, with the same prices, and the same degree of inequality, higher money incomes will increase the size of the market. New products can be introduced via this effect – and old ones phased out.

The normal growth of demand follows the development of a dominant lifestyle and its extension through each of the main social classes. Since a lifestyle is a way of using goods and services to provide the basic social requisites (food, clothing, shelter, education, transportation, communications, etc.) it distinguishes its components as belonging to a certain class in a certain epoch. Goods and services are fitted together in a way which characterizes them as belonging to one another: modern furniture belongs in a modern house, Victorian furniture in a Victorian. Different lifestyles can to some extent be mixed, it is true, and there can be differences of opinion as to what is "correct" within a given lifestlyle. But a lifestyle nevertheless fits together in a coherent and largely determinate way the methods of filling the basic social functions.

As a result, since the social functions are complementary, the dominant methods of filling them will be also. Of course, even within a given lifestyle a social function can normally be filled in a number of alternative ways, substitutes for one another, so that price changes will lead to changes in consumption. But these substitutes are all goods or services of the same *category*; tea and coffee are both beverages, rice and potatoes are both carbohydrates. Within categories, substitution holds, but between categories complementarity tends to be the rule. Pork and beef are substitutes, but meat and vegetables are not (unless, of course, one becomes a vegetarian – but that is a different lifestyle). In production, steel and fiberglass are alternative materials, but energy cannot be substituted for materials. The more specific the good, the greater will be the possibilities of substitution, the more general the category, the more fixed the complementarity. At the most general level, there will be the relationship between output of consumer goods and the coefficient of capital goods. The growth of consumer goods will require equiproportional growth in capital goods.

So if we choose a high enough level of generality in the categories of goods it will be reasonable for the short run to assume fixed proportions in both consumption and production, bearing in mind that our starting point is a social order in which there is a dominant lifestyle, supported by an already existing capitalist industrial system. The implication of fixed proportions, of course, is that all consumer goods will grow at the same rate, and this will translate into the same rate of growth in all capital goods. Over the long term, however, both technical changes and rising incomes will lead to changes in lifestyles.

So we have an irregular but inverse relationship between price and the amount demanded of a commodity. This holds for a representative household from either social class, permitting the formation of aggregate demand functions. There are good reasons to expect it to be rather price-inelastic. Likewise, demand depends on income; when the income of a class or subclass rises we can expect the demand-price relationship to shift out, making it possible to analyze the growth of demand when income distribution or class structure changes, such as the rise of the professional middle class. (Somewhat less reliably, it will shift inwards when income falls – though probably not in the short run.) With these two relationships, we can now develop an approach to pricing and investment that will permit us to consider innovations – introducing new goods, and phasing out old.

"Normal" growth

So can we speak of a "normal" rate of growth of demand? Consider this question in two stages. First, if there were a strong and growing demand for a sufficiently large portion of net output, then the rest of the surplus would have to be invested to produce a growing output to meet that growing demand. Further, the interdependence of industries would require all basic industries to grow in tandem, so that planned capacity could be kept at the level where it would be utilized normally. (In *practice*, of course, the *actual expenditure* carrying out investment plans proceeds unevenly, responding within the framework of the plans, to all sorts of short run and speculative influences.) Hence if a reason to expect a growing demand for a suitably large portion of net output can be found, then the rest of the system will also grow at that rate. This is the second stage – to define a set of forces determining a growing demand for a large portion of net output.

Consider a general increase in the levels of real wages and salaries. A sufficient increase could be expected to lift a section of the upper level of the working class to a level where they could command the resources to invest in "self-betterment." (Only a fraction of the working class can plausibly be assumed to have the potential for self-improvement.) As they are, *ex hypothesi*, economically motivated, they will join the general scramble to rise in the world, investing in their own and their children's future. They will thus add to the market for higher-quality goods and for goods higher on the Engel Curve – social goods, like education and communications (which, in turn, will stimulate secondary investment). They will increase their productivity and their future earnings, thereby improving their present borrowing power, which they will exercise in the attempt to rise. So a one-time, permanent rise in the level of real wages and salaries will set in motion a pattern of *continuing* investment in family self-betterment, which, assuming it is successful, will steadily improve the fortunes of those families. Thus a rise in the real wage will set in motion an increase in the *growth* of family expenditures. (Of course, such a stimulus to the growth of demand cannot be expected to last indefinitely – it will surely peter out after a generation, if not before.)

This suggests a simple theory: a positive relationship between real wages and the growth of demand – with growth on the vertical and wages on the horizontal axis, a curve rising from left to right – could be paired with the inverse relation which is dual to the wage-profit trade-off, and which connects real consumption to the growth of output. Assuming all and only wages were consumed (and interpreting wages to include salaries) the intersection of these two would then determine the real wage and the rate of growth. This can be illustrated with a diagram.

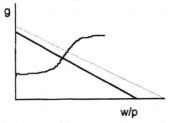

The solid line sloping down from left to right is the consumption-growth trade-off. The rising curve shows the real wage growth of demand relationship. It has a positive intercept, indicating that demand will be growing even when real wages are at the base level. At low levels of the wage, increases will have little effect, and at middle levels the effect will be strong, but it can be expected to flatten out again at high levels, once all the families with the potential to rise have been stimulated to do so.

The dotted line, however, indicates a further point: the very investment in self-improvement that generates growth in demand will also increase productivity, causing the consumption-growth trade-off to shift out and up. A movement *along* the real wage growth of demand curve will tend to cause the other curve to shift. The two curves must be supplemented by considering the effects of higher demand growth on productivity. (In general, under Mass Production, a rise in the growth of demand can be expected to lead to an increase in productivity, since production on a larger scale will permit a superior division of labor and greater specialization.) A function relating productivity growth to demand growth and real wages must be added – bearing in mind, however, that higher productivity growth, in turn, is likely to have a positive effect on real wages. Demand growth will positively affect productivity growth, probably with a diminishing impact, however, and reaching a maximum at some point. Real wages will also have a positive effect (for reasons already given), but only up to a point, after which still higher real wages are likely to have a negative impact on productivity growth.

We could write out three equations with plausible shapes, in three unknowns:

$$g = g\left(\frac{w}{P}, x\right) \quad g'_w < 0, g'_x > 0, \text{ linear}$$

$$\frac{w}{P} = w(g, x), w'_g > 0, w'_x > 0, \text{ but sigmoid in shape}$$

$$x = x\left(g, \frac{w}{P}\right), x'_g > x''_g < 0, x'_w > 0 \text{ up to a point, then } x'_w < 0$$

This (plausibly) assumes that (for a given w/P) there is some level of g beyond which x will no longer increase, and also that (for a given g) at some level of w/P x will reach a maximum and begin to decline. These assumptions effectively bound the level of x, ensuring that the system of equations will have a solution.[3]

But the two central relationships are between the *real* wage and growth; yet in Mass Production there are no forces that work directly on the real wage, and therefore the market will not tend to establish such an equilibrium. Suppose the actual growth rate were too high or too low: what would the effect be on real wages? Suppose the real wage were too high – above the equilibrium; then the growth rate of demand would lie above the growth of capacity. This could be expected to bid up prices. That would lower the real wage – but under Mass Production, a rise in prices would engender a wage-price spiral; but which would rise faster? We shall see that there are circumstances in which, in a boom, wages tend to rise faster than prices, raising the real wage. On the other hand, if the real wage began too low this would certainly *not* lead to a fall in prices, raising it in a stabilizing manner. The idea that real wages induce the

[3] A simple and reasonable linear version of this model can be written

$$g = G - \frac{aw}{P} + hx$$

$$g = \frac{bw}{P} + jx$$

$$x = cg$$

where a, b, c, h, j > 0. A solution is easily found:

$$\frac{w}{P} = \frac{G(1 - jc)}{a(1 - jc) + b(1 - hc)},$$

and it is sufficient for $w/P > 0$ that c, h, j < 1.

growth of demand through self-betterment – the redistributive influence – is important and will come up again.[4] But to develop it further we need an understanding of how prices and wages are actually set in conditions of Mass Production.

Growth of demand scenarios

There are other important causes of the growth of demand. Two scenarios come to mind: one for economies of the last century, more or less, the other for the postwar world. The first depends on colonies and/or the frontier. Population moves to colonize or conquer new territories. There is a steady flow of population to the new area. Then as it grows, the new colony demands growing imports of manufactured goods from the home country. These will be demanded in fixed proportions, reflecting the dominant desired "style of life," and the technologies that the colony wishes to import. (Of course, such proportions are not strictly fixed, any more than they are in the industries the Classical Equations describe. But variations will be limited.) The home country must therefore grow at an appropriate rate in order to satisfy this growing export demand. (The goods exported will be the composite commodity needed to establish the lifestyle desired by the colony. The colony, in turn, will supply the home country with better and cheaper raw materials and foodstuffs, raising productivity and reducing rents on land and mines.) Such export growth is *not* a short-run matter, nor is planning to satisfy such a growing demand a question of animal spirits. Nor does growth depend on interest rates, etc. The ultimate determinant will be the rate at which the colony grows, a matter about which business and government may make informed judgments (and develop policies, e.g., emigration policies). Given this rate of growth, the home country must divide its surplus into exports and investment, so that the investment out of the surplus will be just sufficient to increase the output of exportable goods at the rate the colony is growing. (In the same way, the increase in population must be divided into those that will emigrate, and those that will take the new jobs that home investment creates.)

---

[4] An important practical example: the G.I. Bill of Rights, enacted by Congress following World War II, underwrote self-betterment for a generation of young Americans, providing college educations, job training, mortgages, and medical care, thus giving a strong stimulus to family formation, and purchases of consumer durables, helping to sustain a strongly growing consumer demand for a generation. In turn, this strong demand, especially for housing and consumer durables, stimulated investment. A simple model in which demand growth is generated by policy-based redistribution is presented in Nell (1988), ch. 10.

The second scenario can be built around the Welfare State. With the decline of the extended family and the shift from rural to urban, the training of the young and the care of the aged increasingly devolve on the state. Urbanization and mass production create new kinds of health problems, as well.[5] As the economy shifts to modern industrial methods, spending on education, pensions, and health will have to grow. (Once the economy is fully modernized, such spending will grow in pace with population – or at a higher rate if the age composition of the population is shifting to the elderly.) Again the demands that such growth sets up will be for "composite commodities" consisting of the goods that make up a lifestyle. Subject to the well-known trends in consumption patterns – Engel Curves – these will stand in fixed proportions. As these demands grow, the industries which supply them must grow, hence the means of production must grow, etc. Hence again, the basic sector must expand. Of course, in more modern economies, the "normal" level of capacity utilization may be a range, rather than a point. Hence the connections between industries will not be so tight.

The preceding scenarios define a reasonably uniform balanced rate of growth for the system as a whole. A third scenario can be imagined, however, in which definite expectations of growth can exist, but in which different sectors will be expected to grow or decline at different rates. Consider an economy in which new products and processes are replacing old. Industrial products may displace domestic production – as canned foods drove out mother's preserves, and ready-made dresses those stitched up from patterns. They may displace craft products, as ready-made suits displaced the local tailor, and precision tools the blacksmith. Or whole new sectors may emerge – the railroad, automobiles, airplanes, not only displacing older modes of transport, but vastly increasing productivity and opportunities in the process. The economy as a whole grows partly because the new capital is more productive than that which it displaces, partly because, being cheaper, the new goods are more widely diffused, so that the *scale* of the new is larger than what is displaced, and partly because the new products and processes have "external effects" that increase

---

[5] The development of Mass Production leads the traditional realm of the family to shrink. A century ago, care of the aged was largely handled through the family; two centuries ago, most education – particularly vocational training and the learning of skills – took place within the family. But the nuclear family cannot care for the aged in an urban setting, especially if the family has to move frequently. Nor can parents teach their children skills they do not have – but technical progress regularly changes the skill mix on the job market. Many kinds of production were handled domestically. A large part of the growth of the market for industrial products has taken the form of shifting these activities out of the home and away from the family.

productivity in other sectors. All that is necessary is that the various growth rates of the sectors be reliably forecast, and that expectations be consistent with the interdependence of the sectors. (If a sector is growing rapidly, then its demands for inputs will be growing rapidly, so its suppliers must expect to grow rapidly also – unless they also supply slow-growing sectors.)[6]

The "rate of growth of demand" (or "rate of growth of markets") has not figured prominently in modern growth theory. The reason, skeptics might argue, is that there is no such thing, *ex ante*. Why should new markets open up, systematically, over time, in a definite, normal proportion to current markets? Why should new demand stand in a definite proportion to old, a proportion which remains stable for long periods? Moreover, even if this were so, would businesses know it, and plan investment over several time periods ahead? The answer to the last question is that the first two should be studied by marketing experts. We have already seen cases where an aggregate normal rate of growth of demand makes sense. We can point to phenomena that suggest answers at the sectoral level:

Demonstration effects: new customers are attracted in proportion to the number of existing ones

Complementarities: new "improved" products (made by the same companies using the same processes) sell in proportion to older or ordinary products

Replacement of domestic or nonmarket activities by industry in proportion to general or regional prosperity

These are indeed the stuff of marketing studies. At the macrolevel the growth of demand results from changes in the structure of the economy; for example, earlier in the century each of the following created a new set of markets: the rise of the middle class (resulting from new demands for professionals as mass production brought about the replacement of the family firm by the modern corporation), urbanization, the development of suburbs, the increasing size of government in relation to GNP, and so on. Again the changing relative sizes of different income and social groups is precisely what marketing surveys study.

---

[6] "In the long run" – it may be objected – the slow-growing sector will shrink to a relatively insignificant size, so that the demand for the input will be dominated by the fast-growing sector. No doubt, but in that same "long run" new slow-growing or shrinking sectors will emerge. We *have* to deal with the case where the various sectors of the economy grow (and decline) at different rates, because that is how actual economies behave.

Of course, a normal rate of growth of demand (which, to repeat, *need not* be a steady-state balanced rate)[7] will not last forever. If it is generated by a structural change, then after a time, when the change is complete, it will come to an end – to be replaced, however, by another. Thus the history of capitalism will be the history of growth, driven by market expansions generated by structural changes, where the effect of the market expansion is to bring about further structural changes.

Hence there can be a reasonably defined "normal" rate of growth – in the circumstances stated. When there is such a rate, then prices can be determined (Nell 1992, ch. 17; 1993) so as to provide the profits to finance such growth – hence defining a normal rate of profit. A higher rate of growth will call for increased capacity, so more investment, hence for higher prices in relation to money wages; a lower rate will call for lower prices. Thus, given productivity, a higher rate of growth implies a lower real wage.[8] The associated prices will be those of Chapter 7, or will differ from those because of special conditions in particular markets. When different rates of growth are expected in different sectors, then prices will still be determined in the same way in expanding sectors, but competition will often be inhibited in declining markets.

Normal growth: explaining the trend

Earlier chapters have suggested that the *rate of growth* (of capital and output, assuming a stable and fixed capital/output ratio, with Harrod-neutral technical progress) is the appropriate independent force acting to determine the rate of profit. Given an anticipated rate of growth of *demand* (sales), firms will set prices in order to earn the profits to finance the investments they will have to make to serve their growing markets. Thus the expected rate of growth (of demand) determines both prices (relative to costs) and the rate of profits. But the resulting so-called "Cambridge equation" has been subjected to serious criticism (Garegnani, 1992). It

---

[7] There will usually be declining industries as well as advancing ones. Industries will be at different stages of the "product cycle." The relative importance of different consumption goods will change as income per capita increases – as the Engel Curves show. But the interdependence of basic industries will tend to pull the growth rates of a large part of the economy together. Hence a "core" growth rate can be identified.

[8] The qualification is crucially important. A higher rate of growth of demand leads to higher productivity growth, so that money wages rise. Real wages need not fall, and, empirically, usually rise in periods of high demand growth. Note the suggested sequence: demand growth increases, so investment and planned prices rise. Then productivity begins to rise and money wages increase. The rise in demand leads to increases in both prices and money wages – "creeping inflation."

has been claimed that when growth changes the rate of profit need not adjust to it; nor need consumption change. The larger implication is that there is no consumption-growth trade-off, dual to the wage-profit trade-off.

Normal and actual profit rates must be distinguished. Growth certainly can determine the actual or realized rate of profit, but, it is claimed, it cannot determine the *normal* rate. For the equation $g = s_c r$ cannot relate a *normal* rate of profit to an *actual* (or *ex post*) rate of growth. Actual r can be related to actual g, but such an equation cannot, of course, determine the *normal* or long-period rate of profit (Garegnani, 1992). If investment is determined in the Keynesian way, by animal spirits combined with short run influences, then the resulting rate of growth, equal to current investment divided by the capital stock (valued at normal prices), is not a normal or long-period variable – and hence cannot determine the normal rate of profits.[9]

The claim that investment is unpredictable because governed by animal spirits[10] must be judged implausible prima facie. Taking each in turn: Investment decisions are first and foremost, *long-run* decisions. It is irrational for business to make them on short-run grounds or in response to animal spirits. For business to decide to build a factory or adopt a technology or a product design, they must know that when they have completed the construction and installation, there will actually be a market for the output, at a suitable price. They must have good reasons for thinking this; of course, they can't *know* for certain that the anticipated

[9] Arguably Joan Robinson made this mistake; cf. Garegnani (1992); Vianello (1995). Her defense was that the "normal rate of profit" was an imaginary variable, on a par, methodologically, with neo-Classical equilibrium ideas. She would have been sympathetic to the idea of a "normal" rate of growth only to the extent that it reflects the fact that firms have good reasons for expecting a quantitatively definite expansion in their demand, and for expecting such expansion to continue from period to period. (I think she is wrong to dismiss the "normal rate of profit" for early capitalism; but when it comes to the modern era of Mass Production, some of her points are surely well-taken. But the Classical equations still apply. (See below, and for a full discussion, Nell, 1994.)

[10] This often goes with a related claim that interest rates govern profit rates through the cost effect on prices. Since investment – the rate of growth – cannot explain normal profits, an appeal is made to Sraffa's famous suggestion, that the rate of profit might be determined by the money rate of interest (Sraffa, 1960, p. 33; Nell, 1995). The argument is advanced that interest rates determine the ratio of money prices to money wages, thereby fixing the real wage, and so the rate of profit. But as a mechanism determining the *normal* rate of profit, this is not at all plausible (Nell, 1988, 1992). Monetary policy fluctuates, and the control of the monetary authorities over interest rates is incomplete. More importantly, as we showed earlier, the financial system has serious tendencies to instability. In any case, in Mass Production it is the real not the nominal rate which counts.

market will materialize, but they must have made their marketing studies, sales analyses, etc. So business investment *decisions* – about size of plant, technology, product design, location, etc. – will be made on good, well-studied, long-run grounds. By contrast, having determined to adopt a technology and build, business then has the further problem of deciding how fast to implement those decisions. When should the new plant come on line? Should construction be speeded up or slowed down? At what point should the company enter the financial markets? These are questions about investment *spending*, which is the true Keynesian variable, and decisions on this matter are indeed influenced only by short-run variables, and being somewhat speculative, they are very much subject to guesswork and animal spirits.

Investment decisions, as opposed to investment spending, *can* be given a long-period interpretation; that is, under some circumstances, a *normal* rate of growth, for the economy as a whole, can be defined. Moreover, a look at the evidence tends to support this interpretation: from quarter to quarter and even from year to year, investment fluctuates very dramatically. But when moving averages are taken, the fluctuation smoothes out, and clear trends emerge, which hold steady for considerable periods. We just examined several scenarios giving rise to "normal" growth of demand. When such a rate exists – and is strongly defined – then prices can be determined so as to provide the profits to finance such growth – hence defining a normal rate of profit. The associated prices will be money versions of Classical prices of production, differing from them, however, because of special conditions in particular markets (Semmler, 1984). But if the expected growth of demand is weakly defined or ambiguous both planned prices and planned investment will be uncertain. When different rates of growth are expected in different sectors, then prices will still be determined in the same way in expanding sectors, but competition will often be inhibited in declining markets.

But there is another, interesting case. Suppose that because of innovations, unfocused public policy, or whatever, business expectations are highly diverse, so that no expected "normal" rate of growth exists. Expectations are too diverse; confidence is too low. Investment may become highly volatile; industries and markets may grow for short periods at highly diverse rates, and many will slump altogether, while business withholds spending, waiting to see a new pattern develop. Under these conditions unemployment might rise substantially and remain at a high level; productivity growth would fall off sharply, and money wages could well begin to drift down. Real wages would then tend to fall toward the social subsistence level, as a norm for this condition, which would mean that the normal rate of profit would *rise* to the level corresponding to the social subsistence wage rate. The *actual* rate of profit would sink to a fluctuating and irregular level corresponding to the actual and irregular rate of growth. We would thus expect to find a low rate of profit and a low rate of growth combined with high profit margins, and low utilization rates (as in the United States in recent years).

One point should be stressed, however: when the growth of demand rises, and investment plans are expanded to build capacity at a higher rate, real wages do not fall. In fact, they normally rise. Capacity utilization tends to rise, but more important, *productivity* grows at a higher rate. Prices will be set on the basis of expected costs, but as productivity changes pricing plans will be adjusted. Insofar as operating at higher levels of demand makes economies of scale and scope possible, obviously prices do not have to be raised in order to grow faster. Operating improvements, management practices, scheduling, "critical path" adjustments, and the like, all seem to be stimulated by heavy demand. That productivity grows faster at higher levels of operation seems to be an integral part of Mass Production. So a rise in the expected growth of markets may well not require a decline in the real wage. At a given level of productivity the wage-profit trade-off slopes down; but at higher levels of activity it will shift out. The essential point is that investment and prices are planned together, on the basis of the expected normal growth of markets, which therefore is among the factors determining the normal rate of profits and normal prices.

An objection: capacity utilization rates can differ

It has been objected that a persistent higher rate of growth of output (in one period or economy as opposed to another) need not imply a higher rate of profit, since capacity utilization rates could differ.[11] That is, comparing two economies growing at different rates, the faster growing economy could have a higher degree of capacity utilization, instead of a higher rate of saving. Nor would the faster growing economy necessarily have a lower real wage or higher share of profits. Presumably, the implication is that if the Monetary Authorities in both settled on the same policies, fixing the same interest rate, then both economies could have the same *normal* rate of profits and prices, in spite of the systematic

---

[11] Vianello, for example, asks, "Why should we not suppose that the two economies (one which doubles in 10 years, the other in 15, in response to demand pressures from colonies) have the same propensities to save and to import, and the former (10 year doubling) has a *higher level of exports, investments and national income* than the latter (15 year doubling) – but *not a higher proportion* of its national income devoted to investments and exports?" (F. Vianello, private correspondence, Nov. 1994).

difference in growth rates (and *actual* profit rates). This is the neo-Ricardian picture.

There are three rates of growth to consider – of demand, of output (= income), of capital.[12] For long-run macro equilibrium they must all be equal (assuming no changes in productivity). (They *need not* all be balanced or steady-state. Some sectors may be growing faster than others, etc. But at the macro level the growth of demand must equal the growth of output.) So we have

$$\frac{\Delta D}{D} = \frac{\Delta Y}{Y} = \frac{\Delta K}{K}$$

where D is aggregate demand and we assume that $D = Y$, that current output adjusts to current demand. (Notice that this implies that an increase in Y, due for example to higher capacity utilization, will be matched by a corresponding increase in D. Hence *if* ΔD *is unchanged*, the rate of growth of demand, and so of output, falls.) Continuing,

$$\Delta Y = \Delta K \frac{Y}{K}, \text{ and } \Delta K = I = s_c P, \text{ hence}$$

$$\frac{\Delta Y}{Y} = s_c \frac{P}{K} = s_c r$$

Under these conditions, if the growth of demand rises, the growth of output will have to increase, which will require higher I and so with a given s_c, P will have to rise. So far this differs from the standard neo-Ricardian story only in the emphasis on the growth of demand, g_d (which, however, should *not* be thought of as caused by or generated by the supply process, e.g., the multiplier).

Can the ability of the rate of capacity utilization to vary permit the rate of growth to rise or decline independently of the real wage and the profit margin?

[12] The issue is not steady-state, but "normal" growth – in particular, growth of demand, that is, of markets. Vianello, following Garegnani, contends that "real-world" investment decisions determine the *level* of investment, "whose ratio to the stock of capital will usually be higher or lower than the fully-adjusted rate of accumulation" (Vianello, 1995, p. 13). But the "fully-adjusted rate of accumulation," or the steady-state growth rate, is *not* the relevant concept from which to begin an analysis of long-run adjustments to *demand*. ("Long-run," not "long-period," because the latter carries the connotation of fully adjusted. "Long-run" means capacity – the capital stock – will be adjusted, but not necessarily fully or correctly.) In his contribution to *Beyond the Steady State* Garegnani introduces g^*, a pure supply-side concept. He defines it as the ratio of the savings which take place at normal utilization to the capital stock. Since *this assumes a particular level of demand*, it cannot help us to study the effect of *changes in demand* on the rate of profit.

Why capacity utilization rates differ

Neo-Ricardians contend that a higher level of profit, P, could result from more intensive utilization of existing capital. If production facilities are *elastic*, as Keynes claimed they were, and as they must be when fixed capital is used – as we saw in Chapter 7 – then additional output can be supplied by working them more intensively. If costs are constant or roughly constant, then profits can be increased by more intensive use of existing equipment. If output and costs, including the "user costs" of fixed equipment, rise in the same proportion, profits will increase in that proportion; hence the rate of profit on invested capital (which is given) will rise, also in that proportion.

Output is elastic when it can be expanded at more or less constant costs by adding more labor, such as by recalling layoffs, and running the plant on the weekends; conversely, output can be reduced by laying off labor or by putting workers on short time. But whatever part of the plant is used, and whatever work is done, activities take place in the normal way. There is no difference in the way work is organized, and the coefficients of labor and other inputs are unchanged. (In practice, there may be small variations, but these are considered to offset one another, with some inputs exhibiting a small degree of increasing efficiency and others decreasing as utilization varies. Alternatively, the changes at low levels of utilization may be considered to offset the opposite changes at high levels, so that the coefficients may be treated as constant, on average.) When the plant was built, its ability to operate at higher levels of capacity utilization – including extra shifts – was foreseen and planned for. The transition between different levels of utilization is (ideally) smooth, quick, and costless.

So it is claimed that growth rates can be higher, while real wages and the profit share stay the same – if capacity utilization is higher. The point is well taken, but it is surprising to see it figure in an argument that considers "long-period positions," that is, positions in which the forces making for changes have reached a point of rest. The claim appears to be that two or more significantly different rates of growth could be compatible with the same long-period position. This is very questionable.

Under conditions of Mass Production manufacturing businesses – and quite a number of services also – build plant or service facilities that are designed to operate over a wide range of possible levels, with running costs per unit of output approximately the same at any level within the range. Unlike the Craft-based factories of the nineteenth century, Mass Production plants do not have a single unique level of most efficient operation. Any level within the range is acceptable. Furthermore, as the plant is "broken in," management and workers will learn to use it more effectively, and will see how to improve their use of it, so that its average level of productivity will rise over time – and will rise more rapidly the more intensively it is used. Hence, from the supply-side perspective, it is difficult to define an optimal level at which to operate plant, although, in general, productivity is likely to increase more rapidly when demand pressure is high.[13]

Notice that an investment question may be entangled in the issues here: Should a firm, facing an increase in demand, build a second small plant, so that it operates two plants, each working one shift, or should it reorganize its present plant so that it can be operated for two shifts? The two plants will last longer, but the investment will be larger; the single plant will wear out more quickly, and will operate with higher labor costs, since it will have to pay a premium to the workers on the second shift.[14] Either way they will be able to meet the increase in demand, but they face a trade-off here between operating a *larger* plant for a shorter workday, versus a smaller for a longer. The larger plant wears out more slowly, so lasts longer. But the longer workday requires a wage premium for later shifts. The longer-lasting plant might also become obsolescent before wearing out. The firm must determine which approach will be cheaper, given the current level of the real wage. But this is a question of the *form* in which to construct capacity; the original problem concerned the *amount* of capacity in relation to expected demand.

Firms will build capacity to provide a normal level of operation, which will be cost-minimizing in a broad sense, but other levels of operation may involve no greater running costs. In such cases the normal level of operation will be chosen because it is expected to be the most frequent,

---

[13] Against the effect on productivity must be set the fact that high levels of demand – low unemployment – tend to strengthen unions and worker bargaining, and to erode management's position. Higher levels of operation, involving late shifts, may call for higher wages, and lead to higher breakdown and accident rates. On the other hand, firms may feel that if they operated at such levels for a sufficient time, they could learn to overcome these difficulties.

[14] When formulated strictly, the question is *independent* of demand. It is simply a matter of the best way to utilize equipment, given that more intensive use wears it out faster, but produces output sooner, but at higher labor costs. Kurz and Salvadori formulate the issue as follows: given input coefficients per unit output, per year, and machines that last two years run at one shift, one year when run for two, which will be cheaper – to produce the unit output by operating two shifts, wearing the equipment out in one year and paying a wage premium for second-shift labor, or by operating only the first shift, but using two processes, one with new machinery and one with partly worn out equipment, and paying labor only the normal wage? (Kurz and Salvadori, 1995, pp. 204–7). The answer depends on the wage, the wage premium, the rate of profit, and the price of the new machine. It does not depend on demand.

and because the plant can be operated in ranges above and below that level with approximately the same costs, allowing for an optimal degree of flexibility in adapting to sales variations. This degree of flexibility will be chosen because it permits current production to be adjusted so as to provide the right inventory to minimize selling costs, storage costs, and losses of sales due to delays or quality problems.

So, we can agree that two Mass Production economies (or the same economy at different times), using the same techniques and having the same normal real wages and prices, *could* grow for a long time at different rates because one systematically used its capacity to a higher degree. This would not be possible in a Marshallian world: the more intensive use of Craft-based capacity in the faster-growing economy would imply higher unit running costs. Production systems would be operating above the single most efficient point – the minimum point of the average cost curve. Hence they would have different prices and different rates of profit and would be likely to operate different techniques.

But while the suggested difference in capacity utilization is *feasible* in a Mass Production economy, if the growth rates in the two economies are both *foreseen*, then the situation described in the quote will not be consistent with rational policies by the firms. For capacity is built to a certain size, with a certain carefully designed range of operation, in the expectation that this range will cover the *normal variance of sales*. The ability to adjust output to sales has been built into the production plan, because it is expected to save enough money (in reducing selling costs and sales losses) to cover the cost of designing the plant to have such a range of "constant running costs." (The cost of so designing and building the plant – as compared to a plant with less flexibility – is an investment, which must cover its expense and earn a payoff, within the normal payoff period.)

The range of operating levels will be adjusted to the level and variation of sales, and it will be expanded as the market expands. A certain point in that range, perhaps the midpoint (depending on technical and market considerations), will be some sort of expected average level of utilization. The range will extend on both sides of that point to a distance that reflects the normal pattern of variation of sales (given warehousing costs, etc.). The range is built so that the firm can quickly and easily adjust output to changing market demands, without experiencing serious changes in unit operating costs. Now if in fact capacity is *normally* operated in the upper end of the range, above the midpoint or other point of expected normal use, then the range has been set *too low*. Some fluctuations at the upper end of the scale will fall out of the range. But the range was designed precisely to accommodate such fluctuations; hence capacity should be increased. Conversely, if utilization constantly lies in the lower end of the range, firms are carrying capacity that they do not need. Hence, if these problems are foreseen, firms that are constantly operating at the upper end of the scale will tend to expand their investments and raise the level of normal capacity (thus lowering the degree of capacity utlization), while firms operating persistently in the lower levels will contract theirs (and raise the degree of capacity utilization).

Hence capacity will be constructed on the basis of the demand levels that are foreseen, and their expected variance. Two significantly different rates of growth imply two different levels of demand, and two different patterns of variance. Only the one of these which was foreseen can be consistent with the capacity that is currently in existence.[15]

A complication: the Harrod problem

Of course, just because firms *want*, for example, to increase capacity in relation to normal sales does not necessarily mean they can actually do it. Suppose demand exceeds capacity all around, and all firms try to increase their capacity in relation to sales at the same time, without raising prices. They propose to increase capacity utilization in order to build capacity in proportion to sales, in an amount indicated by the (marginal) capital/output ratio. They will increase their investment spending; but the additional investment spending, in turn, will raise demand according to the multiplier. Hence sales will rise along with capacity – and the two will grow together at a rate given by the multiplier divided into the productivity of capital! The initial excess of demand over capacity will remain. This is the Harrod problem, and, of course, it also holds for attempted reductions of capacity in relation to sales, allowing for problems in relation to scrapping, etc. It is one possible version of multiplier-accelerator interaction.

The Harrod problem implies that there can be *persistent* (but not equilibrium) deviations of actual capacity utilization from that desired. So the case suggested is quite possible, according to this view. But it is *not* an equilibrium, or a normal long-term position; on the contrary, it would be

---

[15] Many issues in capacity utilization involve assessing changes in the way work and plant layout are organized. New shifts are defined, equipment is rearranged, and production tasks are divided up between the shifts. Later shifts receive extra pay or bonuses. This changes the labor coefficients and therefore relative costs, with implications for prices. A change in shift work can be considered analogous to a change in technique and can be analyzed in the same way (Kurz, 1990). Different patterns of shift work may entail different ratios of "labor" to "capital" and hence lead to different normal positions. But *given* the real wage, the normal position will be determined. These issues have nothing to do with degree of demand pressure on existing capacity.

the result of an unstable interaction between capacity construction and demand creation. But instability, or possible instability, in product and labor markets is characteristic of Mass Production.

By contrast, the normal position will be determined by considering the expected *growth of markets* at different prices, and relating this to the ability to finance capacity construction at different prices. As argued above, under various circumstances a "normal" rate of growth of demand can be defined, in which innovating and declining sectors balance, and this will give rise to normal prices and profits. Neither this rate of growth nor the corresponding rate of profits has to be uniform, but they have to be equal as averages at any given time, and over time new sectors must replace old at a normal rate.

Under Mass Production conditions, however, a rate of growth of demand is likely to give rise to (or be accompanied by) steady technical progress. Output per head will rise at a more or less steady rate, but because turnover of inputs increases, capital per head will increase at the same rate, so if wages rise at that rate, the rate of profits will remain steady. So the cross-sectoral wage-bill/capital requirements relation and the Golden Rule will be unaffected.

However, if foreign demand is sluggish, if colonies break away, if state policies falter, there may not exist a clearly defined normal growth of demand. To say that the "normal" rate of growth is not well defined means that the strongly growing sectors are not large enough or growing fast enough to pull the basics into line, so the economy as a whole will be characterized by a distribution of different sectoral growth rates, including declining sectors. If the strongly growing sectors cannot be sustained, and/ or there are few innovators, pulling the basic input producers along, the system may well slide into depression – as happened in the 1930s, and at other times. Or there may be strongly growing sectors, but their growth may be variable and dependent on unreliable exogenous factors. In this case business confidence will be weak, even though growth is expected. When normal growth is weak the volatility of the system can be expected to increase.

(Note that even when there is no well-defined normal rate, if there are innovations, competition will drive investment to take place. This will happen even if, in the long run, the innovations do not increase the rate of profit, so long as those who innovate make a gain, G, relative to those firms who do not, thereby taking a loss, −L. The gain may reflect an increase in market share because of a sleeker, more attractive product, say. The dominant strategy is to innovate, even though in the long run, it will provide no gain. Of course, if innovations do raise the rate of profit, the incentive to invest will be all the stronger. Since innovations have often raised the rate of profit, and it cannot be known in advance whether or not they will, learning through repeated games will not weaken the dominance of the strategy "innovate.")

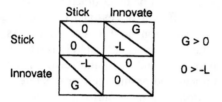

Digression: further objections

This is not the end of the story. There is another arrow in the neo-Ricardian quiver. In the long run, it is argued, capacity itself will adjust, when *capacity utilization* changes. Garegnani (1992) has provided examples showing that a higher rate of utilization need not be permanent, that is, need not go on indefinitely, since the higher utilization will create higher incomes, with consequent higher savings and investment, which will result in the creation of *additional* capacity. Thus a higher rate of utilization implies a *compounding* creation of capacity, moving the system to the desired position faster. Hence, an increased rate of capacity utilization will be able, after a time, to fall back to normal, since the increased capacity will now be able to supply the increased demand.

A *once-for-all* increase in demand can be met with a rise in capacity utilization – and this higher level of utilization will indeed compound, over time, raising capacity. However, there is a problem. If capacity utilization can be expanded to provide the higher income which will generate the savings to match investment – why does there have to be *investment*? Why not simply increase capacity utilization? (This also applies to the compounding argument – the extra savings has to be invested each period, but why build new capacity, if present capacity is elastic?) True, capacity utilization will now be above "normal," but so what? That cannot be a problem for neo-Ricardians. Unit current costs are the same, and average costs are lower. A *further* increase in demand might be difficult to accommodate, but unless this further increase is expected (in which case we are not dealing with the once-for-all situation), why should firms plough back their profits and/or increase their debt, when they could just raise capacity utilization?

If further increases in demand are expected, that is, if demand is expected to increase each period (the rate of increase need not be steady from period to period – if it is foreseen it can be averaged) then it makes sense to invest rather than run up capacity utilization. But now consider, not a once-for all increase in demand, but an increase to a higher *rate of growth of demand*. This means that *each period* demand will increase by the amount required to stand in the new higher ratio to the level of demand of the previous period.

Now consider the argument that higher capacity utilization increases capacity through the compounding effects of higher savings and investment. A rate of growth of demand implies that demand is growing at a compound rate. But a higher rate of capacity utilization will increase the current level of income. With given propensities to spend this will raise the current level of demand. A higher rate of growth of demand will then imply that next period's demand will be increased in proportion to the current level – a level which reflects the higher rate of capacity utilization! Each time capacity is increased through investment, that investment increases current demand, and therefore future demand rises in proportion. Increasing capacity utilization will simply be self-defeating.

Optimal size and output elasticity

Perhaps the argument that a higher permanent rate of accumulation need not imply higher prices (a lower real wage) could be validated in Craft conditions where each firm makes a single once-for-all investment decision, in a context where overall demand is growing. Consider: each period sees a new crop of entrepreneurs, who borrow once only to invest, establishing a firm at its optimal size, at which they operate ever after. Each period demand grows at a "normal" or perhaps a varying rate, but for each crop of entrepreneurs the experience is unique. A larger investment requires a larger group of entrepreneurs, a smaller investment, a smaller group. Entrepreneurs are of varying talents; at higher interest rates only the best will survive, at lower interest rates the less adept will be able to make it. Thus the supply of new firms will be governed by interest rates. The shift in demand for new output each period is the shift of a given demand *function*, showing that additional demand will be larger the lower the price. Prices, however, may (for the sake of argument) be assumed to be determined partly by interest rates. High interest rates require high prices, so the increase in demand will be small. With low interest rates prices can be low, so the demand increase will be large. The number of new firms and the new demand for goods can both be considered governed by the interest rate.

Now suppose there are two such economies, otherwise identical, but one has a higher rate of growth. Must it also have a higher rate of profit and a lower real wage? How else could the savings be generated to offset the higher investment (the new capital goods required by the new firms)? The suggestion is that this can be done by working the existing capacity of the higher growth economy more intensively. If so, then interest rates would govern prices as suggested, and would also determine investment, although not along neo-Classical lines. But this story mixes oil and water.

The system works because firms make a once-for-all decision – they choose an *optimal size* which, once attained, is thereafter maintained. This precludes the elasticity which is essential to the capacity utilization story.[16] For if output is not elastic, then a higher rate of growth will only be attainable if consumption contracts. A higher level of investment will require expansion of the capital goods sector; this will require capital goods industries to produce for investment in themselves; hence they must cut replacement and expansion deliveries to consumer goods industries (cf. Lowe, 1975, on the "traverse"; also Hagemann, 1992). Thus – assuming productivity to be given – to reduce the pressure deriving from the consumer goods sector, the real wage must fall.

Why does investing one time only, to build to "optimal size," preclude output elasticity? Most obviously, if investment is carried out by new firms building to optimal size, it is evident that each plant or organization must have a point of minimum average costs. To vary output will therefore increase costs, and may well be difficult and time-consuming.

---

[16] The textbook "elasticity of supply" does not seem the right notion. (A straight line from the origin always has unitary elasticity – what does that mean?) What we seek is a measure that tells us how the *actual rate of profits* will be affected by increasing supply in response to demand pressure, where supply will be raised by employing more workers in the present plant. The simplest case will be where marginal cost equals average over a wide range, so that supply can be increased by employing more workers at the same wage. Capital is fixed, so employment and the realized rate of profits rise together, in the same proportion. But an increase in employment might yield a smaller than proportionate increase in output – the extra output might be less than the extra wages. Hence the real wage would have to fall. What will then be the effect on the rate of profits? Since K is given in the short period, the effect will take place entirely through changes in the actual rate. This could be expressed by an elasticity formula

$$\frac{W}{P}\frac{dP}{dW} = \frac{dP/dW}{P/W} = \frac{dr/r}{dN/N - dw/w}$$

where $P = rK$ and $W = wN$, and where w is the real wage equal to the marginal utilization product of labor. (Remember, dw and dN move inversely, so have the opposite sign.) This expression shows the change in profits due to a change in employment, with the wage adjusting, divided by relative shares, i.e., the average ratio of profits to the wage bill. When this expression is equal to or greater than one, an increase in employment maintains or raises the current share of profits. When this expression is less than one, an expansion lowers the current share of profits. A negative elasticity means that a decrease of the wage bill raises the profit-wage ratio. (Alternatively, we could consider the inverse of this expression.)

But secondly, output elasticity strongly suggests that firms would follow different patterns of behavior. Specifically, if firms could operate at constant costs over a considerable range, existing firms could always temporarily expand their utilization to sop up any new demand. They could make entry a risky business for new firms, who could not be sure that there would be *room* for a new entrant. Expanding utilization enough to absorb new demand, leaving no room for newcomers, might raise selling costs for existing firms, but such temporary losses might be considered better than a permanent loss of market share to newcomers. Thus when output is elastic, the present firms in an industry have the ability, and the incentive, to shut out new entrants, by absorbing the new demand before the newcomers get their plant running. Existing firms can then plough back their profits and expand to meet the new demand themselves, reducing capacity utilization to normal, and thus maintaining their flexibility.

But if existing firms plough back their profits, then they are planning to adjust capacity to the growth of markets over the long term. If a normal growth of demand is expected, firms will plan accordingly, which requires that they set prices to earn the profits that will make it possible to invest just enough to build the *range* of capacity needed to supply the growing demand, given its expected variability.[17]

Looking to the factors that generate *growth of markets* is surely the right way to extend the Keynesian perspective to the long run. Expected market growth is the basis for investment. But now we need to look at how this relates to pricing.

Benchmark pricing and normal growth

The long-term normal "benchmark" price is the one which will return a normal rate of profit on invested capital, when plant and equipment are used at the planned rate of "full-capacity" operation. (This will not be the maximum possible rate, for firms will normally carry some excess capacity for precautionary reasons, just as they carry cash balances

---

[17] Why doesn't one firm build a huge plant with the most modern technology, and put all the others out of business? First, which firm? If they are all roughly similar, all will be equally well placed to do it, and therefore equally well placed to retaliate if any one tries it. Second, by definition, no one has the capital to do this; it would have to be raised. But who would finance such an operation? Security would be required for such loans, and an equity position would be highly risky. The threatened firms would quickly retaliate by modernizing themselves. There would be a price war. One of the other firms might turn out to be better at the competitive struggle. Moreover, the new technology or better management practices that were supposed to provide the competitive edge might not work out in practice.

in excess of normal transactions needs for such reasons.) Firms may be assumed to use the best technology and face the same wage rate and input prices; hence this benchmark price will be generally known and will be used by financial analysts in calculating the ability of firms to carry debt.

Consider a market for a well-defined product or class of products. Outside of this market there are no very close substitutes; all the products within are similar and are produced by broadly similar technologies, giving rise to similar cost structures. Since the market is defined by the similarity of the goods, other products must be relatively dissimilar. So in the short run there cannot be much in the way of substitution effects. Given more time, of course, consumers could switch in response to price changes, and even in the short run price changes could lead to postponements or stocking up. But under "normal" conditions short-run demand can be assumed to be relatively inelastic – more precisely, the demand curve will be steep. (Therefore, lowering the market price will not create a level of demand that present capacity will be unable to meet.)

Now consider an imaginary firm, operating plant and equipment embodying the average level of technique, as defined by the consensus of the industry. It has built a plant of the optimal size in light of its competitive position in the industry, given the capital that its owners could put up or raise and provide security for. It expects to operate this plant at the "full capacity" level, allowing for normal variations. Therefore it expects a certain share of the market, based on past performance, location, relations with established customers, etc. Its demand, on the assumption that all firms charge the same price (or move their prices together), will therefore always be the same percentage of the market demand that its capacity is of total capacity, assuming, of course, that all other firms have built capacity wisely, neither too little nor too much.

The prices and investment policies of this imaginary firm will be the benchmark prices for the industry. Any firm can make the calculations for this "representative" firm; following its strategy will enable an actual firm to price and invest in line with the growth of the market, keeping its market share intact – so long as others do the same. Deviating from the benchmark position means doing better or worse than the market as a whole. Firms can be expected to differ in the way they implement the benchmark technology; competition will "select" which differences will survive. Successful variations will be incorporated into the benchmark, which will therefore develop over time.

(And what of new entrants? We are considering Mass Production; firms can operate at constant costs over a considerable range. We have just seen that this implies that entry must be a risky business; for existing firms can always expand their utilization to sop up any new demand. In Craft industries, with inelastic production, entrepreneurs will be willing to risk entry, since if expected demand were greater than current output plus planned expansion, they would know that there was room for a new entrant. But in Mass Production, existing firms will be able to expand utilization enough to absorb the new demand for the time being, leaving no room for newcomers.)[18]

Wages, salaries, and fixed costs – debt servicing and normal dividends plus rents and other contractual obligations – are given in money terms. Over the normal range of utilization the firm will use a "standard cost" system, reflecting the technologically determined level of productivity and implying the constant or near-constant marginal costs characteristic of manufacturing under Mass Production. Of course productivity will be growing, and there may well be increasing returns. The following diagrams show constant costs, but continuous or step-function diminishing costs are perfectly compatible with the argument.

The position of the firm can be shown by plotting costs on the vertical axis and output on the horizontal. Here Q_0 is the initial level of normal capacity output and AFC, a rectangular hyperbola, is the curve of average fixed costs. The shaded area in (a) represents fixed costs; any other rectangle under AFC will be the same in area. MC is the level of marginal costs. In (b) required investment funds are added on top of marginal and fixed costs; these are the funds needed to expand capacity by the same percentage that demand is expected to increase – shown here as Q_1, so that $(Q_1 - Q_0)/Q_0$ is the expected rate of growth of demand, g. The problem now is to find a price, p, such that initial capacity (including normal reserves) just balances normal demand at that price and the profits at that price will, when invested, create just the additional capacity needed to supply the growth of demand expected at that price. Such a price is shown in (c), where the initial and the later demand curves are D_0 and D_1, respectively. At every price, revenue is assumed to grow by the same percentage, $1 + g$, where this growth rate is exogenous – the representative firm cannot affect or control it. Consider the firm's calculation now.

[18] When an industry is first being established, neither the product(s) nor the processes are well defined. There cannot be "entry," because there is neither industry nor market as yet. There is a potential market, that is, an opportunity to compete for a portion of household or business income, by serving a particular function in a new way. This leads to fierce competition, but such competition is the struggle to become the defining force, the one who sets the industry standards, e.g., Ford, or in a later era, IBM.

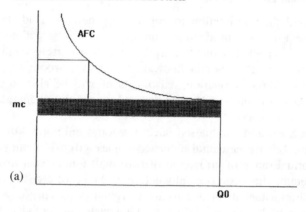

(a)

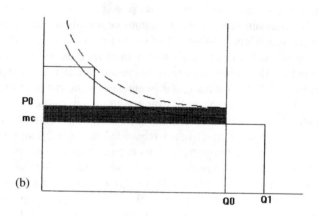

(b)

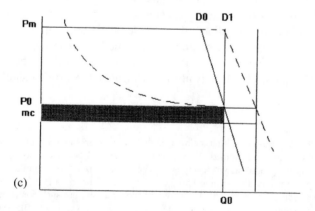

(c)

Figure 10.1

Balancing net revenue and required revenue

The firm has to choose a price that will provide the funds to expand capacity in the same proportion that demand is expected to grow. This price will have to be maintained over a certain period of time; it is not a momentary price. It is the benchmark that guides the day-to-day pricing policies of the firm's sales staff. The relevant period is obviously the time required to construct and install new capacity; the profits from normal operations over this period must be enough to finance new plant sufficient to satisfy the new customers added during this same time. But a longer period must be considered also: over the lifetime of a plant, the profit generated by its operation, plus the recovery of its initial cost, both compounded at a rate of interest equal to the normal growth rate, must finance the construction of new capacity equal to the initial demand times normal (compound) growth over that same period.

The capital-output ratio of the representative firm is given by history. (It is taken as fixed here, but it could be falling from period to period.) The normal growth of the market follows partly from its own conditions and partly from the situation of the economy as a whole. The inelastic demand curve shows alternative prices and corresponding levels of purchase by fully adjusted consumers. However, we also need to think in terms of changing prices. There will be a maximum price such that, if price rose to it from another level, the good would be unaffordable within the short period, and demand would fall to zero. Below that, however, there will be a normal range, in which the good will figure in the budgets and planning of potential users. In this range a lower price will lead to some but not much increased consumption. But over time, as incomes grow, the market will expand.

The normal growth of the market (due to income growth) has to be greater than its potential expansion due to any feasible price cut. If a price cut could expand the market by more than normal growth then firms would concentrate on recovering capital and cutting costs, rather than investing in new capacity, for whoever develops a less costly technique will not only face a huge market, but will have to bring in new capital to service it.

This is shown here as an outward shift of the inelastic (steep) market demand curve. If the initial demand function is D_0, then the new one is $D_1 = (1 + g)D_0$; if D_0 is $q = a - bp$, D_1 is $(1 + g)(a - bp)$. (The p's and q's are prices and quantities as elsewhere; g is the growth rate; the a's and b's and other letters, however, are *not* to be confused with those in other chapters.) This seems to say that the demand curve progressively flattens out from period to period, eventually becoming quite elastic. This would be a misinterpretation. D_1 is the expected demand, at various prices, as judged by the firm, on the assumption that income is growing. Thus D_1 states that if the firm charges the same price from period to period, sales will grow at every price in the same proportion that income increases. So it does not show the various quantities that would be bought at various prices; only D_0 can be so interpreted. (To analyze prices in later periods requires redrawing the shifted demand curve as an "initial" one, that is, as one showing the various quantities the market would buy at various alternative prices.)

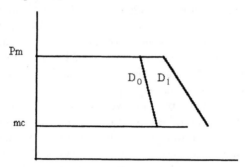

It must now be shown that a unique price always exists balancing the revenue earned from normal operations and the revenue required to finance the additional capacity corresponding to the shift in demand. Intuitively when the price is higher, the required profits will be smaller; since the demand curve has a negative slope, the higher price implies a lower denominator of the (given) growth rate. Hence the numerator must also be lower, so less investment will be needed. This implies a functional relation between price and investible funds, with a positive intercept on the price axis and a negative slope. On the other hand, it is clear that the higher the price the larger will be the investible funds generated, at least over a range, since the demand curve being steeply sloped, is inelastic over most of the relevant range.[19] There is therefore a second relationship between price and investible funds, this time with a positive slope. The price at which these two functions intersect will equate growth of capacity to growth of demand, and the quantity corresponding to that price on the demand curve will be the required initial capacity level output.

In symbols, let Q_1 be capacity output in a given period, where $q = a - bp$ is the demand equation. The maximum price will be p' and the minimum price, m, equal to marginal cost. Hence the slope of the curve will be $-b = (p' - m)/(q'' - q')$.

[19] As we shall see, so long as the demand curve is linear, or approximately linear, neither slope nor elasticity need be specified. The investible revenue curve will always intersect the required funds curve in a stable fashion. In the case of unitary elasticity the investible funds curve will be a vertical straight line.

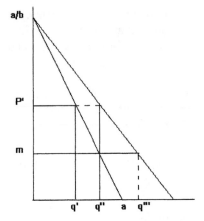

The growth of demand changes the slope to $(1 + g)b$. The theoretically maximal price, the intercept on the vertical axis, will be a/b, which will be unchanged by growth.

Now consider the required increase in capacity at different levels of price. At price p', initial capacity equal to demand would have to be q' and next period's would be $(1 + g)q'$, where $g = (q'' - q')/q'$. At price m an exactly similar calculation can be made, showing a larger difference, divided by a larger initial capacity. Hence we can show the change in capacity, Q, as a function of the price. For $p = p'$,

$$\Delta Q_{p'} = Q_1 - Q_0 = (1 + g)(a - bp') - (a - bp') = g(a - bp')$$
$$\Delta Qm = Q_1 - Q_0 = (1 + g)(a - bm) - (a - bm) = g(a - bm)$$

The difference between $\Delta Q_{p'}$ and ΔQ_m is

$$\Delta \Delta Q = \Delta Q_m - Q_{p'} = g\left[(a - bm) - (a - bp')\right]$$
$$= -gb(p' - m)$$
$$= -gb\Delta p$$

Assuming continuity and integrating, then

$$\Delta Q = -gbp + \text{constant} = ga - gbp,\ p' > p > m$$

The additional required capacity is an inverse linear function of the price. The constant will be the size of ΔQ when $p = 0$, so when p is out of the picture, $Q_1 - Q_0 = (1 + g)a - a = ga$.

To find the required investment funds ΔQ has to be multiplied by the appropriate marginal capital-output ratio. On the assumption that the best-practice technique is given, this creates no difficulties, but unanticipated capital-saving technical progress would imply that a constant price would generate more funds than required. Another adjustment follows from the fact that the capital-output ratios in different sectors reflect the different capital-labor ratios. So the movement of relative prices when the normal expected growth rate of the economy changes will depend on these differences. A sector with a high capital-output ratio, facing an increase in normal growth, will have to raise its price more than a sector with a low capital-output ratio.

Next consider profits, or investible funds, as a function of price. This will be total revenue minus marginal and fixed costs, hence net revenue (remembering actual vintages – if best-practice costs are used net revenue will be overestimated):

$$R = p(a - bp) - m(a - bp) - F = (p - m)(a - bp) - F$$

$$= ap - bp^2 - am + bpm - F$$

$$\frac{dR}{dp} = a - 2bp + bm, \text{ where}$$

$$\frac{d^2R}{dp^2} = -2b, \text{ and}$$

$$\frac{dR}{dp} = 0, \text{ when } p = \frac{a + bm}{2b}$$

Clearly the net revenue or profit function is a parabola which cuts the price axis twice, at $p = m$, where the curve is rising, and again at $p = a/b$, where it is falling.[20] In between it reaches its maximum at $p = (a + bm)/2b$. So when price rises above m, initially, and over most of the relevant range, profits or investible funds increase.

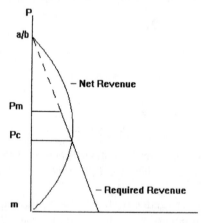

---

[20] The quadratic $R = -bp^2 + (a + bmc)p - amc$ has two positive roots:

$$p = \frac{-(a + bmc) + \text{ or } - [(a + bmc)^2 - 4abmc]^{1/2}}{-2b}$$

Borrowing

The phrase "investible funds" should not obscure the fact that profits need not only be invested directly, but may be used to service borrowed funds. This latter, of course, provides leverage; the profit justifies a larger investment, but the borrowing dilutes equity. The firm must decide on its borrowing policy; the more it borrows the greater the risk and the greater the dilution of equity,[21] but the more it can expand and/or the more capital-intensive it can become. Firms must calculate an optimal policy by balancing these as best they can.

A bond pays a fixed rate of interest; the bondholder does not benefit if the project does exceptionally well, whereas a holder of equity does. But a bondholder is entitled to be paid even if the project fails, whereas the equity investment loses value. A bond or a loan must therefore be secured by independent assets; it cannot be secured simply by the project being constructed with the borrowed funds, for then the bondholder would be in a worse position than a shareholder – equally at risk, but unable to take out excess profit if things go well. Thus the market will not accept bonds unless a firm is able to provide independent security, which restricts the ability to borrow.

Borrowing provides a gain, by increasing leverage; it raises costs by increasing risk. The firm will calculate its optimal ratio of debt to capital, and plan its investments to keep that ratio steady. The principal risk incurred is that of default. This depends on the likelihood of fluctuations around, and/or costly deviations from, the expected path of demand growth. The firm can be presumed to have investigated and ranked all the various possibilities, and to have assessed their costs. The result will be a

---

[21] The traditional view, that in competitive conditions firms can obtain as much finance as they want at the market rate, securing loans with the projects is not consistent with the self-interest of the parties, under conditions of realistic uncertainty. To issue equity is to dilute ownership and risk loss of control; this must be compensated by a *rise* in profitability. Suppose a firm worth 50 shares issued another 50, and built a new project of exactly equal profitability to the first. The earnings per share of the larger firm are exactly the same as those of the original. But now a defection among the original shareholders of a single share could give control of the *entire* firm, the original projects as well as the new, to a new group of shareholders. If control matters, as it surely does for reasons of market position and long-term strategy, the original shareholders will not risk losing control unless the new project is going to *raise* earnings per share sufficiently to warrant the gamble. Of course, the more successful the new project, the greater will be the confidence of shareholders in management and in the control group. These problems are avoided by issuing bonds, which, however, increase the firm's fixed obligations, and thereby raise the risk of default.

probability distribution showing the size and cost, relative to the capital stock, of given sizes of deviations from expected growth, on the horizontal axis, and the frequency or likelihood of such size deviations on the vertical. For simplicity, this can be assumed to be normal, or skewed in a simple way – but in principle it could have almost any shape. The costs, of course, will rise with the real interest rate.

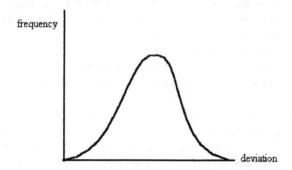

Now consider the corresponding cumulative frequency curve. The horizontal now shows the cumulative cost or losses in relation to capital, the vertical the probability of such deviations, that is, of incurring such costs or losses. If we now superimpose *equity* in relation to capital, $(K - D)/K$, where D is debt, on the horizontal axis, we have a measure of the costs that can be safely incurred without default, together with the corresponding probability. If equity is low, the safe level of losses will be low, and the probability of this being the level will be low. Higher levels of equity will mean higher levels of nondefault losses, with the levels of probability rising at an increasing rate.

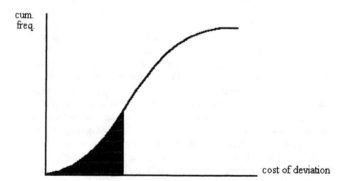

The risk of default therefore increases with D/K. An insurance premium must be calculated to cover this; that is, the firm must consider how much additional growth will be required each year to cover the cost of the additional risk. If this growth premium is less than the anticipated gain from additional leverage, then borrowing should be increased. The point at which the growth premium has risen to exactly equal the gain from borrowing will be the firm's optimal D/K.

Since we have assumed that all firms are similar, and that they follow one of their number as a price leader, all will adopt similar policies.[22]

Determining the benchmark price

These two functions can now be put together to determine the benchmark price. The required revenue function shows the change in needed capacity due to demand growth; to get the required funds it must be multiplied by the capital-output ratio. The generated revenue function shows the net earnings; to find the available funds it must be multiplied by the leverage factor. For simplicity, assume initially that these two cancel each other. For the moment, also neglect fixed costs. Now consider the two functions. From the intercept a/b the net revenue function curves downwards to intersect the vertical axis again at m. The required revenue function also starts from a/b; then as price drops towards m, the revenue required to finance expansion of capacity rises. The two functions will intersect once, determining the benchmark price and the initial capacity output which will enable that price to remain constant while capacity grows in step with demand.

Setting the two functions equal to each other and regrouping yields a simple quadratic:

$$-bp^2 + [b(g + m) + a]p - a(g + m) = 0$$

(where m is still marginal cost). This has the standard form $Ap^2 + Bp + C = 0$. The discriminant, D, is

$$D = [b(g + m) + a]^2 - 4ab(g + m), \text{ which simplified}$$
$$= b^2(g + m)^2 - 2ab(g + m) + a^2, \text{a perfect square, whose root}$$
$$= a - b(g + m)$$

When the demand curve is steep, a will be large compared to b, but the assumptions require that a/b > m and g > 0. Hence, a > or = b(g + m),

---

[22] This calculation concerns the long run: *given* the expected growth rate of demand, and the distribution of likely deviations, firms plan their borrowing policies. In the short-run, however, they will make a different kind of calculation: *given* their debt to capital structure, they will adjust investment spending, as current conditions vary. This will be the subject of Chapters 12 and 13.

so that $D >$ or $= 0$ and there will be either two roots or one. When the discriminant is positive, the roots will be $p = a/b$ and $p = g + m$, and when $D = 0$, $a/b = g + m$, and the single root is $p = a/b$.

But the upper root, a/b, has to be considered irrelevant, since it represents a price set so high that growth, demand and profits are all eliminated. Only the lower root is economically significant, and given m, it depends only on g.

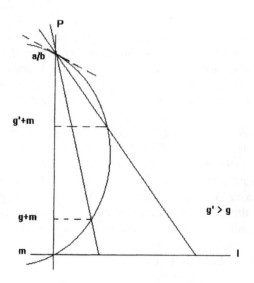

Allowing for fixed costs reduces the available generated revenue, effectively shifting the curve to the left. Allowing for a difference between the capital-output ratio, v, and the financial leverage ratio, f, requires a recalculation. Required revenue must be multiplied by v, generated revenue by f; the resulting equation is

$$-bfp^2 + [b\,(gv + mf) + af\,]\,p - a\,(gv + mf) = 0, \text{ where}$$

$$D = [b\,(gv + mf) + af]^2 - 4afb\,(gv + mf)$$

$$= (af)\,2 - 2afb\,(gv + mf)^2 + b(gv + mf)^2, \text{ with root}$$

$$= af - b\,(gv + mf)$$

When $D > 0$, the two roots will be

$$p = \frac{a}{b}, \text{ and } p = \frac{gv}{f + m}$$

and when $D = 0$, af/b $=$ gv $+$ mf, and the single root is a/b. Again the upper root is irrelevant, and the calculation will always yield a unique price.[23]

The general conclusion has to be that under normal and plausible circumstances,[24] in a steadily growing market with relatively price-inelastic demand, these two relationships will uniquely determine a benchmark price that will hold over a substantial period of time.[25]

[23] Moreover, this price appears to be stable. Suppose that g is high and a/b low, so that the required revenue function intersected generated revenue in the upper region, where demand is elastic. Consider a price above the equilibrium; at that price, generated revenue would exceed required revenue, exactly as it would in the lower region, in which revenue is rising with price. Excess revenue will tend to be translated into additional capacity, leading to a lower price; the movement is stabilizing. Next, suppose the initial capacity level of the representative firm were too low, which is equivalent to saying that at some previous time, demand expanded more than expected. With actual capacity less than "benchmark" capacity, price will tend to be bid up above the benchmark level. At a higher price sales will expand less from period to period, but the profits generated will be larger. When price is above the benchmark level the net revenue generated is above that required; hence extra capacity can be built to offset the initial shortfall. When price is below the benchmark, required revenue exceeds generated; if the initial capacity level were too high, the profit margin would be squeezed and requirements expanded. So the relative movements of demand and capacity will be in the right direction for successful adjustment.

[24] Demand has been assumed to be linear and steeply inelastic, as in fact is characteristic of manufactured products. Suppose the demand curve were of unitary elasticity; then the generated revenue function would be vertical, the required revenue a hyperbola, and the solution simple. Other forms can be imagined, but with substitution limited the functions will have to be similar to those examined.

[25] The famous tangency solution of the theory of imperfect competition is *static* and applies to a single firm; by contrast, here we assume that from period to period the demand curve shifts out, so that the benchmark price is the price that will just enable the firms in the market to invest enough to supply the new demand at such a price. The benchmark price is an *industry* price. But suppose output were set at the point where MC $=$ MR, so that price would follow from the AR curve. Profit would then be *much* greater than needed to build the new capacity required at that price when the demand shifted. The profit rate at this price would be far greater than the interest rate, so it could not be invested in bonds. Nor will other industries be as profitable. If it were distributed as dividends, it would drive up the stock price, lowering the cost of capital to the firm – which would shift the firm's debt-equity ratio. But if invested in expanding, excess capacity would be created, driving the price down. MC $=$ MR cannot be an equilibrium position.

Benchmark prices and market conditions

Actual conditions are seldom "normal," which means that firms will often not be selling their "normal" output. Benchmark prices are guidelines; they can be calculated by anyone familiar with the conditions of the industry; they will be used by bankers to assess creditworthiness and by courts to value assets. Once a benchmark price has become established firms will be reluctant to change it unless the underlying long-run data have changed. But short-term and cyclical fluctuations have to be dealt with. In the face of shortages, surcharges could be added; when there is excess capacity, cuts can be made. Benchmark prices, in short, could well remain stable, while market prices showed considerable flexibility.

Of course, market prices do fluctuate quite a lot. Sales drives, special discounts, "loss leaders," dumping, etc., all lead to variation. But price flexibility has a special meaning: prices will systematically fall when actual demand falls below normal, and they will rise systematically when actual demand reaches or exceeds normal. Demand fluctuates quite markedly, but the expected systematic price behavior does not seem to take place. Why not?

Competitive strategy

Consider a firm whose demand has fallen substantially below capacity. To attract more trade it could cut price – a delicate operation at best; in the ideal case, it would offer bargains throughout the market while keeping its rivals in the dark, but more often it will act without knowing what rivals will do. In any case a preliminary strategy calculation shows that the representative firms must be prepared to act regardless of the behavior of rivals.

Let RF be the representative firm, and AR be all its rivals, meaning all "nearby" firms from whom RF could expect to attract customers or to whom it could lose its market. There are two broad strategies, then, cutting price or sticking fast. The initial price will be p, and the change Δp. The size of a typical firm will be q, and AR is made up of n firms, so will have an output of nq. When RF cuts and AR sticks, a fraction x of AR's demand will potentially switch to RF. Analytically, as a first approximation,

From this it would appear that RF's best strategy is to cut, no matter what AR does, since both $(p - \Delta p) q > -pq$, and $(p - \Delta p) (q + xnq) > 0$. If AR cuts and RF sticks, RF stands to lose everything, since AR, being much larger, will be able to supply RF's whole market. If RF cuts and AR sticks, RF will attract a large clientele – but this potential gain may not be realizable. For RF cannot (except temporarily) supply more than its capacity permits. Moreover, as we shall see, if a large price cut is necessary to attract demand, and the amount of excess capacity is small, it may not be worth it. What is the use of attracting demand that cannot be supplied? We will examine this point in a moment. First, note that AR's calculation is not relevant, although it would appear that cutting would be best for it, too, if p were small and x large, that is, when many customers would move for small price cuts. But such a calculation implies that the firms that make up AR can act in concert. If the market is competitive, however, they will act individually, so the calculation for RF holds for each of the firms in AR, taken separately.

A further reason for firms to favor the strategy of cutting when demand falls short of capacity lies in the fact that a single firm inflicts only a small loss on its competitors, which when spread among them may be hardly noticeable. No one may notice, or bother to retaliate. Thus each firm may be tempted to cut first; which implies that all will cut.

So, at first glance, competitive strategy seems to imply that a representative firm will cut its prices when demand falls below capacity and, by analogy, raise them when it runs over. But it only makes sense for a firm to try to attract demand it can service, and this raises the question of how large a price cut to attract how many customers.

The customer attraction function

Firms each occupy a certain space in the market – a geographical location, and also a position in social space, which together give them a set of habitual customers, who, however, are aware of competitors. Changing suppliers takes time and may disrupt working arrangements between officials of manufacturing and distributing firms or between customers and shopkeepers. Yet if it is worthwhile customers will certainly switch. Firms will try to estimate how much extra trade they could attract by dropping their prices below the benchmark, and also how much they stand to lose if they don't and others do. (Alternatively, anticipating our later discussion, firms may try to estimate how much extra trade they could attract from competitors by introducing *product improvements*. Assuming that such improvements can be graded, the analysis will follow the same course.)

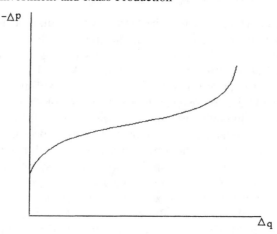

The customer attraction function for the representative firm will show the change in sales, the customers attracted, for price cuts of various sizes (or various degrees of product improvement). In general, it will have a positive intercept; a small price cut would hardly be noticed, nor would it be worthwhile to disrupt established practices for small gains. But as the price cut increases, more and more customers will be attracted, until at a certain level a very large number will shift. Beyond this point, however, it will take very large cuts to attract even a small additional number. All the mobile "nearby" customers have already shifted; those now being attracted are distant, unfamiliar or have significant links with their normal sources of supply. So the general shape will be, starting from a positive intercept, to climb steeply, flatten out, and then climb steeply again. The more "competitive" the market, the smaller will be the required price cuts and the greater the number of mobile customers, so the lower the intercept and the longer the flat portion. In the extreme case, corresponding presumably to the ideal of "perfect" competition, the curve would reduce to the horizontal axis, indicating that an infinitesimal price cut would attract an infinite demand.[26] But in the normal case changing suppliers, or

[26] Textbook definitions of perfect competition are misleading. Samuelson, for example, writes, "The demand curve for the whole industry will look perfectly horizontal [to the firm], with an elasticity of demand of infinity." He adds, "The draftsman will have to train a microscope on the relevant point of the industry curve to show how this sloped curve will reappear as horizontal to the Lilliputian eye of the firm." But a slope is a slope, whatever the units. The point, however, has to be that the demand of the industry is an order of magnitude greater than the demand facing the firm. A quantity is an order of magnitude greater than another if when the first is $f(n)$ and the second $g(n)$, the absolute value of $f(n)/g(n)$ tends to 0 as n tends to 0. Let q be the output of the representative firms and n be the number of firms. The market will then be nq. As the number of firms

(continued)

even seeking out the best bargain, has a cost, which will vary for different customers, giving rise to the functional shape examined above.

Competing for customers

Suppose demand is less than capacity and the representative firm has decided to cut to try to attract customers. It faces a "normal" customer attraction function. How much should it cut its price?

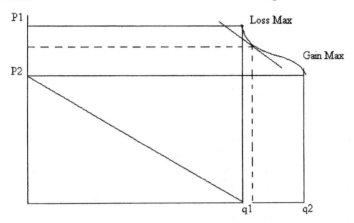

[26] (contd.)

increases, suppose that the industry reaps advantages, economies of scale, which are of benefit to firms, so that q increases. For example, q = an, where a is some positive number. Then the absolute value (AV) nq/an = (AV) nan/an = (AV) n, and the definition is trivially fulfilled. But the fact that the industry's demand is an order of magnitude greater than the firm's tells us nothing about the relation of the industry's demand to its capacity, nor does it say anything about how firms will adjust prices in relation to demand. For this we turn to, for example, Stonier and Hague (1961), who write, "A limiting case will occur when there are so many competitors producing such close substitutes that the demand for the product of each individual firm is infinitely elastic and its average revenue curve is a horizontal straight line. ... [T]he firm can sell as much of its product as it wishes at the ruling market price. If the firm raises its price ... it will lose all its customers. If the firm were to lower its price, it would be swamped by orders" (p. 105). The line is infinitely elastic because an infinitesimal price cut will attract the entire market, an order of magnitude greater than the firm, therefore infinite in relation to it. As defined, it is a customer attraction function, positioned at the level of the initial market price. But it is not an average or a marginal revenue function; if the industry's demand curve is continuous, then if more is to be sold, even a very small amount, price will have to drop. Nor does the limiting case of the customer attraction function justify price-taking behavior; quite the contrary. If demand is below capacity, the representative firm will cut price since an infinitesimally small cut will attract all the additional demand it needs. In general, the firm will push production to capacity so long as price covers marginal cost. Thus price will tend to fall to marginal cost whenever demand is below capacity.

For the price cut to be worthwhile, the gain, the new price times the additional quantity, must outweigh the loss, the change in price times the initial quantity. So the new price must be chosen to make this difference as large as possible. Hence,

Max : $p_2 \Delta q - pq_1$, that is,

$$\frac{d}{dp}(p_2 \Delta q - \Delta pq_1) = 0, \text{ or}$$

$$\Delta q \frac{dp_2}{dp} + p_2 \frac{d\Delta q}{dp} = q_1 \frac{d\Delta q}{dp} + \Delta p \frac{dq_1}{dp}$$

Assuming continuity and manipulating,

$$\frac{p_2}{q_1} = \frac{d\Delta p}{d\Delta q}$$

and this will be a maximum when the second derivative is negative. In words, when p_2 is chosen so that the tangent to the customer attraction function at the point (p_2, q_2) is parallel to the line from p_2 to q_1, the additional revenue reaches an extreme point, which will be a maximum gain at the place marked in the diagram. It seems that the price p_2 should therefore be selected by the representative firm.

But if $q_2 - q_1$ is greater than the difference between the original level of sales and normal capacity it would be a mistake to drop the price to p_2, for part of the demand attracted could not be supplied. The gap between normal capacity and current sales in proportion to the level of sales must be greater than the price cut in relation to the new price to make it worthwhile to cut at all.

Even if this condition is fulfilled for the representative firm it does not follow that there will be a general spate of price cutting. Actual firms differ from the representative firm and from each other, both in relation to costs and in the loyalty of their customers. When effective demand falls off some firms may bear the brunt and begin cutting; but then the rest have to calculate how much they stand to lose if they stick. For instance, a firm with relatively little excess capacity and whose nearby customers are quite loyal to their normal suppliers would not find it worthwhile to cut, unless it stood to lose heavily. But if its customers are reasonably loyal, so that the proportionate loss in sales is less than the proportionate price cut necessary to prevent such losses, the firm should stick to its price and take the loss. By the same token, price cutters will not gain very many customers, so it may not be worthwhile for them. But if firms are fairly similar and the fall in demand substantial, all will calculate that it would be advantageous to cut.

However, if all cut, no one gains. All lose, since the market demand curve is inelastic. The problem has to be reformulated – cutting below the benchmark price is a strategy that will only work if one's rivals do not follow suit. Since such cuts are dangerous it is unlikely that small deviations will be followed; on the other hand large ones, which may attract sizeable numbers of customers, are increasingly likely to be matched. Thus the probability of rivals following suit is an increasing function of the size of the price cut.

Price stability when demand varies

These points can be put together in a diagram that will illustrate the reasons for price stability when demand varies in Mass Production. On the horizontal axis plot the size of the price cut; on the vertical in the positive direction measure the gain from a price cut, in the negative the loss as more and more competing firms follow suit. The shape of the G-function will echo that of the customer attraction function; it shows the gain *expected* from attracting more business. The loss function will show the loss due to the price cut when other firms retaliate, reaching a point when all firms follow suit, so that no customers move – except those brought newly into the market by lower prices. (If there are none the slope will be 45 degrees.) In general the horizontal intercept of the Gain function will be nearer the origin than that of the Loss function – it will take a noticeable price cut to begin to attract customers, but initially the losses to other firms will be small, so they won't retaliate. Once firms begin to feel the pinch they will follow suit, and they will all tend to do so

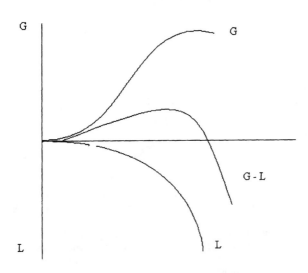

at the same time. So losses will rise steeply, while gain will tend after a point to level out. The net effects are illustrated in the G–L curve – the difference between the two functions – which starts out low, find barely positive, then turns up as larger cuts attract customers without provoking retaliation, rises to a positive but low level as retaliation begins, and then falls sharply to negative values as everyone follows suit and all gains are nullified.

So a small price cut attracts hardly anyone, and merely brings a loss; a somewhat larger price cut may bring a small gain, but if it is large enough to begin to attract customers, it carries a high probability of being matched, and it will "spoil the market." It will lead to retaliation and further cutting will definitely lead to losses. To gain anything by trimming prices requires a delicate strategy – cutting just enough to attract sufficient customers to (more than) offset the cuts, but not enough to be noticed by competitors. Special sales and carefully targeted advertising will be needed – and might be costly. Under the circumstances the best strategy will often be to hold the line. At most the price cuts will be limited and particular.[27]

Price increases

Now suppose demand is above capacity and steady. The firm is in a strong position and its customers should know that alternative sources of supply will be hard to find. This is the time for surcharges. To determine

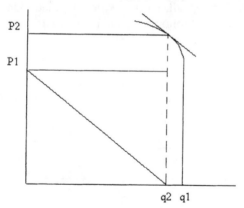

---

[27] To repeat a caveat given earlier: none of this applies to firms engaged in primary production. In the primary sector variable running costs are a low proportion of total current costs, most of which are fixed. Output is difficult to vary, once it has been set. Storage costs are high, so carrying excess output from one sales period to another may not be cost-effective. Hence when demand varies prices tend to vary in both directions. In machine tools prices will vary upwards with demand, but will generally not be flexible down.

how much to add to the normal price, the firm maximizes the additional revenue (see diagram).

$$\text{Max}: \Delta pq_2 - p_1 \Delta q$$

$$\frac{d}{dp}(\Delta pq_2 - p_1\Delta q) = 0, \text{ which becomes}$$

$$q_2 \frac{d\Delta p}{dp} + \Delta p \frac{dq_2}{dp} = p1 \frac{d\Delta q}{dp} + \Delta q \frac{dp_1}{dp}$$

Assuming continuity and rearranging,

$$\frac{q_2}{p_1} = \frac{dq}{dp}$$

The price will be raised until the tangent to the customer loss function at (p_2, q_2) is parallel to the line from p_1 to q_2. Hence at or above capacity there will be a tendency for prices to drift upwards, even though firms may believe that their rivals will not follow suit, so long as they believe their customers will be reluctant to shift, at least for small surcharges. Look at the diagram again: for price increases the G-L curve will tend to be positive. Gains will come from the fact price increases, if not followed by competitors, will not lose customers, while if followed, will lead to higher profits because demand is inelastic.[28] Prices rise more easily than they fall.

The customer attraction matrix

Consider a market consisting of n firms, all producing a product of identical design, using the same technology, but with the possibility of differences in efficiency and the organization of work, and hence in costs. Moreover, firms may also differ in services related to product delivery and in the reliability of quality checking. In the event of a price war, firms will differ in their ability to attract customers from each other. Let c_{ij} indicate the number of customers attracted by the i-th firm from the j-th, as the result of a price cut. Rows will therefore show all the customers the ith firm can attract from each of its j competitors, and the row sums will show the total customers attracted by a unit price cut, on the assumption that all

[28] What have they been waiting for, then? Why were prices not raised initially? Because the benchmark price was chosen as the price that would balance growth of demand with growth of capacity, in competitive circumstances. To raise the price would have disrupted this balance. At this point, *ex hypothesi*, however, demand is above capacity, and the question is whether to deviate from the benchmark, and if so, how much?

firms are operating at the "unit" level, that is, at normal size. Columns will show the customers lost by the j-th firm to each of its i competitors, and column sums will show the total lost. Hence if a firm's (row sum)/ (column sum) > 1, it gains in a price war; if it is less, it loses. We can number the firms according to their (row sum)/(column sum) ratios, putting the highest first:

$$c_{11}\, c_{12} \cdots c_{1n} \rightarrow G_{1j}$$
$$c_{21}\, c_{22} \cdots c_{2n} \rightarrow G_{2j}$$

$$c_{n1}\, c_{n2} \cdots c_{nn} \rightarrow G_{nj}$$
$$\text{---}\ \text{---}\qquad \text{---}$$
$$L_{i1}\, L_{i2}\qquad L_{in}$$

(If we assumed all firms were the same size, the coefficients could be expressed as fractions between 0 and 1. Then the size of the entire market would be n, which would be the maximum gain of customers; the maximum loss would be 1. This simplifies the analysis, but at the price of unrealism.)

If entries are low, few customers move; if high, many move. When entries are low, price cuts are unlikely to be a good strategy, but even when entries are high, price cuts may prove unfortunate if the row sum/ column sum is $\geq$ 1.

Clearly if this matrix is symmetric, $c_{ij} = c_{ji}$, with row sums equal to column sums, price wars will benefit no one. The nearer it is to symmetric, and especially the nearer the majority of row/column ratios to unity, the fewer the gainers and the smaller the gains. But the presence of even a few firms with row/column ratios significantly greater than one will indicate beneficiaries in a price war. The nearer the matrix to a diagonal form the clearer is the hierarchy of market strength and weakness. Over time, we can expect competition to eliminate the weaker firms, moving the matrix toward symmetry – in which case we can reasonably speak of the customer attraction function of a representative firm. But this development also reduces the number of firms.

The instability of perfect competition
The limiting case of a customer attraction function occurs where an infinitesimal price cut would lead to infinite demand – the entire market would shift to the price-cutting firm. In such circumstances the

representative firm will always produce at capacity. When demand fluctuates, however, the response is unstable. Demand above capacity should lead to price increases – but given the inelasticity of demand, these increases may have to become very large before the excess demand is eliminated. When demand falls substantially below capacity, however, price will immediately be driven down to marginal cost. Any one firm cutting will force the others to follow suit; when all cut, however, demand increases only slightly, so the cutting will continue, down to marginal cost. Firms will therefore not be able to meet their fixed obligations, and so will have to go out of business or be reorganized. Thus supply will be drastically curtailed, leading to a very large price increase. So with constant marginal and falling average costs, and inelastic demand (the normal situation of firms in modern industries), the limiting case of perfect competition will be unstable, with price swinging about wildly and firms facing bankruptcy and reorganization.

The result of pushing price down to marginal cost is that profits will fall below the level needed to finance the capacity to prevent entry. New firms will therefore enter, with the latest technology and product design. They will be able to dominate the market and will set about to establish customer loyalty, raising the intercept of the customer attraction function.

Investment in customer loyalty

Loyal customers mean a firm will not have to follow cuts down and will be able to add surcharges in boom times (so long as this does not undermine the loyalty!). Customers are loyal, not as a matter of sentiment, however, but because the firm has made it worthwhile for them to stick by it. To develop a loyal market a firm has to invest in providing customer services – effective sales displays, prompt delivery service, repairs, maintenance and warranties, efficient quality checking, reasonable payment terms, and so on. These are costly and also take up management's time and attention. To invest in them requires building offices and training staff – at the expense of investing such funds in additional productive capacity. The return on the investment is the stream of earnings over and above what they would have been with more mobile customers (minus the current costs of providing the various customer services). These higher earnings reflect higher capacity utilization, steadier prices, and lower risk of default. The rate of return, then, is the rate which discounts the stream of expected earnings, until it just equals the cost of the investment.

This rate must be compared to the rate of return (in terms of earnings growth) anticipated from the construction of new capacity. For, given an amount of investible funds, the firm must decide how much to put into new capacity and how much into the development of "goodwill" or market loyalty. Should the firm sacrifice the construction of some new capacity in order to reduce the downward variance of sales, thereby raising the average rate of capacity utilization and reducing the risk of default? If so, how much?

A very traditional type of calculation can be made. There is a given amount of investible funds. Capacity will be measured on the vertical axis and "goodwill" on the horizontal, both in units of additional revenue. If all the funds are put into capacity, A will be built; if all go into creating goodwill, B will result.

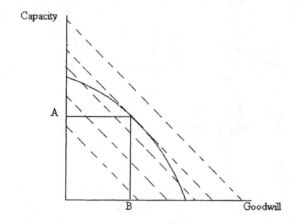

The production-possibility frontier AB then shows the various combinations which could be obtained. This will be concave to the origin, for developing customer loyalty, the relatively amenable and easy-to-reach will be attracted first, and it will take progressively more effort to reach the others. So in units of additional revenue, there will be diminishing returns to investment in "goodwill."

Revenue will be at a maximum when the revenue from additional expenditure on goodwill falls to the same level as that from additional expenditure on capacity. Draw iso-revenue lines with a slope of 45 degrees; the point where the production-possibility frontier is tangent to the highest iso-revenue line will give the optimal division of investment.

This calculation looks very orthodox, but it carries an unsettling implication: firms have a normal incentive *to invest in undoing competitive conditions*. Reducing the mobility of customers – improving the shape of customer loyalty functions – increases average utilization, reduces risk of insolvency and improves competitive position. Over time we can therefore expect markets to change character.[29]

A classification of markets

The limiting case is *pure atomistic competition*, in which the customer loyalty function takes on its extreme form: there is zero loyalty, and so an infinitesimal price change brings an infinite reaction – the entire market moves. But in this case, industrial markets, in which variable costs are constant and average fixed costs fall with increasing output, while demand curves are steep, will be markedly unstable. Moreover, it is contrary to both common sense and common observation to hold that mobility is costless. But if there are costs to switching suppliers, customers will not move unless the price differential is large enough to compensate them.

The normal competitive case can be called *practical atomistic competition*. There are large numbers of similar firms and large numbers of customers are willing to move for relatively small price differentials. But firms calculate their strategies and small amounts of excess capacity will not lead to any price-cutting, nor will small amounts of excess demand bid prices up. Small price cuts will lead to losses; large cuts to retaliation and losses. Firms do not wish to spoil their markets, nor do they propose to erode their goodwill. Prices will therefore normally be more stable than outputs.

But firms will regularly invest in creating goodwill, and as customers develop loyalty the market will tend toward *imperfect competition*. Even large shortfalls in demand will not lead to price cutting; on the other hand firms may be able rather easily to add surcharges when demand is strong. Once customers have become attached, small surcharges will not drive them away. Hence in imperfect markets there will be a tendency for prices to drift upwards over time.

The three degrees of competition are readily illustrated. Q represents full capacity production, mc is marginal cost, and pb is the benchmark price. The curve shows the average price charged as demand varies in relation to capacity, and the shading gives an indication of the variance of

---

[29] Developing product loyalty is not the only way firms can create imperfections to their advantage. Consider a large firm with many small suppliers, e.g., a major auto or appliance company. It could modify the designs of its products, forcing its suppliers to retool in such a way that their equipment would not be adaptable to the specifications of other potential customers (the competitors of the large firm). This will make the suppliers dependent on the firm, enabling it, as a monopsonist, to dictate price and other terms. For the large firm, the calculation whether to do this will pit the cost of product redesign against the stream of gains over time from the resulting monopsony position.

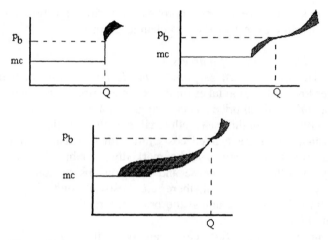

Figure 10.2

prices. In pure competition very slight variations in demand will bring very wide swings in price, with almost no variance among firms. In practical competition there will be a significant range over which price will not change or will change very little, extending both below and above normal capacity. But there will be considerable variance among firms. In imperfect competition prices will rise relatively readily, but it will take a very large shortfall in demand to bring about price-cutting, although the variance among firms will be much larger.

Financial competition and product differentiation

Imperfect competition can develop in another way, in which firms develop a differentiated product, each variant being designed to appeal to specific submarkets. This is the result of pressure arising from stagnation after a period of growth. As lifestyles change with affluence, the growth of expenditure may be concentrated in new fields, so that while established markets may remain large, they will cease to grow. But the capital invested in them will still be under pressure either to grow or to pay dividends equal to growth. A company that fails to grow or to pay an equivalent yield will find that its stock is likely to collapse, making it a target for a takeover. To protect itself management must therefore try to keep up its earnings.

Competitive pricing belongs to the "Ford" phase of industrial capitalism, along with rapid growth, technical progress, and mass production. When growth slows down and stagnation sets in, however, we come to the

"Sloan" period of product differentiation, advertising and marketing (Hounshell, 1984).

Traditional capitalism had just two classes, capitalists and workers. Modern capitalism first created a middle class, and then brought about fragmentation and subdivision in all three classes, ultimately merging capitalists with professionals, and forming distinct styles and patterns of consumption in the various subgroups. In the Ford phase, industry turns out an efficient product which is the same for everyone; in the Sloan phase the product is varied and adapted to the needs and desires (and the unconscious) of each class or subgroup.

The Sloan phase begins when it is clear that the market is stagnant, but the capital is being recovered. Instead of renewing existing plant, the market will be fragmented, and a different product will be designed for each distinct subgroup. The basic production process will remain the same, and the basic underlying product will also. But new plant and equipment will be needed to undertake the modifications necessary to adapt the product to each group – Chevies for the workers, Pontiacs and Buicks for different strata of the middle and upper middle classes, Cadillacs for the rich. Each of the new products is sold to a specialized fraction of the original market, willing to pay a higher price in order to obtain a distinctive version of the basic product. So the new prices will be set by marking up the benchmark price of the original product. Higher prices will mean smaller markets, lower larger. Larger markets will be associated with lower rates of return, unless there are sufficient economies of scale to outweigh the price effect. Normally, then, for each potential product there will be a downward-sloping curve showing the size of the

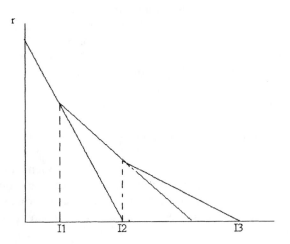

investment (needed to serve the market) in inverse relation to the rate of return.

Arrange the investments in the different products in order of their highest potential rates of return. The first investment will be in the highest-return project, and the amount invested will be the amount that just brings the rate of return in this project down to the highest level obtainable in the next best project. The amount invested in that project will be the amount that brings its rate down to the highest level obtainable in the third project, and so on, until all the funds are invested. Any residual funds, of course, can be invested in the original product, since the new specialized markets, operating at higher prices, will not absorb the whole of the original demand. The ending of a period of growth will thus bring on product differentiation and imperfect competition.

Investment and entry

In traditional theory the entry and exit of firms supposedly correct excessive or inadequate profits, establishing the uniform rate through such movements. By contrast, benchmark pricing is designed to keep new firms out by enabling existing firms to expand enough to supply the (exogenous) growth of demand at a stable price, and the uniformity of the rate of profits is established (if at all) on the basis of normal growth; in general, there will be differentials in profit rates, reflecting different expectations of market growth, and of the evolution of costs. The traditional account is part of the conception of pure atomistic competition, and its account of entry is flawed.

On the analogy of a price cut under perfect competition, when the competitive condition of "free entry" holds, an infinitesimal superprofit will attract an infinite entry and an infinitesimal subprofit will lead to a total exit. An infinite entry – all the firms from all other industries – means oversupply and a collapse in prices; a total exit of all the firms in the industry means radical shortage and a bidding up of prices. Thus perfect conditions of free entry and exit imply severe instability.

To try to limit this instability by placing restrictions on the ability to move implies acknowledging that infinite flexibility is incoherent – but it does not necessarily improve matters. Consider a plausible case: the demand curve facing the market has unitary elasticity, so that revenue in the market is constant. Hence the proportional change in quantities will equal the negative of the proportional change in prices. Assume next that free entry brings new supply, in relation to the equilibrium, in proportion to the price change. In short, quantities change in the same proportion as prices, and prices in the same proportion as quantities. When price rises, new firms will promptly enter, adding their supply to the equilibrium supply, in the same proportion that price rose. But revenue is constant, so this drives down price; when quantity rises *above* equilibrium, price falls *below* equilibrium, in the same proportion. So we have

$$\text{Market elasticity}: \frac{dp}{p} = \frac{-dq}{q}$$

$$\text{Entry elasticity}: \frac{dp}{p} = \frac{dq}{q}$$

Choose coordinates so that the origin represents the normal equilibrium. Plot dp/p vertically, positive and negative, and dq/q horizontally. The demand function then bisects the upper left and lower right quadrants along the 45-degree line, and the supply function cuts the other two quadrants similarly.

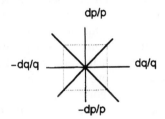

Starting from, let us suppose, a mistaken decision to enter, raising supply above the equilibrium, prices will be driven down below. The mistaken suppliers will exit, but so will other suppliers, bringing output below normal, in the same proportion that price has fallen. This will then attract entry, not only bringing supply back to normal, but overshooting, increasing output, once again, in the same proportion that price rose. And so on.[30]

These assumptions are easily relaxed. Let the two equations be rewritten

$$\text{Market}: \frac{dp}{p} = \frac{-adq}{q}, \, a \geq 1$$

$$\text{Entry}: \frac{dp}{p} = \frac{bdq}{q}, b > \text{ or } < 1$$

Then if b > |a|, the system will converge, but if b < |a|, it will diverge.

[30] Don't they ever learn? Of course they do, but what helps them to learn are the very features of the market that conventional models of "pure" or "perfect" competition exclude, namely indivisibilities, restrictions on mobility, costs of information, costs of shifting lines of trade, customer loyalties, and the like. Formally the model is the same as that used to examine investment-consumption adjustment in Chapter 9.

Of course, "entry" and "exit" are really investment decisions, which brings the discussion back to the relationship of growth in demand to growth of capacity.

In practical competition, firms do enter and exit, though not as often as the textbooks suggest. Oligopoly theorists have argued that firms will adjust their pricing strategies to keep prices low enough to discourage potential entrants. But if prices are kept down, profits will be lower, and less investment can be afforded. Will enough capacity be constructed to supply the growing market, particularly given that the lower prices may stimulate or permit faster expansion of the market? If too little capacity is built, then entry will be virtually impossible to prevent. On the other hand if enough capacity – or more than enough – is built entry will be discouraged even if prices are high, since the firms already in place have a natural advantage (assuming that they are normally efficient and well run).

Entry moves are investment decisions and must be considered on the same footing as the investment decisions of the firms already in an established industry. Leaving aside entry of *capital* through mergers, acquisitions, and lending, firms enter by constructing new plant and equipment. They will do this only if they believe that existing firms are not planning to build enough (or good enough) capacity. Hence, paradoxically, a firm may enter because it believes the present benchmark price too low! Such a price will encourage the growth of more demand than it will generate the profits to build the capacity to supply. So an outsider will enter anticipating a growth of demand in excess of what the industry has expected and prepared for. Outsiders thus will enter when current profits are too low – just the reverse of the traditional doctrine.[31] Similarly, a relatively weak firm would be well advised to exit in a period of high current profits, if it believes that demand growth is likely to be weaker at this high price than anticipated.

There are two cases where the traditional story holds, however. In a new market, the growth of demand is often likely to be both substantial and erratic, so that the firms already producing cannot expect to satisfy demand when it expands, but may find sales drying up unexpectedly from time to time. In these circumstances high prices may indeed be an appropriate signal to enter, and low prices to leave. But when the market settles down and growth becomes predictable, though still not controllable, then the question of entry will depend on outsiders' assessments of the investment plans of the firms already producing.

[31] Asimakopulos (1978, p. 344) cites the case of the Canada Cement Company in the early 1950s, which kept cement prices low and did not expand fast enough, and thus provoked entry to meet the unsatisfied demand, thereby losing heavily, dropping from holding four-fifths of the market in 1946 to one-half in 1957.

The second case concerns a stagnant or declining market. When the market is no longer growing, price cannot be governed by the requirements of investment; with inelastic demand, however, firms could milk the market by raising prices. If entry barriers were weak this could conceivably attract the attention of outsiders. The threat to reduce prices in the event of entry may be a sufficient deterrent; alternatively prices may have to be kept below some temptation level. Much more important, however, is why the market has stagnated; if it is because no potential new customers have been identified, then milking the market by the existing firms may encourage "oblique" entry by outsiders – modifying the product to make it attractive to a new class of customers whom the outsiders proceed to proselytize and supply. On the other hand if stagnation has occurred because a new and significantly more attractive product is taking over the functions of the old, then no one is likely to want to enter – but raising the price will tend to speed up the process of decline.

Even though the growth of the industry cannot govern price, firms will try to earn at least normal profits, so as to have the funds available to invest in other projects. Once a market turns stagnant, the firms in it must diversify; otherwise the capital they represent will cease to expand at the normal rate. Their investors will be disappointed and will tend to pull out, or seek to dislodge the managements. If the possibilities of diversification are good, firms may wish to exit rapidly, and thus will be likely to raise prices even if the higher prices attract entry and speed up the market's decline. Entering firms may buy out the facilities and brand names of existing firms. However, a speed-up of the market's decline will deter entry, or confine it to "oblique entry" – those who believe the old market can be used as a base for developing innovations. Nor can a single firm raise prices unless it is certain that its lead will be followed. The decision whether or not to try to milk a stagnant market will thus depend on the possibilities of diversification, the speed of the market's potential decline and the likelihood of provoking entry.

Rising supply price

When demand is above capacity, prices tend to be bid up; when demand is below, prices tend to fall. Such price movements were held to serve the important economic function of allocating scarce resources to where they were needed most. This traditional view cannot be accepted, but it is true that the market performs an important function. Prices generate profits where they are needed to finance new capacity to service the expected growth of demand. In this sense prices do allocate investible funds, with the important proviso that there is a significant range of variation of demand within which prices do not respond – and for good reason.

Prices here do not reflect scarcities. Traditional theory tried to explain changes in prices, as demand changed, by the movement of costs, requiring the postulate of diminishing returns. We have seen that there is more than a grain of truth in this – for *market prices* in the Craft economy. (Long-run prices in the Craft economy are determined by the interaction of growth with the real wage – which also leads to the Marx-Goodwin cycle.) In modern industry, however, variable costs are constant and average fixed costs fall with increasing output. Costs, therefore, cannot explain the behavior of prices.

The level of capacity utilization reflects an industry's current demand, which in turn is governed by investment. So when capacity utilization is persistently high, pushing prices up, this indicates that investment in general is high, responding to growing demand in many markets. A high price is therefore appropriate, since profits will be needed to finance the capacity needed to meet the growth of demand, and similarly for low-capacity utilization and low prices. Market forces thus tend to push prices in the direction they should move, in light of their function in providing profits to finance new capacity.

Market forces also weed out inefficient – high-cost, low-quality – producers, since they will earn less than normal profits and, with weaker sales and fewer loyal customers, carry more than normal excess capacity. So they will grow more slowly; high-quality, low-cost producers will earn higher than normal profits and will grow faster. If all firms start out at approximately the same size, and if "cost efficiency" and "quality production" are normally distributed, then growth will result in a *lognormal* size distribution of firms, with the larger firms being the more efficient, better-quality producers. The market evolves; new products are developed as part of competitive strategy, and the character of firms changes.

But a great deal of the fluctuation in demand is not long-term, nor reflective of underlying trends. It may be cyclical in nature, or it may reflect temporary aberrations, policy decisions, or speculative excesses. From the point of view of the firm, it would be unwise to continually adapt prices and pricing policies to such changing conditions. Since they operate a Flexible Employment technology, they can adjust output and costs when demand fluctuates. Thus it makes sense to define a normal operating range, and both strategy and customer loyalty will insulate this range from market pressures. Industrial prices are steadier than outputs.

Pricing in administered markets

The effect of investments in customer loyalty, in marketing and in product differentiation will be not only to attach customers to suppliers but also to put firms in a position to affect the growth of markets.

Corporate pricing and investment decisions

When firms are small and the market is new, no one firm will have any control over the growth of demand, nor will any one have a body of established and closely attached customers. But with maturity a few large firms producing somewhat differentiated products will acquire a sufficiently dominant position that they will possess a well-defined and identifiable body of established customers and be able to influence, at a price, the growth of their sales. The analysis which follows concerns a representative firm of such a kind.

Established and new markets

A firm's established market is where it is well-recognized and can count on a loyal and reliable set of customers, either households or businesses, who have incorporated its products into their lifestyles or production processes, so that to switch to an alternative product would entail at least some costs. Switching products thus becomes an investment decision – the stream of gains from the new product, properly discounted, must more than cover the cost of making the switch. Substituting products will therefore not be lightly done, so sales will be predictable, and, within some reasonable range of prices, customers will continue to prefer the firm's products. An established market thus carries itself; the current selling effort is negligible and for all practical purposes capital selling costs are sunk.

Established markets can exist for intangibles – services – as well as for tangible goods, and, in both cases, we can distinguish between durable and nondurable purchases. Durable goods last, removing the customer from the market; nondurables require continuous re-entry into the market. Durable services produce results that last – the surgeon as opposed to the general practitioner, the architect as opposed to the gardener. An established market for a nondurable (an operating input into a household or business) means that there is a regular clientele that repeatedly purchases the good or service for use in its established operating procedures. So it plans for the use of the good, budgets for it, and would have to change its routine to substitute another good or service. In the case of durables, new purchasers similar to those of the past, coming from the same neighborhood, social class, or industry, regularly enter along with replacement purchasers, drawn by the product's reputation and suitability, including the ready availability of ancillary services.

New markets may be markets of the same social class or group, but located in a different geographical region; or they may be in the same region, but involve an appeal to a new social grouping. In the former case new distributional outlets will be needed; in the latter, a new advertising and promotional campaign will have to be devised. So the implications for sales investment will be different in the two cases, but in each investment in a sales campaign will be called for.

Consider a representative firm, or a representative division of a conglomerate corporation, producing a well-defined product with an established clientele. The firm or division has to earn the profits necessary to finance its own growth, either directly by spending internally generated funds, or by showing earnings that will justify the required level of borrowing. In a moment we will see that, given its variable costs, the level at which the firm sets the price will determine the growth rate that it can finance. First, however, we must explore the relationships between the level at which price is set (in relation to money wages) and the rate at which the market can be expected to grow. Over a wide range the large established market will be comparatively insensitive to price, but, by contrast, the smaller new markets which the firm wishes to develop or penetrate will react quite strongly to prices. This can be shown in a pair of diagrams. The left-hand diagram represents the established market, the broken line indicating that it is much larger. The steepness of the demand line indicates that the good is strongly complementary with other goods in the household's normal lifestyle, or in business's production systems. The high and low breaks mark the points where substitution will take place, and customers leave and enter. In the smaller new market, shown on the right-hand side, demand is quite sensitive to changes in price. (The price does not actually have to be the same in the two markets, but if they are different, they must be strictly related and move together.) The current sales drive can be assumed to be concentrated in the new market, and the level of investment in sales and distribution will determine the position and slope of the demand curve confronting the firm in this market. Other things being equal, the higher this level of investment the higher the intercept and the flatter the slope of this demand curve. (Similarly, the position of the demand curve in the established market depends, among other things, on past investments in sales and distribution.) The implication, then, is that a change in price, say from high to lower, will not affect the established market much, but will have a strong impact on the new. At the lower price, the ratio of the new market to the old will be higher. But this ratio represents the growth of sales.

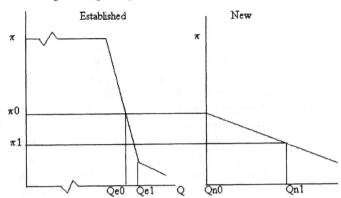

At the price π_0, the rate of growth of sales will be Q_{n0}/Q_{E0}; at $\pi_1 < \pi_0$, it will be $Q_{n1}/Q_{E1} > Q_{n0}/Q_{E0}$. (Again the notation is for this model only.) So as the price falls, the rate of growth of sales increases, for a given investment in sales and distribution. The maximum price for the price – rate of growth function will be the price at which the new market vanishes. The maximum rate of growth, corresponding to the minimum price, will be given by the horizontal intercept of the new market's demand curve.

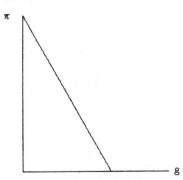

The slope of the growth of demand function then will be

$$\frac{\pi_0 - \pi_1}{Q_{n0}/Q_{E0} - Q_{n1}/Q_{E1}}, \text{ which can be rewritten}$$

$$\frac{Q_{E0}Q_{E1}(\pi_0 - \pi_1)}{Q_{E1}Q_{n0} - Q_{E0}Q_{n1}}$$

In the special case where $Q_{E0} = Q_{E1} = Q_E$, we have

$$\frac{Q_E(\pi_0 - \pi_1)}{Q_{n0} - Q_{n1}} = -Q_E\frac{\pi}{Q_n}.$$

If the demand curve in the established market is quite steep this will be a good approximation to the slope of the price–growth of demand function.[32]

To fix the position of the new market demand function, it is necessary to know the investment in sales effort. There are current costs, both fixed and variable, involved in selling, but the larger part of selling costs constitute an investment.[33] The aim is to capture the loyalty of a set of customers. The time, effort, and expense in selling are to be applied not only to the immediate sale, which may only be a trial order, but to all future sales in that market. A sales campaign requires training and equipping a sales force, providing warehouses, showrooms, office space, personnel, dealers, distributors, delivery systems, servicing arrangements, warranties, and insurance. Independent distributors may be used, but then contracts must be drawn up covering all of the above, and franchises issued. This will involve commitments over various lengths of time. The firm's object is not merely to expand its market, but to develop one that is reliable. It is committing resources to building plant and equipment; it must develop a market corresponding to this new productive capacity that will last as long as the new factories, or at least as long as it takes to write them off. So the expenses in developing the market should be considered an investment; they entail the commitment of capital in construction, training, and the assumption of contractual obligations. These are long-term arrangements, and they are expected to yield long-term benefits.

Investment in market development will shift the growth of demand frontier, but successive investments will not necessarily shift it the same amount. There are good reasons for believing that there will be diminishing returns to investment in sales and marketing. The "pool" of potential customers can be defined geographically and socially. Near the center of the pool it will be comparatively easy and inexpensive to convert

---

[32] This curve shows an *inverse* relation between the markup and the growth rate of demand; earlier we considered a *positive* relationship between the real wage and the growth rate of demand, arguing that a higher real wage (lower markup) would encourage the expansion of demand by inducing a group of households to begin to invest in acquiring the skills that will help them rise in the world. These approaches are complementary.

[33] In the theory of monopolistic competition selling costs are considered, but they are treated exclusively as current costs; the capital costs incurred by developing new markets are never examined (Chamberlin, 1993; Taylor and Weiserbs, 1972; Eichner, 1976; Kaldor, 1950–1). As early as 1931 E. A. G. Robinson referred to the "whole expenditure on wages, buildings, equipment and organization necessary to bring the goods to market" (p. 65) and later, noting the long-term effects of a selling effort, he points out that "once the market has been won, it can be retained at a lower selling cost than necessary to secure it initially" (p. 68). Clearly long-term costs are incurred for long-term gains.

potential customers into actual ones, but the farther from this center the more expensive and problematical the process will become. Transport and transactions costs will rise, and the "fit" between the potential customer and the product will be poorer. This will be important later on.

To summarize the assumed circumstances: The product is well-defined and the established market is given. The productive capacity supplying it has been built, the sales investment has been made and customer loyalties established. The new market has been targeted and projections drawn. It is expected that sales will follow a certain course, depending on the prices charged and the sales investment made. Other things being equal firms will pursue the most rapid possible course of expansion. Their long-term expectations will be assumed to be correct, though allowance will be made at times for unexpected short-term fluctuations in demand. The analysis, then, will hold for the time it takes to carry the project through, that is, to develop the market and build the capacity to service it. After that, the new market becomes part of the established market and attention can be turned to the next investment project. The length of this planning-and-execution period will be longer than the conventional short run, because the new capacity has not only to be built, it has to be operated and the new market consolidated. But the time this takes will itself depend on market conditions and may vary from sector to sector and even from firm to firm.

The supply of finance for growth

Given the foregoing relationships between the corporation's projected growth in sales, its prices, and sales investment, the company's next problem must be to ensure adequate financing to underwrite the investment in sales and to build and equip the new plant to supply the expected new markets. The calculation it must make is relatively straightforward. The company will have a policy, to be explored in a moment, with respect to the burden of debt it wishes to assume relative to its equity. This will depend on the balance between the advantages of leverage and the costs and risk of default. Given this policy the total funds available over the development period will consist of profits plus borrowing, while these funds will be used to cover expenditure on construction of new plant, building up sales and distribution networks, and, of course, meeting existing fixed costs. Thus,

$$P + B = I_F + I_S + F$$

where the symbols stand, respectively, for profit, borrowing, investment in factories, investment in sales development, and fixed expenses. (Again the notation holds only for this model.) So, remembering the earlier symbol

532   Investment and Mass Production

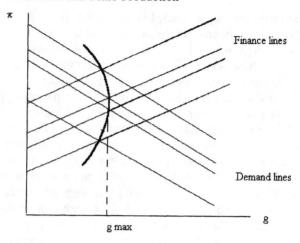

Figure 10.3

for price, and introducing Y for capacity output, W for the capacity wage bill plus materials costs, and g for the growth rate of capacity output, we have

$$\pi Y - W = I_F + I_S + F - B, \text{ and rearranging}$$

$$\frac{\pi Y}{K} = g + \frac{I_S + F + W - B}{K}, \text{ or}$$

$$\pi = gv + \frac{I_S + F + W - B}{Y}, \text{ where } gv = \frac{I_F}{K}\frac{K}{Y} = \frac{I_F}{Y}$$

The rate of growth, g, is defined as $I_F/K = Q_n/Q_E$, so that productive investment is proportioned to the expected size of the new market. The ratio of productive capacity to capacity output, v, is the normal capital-output ratio, and it appears as the slope of the finance relation between price and the rate of growth. This ratio is based on the firm's judgment of its competitive situation, its need for investment spending to keep up with advances in technology, and its expectations of the long-term development of its markets. The intercept will normally be positive, since we can reasonably expect that $(I_F + F + W) > B$. An increase (decrease) in sales investment will cause the finance relation to shift up (down), just as sales investment causes the price-growth of demand function to shift out or in. In addition, the function will shift up with rises in the wage rate, increases in fixed costs, or restrictive monetary policies which reduce borrowing. (Figure 10.3 illustrates these relationships.)

Underlying the definition of g, the rate of growth, is the assumption that the firm's marginal capital-output ratio equals its average. That is, g can be defined as above because $I_F = \alpha Qn$ and $K = \alpha Q_E$. Dividing the second equation into the first, the α's cancel, and the result is g. If the marginal is not equal to the average, but is constant, little is changed. The growth of capacity will simply be proportionate to the growth of demand rather than equal to it. But if there are economies of scale in investment – as there often are – then the relationship between the two rates of growth will vary. As the size of the new market increases, the investment required to service a unit of it declines; hence the finance required per unit of the new market also declines. This can be illustrated readily in a modified version of the preceding diagram. Instead of lines of constant slope, we have curves, starting from the same intercepts and rising, but falling in slope as g rises.

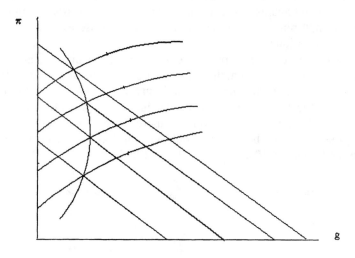

The finance function is defined for capacity output, since the planning period comprises a number of short periods and the relevant concept is the expected norm for the whole period. Nevertheless the model must be capable of analyzing short-period fluctuations, since these are real problems which businesses face. When actual income falls short of the normal operating rate, the finance intercept is increased – the firm will need a higher price to generate the required finance. But the slope of the finance function will remain unaffected, for the firm will not change its judgment of the fraction of its capacity output that it should invest with every shift in the short-run winds.

External finance

The level of the firm's demand for B will be set by increasing risk. Higher levels of B imply higher levels of F in the future; but F/K is set by a balance between the advantages of additional leverage, and the costs of insuring against default, where the risk and the cost of default increase as F/K rises. The risk arises from the probability that a downswing will cut into revenue deeply enough to make it impossible to meet F; clearly the larger F/K, for a given normal rate of profit, the smaller the downswing needed to cause trouble. Given a normal size distribution of fluctuations in sales, risk will increase at a rising rate as F/K rises. The costs of default are the resulting penalties, legal fees, loss of credit rating, and/or reorganization and these clearly increase additively with the severity of the default. But after a point they also interact; when the default is serious, for example, legal fees will be incurred not only to defend the initial default, but also to postpone the penalties, hold off the loss of other credit, defend against reorganization, etc.; in other words, the interaction will be multiplicative. So the costs will tend to rise exponentially, once they begin to interact.

For a given capital-output ratio, a given rate of interest, and a given normal rate of profit, then, the firm's desired level of B/I = F/K will be set at the point where the gain from additional borrowing (a constant) is just offset by the (increasing) cost of the insurance premium to offset the (rising) risk and cost of default. If there were economies of scale the marginal gain from borrowing would increase, as indicated by the dotted line, and a larger B/I ratio would be justified.

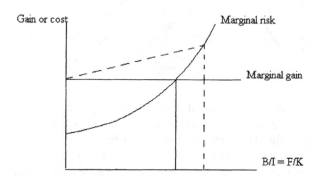

Growth and sales investment

The growth of demand frontier and the finance function together determine the growth rate and the markup, remembering that each relationship is also a function of the amount invested in sales and marketing. Increases in sales investment require the finance function to shift up in a constant proportion, but they shift the growth of demand frontier outward in a diminishing proportion. These shifts can be graphed as above; the intersections trace out the curve indicated. The solution will therefore be to choose the unique price and level of investment in sales that will maximize the rate of growth, g.

Nothing much is changed by the presence of economies of scale in productive investment. As shown earlier, the finance lines will then rise in a progressively shallower curve, but the intersections with the demand lines will still trace out a curve of the same shape, so that there will still be a unique maximizing solution, as in the diagrams above.

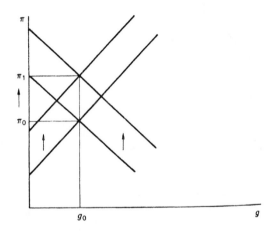

These solutions are predicated on given levels of F, W, B, and Y. If either F or W rises, or if B or Y decreases, then the set of finance lines must shift up, and the equilibrium planned price will be higher. The effects on g, however, require a closer look in at least one case. A rise in F or a decline in B, *ceteris paribus*, will simply shift the set of finance lines, having no effect on the growth of sales frontier. Hence g will fall. But changes in W are more complicated. A rise in W, if it is general and known to be general, implies a general increase in household incomes and so new growth in consumer spending. If the firm's products are consumer goods, the growth of sales lines will shift up, too. Price will increase, but g will be unaffected. The rise in W raises the finance line, and it also shifts up the growth of sales line. If the two are affected in the same proportion, the growth rate will be unchanged, as in the diagram above and the entire effect of the rise in wages will be to increase prices.

But there is another possibility. The increase in wages could increase spending (at every price) in the established markets in the same proportion that spending is increased in the new markets. If at every price spending in the old and new markets has increased in the same proportion, spending in the old and new markets will stand in the same ratio as before the wage increase, so there will be no effect at all on the growth of sales. The rise in W will have shifted up the finance line but will have left the growth of sales line unaffected, resulting in an equilibrium with a higher price and a lower rate of growth, as below.

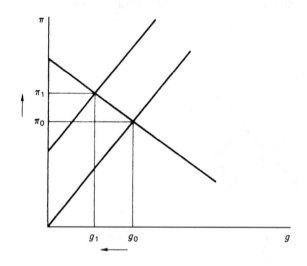

Choice of technique

Modern industrial processes can frequently be computerized and automated, substantially reducing labor costs, raising fixed costs, and increasing the capital-output ratio. We will assume, however, that fixed costs will be set by the calculation of risk above, and that no technique will be considered which implies a rise in F. Alternative techniques therefore change the finance frontier; they cause it to differ in slope and intercept; the technique with the lower intercept – lower W – has the higher capital-output ratio, so is steeper, rising from left to right. Consider two techniques as drawn below: at some point they will cross. If the demand-growth line is steep, cutting the techniques below their inter-section, the relatively capital intensive technique, since it has the lower intercept, will yield the higher rate of growth. But if the demand-growth line is relatively flat and cuts the finance lines above their intersection, then the less capital-intensive technique will give rise to the higher growth rate.

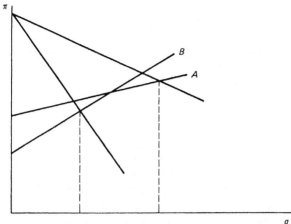

Choice of technique

The choice of technique is normally treated entirely in terms of supply-side considerations. By contrast, the analysis here implies that the possibilities for growth of demand play a significant role in determining technique.

Impact of exchange rate changes

Foreign goods may be imported and sold for domestic currency, either with or without further processing; or foreign capital equipment, energy, or intermediate goods may be used in production. In either case a change in the exchange rate changes the costs which must be covered by the supply of finance; but in the first case, the change increases the cost of sales inventory, in the second it affects the cost of productive capacity. Hence a change in the exchange rate alters the intercept of the finance frontier of a firm importing foreign goods for sale, while it alters the slope in the case of a firm importing foreign capital equipment.

Consider a firm making candy that imports its chocolate and syrup from Europe, processing it with domestic equipment, and compare it to another firm, making scientific instruments, that imports high-grade equipment from Europe but uses domestic materials. For the sake of the argument, suppose that the finance frontiers of both initially have the same intercepts and the same slopes, and further suppose that imported materials are the same percentage of the "intercept" costs that imported equipment is of total capital. The exchange rate change will therefore have the same percentage impact in both cases.

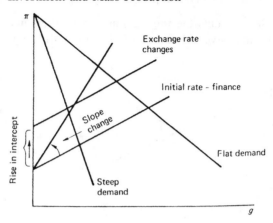

In this figure both the intercept and the slope are shown to double (a very large change, representing a 50 percent decline in the value of the domestic currency, if all the relevant costs are incurred in foreign currency). The two new finance frontiers will intersect at a certain point. If the price-demand lines are similar and both are steep enough to lie below the intersection, we can see that the capital-importing industry will raise its price less and suffer less growth slowdown than the materials-importing industry. If the price-demand lines are similar and both flat enough to lie beyond the intersection, just the reverse will be true. In general, the flatter the price-demand lines, the less the impact on price, and the greater the impact on growth. But in neither case will the full impact of the exchange rate change be passed along as a price increase.

The usual explanations for variations in exhange-rate pass-along, bottleneck and beachhead effects, are based exclusively on supply-side concerns;[34] by contrast, once again, the account here depends crucially on demand, but not on the level of demand – it is the relationship between price and the rate of growth of demand that matters.

[34] "Beachhead" and "bottleneck" models undoubtedly contain a certain amount of good sense (Baldwin, 1988; Dornbusch, 1987; Dixit, 1987). Entry and exit conditions are asymmetrical; if an exchange rate change induces entry, and equivalent change back will not cause all entrants to exit; this is the beachhead effect. Bottlenecks occur at full capacity; an exchange rate change is assumed to increase capacity; when it changes back, however, capacity does not decrease. The result will be a discontinuity in marginal costs, and therefore in the movement of prices. But all such models rely on the framework of monopolistic competition, in which current output and prices are adjusted together. Exchange rate changes, however, take place in the context of a growing economy, and their effects have to be examined in the context of corporate planning for the financing of growth.

Temporary or cyclical variations in demand

The argument so far has been concerned exclusively with expected long-term normal costs and sales, and the variables determined have been planned or "benchmark" prices and the target rate of growth of sales. But this same framework can be used to determine the appropriate responses to short-run or cyclical variations in aggregate demand, affecting the firm's current and immediately future rate of sales. To make the adaptation, we assume that the underlying parameters remain unchanged, and that the firm wishes to maintain its long-run position as well as possible.

Notice, however, that there is an asymmetry between short-term increases and declines in demand. When demand falls, the firm will not necessarily want to cut investment spending, since the long-term pattern of growth is unchanged. But the recession will probably mean that new markets will temporarily dry up. Hence it will be able to exploit its established market by increasing prices enough to maintain the flow of funds required for growth. When demand rises, however, the new rate of growth indicated by the new intersection could not be sustained (since the increase is temporary). So there is no point in moving to it by cutting price and setting up a sales campaign – which could be rudely upset if demand fell back to normal and prices had to be raised. So a rise in demand will not lead to a fall in price; indeed, if the rise is large enough (and regarded as temporary enough) firms might well react by adding surcharges or service fees to ration the demand.

On the diagram, the price-demand growth line is unaffected, since the demand change is only temporary. A fall in demand raises the finance line, a rise lowers it. To maintain normal growth when demand falls firms will have to raise price to the point on the new finance line.[35] When

---

[35] A decline in demand will normally shift the finance function up, but in some cases may leave the sales-growth frontier unaffected, resulting in a rise in price. If this behavior becomes dominant in the capital goods sector, first, the money demand facing that sector will be reduced in real terms, and second, that sector's wage bill will be reduced in real terms. This has two consequences: the new lower level of real demand implies a lower level of output and employment – lowered in proportion to the price rise. Second, with a lower real wage bill, consumer profits, output and employment will be reduced, and the need to raise prices in consumer industries to maintain credit ratings will reduce output and employment still further in that sector. Total profits will remain the same in real terms, equal to the real investment demand facing the capital goods sector, but the *realization* of profits is shifted away from the consumer goods sector toward capital goods, in proportion to the price rise (real wage decline) (Nell, 1992, pp. 584–8).

demand rises, since cutting price will not be a desirable strategy, maintaining price permits a buildup of reserves sufficient, if invested at a later date (to make up for a recession), to raise the growth rate by the amount indicated.

Cyclical downturns, however, can become depressions and last long enough to affect plans. The effects of such a decline can be analyzed in the same way. Consider first the case of a recession, in which (only) new markets will tend to dry up; the sales growth lines will shift down, offsetting the rise in finance requirements. If the shifts are equiproportional, price would remain unchanged, while the growth rate would fall. If the shifts are not perfect offsets, the major effect will still be on g, while the change in price would be relatively minor. Recessions do not tend to lower corporate prices.

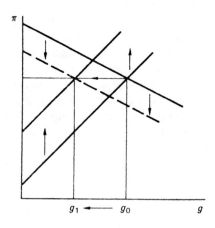

There is another possibility. A decline in incomes, perhaps due to layoffs, may reduce demand in the established markets as well as in the new markets. If demand is reduced in both in the same proportion, then, as in the case of a rise in W, the ratio of spending in new markets to spending in old, for each level of price, will be the same after the change as before. Hence the sales-growth line will not shift, and the entire impact of the decline is brought about they the upward shift of the finance line, resulting in a higher price and a lower rate of growth.

A policy of raising prices in response to variations in effective demand tends to lead not only to inflation, but to sluggishness or even recession in output and employment, while a policy of price cutting in response to demand declines can tend to help to prevent slumps from worsening. This

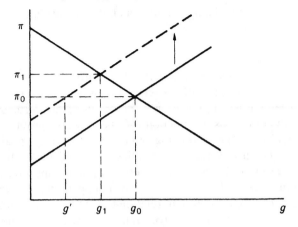

corresponds to empirical findings: in the era of precorporate capitalism, before the rise of the great oligopolies and conglomerates, downturns showed up primarily in falling prices, with relatively minor fluctuations in output. Since the 1920s, however, prices have fallen very little or not at all during downswings, in recent years even rising, while the fluctuations in output and employment have been considerable (Nell, 1988, ch. 4; Sylos-Labini, 1984, part 3; Semmler, 1982, ch.3).

Product cycles

Finally, this framework can be used to analyze product cycles. Start with a given negatively sloped growth of demand curve and a steeply rising finance frontier, indicating that problems are anticipated in expanding scale. with a high intercept, reflecting high sunk costs, e.g. development costs. The markup will be high and the growth rate low. Then as time passes sunk costs will be recovered, so the intercept will fall, and innovation and learning by doing will make it possible to benefit from economies of scale, so the slope of the finance curve will decline. The markup will fall and the growth rate rise. But at some point the new markets will begin to dry up; it will take a larger drop in the markup to bring a given increase in the growth of demand. The growth of demand curve will swing down, becoming steeper, so that, with a given finance frontier, both the markup and the growth rate will decline. But as growth declines still further, or threatens to dry up altogether, firms will decide to wrap up their investment and move on. At this point they will raise price enough to cause the market to decline as they move capital out and into other lines, choosing a price that will generate the optimal rate of decline, in terms of the profit that can be obtained from the remaining customers, the scrap value of the existing capital, and the opportunity costs of investing in new ventures.

Conclusions

The failure to develop a theory of the growth of demand has prevented the full elaboration of the connections between pricing and investment. The Classical equations, describing the structure of capitalism, show prices as a function of the rate of profits. These equations are indispensable in relating prices to the technological interdependence of industries. But a corresponding theory of market behavior is needed, one based on the connection between pricing and investment. The rate of profits is related to the rate of growth; in particular, as we have shown, setting prices appropriately in relation to money wages, yields the rate of profits needed to finance a given rate of growth – the growth of supply needed to match the anticipated growth of demand. (And prices, in turn, can influence the growth of demand.) Yet conventional economics has persisted in trying to determine market behavior by balancing the *levels*, not the *rates of growth*, of supply and demand.

A rise in the growth of demand facing any given company, or group of companies, will lead them initially to plan to raise prices in relation to money wages. But it will also permit a better division of labor, and greater rationalization of production, reducing unit labor costs. A general rise in the growth of markets will enable many firms to take advantage of economies of scale, reducing costs. So while prices will be increased in relation to costs, costs themselves will tend to fall. Under craft conditions a rise in growth normally required a fall in the real wage; under Mass Production, as we shall see, a rise in growth is likely to bring a rise in the real wage, reflecting the rise in productivity.

A theory of the growth of demand must start from an appropriate picture of households, based on class and socio-occupational divisions. As long as demand is thought of in individual terms (or, simply neglected) the marketing plans of firms, and their investment in their marketing divisions, will not be properly appreciated. Consider the usual approach: Rational individuals need only to be provided with technical information about products and market information concerning price and availability. They then make decisions which hold until the parameters change. On the other side of the market firms produce goods to given specifications, choosing the cheapest methods from a preexisting set of techniques. Both consumers and producers are peculiarly passive in this picture – their choices are fully determined by an algorithm over whose terms they have no control. In pure competition firms control only the quantity variables; under various forms of imperfect competition they can, in addition, indirectly influence price, through their control over quantity. But their strategic moves are basically limited to adjustments of output. There is no scope here for imagination, insight, inventiveness, or strategic competitive warfare. By treating a household as a member of a social class, however, subject to the rules of conformity and pressures to "keep up" or advance, that are inherent in its lifestyle, we are able to understand and analyze the way firms try to build customer loyalty, seek out new markets, and develop improved – more marketable – products. Besides price-cutting, there is competition to develop products, services, and technology.

Two cases can be clearly distinguished. In the first, the firm has no control over the growth of its market. (This may be thought of as holding in early Mass Production, perhaps even in the late phases of Craft-based factories, where greater flexibility in employment has been achieved.) It must estimate this growth on the basis of what it knows about its customers – their incomes, the social pressures under which they live, the hopes and expectations they have for themselves and their children, and the way all of these are changing. Firms make their plans on the basis of these data, given their present position on the market and their estimate of their competitive strength vis-à-vis their rivals.

But firms will not accept their fate passively. To improve their competitive position, they will try to bind their customers more closely to them. This, together with the gradual elimination of weaker firms, will lead to a new situation, in which firms, by investing in market development and choosing appropriate pricing policies are able to exercise some control over the growth of their markets. This will be increasingly the case as Mass Production methods spread. As firms gain more control over their markets, their pricing policies come to influence the expected growth of their markets, and the second of the two pricing models comes to apply. To understand the differences between the two cases, and the way they are linked by market development, requires a theory of the social influences affecting household decisions to consume and invest. Many different dynamics are imaginable, but all will draw on an understanding of the way new products enter household budgets and the way new markets can be opened.

This perspective also makes it clear that new products and new processes are being developed and introduced continuously. A new product may be a slight modification of a standard product, designed to make it more attractive to a particular group of potential consumers – witness recent cigarette marketing campaigns; or it may be a new method of performing the same function – compact discs for tapes or records. Whichever, the innovations are part of an ongoing process of adaptation, as are the innovations in production, with which they frequently interact. For both result from and serve to intensify the pressures of competition.

Actual growth fluctuates quite strongly around the norm of expected growth, and as a result, realized profits also deviate sharply from normal levels. But under conditions of advanced technology, actual relative prices, and sometimes even money prices, show considerable insensitivity to these fluctuations, although output and employment adjust rapidly. Traditional theory suggests that prices rather than outputs should adjust rapidly, and in the earlier conditions of fixed employment, they did. Their failure to do so now has been taken as evidence of market imperfection and the growth of oligopoly. But traditional theory analyzes markets in terms of a conception of the firm as an artisan plying a craft; its approach to competition will not transfer to an environment of industrial corporations. Limited liability companies operating industrial technology competitively will set price guidelines on the basis of expected market growth, and have good strategic reasons to keep actual prices steady while adjusting output to variations in sales. Competition will play itself out in ways that leave prices largely stable in the face of changes in demand. The prices set in this way can be related to the theoretical ideal of the prices associated with a uniform rate of profit. But competition also has a way of undermining itself over the long run; the strategies pursued will change the nature of the market, leading to closer ties between firms and customers. When this has become widespread, pricing policies will affect market growth, and prices will respond differently to variations in the market.

CHAPTER 11

Inflation, employment, and market adjustment in Mass Production

Prices are determined along with investment plans. But plans still have to be executed, and the timing and extent of execution depends on *current* variables, especially effective demand and interest rates, a point that will be explored later. But such changes in timing and execution also affect current variables – in the aggregate, demand changes tend to cause further demand changes. In Craft economies such variations in carrying out plans will normally lead to price changes and gravitational adjustments. These effects tend to be movements in the opposite direction to the initial change, and tend to stabilize the system, disposing it to converge on the normal position, as explored in Chapters 8 and 9. But in Mass Production prices will tend to be steady over a wide range of variation. Instead quantities – output and employment – will adjust, but the pattern of changes set in motion will move in the same direction as the initial change; so the adjustments will tend to be destabilizing rather than stabilizing. The differences in the way Craft and Mass Production markets adjust have implications for almost every aspect of the system – and also for theory. Starting from the implications for the idea of equilibrium, we will examine consumption, investment, labor, productivity, and inflation, ending up with a look at a contrasting way to operate a Mass Production system: socialism. In each case we will see that it is difficult to analyze the system adequately on the basis of scarcity and rational choices by representative individuals.

Demand equilibrium in Mass Production

The Keynesian notion of equilibrium differs markedly from the traditional idea of a position which is reached through markets coordinating maximizing activities that are subject to resource constraints. The theory of effective demand permits unemployed resources in equilibrium; the neo-Classical approach holds that, in equilibrium, prices reflect relative scarcities – an unemployed factor would have a zero price. Both start from a conception of equilibrium as a state of rest, reached by some dynamic process. But in the case of effective demand an equilibrium is achieved by arriving at a balance between the forces which encourage

545

expenditure and the forces which discourage it – both sets of forces operate on demand! Equilibrium is achieved by balancing "injections" into the stream of expenditure against "withdrawals" from that stream – the famous hydraulic metaphor.[1]

By contrast in most versions of neo-Classical theory, equilibrium is reached by maximizing an objective subject to a supply constraint; or by finding the optimal allocation, in terms of preferences, of goods produced with scarce factors. A market or multimarket equilibrium is established by considering the simultaneous optimizing of sets of interdependent agents, but the basic problematic remains: to reach the highest preference levels, given the technology, the endowments of factors and their initial distribution. Factor cost is determined by scarcity – "all costs are ultimately opportunity costs" – since opportunities (to achieve preferred states) would not have to be forgone if there were no scarcity. The "state of rest" is imposed by resource constraints. Binding supply constraints figure crucially in the determination of equilibrium. Indeed, the modern interpretation of prices as a signaling system for the coordination of decentralized activities rests on the fact that the information being signaled is precisely where the constraints bind: relative scarcities and opportunity costs.[2]

Of course, many contemporary models do not contain simple fixed constraints; instead this role is filled by convexity assumptions. Action undertaken in a certain direction becomes progressively more difficult or more expensive; more resources must be used to obtain a given effect. The resources, however, are either limited or must themselves be produced at the cost of disutility and/or other resources. Convexity thus expresses the central idea of resource-based limitations on the ability to expand activity.

[1] This approach was compromised by Keynes himself, however, when he declared that the traditional theory held sway at full employment. The idea achieved its clearest statements in the work of Kalecki and, later, the Sraffians. The latter, especially, avoid the neo-Classical setting in which modern economics, following Keynes, has placed the theory of effective demand, thereby forcing it to play second fiddle to scarcity. Setting the problem of effective demand in the context of prices determined by reproduction makes it possible to see the unique features of a demand equilibrium.

[2] Most contemporary discussions of equilibrium focus on either the kinds of behavior – e.g. "quasi non-tatonnement," trading at false prices – or on the nature of the assumptions – tight prior equilibrium, stochastic information, effects of uncertainty, etc. The questions concern the effects the actions of various agents have on each other when these are varied. How do different assumptions about behavior patterns and conditions affect the ability to reach equilibrium? Will the equilibrium position be independent of the path of adjustment? That equilibrium is a relation of desire pressing against constraints is normally taken for granted in these discussions.

It is a more general expression of the idea of scarcity; resource limitations are not absolute, but relative, and thereby give rise to more complicated choice problems.

Rising costs, of course, can inhibit spending as well as reflect scarcity – but such inhibition is not a separate supply-side effect constraining the level of activity; on the contrary, if it happens, it will be a reduction in the flow of demand or spending, comparable to saving or other withdrawals. Rising costs, of course, do reflect supply-side constraints. But these constraints act on demand by changing the distribution of income. Rising costs lead prices to rise; if this leads to an increase in profits and savings, then withdrawals will increase because of rising costs, or because of their influence on prices. This will reduce sales at all possible prices, like any other variation in effective demand. But rising costs may also reflect a more rapid rise in wages than in prices, and such a shift in distribution could well *increase* demand pressure.

More interestingly, as we have previously noted, supply constraints could lead to inverse movements in demand pressure. Consider a Craft-based economy operating a Fixed Employment technology. A rise in investment spending, or in net exports, will lead to price increases; this will reduce the real wage, and with employment fixed, consumption will decrease, offsetting the original increase, and thereby tending to hold aggregate demand steady. By contrast, in Mass Production a change in investment or net exports would tend to produce a further change in demand in the same direction. But both are examples of market adjustment determining the new level of aggregate demand, following a change.

This also shows why the behavior of the economy cannot be studied as if it were the rational responses of a representative individual, reacting to the change. In the Craft economy, different sectors respond differently, and the outcome is the result of their interaction; in the Mass Production economy, different sectors respond in similar ways, but the outcome is the compounded result of their interaction. In each case the final result depends on a process which works itself out through the *circulation*.

In a Mass Production economy the forces which depress or reduce spending on produced goods and services (saving, taxes, hoarding money, speculation, preferences for foreign goods) are balanced against forces which stimulate or increase such spending (business expectations, government needs and political conditions, technical needs or desires for emulation abroad). In the simplest instance there are no effective supply constraints; an inverse Say's Law holds – demand creates its own supply. It is assumed that supply – output and/or employment – adjusts quickly and easily to demand; by contrast, prices are not affected, or not much affected, by changes in the level of demand. Equilibrium here involves only demand: it is a point at which the stimuli to spending just balance the inhibitions, so that the level of expenditure (and, by implication, output and employment) just holds steady, with no tendency to change level or composition. Such a balance, of course, need not determine a fixed level; it could define movement along a path just as well.

In traditional Keynesian theory it is assumed that money prices and wages are somehow "sticky," leading to the view that the problem is one of disequilibrium. As we have seen, this is a misconception. The real difference lies in the shift from a Fixed Employment technology in which most current costs are fixed, to a Flexible Employment technology in which current costs can be varied with output, and output can be adjusted to current sales. In a Fixed Employment system, such as the nineteenth century economy of family firms and family farms operating traditional crafts, it will be hard to adjust employment and output to a decline in sales, especially a sudden one; hence prices will be forced down by oversupply. But when employment and output are flexible, as in a twentieth century Mass Production system, operating mechanized assembly lines, current costs can be adjusted along with output, so there will be no pressure of oversupply and prices can be kept steady.[3]

A demand equilibrium balances flows of revenue injected and withdrawn from circulation. Since these flows are value magnitudes, such a concept must rest on an independent price and distributional equilibrium. This can be expected to reflect deep-rooted, relatively permanent features of the system, whereas spending patterns may be capable of shifting rather easily. But when they do, a new "equilibrium" of demand can be quickly established. Accidental or one-time deviations from normal spending patterns will have limited effects; the spending system itself reestablishes its own pattern – such stability as it has does not rest on supply constraints. Moreover, we will see that shifts in the basic factors determining demand could themselves exhibit fluctuations, so that the successive positions of demand equilibrium could very well trace out a cycle.

Demand and "quantity constraints"
Interpretations of macroeconomics as a "quantity-constrained system" are attempts to force demand analysis into the traditional mold,

[3] Since prices will be determined along with the rate of profit by long-run considerations they will be unaffected by short-run changes in demand. They are independent of endowments, so do not reflect relative scarcities. There is no need to postulate any kind of disequilibrium; prices are determined in one manner, output and employment in another, by quite different forces. Of course, this runs counter to the entire drift of conventional theory, which determines prices and outputs together (Hicks, 1965; Pasinetti, 1975; Sraffa, 1960; Walsh and Gram, 1980).

making supply activities create a constraint on spending. Quantity-constrained systems, such as those of Clower and Malinvaud, suffer from a simple but decisive defect. They represent a somewhat idealized Craft economy, rather than a modern economy of worker and owner households and corporate businesses. In Clower-Malinvaud systems business expenditure plus business demand for money is constrained by current sales plus initial money balances, and household demand for goods and services plus household demand for money is constrained by household factor sales plus initial money balances. This is simply a slightly sophisticated form of the old Robertsonian belief that one period's investment is constrained by the previous period's savings.

In a Fixed Employment economy, in which investment and consumption are inversely related, this is perhaps arguable – but not in a Variable Employment, Mass Production economy, where investment and consumption move together. In such a world there are no direct constraints on business investment; neither current nor past sales, nor current assets, indeed, nothing acts as a direct barrier to investment spending. Under the influence of the profit motive, the banking system can, and (if not prevented by government policy) will, expand finance to accommodate virtually any level of investment that banks and business agree is warranted by the risks and the expectation of future profits – regardless of the level of current or recent sales, savings, profits, or anything. Of course, these will all have an influence on expectations of future profits, and on levels of present risk, so indirectly they will act as inhibitions on spending. But business expenditure on investment is not subject to a budget *constraint* in the sense that there is a precise sum, fixed in advance, composed for example of current sales receipts and initial money holdings, which it cannot exceed. Keynes repeatedly stressed the volatility of investment; this requires that investment spending cannot be precisely constrained. Indeed, the financial system is designed to make it possible to vary investment easily. (In the modern world, households may not be so sharply constrained, either. The development of consumer credit and new forms of near-monies, like credit cards, have freed many households, especially well-to-do ones, from income restraints – as is evident from the many studies of consumption that show the importance of assets in household consumption functions.)

Are there effective aggregate constraints?

Demand analysis is sometimes assimilated to the traditional mold by holding that effective demand will be constrained by finance, or that it runs up against a short-run capacity constraint, or a labor constraint. These cases, however, are fundamentally different. To take the first, a financial constraint is an inhibition on effective demand, on spending, and it applies to individual businesses and households, but not to the system as a whole (except as a matter of government policy, and often not effectively). It is *not* a special kind of supply constraint; finance is not a "scarce resource" – that is the point of the capital controversy. Finance is liquid capital, and a financial constraint is a constraint on the ability to exercise purchasing power, that is, to draw on liquid capital. But liquid capital is capital, and capital, being produced means of production, is not "scarce"; it is not among the "givens" of the economy. In the short run (government policy apart) a profit-driven banking system will always finance the need of production and investment, up to full capacity (and often beyond), regardless of the current level of savings or profits. The aggregate quantity of finance will adjust (Nell, 1984, 1988; Minsky, 1975; Davidson, 1980, Moore, 1988). In the long run capital is wholly variable; in the short run it is variable within limits, depending on the possibilities of over- and underutilization of capacity. But capacity utilization is flexible; underutilization is always possible, while overutilization can generally be managed for a time, though it may involve changes in the way work is organized – a form of technical progress. Given the technology, output and employment – defining the scale of the system – are always, in an important sense, demand-determined, and finance – credit and the money supply – will always adapt to demand.

By contrast to finance, both capacity and labor are supply constraints which are capable of binding the system as a whole. But in operation both are weaker than often supposed.[4] So the notion of a demand equilibrium is not only important theoretically but is the one normally met in practice in modern economics. Keynes wrote that

> it is an outstanding characteristic of the economic system in which we live that, whilst it is subject to severe fluctuations in respect of output and employment, it is not violently unstable. Indeed, it seems capable of remaining in a chronic condition of subnormal activity for a considerable period of time without any marked tendency either towards

[4] Full employment, though it is commonly considered in textbooks to be the point where the labor constraint binds, is ambiguous as between labor or capacity. Moreover, today it is often defined in practice to be the level of employment at which prices begin to rise. Yet this only makes sense if there is no other reason for rising prices than restricted supply in the face of strong and persistent excess demand. But prices may depend more on the relation between rates of growth of demand and supply, than on the relation between their levels – or other factors altogether may be determining; so they may start to rise even when there is excess capacity. There is no justification for the current practice of identifying full employment with the level at which prices begin to rise. Nor is it easy to pinpoint the level of demand at which the elasticity of supply has reached zero.

recovery or towards complete collapse. Moreover, the evidence indicates that full, or even approximately full, employment is of rare and short-lived occurence. Fluctuations may start briskly but seem to wear themselves out before they have proceeded to great extremes, and an intermediate situation which is neither desperate nor satisfactory is our normal lot. (1936, pp. 249–50)

In the interwar period a demand equilibrium with underutilization was the norm. During the war and in the postwar world, however, government policy often brought demand up to or even beyond the level considered full capacity.[5] Yet that did not mean reaching an equilibrium constrained by resources. In 1950–1 and again in 1967, U.S. output exceeded the official estimates of the full capacity level. In advanced economies both the labor supply constraint and the capacity constraint can be shifted relatively easily when there is a strong stimulus of demand, provided there is sufficient practical and political ability to reorganize production.

Shortages and demand

Consider a demand equilibrium that implies a level of output or use of a resource in excess of the amount available. Traditional thinking tells us to expect demand inflation; if demand is pushing against a constraint, the price must rise, cutting back demand. (But some kinds of price rises, e.g., money wages, may also raise demand.) There is another possibility, however; the constraint may give way. If demand is sufficiently strong and persistent, production will take place above capacity; new sources of labor will be found, and new methods of production will be invented or introduced. (By early 1943 the U.S. economy had far exceeded all its capacity limitations and labor constraints.) What is scarce

---

[5] The practical significance of this can be seen by asking what constraint was binding on the American economy in 1944 (or on the German economy in the late 1960s). Clearly it wasn't labor. The American economy drew women into the labor force in unprecedented numbers; the Germans brought in "guest workers." The importance of demand pressure can also be seen in the relation of capacity estimates of the United States economy in 1939 to later performance. In 1939 the United States had a GNP of approximately $91 or $92 billion, and was estimated to be operating under capacity by about 10 percent. By 1944 the GNP of the United States was over $210 billion, with the economy running at almost 150 percent of capacity, by some estimates. During this time the composition of the labor force radically changed as nearly 12 million of the youngest, healthiest and best-trained male workers were withdrawn from the labor force, into the armed services, and were replaced by women who had not had previous training in heavy industry. So the labor constraint was not binding, at all, and the estimates of capacity were wildly wrong in 1939. Of course, the economy had shifted from a market system to an essentially planned mode of operation.

depends on the givens, but in the face of strong and persistent demand, the givens will change.

This is obvious in the case of produced means of production. Capacity limitations can be shifted by investment (Hicks, 1965; Lowe, 1976); but even without investment they can be changed by improved or more intensive methods of operation – reorganizing the flow of production, running additional shifts, speeding up the assembly line, and so on (Nell, 1964, 1982; Kurz, 1990). But it also applies to nonproduced resources. Land use can be improved; methods of production can be altered to avoid reliance on scarce minerals or energy sources. Nor do these changes necessarily increase costs or depend on increases in price. The innovation may not have taken place before, because there was not enough *pressure*. Innovations are always risky; even if they look profitable they are uncertain; when conditions are sluggish, capacity-enhancing and resource-saving innovations may not be implemented, since there is no pressing need.

When demand lies above the level of a scarcity equilibrium, if the demand is strong and persistent, pressures will develop to shift the constraints. Supply adapts to demand, in the long run and in the short; scarcity is itself produced and is always changing in response to demand.[6]

There is, however, a major exception to this: when aggregate demand chronically lies above capacity, not by accident, but because the working of the system has a tendency to generate excess demand, then the system takes on the characteristics of a "shortage economy." Under these circumstances demand pressure will not lead to innovation – far from it: since anything produced can be sold, there will be no pressure at all for improvements or cost-cutting. Product quality and services will tend to deteriorate (Kaldor, 1985; Nell, 1991). This will be examined further in the last section.

Demand pressure and market adjustment

The price adjustments for a Craft economy considered in Chapter 9 have important macroeconomic implications for the relationship between investment and consumption, and for employment and productivity.

---

[6] The Heckscher-Ohlin model tells us that international trade patterns will reflect relative "factor endowments." Repeated tests over a variety of countries and time periods show that, in general, this is not so. Shortages of a factor or of a resource do not lead to *adaptation* to that shortage or to substitution of something else – shortages lead to *innovation*. The best explanation for trade patterns appears to be relative technological advantage (Krugman and Obtsfeld, 1994).

The elasticity of consumption with respect to investment

Even though employment will not vary with changes in investment, output will, although less and with a greater lag than price. But consider the "pure" or extreme case, where the entire effect of a change in investment initially falls on price, that is, investment demand falls by a certain percentage, and price drops by the same percentage. This means that the costs of capital inputs into consumer goods have declined by this percentage, so competition should lower the price of consumer goods in proportion. As a consequence the real wage will rise in proportion, and on the assumption that the whole of real wage income is spent on consumer goods, consumption demand will increase in the same proportion.

Spelling this out formally, with I as investment demand, Y_k as current capital goods output, Y_c the output of consumer goods, p_k the price of capital goods, p_c the price of consumer goods, w/p_c the real wage, and C the demand for consumer goods.

Since $I = p_k Y_k$ in equilibrium, $dI = Y_k dp_k$, when output remains fixed; dividing both sides of the second equation by the first,

$$\frac{dI}{I} = \frac{dp_k}{p_k}.$$

The relation between p_c and p_k can be written in the form used earlier, bearing in mind that these are money prices

$$p_c = RAp_k + wB$$

$$p_k = Rap_k + wb$$

where R is the gross profit rate on capital costs (capital inputs multiplied by p_k), set at the competitive rate, and w is the competitive wage rate needed to cover normal labor costs and provide labor its net earnings. When I falls, p_k will fall in proportion but will continue to cover wage costs; hence the realized rate of profit in capital goods will fall. Competition will therefore tend to push prices down similarly in consumption goods. Three effects will be felt. First, capital equipment and input costs will be lower; second, working capital will be more readily available from the banking system, at lower rates; third, resources and labor will tend to shift in, threatening to expand output. (We may assume that tools and labor's skills are easily shifted, and that many goods serve both purposes.)

On the assumption that rates of profit remain equal as prices fall, and that w remains fixed, solving each equation for R, equating, differentiating, and rearranging, will give

$$\frac{dp_k}{p_k - wb} = \frac{dp_c}{p_c - wB}$$

Suppose first that the capital-labor ratios in the two sectors are the same, for example, in an early stage of capitalism in which the same firms make both tools and equipment and household goods. In these circumstances, when p_k falls, bringing down p_c, so that the rate of profit falls in the same proportion in both activities, since $B/a = b/A$, the above procedure yields

$$\frac{dp_c}{p_c} = \frac{dp_k}{p_k}.$$

But if the capital-labor ratios of the two sectors differ, this equality will not hold. Suppose capital goods is capital-intensive and its price falls because of a decline in I; then with an equiproportional fall in the consumer goods price the resulting revenue would have to cover the labor-intensive wage bill, leaving a rate of profit lower than that in capital goods. However if the current profit rate in consumer goods falls below that in capital goods, resources will no longer tend to move there, nor will working capital be offered preferentially. Once the consumer goods rate of profit has fallen to the level prevailing in capital goods the pressure on consumer prices will cease. In other words, the realized rates of profit will tend to end up the same, so the consumer goods price will not fall proportionally as much as the capital goods price, that is, $dp_c/p_c < dp_k/p_k$. Just the reverse holds when the capital goods sector is labor-intensive; the pressure on prices in consumer goods will continue even after consumer prices have fallen proportionally as much as capital goods prices, since the rate of profit will not have fallen to the level prevailing in the capital goods sector.

When the consumer goods price changes, the real wage changes in the opposite direction:

$$\frac{d\left(w/p_c\right)}{w/p_c} = \frac{-w/p_c^2}{w/p_c} = \frac{-dp_c}{p_c}.$$

Ex hypothesi, all real wage income is spent on consumer goods; since employment is fixed, consumer demand must rise at the same rate that the real wage increases. Yet there seems to be a paradox here – the demand for and output of consumer goods are to rise while the price falls! And this in conditions in which prices are relatively flexible, compared to output and employment. But the paradox dissolves on closer inspection: capital goods demand is falling, which brings down capital goods prices and profits; this is the initiating cause. Consumer goods prices fall only because costs have fallen and competition forces them down. Consumer goods profits are reduced by the threatened or actual movement of capital out of the capital goods sector. But the resulting reduction of consumer goods prices raises real wages and therefore consumer demand. Firms and workers able to move will therefore shift from the capital goods sector, where demand is falling to consumer goods where demand is rising. Hence, in ideal conditions,

$$\frac{d\left(w/p_c\right)}{w/p_c} = \frac{dC}{C} = \frac{dY_c}{Y_c}.$$

Putting all this together, we see that

$$\frac{dI}{I} = \frac{-dC}{C}, \text{ or } \frac{CdI}{dCI} = -1$$

The elasticity of consumption with respect to investment is -1 (assuming equal capital/labor ratios). A proportional rise or fall in investment is exactly offset by the corresponding proportional fall or rise in consumption, brought about by the working of competition and the price system. This can be expressed in a simple diagram of negative unitary elasticity.

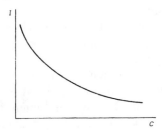

(Note that the price mechanism does not produce a perfect Say's Law offset; that would require the constancy of $I + C$, rather than the product, $I \times C$.[7] However, the adjustment mechanism is significantly stabilizing: since $C > I$, and $dC/C = -dI/I$, $|dC| > |dI|$. So when I falls, adjusted total income, $Y + dY = C + dC + I + dI > Y$, and when I rises, $Y + dY < Y$.

[7] Say's Law has been formulated in many ways: "supply creates its own demand," and "all savings are automatically invested," are the two most common. But as the argument between Ricardo and Malthus showed (Sraffa, 1960; Costabile, 1983; Kurdas, 1988), the first is best interpreted as saying that production distributes income equal to the value of output; it is a separate question whether all income is spent. The second is directed to this point; it has to mean investment will always rise to the level of *potential* (full capacity) savings, and empirically this is obviously false. Theoretically it has been hard to justify; neither interest rate mechanisms nor real balance effects have proved plausible. It is often argued that the postulate of Say's Law began as a wrong but inspired guess, but has now become a part of the free market ideology (Morishima and Catephores, 1985). The suggestion here is that a "Say's Law" economy is a Fixed Employment system, in which movements in consumption and investment spending are offsetting.

In the first case, the increase will tend to stimulate investment, in the second, the decline to diminish it, thus providing a corrective in the right direction.)

(A caveat must be entered. Not all producers will be equally efficient. A fall in price may put some high-cost producers out of business, leaving their labor force unemployed, and thus reduce consumption spending. This reduction in consumer demand may create additional bankruptcies – a bankruptcy multiplier. If the number of high-cost producers is large enough and the fall in price severe enough, the effect of bankruptcy in reducing consumption could rise to the point where it offsets the stimulating effect of the rise in real wages. The result will be a "general glut.")

The pressures of the market, leading firms to seek action to reduce the fluctuations of profits, bring about changes in the technology of production, making employment and output more readily adjustable. But when these innovations have spread, this, in turn, changes the way the market adjusts. We can start with Keynes.

Adjustment in the simple case: alternative theories

Keynes developed an account of adjustment in which consumption varies in the same direction as investment. But he sought to keep his account close to the conventional approach.

On this view, income adjusts to equate savings, which varies with income, to investment, which, as a first approximation, is taken as exogenous (but later is considered to depend on income and interest rates). Investment is an addition to the stream of spending, saving is a withdrawal, and income settles at the point where they offset one another. This is definitely a demand equilibrium. But according to Keynes it is set in a context in which full employment prices and quantities are determined by supply and demand, that is, by the scarcity principle.

Keynes, 1936, wrote:

> Our criticism of the accepted ... theory ... has consisted not so much in finding logical flaws ... as in pointing out that its tacit assumptions are seldom or never satisfied. ... But if our ... controls succeed in establishing an aggregate volume of output corresponding to full employment ... the [neo-]classical theory comes into its own from this point onwards. If we suppose the volume of output to be given, i.e. to be determined by forces outside the [neo-]classical scheme of thought, then there is no objection to be raised against the [neo-]classical analysis of the manner in which private self-interest will determine what in particular is produced, in what proportions the factors will be combined to produce it, and how the value of the final product will be distributed between them. (p. 378)

Prices *and quantities* are therefore determined at full employment by the equilibrium of supply and demand. How is this to be reconciled with the idea of a demand equilibrium, which may occur at virtually any level of output and employment?[8]

Arguably, Keynes never provided an answer, but the conventional approach is simple and ingenious. As a matter of pure theory the only *equilibrium* is that of supply and demand at full employment. Aggregate demand may settle for longer or shorter periods at other levels – but these will be disequilibria ultimately due to wage or price rigidities. When all variables are perfectly flexible the only equilibrium position is full employment. Hence the *level of spending* must adjust – full employment supply must create the necessary demand. Below the full employment level, for example, prices must fall, raising the value of real wealth and thereby stimulating spending. Demand variables may influence the determination of the level of full employment, but the main feature is the adjustment of demand to the scarcity-determined position of full employment.

However, in this approach "income," sometimes means payments to factors of production – wages plus profits, sometimes total spending – consumption plus investment, and sometimes total output – consumer goods plus capital goods. The relationships between employment, productivity, and output have not been clearly specified.

Consider a conventional Keynesian Savings-Investment diagram, with Y, "income," plotted along the horizontal axis, and S and I along the vertical. The I-function rises slightly – higher Y encourages investment – and the S-function rises steeply starting from a negative value. The intersection shows the level of Y where withdrawals and injections balance. The problem here (concealed when I is taken as purely autonomous) is that Y has two different meanings, and both cannot be measured on the same axis. In the S-function, Y means incomes received by households, or W + P, wages plus profits, but in the I-function, Y stands for total

---

[8] Keynes only arrived at the view that variations in investment determined output and employment in the spring of 1932. It was in his Easter Term Lectures that he first set out this view clearly, provoking the "manifesto" by Joan and Austin Robinson and Richard Kahn, trying to clarify and sharpen his argument. The argument, however, never did get straightened out. But consider a production/utilization function with such sharply diminishing returns at the initial point that the proportional fall in the marginal product is greater than the proportional rise in employment. Expansion due to a rise in investment would thus imply a decline in the wage bill, and with a Classical saving function, in consumption (Nell, 1992, ch. 16). Conversely, if returns diminish only slightly, a rise in investment will lead to a proportional increase in employment that is larger than the proportional decline in the real wage. In this case, consumption will increase with investment.

expenditure, C + I, consumption plus investment. Investment rises when aggregate spending increases. Notice that this is not arbitrary; it would make no sense to think of saving as a positive function of total spending, or of investment as an increasing function of the costs of production, that is, the incomes business is currently paying out. Since workers can save and capitalists consume, the two meanings of Y are not identical and need not be equal, except in equilibrium.[9]

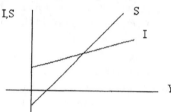

In reality there are *three* separate and distinct ideas involved in the determination of demand equilibrium.[10] First, there is output, which with

---

[9] It has sometimes been argued that sales receipts, which necessarily equal expenditure, are automatically passed along as income, so that income does always equal expenditure. This is simply wrong; income includes adjustments for inventory changes and for capital gains and losses. It is certainly not identical to, and does not necessarily even equal, sales receipts. Net income paid out, properly defined, does have to equal net output produced, but this will only equal net sales receipts if output adjusts quickly and accurately to sales. In equilibrium, of course, income, output and expenditure (sales receipts) will all be equal. But otherwise, and most of the time in reality, there will be differences. Income and output have to be equal in the final accounting, because income consists of claims to the value of output. Wages and salaries are deducted, then a portion goes to replace used up or depreciated means of production – and whatever is left belongs to the producers, either as realized or unrealized profits (inventory adjustments). But neither income nor output has to equal sales receipts; there is no reason to expect all output to be sold (except in a Say's Law economy). However, when output can be rapidly adjusted to sales, the differences will be slight – but this is because there is a strong tendency to equilibrium.

[10] Once income and expenditure are distinguished the conventional Saving-Investment diagram becomes hard to interpret. At a point on the horizontal axis, Y_1, for instance, conventional theory would say $I(Y) > S(Y)$, leading Y to increase. But $I(C + I) > S(W + P)$ is harder to understand; if $C + I = W + P$ at Y_1 will they still be equal at a higher level? Why? Which components will increase? What are the causal relations between expenditure and income? Analytically we have:

$$S = S(Y)$$
$$I = I(E)$$
$$S = I$$
$$Y = E$$

(*continued*)

given plant and equipment, operating a Flexible Employment technology at a constant level of productivity, will vary regularly with employment, $Y = Y(N)$. Second, there is expenditure (identically equal to revenue), $C + I$, which depends on income, $W + P$, and on expenditure itself. $C = W$, where W determines C, but, as we shall see, $P = I$, and the causality runs from I to P. Income, in turn, depends on employment and on the level of the real wage, $W = wN$, and $P = Y - wN$. These relationships can be plotted on a diagram (similar to that of the aggregate Marshallian system, except here the output-employment function is a straight line).

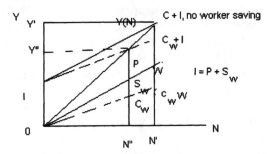

For a given I, worker saving reduces profit, employment and output: $Y''' < Y'$, $N'' < N'$.

Measure employment, N, along the horizontal axis, and output, expenditure, and income, expressed in monetary units, on the vertical. The first relationship above will be the line Y(N). Here employment varies with output, while productivity remains constant. Output is adjusted to changing sales by varying employment, rather than productivity – labor is a variable cost. The given wage rate will be the angle WON, and the wage

10 (contd)
where $Y = W + P$ and $E = C + I$. If $I = I(Y)$ replaced the second equation, the system would make sense; $S = I$ is the balance between injection and withdrawal. But what is the rationale for $Y = E$? The familiar argument that since income is either consumed or saved, we can write $Y = C + S$, so that whatever Y is, given $E = C + I$, $Y = E \longleftrightarrow S = I$, cannot be accepted. The saving function states that saving (consumption) varies with *income*, $W + P$; $S = S(C + S)$ does not provide the required information, or even make sense. So what does it mean to claim Y=E? What kind of adjustment process is involved? How do we know such adjustments will not cause shifts in the saving or investment functions? To answer these questions it is necessary to turn to the relationship between employment and output.

bill will be given by the line W. Potential profit at any level of employment is therefore $Y(N) - W$. For simplicity assume that all profits are saved; if workers do not save, consumption will equal the wage bill, if they do then it will be given by the dotted line, c_wW, where c_w is the worker propensity to consume. (Note the implications of worker saving.)

Total expenditure can now be shown by marking off autonomous investment on the vertical axis and drawing from it the upward-sloping line labeled E, which represents $C + I$. If all investment is autonomous – higher levels of activity do not stimulate investment – then this line will be parallel to c_wW; but if higher activity has a positive influence on investment then E will have a slope steeper than that of c_wW. Demand equilibrium is reached at the point of intersection between E and Y, where profit withdrawals just balance investment injections. The relationship between Y and E is spelled out clearly and depends on employment and output. The conventional presentation was more inadequate than incorrect; it failed to show how the adjustment actually came about. Here it is clear that *employment* adjusts to equate the surplus to investment; the higher the real wage the larger the employment adjustment required for a given change in investment.

Even though Y is output and $Y(N)$ an output-employment function, this is not a conventional supply-demand equilibrium, for supply exercises no restraining force. Output and employment vary with sales, productivity at the margin remains constant. Supply thus adjusts, and the effects of such adjustment are felt as influences on demand. Supply conditions – capacity and productivity – establish boundaries, so to speak, for they define the potential profit and determine the size of the required adjustment in employment per unit change in autonomous spending. But the equilibrium point itself is a balance between influences furthering expenditure and those reducing it. There is no hint of scarcity to be found here.

Multiplier analysis of employment

Plant and equipment, then, will be designed to run at various levels with constant productivity, making it possible to adjust output to sales easily, while avoiding pressures to alter prices, which can be kept at levels appropriate for long-run development. On this view, in the long-run, "normal" prices are considered determined by capital accumulation, seeking the most profitable outlets, in the light of growing markets. Given such prices, as determined in Chapter 10, variations in demand govern employment. But a central element in demand will be the real wage.

To study this we must look again at the "utilization" function, showing the degree of capacity utilized in response to demand. In conditions of inflexible employment and diminishing returns, as we have seen, changes in the real wage might well be required for changes in employment.[11]

But Mass Production is designed for flexibility; start-up and shut-down costs are slight, work is designed to require minimal skills, and the pace of work is controlled by the machinery. Productivity is therefore fixed, but output can be varied in line with the rate of sales. When demand falls, production lines can be shut down or put on short time, and workers laid off; when demand recovers, workers will be recalled as production is raised. (Laid-off workers retain a connection with their firms; moreover, they still belong to the union.)[12] There is no need to reorganize work crews, or reactivate whole plants. The system is designed for flexibility. Since both output and variable costs can be adjusted easily in the short run, changes in demand need not result in price changes. With both prices and money wages given, the real wage is fixed. In conjunction with the real wage, then, the level of employment determines the total wage bill. (Money wages and prices can, of course, deviate from their long-period levels during the short run, but these changes should be considered second-order modifications of the analysis.)

Since the great bulk of wages are consumed, by determining employment a substantial step is taken toward determining short-run consumption. If we further assume that profits are largely retained, and temporarily take investment as exogenous, we have the makings of a simple macro-system, illustrated above. We remember that the vertical axis measured output, the horizontal employment. We made the provisional assumption that the productivity of labor would be unaffected by changes in utilization – for the very good reason that mass production plants are designed to ensure just that (Barnes, 1958; Nell, 1992, ch. 16). Given the real wage, the wage bill is expressed by a ray from the origin, as shown, and on the assumption that all and only wages are consumed, this also gives us house-

[11] Hicks (1989), commenting on Pigou and Keynes, remarked, "Pigou was arguing from a fully Marshallian position, on the formation of the prices of manufactures, that in the 'short run' an increase of demand must raise their prices. So if money wages are given, an increase in 'effective demand' must lower real wages." As we saw in Chapter 9, both Marshall and Pigou discuss short run production functions in terms of "idle equipment" (Pigou, 1944, pp. 51–2) and consider "How actively ... to work ... appliances" (Marshall, 1961, p. 374). The conception is clearly that of a utilization function, rather than a production set from which techniques are chosen. (See p. 450 below.)

[12] The layoff *system* did not develop in the United States until after World War I and did not become general until after World War II. The layoff principle states that workers have a vested right in their jobs, implying that when laid off they have a right to be rehired in a definite order when business conditions improve. Early discussions took place in Massachusetts in the 1890s.

hold consumption. Then adding investment, and perhaps a fixed level of consumption for managers and property owners, we get aggregate demand. Of course, worker savings will reduce the angle of the consumption function; unemployment compensation can be expressed by a line of payments inversely related to employment. But ignoring these complications, in the basic case the multiplier depends only on the real wage and the productivity of labor, so that

$$\text{multiplier} = \frac{1}{1 - wn}$$

where w is the real wage and n is the productivity of labor. Hence, assuming a closed system, and neglecting government, $Y = mI$, and $\Delta Y = m\,\Delta I$.

The multiplier requires that firms be able to adjust their current spending on wages and materials quickly and easily. Banks will find it profitable to provide funds by offering lines of credit. (Compare Chapters 6 and 13: For this system to work a lender of last resort will be required. The result will be to make the supply of bank money endogenous in the sense of being strictly and rapidly responsive to changes in demand; Moore, 1988; Deleplace and Nell, 1996; Lavoie, 1992, ch. 4.)

The multiplier and the elasticity of consumption

In a Mass Production economy employment reflects the degree to which capacity is utilized. Variations in demand are met by varying utilization, keeping prices and productivity constant. When demand falls, workers are laid off, reducing their incomes and so also their consumption; when it increases they are reemployed, up to and even beyond rated capacity.[13] Investment constitutes the demand for capital goods; neglecting other forms of consumer spending, wages provide the demand for consumer goods. Gross profits are withdrawn at each stage in both sectors. Given a variation in investment spending, the multiplier sequence will be

$$\Delta C_1 = wn_k\,\Delta I$$

$$\Delta C_2 = wn_c\,\Delta C_1 = wn_k\,wn_c\,\Delta I$$

$$\Delta C_3 = wn_c\,\Delta C_2 = wn_k\,(wn_c)^2\,I, \text{ etc.}$$

---

[13] A popular textbook and "New-Classical" argument holds that the emergence of unemployment will bid money wages down, whereupon competition will force money prices down in proportion. Unemployment will certainly have some effect on the level of wages in new hirings; but it is difficult to see how a rise in unemployment, even over a long period of time, can affect the wages and salaries of those already in stable employment. Traditionally, the relation between wages and employment has been a *long-run* matter, as with other factor markets. But the discussion concerns short-term adjustments.

Hence, the sequence converges to

$$\Delta C = \frac{wn_k}{1 - wn_c} \, \Delta I.$$

This is the multiplier; a change in investment generates an accompanying change in the same direction in consumption. In this form the multiplier depends on the share of variable costs in revenue, and on the real wage. The psychological propensities of households are not significant (Nell, 1977; 1992, chs. 16, 20, esp. pp. 504–7).

In a Mass Production economy, money wages and money prices are stable or move together, so that with fixed productivity, output is adjusted to sales. Fluctuations in investment will then lead to changes in employment, as workers are either laid off or rehired or find themselves working short-time or overtime, resulting in changes in their pay and so in consumption spending. Rises and falls in investment will be accompanied by changes in the same direction in consumption. This can be illustrated on the previous diagram. The aggregate output-employment function will be a straight line, since productivity is fixed by the nature of the equipment. Consumption, investment, and the wage will be the same as before, and the equilibrium will be as shown, with a multiplier of $1/(1-wn)$.

To see its full significance, rewrite the multiplier formula. In the aggregate integrated form it says that $I = Y(1 - wn) = P$. Business savings are with-drawn each round, until withdrawals balance the injections of investment. This is correct, but by aggregating we have lost information. In the two sector, integrated form the formula reads

$$Iwn_k = C(1 - wn_c).$$

This shows that the expansion takes place in the consumer goods sector, and it states the familiar result that the wage bill of the capital goods sector will be rendered equal to the gross profits of the consumer goods sector by the process of respending. This is the same relationship that played a central role in the theory of circulation, where we examined the process of "secondary circulation" of the wage bill in the capital goods sector. In that circulation the proceeds of the sale of capital goods to the consumer goods sector moved successively through the capital goods sector, enabling the producers to pay their wages. The focus then was on the movement of funds within the capital goods sector. The *spending* of those wages on consumer goods, producing the profits of the consumer goods sector, is what we see above in the multiplier formula. In the discussion of mercan-tilism and early capitalism – Fixed Employment systems – the spending of the wage bill of the capital goods sector returned the funds to the merchants, but this time – with Mass Production technology – the spending leads to changes in employment and output. The multiplier here is the obverse of the theory of circulation, in the context of a variable employment technology.

Finally, let the term in brackets in the two sector multiplier formula be M, so that $\Delta C = M\Delta I$; it then follows, by integration, that $C = MI$. Dividing the former by the latter, it follows that

$$\frac{dC}{C} = \frac{dI}{I}, \text{ and } \frac{IdC}{CdI} = 1.$$

A proportional change in investment generates an equiproportional change in the same direction in consumption. The elasticity of consumption with respect to investment is unity. Under Mass Production the market maintains the constancy of the *ratio*, I/C, rather than the product of the two, as in the Fixed Employment economy.

Expenditure and the "terms of participation"
The change from negative to positive elasticity of consumption with respect to investment has an important significance for method. When the elasticity is negative, the economy is stable; barring financial or political disruptions agents hold their positions. Hence the working of the system as a whole depends on the reactions of the agents, and the agents of each class or sector can often be represented by a typical or average figure. But when the elasticity is positive this will no longer be so.

In a Craft economy the terms according to which an agent participates in the system are largely set by birth, modified by accident. Agents are born into families with or without land, tenure rights or other property, with or without craft skills and trade secrets. Primogeniture or other rules determine how a family's assets are passed on to the next generation. Some will be landowners, others capitalists, still others skilled workers and craftsmen. Many are born into families with no assets, however, and these must take their chances in the job market. Here luck and accident will play a major role. A large lower class will have to rely on day labor, with no security and no prospect of advancement. Such labor will be drawn on as needed in good times and discarded in bad. The very poor and a relatively few others, such as war pensioners, are entitled to state grants. But once the way a person participates in the economy is set, this status will tend to be stable. Once a worker or craftsman is employed full time, the job will be secure, unless the times are hard enough to cause the firm to shut its doors. Workers will not lose their status unless employers also find theirs threatened. However, when the monetary rewards fluctuate, the two main classes tend to prosper at each other's expense, high real wages tending to be associated with low profitability, and vice versa.

This picture changes in three ways in Mass Production. First, basic status is somewhat less rigidly fixed by birth; there tends to be greater social mobility. Second, state entitlements are very much larger. Third, of most relevance here, the way the terms of participation vary have changed. Real wages and actual profitability tend to move together, rather than inversely, and worker status can be varied easily – large firms have developed a layoff policy. When sales are slow, production will be cut back by laying off workers. This procedure developed with Mass Production. Workers can be temporarily dismissed but have a right to return to their jobs when sales pick up. The layoff policy is negotiated with the union, and generally follows seniority; recall is often a matter of right written into the union contract. Hence worker status can vary between being fully employed, being employed part-time, being laid off according to seniority, with the expectation (or right) of reemployment, and, finally, being fully unemployed. Within the category of employed labor, workers may be promoted or demoted, or find their jobs upgraded or downgraded.

In each of these circumstances workers will have a different income, and a different normal pattern of consumption. Consumption patterns may provisionally be understood as being what is required in order to perform the job properly, and bring up a family to provide replacement workers of the same skill and education. Unemployed workers will have an income provided by the state, through welfare and/or unemployment compensation. This will tend to be substantially lower than their normal earnings, but enough to tide them over until they can be reemployed. Hence aggregate consumption will change as the pattern of employment changes.

These differences can be summarized in a table:

	Status	Income	Pattern
Craft	Fixed	Variable	Offsetting
Mass	Variable	Fixed per status	Compounding

In Craft conditions status tends to be stable, but the income earned in a given status will vary. The pattern of variation, however, will be such that overall, the changes will be offsetting. In Mass Production status will be variable, but the incomes earned in a given status will not vary, while overall, the changes will tend to compound.

Consider aggregate consumption as a function of income from this perspective. In the Craft economy, in the short run, money income does not change (or changes little), but prices do. Real consumption varies because of the price changes. Under Mass Production, the situation is very different. When aggregate demand changes, say because investment falls, some workers will be laid off. Their status changes from employed to unemployed. Others will be shifted to part time. The result is a further decline in aggregate demand, since household spending is determined by status. This will then lead to still further layoffs, followed by further reductions in spending.

Simple rules can be developed to determine the patterns of spending that will be followed by agents as they change from one to another of the different status positions. These rules might follow from solving one or another maximizing problem. Alternatively, they may be developed from studies of institutions. Moreover, it is likely that they will change over time, as agents learn and innovate. But it is not possible to explain changes in consumption by considering the reactions of a "representative agent," or to suppose that either sectors, or the economy as a whole, will behave like such agents. On the contrary, the changes in consumption are largely due to the changes in the *status* of groups of agents, where each status has associated with it a well-defined, socially enforced, normal pattern of consumption.

In early Craft conditions – AS/TA – short-run labor demand will call for varying hours and/or intensity of work from existing workers; the labor supply function will indicate their willingness to increase or reduce hours. (A change in intensity can be expressed as a variation in hours of constant intensity.) In CBF/MA, however, in addition to hours and intensity, the short-run demand for labor may also call for varying the number of workers. Workers can shift from declining industries to expanding ones; so demand and supply of labor must consider movements of workers. But neither case poses the question of workers changing *status*, from employed to unemployed or vice versa, in the short run. Yet just this is the characteristic issue under Mass Production.

A simple – overly simple, perhaps – short-run Theory of Consumption can be sketched on a diagram. Plot C along the vertical axis, N on the horizontal. (In Mass Production output will vary directly with employment, so C will also be a function of Y; in this simple example it is reasonable to take both functions to be linear.) At the minimum level of employment, state welfare payments and unemployment compensation will be at a maximum, and wages will be at a minimum. As employment increases, wages will increase, while welfare payments and unemployment compensation decrease, falling to a minimum at full employment, N_f.

Here C_a is autonomous consumption. The line rising from left to right is the wage bill, and its slope is the real wage rate, adjusted for taxes. Workers are assumed not to save (worker saving is spent on consumer

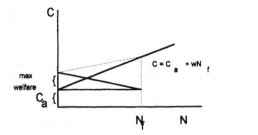

durables). As employment increases, the wage bill (disposable worker income) rises and with it, worker consumption spending. But welfare payments decrease, and consumption at any level of employment is the sum of the two effects, indicated by the rising dotted line. Income will fluctuate with N, but because of C_a, the proportional fluctuations in consumption will be less than those of income.

Besides being employed or unemployed, there is a third "status," that of wealth-owner. Consumption based on disposable income from human and nonhuman wealth is subsumed in C_a, on the grounds that wealth-owners will be able to sustain their normal patterns of consumption in the face of short-term fluctuations through borrowing or spending out of wealth. ("Human wealth" refers to capital invested in education and the acquisition of skills, which raises earning power above the norm for wages.) A reasonable hypothesis is that disposable income from wealth is largely saved, that from wage income largely spent. A larger proportion of the household income of higher-income groups comes from wealth; hence we should expect to find their household savings propensities to be higher. This is well known from cross-section studies. On the other hand, "longitudinally," as income increases over the long term, the overall saving propensity will vary with the aggregate ratio of wealth-based income (Y_k) to wage income, Y_k/Y_w. If this ratio does not vary much, then aggregate consumption will closely track aggregate income. Further, if, as seems true in many cases, the variations in Y_k/Y_w are procyclical, then saving will tend to vary procyclically. Again this is a familiar result.

Although this approach is suggestive, for empirical work it is seriously inadequate in several respects. First, expenditures on durables, nondurables and services should be distinguished. (Interest rates and liquidity affect durable goods purchases, but are not significant for the other categories. Durables, in turn, can usefully be subdivided into automobiles and other durables.) Second, households can usefully be divided into three groups: wealth-owning, non-wealth-owning workers, and welfare-dependent. Third, an effort should be made to discriminate between fundamental and discretionary household spending. Households will attempt to maintain the former even at considerable cost; the latter will vary more readily. Fourth, besides the three types of income above, other variables affecting consumer spending need to be considered – interest rates and liquidity, household financial assets, stocks of goods already acquired, and consumer attitudes and expectations (Evans, 1969; Eichner, 1991, ch. 9; Deaton, 1992).

This approach can also be used, with some modifications, to examine the behavior of aggregate investment, as income and interest vary. For example, firms have as possible status positions: good credit risk, poor credit risk, or bankrupt. As firms change status from good credit rating through poor to bankrupt, their ability to sustain investment spending declines, in the latter case to zero. An aggregate investment spending function thus need not be derived from the optimizing behavior of a representative firm. It could be constructed by aggregating the changing status positions of a variety of firms, as relevant variables change.

Investment and capacity utilization

Investment is notoriously difficult to explain. Not only is it volatile, but more than a half century after Keynes, there seems to be little agreement on basics. Technical changes, product developments, location in relation to markets, potential economies of scale, and competitive pressures all appear to be a part of the decision to commit resources to expanding. However, the first step in disentangling the problem may be to separate such investment decisions from investment spending.[14] We have explored investment decisions in Chapter 10. But one matter remains: under Mass Production, given widespread economies of scale and processes that tend to exhibit increasing returns, there is a general competitive pressure to build ahead of demand, compounding the necessity already inherent in the use of fixed capital, as explained in Chapter 7.

A tendency to overbuild

Each firm can calculate the pricing and investment plans of the representative, average firm, and can thereby determine the amount of additional capacity it should build to maintain its market share. But this determination is subject to two kinds of uncertainty: the future expansion of the market can only be estimated within a range, and the exact impact of technological improvements on costs and output can only be approxi-

---

[14] Most modern analyses, like Keynes's, concern investment decisions – the decision to commit capital to a project, or keep it invested in a certain line, by committing to replacements. But once the decision is made, as Kalecki made clear, there remains a further question, the timing and pace of expenditure to carry out the decision (Kalecki, 1937).

mately judged. But it is known that economies of scale exist. Let us suppose that no firm has any definite technological or managerial advantage, so that no one has an incentive to try to improve their market position. Nevertheless, they will have to decide whether to build for the high demand and low technical progress estimates, or for low demand/high technical progress, that is, whether to err on the side of overbuilding, or on the side of underbuilding.

Consider the strategic position of a representative firm vis-à-vis the rest of the firms:

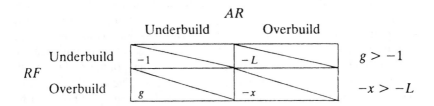

If "all the rest," AR, underbuild, and RF also underbuilds, the result will be unsatisfied demand, creating room for entry, so that both RF and AR will suffer losses, -1, due to new competitors.

If AR underbuilds and RF overbuilds, then RF makes a gain, g, since it increases its share. If AR overbuilds and RF underbuilds, RF suffers a substantial loss in market share, $-L$, since the overbuilders will both enjoy economies of scale and be under pressure to press their competitive advantage. If AR overbuilds and RF also overbuilds, RF suffers only a small and uncertain loss, $-x$, due to carrying excess capacity, which will be partly offset by economies of scale.

The best strategy for RF therefore, is to overbuild, or build ahead of demand. This applies to every firm taken successively in isolation, and considered against all the rest. Hence there will be a tendency to excess capacity. (By contrast, in a Craft economy, with little technical progress and few economies of scale, new competitors would create no necessary problems, while overbuilding will carry a heavy penalty in risk. The safest strategy will be to underbuild.)

Re-organizational investment

Investment means adding to productive capacity, and this in turn usually entails the purchase and installation of capital goods – with consequent multiplier effects. But there is another kind of investment, which may be critically important at certain times. In this case, the firm incurs investment costs, not by laying out funds for the purchase of equipment, but by shutting down its normal operations. Its investment costs are its forgone opportunities of producing and selling its product. It then deploys its normal labor force (or, perhaps, lays them off and hires outside specialists) in *reorganizing* its production system, changing the layout of the plant, or the sequencing of operations, altering the division of labor and changing the distribution of responsibilities in the work force. As a consequence, the plant and equipment may be completely unchanged, but the production process and the definitions of tasks in work activities will be different, resulting in an increase in productivity. From the point of view of the firm, the forgone earnings due to the shutdown are an investment, to be compared with the discounted value of the stream of earnings from higher productivity; but from the point of view of the economy, no new capital has been put in place. The existing capital is simply more productive, as is the labor force. The wage-profit frontier has shifted outward.

The importance of this kind of investment arises from two related facts. First, it is an *alternative* to adding to capacity by purchasing capital goods; by increasing the speed of throughput and reducing waste and losses during operations it raises the productivity of existing plant and equipment – but it has no or very few multiplier effects. It involves little or no additional spending; a firm's own labor force may carry out the reorganization, or its normal workers may be laid off in favor of specialists. In either case the flow of spending will be approximately the same and will chiefly be directed to labor costs. Second, this type of investment may become competitive with normal investment at crucial points in the business cycle. One such time is when risk has become high. Adding to capacity by acquiring new equipment involves laying out funds; incurring opportunity costs does not. After a long expansion, many ideas for reorganization will have developed through learning by doing. Hence, as the boom is drawing to a close, risk may be expected to rise, and it may seem prudent to expand capacity by reorganizing existing plant and equipment rather than building new. Another example: such reorganization is clearly cheaper when the economy is in a slump. Hence there may be a tendency to increase capacity by this method when conditions are depressed, thereby intensifying the depressed conditions. Thus this form of investment may contribute to the downturn and may tend to prolong slumps.

(Szostak, 1996, considers this an important factor in the weakness of the recovery from the Great Depression.)

Investment decisions and investment spending

In a Mass Production economy the level of investment will tend to adapt to changes in demand; that is, investment will be adjusted so that output will be produced at the proper or desired capacity-output ratio. That such a tendency exists is perhaps the best-established empirical finding in connection with investment. But most presentations of the "capital-stock adjustment principle" do not distinguish between investment plans, and investment spending, the implementation of those plans. The capital stock adjustment principle is chiefly relevant to the formation of plans, not to their implementation. (In Chapter 10 investment plans were determined in proportion to the expected growth of demand, with prices planned to provide the required finance.)

The two phases of investment – planning and implementation – depend on different variables and have different consequences. Investment decisions concern the expansion of capacity, the choice of technique and product design, location and desired labor force – but above all a decision has to balance the plan to build (or maintain) capacity with the need for it. (Nickell, 1978, ch. 12, summarizes the evidence that expected demand is the chief influence on investment decisions.) Investment decisions therefore depend on the overall growth of demand in the economy; investment plans, including planned prices, reflect the expected normal growth of markets. But the expected demand in any particular market will not be fully determined until a decision is made about price. So the investment decision has to balance the expected expansion of the market at possible prices with the new capacity that the profits from such prices will finance. As we saw in Chapter 10, investment decisions are also pricing decisions (Eichner, 1976; Nell, 1992). In what follows, it will be assumed that a well-defined normal growth of demand exists.

Once the decisions are made, however, new problems emerge, for carrying out the decisions presents a new set of questions. First, there is the matter of timing – when and how fast should the new capacity be constructed? (For Haavelmo [1960] this is the essence of the Keynesian conception of investment, and it is the issue that neo-Classical theory neglected.) Secondly, how should it be financed? These are the decisions that will have multiplier effects on the short-run levels of employment and output – and they will, in turn, be infuenced by current activity and interest rates. Investment decisions look to the comparatively distant future; uncertainty is inescapable, and decisions have to be made on woefully inadequate grounds. But investment spending is concerned with more immediate problems. The future is still uncertain, but it is not necessary to look so far ahead. Expectations can be based on better foundations.

The adjustment of investment spending, therefore, is not based so much on the accelerator, as on the changing degree of risk. When sales fall, it is more difficult for firms to meet their fixed obligations; to continue to spend on investment may put them at risk, in the event of a further drop in sales. Thus cutting back, at least temporarily, may be the best strategy. Moreover, with lower sales in relation to fixed obligations, banks will lower their credit ratings, and the bond market will downgrade their bonds. In the same way a higher rate of interest will lead banks to view prospective borrowers more skeptically – and will lead borrowers to reassess their current needs, postponing the implementation of plans. (These reactions will be examined in more detail in Chapter 13.)

In Keynesian theory the current level of output influences investment through its effect on the MEC schedule, and investment will be pursued up to the point where the interest rate equals the MEC, acting as a cutoff. The conception is significantly different here; the importance of current activity lies in the revenue received by firms, while the interest rate is significant as a measure of cost and risk. The factors that, in traditional theory, enter into the MEC calculation have already been accounted for in the investment decision; they have nothing to do with the questions of implementation.[15]

Investment functions

At this point one might ask for the development of an investment spending function. Roughly, such spending will be shown in Chapter 13 to depend on the current state of capacity utilization, and on the real rate of interest in relation to the rate of growth. But an investment spending function is more difficult to develop than, for example, a consumption function. Investment spending responds strongly to changes in expectations, and even when expectations are unchanged, to variations in the state of confidence. This last is affected by many noneconomic factors, and may be affected in different ways at different times by economic factors. As a result, it is highly volatile.

---

[15] In particular, the MEC calculation concentrates on recouping costs, including opportunity costs in interest. This is implied in the pricing-investment analysis of Chapter 10, however: prices are set so as to maintain a flow of gross investment that, given the costs, will replace capital and expand at the pace at which sales are expected to grow at these prices. Since this flow is continuous, costs are continually recouped.

In order to deal with these complications the problem of investment must be situated in the context of a full model of the macroeconomy. Hence the question will be postponed until Chapter 13.

Labor and productivity

The labor market

It can easily be shown that under Mass Production the real wage and the amount of employment will tend to vary together in the short run (Kalecki, 1933; Nell, 1978, 1984, 1988, 1992; Lavoie, 1992). Rewrite the multiplier equation as an employment multiplier. The initial employment will be nI, where n is the labor per unit of output (assumed for simplicity to be the same in both sectors). Then employment will be given by the multiplier equation: $N = m(nI)$, and $\Delta N = mn \, \Delta I$, the real wage and productivity being held constant.

Employment will vary with the real wage according to the partial derivative:

$$\frac{\delta N}{\delta w} = \frac{In^2}{1 - wn^2} > 0, \text{ where}$$

$$\frac{\delta^2 N}{\delta w^2} = \frac{2\,In^3}{1 - wn^3} > 0$$

The higher the level of the real wage, the greater the increase in N associated with a given further rise in the real wage. This curve can be drawn as below; the intercept on the horizontal N axis will be nI. This should be modified to take into account all forms of autonomous expenditure. In particular, it should include the spending of managerial incomes, C_m. From this intercept the curve rises to the right with a declining slope, approaching the line 1/n asymptotically. The real wage and employment – the demand for labor – are positively related, exactly the opposite of the relationship in the Old Trade Cycle. A cut in wages will diminish employment, not increase it.

Now consider the question of the supply of labor. It is common to assume that higher real wages will lead to the offer of more labor – but not much more. The function might then be a steep line rising from a point representing the basic pool of labor, as in the second diagram below. This might be plausible in a prosperous era, and might be especially relevant to cases where the real wage is rising. But there is another possibility: households have commitments which must be met if they are to maintain their customary style of life. These commitments and basic expectations add up to a certain level of real income, which may be taken as the standard of living, given in the short run. When real wages fall, the breadwinners must work harder or longer; alternatively, more members of the household must enter the labor market, perhaps in search of part-time work. When real wages rise, such work will not be necessary, and labor force participation may be cut back. This approach might be most relevant in difficult times, especially when real wages are falling. In the simplest case, such a curve will be a rectangular hyperbola, sloping downwards from left to right.

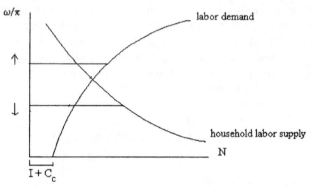

Finally, put the employment curve and the household labor supply together – whichever labor supply function is considered more plausible, a real wage above the equilibrium will be associated with excess demand for labor, and a real wage below, with excess supply.

In each case the production system will be able to adjust output. But when there is excess demand for labor, the goods cannot be produced because of the labor shortage. Prices and money wages both will therefore tend to drift up. When there is an excess supply of labor, output will be cut back and there will be no pressure on prices. Money wages, however, will tend to fall, moving the real wage in the wrong direction. When the labor market is out of balance, market forces will tend to leave the disequilibrium unchanged, or move it further in the wrong direction (Nell, 1988, 1992; Lavoie, 1992).

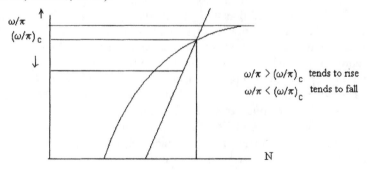

Clearly the working of this labor market, corresponding in many ways to post–World War II conditions, contrasts sharply with the labor market of earlier eras. Downward flexibility of the real wage would reduce rather than increase employment. No market forces work to adjust the level of employment; no market-clearing forces exist for labor. The change reflects the emergence of the multiplier, which, in turn, itself is a consequence of the evolution of technology and the associated developments in the organization of business.

The neo-Classical synthesis joins a demand function reflecting marginal productivity, derived from the firms' choices of technique, and a supply function derived from households' work-leisure choices, to the IS-LM system. Aggregate supply is determined by the labor market, which gives the equilibrium level of employment, and the production function; aggregate demand is determined by the IS-LM interaction. The price level supposedly ensures that demand adjusts to the equilibrium supply, by changing the value of real balances and thus consumption, or working indirectly through the interest rate. The model is thus partitioned, with the Keynesian demand side adjusting to the neo-Classical supply side. By contrast, the multiplier approach unites aggregate demand and employment; the labor market directly reflects the multiplier. There is no partitioning, and, in the absence of policy or "built-in" government response, no stabilizing adjustment.

Mass Production and productivity bargaining over money wages

A commodity-based currency is not flexible enough for Mass Production and is too expensive. But with the emergence of bank deposits as the principal currency, money wages are no longer tied to real wages. When money wages rise, prices can also be raised, defending the profit rate. The labor market no longer adjusts real wages to reflect shortages and surpluses. Instead, changes in employment will primarily reflect variations in aggregate demand, while changes in wages will reflect productivity.

From capital's point of view productivity bargaining is dangerous. What workers give freely they can take away. Early in the history of capitalist industry, it became apparent that a cooperative arrangement with workers could bring great advances in productivity. But these had to be bought at the price of control. The security of capital and the prerogatives of management could not be safeguarded.

With Mass Production, however, productivity bargaining comes into its own – precisely because output does *not* depend heavily on worker morale and effort. Productivity depends on the speed of the machinery; it is built into the equipment and the organization of work. So long as workers make a normal effort, so long as they are not actually dragging their feet, productivity standards will be met. But even if they make an exceptional effort, unless the pacing of the line and the organization of work are changed, the speed of output will be the same. Work norms are set for a range of normal effort. A little bit of slacking should not make much difference; exceptional effort should not be necessary. Workers can slow down the process, but only if they deliberately set out to do so. They can speed it up – but only if they cooperate with the technical experts and engineers.

Such cooperation, in fact, is the key to continuous productivity change: labor must agree to changes in line speed and organization, and actively help in implementing them. If labor resists reorganization and changes in the division of labor, the improvements will not work. A constant battle over norms and job definitions will impede progress. But if labor agrees to the changes, and allows regular time and motion studies, leading to regular reorganization and redefinition of jobs, then productivity can increase year in and year out. To win labor's acquiescence pay rises are tied to productivity increases. It is not greater effort that is rewarded – it is the willingness to accept changes in work norms, job definitions, and plant layout, to accept operations research and programming improvements as a regular part of normal life. An important role of unions is to ensure labor's willingness to accept this state of affairs.

Once this happens, money wages are no longer determined by the state of the labor market, and the cycle of rising and falling wages ceases to exist. Henceforth money wages will only rise, and their rate of increase, though correlated with the labor market, will be determined along with that market by the pressure of effective demand. Even so, the contrast is striking: in Craft economies, the *level* of the money wage varies with unemployment/excess demand for labor, while under Mass Production, its *rate of change* varies with the state of the labor market.

Cutting money wages
This sheds light on an important characteristic of Mass Production economies. Even when demand has fallen sharply and unemployment is high, money wages are rarely cut back.

In an earlier era such unemployment would bid down money wages, with competition then pushing down prices in proportion, so that prices would fall with demand, just as the traditional theory has argued. (It would be harder to argue, however, that a recovery leading to reemployment would bid up money wages, leading prices to rise with demand.)

Wage-cutting is dangerous, however; it upsets relationships between the firm and its employees, endangering technical progress. Worse, it can provoke slowdowns and sabotage – direct retaliation. Moreover, if rivals do not follow suit, the best workers will leave for greener pastures, and the firm will face turnover and training costs. Consider the strategic position of a firm contemplating a cut in the money wages it pays its workers.

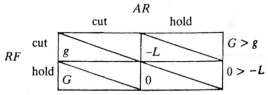

Suppose all the rest are cutting; if RF also cuts it will make a small gain, g, due to the lower wages, but it will have to deal with disgruntled employees and low morale. If it holds the line, however, it will attract the best workers from everywhere else, and will reap the rewards of higher morale, which it can cash in the form of worker willingness to accept speed-ups and cost-cutting job redefinitions. These productivity gains, G, could easily outweigh the gains from lowering wages.

Suppose the other firms are holding wages steady. Then the firm that cuts would become a pariah in the eyes of workers. The best would leave, the rest would turn sullen and obstinate, and the result would be a loss, −L, in productivity, outweighing the gain from the cut. To hold the line will leave the situation unchanged. So, if workers can speed up or hold back technical progress, so that G and L are large, it will pay the representative firm, and therefore every firm considered in succession, to hold the line on wages, rather than cut, in the face of unemployment and deficient demand.

"Downsizing" and productivity changes: blue- and white-collar labor

Given a level of demand, a rise of productivity implies a reduced demand for labor; regular increases in productivity thus raise the spectre of potential unemployment. Under Mass Production this was obviated by a double arrangement. First, productivity increases themselves were the result of investment spending; productivity rose at a rate which reflected the growth of capital, which in turn was maintained by a regular growth of investment. Second, productivity bargaining saw to it that the real wages of labor, and therefore household spending, rose at the same rate that productivity grew. As a consequence the lower need for labor implied by higher productivity was regularly offset by the growth of demand.[16]

---

[16] Suppose productivity growth is greater than demand growth; excess capacity and unemployment will emerge. This will dampen investment, and reduce productivity growth. But it will also further reduce demand. For stability, productivity growth would have to fall more than demand growth, a difficult condition to meet, given the volatility of investment.

Though no doubt made in heaven, this arrangement is not inevitable. Other forms of technical progress can take place, and have been much in evidence. In particular, "company restructuring," with a particular focus on office work, takes advantage of computerization to rationalize white-collar and managerial labor, reducing overhead costs. This may have little or no effect on production work and throughput, but it may significantly reduce overall costs.

The effect of this kind of technical progress can be analyzed with the help of the preceding diagrams. By reducing overhead costs and "down-sizing" office work at the managerial level, management incomes will be reduced. If this takes place on an economy-wide scale, management-level household spending will be reduced. This will lower the lavel of aggregate demand. The real wages of production workers are unaffected; at any wage, therefore, employment will be lower. Alternatively, a *higher* wage will be required to provide the spending to sustain any given level of employment.

Of course, "perestroika" itself is not cost-free. It must itself be the result of some form of investment expenditure, which would tend to shift the intercept in the opposite direction. But presumably companies know what they are doing; they expect to save money by restructuring. Hence the *net* movement must be in towards the origin. (Alternatively, we could consider a one-time expenditure of additional funds, followed by I falling back to its normal level, and a permanent reduction of C_m.)

A striking feature of "perestroika" is that its effects on productivity will not be fully revealed in the current aggregate figures. For productivity is measured as average current output per currently employed worker. In the case of normal or blue-collar productivity increases, based on the double arrangement above, both real wages and investment increase *pari passu*, and the same number of workers can be seen to produce more output.

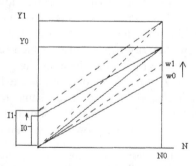

Measure Y on the vertical axis, and N on the horizontal. Output is initially Y_0, and employment, N_0. Then an increase in productivity swings the production line upwards by x percent; the real wage line rotates by the same percentage. If to achieve the productivity increase I has also risen by the same percent, the new level of output will be Y_1, with employment remaining at N_0. Measured productivity was initially Y_0/N_0; it then rose to Y_1/N_0. By contrast when managerial workers are dismissed, they reduce their household spending; moreover, there are no compensating wage increases, and, as we just saw, there will be no net rise in investment spending.

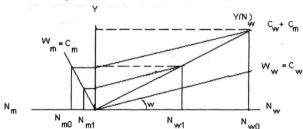

The vertical axis measures Y, the horizontal N, N_m to the left, N_w to the right. Output varies with the employment of production workers, as shown by the function $Y(N)$, rising from the origin to the right. Investment is ignored here, so output consists only of consumer goods, demanded by the two categories of employees, managers and production workers respectively, C_m and C_w. Both managers and production workers spend their entire incomes; when either loses job, the family emigrates, so their household spending falls to zero. The earnings and spending of each class are given by a line rising from the origin – to the left for managers, to the right for production workers. Manager spending is mapped onto the Y-axis, and demand by both categories of labor sets the level of production worker employment, N_{w0}.

Managerial workers are necessary as overhead labor. Productivity is therefore output divided by the sum of both types of labor.

Now consider a downsizing, that reduces the managerial labor force from N_{m0} to N_{m1}. Let this cut be a fraction, a, so that $N_{m1} = aN_{m0}$. Then $C_{m1} = aC_{m0}$. But it also follows, from similar triangles, that $N_{w1} = aN_{w0}$, and therefore likewise that $C_{w1} = aC_{w0}$.

Aggregate productivity is Y/N, total output over the total labor force, workers plus managers. Hence, initially, (ignoring investment) it will be

$$\frac{C_{m0} + C_{w0}}{N_{m0} + N_{w0}}.$$

the new level of productivity will be

$$\frac{C_{m1} + C_{w1}}{N_{m1} + N_{w1}}.$$

But the latter can be written

$$\frac{a(C_{m0} + C_{w0})}{a(N_{m0} + N_{w0})}$$

and the a's cancel out. Hence the two levels of productivity are equal. Measured average productivity has not changed! Yet clearly potential productivity has risen.

Suppose the real wages of production workers had been raised. Then the wage line would have rotated upwards, so the total demand function, $C'_m + C'_w$, would also have swung upwards, cutting the prodution line at a higher point, on a ray from the origin lying *above* the initial average productivity line. Hence measured productivity would have been increased by paying higher real wages! Similarly, if real wages had been reduced in the process of perestroika, the same reasoning would lead to the conclusion that productivity had fallen. Measured productivity is not a reliable guide.

Inflation

Low-level inflation and productivity bargaining
So long as all work is productive, and all industries coordinate, either speeding up together, or altering size appropriately, a rise in output per head will be matched by a rise in capital per head, so that paying workers in proportion to the increase in productivity will leave the rate of profit unchanged. By the same token, if workers are paid a little less, or if the wage payment lags a period or so, profits will be increased. In Craft economies and early Mass Production it might be reasonable to expect these conditions to be met.

But as Mass Production developed, white-collar work increased in relation to blue-collar work. Speeding up the process of production affected only the output of blue-collar workers; managers, office workers, and sales personnel were not affected. Such workers would normally be employed in proportion to the capital stock. A speed-up in turnover, however, brought about by *investment* in productivity-enhancing methods or equipment, which results in a corresponding increase in circulating capital, would also require a proportionate rise in the numbers of white-collar employees. But, when output per blue-collar worker increases, and both capital per blue-collar worker and white-collar employment also increase in the same proportion, the rate of profit will be unchanged if blue-collar wages rise in that proportion.

There is a catch, however. White-collar workers are higher-status workers, and their salaries must reflect this. Hence their pay must maintain a traditional differential with respect to blue-collar work. Thus their pay must also rise at the same rate. If it does, the rate of profit will fall – unless prices are raised. Hence these productivity increases create inflationary pressure.

Let x be the percentage change in productivity, with N_b and N_m the numbers of blue-collar and managerial workers respectively, where w is the wage and w' the salary rate. Then Y and K are output and capital, and P the price level. So,

$$PY = wN_b + w'N_m + (1+r)KP, \text{ and rearranging,}$$

$$\frac{(1+x)PY}{N_b} = (1+x)w + \frac{(1+x)^2 w'N_m}{N_b} + \frac{(1+x)KP}{N_b}.$$

Investment increases the capital stock in the proportion x, which makes it possible to raise productivity by the same factor. (Kaldor's laws support a correlation; to simplify, we have gone a step further and assumed that investment brings productivity growth in the same proportion.) Clearly if all terms are multiplied by $(1+x)$, no relative changes take place. But if managerial workers increase by $(1+x)$ and at the same time, to maintain the status differential, salaries are also raised by $(1+x)$, white-collar costs will rise and profits fall, so that

$$\Delta r = r_2 - r_1 = \frac{-(1+x)w'N_m}{N_b}.$$

To offset this, prices must be raised. Initially, ignoring the effect of higher prices on capital, firms will increase prices at the rate

$$\frac{\Delta P}{P} = \frac{xw'Nm}{PY}$$

This will cover the additional white-collar costs and restore the original rate of profit.[17] However, in the longer term this rate of increase will prove incorrect since it does not allow for the fact that capital costs rise when P

---

[17] Consider a simple numerical example:

pY	wN_b	w'N_m	K_p
100	2 × 10	4 × 5	50

Blue-collar workers are paid a wage of 2; managerial employees receive a salary of 4. Total expenses are 90; the rate of profit on capital is therefore 10/50 = 20 percent. Investment now takes place at a rate of 10 percent; the effect is to increase blue-collar productivity by 10 percent, and to require 10 percent more managers. The new numbers are:

110   2.2 × 10 4 × 5.5   55

Output is 110, costs are 99, and the rate of profit, 11/55 = 20 percent. But the managers have lost ground relative to production workers. To maintain the normal differential, managerial salaries must rise to 4.4; managerial costs will rise to 24.2, and the rate of profit will now fall to 8.8/55 = 15.9 percent. To counteract this, firms will raise prices in proportion to the share of managerial costs in output times the initial increase in productivity, 20/100 × 10 percent = 2 percent. Then 112.2 − 101.2 = 11, and 11/55 restores the original rate of profit. (If the higher price must be applied to the capital costs, an additional complication is added.)

rises. Taking this into account, for a rate of productivity growth of x, the rate of inflation required to keep the profit rate steady when white-collar pay is raised to maintain status differentials will be

$$\frac{\Delta P}{P} = \frac{w N_b + (1 + x) w' N_m}{(w N_b + w' N_m) - 1}.$$

In practice, status differentials are not wholly inflexible, nor is it necessary for white-collar employment to rise exactly in step with investment. The rise of costs will therefore be somewhat less than this indicates, so the rate of inflation will also be lower. But this type of inflation will normally be endemic in Mass Production.

Burden-shifting inflation

In a demand-constrained economy, inflation originates in changes in costs. Two qualifications should be made immediately, however. First, while prices of manufactured goods are generally insensitive to changes in demand, primary products may be quite sensitive, especially to large or sudden changes. Thus a sudden, unexpected, or large demand increase may lead to a rise in primary products, which is then passed along as cost inflation, setting off a wage-price spiral. When demand declines, especially if it collapses rapidly, primary products may decline in price, and this will help to dampen a wage-price spiral.

(Financial innovation may have an important effect here. Futures markets in commodities will be sensitive to the effects of demand pressures on primary products. When world demand begins to rise, commodity futures will rise. But speculators will buy now for delivery at the higher future price, thus driving up the current price. Hence as markets become more sophisticated, anticipated future pressures of demand on primary prices are more swiftly translated into present price increases.)

Second, sharp increases in demand will drive up prices in machine tools; these price increases will then be passed along as cost increases in all other sectors, since machine tools enter directly or indirectly into the production of all goods. This will tend to set off a wage-price spiral. Decreases in demand, however, will generally not lead to declining prices in machine tools; because production is to order, and is customer-specific there is little direct price competition. Moreover, when times are slack, machine tools firms will shift into other activities, to maintain their revenues.

In the case of machine tools the rise in price may be temporary, since ideally, it helps in adjustment. Because of the price rise, fewer projects will be ordered. Thus more of machine tool capacity can be devoted to the construction of machine tools to increase the sector's own capacity. Once this capacity is built up, the sector can deliver a larger output, and hence can meet demand at normal prices. However, in the meantime, the rise in machine tool prices will have driven up costs throughout the economy, setting off a wage-price spiral. The fact that the initiating increases turn out to be temporary will not stop the inflationary process.

Generally, in a demand-constrained economy inflation is the market process by which it is determined which groups shall bear the burden of increased costs. This process takes the form of a wage-price spiral, in which groups try to pass along the cost increases they experience. (In a resource-constrained economy inflation is the market process by which it is determined which groups shall bear the burden of the shortages.) Let us spell this out.

When a cost increases (say oil imports rise in price, although the story would be much the same if a sharp rise in demand drove up primary prices) the various industries using oil pass along as much as they can in higher prices. Consumers thus face a rising cost of living, and so demand higher wages and salaries, further raising costs to business, which in turn are passed along again in price increases, to the extent possible. But while business will try to pass cost increases along, and workers will try to recoup cost of living increases in higher pay, their ability to do so depends on their respective market positions. Not all businesses and not all groups of workers will succeed; indeed, it could even happen that none were wholly successful, but in general some will do better than others. Those who are relatively most successful, round after round, will escape most of the costs, which the least successful will have to bear. As prices and wages rise, however, the burden is lessened in real terms, and the wage price spiral peters out when the reduced burden has been distributed between business and labor in proportion to their inability to pass it along.

The process can be illustrated with a single-equation model. Let k stand for means of production per unit (aggregate) output, and n for labor per unit output, with m as the aggregate markup. $ will be the price of capital goods, w, the money wage rate, and P the money price index of output. Initially,

$$mk\$_{(t-1)} + mnw_{(t-1)} = P_{(t-1)}$$

$ then increases, P is increased accordingly, and w remains fixed:

$$mk(\$_t - \$_{(t-1)}) = P_t - P_{(t-1)}$$

However, once prices go up, households respond by demanding wage increases to compensate:

$$\frac{w_t - w_{(t-1)}}{w_{(t-1)}} = \frac{x(P_{(t-1)} - P_{(t-2)})}{P_{(t-2)}}, \text{ where } 0 < \text{ or } = x < \text{ or } = 1$$

The parameter x indicates wage-earners' market power; if x = 0 they are not able to raise the money wage at all, and the full burden of the cost increase will fall on them; if x = 1 they are able to keep pace fully with price increases, and the wage-price spiral will continue until the original ratio $\$/w/P$ is reestablished. Any value in between means that workers can keep up partially but will end up bearing the larger share of the burden. (In a labor-dominated system workers might be able to keep up fully with any cost increases, but business would be able to raise prices only a fraction. Interchange the w's and P's in the equation.) In any event the wage-price spiral comes to an end when the burden, reduced by inflation, is distributed.[18]

Inflation is a market response to an external shock, whose function is to determine who will bear the burden – of the cost increase in capitalism, of the increase in shortages in socialism. The rise in prices and wages reduces the burden to be distributed, while shifting it to the weakest, those least able to pass along or keep up with the increases. The more evenly matched the market positions of the various players, the longer the process will continue, and the lower the final burden to be distributed.

---

[18] In a *resource-constrained* context – see the discussion on socialism below – inflation will result from the impact of an increase in excess demand, e.g., a rise in investment; prices will be bid up by the competition for the scarce goods as consumers and enterprises try to shift the burden of the shortage to those who cannot afford higher prices. But as prices rise workers will demand pay increases, and enterprises in turn will increase output prices as their costs rise. Some groups of workers and some enterprises will be relatively successful; but those in weaker market positions will do poorly, and will end up bearing the burden of the shortages, reduced by the effects of the general price increases. Here, however, the Kaleckian dictum, "workers spend what they get, capitalists get what they spend," must be adapted and considered. Workers can only spend more if they receive raises; enterprises, however, will collectively get back whatever they collectively spend – from each other for capital and intermediate goods, from consumers spending their wages on consumer goods. In the nature of things, then, enterprises will keep up with demand pressure.

Let us suppose that some input in short supply is bid up in price, to ration supplies to those who can afford them. Enterprises using the input then try to pass the costs along; enterprise spending in the aggregate returns to them. Households respond to the higher prices by demanding wage increases. If they get them, their wages return to enterprises in the form of receipts from consumer goods sales. To the extent they fail to keep up, real wages are reduced, and workers bear the burden of the shortages. A corollary is that real supply and effort will tend to shift away from the consumer goods sector to production for interenterprise transactions. (Trying to reduce demand pressure by cutting back money wages could backfire if, in anticipation, enterprises intensified this shift.) Such processes may be open or suppressed.

Inflation and unemployment

Traditionally it has been claimed that inflation moves inversely to unemployment, but whether wage or price inflation was meant has not always been clear. Indeed, it has been typical of conventional economic analysis to presume that wage inflation and price inflation proceeded at the same rate, when account was taken of productivity changes. That is, the point at which the elasticity of wages to prices is unity has been taken as a *center of gravitation*, a point toward which macroeconomic pressures will drive the system. The grounds for this belief – belied by all evidence – lay in the view that the real wage was determined by stable processes in the labor market. This is unacceptable for Mass Production economies. As a result the unitary elasticity of wages to prices must be considered a special, not particularly likely case, but important because it marks the point at which the real wage is constant. Below this point, the real wage is falling, above, it is rising. And, we shall suggest, this elasticity can reasonably be considered a function of the level of unemployment.

Here we shall take "inflation" to mean a wage-price, or price-wage, spiral. The causes setting off such spirals are those just discussed – rises, especially sharp rises in demand for primary products, or for machine tools. (It could also be set in motion by a rise in money wages above productivity growth, resulting from successful bargaining.) But any such spiral can be one in which wages rise faster or slower than, or at the same rate as, prices. This can be expressed by the elasticity of wages with respect to prices, the ratio of dw/w to dP/P, or $P/w(dw/dP)$. Plausible relationships can be set out in a diagram. On the vertical axis we plot the rate of unemployment, u; on the horizontal, moving from the origin to the right, we measure the wage-price spiral; moving to the left, the elasticity. In each quadrant there is a downward sloping line – drawn here as straight lines, not because there are good theoretical reasons to do so, but because there are no grounds for supposing that the curves bend inwards or outwards – or alternate in curvature. The lines start from the same point, which is a level of unemployment so high that the money wage may begin to fall. That would turn both wage inflation, dw/w, and the elasticity, P/w (dw/dP), negative. But the axes only show positive values of the two variables. The functions show that at high levels of unemployment both the wage-price spiral and the elasticity of wages to prices will be low. That is, the level of inflation will be low, but real wages will be drifting down. Demand is strong enough that business can pass on price increases, but the labor market is sufficiently slack that workers cannot successfully push to keep money wages rising in pace. At lower levels of unemployment, the wage-price spiral will increase, and after a certain point, marked by the dotted lines, the elasticity of wages to prices will rise above unity, indicating that real wages are rising in the more rapid inflation. Part of the reason for this lies in the tightening of the labor market; workers will be better able to push for money wages to keep up – or even to demand raises. But a crucial aspect concerns the *growth of demand*: at the point where the actual growth of output comes to exceed the normal (expected) growth of demand, businesses (as we saw in Chapter 10) will have to adjust prices to encourage the expansion of new markets. And demand in such developing markets can be expected to be price-elastic – which will inhibit business in raising prices. When low unemployment reflects high growth we can expect profits to be squeezed. So this picture suggests that the inflationary process will typically move the real wage procyclically.

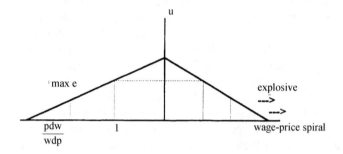

Perhaps less plausibly, it also suggests that at "full employment" – the minimum level of unemployment – the wage-price spiral will reach a maximum with a rising real wage. This may be so; but at some point the spiral may begin to accelerate, meaning that the longer the system rests at or below that level of unemployment the greater the rate of inflation.

This deserves a moment's reflection. The suggestion is that the wage-price spiral might become *explosive*, at a certain level of unemployment. Even though inhibited by the need to encourage new markets, with a rising real wage – especially with expectations that this rise will continue – demand pressures may be strong enough to allow business to raise prices in line with costs; while, at the same time, with unemployment low workers may find it possible to drive up wages by *more* than the cost of living. Business is able to match wage increases with price increases, labor is able to more than match cost-of-living increases, and so the wage-price spiral must begin to increase. Instead of the line continuing down to meet the horizontal, at any level of u it would tend to shift outward, as indicated by the arrows. Although dw/w and dP/P would both increase, their ratio would remain steady, so this would be the maximum level of the elasticity, as indicated by the dotted line.

Technology and social organization:
capitalism and socialism

To see what "demand-driven" means we need to contrast it with its opposite. So we will compare two modes of operation of Mass Production economies, one in which demand is perpetually in short supply, the other in which there is always excess demand. (Very loosely, the first can be identified with "capitalism," the second, with "socialism.") We will also see that the dividing line between them is what Harrod and Domar took to be the "warranted rate of growth," which, however can be shown to be neither warranted nor a rate of growth. Harrod-Domar has been taken to be the extension of Keynesian thinking to the long run. This has been a mistake; the so-called warranted rate simply balances aggregate demand against productive capacity. It has no inherent connection with growth. The instability property attributed to the warranted rate turns out to be the *stability* of the mode of operation, the force which prevents the system from crossing the barrier to the other mode.

Our central theme has been that the working of the market generates pressures for the development and diffusion of technical innovations. It has been argued that uncertainty and the possibility of shortfalls in sales – presence of excess capacity – will intensify these pressures. On the other hand, it has also been noted that increases in demand lead to faster growth of productivity. So it appears that excess capacity and increases in demand both stimulate technical progress! Is there a contradiction here?

Incentives and the mode of operation

To explore what is at issue, we need to examine the way the market creates pressures to innovate and to sell. One way of doing this is to contrast a market-driven system – one with "demand shortage" – with a system operating under chronic excess demand – a "shortage" economy. Capitalism is market-driven; socialism means many things to many people, but in practice, real-world socialist systems have always been shortage economies. We will explore the differences, especially in market incentives, but this is *not* an essay in comparative systems. What actually happened in Eastern Europe and the Soviet Union is not the issue (Nell, 1992, 1991).

The two economic systems – capitalism and socialism – will be considered as abstract, idealized forms. Our aim is to understand the role of excess capacity – unemployment! – in capitalism. Shortage economies will only be brought in by way of contrast. A capitalist economy will be understood as we have already presented it: some (families? institutions? individuals?) own the means of production while others do not; capitalist production generates a surplus through the employment of wage labor, and competition establishes a common ratio of surplus to the value of the means of production used. This is the rate of profit, and every capitalist system is characterized by a "normal" rate of profit (expressed in the rate of interest on money), which makes it possible to calculate the "amount of capital" in any sector or industry by capitalizing its net income stream. On this basis, therefore, economic activities can be bought and sold. Capitalist enterprises compete with one another; and liquid capital, funds conferring ownership of or claims against such enterprises, will actively seek out those with the highest rates of return. Hence there is constant pressure to increase the surplus, that is, to raise productivity.

Under capitalism the ownership and distribution of wealth is given, and the system generates pressure to operate the means of production most efficiently (productively). By contrast, under socialism the efficiency and productivity of the means of production are assumed, but private ownership is abolished and the system seeks to distribute the gains most fairly, taking account of both the general interest and the interests of all. Ownership is vested in bureaucracies supposedly representing the general interest,[19] run in accordance with an overall plan, and income is distributed in proportion to productive contribution, modified by subsidies to those with special needs. Investment is planned to bring about balanced growth at the highest rate consistent with planned consumption. Job security and a basic standard of living are guaranteed to all. Capitalism is regulated by prices and the rate of profits, socialism by quantities and the rate of growth.

Modern economies are essentially monetary systems: capitalist profits do not count until they are realized in money (and socialist incomes must be both paid and spent, before any judgment of fairness can be rendered). Since prices are realized in money, in neither system, therefore, does the money wage – fixed by the wage bargain in the labor market – determine the real wage. Production and distribution are carried out at least partly through market processes – wages are paid and spent, accounts are kept of purchases and sales – although the markets work differently, and the socialist markets are not competitive. Further, in both systems production is largely concentrated in the hands of giant bureaucratic organizations, with easy access to funds and well placed to lobby the government. And in each a privileged class or stratum can be identified.

---

[19] Of course, in practice socialist bureaucrats will develop interests of their own, sometimes conflicting with the Plan, just as capitalist managers develop interests separate from, and sometimes opposed to, those of the firms they manage. (And in either system the interests of the firm, as a particular institution, may clash with more general interests, as embodied in the shareholders, or the Plan.) These are important questions but not central to the issues here.

Long-term prices are designed, ideally, to recover costs and provide the wherewithal for expansion. (Socialist prices are "distorted" by social policies abetted by entrenched interests; capitalist prices are distorted by monopoly power and special interest.) An equation for prices that would provide the profits that would underwrite growth can be defined, which, in each case, must be modified by a vector expressing the influences causing deviations from the ideal prices.

Each system employs Mass Production technologies, but one mode of operation is capitalist, the other socialist. Products and productive equipment are standardized. Worker skills are required, though jobs are also standardized, and the pace of work is set by the machinery; costs are kept down and economies of scale are realized by large plants and long production runs. Prices in both systems have to cover current production costs and contribute to meeting the fixed monetary costs incurred in setting up the mass production plant. In both systems normal or long-run prices of manufactured goods will be inflexible, determined by reproduction costs and a markup. The differences will lie in the way the markup is set, and the extent to which cost overruns are permitted. Costs and outputs of primary goods (nonproduced means of production and basic consumption – farm and fish products, minerals, raw materials, oil, etc.) will fluctuate in both systems, leading to temporary market price changes in capitalism and variations in subsidies in socialism. In neither system, however, do prices reflect relative scarcities.

Prices serve the same function in both systems. At their normal levels they reflect the requirements of reproduction and distribution; when exchanges take place at the correct long-term prices, distribution will be accomplished and reproduction will be made possible. Prices do not reflect relative scarcities – they cannot in capitalism, because with excess capacity and unemployment, factors are not scarce, and they do not in socialism, because planning must ensure that exchanges will accomplish the desired reproduction and expansion.

Prices cover costs and earn normal profits from the operation of presently existing equipment, as we saw in Chapters 7 and 10. Choice of technique is therefore irrelevant; a new technique may indeed be more profitable, but it will take time to build new factories, and in the meantime demand will have to be served by current, less efficient plants. Normal prices will be based on normal levels of operation, reflecting normal costs. These will not generally be affected by aggregate shortages or excesses; indeed, within a wide range, variations in demand will tend to have little effect on prices. "Benchmarks" for normal prices will be established at the time investment decisions are made, based on the prevailing levels of the wage and other costs, on the one hand, and expected market growth on the other. Market prices can be expected to vary around these norms, but in manufacturing sectors even considerable swings in demand may have comparatively little impact (Kalecki, 1952; Eichner, 1976; Nell, 1991, esp. chs. 16 and 17; for a different route to a similar conclusion, cf. Blinder, 1988, and Gordon, 1990). Nor would variations in prices tend to correct aggregate imbalances between demand and capacity (Nell, 1991, ch. 20; Tobin, 1984).[20]

Modes of operation

Capitalism and socialism have traditionally been defined as modes of production, meaning ways of organizing and controlling the means and processes of production, so as to appropriate the resulting surpluses. This traditional approach does not explain either capitalism's combination of wasted capacity and unnecessary products with innovative dynamism or the corresponding mix of high capital construction, shortages and frustration apparent in socialism. To understand these problems we shall examine the characteristic modes of operation of the two systems, which following Kornai and Kaldor, may be called "demand-constrained" and "resource-constrained," respectively.

The characteristic mode of operation pervades the economic sphere and colors all aspects of it – and much beyond as well. Expectations of enterprises as to prices, quantities, revenues, and capital values, all will be formed on the assumption of normal demand scarcity or normal shortage. Households will likewise plan careers and education of children with an eye to the normal state of the labor market. Public bodies will shape their expenditure and capital construction plans on the basis of the normal conditions of operation. Even the agenda of public policy and the issues in political debate may be shaped more by the mode of operation than by the mode of production.

Characteristic incentive patterns

Each system's mode of operation sets up characteristic incentive patterns, which fit together into a definite style. Under capitalism the presence of near-universal excess capacity, dampens the inducement to invest, in the absence of technological improvements. Capitalist eco-

---

[20] This leads to an important conclusion in regard to the debate over "socialist reforms": if prices reflect the requirements for reproduction and distribution, and if pricing benchmarks are set in connection with investment decisions, the claim that aggregate demand imbalances are due to price or wage "rigidities," and can be corrected by restoring "free markets," cannot even be entertained (cf. Eichner, 1976, 1980; Wood, 1976; and Nell, 1991, ch. 17, for the relationships between pricing and investment).

nomies tend to build capacity sluggishly, punctuated by strong bursts of expansion, usually stimulated by innovation. Weak and/or uncertain investment, in turn, tends to keep capacity utilization low and to create a shortage of jobs. By contrast, under socialism, near-universal shortages of goods, engendered by the attempt to run all productive processes at full potential, strengthen the inducement to invest, which in turn further intensifies the pressure of demand on capacity. Socialist economies build capacity rapidly and regularly, but fail to innovate or to produce high quality. Output growth in capitalism chiefly comes from technical progress, in socialism from adding capacity.

A shortage of demand in relation to capacity tends to intensify competition; sales are uncertain – a firm's market could always be lost to competitors. Hence cost-cutting and quality enhancement will be important, perhaps competitively necessary, to attract and keep a share of the limited market. Technical progress in regard to both products and processes is therefore stimulated by the characteristic situation of capitalism and accounts for a large part of the growth of output.

Such technical development will be of the kind analyzed by Adam Smith and Charles Babbage – separation of function and division of labor. Tasks and designs are simplified, clarified, broken down, and made more precise, so that tasks and skills are carefully matched and products fit proposed uses. Expensive skilled labor/equipment will not be used for tasks that unskilled workers can perform.

By contrast, chronic excess demand means that neither product improvement nor cost-cutting is necessary to make sales; indeed, sometimes good quality is not even required. When shortages are severe enough practically anything will be absorbed by the market. But generalized shortage sets up pressure for innovations that can meet several needs or perform several functions at the same time – two birds with one stone. In the face of chronic shortages, jobs must be accomplished without the proper tools or materials, which provides an incentive for redesigning products and equipment, and redefining jobs; equipment and work teams must be adapted to multiple functions. So technical progress takes the form associated with the Pentagon:[21] functions are combined, rather than

---

[21] Perhaps the most familiar and striking examples of such baroque technological innovation are to be found in the U.S. military, e.g., the Multi Role Combat Aircraft or nuclear submarines. But the space program also provides fine examples – not least the shuttle – and a study of Soviet military and space technology will also provide specimens, to say nothing of Soviet tractors – two models to do everything. Incidentally, this illustrates the point that actual systems have usually been a mixture of corporate planning and competitive markets. The U.S. military industrial complex is a planning system embedded in market arrangements, just as Soviet agriculture embedded a limited market system in a planning regime.

separated, and tasks are multiplied instead of divided. These innovations are often admirable – Swiss army knives, vegematics – but they seldom reduce costs in the long run, for a breakdown in any one function usually incapacitates the whole, so that all functions must be scrapped or shut down for repairs. Thus as functions are *added*, breakdown/repair costs are *multiplied*.

Similarly, since a shortage of demand means competition for sales, costs must be kept down by driving hard bargains. Companies will therefore ride herd on money wages; for the same reasons they will try to keep other material and input costs down. Moreover, they will insist on quality for money, since sloppy work or poor quality inputs can mean uncompetitive, unsaleable products. Socialist enterprises, on the other hand, do not feel pressures to keep costs down and quality high. Even with declining quality they can sell their products, and rising costs, though a nuisance, will seldom interfere with the enterprise's plans for expansion. Given the widespread shortages, virtually any reasonable expansion plan will be approved; neither prospective nor realized profitability governs or constrains investment. Capitalism hands out harsh penalties – too liberally, for they fall on many who do not deserve them; socialism hands out easy rewards, also too liberally, for they accrue to many who have done nothing to deserve reward.

These arguments must be treated carefully; it does not follow that capitalism will generate progress and turn out high-quality goods, while socialism will stagnate, drowning in junk. Producing high-quality goods is one important way of competing; introducing marketable innovations is another. But producing cheap goods with hard-to-detect flaws is also a good strategy, as is covering up dangerous defects, pandering to unhealthy desires, building in obsolescence, and distributing advantageous misinformation through advertising. Socialist enterprises must meet Plan requirements and deadlines, but are under no competitive pressures to sell. Hence, although they may let quality decline and costs rise, for example, socialist publishers can concentrate on culturally significant works, rather than best-sellers. Socialist medical care could be delivered to those who need it, rather than those who can pay for it – although it might arrive too late. The contrast may be less between high versus poor quality goods than between, say, classics that fall apart and are delivered late, and swiftly produced, elegantly marketed trash.

At the risk of generalizing too easily, the argument can be put schematically: under capitalism waste is generated by "commission," by actions deliberately undertaken – to produce unnecessary or harmful goods, to add unnecessary features to products, to take expensive but socially wasteful actions to sell, market or promote. By contrast under socialism, waste is generated by "omission," by actions deliberately left undone or overlooked or neglected – failing to control costs, keep discipline in production, keep a check on quality, distribute effectively, inform the market adequately, and so on. Socially wasteful goods that sell, or activities that promote sales, are not penalized under capitalism – but failure to sell is; omitting to control costs and quality is not penalized under socialism – but failing to meet the production quota is.

Market and bureaucracy are seen as two opposed and incompatible forms of organizing economic activity. Nothing could be further from the truth; modern capitalism is highly bureaucratic, and contemporary socialism is equally obviously a market economy, though a market operated with *excess demand*. The production units of all mass production economies so far have been run by bureaucracies; no alternatives have yet proven workable on a large scale. *Both* systems are bureaucratic and *both* are planned through state agencies, although the scope, nature, and objectives of the planning are different. Moreover, all modern economies are market economies; the market may be planned by the State or administered privately or through some mixture of state and private, but it is still a market – goods are produced for sale; ownership changes hands through monetary transactions; monetary income, arising from property or from work, confers the power to consume. But the mode of operation of a market system can be demand-constrained or resource-constrained, and that, we shall argue, makes more difference than whether production is run by bureaucracies professing to represent the citizenry as collective owners or representing shareholders as collective owners.[22]

[22] That economies characteristically operate either as demand or as resource constrained runs counter to most current economic thinking. On the one hand it is assumed that aggregate demand in capitalist societies can and often does reach or surpass the level of full employment. This never happens except in wartime, and seldom then. "Full employment" has been redefined – as an upward-drifting "natural rate." In World War II, the only time the United States ever exceeded full capacity, it began to behave as a shortage economy. On the other, shortages in socialism are widely held to be due to systemic inefficiency and slackness in production – "soft budget constraints," in Kornai's phrase (Kornai, 1986; Davis and Charemza, 1989). The argument is made that for political and administrative reasons, lazy workers cannot be fired, and inefficient firms cannot be shut down, so bureaucratic socialism is unable to enforce budget constraints. Consequently, enterprises feel no compulsion to cut costs or produce efficiently; so long as they meet their quotas, they will suffer no penalty for being unprofitable or for making costly and unwise investments. But managers and bureaucrats will regularly try to expand their territory; hence, careless of costs, they will bring about inefficiency and general shortages. (An extension of the argument holds that markets and private property, bureaucracy and public property, are strongly linked; efficiency is not, in general, consistent with public ownership; Kornai, 1990).

(*continued*)

The "mode of operation" thus refers to the system as a whole; it determines the character of the system, and, in particular, the incentives which govern market behavior. It follows that a system must be one or the other; demand scarcity and supply shortage cannot easily be mixed without losing the distinctive virtues of each. Capitalism is "demand-constrained," that is, productive capacity will normally exceed aggregate demand; whereas socialism is "resource-constrained," meaning that aggregate demand will normally exceed productive capacity. Capitalist firms face a buyer's market, socialist enterprises, a seller's market.

Multiplier analysis

In a capitalist industrial economy additional investment spending increases employment in the capital goods sector, leading to an increased wage bill, the proceeds of which are then spent on consumer goods, leading to increased activity in that sector. Investment spending thus causes consumption spending to move in the same direction. (But the reverse does not hold; a decline in consumption spending need not always have the same, or indeed any, general effect on investment.) It also causes energy and materials production to vary directly in the same proportion, and it stimulates replacement activity. Each of these in turn leads to increased activity among its suppliers, as expressed in the matrix multiplier.

In a socialist industrial economy additional investment spending means intensifying the excess demand for capital goods. Since in general changes in the intensity of excess demand lead to attempts to change output in the same direction, when excess demand for capital goods increases, overtime work will rise, equipment will be overworked more, breakdowns and accidents will rise, etc. Any of these effects may result in additional wage income, the spending of which will further increase the demand pressure on consumer goods. Changes in excess investment demand thus generally cause changes in the same direction in excess consumer demand and may cause further pressure on suppliers of materials

[22] (contd.)

The facts are correct, but the causality is exactly backwards. Shortages *result* from excess demand, which in turn leads to inefficiency, since everything produced can easily be sold. Budget constraints are soft *because* the incentives to expand are strong; not the other way around. Costs are ignored because of the intensity of demand. (As for the idea that soft budget constraints in themselves engender inefficiency, what could be "softer" than the budget constraint of a large American corporation? Perhaps only the budget constraint of a Savings and Loan! Many U.S. government-owned firms, such as TVA, have long been models of efficient operation.) It is competition for scarce demand, not restrictions on current or capital spending, that stimulates cost-cutting.

and replacements. But as in capitalism, a decrease in excess consumer demand need have no effect on excess investment demand. Suppose, for example, that a rise in consumer prices relative to fixed money wages caused excess consumer demand to fall to zero; no productive capacity would thereby be released which could be transferred to the capital goods sector. (This point will be important when we come to the question of reform in socialism.)

In both capitalist and socialist economies the multiplier reflects the turnover of funds, which passes along the stimulus to activity. "Injections" set off activity, and variable costs are passed along in the current period, transmitting the stimulus to further industries or sectors. Funds representing capital charges, depreciation, and fixed costs are withdrawn, or turn over more slowly. In simplified form, then, the multiplier rests on the ratio of variable costs to total revenue, modified, if necessary, to take account of worker saving. (The secondary effects on produced means of production follow from the matrix multiplier – if $\mathbf{i}$ is a vector of injections, and $\mathbf{y}$ one of outputs, then $\mathbf{y} = \mathbf{i}(\mathbf{I} - \mathbf{A})^{-1}$, where $\mathbf{A}$ is the input-output matrix – but will be neglected here to concentrate on the aggregate relationship between demand and capacity.)

The principal injections into aggregate demand are gross investment, I, current business spending (energy, consumption by overhead labor, office expenses), B, government spending, G, and exports, E. To get total demand these injections (measured in normal prices) must be multiplied by an expression which takes account of taxes, imports, saving out of wages, the wage rate and the productivity of labor (Nell, 1988, ch. 5, Appendix). Let the coefficients be $t = t(\mathrm{w})$, $m = m(\mathrm{w})$, and $\mathrm{s} = \mathrm{s}(\mathrm{w})$, where these show the additional taxes, imports and savings that take place when aggregate income (output) increases as the result of additional employment, prompted by additional demand. Hence they are each positive functions of the real wage; even if the marginal tax (import, saving) ratio to individual income were constant, a higher income would mean higher taxes (imports, savings) when an individual changes from unemployed to employed. Moreover, there are good reasons to think that all three may be progressive in both systems. Hence aggregate demand can be written (Nell, 1991, IV):

$$[\mathrm{I} + \mathrm{B} + \mathrm{G} + \mathrm{E}] \, \frac{1}{1 + t + m - (1 - \mathrm{s})\,\mathrm{wn}}$$

where t, m, and s are all increasing functions of the real wage, w.

Aggregate productive capacity is given by the capital stock, measured at the given normal prices, multiplied by the productivity of the system. This last depends on the normal average ratio of capital stock to the labor force, and on the number of workers required per unit of output, on average. Aggregate capacity can therefore be written very simply:

$$K \cdot \frac{1}{kn}$$

where K is the total capital stock, k is required capital per worker, and n is labor force per unit of output. Both these coefficients must be measured at established or normal prices.

Now consider these expressions in the light of the earlier discussion. Characteristically, capitalism will find itself with excess capacity, socialism with excess demand (Nell, 1988, chs. 5, 8). Hence for capitalism:

$$\frac{[I + B + G + E]}{1 + t + m - (1 - s)\, wn} < \frac{K}{kn}$$

For socialism:

$$\frac{[I + B + G + E]}{1 + t + m - (1 - s)\, wn} > \frac{K}{kn}$$

However, care must be taken interpreting these, for they are not the same. When demand exceeds capacity, the multiplier cannot work properly because additional workers can't so readily be hired; however, existing workers can work overtime, and sometimes additional shifts can be added. So the rate at which wages are paid and respent is likely to change as output rises above capacity. An increase in demand pressure will tend to raise w and n, thereby increasing the multiplier, intensifying the pressure, even though employment may not have risen. With this in mind let's compare the two.

Under capitalism, the existence of excess capacity requires firms to compete for the scarce demand, by cutting costs and improving products. Hence n will tend to decline, increasing the expression for aggregate capacity, while reducing the multiplier. Thus the gap between capacity and demand tends to widen. However, competition may force firms to increase w in proportion to the decline in n, offsetting the impact of increased productivity on the multiplier. But t, m, and s are all increasing functions of w; hence the multiplier will still tend to decline and the gap widen. In any case, however, if overall productivity increases by x percent, the new level of income is $(1+x)Y$; if wages rise in proportion and are wholly spent on consumption, its new level will be $(1+x)C$. So the new level of demand will be $I + (1+x)C < (1+x)Y$; excess capacity increases.

The competitive pressures arising from demand scarcity will tend to reduce normal investment and business spending, or at least increase their variability. Rising productivity increases capacity under conditions in which excess capacity already exists; this will dampen I. Increased efficiency in the use of energy, labor and materials will cut into B, and as superior or more cost-effective equipment designs become available, so that k falls, the reductions will affect I and G as well. Only exports are affected in the reverse way; if product or equipment designs improve, and costs are cut, then exports become more competitive and may increase. Otherwise, the pressures tend to reduce each of the major injections, intensifying stagnation.

This pattern is reversed under socialism. Excess demand – a state of generalized shortage – creates incentives to push production to the extreme. The basic ambitions of the system require pushing production to the limit, and there are built-in tendencies leading to further excess. Demand pressure can arise from the attempt to establish fair levels of pay and appropriate differentials, especially between different ranks in both enterprise and state hierarchies. Fairness requires granting regular pay increases when productivity permanently improves as a result of worker efforts. But if a certain kind of blue-collar pay increases in pace with productivity growth, relativities and hierarchical differentials will be eroded; to preserve them the pay of other workers, including management and white-collar pay, must rise. Thus localized increases in productivity can give rise to generalized increases in pay, and consequently in consumer demand.

This can take other forms. New capital goods are normally more productive than old. Thus productivity rises as a function of investment; however, workers using the new and more productive goods are normally exercising the same skills, often in the same jobs, as workers in the old. Fairness therefore demands that they be paid the same. If pay rises with productivity for workers using the new goods, and then, out of concern for fairness, rises for workers using the old, demand will increase more than productivity.

As a consequence of demand pressure, bottlenecks develop, older and outmoded facilities are utilized, workers put in longer hours and make more mistakes, so that productivity falls, that is, n rises. Moreover, demand pressure will tend to call forth basic productive inputs of poor quality, which often only become available in the wrong proportions or at the wrong times. As facilities are pushed harder, previously retired equipment will be brought back into production, and inappropriate equipment will be adapted, all of which will tend to raise capital used per worker, k. (This is very much in line with the traditional view that costs rise as production facilities are pushed beyond a certain limit.) Hence, as k and n rise, even though K rises, the addition to capacity will be less than is needed, and the general downward pressure on productivity in all facilities may even reduce aggregate capacity, while the increases in n and w will raise the multiplier, expanding aggregate demand; both effects tend to widen the gap.

Scarcity of demand in capitalism promotes product improvement; a better mousetrap attracts the market. Excess demand – generalized shortage – on the other hand, implies a seller's market; leading, after a time, to product deterioration and to delays and inefficiency in services. Product improvement/deterioration is often represented as an increase/decrease in the productivity of inputs, which here would be a further decline/rise in k and n, compounding the effects already noted.

As in the capitalist case, both the presence of the gap and the tendency for it to widen, due to its effects on productivity, will lead to pressures on the spending plans of enterprises. Shortages of capacity in relation to demand are a signal to increase the pace of investment spending, to try to bring new capacity on line as fast as possible. Shortages of inputs will lead enterprises to stockpile inventory. Inefficient operation will lead to larger than necessary business expenses. Hence both I and B will increase. The same will hold for G; in the face of generalized shortages and inefficient operation, the government will have to increase its activities, expand its facilities and stockpile scarce items. Again, the impact on exports will be different. If costs rise and the quality of goods declines, exports will tend to fall. By the same token the propensity to import is likely to rise. Moreover, if selling is easy in the domestic market, but competitive internationally, enterprises will prefer to focus on the domestic scene.

For closed economies, then, in capitalism the tendency to stagnation is reinforced by competition, while socialist markets tend to intensify shortages. Capitalist pressures tend to stimulate technical progress in the form of cost-cutting and product improvement; socialist pressures tend to foster inefficiency, cost overruns, and quality deterioration. Capitalist economies deliver services to those with money – which tends to be a buyer's market; socialist economies to those with need – a seller's market. So again the quality is better under capitalism.

For open economies, these conclusions must be modified by noting that the effect on exports (and perhaps imports) will tend to run in the opposite direction in each case; capitalist incentives stimulate exports, socialist ones weaken them.[23] Neither capitalist nor socialist systems are *radically*

---

[23] The respective systems of international trade work the same way. The Western system puts the burden of adjustment on the weaker nations that run deficits; surplus nations do not have to adjust. To restore balance of payments equilibrium basically requires austerity and unemployment, thereby lowering imports. Thus demand will be lowered throughout the system, until the deficit nations are all either in balance, or at an acceptable level of imbalance. The Comecon system financed deficits; planners tried to achieve balance, but if an imbalance arose the Soviet Union would finance it. No austerity measures were required. Hence the system tended to augment domestic demand pressure.

unstable. Capitalism tends to stagnate, socialism to run shortages, but both tendencies meet countervailing pressures and stay within limits. One source of such pressures is external trade, but others can be found within the domestic economy itself.

Relation to the "warranted" rate of growth

This provides a new insight into the famous Harrod-Domar definition of the warranted rate of growth. That rate just balances aggregate demand and aggregate capacity: aggregate demand, in simplified form, is investment times the multiplier, while total capacity is the capital stock times its productivity. Equating the two gives us the rate of growth (investment divided by total capital) that will keep entrepreneurs just satisfied, namely a rate equal to the productivity of capital times the ratio of withdrawals to income. But a small deviation from this rate in either direction will be self-augmenting. For a rate that is too low will imply that aggregate demand is less than total capacity – which gives business the signal to cut back investment spending, thereby reducing demand even further. Too high a rate in turn signals a shortage of capacity, and thus the need for speeding up investment. In each case, responding to the market signal worsens the initial imbalance, apparently showing that capitalist growth is seriously unstable (Morishima, 1975; Kregel, 1980).[24]

The warranted rate is only *incidentally* a rate of growth; it is actually defined as a *level of full utilization*. This degree of utilization is brought about by a certain level of investment, which, divided by the capital stock, identifies a growth rate. But exactly the same argument could be made with respect to government spending, G, in a stationary system. Suppose higher levels of G increase the productivity of capital, both private and public (as in fact they do; Aschauer, 1992; Munnell 1990; Uimonen 1993). Then

$$\frac{G}{z} = \frac{K}{v} \text{ which implies } \frac{G}{K} = \frac{z}{v}$$

---

[24] This, of course, has led to an enormous literature. Two problems are usually identified. First, there appear to be no market forces to bring the warranted rate into line with the natural rate. Second, the warranted rate is unstable. To rectify the first, the neo-Classical model makes v variable drawing on an aggregate production function. However, investment is assumed to be governed by savings. This approach is undermined by the "capital theory controversies." The neo-Keynesian model makes s variable, through the pressures of demand on the distribution of income. But it is easily shown that these forces are not in general stabilizing (Nell, 1982, 1992). In any case, none of the models proposed in this literature explains the *growth of demand*, in the sense of giving reasons that would lead firms to expect that their markets would *expand* in the foreseeable future at a certain rate.

This is our version of the multiplier, where $z = 1 - wn = P/Y$. The formula states that the ratio of government spending to total capital must equal the ratio of profit's share to the productivity of total capital, in order for capacity to be fully utilized. This would appear to suggest a *policy* goal, rather than a growth target. Moreover, if $G/z > K/v$, capacity appears to be too small; hence productivity should be increased. Since productivity is an increasing function of G, the state of the market calls for more government spending. Similarly, if there were excess capacity, the market signal would call for reducing G. There is no growth here at all, yet we find both the Harrod-Domar formula and its instability property!

Nor is the warranted rate a potential, satisfactory balanced growth path for an actual economic system. No capitalist system has ever grown for any time at the warranted rate. Capitalism *always* operates with a margin of *excess*, not just reserve, capacity. (World War II is the exception that proves the rule: the Allied economies were planned and developed shortages.) Socialist economies, that is the economies of the former Eastern bloc, always tended to operate with a level of aggregate demand above full capacity. In short, the warranted rate is not an achievable target (although it may be approachable, as a matter of policy); instead, it is a *dividing line*, separating two contrasting *modes of operation*. The same economic system cannot cycle around the warranted rate, first below, then at, then above it, and so on. Below the balancing point, the system operates one way, generating one pattern of incentives and results; above, an altogether different pattern holds. The original problem was misstated.

Implications

The inherent dynamism of the capitalist market system turns out to be a *macroeconomic* phenomenon; it results from a systematic scarcity of aggregate demand. This provides the setting in which innovation and technical progress occur. Within that framework, both will tend to be greater as investment is greater. Similarly, the shortages of socialism are not due to inefficiency or to soft budget constraints or any other microeconomic factors; shortages, inefficiency, and soft budgets all are the results of fundamental *macroeconomic* pressures. *The characteristic micro behavior of each type of economy has its foundation in that system's macroeconomic mode of operation.*

This in turn implies that a dynamic market economy cannot, in principle, allocate scarce resources optimally. For if it did, it would be resource-constrained, and so not dynamically competitive. Optimal allocation, it seems, is not consistent with a competitive capitalist system. Uncertainty, on the other hand, is necessarily pervasive, since demand scarcity implies that firms can never be sure of their markets. Conventional price theory, concerned with optimal allocation under conditions of market certainty, would thus, paradoxically, seem more at home in socialism.

Conclusions

The preceding has tried to present a survey of various aspects of the working of the system of market adjustment – commodity and labor markets – in conditions of Mass Production, emphasizing the differences between it and the earlier system of adjustment in markets based on Craft technology.

Mass Production economies are demand-driven; the level of demand pressure is determined by finding a *demand equilibrium*, in which market forces acting to increase demand just balance those acting to reduce it. Even Craft economies are demand-driven, though not in the same way. In Craft economies the inflexibility of employment and output leads to *prices* adjusting to offset the demand pressures. These adjustments ensure that fluctuations in investment will be offset by opposite movements in consumption. However, when production processes become more flexible, so that employment can be adjusted, keeping productivity constant, then *quantities* will adjust. Relative prices will tend to remain steady in the face of changing demand. Under these conditions fluctuations in investment will be augmented, rather than offset, so that at the macrolevel the system will be unstable.

Other features of the system differ, too, and these differences are not due to "imperfections." They are part of the system's normal pattern of operation. For example, the labor market under Mass Production will not ordinarily clear. At times, it may adjust in a direction away from equilibrium. Real wages may move procyclically. When wages are tied to productivity growth in the context of a status hierarchy there will be a tendency to low-level inflation. Inflation will also reflect conflict over relative shares. In general, it will not be neutral in its effect on distribution. On the contrary, typically, inflation at high levels of unemployment will work to labor's disadvantage, while in strong or prolonged booms real wages will tend to rise. But when productivity growth leads to managerial "downsizing," the increase in productivity may not appear in the aggregate statistics, but the depressing effect on wages and salaries may be all too evident.

Next we must examine money and financial markets in Mass Production, and how they interact with commodity and labor markets. We will see that the system functions in a distinctive pattern – the new business cycle – which, in turn, leads to a distinctive set of pressures for innovation and change.

Appendix: Stability of the modes of operation

Both systems have built-in tendencies to exacerbate their characteristic condition – stagnation and shortage, respectively. (Both also tend to generate offsetting influences in international trade and in the informal sector – but that is not at issue here.) The interaction between demand pressure and the building of capacity, which determines the extent of excess capacity of shortage, itself tends to preserve the gap between demand and capacity, allowing it to fluctuate, but keeping it within limits. This can be shown by examining simple interactions between two variables – investment and excess capacity for capitalism, investment and shortage for socialism. In each case we will find a cyclical pattern, confining the variation within limits, but always remaining on the same side of the "warranted" or balanced position, which must therefore be considered a dividing line, rather than a target for practical policy. (Admittedly, these models are too simple and abstract to be realistic, but the forces portrayed are present in each system.)

First we must explore a crucial difference in flexibility.

Output, capacity, and demand

When excess capacity exists, whether in a particular sector or in general, changes in demand can be met simply by changing the degree of capacity utilization, which will usually require corresponding changes in employment. When the level of activity is very near the rated maximum, reorganization may be required, for shortages may develop in specific inputs or labor skills, and unit costs may rise – but if the demand pressure is regarded as likely to persist, these temporary shortages can often be overcome by a once-for-all effort, so that costs will fall back to normal. An important implication of this is that changes in the proportions of demand can easily be met, so long as the changes remain within the bounds set by existing capacity. This is extremely important for capitalism, since the incentives to innovation and productivity growth mean that different sectors will be growing at different rates.

By contrast, in an economy in which every sector is operating at full blast, getting the proportions of output correct is both important and difficult. The difficulties are obvious – if there is no reserve capacity, neither aggregate output nor the proportions of output can be adjusted with any ease. There is no room to correct mistakes. It is important because if the proportions are not correct, investment cannot be carried out as planned, with the results not only that growth will be slower than expected, but that some output will be wasted – it may even happen, paradoxically, that excess capacity may emerge!

The maximum rate of growth of the system (with given technique, assumed to be embodied in its plant and equipment) will equal the maximum rate of profit, but will only be attainable if outputs are in the proportions that will produce a physical net surplus consisting of the same goods in the same proportions as the aggregate means of production (Pasinetti, 1977, pp. 208–12; Abraham-Frois and Berrebi, 1979). For any wage rate above the basic standard of living, there will be a corresponding rate of profit lying below the maximum, according to the inverse wage rate – rate of profit function. If wages are consumed and profits invested, there will be a maximum balanced growth rate corresponding to each level of the wage – but this growth rate can only be attained if the outputs are in the correct proportions. (This has been spelled out in Chapter 7.)

Growth depends on the structure of the economy. A certain group of industries can be identified which form a special sector – Lowe termed it "machine tools"; for Sraffa it would be the basics. This sector produces its own means of production, as well as producing basic capital goods for use in other sectors which produce capital goods for use in sectors producing final output. The capacity of this sector not only forms an upper limit to growth, but to increase the growth rate it will be necessary first to increase the capacity of the machine tool sector, which means reducing its sales to other sectors and thus lowering the growth rates of those it supplies. If there is constant pressure on machine tools to deliver, it can never augment its own capacity.

For growth to take place at the maximum rate the output of the basics – machine tools – has to be balanced. If basic output is produced in any other proportions, the system will have to grow more slowly; some goods will be in short supply, others in excess. For each basic good a ratio can be formed, the numerator being its total production minus the amount of it used as input in all the various industries, the denominator being the amount used as input by the various industries. These ratios can be called the "own-rates of surplus"; when they are all equal, then outputs are produced in the proportions that will enable investment of the surplus to reach the maximum rate of growth consistent with the given consumption/wage level. But when the own-rates are not equal, some will be higher and some lower, meaning that some goods will be in surplus and others in deficit, compared to what is needed for investment. The investment which can actually be carried out will therefore be limited by the amounts available, that is, the limit will be set by the good most in deficit. The surplus goods are simply excess. So the sustainable rate of growth for an economy operating in given proportions will be set by the *smallest* of the physical own-rates of surplus of the basic commodities generated by production in those proportions. In other words, if growth has become un-balanced, in an economy with no excess capacity, the lowest of the "own-rate" sets the sustainable rate of growth. The more unbalanced growth becomes, the slower it becomes.

Growth cycles

Capitalism: The mode of operation is demand-constrained, which implies that $K/v - I/z > 0$, that is $I/K > z/v$, where $z = 1 - wn$, w being the real the wage, n the labor requirements per unit output, and v the capital-output ratio. Growth will always be below (or at most equal to) the level that would just balance aggregate demand and total capacity. Excess capacity is always present and exerts a dampening influence on the ability of investment to grow, where investment should be understood here to mean investment spending. Why build more capacity when what already exists is underutilized? On the other hand, investment spending generates technical progress, which increases potential output, and thus increases excess capacity. So the growth of excess capacity depends positively on the level of investment, while the growth of investment depends nega-tively on the level of excess capacity. This looks like foxes and rabbits; let's set it out.

First consider the inhibiting effect of excess capacity on the growth of investment. Let excess capacity, x, be defined in percentage terms as

$$\frac{K/v - I/z}{K/v} = \frac{1 - g_a}{g_w} = \frac{g_w - g_a}{g_w}.$$

Each firm will choose its target capacity in light of its expectations as to the growth of its markets; since the markets are interconnected, these choices can be combined to imply a rate of expected market growth, which we may call g, and which for reasons outlined earlier lies below (at most equals) z/v. But because of overbuilding and the need for reserves, the firms' choices taken together will imply normal excess capacity, which will tend to dampen investment growth, reducing it below the collectively expected rate of market growth. To maintain long-run balance, the growth of investment must match g, but firms will tend to reduce their commitment to expand their construction plans in proportion to the excess they already have. The current level of investment spending has been adjusted to expected market conditions; but when this gives rise to excess capacity, firms will trim the expansion of investment in proportion to their unused plant and equipment. Capital construction will continue to be governed by normal expectations, but its *acceleration* will be adjusted in proportion to excess capacity. Thus we can write

$$dI = gI - axI, \text{ where } 0 \le a \le 1$$

If a = 1, then the growth of investment will be reduced in proportion to the full amount of excess capacity; if a = 0 excess capacity will have no effect – investment will grow at the same rate as capital. In between, investment will grow more slowly than capital, but not as slowly as the excess capacity would warrant. Hence,

$$dI = [g - ax]I$$

The growth of investment varies inversely with the level of excess capacity.

Now consider the growth of excess capacity in relation to the level of investment. Technical progress will increase in proportion to the level of investment. If wages rise with productivity, technical progress will leave the multiplier unchanged but will increase output per unit of capital, that is, reduce v, and so raise g_w. Hence $dg_w = tg_w = bg_a$, where t is the rate of technical progress (that is, $dY = tY$; since $v = K/Y$, and $g_w = z/v = zY/K$, $dg_w = (z/K)dY$), which can be taken as proportional to the level of investment, that is, $t = bI/K = bg_a$. Next, note that $dg_a = g_a(dI/I - g_a)$, and that when $I = 0$, $dx = -hx$, where $-h$ is the proportional rate of scrapping underutilized capacity, due to age or obsolescence; then, given x as above, after manipulating,

$$dx = \frac{g_a}{g_w} bg_a - \frac{dI}{I - g_a} - hx$$

$$dx = \frac{g_a}{g_w} g_a(1 + b) - \frac{dI}{I} - hx$$

Substituting, this becomes

$$\frac{dx}{x} = \frac{g_a}{g_w - g_a} \left[g_a(1 + b) - \frac{dI}{I} - h \right.$$

So the rate of growth of excess capacity depends directly on the level of actual growth, multiplied by the ratio of the actual rate of growth to the difference between the warranted and actual rates; and from this must be subtracted the desired rate of growth minus the adjustment for excess capacity, multiplied by the same ratio. And the whole must be adjusted for scrapping.

The dynamic system is completed by noting the two additional equations, already implied in the discussion,

$$\frac{dg_w}{g_w} = bg_a, \text{ and}$$

$$\frac{dg_a}{g_a} = \frac{dI}{I} - ga$$

These equations define a dynamic system with some affinities to a Goodwin-Volterra cycle, except that it has four variables, dx/x, dI/I, g_a, and g_w, instead of two. An analytic solution is not easily found, but computer simulation shows that cycles of various kinds will be generated with plausible values of the constants.

This confirms that a system of capitalist demand scarcity will normally contain inherent limits. Even if the economy is closed, and no opportunities exist for a black market, the system itself will tend to see that the development of stagnation will reverse itself – and also that there will be no escape from stagnation. As the system approaches the point of full utilization, pressures will develop driving it away; as it moves further and further away, forces will arise pulling it back. The system both creates and limits the amount of excess capacity, generating a cycle that takes place entirely in the realm of demand scarcity. The demand-constrained mode of operation is stable. Now consider the other system.

Socialism: In general, $I/z - K/v > 0$, that is, $I/K > z/v$, intended growth exceeds or just equals warranted. Excess demand or shortage is the general condition. Output is already at full capacity, and demand is still unsatisfied – *in general*. Effective demand for consumer goods exceeds the available supplies, so that queues form in all shops. Order books for capital goods lengthen. But output can only be further increased by operating above the normal level, which in turn can only be done at the expense of higher costs and lower productivity. A given level of shortage implies a proportional level of such extra costs, delays, etc. Hence, although a condition of shortage implies an *incentive* to expand investment more rapidly, it implies an *ability* to do so less rapidly, and the greater the level of shortage, the more pressure there will be on productive facilities.

As in the case of excess capacity, shortage can be defined as a ratio,

$$s = \frac{I/z - K/v}{(K/v)} = \frac{g_a}{g_w} - 1 = \frac{g_a - g_w}{g_w}.$$

Shortage is the difference between the income pressure in the system and its productive capacity, considered in relation to productive capacity, which can be expressed as the difference between intended growth minus warranted growth over warranted growth.

Investment must be understood not just as spending, but as output of capital goods. When the system is already operating at normal full capacity, the multiplier cannot generate secondary incomes in the usual way, through hiring more workers and using idle capacity; instead, overtime, overutilization, running down inventory, and extra shifts, possibly at the expense of maintenance, will be required – and these will generate costs that will reappear as income and spending. But this will only happen if the output is actually being produced. In the capitalist case, investment meant spending; here it must be understood as output.

In the absence of shortage, investment would grow at the same rate as the capital stock, the growth of which in turn will be governed by the Plan. If the Plan's proposed rate of growth (not to be confused with the earlier symbol for government spending) is G (which for reasons already given, will exceed or at best equal, g_w), then dI would equal GI. The presence of generalized shortage, however, will cause investment growth to fall below that rate as capacity is overworked and bottlenecks accumulate. The capital goods sector will be unable to expand its output in accordance with the Plan, and the bottlenecks and shortages will make it difficult to complete projects on schedule. Demand pressures will lead to the production of outputs in the wrong proportions for investment; production can be expected to respond both to demand and to political or bureaucratic pressure – and it would be remarkable if the ability to muster such pressure were distributed in the exact proportions required for growth at the highest attainable rate, given the standard of living. But if goods are produced in the wrong proportions in response to demand and bureaucratic pressure, some goods will not be usable because complementary goods are not available in the correct proportions.

We can spell out an exact relationship here. With a given technique, there is a unique maximum growth rate; when the system is not producing the output corresponding to this, it can only grow at a lower rate. Putting this another way, we know from growth theory that the maximum rate of growth is reached when the "own-rates" of surplus (output minus direct plus indirect use as input, divided by direct plus indirect use as input) for every basic good are equal. Let the q's indicate output levels, and the a's be input coefficients; then

$$\frac{q_1 - [a_{11}q_1 + a_{21}q_2 + \cdots + a_{n1}q_n]}{[a_{11}q_1 + a_{21}q_2 + \cdots + a_{n1}q_n]} =$$
$$\frac{q_2 - [a_{12}q_1 + a_{22}q_2 + \cdots + a_{n2}q_n]}{[a_{12}q_1 + a_{22}q_2 + \cdots + a_{n2}q_n]} = \text{etc.}$$

But the outputs, and embodied productive capacities, of the various basic goods will respond differentially to demand pressure. Hence as shortage intensifies, demand pressure will move the system away from the "turnpike" proportions, in which the own-rates are equal. Some will rise, responding to demand pressure, which necessarily causes others to fall, since the available labor force will provide a constraint. But the sustainable rate of growth is set by the *lowest* of the own-rates among the basic goods. Hence the effective increase in investment will be reduced by demand pressure. These difficulties can reasonably be assumed to be proportionate to the level of s; hence investment output will be reduced in proportion to s. If the proportionality factor is A, then AsI will be the shortfall, which must be subtracted from GI. Hence,

$$dI = (G - As)I$$

The rate of growth of investment is inversely related to the level of shortage. Given s as defined above, and defining t' as technical regress, the percentage decline in productivity as a function of the level of investment, where $dY = t'Y$ and $dg_w = zdY/Y = t'zY/K = t'g_w$. As before, $dg_a = g_a(dI/I - g_a)$, so that as a preliminary we see

$$ds = \frac{g_a}{g_w}\left(\frac{dI}{I} - g_a - t'\right)$$

where, of course, t' is negative. Then, assuming the decline in productivity to be directly proportional to I, $t' = BI/K = Bg_a$, where B is negative, and bearing in mind that when I falls to zero, shortages will decrease due to the absence of demand pressure (since projects will be completed, and productivity will rise to the planned levels, so that capacity will increase) at a rate H:

$$ds = \frac{g_a}{g_w}\left[\frac{dI}{I} - (1 + B)g_a\right] - Hs, \text{ and substituting}$$

$$\frac{ds}{s} = [G - As - (1 + B)g_a]\frac{g_a}{g_a - g_w} - H.$$

So the rate of growth of shortage equals the desired rate of growth, adjusted for the pressures created by shortage, and reduced by the rate of increase of capacity, all multiplied by the ratio of the actual growth rate to the difference between actual and warranted, and finally reduced by the normal rate of expansion of capacity.

To complete the dynamic system we note that

$$\frac{dg_w}{g_w} = Bg_a, \text{ and}$$

$$\frac{dg_a}{g_a} = \frac{dI}{I} - g_a.$$

These equations now define a cycle which lies entirely in the realm of excess demand. The general format can easily be seen. High investment will cause rising shortages, and higher shortages will lower the growth of investment; and when investment reaches an average level, shortages will have risen to their maximum. But the declining level of investment will now bring shortages down, to their average level at the point where investment hits its minimum. Shortages will continue to decline, permitting investment to turn up, so the cycle will repeat. An analytic solution is not available, but computer simulation shows various cycles of the sort described.

As in the case of capitalism and demand scarcity, socialism sets limits on its characteristic excess demand; it will never disappear, but it will never reach unmanageable levels either. Each system works within definite boundaries, even though there are no "floors" or "ceilings." Nor does either system ever operate at normal full employment, or grow at the warranted rate. Of course, these two models are greatly oversimplified; but they deal with a central issue – the relation between the capacity-creating and demand-generating aspects of investment – and they show that the two modes of operation are each stable. Each creates forces repelling the system when it comes to close to the balanced position, and pulling it back when it moves too far away. Each therefore can generate cyclical behavior while remaining entirely on one side or the other of the balanced or warranted rate of growth.

PART VI

Money and fluctuations in the modern economy

Modern monetary theory has chiefly concerned itself with money as an asset, deriving this focus from Keynes. But circulation has been neglected, subsuming the relevant concerns under the "transactions demand." This is not justifiable. Moreover, in spite of the attention to assets, capital arbitrage has not been fully explored. Capital can be shifted into and out of real assets very quickly in modern conditions; when this is taken into account, the Keynesian approach to money and financial markets can be shown to result in the "Classical Dichotomy." In addition, it provides an inadequate analysis of the stability relationships between money and finance, on the one side, and commodities and labor, on the other. The neo-Classical synthesis fares even worse, however: it turns out to be inconsistent.

Chapter 13 presents our picture of market adjustment in Mass Production. The real side of the economy is unstable; but this instability is partly constrained by the financial system under appropriate conditions. The economy will tend to expand, but the boom will lead to an increase in fixed costs and under extreme conditions to profit squeezing inflation. However, declining inflation raises real interest rates again, holding the economy in a slump, until a fall in nominal rates, coupled with the reduction in fixed costs due to reorganization, allows for expansion again.

Interestingly this cycle tends to settle either into one or another of two regimes, where the real interest rate lies either below or above the growth rate. These regimes have different characteristics, but business faces a dilemma. The low interest regime will tend to strengthen the position of labor and raise the danger of a squeeze on profits. The high-interest regime avoids this but benefits financial capital at the expense of industry, while generally increasing the burden of excess capacity.

These problems have led to innovations developed to increase the flexibility of production, shifting productivity increases to reduce the dependence on organized and high-risk labor, and outsourcing to reduce the burden of capacity. These innovations, in turn, have tended in a number of complex ways, to raise real interest rates and reduce growth.

A variety of policies can be developed to manage the cycle and reduce its burdens on the different parties. There are good reasons to believe that such policies would work, if unanimously pursued. But such activist policies depend on political consensus, and on a strong and stable state. At present neither seems much in evidence.

CHAPTER 12

Money and interest in the Keynesian system

The Keynesian system emerged at a critical juncture when the economy was shifting from Craft-based factories to Mass Production. It embodies aspects of both systems. Diminishing returns are assumed; the real wage is supposed to move inversely to output and employment. Prices are flexible – but so are employment and output. Demand changes have multiplier effects. Investment can be volatile. The system may not adjust properly and certainly need not adjust to full employment. Money is nonneutral and monetary factors may prove to be destabilizing.

Keynes offered two new features in the *General Theory*: one was its theory of output – that output adjusts to bring savings into line with investment. The central insight is surely correct. What the preceding chapters have done is first to relate it to *income* – that is, wages and profits, and secondly, to dovetail it with the working of demand pressure in the Marshallian system, in order to show the changing character of market adjustment to variations in demand pressure as technology and institutions developed. In the world of Craft-based factories markets adjust chiefly through prices, after the development of Mass Production, through quantities.

The other was his theory of money. Its value is anchored in money wages, and variations in its quantity affect not prices, but interest rates. The centerpiece of the theory is liquidity preference, the idea that the rate of interest is determined by the supply and demand for money as an asset. By affecting the rate of interest, monetary forces would impact investment, which in turn would have consequences for output and employment. For Keynes, then, money was nonneutral.

But the theory of liquidity preference is problematical, to say the least. The "transactions" demand, although not clearly defined, appears to be money needed for or in proportion to current activity – money "on the wing" – yet is added together with money held for capital asset transactions – money in the vault (Robertson, 1925). At the very least these two respond to different economic forces. In Chapter 6 we argued that the relationship between the two was potentially unstable. Moreover, the theory presents an oversimplified and arguably misleading account of financial markets – for example, it neglects the market for corporate ownership. It considers arbitrage with "real" projects only at the "margin," neglecting the many ways capital can flow into presently operating ventures.

The theory of monetary circulation developed here has shown that the quantity of money in circulation adapts to the pressures of demand in the short run and grows with income in the long. The pattern of circulation remains while the medium changes from metal to paper to bank deposits. But such money is *active*; it is money on the move. Idle or slow-moving funds depend on capital-deepening and only make their appearance as capital-intensity grows. Unlike active money, however, these funds can be shifted into financial markets. But it is not the rate of interest, per se, which matters here, but the relation of the rate of interest to the rate of growth. Capital flows to where the rate of return is highest; moreover financial capital can move in the short run. But the relationship between rates of return turns out not to be stable.

Even without considering this last point, when the shifting of liquid capital is taken into account, the monetary side of the Keynesian system proves to be unsatisfactory. This can be shown both for models based closely on Keynes, in which the "Classical Dichotomy" reappears, and for the "neo-Classical synthesis," which becomes contradictory.

The later Craft-based economy:
Keynes on Marshallian foundations

As we have seen, Marshallian microeconomics describes the working of the Craft-based economy. Diminishing marginal returns and price flexibility are tied to technology and institutions and are empirically reasonable.[1] Keynes based his analysis firmly on Marshall and accepted the elements of marginal productivity theory, understood in Marshallian terms, as developed by Pigou.[2] First, a review of our earlier discussion.

---

[1] Neo-Classical theory, as we argued earlier, confused this descriptive validity with the prescriptive claims of "rationality," thus losing sight of the empirical and historically specific character of the price mechanism.

[2] Since we are basing Keynes on the Classics, and invoking the authority of Pigou, we should consider his (and Patinkin's) "long-period" interpretation, namely that the *General Theory* presents a reasonable "short-period" theory, but when sufficient time is allowed for the price level to adjust, the resulting "real balance effects" will restore full employment. This supposedly holds in theory, but in practice should be considered unlikely to work unless helped out by policy. Such a view conveniently enables one to continue to believe in neo-Classical theory, while supporting Keynesian policies. However, the idea underlying "real-balance effects" is incoherent (Tobin, 1981; Nell, 1984; 1992, chs. 18, 25, esp.

(*continued*)

Marginal productivity in an economy with
slowly adjusting employment

The world is uncertain, and investment volatile. Sales will fluctuate, but employment (and output) are "sticky" in the short run. Consider the case when demand increases, and business comes to expect, correctly, that the new higher level of demand will continue indefinitely. Suppose that this takes place in conditions in which there are unemployed workers. The extra demand will bid up prices, but, since it is difficult to expand employment, there will be no immediate additional demand for labor. So the real wage will fall. But household consumption expenditure depends on the real wage; consumption spending will therefore decline, partially offsetting the increase in demand. The lower real wage now makes it profitable to expand employment, bringing in new workers to increase output to meet the remaining part of the increase in demand.[3]

For the economy as a whole, we form an aggregate utilization function for all firms, summing output on the basis of normal prices. (Assume that households spend the entire wage bill; for simplicity ignore any other forms of consumption spending. Hence $W = C$, as in Chapter 9 and elsewhere.)

To get the total level of demand, to consumption, C, at every level of N, there must be added investment, I, which as a first approximation, may be

---

[2] (contd.)
   pp. 599–614). A different "long-period" interpretation of Keynes has been offered by Eatwell and Milgate, namely that he proposed a determination of relative *quantities*, independently of prices, thus capable of being conjoined to the Classical Equations, which determined prices independently of quantities. All the objections of chapter 8 apply to this view. Moreover, there is little textual evidence for it (except in the contribution to the festschrift for Irving Fisher which they do not mention, since it offers an apparently different theory (but cf. Nell, 1983, and comments by Kregel, 1983), and it is contrary to Keynes's evident interest in dynamics and the processes of market adjustment. In any event, his repeated assertions regarding the volatility of investment would seem to preclude the view that the system could be expected to converge to, or usefully be considered as tending towards, a definite long-period position.

[3] In commenting on chapter 2 of the *General Theory*, Hicks recently (1989) remarked, "Pigou was arguing from a fully Marshallian position, on the formation of the prices of manufactures, that in the 'short period' an increase of demand must raise their prices. So if money wages are given, an increase in 'effective demand' must lower real wages. Pigou maintained that it was this reduction in real wages which raised employment. Keynes, accepting that this would happen, claimed that Pigou had got the chain of cause and effect the wrong way round" (p. 36, n. 14).

taken as exogenous, and volatile. Consider first a level of investment somewhat below the "normal" level, that is, the level expected when plant and equipment was constructed. At normal employment levels output will be excessive, and prices will fall. The real wage will therefore rise, so that spending on C will rise to make up for the deficiency in I, and the aggregate demand line will swing upwards, until it becomes tangent. Employment and output will be slightly reduced, but demand will have been increased. Starting from normal employment, a high level of investment demand, above the normal, will bid up prices, lowering the real wage, lowering C, swinging down the aggregate demand line, until it becomes tanget at a point above the normal level, thus leading to a slight expansion of output and employment. In this way, flexible prices, in conditions of sticky employment and given money wages, will lead to the equality of the real wage and the marginal product of labor, and the adjustment to that point will determine the levels of aggregate output and employment.[4]

Effective demand and the elasticity of employment

A system of Craft-based production with relatively inflexible output and employment seems to correspond to Keynes's idea of a Classical system. If an increase in demand cannot easily be met by increasing output, prices will be driven up substantially, lowering real wages substantially. Employment then will rise only slightly, as will output. The proportional fall in the real wage will be greater than the proportional rise in employment. Hence, in such a system, on the plausible assumption that wages are consumed and profits saved, consumption will fall when the fluctuating component of aggregate demand, investment, rises, and vice

---

[4] On pp. 139–40 of *The General Theory* Keynes discusses Marshall's treatment of interest and marginal productivity, showing that "Marshall was well aware that we are involved in a circular argument" in trying to determine the rate of interest from the marginal productivity of capital – since the use of capital will be increased until its marginal productivity equals the independently determined level of the interest rate. The argument sets up the point that (since a saving-interest rate schedule is not independent of income) there is a need for a separate theory of interest–liquidity preference. But in a footnote Keynes asks about Marshall, "was he not wrong in supposing that the marginal productivity of wages is equally circular?" That is, employment will not be increased or diminished until the marginal product equals an independently determined real wage; rather, the real wage and the marginal product will be determined together, by aggregate demand and supply.

versa. This appears to be similar to the "Classical" system, as Keynes described it.[5]

Keynes wrote that "if our central controls succeed in establishing an aggregate volume of output corresponding to full employment as nearly as is practicable, the classical theory comes into its own again from this point onwards. If we suppose the volume of output to be given, i.e. to be determined by forces outside the classical scheme of thought, then there is no objection to be raised against the classical analysis" (Keynes, 1936, pp. 378–9).

This need not mean that at full employment the *neo*-Classical analysis will hold. Keynes was familiar with Sraffa's equations. If we interpret "full employment" as the labor employed when the system is operating at "normal" capacity, that is, at the lowest level of average variable costs, then we may take those costs as defining the coefficients for the Classical Equations. With the normal wage given, then, normal prices and the rate of profit will be determined. The short-run "marginal productivity" conditions hold, but they are conditions for the proper utilization of capacity under normal conditions.

Inflexible production in the face of fluctuating demand is a recipe for disaster; we can assume that effort would be devoted to overcoming such inflexibility. If it succeeds, then prices will not be driven up so far by a rise in investment demand, employment will become adjustable more easily, and the proportional increase in employment will eventually come to outweigh the proportional decline in the real wage. At this point a rise in investment spending would bid up prices, reduce the real wage, employment would expand more than proportionally, and consumption would rise. Thus the initial increase in demand would be multiplied, rather than offset. In reverse, a fall in investment will be multiplied; that is, a collapse in demand will spread and be intensified. This is clearly relevant to

---

[5] In a letter to Hicks, Keynes writes, "the classical theory assumes that the supply of output as a whole is wholly inelastic; increase in one direction being necessarily offset by a decrease in another" (Keynes, 1973, p. 71). An inflexible version of a Craft economy was explored in Ch. 8, as "Classical gravitation." In Chapter 9, more flexible, Marshallian-type simple "supply-and-demand" relationships were derived as deviations from normal (Sraffian) positions, in response to variations in aggregate demand, caused by autonomous fluctuations in investment. In this system, also, deviations in one direction were offset by opposite deviations in another. Such systems seem to correspond to Keynes's concept of the "Classical" theory.

explaining the Depression, and arriving at this position was an important step in formulating the Keynesian approach.[6]

The emergence of the multiplier, therefore, is not simply a development in economic theory – another example of how we have come to understand the economy more adequately than our forebears. Rather, it is the consequence of an historic change in the nature of technology, making it possible for firms to adjust employment and output more quickly and with fewer costs to changes in demand. Technical progress has usually been represented as a *shift* in the "production function," in which given inputs are enabled to produce higher levels of output. This is certainly important, but in addition, we have argued, the utilization function loses its curvature, so that average variable cost functions acquire a long, flat section. The inflexible system of production put a heavy burden on firms, but for small variations in demand, provided a stabilizing response – variations in investment would be met by opposite variations in consumption. With the growth of flexibility in the response of *individual* producers, however, the *system's* response became destabilizing, as variations in demand became subject to the multiplier.

Digression: a critique of Keynes's aggregate demand and supply

Many commentators on Keynes have sought to represent the Marshallian aspects of his argument by means of the "aggregate supply and demand functions" proposed in the *General Theory*. This approach can be shown to be inferior to the "tangency diagram" suggested earlier. Let us suppose that the aggregate employment and output in the economy can be represented by a constant-returns Cobb-Douglas utilization

---

[6] Keynes first formulated this position in his second lecture of Easter Term 1932, and he developed it further in the second and third lectures of Michaelmas Term 1932, following the important arguments over the "manifesto" of Joan and Austin Robinson and Richard Kahn, who had attended the lecture of May 2, 1932, and wrote a critique of it (Rymes, 1989, pp. 30–43, 55–63; Keynes, 1972, pp. 373–80; 1949, 39–48). The central proposition is "the remarkable generalization that, in all ordinary circumstances, the volume of employment depends on the amount of investment, and that anything which increases or decreases the latter will increase or decrease the former" (Keynes, 1979, p. 40; Rymes, p. 31). There followed an attempt to find conditions under which this would not be true, and to show that such circumstances were highly improbable or particular. The "manifesto" challenged parts of this analysis, with the aim of simplifying the conditions and extending the generality of the argument. Neither Keynes nor the authors of the "manifesto" seem to have succeeded in formulating these conditions adequately, but the argument opened the way to the study of quantity adjustments.

function, and compare Keynes's aggregate supply and demand functions to the tangency diagram.[7]

Let us translate the previous analysis into Keynes' terminology. With a Cobb-Douglas function, relative shares will be constant as employment and output rise. Assume that all profits (or a given fraction) and a constant fraction of wages are saved; since shares are constant, this translates into a "Keynesian" saving function – saving, and so consumption also, will be a given proportion of total income. Aggregate demand will be composed of consumption as a function of income added to the given or autonomous level of investment. Aggregate supply will be determined by the entrepreneur's "expectation of proceeds." If the marginal productivity conditions are met, the share of profits will remain constant, so profits will rise with employment. Output will increase with employment, also, in conformity with the utilization function, thus exhibiting diminishing returns at the margin. Hence there will be a straight line aggregate demand curve, and a concave aggregate supply function. As explained above and in Chapter 9, they will touch once, at a point of tangency, where the marginal product of labor equals the real rate of consumption by workers. An increase in investment, then, will lead to a (comparatively small) rise in prices, lowering the real wage, and therefore lowering the slope of the consumption, and thus the aggregate demand, function. The lower real wage will permit a rise in employment, and so in output. If the change in prices and the real wage is small, that is, if the curvature of the utilization function is slight, then the change in output will be approximately one over the propensity to save (approximately equal to one over the share of profits) times the change in investment (Nell, 1992, ch. 16).

In fact, this approach does not require the function to be Cobb-Douglas; it can take any form. Further, it encourages an analysis of the effects of increasing flexibility in production – as the utilization function straightens and the cost curves flatten with technological development, employment ceases to depend on changes in the price/wage ratio, and the tangency solution gives way to a "Keynesian cross." The "marginal productivity" principle, which in any case does not enter into the determination of the normal wage, ceases to play a role. Prices relative to money wages become more stable in the short period, while output and employment become more flexible.

By contrast, Keynes's account of "aggregate demand and aggregate supply" seems confusing. He offers two different definitions of the aggregate demand function in chapter 3 of the *General Theory* (Asimakopulos,

---

[7] R. F. Kahn, in a letter to Keynes 24 Sept. 1931, observes, "You are assuming all the time conditions of constant cost" (Keynes, 1972, p. 375).

1991; Robertson, 1991; Wells, 1962; Weintraub, 1988; Davidson and Smolensky, 1964). One depends on "expected proceeds" and is dismissed by all as illegitimate; the other combines the propensity to consume and the inducement to invest and is simply a C+I function – where in the simplest case I is taken as given. C, however, is a function of income, but since income is a function of employment (with diminishing marginal output) it can be rewritten as a function of employment. (A linear C, Y function becomes a concave C, N function.) But this is not analogous to the normal Keynesian function; unlike the latter, as employment changes labor's share may change, which leads to consumption changes. The standard assumption (cf. Asimakopulos, 1991) appears to be that as employment rises, labor's share will fall, at least after a point. As we shall see, this creates problems.

The aggregate supply function is harder to define; it is not obvious how to aggregate the supply functions of firms. Asimakopulos, for example, examines four proposals and finds none uniquely compelling, but concludes that all will yield upward-sloping curves of the same convex shape. The explanation of this shape is problematical: "The aggregate supply curve is generally drawn to be convex ... and certainly it must become convex at least after some point, as marginal costs increase sharply and much higher prices are required to bring forth increased output (and employment) thus lowering labor's income share" (Asimakopulos, 1989, p. 110).

This must mean *net* proceeds – profits – otherwise it would imply *increasing* returns. The curve must show that a given increment in employment generates progressively greater net proceeds as employment increases. In *real* terms output at the margin is assumed to be diminishing; the real wage is falling. So prices – equal to marginal cost – must rise, and the implication is that profit increases. This creates a difficulty: Keynes's central insight in the early lectures was that employment, output and investment vary together. But under these assumptions, a rise in investment could drive up prices sufficiently to *lower the real wage more than in proportion to the increase in employment*. As is easily seen on the tangency diagram, the expenditure on consumption would therefore *fall*, and if returns diminish sharply, the effect will be to return output and employment to almost the normal level, even for a large increase in investment.

In short if (following the Classical tradition), we allow that wages play a dominant role in determining current consumption, the validity of the aggregate demand/supply approach depends crucially on restrictions limiting the utilization function. For if relative shares are not constant, then changes in the level of employment will imply changes in the share of wages in income, and therefore in the propensity to consume. But the aggregate demand function is supposed to show the level of C+I associated with each level of employment, on the assumption of a *given* propensity to consume. Moving along the aggregate supply curve cannot cause the aggregate demand curve to shift or change shape.

The problem can be stated another way. Marshallian conditions have been assumed. Hence when demand varies, prices will vary in the same direction, and real wages in the opposite. If these variations are large, changes in I could lead to opposite changes in labor's share, and so in consumption. That is, if labor's share falls when employment increases, a rise in I could lead to an offsetting fall in C. This is inconsistent with a Keynesian multiplier, and has to be ruled out explicitly.

The "aggregate demand/aggregate supply" approach does not explain why prices change, and it does not relate such changes to the level of employment and output. The shapes and positions of the curves are assumed, but they are not related to the core economic principles. By contrast, the "tangency" diagram shows exactly how and why price must change relative to the money wage, in order to reach a new short-term equilibrium, and the role of diminishing returns – the curvature of the utilization function – can be seen visually.

A Keynesian model

These ideas can be brought together in a simple model. The analysis will be comparatively static, although some simple stability questions will be addressed. (Again the notation applies only to this chapter.) To the production system which fixes employment and output through flexible prices and marginal productivity, we add the Keynesian liquidity preference theory of interest and the interest rate–marginal efficiency determination of investment. Since the real wage depends on prices, which in turn reflect the demand for goods, it is not set in the labor market; "supply and demand" in the labor market do not determine either real wages or the level of employment – although the demand for labor is governed by the marginal (utilization) product. But the labor supply plays no role in short-period equilibrium and only affects money wages in the long. Unlike the "neo-Classical synthesis" model, in which output, employment and the real wage are determined by the production function and the labor market, while the demand system adjusts to produce the required level of demand, this version of the Keynesian scheme is not block recursive; all the parts interact, and an equilibirum at less than full employment is distinctly possible.

It will be assumed that the utilization function is flat enough that increases in I generate increase in C; the shift to Mass Production has already begun. This suggests that firms are also tending to adopt a corporate structure, and that employment has become variable. We will see that these points are important.

Equations

$$Y = Y(N, K^*)$$ (1)

This is a Marshallian function with a positive first and, at least over some range, a negative second partial derivative with respect to labor. Nothing, however, has been said about the marginal product of the existing capital stock; that partial derivative may not even be defined, and profits may simply be determined as a residual. Of course, profits plus wages must add up to output and the condition that the real wage equal the marginal (utilization) product of labor will maximize short run profits. Moreover, they will equal the marginal (utilization) product of capital, if that derivative is defined as a function of the capital/labor ratio. Capital – plant and equipment is fixed, although its degree of utilization varies in response to demand. (Embodied capital and embodied technique remain the same, but the amount of employment, and therefore the organization of work, change.) When demand changes, prices change relative to money wages, and the real wage changes, affecting profits. So the level of employment and output changes.[8]

On the other hand, a form such as a Cobb-Douglas *could* be assumed. The capital stock is fixed, but the rate of profit – the "marginal product" – would depend only on the ratio of the fixed capital stock to the variable amount of labor employed;

$$\frac{w}{\pi} = \frac{\delta Y}{\delta N} = Y'(N)$$ (2)

The real wage equals the marginal product of labor:

$$C = \frac{w}{\pi} N$$ (3)

---

[8] Since labor intensity will normally differ in the two sectors, the impact of a change in prices relative to money wages will also differ; relative prices will have to change to maintain a uniform profit rate. But this will only happen if the change is deemed to be a permanent change in the normal price-wage ratio. Short-term fluctuations will be assumed to cancel out over the course of booms and slumps. Given the conditions of production, aggregate demand is the ultimate determinant of both the real wage and, as a residual, the realized rate of profit.

In line with our earlier discussion it is assumed that production technology has become sufficiently flexible that C increases with N; that is, in an expansion the proportional fall in the real wage is less than the proportional rise in employment. The wage bill is wholly spent on consumption by households; for simplicity we ignore any other consumption spending. (A fixed level of capitalist consumption would simply be added on to the wage, leaving the slope unchanged. But saving by worker households would imply that the slope of the aggregate demand line was no longer parallel to the real wage – unless capitalist spending rose with profits just so as to offset worker saving. Assuming that capitalist consumption is fixed, and that workers save, the equilibrium condition for employment – the tangency point – becomes the equality of the marginal product of labor with real household per capita consumption).

$$I = I(i, C), \quad I_i < 0, I_C > 0 \tag{4}$$

Investment, understood as current spending on investment projects, depends on the rate of interest through the marginal efficiency of capital schedule, following Keynes. Higher levels of consumption will shift the MEC schedule out. I therefore depends negatively on i, and positively on C. (Or, in line with most Keynesian models, Y. The central relationship is between I and i, through the MEC. The influence of current variables such as C or Y runs entirely through their supposed – and as Hicks admitted, "unreliable" – effect on future levels of spending.)[9]

The MEC schedule is constructed by ranking projects on the basis of long-term or normal prices, based on an expected normal real wage. All such projects are assumed to be known; new firms are waiting in the wings, so to speak. Changes in prices relative to money wages, which even though not permanent, are expected to last for several periods, will nevertheless not affect the schedule, because the implied change in costs will be offset by the implied change in demand. (Yields are increased by the higher prices, but reduced by the lower level of C.) Strong demand, however, may favor investment by creating a climate of confidence, which could lead to cumulative expansion, even though above normal wages

---

[9] Hicks introduced the influence of Y on investment in his famous article. Keynes, acknowledging that he had also tried writing the function the same way, objected that to do so overemphasized the influence of current income on investment. Any effect it might have was already accounted for in the expected yields. Hicks replied that the effect of current income on expected income was precisely the point at issue, that such an effect would normally be significant, and that it was important to make it explicit (Keynes, 1973, pp. 80–82). A rise in current income will raise expected incomes shifting the MEC schedule out, for example. Hicks agreed with Keynes, however, that this effect would be "unreliable."

(which may partly underlie a high level of demand) might continue to be seen as temporary (Nell, 1992, ch. 16).

The institutional background is that described in Chapter 9. Profits are distributed and constitute the main source of savings. New firms borrow the savings to set up operations, which are not significantly different from those of existing firms, but serve new markets. These new markets and the projects that will serve them are taken for granted; they are not explained. ("Normal growth" in markets is assumed; what is to be explained is the volatility of investment.) Projects can be evaluated by discounting the flows of expected returns, calculated in "normal" prices, and comparing the ratio of such returns to set-up costs with the current rate of interest. Investments that pay better than bonds are clearly worthwhile. It cannot be objected that changes in the rate of interest would lead to changes in prices; only normal prices are considered, and the projects are evaluated at the normal real wage. Changes in the current rate of interest will not affect these normal values.

Strong demand will outweigh high costs in a boom at least initially; in a slump weak demand will outweigh lower costs.

$$\frac{M}{\pi} = L(i, Y), Li < 0, L_Y > 0 \tag{5}$$

Following Keynes, M is taken as given, inherited from history, or set by the Central Bank. (Later we will consider making M an endogenous money supply, setting it equal to liquidity demand as represented by L( ).) M is made up of holdings of coin and bullion, paper and bank deposits, the short-run endogenous nature of which will be ignored for now. The real balance liquidity demand function depends inversely on interest, and positively on the level of real income. But the Keynesian reasons must be carefully reinterpreted. The "transactions" demand must be understood as demand for money as an asset. This is a stock of money that is *held*, not spent. It does not include the circulating funds used in production, or for household expense. Such active funds are part of the stream of revenue; they are not capital. But the transactions demand for money as an asset, as explained in Chapters 5 and 6, is important in coin and paper currency regimes but loses significance when bank deposits and credit cards are the dominant forms of currency – except to the extent that banks and credit card companies must hold cash for change-making and payments purposes.

The other major asset demand for money will be the speculative demand – money that is held in anticipation of more favorable developments in capital markets. When interest rates are high the opportunity cost of idle funds is high, and speculative possibilities are favorable; when interest rates are low, the opportunity cost of holding money is low, and speculative dangers are significant. Hence the speculative demand will vary inversely with interest rates. (The same reasoning applies to current profit rates. When current profit rates are high, the opportunity cost of holding idle funds, as opposed to buying shares in existing businesses, will be high. Of course, shares are somewhat riskier than bonds, and the market may not be as well developed. But the logic is exactly the same).

$$Y = C + I \tag{6}$$

Income equals expenditure on consumption and on investment.

$$Y = \frac{w}{\pi}N + P \tag{7}$$

Income equals wages plus profits.

$$w = w* \tag{8a}$$

The money wage is fixed in the short period. Keynes, as did Pigou, held that the money wage was slow to adjust; but they both argued that it did move in response to fluctuations in demand, although slowly. In a strictly Fixed Employment Craft economy, it would be reasonable to treat money wages as unaffected by changes in demand. But the Keynesian system is designed precisely to analyze *variable* employment, and if employment varies in response to changes in demand, it is likely that money wages will too.

As the system develops a more sophisticated technology, it also adopts the managerial form, giving rise, if not yet to a full market in equity, at least to ways that capital can take positions in existing enterprises. So there is another possible closure of the system:

$$\frac{P}{K} = \frac{Y - (w/\pi)N}{K} = i \tag{8b}$$

Instead of fixing the money wage, it may be allowed to vary consistently with the requirement that capital arbitrage result in a profit rate equal to the interest rate.

Solutions: A solution may be found by constructing two functions in i and N, and solving them simultaneously. One function will be based on equations 4 and 7, drawing on equation 1 through equations 3 and 6; this sets I = P, and is therefore analogous to the IS. The other is based on equation 5, drawing on equations 1 and 2, in order to write i as a function of N, such that the demand for money equals the given supply. This is clearly analogous to the LM. But both functions draw on equations 1 and 2; the labor market employment and output functions are fully integrated with the demand system. To complete the system, equation 8a must be substituted in both equations. Now set the two expressions for i equal to each other. This will yield a polynomial in N, with one or more economically meaningful solutions.

With a solution for N, π follows, and then from equation 5, i can be found. Given i, from equation 4 we find I, and given N, from 1 we find Y. To see the solution visually, draw in I on the diagram for Y(N); then the real wage line will be parallel to the line for C + I that is tangent to Y(N) at the equilibrium N, which is to say, given the money wage we can find π. From this it is easy to calculate C and P.

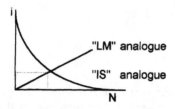

An example: Suppose the utilization function were Cobb-Douglas, and the investment and liquidity functions were linear, as in the example of the neo-Classical "Synthesis" model later in this chapter. We would have

$$Y = xN^{\alpha}K^{(1-\alpha)} \quad \alpha < 1$$

$$I = d + eC + fi, \quad d > 0, 1 > e > 0, f < 0$$

$$\frac{M}{\pi} = \beta Y + \gamma i \quad \beta > 0, \gamma < 0$$

and the rest of the equations would be the same. The IS analogue would then be

$$i_{IS} = \frac{1}{f}\left[\frac{\chi}{\alpha}(1-e)N^{\alpha} - d\right], \text{ and the LM analogue comes out}$$

$$i_{LM} = \frac{1}{\gamma}\left(\frac{(M\chi N^{(\alpha-1)}}{\alpha w * -\beta xN^{\alpha}}\right)$$

There is no separation here between the supply and demand systems; the supply functions enter into both the investment-savings and the monetary demand equations. Plotting i on the vertical axis, and N on the horizontal, the IS analogue has a positive vertical intercept, $-d/f$, and a negative first and positive second derivative. The LM analogue rises from the origin with a positive first derivative. This is sufficient to show that a unique positive solution will always exist, under plausible values of the coefficients. (There may also be a solution in another quadrant.)

Unemployment and money wages

The simplest course is to take the supply of labor as given in the short run, independently of either the real or the money wage. Clearly there is no reason to suppose that the equilibrium, assuming it exists and is unique, will coincide with full employment. Hence this model supports the Keynesian argument that in the short run, unemployment equilibrium is possible. But what happens if, over time, such unemployment causes money wages to fall? (In comparative static terms: to be lower.) If a short-run equilibrium with unemployment leads to falling (or, implies lower) wages and prices, thereby reducing the transactions demand for money, it would be reasonable to expect a fall in (or, a lower level of) interest rates. It seems that the result should be a rise in investment demand, leading prices to fall by less than money wages, so that real wages fall, and employment and output increase. Over a longer time period, then, or allowing for fuller adjustment (but still holding to the assumption of a given capital stock), this approach would appear to support the contention that, given flexible money wages, the system would reach equilibrium at full employment, except when interest rates are stuck in the liquidity trap and/or the MEC schedule is inelastic.

Keynes, of course, objected that competition would not necessarily bring about a fall of money wages and prices. Relatively fixed money wages, in his view, were necessary for a stable monetary system (Chick, 1983, ch. 7). More significantly for our purposes, however, he argued that a decline in money wages and prices would increase the burden of debt, and could well set up expectations of further declines. Both would have a deleterious effect on the MEC (Chick, 1983; Davidson, 1979; Keynes 1936, ch. 19). The reference to the burden of debt is interesting because, following Keynes, and most Keynesians, this model has not specified the structure of debt; what the rate of interest is paid on remains mysterious. Yet P must be paid out as income, and must be borrowed to finance investment. By implication the interest rate is either actually paid, or figures as opportunity cost of the funds to finance investment; hence presently existing capital must have been financed at past equilibrium interest rates. So there must be previously issued bonds paying an interest rate – and that rate on existing capital must yield $iK = P = Y - (w/\pi)N$.

This poses a dilemma. If we take the money wage as fixed, our Marshallian-based Keynesian model will be open to objection that the capital market has been left in disequilibrium (Nell, 1992), for the condition that $i = P/K$ has not been imposed. The behavior of investors has been left underdetermined in a crucial respect: Why should savings be invested in new projects, if buying previously issued bonds or buying into existing firms might be more profitable? And why should existing capital be held, rather than replaced, if new capital is more profitable? But if we add the condition that the rate of interest equal the rate of profit, the model will be overdetermined and inconsistent.

If the money wage is allowed to vary, however, we have the answer. We drop equation 8a, and replace it with equation 8b. The money wage will move gradually, as workers change jobs, or firms go out of business, or reorganize and enter a new industry. The money wage will be variable, but slow-moving. The change in the money wage can be assumed to be inversely related to the level of unemployment, in line with Keynes's discussion in chapter 18 of the *General Theory*. But this assumption will not be used directly to determine the level of the money wage. No behavioral functions will be written showing the offers (or acceptances) of money wages as a response to labor market variables. With the development of the factory system, money wages will not be an objective of policy in the short run (and will be adjusted in productivity bargaining in the long). But they may still be adjusted when market pressures mount. These will follow when the condition is added that $[Y - (w/\pi)N]/K = i$ (which must also represent a slow adjustment).

Profit adjustments: intuitive dynamics

To see what is at stake, let us consider the intuitive economics of these relationships. The adjustment of money wages paid by currently operating firms will see w bid up, in an expansion, and tending to fall in a contraction. Firms will be slow to change what they pay their current employees – but if they have to bid for labor, they will adjust their pay to present workers in line with what they have to pay to attract new workers (Okun, 1980). Similarly, in a contraction when laying off workers, pay cuts can be offered in place of short time. In each case the movement will be comparatively slow, and the changes will likely be small.

As for the relation between profits and interest, funds will shift between new investment and the operation of existing capital. Financial markets develop in line with the corporate form precisely to facilitate the rapid movement of capital. A divergence between P/K and i means that existing capital pays more or less *than new investment, and the bonds financing it.* The rate of interest and the MEC will be assumed equal, any divergence being quickly corrected by an adjustment of investment.

To simplify, let us assume that the oldest vintage of existing capital is located on the East Coast and newer vintages are built on the frontier, moving progressively further West. New investments are therefore physically separate from older capital. Remember, in CBF/MA firms build plants of optimal size, which they operate indefinitely; they do not invest to expand their own facilities. They distribute their profits, which are saved and loaned to new firms. Once a firm is established, the only question is how intensively it will be operated.

Now consider an inequality between P/K and i: The capital stock in the East does not change, but the intensity with which it is operated does. If P/K< i, current operations in the East are less profitable than investment in the West (since projects in the West can afford bonds paying a higher interest rate than the profit rate currently earned by Eastern firms). So activity in the East will be reduced, cutting output and employment, in order to bid for higher paying bonds. This will drive down i, and thereby expand I. In effect, operating funds will be shifted from the East in order to finance new investment in the West. But if P/K > i, current operations in the East will be expanded as potential bondholders would prefer to offer funds to existing firms, lowering bond prices and reducing I.

Taking a closer look, suppose $[Y - (w/\pi)N]/K > i$. Potential bondholders will prefer to lend to or buy into Eastern firms, joining existing entrepreneurs and reorganizing, expanding operations with existing equipment by adding to established working crews. Since labor currently receives its marginal product, and spends the entire wage, output and demand will increase by exactly the same amount. But the pressures to expand will tend to bid up money wages, at the same time that new investment will be falling off, owing to the shift of funds, and the rise in interest rates. On the tangency diagram we can see that the decline in investment will have to be offset by a rise in the real wage; this will partly come through a rise in the money wage, caused by the shift of funds and consequent pressure to expand operations, and partly through a fall in prices due to the lower investment demand. In any case profits will decline, while the shift out of bonds and new capital will tend to raise the rate of interest, reducing I. The movement will be toward establishing equality.

Similarly, when $[Y - (w/\pi)N]/K < i$, entrepreneurs at first will tend to reduce current activity, cutting back the utilization of present plant and equipment, freeing funds for buying bonds, which in turn will increase the investment in new capital goods. As a result workers will be let go, and money wages can be expected to drift down. As workers are let go, current production will fall off at the margin; since workers both receive and spend their marginal products, output and demand will initially fall by the same amount. The decline in money wage rates will lower the real wage, easing the way for the expansion of employment and output to meet the new demand for investment, which will tend to increase profits. And, of course, this new demand results from the shift of funds to bonds, to finance new capital formation, which reduces the rate of interest and thereby raises I.

So in each case, the outcome for business as a whole depends on what it collectively spends. When funds are shifted from lending for new investment to financing current production, the unintended result is a contraction of current activity and a reduced level of profits for existing industries. When funds are shifted to investment, the result is expansion and higher profits. Kalecki's dictum that "workers spend what they get, capitalists get what they spend" holds here, too. In short, both prices and money wages can be flexible, with the real wage equated to the marginal product of labor, while the rate of interest is brought into line both with the rate of return on existing capital, and the marginal efficiency of new, and yet the system will be demand-driven and will exhibit Keynesian unemployment and Kaleckian profits.

The question of labor supply

Suppose, however, that we now consider the labor supply. In a Fixed Employment system the "labor supply" tends to refer to the *hours* or *effort* which the already employed are willing to put forth. But the Keynesian problem arises precisely because we are no longer in such a system. It is now possible in the short run to shift status from employed to unemployed. The question is, what will be the division of the potential workforce between the two possibilities. Let us take it that there will be a certain labor supply, in the short run, at any real wage above a minimal level – a vertical line. Suppose further that this intersects with a descending marginal productivity curve, derived from the utilization function. If the level of employment is below this intersection, so that the real wage will be above it, surely if the money wage is flexible, the resulting unemployment will drive it down. Will this not lower the real wage, and raise employment?

First, in CBF/MA firms are concerned with productivity and productivity bargaining. Hence they are not going to break up work teams in order to bring in new workers at lower pay. The most they can do is replace retiring workers and those who quit at lower rates – but the effect of this on morale must also be considered. Workers are notoriously sensitive to changes in differentials. Unemployment cannot have any very direct or rapid effect on current levels of money wages.

However, suppose the money wage did fall and brought down the real wage as well. Further suppose that this tended to reduce unemployment. Then the rate of profit on existing capital would be higher – but since this represents a short-term adjustment it would not necessarily raise expectations of future profits. Arbitrage, however, would imply a higher rate of interest, requiring a higher MEC on investment, thus indicating a lower level of investment. This would imply a reduced effective demand, with lower prices again, so that the real wage would be higher, re-creating unemployment.

The preceding argument only considered the comparative static implications. When money wages fall: what will be the dynamics? Why will the real wage fall – and if it does, how fast will it fall? Lower money wages will mean lower costs, but also lower consumption spending (Keynes, 1936, p. 261). The two effects will counteract one another, leaving investment unchanged. Hence profits cannot increase, and the lower costs will translate into lower prices and the real wage will remain the same. The lower transactions demand cannot bring down the rate of interest, since arbitrage will keep it equal to the rate of return on existing capital implied by the real wage.[10]

As Keynes argued in chapter 2, the real wage cannot be set in the labor market, since flexible prices depend on effective demand. The level of effective demand depends on, and in part determines, the rate of return. But given the capital stock and the rate of return, the real wage also follows. We can draw the curves for the labor market, showing the level of the real wage at which the supply and demand for labor will be equal. But the analysis has shown that there are no market forces which work to bring about that equality. Since the labor market cannot determine the real wage, it cannot affect the level of employment, and, at most, it can exercise only residual influence over the money wage.

The Classical dichotomy

When the money wage is taken as fixed, the markets for Savings-Investment, Output and Employment, and Money and Interest all interact. But when the money wage is left free to drift, and the equality of the rates of interest and current profits is imposed, the "Classical dichotomy" appears quite strikingly. That is, aggregate demand in real terms, together with the equations for production and employment, determine the real

---

[10] Here the presumption of a fixed money supply creates an unnecessary problem. If "the money supply" consisted of bank balances, then the lower money-wage and price level would translate immediately into lower money bank balances – the money supply would adapt to the demand, as the theory of monetary circulation tells us. But in the Keynesian framework, if "transactions demand" is down, with a *fixed* money supply, there will be excess transactions balances. These may at first enter the market, lowering short-term interest rates, only to find arbitrage pulling them back up, whereupon, facing falling bond prices, the funds might take refuge in speculative balances. (The lower money wages and prices, of course, imply a higher value of real balances. They also imply a higher burden of debt [*General Theory*, p. 264], a matter of much greater significance; cf. n. 3. Part I.) Once we focus on circulation, however, it is clear that money adapts to activity.

values of all the variables, including the real wage and the rates of profit and interest, while the monetary equations determine only nominal values – the price level and the money wage.

This can be seen as follows: Start with equation 4, for investment, and substitute into it using the production function, equation 1, the real wage equation, 2, consumption, 3, and the income equation, 7, and finally, the profit rate-interest rate equation, 8b. By this means C and i on the left-hand side can be eliminated; through regrouping and repeated elimination, the other variables can be removed until only N and coefficients or constants appear. (This cannot be done without equation 8b, the profit rate-interest rate equation; when the model contains equation 8a, the fixed money wage, i cannot be eliminated from the investment equation, 4, without drawing on the monetary equation. The monetary system has real effects, and vice versa; the monetary and the real interact.) Then use equation 6, $Y = C + I$, first to eliminate I, then using equations 1, 2, and 3, to eliminate the other variables on the right-hand side, leaving only N and coefficients or constants. In the example earlier, the result will be

$$N^{\alpha} = \frac{d}{x[(1 - f/K)(K^{1-\alpha} - \alpha) - e\alpha]}, \text{ so that}$$

$$N = \left\{ \frac{d}{x[(1 - f/K)(K^{1-\alpha} - \alpha) - e\alpha]} \right\}^{1/\alpha}$$

and for plausible values of the parameters this will have a single positive solution for N. Drawing on the solution for N, from equation 1 we can obtain Y; from equation 2, w/π; from equation 3, C; from equation 6, I; then with I and C, from equation 4, i will be determined. Finally, from equation 8, P follows. Thus, the seven equations pertaining to Savings-Investment and Output-Employment determine the seven *real* variables: N, Y, C, I, P, i, and w/π. Equation 5, for the money market, is left to determine π, the price level, and w, the money wage, then follows from equations 2 and 3. "Real" relationships determine the real variables; monetary forces, the actions of the money market, only determine nominal variables!

This result continues to hold even if unemployment influences money wages, provided the money supply (or part of the money supply) is allowed to be endogenous, in the sense of adapting to the needs of circulation. Add two further equations:

$$N^* = N\frac{w}{\pi}, N' > 0$$

$$w = w(N^* - N), w' < 0 \quad [\text{or:} \bar{w} - w = w(N^* - N),$$

where $\bar{w}$ is the normal money wage.]

where N^* is the labor force and N the level of employment. The labor supply is shown as a positive function of the real wage, as in conventional theory. Many labor-force participation studies suggest this – although, as noted earlier, some studies show an inverse relationship. The unemployment-money wage equation takes the place of the market-clearing condition; it shows that higher unemployment will lower the money wage, driving it below the normal level. Solving the equations as before determines the seven real variables. Given the real wage, N^* is determined. The unemployment equation will set w, the money wage, then the monetary equation will determine M/π, real balances. Finally, M will adapt to ensure that π will be consistent, given the unemployment-determined w, with the real wage, as determined by the "real forces." M is determined as a residual, passively adapting to the needs of the market.

These startlingly "Classical" results are obtained in a model which is demand-driven, reaches equilibrium with Keynesian unemployment, and in which the Kaleckian relationship holds between investment and profits, even though the wage equals the marginal product of labor. And the money supply may be endogenous. Clearly, the Classical dichotomy does not rest on full employment, or Say's Law. Equally, unemployment does not necessarily depend on monetary factors. Liquidity preference – the transactions and speculative demands for money – and the effects of interest rates on investment are not sufficient to establish the non-neutrality of money.

Indeed, the supply and demand for money as an asset considers only one aspect of the monetary system. Circulation, especially the interaction between money in circulation and money held as a store of value, is wholly neglected in this approach. The effects on investment and output of a collapse of the banking system, for example, due to overissue or overlending, cannot be adequately studied in this framework. The risks of default on money-denominated obligations do not appear. The effects of changes in the ratio of fixed obligations to current earnings are not considered; the "burden of debt" is not modeled at all. Inflation is not examined, so the effects of inflation on the real level of the interest rate cannot be studied. And while equation 8b does relate the interest rate to the rate of return, the possibility of a *systematic inequality* between them, one which sets a dynamic process into movement, cannot be dealt with in this context. In a word, there is not enough money here – the monetary-financial side of the economy has been short-changed.

Conclusions

Keynesian conclusions, then, can be established on Marshallian foundations, provided the Marshallian production/utilization function is clearly distinguished from its neo-Classical namesake. When the money wage is taken as given, the resulting model is not block recursive; the demand system does not adapt to the supply side, as in the neo-Classical synthesis (to be considered next) and the model gives rise to under-employment equilibria. The real and the monetary aspects of the economy interact.

The analysis can be criticized, however, for leaving the capital market in disequilibrium. Liquidity preference theory tries to account for the positive relation between activity levels and interest rates through the effect of higher activity on transactions demand. But the supply-and-demand approach to money and interest is deeply flawed. It overlooks capital arbitrage, in which interest rates are pulled up when demand is strong, because realized profits are high. But when such arbitrage between bonds and existing capital is incorporated, allowing for pressures tending to bring rates of return on financial assets into line with one another, the structure of the model changes, and the Classical dichotomy reemerges. The awkward results continue to hold even when the labor supply is shown as a function of the real wage, and unemployment influences money wages. There is no link between the asset-money market and the real system, because the real side of the economy chiefly relates to money through circulation. Real-money interaction, then, takes place through the interaction between money as an asset and money as a medium of circulation – but here the theory of circulation is missing.

The Marshallian picture of diminishing returns to utilization represents a particular point in the development of modern technology. It refers to a period in which the factory system had developed, but was still being operated by skilled work teams more characteristic of the Craft technologies that had preceded modern industry. As Mass Production technology matured, the pace and control over work came to be embodied in the equipment itself, so that the curvature characteristic of the Marshallian utilization function tended to disappear. In other words average variable cost curves flattened, and by the end of World War II modern industry had become more flexible in its ability to adapt output and employment to variations in demand. The financial structure evolved *pari passu*. A new macroeconomics, without marginal productivity, came on stage, prepared to present its dramas in a dynamic mode. That will be the subject of the next chapter.

Investment and production in the neo-Classical synthesis

The Keynesian model examined above accepted Marshallian marginal productivity, yet nevertheless generated unemployment. When arbitrage between real and financial rates of return was added, however, the structure of the model changed. The "neo-Classical synthesis" is likewise based on marginal productivity theory, but reaches equilibrium at full employment, allegedly showing that Keynes was wrong. Awkwardly, however, the neo-Classical version accepts marginal productivity theory in the labor market, yet overlooks its implications for the capital market. When arbitrage between rates of return on claims to the existing capital stock and financial assets is taken into account, an inconsistency emerges, which reveals the crucial differences between the Keynesian and neo-Classical approaches.

Traditional macroeconomic theory

To see this let us write a simple linear version of the conventional model of the IS-LM plus labor market (Modigliani, 1944). (Linearity will not affect the model; nothing in conventional macroeconomics depends on second derivatives.) The variables will be: S, for saving, I, for investment, Y, for aggregate demand, i, for the rate of interest, L, for the demand for nominal money, π, for the price level, N, for the amount of labor, and w for the money wage - eight in all. The system can be represented by the following equations:[11]

$$S = a + bY + ci \quad a < 0; b, c > 0 \tag{1}$$

where a is autonomous dissaving, b is the marginal propensity to save out of income, and c indicates the (probably weak) influence of interest on

[11] The model to be examined does not contain a "Pigou" effect. Patinkin's suggestion (1965, 1991) that money should be included in the utility function suffers from the defect that the resulting real balance effect must be considered along with the substitution and income effects in deriving demand functions, and can easily create perversities (Lloyd, 1965; Nell, 1984, 1992). But if this is not done, the proposed "effect" has neither foundation nor intuitive plausibility; changes in the burden of debt – which run in the opposite direction – are more readily observed, and much more significant. For the Pigou effect to work therefore, it must be assumed that when wages and prices change, the entire capital stock is rebuilt at the new price level, financed by bonds at the new nominal level. Why? Where does this appear in the investment function? (Patinkin, 1965, p. 217, esp. note 13; thanks to Wynne Godley for the reference). Alternatively, when prices rise firms must pass on capital gains to entrepreneurs and be recompensed by them for capital losses when they fall. That is, an increase in the burden of debt will be assumed to be offset by "entrepreneurs"! Why? What motive do they have? There is an implicit and implausible equity market hiding here. In any event, the Pigou effect cannot plausibly be considered a short- or medium-run adjustment. But if it is long-run, or even medium-term, on what grounds do we suppose the supply of money to remain constant, while the price level and real balances are changing?

saving (the notation is not to be confused with that elsewhere);

$$I = d + eY + fi \quad d > 0; e > 0; f < 0 \tag{2}$$

where d captures the various autonomous influences on investment, e is the marginal propensity to invest in response to aggregate demand, and f represents the slope of the marginal efficiency of capital schedule. The latter, of course, is the principal concept in Keynes's own theory; e, however, can also be justified on the grounds that higher levels of activity today give rise to expectations of higher revenue in the future, thus raising the prospective yields.[12] Modigliani and Patinkin follow Hicks in making investment depend on current income, Y. (The interpretation of this function will come up again later.)

A different interpretation can be suggested, following Kalecki: Risk increases as investment spending rises in relation to the existing capital stock (or level of output) because the danger of being unable to service fixed obligations rises as spending commitments increase. (Not only that, risk would likely rise at an increasing rate with investment spending, since the costs of defending against default interact multiplicatively.) Higher Y will therefore encourage investment, and higher i discourage it.

$$\frac{L}{\pi} = \alpha + \beta Y + \gamma i \quad \alpha, \beta > 0, \gamma < 0 \tag{3}$$

L is the nominal demand for money, L/π is the real; βY represents the Quantity Theory, while γi is the Keynesian speculative demand.

Next we have the two standard equilibrium conditions and one side condition, representing the liquidity trap:

$$S = I \tag{4}$$

$$M = L \tag{5}$$

$$i > i_T \tag{5'}$$

---

[12] As noted earlier in footnote 9 Hicks introduced this idea in his famous article, while Keynes remained unpersuaded. Both agreed, however, that the effect in question was unreliable. From an empirical point of view the best established and most reliable relationship between investment and income is the modified accelerator.

From these we have

$$i_{IS} = \frac{d-a}{c-f} + \frac{e-b}{c-fy}$$

$$i_{LM} = \frac{(M/\pi - \alpha)}{\gamma - \beta/\gamma Y}$$

which solve for the equilibrium values of Y and i

$$Y = \frac{(c-f)[(M/\pi) - \alpha] - \gamma(d-a)}{D} < \text{ or } = Y_F$$

$$i = \frac{(b-e)[(M/\pi) - \alpha] - \beta(d-a)}{-D} > \text{ or } = i_T$$

Here $D = \beta$ (c–f) + γ(e–b). For unique and plausible equilibria, under normal circumstances, and for stability, b > e must be assumed, although there is no economic justification for this (Kaldor, 1940; Nell, 1992).

These equations make up the monetary aggregate demand system, determining saving, investment, and the real demand for money as functions of income and the interest rate. To this we now append the real supply side:

$$Y = Y(N, K^*) \tag{6}$$

This is a well-behaved neo-Classical function, with positive first and negative second derivatives, constant returns to scale, with the value of the function, in ratio form, going to zero as the first derivative tends to infinity, and tending to infinity as the first derivative tends to zero. K^* refers to the fixed stock of capital, the real means of production, and its earnings are not differentiated between debt and equity.

$$w = \frac{\pi(\delta Y)}{\delta N} = \pi Y'(N^D, K^*), \text{ where } Y'(N^D, K^*) < 0 \tag{7}$$

This is the traditional demand curve for labor, the descending marginal product curve. It depends only on the real wage; this is because the labor market here is a simplified "parable" version of the neo-Classical price mechanism, in which output and prices are determined simultaneously on the basis of given preferences and endowments, with a given technology.[13]

$$w = \pi S(N^S, K^*), \text{ where } S'(N^S, K^*) > 0'' \tag{8}$$

This is a rising supply curve of labor, usually, though not necessarily, derived from household utility maximizing. Households respond to the real wage by changing the hours they offer; this function therefore shows variable hours offered by a fixed number of employed workers. This is appropriate for a Craft economy, in which employment is fixed in the short run. But for a Keynesian system, demand for labor must be considered demand for a variable number of workers for a fixed set of hours. This is not just a dimensional quibble; the labor supply function shows the willingness to work longer or shorter hours on the part of those whose *status* as employed is already settled, whereas the demand function is calling for workers to *change status*, between employed and unemployed. Finally, the equilibrium condition,

$$N^D = N^S \tag{9}$$

The labor market must clear, which implies that excess demand for labor must bid up the real wage, excess supply must drive it down. This is what Keynes denied; the real wage is not set in the labor market. There are no market forces to bring about this adjustment.

These last four equations (three independent, when we consider N as one variable) comprise the supply system, determining the real wage and full employment, and the full employment level of output. The system is block recursive; the supply side determines output and employment, and the demand side will generate the required demand.

From the solution of the supply system, we get the full employment level of output, while the demand system solves for the IS and LM loci respectively. These will intersect at some combination of Y and i. The LM locus can then be adjusted by varying the price level, until the intersection coincides with the full employment level of Y. This then determines π and i, and the price level in conjunction with the marginal product of labor

---

[13] Even Patinkin, 1965, comments, "though it is obvious that there must be some connection between the firm's output of commodities and their input of labor, this connection is not explicit in our demand function for labor ... to all outward appearances this function depends only on the real wage and not on the volume of output ... this absence of an express dependence on ... output ... holds for any labor demand function derived in the standard way from the principle of profit maximization" (p. 319).

fixes w, the money wage. Eight independent equations determine the eight unknowns,[14] and the model is usually interpreted to mean that the monetary demand system adjusts to the level of output determined by the supply system. If Y is below its full employment level, a lower price level would reduce transactions demand, and so imply a lower interest rate, resulting in higher investment and a higher level of income. Assuming a sufficiently elastic marginal efficiency schedule, and a low enough liquidity trap side condition, a low enough price level will ensure full employment.[15]

Marginal productivity theory

But the supply side has another implication, which has been ignored in this story. Consider an ordinary supply and demand for labor diagram. The demand curve falls from left to right, intersecting a rising supply curve. The area under the marginal product curve is equal to total output. The vertically shaded area is the wage bill. By definition the remaining area, horizontally shaded in the diagram below, represents profits. If exactly this amount of profit is not paid out, income will not have been properly distributed, and we cannot say anything about savings or consumption. (Hence, we cannot be sure about investment or the transactions demand for money.) Moreover, the profit must be paid at a uniform rate on capital, allowing for various imperfections, or the system will be in disequilibrium.

[14] It is usually assumed that the IS locus slopes down from above the LM, which slopes upwards, so they intersect just once. The IS, however, could have an intercept below the liquidity trap level of interest, in which case there would be no intersection in the positive quadrant. If e > b, the IS will have a positive slope; if this is steeper than that of the LM, then if the intercept lies above the liquidity trap there will be no intersection in the positive quadrant; if below, one – in the liquidity trap. If the IS is shallower than the LM, there will be one intersection if the intercept is above the liquidity trap, and two if it is below. Moreover, e > b implies that the IS is unstable. Since changes in the price level in this model shift the LM locus only, they will do little to "correct" these problems (see Harcourt, 1980; Nell, 1992, chs. 18 and 25 for further discussion).

[15] Suppose the money supply were "endogenous." Less change in the price level would be required to reduce the interest rate. If the money supply adapted fully to transactions demand without requiring a rise in interest, no price level changes would be required; if the pegged rate, or the liquidity trap rate, were low enough and the marginal efficiency schedule elastic enough, the model would always produce full employment. Moreover, changes in activity would cause no changes in interest rates. These are implausible conclusions; but it is quite realistic to claim that in the postwar world the money supply has been highly adaptable. It is the "synthesis" model that seems to be at fault.

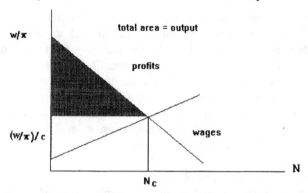

Curiously, the neo-Classical synthesis fails to tell us how this profit is to be paid. It must be a monetary flow from firms to households, either interest on bonds or dividends on equity. Equity, however, is never mentioned; it is presumed and sometimes stated explicitly (Patinkin, 1965, ch. ix) that all capital is financed by bonds. The only assets, other than real capital goods, are money and bonds, and bonds are floated to raise money for investment in real capital. But no expression for the payment of income appears in the model, and neither the earning of profit by firms nor the payment of interest on bonds is shown anywhere.[16]

(The justification advanced – if the matter is mentioned at all – is that owing to Walras's Law the bond market may be "dropped.")

At any given time, therefore, real capital goods will be purchased with the funds raised from selling bonds of equal value. So the accumulated capital goods and outstanding bonds must be equivalent, and the marginal product of capital, the earnings from real capital, must equal the rate of interest. Otherwise, the total income paid out would not equal the value of output.

A simple diagram

Since the production function is well behaved, it can be written in ratio form. Instead of the more usual practice of writing it per capita, here let us divide through by the fixed capital stock, K^*, and write all

---

[16] In an early presentation of the argument of the *General Theory*, strongly endorsed by Keynes, R. B. Bryce defined income "as the money receipts of all individuals ... in the given period ... for their productive services or rights used during that period. ... These receipts are considered to include those not actually paid over – for example wages earned but not paid till the end of the week, or profits earned but not paid till the end of the year" (Bryce, 1935, p. 134). Income, in short, was explicitly understood as wages plus profits.

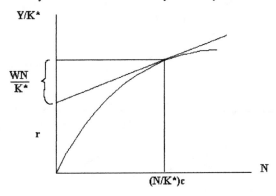

variables in per-unit-wealth terms. Hence in this diagram the horizontal axis will be N/K^*, and the vertical, Y/K^*. If the labor market equations are similarly rewritten, we can determine the full employment level of N/K^*. Draw this in and determine the full employment level of Y/K^*. At this point on the production function draw the tangent and extend it to cut the vertical axis. The slope of this tangent is the marginal product of labor, equal under competitive conditions to the real wage. In the triangle formed by the tangent, the horizontal stretch is N/K^*, so the vertical must be $wN/\pi K^*$, the wage bill over the capital stock. Since the vertical height measures full employment output, the remaining portion must be profits over capital, that is, the rate of profit, equal to the marginal product of capital. As seen above, this must equal the rate of interest paid on existing bonds. But the question is, will this equal the rate of interest determined by the supply and demand for money interacting with savings and investment?

Revising the A, B, C model

To make use of this diagram, we rewrite the IS-LM equations, putting them in per unit wealth terms. This can be done very simply by multiplying and dividing the Y terms by K^* and regrouping. The results are

$$i_{IS} = \frac{d - a}{c - f} + \frac{K^*(e - b)}{(c - f)Y/K^*}$$

$$i_{LM} = \frac{(M/\pi) - \alpha}{\gamma - K^*(\beta Y)/(\gamma K^*)}$$

Nothing is changed, but we can now map the equations onto a diagram with Y/K^* on the horizontal axis. This makes possible a three quadrant diagram of the whole system (see Figure 12.1). In the lower left, we have the labor market, with the real wage along the horizontal axis, and N/K^*

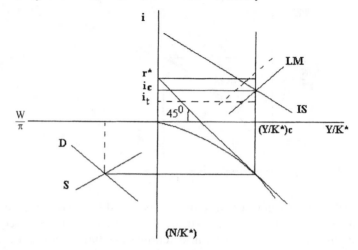

Figure 12.1

on the lower vertical. The production function is then drawn in the lower right, with the horizontal axis measuring Y/K^*. The upper right quadrant shows the revised IS-LM equations, which intersect at the full employment level of Y/K^*, determining the price level and the rate of interest, as shown.

An inconsistency

Now draw in the tangent, as in the preceding diagrams. The point where it intersects the horizontal axis divides the full employment level of income per unit wealth into the wage bill per unit wealth and the rate of profit. This rate of profit can then be mapped onto the vertical axis by a 45-degree line, as shown. It can thus be compared directly with the rate of interest determined by the IS-LM system. There is no reason why they should be the same.

There is, therefore, a flaw in the heart of the traditional model: it implies a distribution of profit from real capital as interest on bonds at a rate that need not equal the rate of interest on money, as determined by the money market interacting with savings and investment. Yet the two must be equated by market arbitrage, for the same reasons that the rate of interest must equal the marginal efficiency of capital. Old and new capital are in competition, both with one another, and for funds, so that old bonds and new are also in competition.

It could be argued that, in a short-run model, a discrepancy between the marginal product and the marginal efficiency of capital is not serious. In the long run it will be resolved – for that is the function of investment! This is a misunderstanding; such an argument might be made in "real" neo-Classical models, but cannot be countenanced in a monetary, Keynesian context. The arbitrage between the outstanding bonds previously floated to finance existing capital and those for current investment will be immediate – and will result in a discrepancy between the value of the old bonds and the value of the capital stock securing them. What then happens as existing capital is worn out and replaced? How can the bonds be rolled over without capital gains or losses?

Nor is the time dimension of the "short-run/long-run" distinction adequate. The model provides for enough time for unemployment to force money wages down, and then for competition to reduce prices in proportion. Both of these processes will face opposition – from workers and firms, respectively; so for money wages and prices to adjust we must assume both that agents are strongly sensitive to their interests, and that market conditions are highly flexible. But then how can it be that investors will not move their liquid capital into those assets offering the highest returns?[17]

---

[17] Most textbooks today present the neo-Classical Keynesian system in the form of "Aggregate Demand and Aggregate Supply" curves. The diagram plots the price level against output, and is (misleadingly!) analogous to ordinary supply-and-demand curves. (In conventional supply and demand, the price is what the demand side pays and the supply side gets. By contrast, the macro *price level* is a construct, an index. No one pays the price level for a unit of aggregate output; no one charges the price level for such a unit.) The AD slopes downwards, on the grounds that at lower price levels "wealth" (e.g., money balances) will be greater, at higher, smaller, and wealth effects are presumed to have a significant impact on demand – contrary to all evidence. No account is taken of the empirically significant effects, running in the opposite direction, of price level changes on the burden of debt (e.g., Tobin, 1984; Bernanke, 1993). (Alternatively: it is assumed that nominal debt can be easily and quickly adjusted, so that the debt/price level ratio stays constant.) The AS would be vertical in the model above. The simplest argument for a positive slope is that the price level moves faster than money wages. Hence as prices rise, the real wage falls, and firms hire more labor, so produce more output. Unfortunately, if the production function is anything but Cobb-Douglas, this also implies that relative shares will change as output changes. A change in relative shares is likely to affect both consumption and investment, and unlikely to affect them in exactly offsetting ways. Hence moving *along* the AS function will *shift* the AD function. If the production function is Cobb-Douglas relative shares will not change, but the marginal product of capital is an increasing function of output (since the capital stock is fixed), and, of course, the difficulties discussed in the text will have to be faced. To explain the positive slope of AS – the sluggish reaction of money wages – hypotheses about asymmetric information have been developed. Workers are assumed to learn less quickly, or react more slowly, when market variables change. Initially these assumptions were largely ad hoc (Cherry, Clawson, and Dean, 1984), but the work of the "New Keynesians" has provided more secure foundations.

The rate of profit and the rate of interest

The "synthesis" model contains no explicit equity market; nevertheless firms have owners – "entrepreneurs" – and if a firm or group of firms earns more or less than the going yield on bonds, entrepreneurs will try to buy in, or sell out, respectively. Even if there is no formal market, an informal equity market, for example, a market in partnership shares, will tend to develop. Moreover, each period the capital stock depreciates, and the depreciation allowances must be placed. If real capital is earning less than bonds, firms would be foolish to purchase replacement equipment; if real capital is earning more, firms would be well advised to use replacement funds to expand.

Suppose, for the sake of argument, that the rate of profit, r^*, equal to the marginal product of existing capital, does not equal the equilibrium rate of interest, i^*, where we accept the "synthesis" model's assumption that the latter will be equated to the marginal efficiency of capital.

Suppose, first, that $r^* > i^*$. But why would anyone invest in a new project that paid less than existing activities? Surely, given that the rate of interest is lower than the earning potential of existing firms, it would pay to borrow to buy the bonds of those firms, which would be a better investment than new projects, since these pay less. Hence the price of existing bonds will rise, creating a discrepancy between the value of the bonds and the value of the capital securing them. At this point existing capital equipment is a bargain, and investors will either try to buy into companies, or bid for the equipment itself, driving up the price of equipment. New investment will fall off, raising the marginal efficiency; it might seem that the rising price of equipment would lower r^* until the two are equal. This would be to misunderstand marginal productivity theory, however; r^* is derived from the production function and, given the assumptions of the theory, depends *only* on employment. (If this seems absurd, it is the fault of the theory.) But the lower level of activity will mean a lower transactions demand, and hence would indicate an even lower level of the interest rate. Further, since investment will decline, Y/K^* will fall, which should imply a lower price level. In other words, $r^* > i^*$ implies a general deflation, together with contradictory pressures on investment.

Next consider $r^* < i^*$. If new capital has greater earning power than old, why should anyone be willing to continue holding the old stock? It should be scrapped and replaced! Moreover, no one will wish to hold the old bonds. They will be dumped, and their price will fall, until the effective rate on them rises to i^*. Further, as debt becomes due, it will not be possible to roll it over – the interest rate is now higher than old capital's earning power. Hence there must be a rush to scrap the old stock and rebuild it. This appears to violate the model's assumption that K^* is fixed! Of course, all the investment activity should tend to lower the marginal efficiency, but this will not bring equilibrium. For there will be general inflationary pressure, which will tend to raise the transactions demand, and therefore further increase the rate of interest.

Of course, by accident the two rates could be equal. But even if they were, a change in any of the parameters of the system would make them unequal, and no forces would tend to restore the equality. And if they are not equal, the model is inconsistent, in a manner that affects decisions and markets essential to the core of traditional macroeconomics – investment decisions and the bond market. The general implication is that new capital competes with old, the cutoff level of the rate of return on new capital being set by the own-rate of interest of money. But the rate of return on existing capital is set by the marginal productivity conditions, which are independent of the arbitrage between money, new capital and old.

In fact, on a closer look, the money rate of interest does not seem to be the crucial variable. Investment will take place if the rate of return on new capital projects is higher than the marginal product of existing capital, and not otherwise. New projects will be developed, or existing equipment scrapped and replaced, until the marginal efficiency falls to the level of the marginal product. But once these two rates are equal, a lower money rate of interest will not be sustainable, and will not lead to additional investment. Further investment would lower the MEC, but further purchases of equity will not affect the marginal product, which depends only on the capital/labor ratio. Why would anyone hold bonds that paid less than new capital projects? So an attempt to lower the rate of interest would pump money into purchases of equity positions (which the model does not explicitly recognize) but will have no effect on real variables. (On traditional liquidity preference grounds, however, the rise in equity prices could raise the demand for money for financial transactions enough to pull up the rate of interest.)

Alternative approaches to the problem

The difficulty arises from a conflict between the Keynesian investment function and the neo-Classical production function. Each independently determines a rate of return, which, in equilibrium, must equal the rate of interest. But there is no mechanism to bring the marginal product of capital into line with the marginal efficiency of capital.[18] One response is to drop the independent investment function and work out a fully consistent neo-Classical static model.[19]

This suggests a return to the neo-Classical idea that investment consists of two parts: replacement, which simply maintains the capital stock intact, and net investment, which is the currently required change of the capital stock needed to move the economy nearer to its long-run equilibrium in relation to labor.[20] (Such an equilibrium, of course, need not be stationary, if the labor force is growing.)

The investment function could be defined so that it will not conflict with marginal productivity theory. Suppose that the MEC curve were defined to be identical to the marginal product curve derived from the production function. There is still no guarantee that the savings-interest curve will intersect the MEC schedule at the level of the rate of interest equal to the

[18] Keynes notes the conflict on p. 139 of *The General Theory*, where he observes, "The ordinary theory of distribution, where it is assumed that capital is getting now its marginal productivity..., is only valid in a stationary state. The aggregate current return to capital has no direct relationship to its marginal efficiency." Later, however, he observes (p. 151) that "there is no sense in building up a new enterprise at a cost greater than that at which a similar existing enterprise can be purchased (on the Stock Exchange)." This implies arbitrage between old capital and new. Earlier, Lionel Robbins noted a similar difficulty in connection with Schumpeter's theory of the interest rate (Robbins, 1930). Schumpeter argued that interest was a dynamic phenomenon; the interest rate could not be positive in stationary conditions, since it was determined by the interaction of savings and investment. Robbins retorted that the existing stock of capital generated a net yield, which was the reason for abstaining from consuming it in the present. Though Robbins does not appear to see it, this puts the problem in plain view: the existing stock of capital generates a net yield, that is, a rate of return, and saving and investment determine another. How are they to be reconciled? Even earlier, as Hayek pointed out, Wicksell had given inconsistent definitions of the "natural rate," one resting on marginal productivity and loanable funds, the other on average productivity (Wicksell, 1935, vol. 2; 1936; Hayek, 1932; for a commentary see Nell, 1967, esp. pp. 389–91 and notes 10, 11, 13).

[19] This is the approach taken by Wynne Godley (Godley, 1992). Godley has developed a consistent neo-Classical model, but besides eliminating the investment function, it proved necessary to significantly rewrite the consumption function. The only Keynesian feature left is the demand for money, and that is indistinguishable from a modified Quantity Theory.

[20] Hayek (1941) is particularly clear that net investment – capital accumulation – is simply the process of moving towards the final, stationary equilibrium (chs. 19, 20, esp. pp. 263, 268). During the process the stock of capital may have a rate of return different from that expected on investment – but the equilibrium rate, towards which all will move, will be that on the stock of capital in the stationary state. Haavelmo (1960, pp. 3–7) also contrasts the idea of investment as movement to equilibrium with "Keynesian" investment – and finds that it is difficult to portray investment as a continuous function of the interest rate consistently with the rest of neo-Classical theory.

marginal product of capital implied by the equilibrium of the labor market. The latter depends only on the capital-labor ratio, which has no influence on, and does not interact with, savings. And even if the intersection were at the correct interest rate, that rate still might not be consistent with the IS-LM equilibrium.

Sargent has developed q-theory in a manner that may appear to offer a way out. Investment depends not on the rate of interest, but on the difference between the rate of interest and the marginal product of capital (Sargent, 1979, ch. 1), that is, between the rate of interest and the rate of return on the existing capital stock. The larger the difference the larger will be investment, while any investment will raise the capital stock and lower its marginal product – in the long run. (The MEC is assumed to be the same as the marginal product of capital.) But if arbitrage or accident brought the rate of interest and the marginal product of capital together, net investment would cease, although net savings could still be positive. The system would have to contract until savings disappeared. This is inconsistent with equilibrium in the labor market (Keynes, 1936, ch. 16; Dutt and Amadeo, 1990, pp. 63–4).

With full information and well-developed financial markets, arbitrage will tend to eliminate any differences between the rate of interest on financial assets and the rate of return obtainable from existing capital. To make q-theory work, adjustment costs must be assumed, sufficient to permit a discrepancy between r and i. Otherwise investment will be driven to zero, a conclusion that suggests something wrong with the theory. But the fixity of the capital stock is not enough in itself to prevent bringing r and i into alignment in the short run. For example, when the marginal product of capital is greater than the rate of interest, r > i, bond holders will sell bonds and acquire equity in firms (or buy into partnerships, etc.). They can also buy second-hand capital goods. Imperfections in the markets for capital goods will not prevent this. Even if firms have difficulty adjusting their level of investment, because they face severe or rising costs of adjustment, the rate of interest will be pulled up by arbitrage. On the other hand, if i > r firms could shift their capital into bonds to the extent of their annual depreciation. They could also shift working capital. The effect would be to drive bonds up and lower the rate of interest.[21]

---

[21] Sargent appeals to the assumption that capital is fixed to each firm at each moment of time, thereby ruling out the existence of a perfect market in the existing stock of capital. On the other hand, he assumes that all and only the profits of the capital stock are distributed as dividends on equity, and that equity and government bonds (the only bonds) are perfect substitutes (Sargent, 1979,

(continued)

None of these movements are difficult or time-consuming. It is difficult to see how adjustment costs could be very great. Arbitrage will therefore tend to move the rate of interest to equality (allowing for risk, etc.) with the rate of return on capital, reducing investment to zero. The q-theory provides no escape.

Could the neo-Classical approach provide an adjustment process that would establish $r = i$, consistently with the rest of the model? If $r < i$, investment falls off; then contraction tends to reduce real income, so equilibrium in the labor market would employ too much labor; hence money wages and prices would tend to fall. This would reduce the transactions demand for money, and therefore bring down the rate of interest. This should reduce saving and stimulate investment, strengthening profits.

But the influence of interest on saving is ambiguous. Those households saving to accumulate a definite sum at a definite future time will have to save more with a lower rate of interest. The effect of interest on investment is likewise problematical when wages and prices are falling. In depressed conditions a low rate of interest may not stimulate investment, because the favorable implications for the firm's cost position may not outweigh the weakness of the prospects for sales. It is no use building at low cost, if the product cannot be sold in sufficient amounts to warrant the new capacity. Further, as Keynes also noted, a move to a lower general level of wages and prices implies a higher burden of

[21] (contd.)

pp. 7–8, 10–11). He also assumes a given rate of depreciation. Hence, contrary to what he says, firms can at any moment change their capital stock, by shifting their depreciation funds. The marginal product of capital governs the rate of return on equity; hence arbitrage will adjust the price of equity to make the dividend rate equal the interest rate. But this arbitrage will also affect the interest rate. The difficulty facing the "neo-Classical synthesis" consists in trying to determine the rate of interest by the supply and demand for money interacting with the supply and demand for savings – the bond market. The assumption is that there are only two assets: money and bonds. Hence by Walras's Law one market can be dropped. Sargent tries to evade the problem by claiming that bonds and equities are perfect substitutes; but his assumptions imply that the return on equities is determined by the marginal product of capital. If this is not equal to the rate of interest, 1 $ of bonds will not be a substitute for 1 $ of equity. Also since employment and the real wage are changing, this may cause the marginal product of capital to change (being a function of N/K, with K fixed), unless the production function is arbitrarily restricted.

debt, and therefore a squeeze on profits; a sufficient decline must lead to bankruptcy.[22]

Perhaps the most serious difficulty facing this suggestion is that, if contraction reduces income, and so employment, then capital intensity will rise and from the production function, the marginal product of capital will decline. So even though falling wages and prices will reduce the rate of interest, contraction will reduce the rate of return – the two rates will move in the same direction. (Moreover, shifts in depreciation funds would tend to keep the rate of interest in line with the marginal product of existing capital.) Hence there will be no forces pulling the rates together and no stimulus to investment!

Perhaps the problem could be overcome by writing a compromise function? For example, $I = d + eY + f(i^* - r^*)$. When $r^* = i^*$, investment due to the difference between the marginal product of capital and the rate of interest would cease, but Y, or perhaps, ΔY, or even $\Delta Y/Y$, could still exert influence. So investment need not fall off completely. But the two parts of this function are inconsistent. The level of income exerts its influence on expectations of future yields. But when $r = i$, expected future yields must be consistent with present; if not, this would mean that the final equilibrium had not been reached – either r or i will be changing. Making investment depend on dY or dY/Y is slightly different; it is saying that there is a desired capital/output ratio, and that investment will be necessary to adjust to that. But if the production function is well behaved, each capital/output ratio is uniquely associated with a rate of return – so there is no difference between achieving the desired rate of return and the desired capital/output ratio. Hence dY and dY/Y do not provide a separate motive to investment. Of course, as Marshall pointed out, and Solow proved, the equilibrium need not be stationary, so long as capital, labor (measured in efficiency units) and output are growing at the same rate. But the argument only holds because there is no independent investment function in Solow's model.

---

[22] As noted above, Patinkin assumes that the capital stock is rebuilt at the new price level, financed by new nominal bonds. Other writers have assumed that the stock of debt might simply be "rolled over" at the new level of prices and the new interest rate. This would imply that lenders would exchange lower nominal bonds, earning lower interest, for higher nominal bonds, earning higher interest! A more general approach waives all detailed discussion of real balance effects, and derives a general demand for "wealth," based on preferences and endowments, as a function of many variables, including the general price level (Niehans, 1978, pp. 237–9). But such a discussion simply evades the issue of what happens to existing nominal debt contracts when money wages and prices change.

To reconcile the investment function with marginal productivity theory calls for mixing two different approaches. The central insight of the Keynesian investment function is that a capitalist, industrial economy invests, not because it is moving toward a predetermined stationary state, or state of steady growth, but because expansion of capital is inherent in its nature – it has to grow to stand still, so to speak. Such growth is inherently volatile, and is the source of the system's fluctuations.[23] By contrast, the neo-Classical production function and labor market analysis assumes full information, easy adjustment to equilibrium and determines a stable position to which the demand side adapts.

Investment in Keynesian and post-Keynesian growth theories is *not* a movement to some envisioned final position, but a stage in an activity that could be expected to continue indefinitely. Investment, like progress itself, has no final destination, and does not adjust to equilibrium. Capitalists invest, just as consumers consume – because it is their nature. They compete for wealth and power, and investment is the means to both. Other economic variables adjust to investment.[24]

Investment, in Keynes's famous phrase, depends on "the animal spirits of entrepreneurs," and is inherently volatile, it is the driving force behind the system's development – as well as the cause of its depressions. Instead of giving up the Keynesian approach to investment, it might be better to reconsider the labor market and marginal productivity theory.

Toward a new approach

The neo-Classical synthesis is composed of two very different parts. The IS-LM system is demand-determined, whereas the labor market and production function make up a supply system in which the real wage supposedly measures the relative scarcity of labor to capital. It is admitted that this is "only a parable"; but nevertheless every effort has been made to show that the demand part of the system will adapt to the supply part. The model examined relied on interest rates; the Pigou effect has also been suggested, but it is not very plausible, and very little evidence of it

---

[23] In *Industrial Fluctuations* (1927) Pigou observed that capital goods industries had fluctuations of wider amplitude than consumption goods industries, and that their fluctuations normally preceded those of the rest of the economy. But he refused to draw any causal conclusions (pp. 13–14, passim).

[24] To be sure, Keynes thought it possible that, in a few generations, the marginal efficiency of capital might fall to zero – bringing "the euthanasia of the rentier." But this was by no means certain, nor was it the destination towards which the system was necessarily heading. Quite the contrary; were this to happen, it would be a new kind of world, requiring new patterns of behavior.

has ever been found. Yet logic and facts both argue that the adjustment should run the other way, from the level of aggregate demand – as determined by saving-investment interacting with the money market – to the labor market. If goods cannot be sold, it seems reasonable that workers should not be employed to make them.[25]

Even Patinkin admits that the interest rate and real-balance effects must be assumed to work very quickly, otherwise, "the adjustment process becomes a long drawn-out one. It cannot then realistically be assumed that firms will continue producing at an unchanged level, for this would require them to accumulate inventories at ever increasing levels." Hence they would have to curtail production, and so reduce employment. In other words, even from the pespective of the "synthesis" model, it is possible to argue that the adjustment will run the other way. But, of course, this means abandoning the doctrine that the real wage measures relative scarcity.

In reality markets have developed historically in such a way as to compel the adjustment to run from demand to output and employment. There is little evidence of the multiplier (or the accelerator) in the nineteenth century; on the other hand, fluctuations in prices and money wages were much more severe in both directions in that century than they were later. The evidence from technological history suggests that it was much more difficult to adjust employment and output under Craft conditions. The development of Mass Production made such adjustments easier, while keeping the productivity of workers on the line constant. But if a fall in demand leads to layoffs, then the households of laid-off workers will, in turn, reduce their consumption, leading to further layoffs, and further reductions in consumption (Nell, 1992, ch. 16).

Keynes employed a Marshallian version of the production function, in which utilization varied, while "scarcity" played no role. He then

---

[25] Patinkin recognized this, and developed an "out-of-equilibrium" analysis to deal with it. When aggregate demand failed to adjust, the actual demand for labor would be "off" the marginal product curve, and would depend on the demand for goods, although the exact functional form of this dependence is not spelled out (Patinkin, 1965, p. 319). "The involuntary departure of firms from their labor demand curve ... is the simple counterpart of their involuntary departure from their commodity supply curve" (p. 322). By contrast, in Bryce's early presentation of the *General Theory* the argument was straightforward (Bryce, 1935). The supply function of consumption goods "relate(s) the quantity of labor which will be employed in making consumption goods, N, to the expenditure on consumption goods, C," a similar definition being given for the supply function of investment goods. He adds, "it seems quite reasonable – that higher quantities of employment are associated with higher expenditure" (p. 136).

dropped the supply curve of labor. Employment was adjusted in the light of the real wage, but ultimately determined by demand, which reflected an inherently volatile propensity to invest. In this way the two parts of the model could interact, as shown in the text. And he was surely right. But his theory of money was inadequate. He had no theory of circulation and no way of relating circulation to productivity.

Afterword: the New Keynesians

A new variant of neo-Classical Keynesian thinking has flourished in recent years. "New Keynesians" have sought to show that typically Keynesian conclusions – the existence of underemployment and/or inflationary equilibria, the need for and effectiveness of government intervention, etc. – can be derived within a framework of competitive markets peopled with rational optimizing agents. This work tends to be ingenious and often provides suggestive insights into the way markets can fail. But it provides no help with the problems we are considering, for three reasons.

First, it draws on ideas which we have shown to be historically limited, when not actually invalid. For example, most New Keynesian works assume diminishing marginal returns and diminishing marginal utility. The "Capital Theory" debates demonstrated that the former is invalid in general; we have suggested a limited validity for the *Marshallian* version in the era of Craft-based factories. Similarly, demand theory requires an understanding of *households* and their changing needs. Diminishing marginal utility – or convex preferences – will not provide this. Finally, New Keynesians, following the neo-Classical approach, consider prices to be determined by "market-clearing" which they take to be governed by the equality of supply and demand. We have shown that prices should be understood to be set by the balance between the *rates of growth* of supply and demand.

Second, New Keynesians do not draw on Classical theory at all. They do not make use of the Classical equations for prices, nor do they consider what has been called the "grand dynamics" of the Classics – the growth and change of the economy in historical terms.

Third, as mentioned earlier, a large number of their best and most interesting ideas are caught in an historical paradox. They have advanced these ideas to explain various kinds of "market failures" – failures, that is, of the price mechanism to bring about market adjustment. Yet the conditions indicated were invariably stronger and more widespread in the historical era when the price mechanism appeared to function, than in the post-war era, when it is no longer visible. Two groups of examples will illustrate this:

First, the "economics of information":

> Asymmetric information: This has become one of the most general and wide-ranging New Keynesian hypotheses. It has been used in many ways to account for the stickiness of prices and wages, and more specifically for the negative slope of the short-run Phillips Curve, and the positive slope of the Aggregate Supply function.
>
> Search costs: These are often a consequence of asymmetric information. They are used to explain sluggish market adaptation.
>
> Menu costs: Again, menus would not be needed if information were costless, and evenly distributed.

In the era when telephones were few and far between, before the radio, let alone the TV, when the Sears catalog had to be delivered by the Wells Fargo wagon, all these informational costs must have been more serious. But that was the era of price flexibility.

Second, more traditional obstacles to market adjustment:

> Efficiency wages: Workers will be paid more in effect to prevent them from sabotaging productivity.
>
> Market imperfections: essentially developments of the model of imperfect competition.
>
> Limited rationality: generalized optimizing can be too complicated, require too much or too costly information, might take too long, etc. Hence rules of thumb or limited optimizing will yield superior results over a period of time.

Each of these has been advanced as a reason for "sticky" prices or wages, or as creating a rigidity that might lead to adjustment failure. But in each case it is easy to see that the condition would have been more prominent in the era preceding the present. In Craft conditions workers had more control over work processes. Market imperfections were greater when transport was slower and mobility less. Limitations on optimizing must have been greater prior to the invention of the computer and the development of computerized databanks.

The Keynesian insights were brilliant, but Keynesian modeling was flawed, a condition reproduced in the New Keynesians. It's time to see if we can do better. The next chapter will set out the theory of effective demand, and the relations between commodity, labor, and financial markets, in conditions of Mass Production, based on the Classical equations and the theory of circulation.

CHAPTER 13

Growth and cycles: financially constrained instability under Mass Production

One basic Keynesian insight is that the economy is demand-driven. Another is that it runs on money. Both are correct. But the first had to be recast in the light of transformational growth, and the second reformulated in accordance with the theory of monetary circulation. Now we can draw on both to develop an account of the cyclical pattern of growth of Mass Production.

Actual growth cycles around the trend. Booms and busts follow a more or less regular pattern, generated by the interaction of financial markets with the multiplier. The Mass Production economy is unstable, although not wildly so. But unlike the Craft economy, it is not a stable system destabilized by the monetary/financial system; on the contrary, the labor and commodity markets are highly volatile, and are partly *constrained* by the financial system. Moreover, the pattern of movement in the Mass Production economy is quite unlike the sawtooth pattern of the Old Growth Cycle; it is more like a succession of hills and valleys, and the reasons for this need to be explained.

Financial markets both constrain, and on occasion, exacerbate the instability that underlies the cycle. The cycle sets up systematic pressures on firms, which inhibit their ability to compete. In particular, firms find themselves carrying excess capacity, and excess inventory. They experience a rise of fixed costs associated with the development of bureaucracy. Their fixed equipment tends to be inflexible. It is intended to carry out long runs producing one particular design; orders of a different kind require expensive retooling and down time. Finally, strong upswings – which will reduce the burden of capacity and spread out the fixed costs – tend to favor labor, and lead to wage-price spirals that squeeze profits.

Moreover, the interaction between the financial and real aspects of the economy tends to promote financial innovation and the evolution of financial institutions. On the other side, it also promotes the development of new and competitively superior forms of organization of firms.

To curb the potential swings in output and growth, however, calls for countercyclical policy. Such policy has been indispensable in keeping the cycle under control – but it has also meant that capital has ceded its

654

dominion, has given over some of its prerogatives, to governments repre-senting the general public. Regulation in the public interest – curbing excesses and providing well-timed stimulation – keeps the system running within acceptable bounds, but it represents the abdication of power on the part of capital.

In the fully developed Mass Production economy there has generally been a strong and well-defined growth of demand, partly generated and managed by the State. This has tended to lead to clear-cut investment decisions. But a combination of market forces has nevertheless led to systematic fluctuations in investment spending, holding capacity utilization down, and leading to periodic crises.

The resulting pressures have led business to innovate, both technolog-ically, and organizationally. The technological innovation – the high-tech revolution – has eliminated the need for many levels of management, while at the same time, making it possible to control production from a distance, thus allowing for widespread outsourcing, and for the reorganization of firms. This, in turn, has led to new forms of corporate control, and new systems of management.

The relationship between the financial and the real aspects of the economy can be established in either of two "regimes": one in which $i < g$, the other in which $i > g$.

The changing character of the business cycle

By the mid–twentieth century methods of Mass Production had been widely adopted among the advanced countries. As a result firms were able to adjust to variations in demand much more easily. In the face of a drop in sales, assembly lines could be temporarily shut down, or run on short time. Workers could be laid off, or put on short workweeks. An unusual rise in sales could be met by continuous operation over weekends, a full third shift, or other forms of intensification. Mass Production methods allow for considerable expansion and contraction of current output with-out much variation in unit costs. The typical average and marginal cost curves for a Craft economy first fall and then rise, with a well-defined minimum point. The typical average and marginal cost curves of a Mass Production system may fall a little at very low levels of output, and rise at very high, but in between, in the large normal operating range, the curves will be flat. There is no minimum point; instead there is the normal operating range (Lavoie, 1992, pp. 126–8). Variations in demand no longer necessarily brought variations in prices.

One consequence is that prices can be held steady at the benchmark level for variations in demand that fall within this range of constant variable costs (Carlton, 1986). Previously a rise in demand brought a fall in the real wage in the short run; this tended to have a stabilizing influence. Under Mass Production variations in demand are magnified. Previously a rise in the growth of demand called for a rise in the rate of profit, through a lower real wage. But under Mass Production a rise in the growth of demand will tend to lead to a rise in productivity, and consequently to a rise in the real wage.

The advent of Mass Production changed the economy in other ways. Firms had to be much larger in relation to their markets; moreover, technical improvements became a matter of competitive effort, and hence best-practice methods changed regularly. Firms no longer planned to achieve an optimal size. They retained earnings and reinvested them in improvements and new and superior equipment. As a result firms ceased to be family affairs and became hierarchical corporations, managed by professionals, with ownership determined in the "market for corporate control."

Once technological advance became a part of competition, firms could no longer continue to operate their initial plant unchanged. Nor could they permit new firms to enter, with all-new and superior plant and equipment. Existing firms plan to expand in pace with the growth of the market, adding enough capacity to service the new customers each period. Thus each firm tended to operate a mix of equipment of all vintages.

Expectations and confidence

In the Craft economy expectations pretty much had to be adaptive. Not enough was known about how markets adjusted, or about the causes of booms and crises, or the pattern of cyclical behavior (which was not studied in statistical detail until the early twentieth century; cf. Burns and Mitchell, 1946), for businesses or households to form their expectations guided by statistical regularities, let alone forming them in accordance with a theory. The best guide to the immediate future of the market appeared to be the recent past, modified in the light of long-term experience, and perhaps the current direction of change of key variables. In short, adaptive expectations.

But sometimes there is no good ground for changing expectations, for example, of sales or price, from what they have been to some other specific values, yet, on the other hand, there are signs either that all is not well, or that a groundswell is building for an upswing. *Expectations* will not change, since there are as yet no grounds on which to base a determination of new values of the variables, but the *confidence* with which they are held will decline, since in each case premonitions of change are felt. In Craft conditions such premonitions must remain vague; when conditions do change, in a way evident to all, then the shift in expectations will be sudden and general. This will help to bring about the rapid collapses and booming upswings characteristic of the era.

By contrast in Mass Production there will be a great deal more information, and detailed theories on which to base planning. Expectations can be formed in greater detail, and the emerging signs of change – leading, current, and lagging indicators – will provide statistically precise reasons for the confidence with which these expectations are held. On the other hand, the greater flow of information has also made it clear that the cycle is underdetermined in many respects. In particular, exactly when it will change direction cannot be foreseen with reliability. Apparently equally good reasons can be advanced for different views. A distribution of beliefs can be expected, of more or less normal shape, centered around the correct turning point. Those who are too early in anticipating a downturn, for example, will lose profits they could have made; those who are late will take losses they could have avoided. (However, if there is a general skewing to the early side, this could tend to speed up the downturn; if to the late, it might slow it down.)

Procyclical movements

The real side of the Craft economy tends to stabilize itself. That is, a movement in one direction or another, if it comes to overshoot the normal position, will tend to generate a correcting movement in the opposite direction. By contrast the multiplier – especially in conjunction with the accelerator – implies a procyclical pattern of movement of employment and output. Movements upward or downward generate further movements in the same direction. One striking aspect of the modern economy is how widespread this procyclicality is.

Chapter 11 examined the procyclical behavior of output, employment, and consumption, as a result of the multiplier. The same pressures also extend both to investment spending and, if somewhat less directly, to investment plans. But procyclical forces also operate in regard to productivity growth and real wages. And they act on inflation. And we shall see that important financial variables are also procyclical. Risk, for example, tends to rise at higher levels of activity, pulling up interest rates. As firms earn more and grow faster, stock prices appreciate faster, attracting funds from the bond market, weakening bond prices and raising interest rates. As interest rates rise, fixed costs rise; we shall see that real fixed costs also tend to rise, accelerating at a certain point.

Two consequences should be noted. First, such widespread procyclicality suggests that economic activity could be raised to a high level of activity and maintained there, provided policy ensures an initial movement in the right direction. By the same token, even a small initial downward movement could set all forces working to establish and maintain a depressed state of activity. The second point, however, offers a corrective. The procyclical movement of financial variables – and others, like real wages – implies that important *costs* move procyclically. Such cost variations suggest that procyclical fluctuation may generate countervailing forces, which might eventually undermine or reverse the movement. There is also the possibility – likelihood? – that the strong movements of output and employment will *overshoot* the normal levels implied by the normal rate of growth of demand. As we saw in Chapter 11, when actual growth tends to outrun normal, the wage-price spiral will tend to squeeze profits. Taken together, these provide the outline of a cyclical mechanism: an upward movement first reinforces itself, then overshoots normal levels, while at the same time driving up costs, until these form a barrier to further expansion, leading to a downswing, which then reinforces itself, overshooting again, but bringing a decline in costs, until a lower turning point is reached.

To develop this idea will require spelling out a number of behavioral functions, which, in turn, are set in the context of the structure of the system as described by the Classical Equations. However, it will be useful to recall that these behavioral functions are not, in general, based on presumed "maximizing decisions" made by "rational agents." Instead the changes in behavior come about because changes in macroeconomic variables lead to changes in the status of many agents – with consequences for their behavior.

Changes in the status positions of decision-makers

The basic change comes with the working of the multiplier: workers move from employed to unemployed, or vice versa. As a result their income changes from welfare payments/unemployment compensation to normal wages, leading to a corresponding change in the spending of their household. This change in spending then shows up as altered revenue for business. This, in turn, may affect the credit ratings of various businesses, changing their status with respect to banks and financial markets. In response to these developments, financial markets can be expected to react, leading to adjustments of portfolios, which, in turn, may lead to changes in the values of financial variables – ultimately reacting back on the level of activity.

Circulation and financial markets

For example, a downswing will lead to reduced earnings in business; marginal firms, and even otherwise strong firms with high debt-equity ratios, will find their credit rating reduced. As a result they will experience difficulties and higher costs in raising funds; they will have to cut back on their expenditures. On the other hand, in an upswing many firms will find their credit ratings improved, so that it will be easier and less expensive for them to obtain funds.

Components of a model of the new business cycle

Investment: decisions versus spending

The business firm is the central player in the story of the cycle. Business decisions as to prices and investment set the stage; business spending implementing those plans determines the I component of aggregate demand, while the firm's current employment and output decisions establish the wage bill, and thus largely fix household consumption, the C component of aggregate demand. By contrast to business, households and banks are comparatively passive.

The spending on investment, to implement the decisions, will be determined by balancing the earnings firms can reasonably expect, given the expected level of activity, against the earnings required to safely carry out a level of investment. Two economy-wide or aggregate functions can be constructed, relating the current (realized) rate of return on present capital to investment spending on the acquisition of new capital, in relation to existing capital. (The axes are therefore r and g.) Capital is valued at the supply price, but we assume that the investment being carried out reflects implementation of investment decisions made by balancing capital asset demand prices – the expected streams of quasi rents – with the corresponding supply prices. Hence the value of the capital stock is well defined.

One aggregate function will show the rate of return on the current capital stock generated by investment spending, through the multiplier; the other, the rate of return required on current capital in order to support a level of investment, consistently with firms' other obligations. These two are differently constructed. The first shows the results of an aggregate process – sales resulting from the multiplier – distributed over the population of firms. The second, however, is the aggregation of the individual calculations of firms – each firm considers what rate of return it would currently require in order to feel safe investing a certain amount. (Conversely, given a current rate of return, how much investment would it feel comfortable undertaking?)

More specifically, in the first of these relationships, which we can call the rE function, the rate of return generated through the multiplier by different levels of investment expenditure, will rise from left to right. It will begin from a positive intercept expressing fixed income consumption and autonomous investment, including government spending on infrastructure. Note, however, that autonomous investment and government spending will tend to generate growth, especially productivity growth.

This must be marked off on the horizontal axis. Autonomous spending, measured as r_0 on the r-axis, gives rise to productivity growth, g_0, on the g-axis; but thereafter, investment spending, measured as g, gives rise to realized profits, r. The direction of causality is different in the two parts of the rE function.

The slope of the rE rises from left to right, reflecting the generation of profits by investment spending. It would be reasonable to take this slope to be unity, although new investment when completed will outcompete some existing firms, and can be expected to cause losses to them. (As the spending on new competing projects takes place, the losses will be anticipated, and the capital stock of firms using inferior methods will be devalued. This should be subtracted from profits.) Hence the rE will start from a positive intercept, and rise from left to right with a slope of 45 degrees, or perhaps a little less. (The growth rate on the horizontal axis is investment spending divided by the capital stock; in the long run the losses will be reflected in slower growth or decline in the industries in question, reducing the realized growth rate. But that has no effect on spending, whereas the anticipated loss will lead to profits being marked down.)[1]

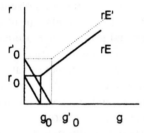

The effect on productivity growth of the spending that gives rise to r_0 has to be marked out on the g-axis, g_0. The rE then rises from the point

---

[1] When g is adjusted for the declines in output that correspond to the projected losses then it will tend to move closely with r. They will differ because of the intercept of the rE, which represents various kinds of "capitalist consumption," supported by private portfolios – foundations, universities, hospitals – sometimes referred to as the "Grants Economy." (Salaries, management bonuses and plush offices must be subtracted as costs of doing business, or as expenditure out of monopoly rents.) As mentioned earlier these represent nonbasic consumption from the point of view of individual agents, but they are basic collective investments from the point of view of the system as a whole. They are not purchased to be entered, directly or indirectly, into the production of any single good, but the outputs of these activities do enter indirectly into the production of all. In addition, pension fund earnings should be removed, since they are part of deferred compensation. Making these adjustments would reduce the intercept (nearly) to the origin.

(r_0, g_0) with a slope reflecting the generation of profits by investment spending. An increase in autonomous spending, for example, by the government, giving rise to r_0', would require marking out the impact on productivity growth, g_0', before drawing in the new, higher, rE line, shown as rE'.

The second function, IF, for "investment finance," shows the rate of return which must be currently earned for firms to feel justified and secure when carrying out such investment spending, while meeting their various fixed-cost obligations. Any spending commitments carry risk, and have opportunity costs. To justify such commitments, firms must feel that their current cash flow will be sufficient to cover the risks they imply, *given their importance*. That is, firms can be expected to attach a degree of importance to the various stages in implementing their investment plans. Each level of investment spending represents a different level of implementation of investment plans. The different stages may all be considered equally important, or they may carry very different weights, some urgent, some less so. For example, all components of investment might be considered equally important; higher levels of investment would then carry *proportionally* higher weight. On the other hand, low levels of investment might cover spending that the firm felt absolutely must be done, while higher levels might include projects whose urgency is less immediate. That is, there would be a large risk or anticipated loss from *not* carrying out certain spending, but delays in other projects would be less costly, or not costly at all. Thus as the level of investment spending rose, the urgency at the margin would decline. For low levels of investment, with high urgency, only the minimal current cash flow would be required to cover them. But as marginal urgency declines higher levels of investment spending would require greater marginal coverage – the extra risks would be less worth taking. In the first case, the cash flow required, represented by r, would rise at a constant rate with the level of investment, whereas in the second case it would rise at an increasing rate.

Let us take the case of constant importance first. The function will begin from a positive intercept, reflecting fixed costs, and then rise with a constant slope, showing the rate of return minimally required to support each level of investment spending. This rate of return will not have to cover the full costs of the investment – which would imply a slope equal to that of the rE. The possibility of external finance permits it (initially) to have a shallower slope than the rE. A rise in investment spending requires an increase in current earnings to underwrite it, but the increase is less than the rise in spending. Only a fraction (usually a majority) of current investment spending comes from retained earnings; the rest can be borrowed, usually with a minimal outlay in expenses. Given a debt/equity ratio, the slope will be constant, and if the ratio is maintained (and the interest rate is unchanged), the intercept will remain the same from period to period, as fixed costs will rise at the same rate as capital.

These two together form an unstable system, as can be seen in the diagram. (If IF lies wholly below rE, the system will be unstable upwards; if they cross, the intersection will be unstable in both directions.) However, the effects of fixed-cost obligations must be brought into the picture; as current spending on investment projects rises, after a point, the possibility must be considered that current sales might falter. Risks would be greater. Moreover, as noted earlier, the urgency attached to some of the components of higher levels of investment will be lower. Hence the IF curve will turn up and rise with increasing steepness, indicating the extent to which the rate of return must rise to offset the increasing risk attached to progressively higher levels of investment spending.

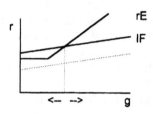

The intercept of IF will change as fixed costs – financial obligations and contractual managerial costs – are adjusted to reflect current market conditions. As productivity changes, affecting inflation (as outlined in Chapter 11), the position of the rE function will change. Productivity changes may also affect the IF. These shifts in the two curves determine the movements of output and employment, and the interaction traces out a simple cycle. But first we need to explore the curves in more detail.

Investment: increasing risk

It is common to define two kinds of "increasing risks," borrower's and lender's. Borrower's risk is a subjective judgment that reduces the expected value of a stream of quasi rents and rises with investment. Lender's risk manifests itself in bankers' demands for higher rates or shorter maturity dates; it also rises with investment and will shift with changes in debt/equity ratios. There are two problems with this approach. First, it is usual to apply both to the calculations involved in the plan – the Investment Decision. Lender's risk, certainly, and a practical version of borrower's risk should be considered at the stage of implementation, as part of the cost of investment spending – since the risk is the risk of failing to meet obligations on time. Second, though in a general way reasons for increasing risk are given, few writers have explained the rate at which the risks increase, or whether this will stay constant. Moreover, many draw diagrams with risk rising at an increasing rate! This is never explained at all. Yet it is quite plausible, provided we shift the focus from plans to implementation.

Increasing risk can be explained by two factors. First, there is the probability of failure, that is, that the firm will not be able both to meet its continuing obligations and carry out the program of investment spending. This depends on the expected level of activity in its market, and the distribution of sales among the firms in that market. If this distribution were reasonably random we would expect a (no doubt skewed but still) bell-shaped curve, relating levels of sales to the number of firms achieving that level. Now consider the level of sales that will just cover the firm's fixed obligations plus its projected level of investment spending – given its leverage ratio. The cumulative frequency of firms failing to reach that level of revenue, as a percentage of all firms, is then a measure of the probability of failure. Since the cumulative frequency curve has an S-shape, it is apparent that this probability will rise, and rise at an increasing rate, over a considerable range of investment levels. Firms' expectations will be based on this, since it can be assumed that they have a good idea of how fluctuations in demand affect their market.

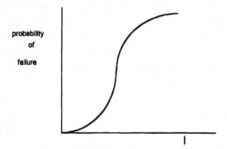

The lower the level of sales, the higher the frequency of firms failing to reach the revenues required to meet their obligations. Since the problem is to determine the rate of return required to sustain a *current* level of investment spending consistently with meeting fixed obligations, the level of sales in question is that immediately ahead. Sales are running today, this week, at a certain level, which is expected to hold for the rest of the quarter, or year, unless firms have good reasons to think otherwise. Firms must be assumed to use all the information their sales and marketing staff can come up with to determine their expectations. In this sense their expectations will be "rational."

But it does not follow that errors will be random, for the level of sales will not be the result of a deterministic process subject to random shocks. On the contrary, the model here is subject to sudden and partially – but not fully – predictable switches in the direction of movement of major variables. Many changes will not be random, but they will not be altogether predictable, either. Expectations will therefore be subject to uncertainty, and different agents can reasonably come to different conclusions.

In these circumstances the "state of confidence" becomes important. All available information may indicate that sales should continue strong in the immediate future; but the bloom is fading from the boom. There is no hard evidence of troubles ahead, but the time is ripening, and the downswing could begin any moment. So the expectation of strong sales will be held, but with weakened confidence. Conversely, when the slump is over and the upswing begins, confidence will improve. The effect of changes in the state of confidence is to increase or decrease the curvature of the IF. When confidence falls, a higher *subjective* risk premium must be added, increasing the upward curvature; when it rises a subjective factor can be subtracted from risk, and the curvature will be flattened. (Any weakening in the normal rate of growth of demand, any weakening of confidence in it, any widening of the variance in expectations, will be reflected both in an increase in the curvature of the IF, and in its propensity to shift with changes in the rate of interest.)

The other factor is the cost of failing to meet obligations. If revenues fall short, either fixed obligations or contracted payments for the investment project will have to be postponed or renegotiated. In either case penalties, legal fees and/or emergency borrowing will be required. These costs will be larger, the larger the sums involved; hence they can be expected to rise with the planned level of investment spending.[2]

The sum measuring risk, then, is the product of the probability of failure times the anticipated cost of failure, and the rate of return that would compensate for risk, is that sum divided by the capital presently in the market. Notice that even if both the probability and the cost of failure rose at a constant rate with planned investment spending, the sum at risk, their product, would rise at an increasing rate. An example: Let prob-

---

[2] At low levels of investment the disruption to a firm of having to halt spending on an investment project may not be very great. But larger projects will involve more of a firm's management and affect more of its current operations; a break will therefore disrupt a larger *proportion* of the firm's activity. The example below in the text can be reinterpreted: the impact on the firm rises linearly with I (100, 200, 300, 400), but the proportion of the firm's activity affected also rises linearly (10%, 20%, 30%, 40%). Then at higher levels of I the cost of disruption will increase more than proportionally – 10, 40, 90, 160.

ability of failure rise linearly with I; as I rises: 1, 2, 3, 4, the probability rises: 10%, 20%, 30%, 40%. Let cost rise linearly with I; as I rises: 1, 2, 3, 4, the cost rises 100, 200, 300, 400. Then "risk," meaning the product of probability of failure multiplied by the anticipated cost of failure, rises with I (1, 2, 3, 4): 10, 40, 90, 160 (cf. Nell, 1992).

Risk can now be added to the IF curve, causing it to turn up and rise at an increasing rate with investment spending. Moreover, as noted above, the marginal urgency of higher levels of investment will be lower, so that the cash flow coverage required will be proportionally higher. Putting the curves together – taking the origin as (r_0, g_0) to simplify the diagram – there are three possibilities, depending on the position of the rE with respect to the IF. Taking the top rE line first: the IF could cut rE from below; the second rE line illustrates that IF could intersect rE twice; while the third shows that it could be just tangent from above.

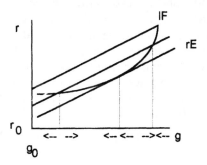

When the required rate is above the market-generated rate, spending will be cut back; in the reverse case, spending will be expanded. Hence when the IF cuts the rE from below, investment spending is stable. In this case, it is clear that financial markets constrain an otherwise unstable system. When it intersects from above, however, investment will be unstable, and the tangency point is unstable downwards.

Notice that when the IF cuts the rE from below, the implication is that if growth is low, the system will expand in a boom, whereas if g is high and the IF is tangent, there will be a downswing. These movements have implications for inventory policy. If there is a possibility of an upswing, firms must have inventory at the ready. For as the system swings up shortages will develop. If a given firm, waiting to be sure, has failed to stock up, and so, as the boom takes off, lacks inputs where its competitors have them, it will lose markets. Similarly, when a downswing looms, firms must be careful about cutting inventory. If, anticipating a slump, a firm tried to run lean and turned out to be wrong, they would lose markets to those who didn't. On the other hand, if the slump does come, those who cut back early may weather it better, but they won't gain any ground on competitors. Inventory mistakes are likely to be more costly, or costly in a more permanent way, than is justified by the gain from being right. On the whole firms are better off carrying inventory that is likely to prove excessive.

So far we have considered investment on the basis of expectations of the revenue generated by the anticipated level of current spending. Clearly the higher this is, the lower the probability of failure and so the lower the risk; hence the higher will be investment spending. But there is another element to consider: the rate of interest, for this will affect both the level of fixed costs – the intercept of the IF – and the costs of failure, a determinant of its slope.

The rate of growth and the rate of interest

Interest rates will come under pressure from the equity market. This is a consequence of the Keynesian arbitrage outlined earlier. Suppose firms invest their retained earnings, and that the market considers the new capital equal to the old in profit potential. If no new shares have been issued, then the old shares should rise in price in proportion. Let N be the number of outstanding shares, $\$$ the price of one share, K the value of capital stock, F fixed obligations, I new investment, and D new indebtedness. Then $N\$ = K - F$ and $Nd\$ = dK - dF = I - D$. Hence $(\$N/K) d\$/\$ = I/K - D/K$. Define v, the valuation ratio, as $N\$/K = 1 - F/K$. Then

$$\frac{d\$}{\$} = \frac{g - D/K}{v} = \frac{I - D}{K - F}$$

But if the ratio of new indebtedness to investment is the same as the ratio of fixed obligations to the stock of existing capital, $D/I = F/K$, then $D/F = I/K$, and this in turn equals $(I-D)/(K-F) = g$. That is,

$$\frac{d\$}{\$} = g \quad \text{(cf. Chapter 6)}$$

So equity price appreciation will reflect the growth rate resulting from investment spending. Holding equity is an alternative to holding bonds; bonds pay interest, equity yields dividends plus appreciation – or in the case of pure "growth stocks," just price appreciation, in proportion to investment, and the market's judgment of that investment's wisdom.

If $d\$/\$$ is greater or less than i, it will make sense for portfolio managers to shift funds accordingly, within the bounds of normal risk. But such arbitrage may not bring d$/$ and i into equality with one another. For while a shift of funds out of bonds into growth stocks, when d$/$ > i, will lead bond prices to fall, and so i to rise, it will also contribute to faster appreciation of the growth stocks. The two rates will move together; they will tend to track one another, but they need not move closer together. The disparity may be perpetuated, or the stock prices might even accelerate. In the same way a shift of funds from growth stocks to bonds may lower i, but may also lead to a serious collapse of the stocks, so that d$/$ might even turn negative. Such speculative indeterminacy implies the possibility of overshooting, and stands in the way of a simple link between interest and investment.

In sum, a given level of investment spending, once established, will result in share price appreciation; this in turn will affect the rate of interest through arbitrage. There is, therefore, a two-way interaction between the rate of interest and investment. Interest affects both the intercept and the slope (curvature) of the IF curve; in this way it affects its intersections with the rE line. Higher rates of interest will reduce investment spending, lower will increase it. On the other hand, after a lag, the rate of interest will adjust to the rate of growth, although speculation may substantially delay and distort this adjustment.

This has a double significance. On the one hand it confirms the view that the Mass Production economy contains important tendencies to instability. On the other, however, it confirms that market pressures exist which tend to move the interest, profit and growth rates in line with one another. In Chapter 5, we saw that this was a condition for the working of the circulation system; in Chapter 6 we saw that arbitrage led to tracking but not to equality, and we shall now see that arbitrage plays an important role in the Mass Production cycle.

Investment: money and the banking system

Let us draw on the theory developed earlier. Consider a simple banking system financing current operations, with, initially, no assumptions a Central Bank. (When it enters we shall treat it as a lender of last resort, with responsibility for preventing runaway speculation – which may require "leaning against the wind"; see L. Currie, 1934, 1968.)

Commercial banks finance production and household spending, while investment banks provide the funds to underwrite the portion of investment spending that is not financed by retained earnings, in particular, the finance fixed capital (Chapter 7). As we showed earlier a "free banking" system, a system of competitive private banks, can always provide the circulating medium needed, and can finance any feasible level of investment spending. But such a system will tend to be unstable, so it needs to be supported by a Central Bank, acting as lender of last resort.

Commercial banks accept deposits from firms and households and make short-term loans to firms to finance the wage bill, and to households to finance consumer spending, such as on durables. Investment banks make long-term loans (buy bonds) to underwrite fixed capital construction. Commercial banks deposit their reserves with investment banks. Investment banks lend a portion of these deposits and hold the remainder as reserves (or deposit them with a Central Bank/lender of last resort).

Commercial banks provide the wage funds equal to the wage bill of the capital goods sector. These funds complete the circuit, including the "short circuit," as described in Chapter 5. Investment banks underwrite the sale of securities for fixed capital goods; these funds are circulated as outlined at the end of Chapter 5 and further explained in Chapter 7. Earnings during the period from banking and financial activity will be added to financial capital, permitting a larger volume of lending. Hence additional funds will enter the stock market as portfolios are updated in the financial circulation, increasing stock prices in proportion to investment.

A bank-dependent system reduces the need for actual money, paper or metal. Checks serve in place; a small amount of high-powered money must be kept in reserve, as a precaution against a run. The amount of such reserves can be quite small, if the system is efficient in making it quietly available in case of need. But reserves are created by deposits; not the other way around. It is activity – circulation – that generates deposits. Banks make loans, deposits flow in, and banks retain reserves. The important point is that the *system* should have adequate reserves and proper oversight by the lender of last resort.

In short, given the level of interest rates, the banking system will supply credit-money to business (and households) as they need it to carry out the level of activity mandated by the investment spending plans of business. Changes in the money supply neither cause nor require changes in interest rates, in the short run, contrary to conventional theory. Changes in the money supply *reflect* spending; they do not cause it. Moreover – on these assumptions – money simply accommodates.

By the same token, however, a solvent banking system is essential to the smooth working of the system. For if banks fail, the supply of working capital dries up, and production will be constricted (as in the Great Depression, Bernanke, 1993). Speculation in financial markets and over-lending are both dangers which the Central Bank must guard against. Further, when the banking system is weakened, it will be advantageous to keep interest rates high, to provide a cushion of earnings.

The fragility of financial institutions

It has been argued, by Minsky and others, that the instability of the economy is a reflection of the instability of its financial institutions. Banks and other financial centers grow more fragile in booms, as they optimistically take on excessive debt, leading to failures and disorder during the downswing. This was certainly common during the Craft era. But it is important to understand the nature of that fragility. There is an older tradition, which has always held that "banks" were less secure than "firms," meaning they were less able to survive a downturn.

This can be illustrated with a diagram. On the vertical axis plot the percentage change in net earnings, and on the horizontal the percentage fall in sales due to the downswing of the business cycle. Draw in a 45-degree line; along this line the decline in earnings exactly reflects the decline in sales. Above it, however, the effects of the decline have been compounded – a drop in sales causes a larger drop in net earnings. Below it, by contrast, the drop in sales has been partially offset, so that the decline in earnings is less. At the appropriate level on the vertical axis, mark the bankruptcy point, and draw a horizontal to the 45-degree line. Reading down to the axis, that will give the bankruptcy level of sales decline, if there are no offsets.

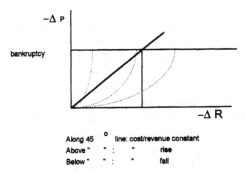

We can now define "robust" and "fragile" institutions. A fragile institution works in such a way that the effects of a decline are compounded, so that its earnings-loss curve lies above the 45-degree line. A robust institution is one capable of carrying out offsetting strategies, so that its earning-loss curve typically lies along, or even below the 45-degree line. These cases are illustrated by the dotted lines.

Firms in the era of Mass Production will tend to be robust, lying along or even below the 45-degree line; in a downswing they can adjust output and lay off labor, reducing their costs in proportion to the decline in sales. If they have market power, they can raise their markups to their established customers (since in a downswing they will most likely not be trying to break into new markets). They can delay replacements, since business is poor, anyway. By contrast, firms operating Craft technologies will be less able to adjust output and lay off labor, since that would entail breaking up work crews. Hence they will have to absorb most of the loss in their profits and thus would lie well above the 45-degree line. However, by the same token, in boom times, most of the revenue from above average sales goes into profit. Hence they could accumulate funds in good times to tide them over bad (Broehl, 1959).

Banks and financial institutions, however, are vulnerable on both counts. They cannot easily lay off workers; virtually all their running costs are fixed. They have little market power and cannot easily impose higher interest rates in a slump. Moreover, they will have to be especially careful about lending in hard times, since risks are higher. Yet at the same time macroeconomic forces and policy considerations are likely to combine to reduce interest rates, squeezing their margins. To maintain reserves, as the recession reduces activity, they may be forced to call in loans, further reducing their earnings. If word spreads that they are in trouble, they may face a run, which will force them to liquidate their earning assets, or take on heavy costs to meet withdrawal demands. However, unlike Craft firms, it is not so easy for banks to build up precautionary reserves in boom times. The Craft economy increases sales somewhat in a boom but chiefly faces higher prices, producing greater profits. It does not necessarily invest more; hence the profits can be accumulated as a reserve fund. But in a boom, a bank is called on to make more loans; these will bring more profits, because rates will be higher, but the profits, in turn, will be required as insurance premiums since the increased volume of loans also embodies higher risk – as the downswing will shortly prove.

Thus banks and financial institutions will tend to lie above the 45-degree line and will be more vulnerable in a downswing than firms. This is an important reason for having a Lender of Last Resort. Even so, as we shall see, this vulnerability will tend to intensify the downswing and delay recovery.

Inflation

In a demand-constrained economy, inflation originates in changes in costs. Two types of inflation were examined earlier. There is, first, a persistent low-level upward drift in prices, which stems from the pressures to maintain socially mandated wage and salary differentials. Secondly, there are wage-price spirals that arise from determining through the market which groups will bear the burden of a once-for-all cost increase, such as an oil price shock, or a devaluation. Like many dynamic processes these may gradually peter out, go on indefinitely, or accelerate.

While prices of manufactured goods are normally insensitive to changes in demand, primary products may respond to large or sudden changes. An unexpected, sudden or abnormally large demand increase may drive up primary prices, and this increase will then be passed along as cost inflation, setting off a wage-price spiral. When demand declines, especially if it collapses rapidly, primary products prices may sink, tending to dampen inflation, but seldom leading to an actual fall in the price level.

(Financial innovation may increase the sensitivity of futures markets to the effects of demand pressures on primary products. A rise in world demand will tend to pull up commodity futures, and speculators will buy now for delivery at the higher future price, thus driving up the current price.)

As we saw earlier, one area of manufacturing, machine tools, retains Craft-like characteristics. Here sharp increases in demand will be difficult to accommodate without a rise in unit costs. Hence prices will rise and these price increases will then be passed along as cost increases in all other sectors, since machine tools enter directly or indirectly into the production of all goods. A sharp increase in demand affecting machine tools will therefore tend to set off a wage-price spiral.[3] But the effects of increases and decreases in demand are asymmetrical: a fall in demand will generally not bring prices down. In machine tools, production is to order, and is customer-specific, so that there is little direct price competition. Moreover, when times are slack, machine tools firms will shift into other activities, to maintain their revenues.

Inflation also serves as the market process determining which groups shall bear the burden of increased costs. This process takes the form of a wage-price spiral, in which groups try to pass along the cost increases they experience. (In a resource-constrained economy inflation is the market process by which it is determined which groups shall bear the burden of the shortages.) When a cost increases (say primary products rise in price) the affected industries pass along as much as they can in higher prices. Consumers thus face a rising cost of living, and so demand higher wages and salaries, further raising costs to business, which in turn are passed along again in price increases, to the extent possible. But the ability to raise prices or wages depends on the respective market positions of business and labor. Those who are relatively most successful, round

---

[3] This price increase may be temporary. Fewer projects will be ordered, so more machine tool capacity can be devoted to the construction of machine tools to increase the sector's own capacity. Once this capacity is built up, the sector can deliver a larger output and hence can meet demand at normal prices. In the meantime, however, the higher machine tool prices will have driven up costs throughout the economy, setting off a wage-price spiral.

after round, will escape most of the costs, which the least successful will have to bear. As prices and wages rise, however, the burden is lessened in real terms, and the wage price spiral peters out when the reduced burden has been distributed between business and labor in proportion to their inability to pass it along. The more evenly matched the market positions of the various players, the longer the process will continue, and the lower the final burden to be distributed.

General inflation lowers the burden of nominal fixed costs; it also tends to lower real interest rates. Once it comes to be expected, projects, especially large ones, may be speeded up to take advantage of current prices, anticipating higher prices later. Both effects cause the IF to shift down and flatten. Profit inflation will intensify this. Both general inflation and profit inflation thus lead to expansionary changes in the IF – but if profit inflation lasts for long, it may have an inhibiting effect on investment decisions, since the corresponding squeeze on wages will make it harder to develop new markets. Wage inflation will squeeze profits, and cause an upward shift, and/or increase in the curvature, of the IF, but the rise in real wages will increase *employment* per unit of investment spending, and over a long period, the rise in working class prosperity will encourage the development of new markets. (If there is a tax on wages – see below – the rise in W per unit I will tend to raise the IF and swing the RE down.)

In general, inflation will develop with expansion, increasing as during the upswing, but in the early stages, it will tend either to be broadly neutral or to benefit capital. That is, prices will tend to rise as fast or faster than money wages. But in the boom, when actual growth is close to the "normal" growth of demand, and especially when it comes to overshoot normal growth, the ability of business to raise prices will be inhibited, and wage inflation will tend to outrun price inflation. For when actual growth exceeds the growth of demand, supply will tend to outpace the development of new markets, and businesses will be forced to compete for new customers. They will have to curb price increases. By contrast, demand for labor will tend to outrun supply, putting both unions and unorganized labor in a position to demand wage increases that fully match – and perhaps exceed – the rise in the cost of living.

Interest rates and growth: two regimes

Let us now reexamine the financial relationships here. So far we have considered r and g, on the understanding that the meaning of r on the IF was that it provided the funds that would support the level of investment spending expressed by g, which in turn would generate the level of r on the rE. But there is more to the support for g than the level of current earning; the terms of new borrowing and the levels and cost of current indebtedness must also be considered, and so far this has been left to one side.

To study these questions we need to examine the relationships between i and g, where i is understood to be the *real* rate of interest, the nominal market rate adjusted for inflation.[4] Two functions, similar to the rE and the IF, respectively, can be defined. In effect, instead of earnings generated or required on capital, we shall consider interest on debt, and one function will show the interest business can afford, for each level of g, while the other will show the interest banks will feel they need to charge, to fund each level of g. Both will be defined in relation to the basic rate of interest expected to be "normal" in this period – or if there is no such rate, in terms of the current prime rate.

In the first, loosely analogous to the rE, we have a function showing the interest rate firms would be able to pay in order to fund each level of g, given the profit generated by that level of g. This function – call it the iFP, "interest firms pay" – will consist of two parts: First, a constant term showing the interest rate for which the firms planned, in the light of the expected normal rate of growth of demand. Second, to this will be added the difference (positive or negative) between the rE and the IF, for each level of g. Suppose the rE rises at (near) 45 degrees, and that the IF begins below it, rising slowly, and then in view of increasing risk turns up sharply to cut the rE at a high level of g. Adding this to the normal rate expected when investment plans were drawn up (and which is therefore covered by the pricing policy), gives a function that first rises slowly from a positive intercept, then falls sharply.

In place of the IF we have a function showing the interest rate banks and financial corporations would require to finance each level of g. Call this the iBC, "interest banks charge." This will likewise consist of two components. First there will be the basic "normal" rate, based on a markup over the Central Bank's lending rate. To this will be added a further

---

[4] The *ex post* adjustment is simple: we subtract the inflation rate of the corresponding period. *Ex ante* is more complicated. The appropriate rate to subtract is the *expected* inflation rate. But what do expectations depend on? The simplest answer is the current and immediately preceding rates – adaptive expectations. To get the real rate, then, we would subtract a backward moving average of the inflation rate, with lower weights on earlier years. This cannot be correct, however. In the upswing, agents will expect inflation to *rise* in the future – a backward moving average would pull it down. In the downswing agents will expect inflation to fall. Hence expectations must be based on the position in the cycle – the expected rate of inflation will be the current rate plus the normal rise (upswing) or fall (downswing), as computed, for example, from a reference cycle.

markup reflecting "lender's risk," the risks that banks think they see in what firms are doing.

Given the desired debt-equity ratio of firms, the intercept of the iFP must reflect the interest rate expected to result from capital arbitrage when g centers on normal. The intercept of the interest-banks-pay function is what banks and financial institutions will require to finance basic operations – it is what they need to cover their own fixed costs of doing business. The slope of iFP rises with the slope of the rE, until the IF begins to turn up; then it flattens out and eventually turns down. The iBC is based on the prime rate. When activity is low, however, it will lie above the prime, both because of lender's risk, and because the banking system has to spread its largely fixed operating costs over a smaller volume. At middle levels of activity it will coincide with the prime rate, and then it curves up. This may partly reflect banks' understanding that firms are facing increasing risk. But it will chiefly be due to the fact that at high levels of g stock and shares will be appreciating, making lending to the stock market attractive. If the intercept of the iFP lies below that of the iBC (see dotted line), so that there are two intersections, the lower one will be unstable, the upper stable.

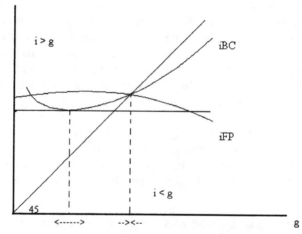

Here the intersection of the iFP and the iBC has been drawn so that it lies exactly on the 45-degree line. This should help to make clear how unlikely it is that the equilibrium would generally occur along that line, even though a discrepancy between i and g sets capital arbitrage in motion. For example, had banks not felt concerned over risk, the intersection would have taken place at a level of $i < g$.

Suppose that were the case; $i < g$ would then lead to tend to shift funds from bonds to real investment, and/or to shares (as we saw in Chapter 6). This would reduce the IF, so raise the iFP. Debt/equity ratios will tend to fall, reinforcing this movement, resulting in a rise in both i and g. This, in turn, might also raise inflation, which would further reduce the burden of debt and tend to diminish risk, leading to an additional rise in iFP. In the absence of adjustment by the Central Bank, inflation lowers the real base rate of interest, so that the iBC shifts down. This, however, might well lead banks to demand a rise in the nominal base rate. In short g would tend to rise, while i would be the subject of conflicting forces, some leading to a rise, others, at least in the short term, to a decline. There would be *no* tendency to move back towards the 45-degree line.

The same is true if the intersection occurs above the 45-degree line, so that i > g. The resulting capital movements – and growth of debt – will create pressures for the IF to rise, bringing the iFP down, which will tend to reduce g, moving it in and away from the 45-degree line. It will also tend to reduce inflation. In the absence of Central Bank nominal adjustments (for which the banking community will not press), the effect will be to shift the iBC up. The low level of activity and low inflation will combine to keep risk high, perhaps increasing it, further tending to push i up. On the other hand funds will tend to flow into the bond market, tending to bring i down. Again there are conflicting pressures on i, but a clear tendency of g to move away from the 45-degree line.

We can now explore further relations between g and i. Earlier we noted the possibility of two distinct "regimes," which tended to be self-maintaining, one in which i < g, and one where i > g. We have just seen that in such regimes movements of i and g tend to preserve the relationship between them. These regimes can be indicated on the diagram in i,g space, with the 45-degree line, and the iFP and iBC curves, a pair in each of the two partitions.

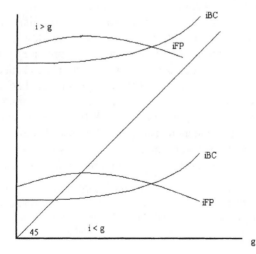

Above the 45-degree line Austerity rules, below, Expansion. The way the economy works will be significantly different in the two cases. Debt in all forms will be rising in the upper regime; below it will be falling. Above the 45-degree line, g will be lower and i higher. Risk will be higher. Fixed costs and wealth-based consumption will be higher, implying a more unequal distribution of incomes. The costs of investing in "human capital" will be higher, requiring wider differentials between professional pay and basic labor. All these combine to raise the markup so that the share of profits in income will be higher, implying a lower multiplier and a weaker accelerator. We can expect unemployment to be higher in the upper regime.

By contrast, in the lower regime all these will be reversed. Growth will be strong, debt falling. Inflation will be strong and will vary with growth, as will productivity growth. Unemployment will be low, the income distribution more equal. The upper regime favors financial capital, the lower labor. In the upper, industrial capital benefits from a high profit margin, but faces weak and volatile demand; in the lower it benefits from strong demand, but will have lower margins, as it faces strong labor.

When i > g, and debt is rising, IF will tend to drift upwards, and inflationary pressures will be weak, providing little or no offset to the upward movement of nominal rates. Since g will lie in a relatively low range, productivity growth will tend to be weak, so r_E will not swing down very markedly. The downward movement of both curves will be limited. The accelerator will be weak, and growth will remain sluggish.

By contrast, when i < g, and debt is falling, the IF will only shift upwards in the boom, and not strongly then. Inflation will tend to offset upward movements of nominal i, and productivity growth will be strong.[5] Fixed costs will stay low, the wealth-based spending will be low; income distribution will tend to be more equal, and both the multiplier and the accelerator will work strongly, tending to keep both curves low.

One feature of the i > g regime deserves special mention: an expansion is likely to generate a runaway stock market boom. Earlier, we argued that the ability of the banking system to create adequately backed funds expands at the rate i, whereas the real side of the economy's demand for funds rises with g. Consider an equilibrium case: When g = i new bonds and new funds will be evenly matched. These new funds will be divided in the same ratio as previously between bonds and equity. Some new equity will be created (in the same proportion as in previous years), but the bulk

---

[5] More precisely, when i < g, and g varies, inflation, dP/P, will vary strongly in the same direction, but when i > g, and g varies, dP/P will not vary much. (In the first case, it might be said that a normal Phillips Curve exists, in the second case, inflation has other causes, and does not behave "normally.")

of funds destined for the equity market will go toward raising the prices of existing shares, reflecting the reinvestment of profits, so that $d\$/\$ = i = g$. Now suppose that $i > g$ and that both are expanding, so confidence is strong: new funds will be created at rate i and will tend to flow in the established proportions into the bond and equity markets. But new bonds and new shares will only be offered at rate g, reflecting the rate of expansion of business. Hence both bond prices and share prices will be driven up. The effect will be to raise share prices at a greater rate than warranted by real expansion, and it will also initially tend to lower the real interest rate. Hence the relationship $d\$/\$ > i > g$ will tend to emerge; when it is clear that $d\$/\$ > i$, funds will tend to flow from the bond market into the stock market, raising interest rates and driving up stock prices even faster. The resulting boom will tend to pull idle funds and "near monies," M_3-M_1, into accounts that are active in the financial markets, adding further fuel to the flames. When g falters or turns down, however, ... Let's look at the whole story.

The cycle in a Mass Production economy

The basic pattern of the cycle

Putting these ideas together then, we can describe a typical cycle in a Mass Production economy, starting with a very general and simplified account, without specifically considering the interest rate or the two regimes. This will be a *typical* cycle, an idealized portrait, useful for organizing our thoughts about the essential relationships. No actual cycle will look exactly like it; nor are the relationships very precise. All we can legitimately claim to know, and that only roughly, is the general shapes of the functions, and the relative speeds with which variables adjust.

We take it that investment *decisions* have been made; they are given, along with benchmark prices. Investment spending, however, will vary, and the cycle will be driven by its movement. Employment will adapt through the multiplier, with real wages changing only marginally, due to differences in the rates of change of money wages, prices and productivity. The labor market tends to move procyclically and generates no pressures to adjust, although its destabilizing tendencies are weak compared to those affecting investment. When the rE lies above the IF, the economy will *expand*; when the reverse, it will *contract*. These movements "along" the curves will be relatively rapid. In addition, the intercepts will change through slower processes, shifting the curves. (Moreover, processes will be identified which change the curvature of the IF and the slope of the rE; these will also be "slow," but they will not play a major role in the story.) Shifts in the curves bring about the changes in the direction of movement.

Shift factors affecting the rE and IF

The rE equation, showing the rate of return generated by current I spending, can be written

$$r_{rE} = r_0 + (1 - d)\, g_a, \quad 1 > d \geq 0$$

where r_0 is the intercept, reflecting private consumption by the wealthy and the net government stimulus,[6] and $1 - d$ is the slope, where d represents the markdown of the value of existing capital due to competition expected from new investments, expressed here as a fraction of g_a.[7] The rE equation will shift up or down with variations in the government stimulus and with changes in private wealth-based spending. Its slope will change little, but may be affected by variations in the costs of, or losses due to, the collections and payments systems, for example, in the slump.

The IF equation shows the rate of return required to sustain a given level of I spending, and it can be written

$$r_{IF} = r_F + eg_a + \chi(g_a), \quad 1 > e > 0, \; \chi', \chi'' > 0$$

where r_F is the rate of profit required just to cover fixed costs, e is the fraction of the value of capital in equity (the portion of I that must be covered by retained earnings), and $\chi(g_a)$ is the increasing risk term, imparting upward curvature.

The two intercepts are closely related, and will tend to move in the same direction, but not necessarily by the same amount or proportion. Assuming no government stimulus for a moment, the fixed costs of the IF intercept are the source of the private consumption spending that generates the rE intercept. These fixed costs represent interest income, managerial salaries, and various kinds of legal and consulting contracts. The two intercepts will therefore differ by the amount of saving out of wealth-based income, and that of the IF will normally lie above the rE. A rise in fixed costs would not necessarily raise the intercept of rE by the

---

[6] This is government injections minus government withdrawals. Government spending on goods and services plus transfers that are spent minus taxes on active funds. Government payments of interest or bailouts of capital losses are not injections; taxes on savings and on profits fall on funds already withdrawn.

[7] For simplicity it is assumed that d is constant. At higher levels of g, however, productivity growth and innovation will generally be higher, and diffusion more rapid. This suggests that d would rise with g, in which case the rE would have a falling rather than a constant slope. This will not affect the argument.

same amount, because of wealth-based saving. If the net government stimulus exceeds the amount of wealth-based saving the rE intercept could lie above that of the IF. A change in government spending could shift the rE but leave the IF unaffected.

The IF intercept depends in part on the real interest rate. Variations in i may shift the IF and may also change its curvature. But while nominal interest rates rise with activity levels, under Mass Production, so does inflation, keeping pace quite closely. Hence real rates move less and move with a lag. Other factors may have a more immediate impact on the curvature and position of the IF.

Consider procyclical movements in productivity. Higher levels of activity lead to small innovations and make it worthwhile to improve organization; there are also small economies of scale. All these lead to higher productivity; a higher rate of growth will lead to higher productivity growth. This tends to move g_0 out; altering the position of the rE, moving it to the right and down. But there will be an exactly similar effect on the IF. However, that is not the end of the story: a rise in productivity in effect increases the firm's *productive capacity*. But that is what investment is intended to do. Hence when productivity has risen, investment is no longer so urgent. But the urgency of investment is a determinant of the level of cash flow required to cover it. When the urgency declines, a higher and safer level can be required. The IF shifts up.

In regard to the intercept of the IF: when the real wages of production workers are adjusted to their productivity, there is a significant effect on managerial costs. Managerial staff must be paid at a suitably higher rate than production workers; the differential is set by social convention. When wages rise, therefore, managerial salaries must rise also, even though managerial productivity has not changed (and may not even be easily measurable). But managers, office staff, equipment, and other organizational needs will rise in step with new investment. Putting the two together, then, if labor's productivity is increasing, overhead costs must rise *faster* than investment. Hence F will tend to rise during the upswing.

There is a further cause for the increase in F during the upswing. During a period of increasing I, the administrative and contractual costs of integrating new projects into the table of organization may tend to rise, not additively, but multiplicatively. When a new project is added to an existing organization, it has to be coordinated with each of the existing divisions or units. Setting up and operating each such channels of coordination takes time and adds to fixed costs. But the total increment of fixed costs is a multiple of the cost of coordinating one unit with another, the number depending on how many units there are in the organization. Of course, efficient management design may reduce the problem, but it is unlikely to eliminate it.[8]

Overhead costs will rise for another, quite separate, reason. As the boom develops financial markets will flourish, and speculation will begin. Prices of assets of all kinds will rise, and firms will be increasingly tempted to take positions in them, anticipating a further increase. Competitive pressures reinforce this temptation. If a given firm does not speculate, and its rivals do, successfully, the rivals will have increased their asset base, improving their competitive position. On the other hand, if a firm speculates, successfully, and its rivals do not, it will gain a leg up on them. If neither speculate, or if both speculate successfully, their relative positions will be unchanged, but in the latter case, both gain. So the dominant strategy will be to speculate – always assuming that the dip into the market will bring a gain. This attitude can only be expected to prevail in the heady days of the boom.

Once firms begin to compete in speculative markets, they will run up their financial costs – brokerage fees, commissions, margin costs, etc. Their risk position will also increase, though with confidence at a high level, they may initially discount this. But it will become increasingly apparent as the boom begins to fade.

When the market is weak another factor enters. As activity rises, with Investment Spending, the real assets of companies increase. In a weak market, their stock prices may lag; the market may not know enough, or it may not be in a position to judge whether the investment projects are in fact well chosen. Or it may be in a slump. If stock prices do not keep up, it will be expensive in terms of dilution of control to raise funds by issuing new stock. Moreover, if the newly acquired assets, or new assets under construction, are in fact well chosen, and potentially profitable, and the stock does not rise, then the company is a good target for takeover. In such circumstances the company must either buy its own stock, or distribute dividends, to raise its stock price. The higher the level of investment and the longer the boom has lasted (and the better the investments), the more serious this problem is likely to be. Hence the cost of raising the stock price will rise with the level of activity (level of investment spending), which will shift the IF upwards, and impart an increased upward curvature to it.

---

[8] The current interest in "network" and "cell" arrangements of production reflect a desire to simplify coordination and reduce organizational and informational costs, drawing on computers.

Phases of the cycle

Starting from the bottom of the slump, once the IF lies below the rE, investment *spending*, increases; this leads to an expansion. Output and employment increase according to the multiplier. As the expansion progresses, however, risk increases; the IF curve rises to an intersection with rE. As it rises, however, confidence increases and the curvature flattens, so that the intersection is at a higher point than would be justified by the objective situation alone. This intersection is a stable high employment boom. Moreover, it is likely to lie above the "normal" level of g_d; that is, the boom will have led to "overshooting." This may be temporarily corrected by tendencies of firms to revise their expectations of demand growth upward, and hence to enlarge their investment plans. But such an upward revision would very likely lead to overconfidence and reduced assessments of risk, further encouraging investment spending.

But this boom will eventually undermine itself. First, the high level of activity will lead to high productivity growth and therefore to creeping inflation. Initially this will slightly shift the IF down, and widen its curvature, intensifying the boom. But as the boom progresses the creeping inflation will develop into expansionary inflation. Moreover, the high demand will create pressure in the *machine tool* sector (which retains elements of Craft organization). Machine tool prices will begin to rise, creating additional inflationary pressure. But as the boom drives actual growth up to or beyond the level of the normal anticipated growth of demand, two things may be expected to happen. First, the ability to raise prices, relative to the rise of wages, will weaken, leading to a squeeze on profits, and second, uncertainty will increase. New markets simply may not open up as expected; new products may not catch on. The best innovations will have been put into play; the next best may not work so well. Hence perceived risk will increase, and curvature of the IF will rise.

The high productivity growth of the boom will also have been accompanied by the accumulation of ideas and proposed innovations in organization. Given an imminent profit squeeze and an increase in perceived risk, businesses will be reluctant to lay out funds for new plant and equipment. At least some portion of new investment will normally require shutting down operations, for example, to replace or upgrade equipment, or to integrate old operations with new. Particularly where a shutdown would have been required in any event, firms may now choose, in the light of the increased risk, to substitute *reorganization* for expenditure on additional equipment. The reorganizational investment will increase capacity without requiring firms to take on additional indebtedness; so it will tend to reduce the ratio of debt to capacity output. But there will be little in the way of multiplier effects. As a result the rE line will swing down.

Inflation and the high level of investment eventually will drag up i, a period later. The inflation leads the banking system to raise nominal rates. The high level of investment leads to capital arbitrage; higher g drives up dp/p, which pulls up i. Capital arbitrage also raises the spectre of take-overs, defending against which requires firms to bid up the prices of their shares. The salary increases required to maintain income differentials lead to rises in managerial fixed costs of all kinds. (But because of high saving out of wealth this does not tend to raise the rE much; saving out of wealth-based income is procyclical. Hence, a rise in the IF intercept will not shift rE proportionally.) These two effects combine to raise the intercept of the IF curve. As the IF drifts up, while the rE swings down, the intersection moves down, perhaps only slightly at first. But this signals the fading of the boom, and confidence begins to weaken, increasing the curvature of the IF, and further lowering the inter-section point. Eventually the upward shift combined with the increasing curvature will reach a point of tangency with the rE. This point is unstable downwards, and leads to a

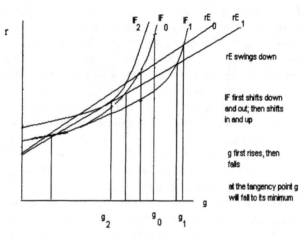

At the outset: rE$_0$ and IF$_0$ intersect at g$_0$ close to g$_d$, triggering boom.
Boom conditions lead IF to shift out and down to IF$_1$, so g rises above g$_d$.
Reorganizational investment and profit-squeezing inflation lead rE to swing down to rE$_1$, so g falls back slightly.
Rising overhead costs and rising risk – due to falling g – lead IF to begin rising and increasing in curvature, so g continues drifting down.
When IF reaches a point of tangency with rE$_1$ g will enter the recession phase – downward instability – and fall to its minimum level.

collapse in I. Output and employment will decline, perhaps all the way to some minimal level, maintained by govern-ment spending and "automatic" stabilizers. At this point, however, firms will cut back on fixed managerial costs, bankruptcies will renegotiate fixed obligations, and interest rates will decline. Hence the IF will shift back down, and the cycle can begin again.

Remarks

That is the general outline. The cycle is based on the interaction of commodity and labor markets with financial markets. The former are unstable; the latter, somewhat surprisingly, tend to provide a weak stabilizing effect, shaping the system into a cycle. Speculative financial excesses can convert the upper turning point into a crash, but the downturn does not depend on financial instability.

The expected normal growth of markets has been taken as given – if this weakens or becomes less well defined, risk will increase. The curvature of the IF will become more pronounced, and its shift with changes in the rate of interest will become larger. The effect will be to increase the volatility of the system. Conversely if the expected normal growth of markets becomes more definite, and confidence rises, the IF will tend to flatten and shift less; volatility will diminish.

But the increases in administrative and overhead costs take time to mature; F will rise, but only with a lag. When I falls, and especially when it collapses, these costs will be swept away by reorganization – but again only over a period of time. Of course the movement of I generates a cor-responding movement of Y through the multiplier. The impact of changes in Y on F can be assumed to take place in accordance with the following:

$$\Delta F_t = \phi(Y_{(t-1)}, \Delta Y_{(t-1)}),$$

where the sign of $\Delta F_t = $ sign of $\Delta Y_{(t-1)}$, and the size of the change in F is proportional to the size of $Y_{(t-1)}$, depending on the sign of $\Delta Y_{(t-1)}$. We have

$$\Delta Y : \quad + \quad -$$

High Y :   L   S   L = large

Low Y :   S   L   S = small

When ΔY is positive, the change in F will be large when Y is high, and small when Y is low; when it is negative, just the reverse. Thus in the upper stage of a boom fixed costs will rise steeply, in the early recovery period, they will rise weakly; at the beginning of the downturn they will begin to fall, and as the slump deepens they will fall sharply.

(We could be even more subtle, and have the largest changes in F occur when the change in Y has fallen to zero, but is changing sign. At the peak of the boom, just before the downturn, the change in Y turns from positive to negative – and this is when the rise in F could be considered largest. Similarly at the bottom of the slump, just before the upturn, the fall in F would be greatest.)

Once the IF shifts up, however, investment will be cut back, and when it has shifted up enough to become tangent, I will enter the downswing and fall precipitously. But as I falls the banking and financial community comes under pressure. Bank earnings will be squeezed, and efforts will be made to keep interest rates up, in order to prevent banks collapsing. Hence interest rates will adjust more slowly. Similarly managerial costs tend to be contractual, so will also adjust only slowly. Thus the position of the IF will tend to remain fixed during the downswing.

Eventually the decline will lead to cost-cutting, bankruptcies and reorganization, at which point the intercept of the IF will shift down. The intercept, F, reflects not only the interest costs of the debt incurred in the construction of the present capital stock; it also includes contracts, administration costs, consulting and legal fees, etc. The changes in this intercept are "slow" changes, slower even than the changes in i, which lag behind the movements in I. When the intercept shifts down sufficiently, a portion of the IF curve will lie below the rE line, and a rise in investment spending will be called for. However, the upswing cannot begin until order has been restored in the banking system. There may be a further delay as bank portfolios are readjusted and banks recapitalized; only when credit for working capital can be provided easily can the recovery begin.

So the movement of F, lagging behind the movement in I (and Y) is central to the cycle. When I rises, leading to a boom, F will be dragged up after a time, leading to a downturn in I (and Y), which will eventually bring a reduction in F, permitting an upturn. Changes in the state of confidence intensify this pattern, by increasing the upward curvature of the IF as the boom ends, and reducing it during the upswing.

This pattern will be further reinforced by the likelihood of a shift to reorganizational investment as the boom wears on. This will swing the rE down at or just past the peak of the peak of the boom, and tend to hold it low at the bottom of the slump. This downward drift of the rE can also be intensified by corporate pricing practices.

As we saw in Chapter 10, the corporate response to cyclical variations in sales will be to raise prices in a downturn, while maintaining them steady or allowing them to drift up, during the upturn. Price increases in the downturn, lowering the real wage, will reduce the multiplier, so that a given level of I will generate less employment, weakening the "accelerator." That is, a reduced multiplier will hinder the development of new markets, and so reduce expected growth, leading to a reduction in *planned investment*. A reduction of planned I, in turn will very likely lead to cutbacks in infrastructure spending, tending to reduce the government stimulus, shifting the rE line down. Declining prices in the boom would tend to have just the opposite effects.

These movements in rE are matched by corresponding shifts in IF. At the peak of the boom, pressure to maintain wage-salary differentials creates cost-increasing inflation, which will tend to shift IF up; in the slump, the pressure of unemployment on money wages will allow productivity to run ahead of real wages, tending to bring IF down.

So the rE line will tend to swing down just past the peak of the boom as F is rising, leading to the downswing, during which it may shift down; it will swing back up at the end of the slump, at the same time that fixed costs are reduced, until the IF once again lies below rE over a long stretch. This then brings on a recovery and a rapid expansion. I will rise to a boom level, bringing up the rate of interest in its wake, raising the rate of risk, shifting the IF up, leading to the gradual constriction of investment spending again, and eventually, once again to collapse.

The role of financial markets

Financial markets, however, play a special role here, which needs to be explored further. We saw in Chapter 6 that the arbitrage between the rate of profit, the rate of growth and the rate of interest could become unstable, if share price appreciation became an object of speculation. This could happen in a boom, where the effect would be to attract funds away from real investment, creating a bubble that would break as I turned down. In general, financial speculation can draw funds from the circulation, creating downward pressure on goods and labor markets, while fueling the speculative frenzy. But what is notable here, is that there exists a cycle in the Mass Production economy even without such speculative excesses.

Clearly this is a cycle based in large part on changes in costs that come about because of changes in financial conditions. F rises and falls with changes in the rate of interest, and through changes in other fixed money contracts. The cycle also exhibits a procyclical pattern of real wage and productivity movements. The effects on rE in the boom help to bring the expansion phase to an end, and the movement in the slump promotes recovery. The downswing in rE results from the effects of productivity growth; the upswing because of a divergence between money wage and productivity changes. The turning points in the cycle depend on *monetary* variables. Even increasing risk must be understood as based on finance, since the risk is the risk of default on fixed obligations denominated in monetary terms. Taken by itself, the "real system" is simply unstable. It is the interaction with monetary and financial variables that converts this instability into a cycle. The contrast with the Craft economy – as outlined in the Hayekian story at the end of Chapter 9 – could hardly be greater. There an essentially stable real system is destabilized by interaction with the monetary/financial side of the economy; here interaction with financial markets turns around an unstable real system, preventing runaway movements.

But the connections are not immediate – or mechanical. A boom could be maintained for a long time, especially if sustained by policy, and if speculative excesses can be kept under control. Recovery from a slump could be inhibited by weaknesses in investment decisions, which depend on long-term matters, such as technological innovation and the anticipated growth of markets, matters discussed elsewhere. Recovery could be delayed by financial weakness and disorganization. If the banking system were inadequately capitalized, for example, it would not be able to provide the working funds necessary for expansion. The cycle depends on institutional arrangements.

Over time, it could be expected that business would realize that running up managerial and fixed costs in the boom would lead to problems, especially when booms may bring profit-squeezing inflation. Nor should business allow itself to accept the burden of excess capacity. But business cannot escape the cycle just by "learning." These two points are sharp and dangerous; they are the horns of a dilemma! Rapid growth leads to profit-squeezing inflation and the runup of managerial fixed costs, but stagnating growth imposes the burden of excess capacity. Trying to keep managerial costs down in the boom may lead to losses of market to competitors; building smaller plant to avoid excess capacity may mean failing to capture economies of scale – which competitors have.

Here is where technology can help. Controlling managerial costs and reducing excess capacity figure among the chief aims in deploying computers for office work, leading to downsizing and "perestroika." In the same way, variable interest rates have added to flexibility in adjusting to fluctuations.

The cycle in the two regimes

Such is the anatomy of the cycle – with the qualification that such a picture can never be complete, nor can the relationships be known with precision. Even so, more could be said about financial markets and interest rates. That story was left largely implicit. It is time now to examine this explicitly, and, in particular to look at the cycle in relation to the two regimes, expansion and austerity, discussed earlier. That discussion centered on a diagram with the real interest rate and the growth rate of output on the axes, showing the interaction of the real and financial sides of the economy. We need a clearer picture of this interaction. Moreover, in spite of capital arbitrage, we found no tendency for the economy to gravitate to the 45-degree line. But we need to understand better why it tends to cycle either above or below. It is necessary to show that if policy established i = g, the economy would not tend to remain there. Let us consider this further.

Interaction between i and g

First, a review of the cycle, bringing out the role of financial-real interaction. Consider the expansionist case where i < g. Begin with the phase of expansion, where the rE lies above the IF: this implies that investment spending creates a cash flow in excess of that required to sustain such spending. This will bring an upswing, as investment spending rises, carrying to the point where increasing risk turns the IF up sufficiently sharply to intersect with the rE. In its enthusiasm, the boom may carry investment spending beyond the level called for by investment plans. And it will give a strong boost to productivity growth. Both of these will encourage the stock market. Stock prices will appreciate, pulling up interest rates. (Stock prices will *rise*, at a rate d$/$ = g, but will not *boom*, taking off into a self-reinforcing expansion.)

The implications for interest rates can be seen on the i,g diagram. The iBC is based on the horizontal line indicating the level of the prime rate, that is, at a rate marked up above the Central Bank's rate by an amount that would cover the fixed costs and normal profits of the banking system, if realized on a normal volume of low-risk loans. At low levels of g, however, lender's risk will require a rate above prime; at high levels the stock market will be expanding, and opportunity cost dictates a rate above prime. The iFP starts from an intercept either above or below this rate – depending on the relative position of the intercepts of the rE and IF. Then it rises, reflecting the widening gap between the rE and the IF, after which it begins to fall as this gap closes, and it finally cuts the iBC at the level of g where the rE and IF intersect – *provided*, that is, that the iBC has not risen above the prime rate because of the pull of the stock market (or because lenders fear that the boom is fading, and/or that currently projected investment is risky). Remember, the IF exhibits risk as seen by firms, whereas the iBC shows risk as seen by the banks. The two calculations need not coincide, and will normally involve different considerations. If the iBC does rise, so that its intersection with iFP takes place at a lower level of g, then firms will find that their borrowing costs have risen. This will shift the IF curve up, lowering g, until the intersection points are the same.

Moving on: the expansion phase will stop when the system has reached the upper equilibrium, at which point growth will be near to, and may exceed g_d. Productivity growth will now be high, and inflation will have increased. The initial effect of inflation – so long as it is favorable to profits or neutral – was to shift IF and iBC down, at least temporarily, which tended to intensify the expansion. This stimulus, however, will be counteracted by the reaction of banks, who will push for a rise in nominal interest rates, leaving the overall impact uncertain. But if and when growth comes to surpass g_d, inflation will begin to squeeze profits, and this will raise the IF.

The effect of productivity growth, on the other hand, will be complex, and will affect both curves. The rE will shift out and down. That is, every level of cash flow, represented by r, the current rate of return on existing capital, will be associated with a higher rate of growth of output. In causal terms: a higher rate of growth of output will be needed to generate a given level of cash flow. As noted earlier, this movement will tend to be matched by the IF, but productivity will have further effect, that of diminishing the marginal urgency – as defined earlier – of investment spending. And, as we noted then, this will tend to shift the IF upwards. The result will be a slight decline in g. This will not, however, affect inflation; the wage-price spiral will by now have acquired a momentum of its own. But such a decline signals a rise in risk, which will tend to trigger a shift on the part of firms away from purchases of plant and equipment to re-organizational investment, and this will swing the rE line down.

Thus the downturn begins, slowly at first, then gathering force, as banks become suspicious and firms cautious. Firms may begin to experience difficulties in collection, reducing their levels of cash flow. This will reduce the slope of rE, further lowering g. The IF will continue to rise reflecting the continued growth of managerial costs, and it will now increase in curvature, reflecting a fall in confidence, due to the decline in the growth rate. Inflation, however, will tend to lower interest rates, but not enough to offset the upward drift of the IF. When the IF rises to become tangent to the rE, the iFP will have drifted to a point of tangency below the iBC, and the system will careen downhill full speed. At this point the stock market will stop expanding and confidence will flag, leading to a rush to liquidity.

Real interest rates will fall during the period of the turning point, but as the downswing gets under way, inflation will slow down, allowing rates of interest to begin to recover. Moreover, banks will push for higher rates, since they need the funds to meet their inflexible operating costs. Thus at some point in the downswing, i will actually rise, further intensifying the downward pressure.

At the bottom, firms and banks will reorganize, cutting back on fixed costs, both voluntarily and following court orders resulting from bankruptcy proceedings. F will therefore decline, and the IF shift down again. But interest rates will be slow to fall, and collection difficulties will persist for firms. Eventually, however, rE will recover, and IF will decline, to the point where expansion can begin again.

The shape of the cycle

Let's consider the shape of this cycle – bearing in mind that our analysis cannot be considered more than suggestive. As the upswing gathers steam, g rises while i first stays low, and then rises as banks begin to "lean against the wind," shifting up the base rate. Also as g moves out, the slope of the iBC will begin to rise. The expansion continues at the higher rate of interest, but as inflation picks up, i begins to fall, with g continuing to rise. Then g falters and the downturn begins in earnest, so g and i decline together. At this point inflation cools off, and i begins to rise during the downswing, a rise reinforced by the needs of bankers for higher rates to cover their expenses in the face of a lower volume. At the bottom of the slump, banks as well as business will reorganize and i will drop, helping to turn the system around, so that expansion will begin and g will rise in conjunction with a low level of i. To summarize:

The upswing:
Phase I – g rises; Phase II – i steady; g and i rise together; Phase III – g rises, i steady at a higher level.

The upper turning point:
g steady, i falls; then g falls with i still falling

The downswing:
Phase IV – g falls, i steady; Phase V – g falls, i rises; Phase VI – g falls, i steady at a higher level

The lower turning point:
g steady, i falls; then g begins to rise, i still falling (see diagram).

THE CYCLE IN (G,I) SPACE

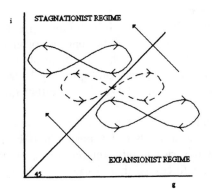

The case where i > g follows essentially the same pattern, with a few important differences. As a result the figure is drawn illustrating both cycles, each represented as a figure eight, respectively above and below the 45-degree line. However, the differences deserve discussion. Inflation will not necessarily be weaker in the upper cycle, but it will not vary reliably with g, or at least it will not vary with g for the same reasons. As a consequence, neither the fall in i at the end of the upswing, nor the rise during the downswing, can be attributed to inflation. Productivity growth will generally be lower, and/or due to different cause, so will also not vary reliably with g. Hence the account of the upper turning point needs to be modified.

Taking these in order. The rise in i during the upswing will be due partly to banks "leaning against the wind," but it will be strongly supported, as we saw earlier, by the emergence of energetic financial speculation, fueled by money creation, on the one hand, and by the use of idle funds and near money. This will drive up stock prices, drawing in funds from the bond market, so pulling up interest rates. The turning point will come when IF has risen, first, stopping the expansion, and then leading to a perhaps small decline in g. Small, but with g slightly falling, the real situation no longer justifies the levels which stock prices are reaching. The discrepancy will bring a crash, and the collapse of the stocks will send funds back into the bond market driving up bond prices. Interest rates will fall and so will the growth rate. As the downswing gets under way banks will try to stabilize rates, and as it proceeds, they will seek to raise them so as to compensate for reduced volume and higher lender's risk.

So the shape of the cycle is the same in the two cases – and the dilemma posed for business is the same, also. In each case, low growth imposes the burden of excess capacity (and forgone opportunities), but high growth brings a buildup of fixed costs and the risk of an unfavorable inflation.

The instability of i = g

Consider a case where policy has carefully set the normal rate at a level very close to equality with the expected normal rate of growth, g_d, with the objective that the system should then oscillate about the 45-degree line position. Could we then expect i and g to move together, so that both would remain close to equality? We shall see that this is not likely.

Note that, in general, it can be expected that g will prove more volatile. In the Craft economy where output tends to be fixed, g may be limited; but in Mass Production g will vary strongly. Interest rates, as we have seen are subject to conflicting forces so their variance will tend to be less.[9] In general lower interest rates encourage expansion, while higher discourage it. It follows that in the slump we might expect i > g, while in the boom, i < g.

As a preliminary, it is immediately apparent that there are factors which will tend to preserve, rather than correct, the inequalities. Interest rates lower than the growth rates tend to encourage investment, thus tending to raise g, which is likely to stimulate inflation. Inflation will then tend to offer further support for the case i < g, for it will tend to encourage spending and expansion, on the one hand, raising g, while it will reduce real interest. Interest rates greater than growth will discourage investment, and dampen inflation, which will tend to have the opposite effect, although perhaps less strongly. When i > g, that is, low or declining inflation supports interest rates, while discouraging spending.

In the same way, high-productivity growth will tend to support the inequality i < g. It stimulates investment, and tends to reduce risk (although we have also noted that productivity growth reduces the urgency of investment, especially when the boom has peaked). Low-productivity growth can be expected to provide support for i > g; risk will tend to be high, and stimulus to spend low.

Most important, when i and g are unequal, if this inequality is prolonged, four significant long-term changes take place, which make it more difficult to bring them back together.

First, when i < g, corporate debt will tend to be falling as a ratio to total output, and when i > g, it will tend to rise. We remember that in the steady state $i = D/F$ and $g = I/K$, which implies $i = g = (I - D)/ (K - F)$. But profit $P = I = D + E$. Earnings attributable to equity can be reinvested. $D = iF$ must be paid as debt servicing. If $P = I$ is to be invested, then, and if capital is to grow at rate g, with an unchanged debt-equity ration, then $P-E$ must be borrowed afresh. If $g = i$, all is well. But if $g > i$ each period $D/F < I/K$; hence E/P will rise, and so will E/K. The ratio of debt to capital will tend to fall, in the aggregate. By the same token, then, when i > g, it will tend to rise. A falling ratio of debt to capital is generally stimulative, while a rising ratio calls for austerity. The former tends to raise g and lower i; the latter brings just the reverse movement.

Second, the faster growth of earnings in the financial sector will be reflected in the relative expansion and greater robustness of that sector.[10]

---

[9] Casual inspection supports this observation for both the United States and the United Kingdom, 1950–95.

[10] The ratio of FIRE (finance, insurance and real estate) to U.S. GDP was approximately constant at slightly over 14% from 1959 to 1979, during which time i stood consistently below g. Then in 1980 i rose above g and has remained above ever since. The FIRE/GDP ratio has risen to over 18%. A similar pattern can be observed in the United Kingdom (Felix, 1995).

But as we saw earlier financial institutions are inherently "fragile" in the sense that a high proportion of their operating costs tend to be fixed. Thus an expansion of the financial sector increases the costs that have to be covered by financial earnings, which in turn will tend to limit the downward flexibility of i. On the assumption that all earnings are ploughed back (or that financial and real side businesses plough back the same proportion) the financial sector will grow at rate i. However, the *volume* of business that it does will only grow at the rate g. Hence there will be pressure to raise i to insure against lender's risk.

Third, as we saw in Chapter 6, the relation between g and i determines the ratio of idle funds to active funds, roughly $(M_3 - M_1)/M_1$. When i < g, this ratio will be high and will tend to rise; when i > g it will be low and tend to fall. That is, the opportunity cost of idle funds is low when i lies below g. It will be preferable to be relatively liquid, and able to take advantage of investment opportunities. But when interest rates are high relative to real expansion, liquidity has a high opportunity cost. When $(M_3 - M_1)/M_1$ is high and growing, the banking and financial system will be relatively stagnant; but when it is low and falling, financial markets will be booming. Indeed, the rise in M_1 will be chiefly due to the influx of funds into accounts that turn over rapidly because they are used in financial transactions.

Finally, i > g creates pressures tending to increase inequality. Recall the discussion in Chapter 10 of investment by households in skills: the long-term calculation was based on an equality between i and g, for the *earnings* could be expected to rise at rate g, while i represented the normal discount factor. If g < i, then the ratio $(w^* - w)/w$ would have to be increased, in order to make up for the slower growth in earnings. Otherwise investment in human capital will fall, and those who have already invested will find that they are not earning the returns they expected. But w^* is an index of management salaries, professional pay, and high-skill work; to increase this ratio means widening the disparities in pay, making the earned income distribution more unequal.

When i < g the real side of the economy will tend to expand, but financial markets will be relatively sluggish, responding only to strong movements in growth and profitability. By contrast when i > g, the financial markets will tend to boom, while the real side of the economy will tend to be relatively stagnant. To shift from the first to the second, then, will require a prolonged recession in the real economy, with high unemployment, low growth, and lost productivity. To shift from the second to the first will require weathering a financial crash and a shakeout of the banking system, with the strong likelihood of a disastrous impact on pensions and insurance. In each case care will be needed to see that the contracting side of the economy does not bring the other side down as well – that high unemployment does not cause a stock market crash, and that a financial collapse does not throw the real economy into recession. But policies would also need to be designed in each case to help the ordinary public through the transition. (Such policies were not prominent in the transition that took place in 1979–81.) This would require careful planning.

But policy is not the only force behind the shift from the high growth to the stagnationist regime. The characteristic dilemma which the Mass Production cycle poses for business has brought forth responses that themselves tend to lead through market pressures to a regime in which $i > g$. Computerization has been significant in several respects. It has helped to reorganize plant and equipment to permit "just-in-time" production, reducing the need for inventory. Production lines can be preprogrammed in multiple ways, so that they can be shifted from fabricating one design to another, even permitting custom designs. This cuts down on idle machine time. Computerized control over production makes it possible to separate tasks, so as to locate those processes requiring only unskilled labor in low-wage areas. This cuts down the need to carry capacity. Moreover, such a move offers business the double benefit of low wages in the new location, and a weakened labor movement back home. The new technologies have been a major factor in the *globalization* of modern business.

Globalization and stagnation

The new world order

The term "globalization" will be taken to indicate three phenomena: first, free trade; second, the removal of restrictions on the mobility of financial capital; and third, the technological ability to disperse production facilities around the world, while maintaining detailed control of the production process.

This technological ability has three important features: first, computerized control, making it possible to monitor processes in detail; second, cheaper, faster, more comprehensive communications, making it possible to transmit information about productive processes rapidly and completely (pictures, colors, numerical data, mathematical commands, etc.); and third, reduced transportation costs, together with higher transportation speeds.

"Stagnation" will be taken to mean the combination of a low growth rate, low productivity growth, and low capacity utilization. It represents the failure to use and improve the productive facilities available at the optimum level.

The suggestion is that the development of globalization brought a rise of stagnation in its wake. The context is that of an international system in which there is no "public authority"; that is, there is no sovereign power which can regulate, control or stimulate or restrain the activity level of the international system. The argument is that under such conditions globalization will set up forces that reduce capacity utilization and therefore growth, while at the same time driving up interest rates.

At the outset, however, a number of countervailing tendencies should be noted:

> Removing restrictions on trade should have the effect of increasing specialization, and thus efficiency. Hence productivity should rise among the affected nations.
>
> Removing restrictions on capital mobility should better permit capital to be allocated to its most productive uses, again increasing productivity. As capital leaves less productive employments, this will temporarily reduce activity, but the reduction will be offset by the rise in the new area of investment – and there will be a net gain in overall potential GNP resulting from the increased efficiency.
>
> The flow of capital to low-wage areas may lead to reductions in wages or to stagnation in high-wage areas, but it should bid up wages in the low-wage areas.
>
> Similarly the flow of short-run financial capital to high-interest areas will raise interest rates in the regions such capital is leaving, but lower them in those to which it is flowing.
>
> These countervailing tendencies should be borne in mind. The working hypothesis is that they will not prove sufficient to offset the market tendencies outlined below, but the case must be proven.

The effects of free trade

In regard to trade (in conjunction with capital mobility) the argument is based on the view that competitive advantage in the postwar world has depended and still depends importantly on technological superiority. Such superiority not only provides a cost advantage, but also enables production of a better product, sometimes a product that is indispensable to buyers over a wide range of prices. This superiority often arises from economies of scale. The larger advanced countries are able to produce more cheaply and to turn out better goods.

In abbreviated form the argument can be summarized as follows:

> A rise in free trade – removal of trade restrictions – strips away the protection from many smaller markets, enabling import penetration by large-scale advanced firms.

This creates higher foreign deficits for countries with weaker and/
or smaller – and so less competitive – firms, putting pressure
on their currencies.

To defend their currencies they must raise interest rates and
reduce capacity utilization, which lowers inflation, raising real
interest rates.

The higher interest rates attract short term capital from the
stronger countries, while the reduced capacity utilization reduces their import purchases from the stronger countries.

Interest rates will be driven up and demand will be reduced in the
stronger economies, leading to a decline in activity levels.
Potential growth and GNP may have increased as a result of
the superior goods being imported, but actual levels will be
reduced because of the effects on interest rates and capacity
utilization.

Free trade and unrestricted capital mobility make it difficult to maintain
relative currency values; that is, a regime of fixed exchange rates cannot
easily be sustained.[11] The argument can be sketched:

Trade will increase as technologically advanced firms penetrate
the markets of weaker and more backward nations, and/or
markets in other advanced nations that, prior to the removal of
trade restrictions, were dominated by relatively backward firms.

In these nations imports will rise, but, being relatively backward
– uncompetitive – technologically, they will be unable to

---

[11] The Bretton Woods fixed exchange rate system was not sustainable, not only
because of the fixed rates, but also because of the reliance on the dollar as the
reserve currency. Supplies of dollars to trading nations came principally through
the excess of U.S. spending abroad over earnings abroad, in the end such a
system had to fail. For the growth of trade depended on the growth of an
adequate supply of the reserve currency, the dollar. But the supply of the dollar
depended on the U.S. deficit. The size of the deficit could be expected to grow in
keeping with the growth of output. But for world trade to expand relative to
world output the U.S. foreign deficit would have to expand relative to U.S.
output.

If the increase in the deficit came on capital account, then "Gaullists"
everywhere would object that the United States had become able to
buy up more of the world's best assets essentially free – by
supplying money that no one will or can use.

If the expansion took place on current account then confidence in the
dollar would be weakened.

Hence for T/Y to rise, the U.S. foreign deficit must rise in relation to
U.S. GNP, and there will be a point beyond which the U.S. deficit
becomes unsustainable.

increase exports pari passu; hence their balance of payments will move into deficit.

The supply of the currency of such a nation on world markets will exceed the demand for it, for use in trade; conversely their demand for other currencies will exceed what they can earn, and hence their borrowing requirements will rise.

As a consequence their currency will tend to fall, and the interest rate they must pay will tend to rise – since they are obviously risky, being competitively unable to earn their keep.

To keep the exchange rate fixed would require lending the foreign currencies needed to support the weak nation's imports.

But the supply of these currencies will be limited (since *ex hypothesi*) they are the currencies of strong nations running surpluses, whereas no one will be interested in holding the weak currency.

Given free capital mobility speculators will bet that, eventually, the value of the weak currency will fall, the strong rise; they will therefore sell the weak and buy the strong.

To counteract this Central Banks and the strong currency countries must be prepared to issue the strong currency to the extent needed and absorb the weak one – in spite of the evident fact that the latter is useless.

Thus pressure will be exerted, through policy, to make the strong currency tend to circulate, while Central Banks and financial institutions will be encouraged to hold the weaker as reserve. This runs directly counter to Gresham's Law and is not sustainable.

The implication is that a fixed exchange rate system will have to be abandoned, so that currency values will become variable. But in circumstances in which competitiveness reflects technological development, flexible exchange rates will very likely make the problem of foreign deficits worse. If trade deficits exist because the competitive strength and weakness of different countries reflects their relative technological development, some sizable fraction of imports will be likely to reflect technological weakness. Such imports will tend to be price inelastic, so that devaluation will raise the import bill. Advanced countries are likely to have exports that are also price-inelastic, so that cheapening them will lose revenue. Such circumstances will make it difficult to meet the Marshall-Lerner conditions. Hence flexible exchange rates are likely to require even greater contraction to bring deficits to an acceptable level.

Summarizing the implications, then

It can be assumed that the international financial system will be willing to fund only a certain level of deficits in relation to world GNP. This willingness will be broadly proportional to world assets, which in turn will reflect world potential GNP.[12]

The U.S. foreign deficit is a special case – but even so, a run on the dollar is possible. The problems of the United Kingdom are well-known, and contractionary conditions for finance have been imposed throughout Latin America and Africa. The reasons that world financial markets will not underwrite more than a certain level of deficits need to be explored; also important is whether and why some countries are privileged.

The acceptable level of world deficits depends on assessments of risk, which in turn depend on judgments as to competitive strength, which will determine the ability of a country to earn the foreign exchange necessary to service its foreign debt, and to repay (or roll it over).

The higher the volume of trade in relation to GNP, the higher the risk, moreover, risk may be increasing at a faster rate than GNP.

Under these conditions a higher ratio of trade to GNP will have to be accommodated at a lower level of GNP. That is, a rise in trade/GNP – T/Y – would imply a larger volume of trade if GNP continued to expand at the same rate. But if deficits rise

---

[12] This appears to suggest that there is some sort of "world financial constraint," an odd position to take if one also believes that "money is endogenous." But the world monetary system is not fully organized; that is, the system has no lender of last resort, nor does it have a regulator with powers of enforcement. Hence risk and uncertainty are bound to be pervasive. Central Banks will be pressured by bond-holders to defend the values of their currencies; even if there is little actual risk of default, a balance of payments deficit may be a signal to speculators. Moreover, financial innovation, creating new types of securities, pyramided on other securities and commitments, raises the cost of financial activities. The ratio of FIRE/GDP in all advanced nations has risen – particularly in Germany and the United States. But financial institutions have few variable costs; hence a downturn in financial activity or in prices affects earning strongly – operating costs cannot easily be adjusted. Risk therefore rises. But it rises for another, probably more important reason: pyramiding securities means that a default in the underlying commitments may bring down the whole house of cards. The impact of a given default is greatly increased. Derivatives also, arguably, lead to higher risk (although the opposite is sometimes contended). These various reasons for higher risk mean that basic real interest rates must be higher to compensate for the greater risks.

in proportion to trade, reflecting the given competitive conditions between nations and firms (based on degrees of technological development), then since only a certain level of deficits can be supported (any higher level will be too risky), the higher ratio of T/Y will require a lower level of Y to bring T down to the point where the implied trade deficits can be financed.

The austerity dilemma

The resulting strategic trading position can be called the "Austerity Dilemma." It is easily expressed in the form of a simple "game" between any given country and its trading partners, considered together or separately. Each country can operate a policy of Expansion or one of Austerity. If both choose Austerity, both make small losses; if both Expand, then both make small gains. But if one country chooses Austerity, and the other chooses to Expand, then the Austere country will have low imports, but its exports will boom. It will experience an inflow of short-term capital and a strengthening of its currency. The Expanding country will suffer from weak exports (since its trading partner is Austere), while undergoing an import boom, at a time when short-term capital will be fleeing, as the currency drops. Austerity will produce a large gain, and Expansion a large loss. The game looks like this:

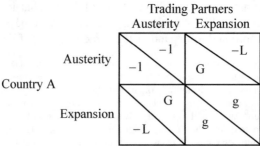

It is clear that Austerity will be the favored policy for both sides. Repeated games might lead to an understanding that Expansion should be favored – but who are the decision-makers: countries or governments? Countries repeat, but governments or regimes are the ones who make the choices. It may not be so easy to define a repeated game.

The impact of new production technologies

In regard to wages and production, the argument is that the new technologies permit production processes to be broken up, so that parts can be contracted out. They can also be picked up and shifted around the world, because the new technologies have dramatically reduced costs of transportation and communication, while permitting through computers the precise monitoring of technological processes from a distance. The effects on macroeconomic performance are

> Labor markets with the lowest wages, adjusted for productivity, risk – including political risks – and transportation costs, will attract direct foreign investment: that is, these markets will be chosen for the placing of production facilities, by advanced corporations; and in addition to being attracted by low wages, capital may be attracted by the prospect of nonunion labor, or by local regimes capable of controlling and policing labor.
>
> Because of the improved ability to control production, processes may be broken down into component parts, many of which can be contracted out to low-wage low-cost (and/or nonunion) producers in underdeveloped regions of the world, or in backward regions at home.
>
> Corporations will not need to invest in facilities to produce these components, and may dispose of facilities they already have.
>
> Investment will therefore be reduced in high-wage economies, or high-wage regions, and increased in low-wage ones.
>
> Capacity utilization and employment levels will fall in high-wage economies and regions, but will not necessarily rise in low-wage ones, if they are also labor-surplus economies.
>
> Wage stagnation, leading to a reduction of the wage share, in high-wage economies and regions, will tend to reduce the growth of demand, which will contribute to the stagnation of investment.
>
> Stagnation and sluggish demand tends to raise risk, leading to higher real interest rates.

The capital invested in production processes may usefully be divided into "low mobility capital" (LMK) and "high mobility capital" (HMK). The former is not readily movable; the latter can pull up stakes easily. LMK is capital invested, for example, in traditional Mass Production processes – heavy long-lasting plant and equipment – which require substantial vertical integration, achieve economies of scale, and cannot easily be broken up. To change location, the whole plant would have to be dismantled. It would be difficult or impossible to move it physically and it would be hard to find a buyer. Such capital is immobile. Likewise, capital less physically encumbered might nevertheless be immobile, because strongly dependent on infrastructure provided by the local community. Or the production processes might be tied into local or national government activities – as with firms using defense-related technologies. HMK is capital invested in processes that are easily separated, so parts or processes can be contracted out. Also such plants must be possible to dismantle, and/or the equipment be quickly written off or easily sold.

> The new technologies have raised the proportion of HMK to LMK in the advanced economies.
>
> HMK has little or no long-term interest in the region in which it settles; hence will not spend in that region, nor will it be concerned with community development or improvement. LMK, by contrast, will be concerned with the region, and its managers will be concerned with community development, since they and their families will benefit from the public goods in that region.
>
> LMK will develop a regionally based organization; that is, the administrative hierarchy that operates the production processes will be regionally based and will spend locally. HMK will tend to administer the production processes as far as possible from its center, so will not develop a regional base. Nor will the system of administration necessarily take the form of a traditional corporate hierarchy. Hence HMK will tend to spend very little locally.
>
> LMK therefore will have a greater multiplier effect, and will be more likely to generate local accelerator effects, than HMK.
>
> Raising the ratio of HMK/LMK will therefore weaken the multiplier and accelerator effects of investment and business spending on the economy, tending to accentuate stagnation.

The conditions for capital to move to a new low-wage location can be expressed in a formula. Here T stands for the transportation costs, C for communications, S for software and computing charges, M for management costs, all referring to the excess over such costs required for production at home (so all are in real terms), and R stands for political and other risks associated with the proposed new location. Then we have

$$T + C + S + M + R < \frac{w_a - w_u \, (u/\$)L}{P}$$

where the RHS shows the difference between the wage in the advanced area and that in the underdeveloped area, expressed in terms of the advanced area's currency, multiplied by the necessary labor, L, and divided by the advanced area's price level, P. On the RHS the first four variables here are all reduced by the new technologies. On the LHS the effect of capital mobility and greater free trade will tend to reduce the exchange rate u/$. Hence the movement will be to encourage the mobility of real capital.

Implications

The interesting point is that all these pressures tend to raise real interest rates, and to lower the growth rate.[13] That is, they tend to generate i > g, where i stands for the real interest rate. Earlier we saw that i > g, and its opposite, i < g, are stable "regimes" in the sense that market caused variations in g will not change the sign of the difference between i and g. We also saw that each regime brought in its wake a variety of economy-wide effects, including social consequences. The movement toward globalization has tended to establish the regime i > g, worldwide, bringing stagnation, austerity, and widespread misery in its wake.

Toward the information economy

The normal working of the new business cycle, reinforced by the effects of the i > g regime, creates pressures for new technologies. As we saw, the cycle forces business to carry excess capacity and, at times, high inventory; it leads to the building up of fixed managerial costs, and it confronts firms with profit-squeezing inflation in the boom, when capacity is finally being fully used.

Each of these can be helped by the new technologies. Just-in-time production reduces the need to carry inventory. Breaking up large production processes permits outsourcing and more flexible product mixes; this both reduces the need to carry large capacity, and allows for targeting specialized markets with custom-tailored production, reducing idle time. Combined with computerized monitoring and control over production processes, it also allows business to take advantage of cheap overseas labor, while dealing a damaging blow to unions at home. Downsizing reduces white-collar and managerial costs.

The new technologies allow products to do more, but there is a cost to the user. Users must learn how they work, and how to use them to best advantage. This, in turn, requires support services. As a result, companies are tending, more and more, to sell, not simply products, but *systems*,

---

[13] In the early postwar period, the real interest rate tended to lie below the growth rate, in the later years, however, this is reversed, the real interest rate generally lies above the growth rate. Recall the charts for the U.S. and the U.K. at the end of Chapter 6 (Figures 6.1 and 6.2, pp. 284–5). In both countries the first twenty years of the postwar period are generally characterized by g > i. The brief exceptions are caused by the Eisenhower and Nixon recessions in the U.S. and by balance of payments crises in the U.K. In both countries from the 1960s on, i and g tend to track one another. Prior to 1960 in the U.S., i and g tended to move inversely. In the 1970s volatility becomes greater in both countries, compared to the late 1950s and the 1960s; moreover, real interest rates go negative for a time. Then at the same time in both countries, 1979–80, the real rate of interest rises above the growth rate and stays there.

mixes of products and services. This is evident not only in new fields, like computers, but also in traditional areas, such as consumer durables, where warranties and service contracts have become more prominent.

A significant feature of systems, however, such as software, is that they are more useful and provide greater benefits, the more widely they are used; that is, they are subject to *increasing returns*. This is true both for consumer products and for capital goods; nor is this necessarily a *long-run* phenomenon. The more intensively processes are used, the better they are understood, and the more productive they are. The more widely they are used the greater the benefits, particularly, of course, if there are positive externalities, as in communications and transportation.

It is difficult to predict at this stage how widespread increasing returns will become. We have argued that it makes sense to describe Craft economies as systems with diminishing returns to utilization, whereas Mass Production economies have constant returns. But while it may be premature to suggest that Information economies are characterized by increasing returns, it nevertheless, might be worth sketching a possible pattern of short-run adjustment.

On the vertical axis we measure output and expenditure. On the horizontal, to the right we have employment. To the left, profits. In the right-hand quadrant, we draw a utilization function showing increasing returns, and below it a straight line rising from the origin, representing the wage bill, the angle of which will be the real wage. As before we will assume that worker households spend their entire income; hence $W = C_w$, the wage bill equals worker consumption.

In the left quadrant we will consider what happens to profits. They are used to underwrite investment, of course, but before they do, they are thrown into the financial markets. A certain proportion will be deducted to pay brokers' fees and commissions; another portion will go to pensions, still another to hospitals and universities. In each case that part of profits will be spent on consumption. Profits will be indicated on the horizontal axis by drawing a 45-degree line from the origin, rising to the left. The surplus of expenditure over wage costs will then be mapped onto the profit axis. Below the 45-degree line will be a straight line rising at a lesser angle. The angle of this line will represent the propensity to spend out of profits, its height at any point will be C_p, consumption out of profits.

Now mark off the level of investment spending on the vertical axis. Such spending will create employment, and the further spending of wages by the employed workers, will generate further demand, leading to still further employment, resulting in the intersection of the $C_w + I$ line with the rising curve, $Y(N)$. This intersection takes place at N_I, the level of

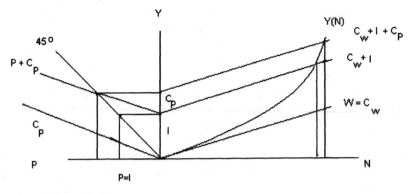

Figure 13.1

employment corresponding to the investment spending. But this gives rise
to profits equal to the level of investment, as indicated on the profit axis by
reference to the 45-degree line. At this level of profits, however,
consumption out of profits will be that indicated by the C_P line. This must
be added to investment, and a new level of employment calculated, to
which will correspond a higher level of profits – and, once again, there
will be spending out of such profits. That is, the total spending will be I
plus the total spending out of profits. Hence we must draw I + C_p, running
from I on the vertical axis to intersect the 45-degree line. This will give
the total spending on the vertical axis to combine with worker consumption to determine total employment, N'. What we see here is a multiplier
operating in financial markets.

Increasing returns and low labor costs, however, do not mean that all
variable costs are low; capital charges can become variable. Interest rates
can be made changeable, and dividends are certainly variable. Raw
materials, intermediate products, and energy vary with adjustments of output
to demand and have not been considered in the preceding discussion. A
more comprehensive multiplier can be defined, in which expenditures for
energy, materials, etc., are related to output, dividends are tied to sales
revenues, while retained earnings and depreciation allowances are the
chief withdrawals. As in the capital market multiplier above, dividends
will be partly saved, and partly paid to foundations, universities, hospitals,
and other bodies that pass the funds along in further, mostly consumer

expenditures. Investment will be the volatile factor, while overhead labor and managerial salaries will be fixed expenditures which underwrite stable patterns of consumption. Thus a balance can be struck between withdrawals and injections, through the adjustment of expenditures that vary with sales and output.

Consider a case of *fixed employment*, for example at the lavel where $Y(N)$ in the diagram has become vertical: let I be net investment, C be consumption of workers and managers $= wL$, where w is an average of wages and salaries and L is total fixed employment, $E = eY$ is energy expenditures per unit output, $M = mY$ is materials per unit output, and $D = dY$ (not to be confused with the calculus operator) is dividends paid out per unit sales revenue, where the receipt of the dividends finances consumer or other spending. Then $Y = [E + M + D] + C + I$, where C is stable and tends to be constant, but I will be volatile. Then,

$$Y = \frac{C + I}{1 - (e + m + d)}, \quad \text{and} \quad \frac{1}{1 - (e + m + d)}$$

is the multiplier, based partly on variable costs of production, and partly on the passing along of earnings in proportion to sales. Employment is fixed, but output is variable, and returns may be constant or, as above, increasing. If firms are successful in moving prices countercyclically, the impact of demand fluctuations on profitability will be dampened, but successive rounds of inflation may be triggered.

If these models capture some elements of the emerging system, we can see that it will adjust differently. Prices and wages will still tend to be inflexible downwards, but quantities will not behave in the accustomed ways. For one thing increasing returns will make it more difficult to increase employment – and, indeed, employment may become inflexible. Moreover, expansion may tend to reduce labor's share. For another, financial markets actually generate a multiplier process, operating in those markets. To this multiplier we can add further elements of variable costs, such as energy and raw materials. But it will clearly lead to different adjustment patterns. However, all this has to be considered speculative; it may still be premature to model the adjustment of the Information economy.

Countercyclical policies

The preceding suggests that the traditional approaches to macroeconomic policy are not properly focused. They tend to be directed toward specific issues – reducing unemployment, curbing inflation, promoting productivity growth, alleviating poverty, maintaining a favorable trade position, and so on.[14] But we see that governments face several different *kinds* of policy issues. Cyclical issues intertwine with questions concerning growth. With regard to cyclical policies, it must first be determined what policies will move the system to the desired regime, a matter that cannot be separated from the desired rate of growth. When i < g, the system has a built-in expansionary bias, working to the benefit of labor, certainly, but also providing high profits and strong growth for industrial capital, although profit margins will be relatively low. It is not so favorable to financial capital, however, with falling debt and strong inflation. By contrast, when i > g, banks and financial corporations prosper. Debt will be increasing and inflation weak. Industrial capital benefits from high profit margins; labor finds itself in a weak bargaining position. The choice between regimes will clearly depend on politics. Popular and labor-based movements will tend to favor the lower regime; wealth-based rentier-funded movements will favor the upper.

Given the choice of regime, however, there will still be a need for policies to limit and control the cycle. Regardless of which regime is preferred, governments will wish to put limits on output volatility, prevent excessive unemployment and inflation, encourage growth, and prevent financial crashes. But we will find an asymmetry here: the policy instruments for controlling the cycle also tend to establish the lower regime. In order to move the system to the upper, it appears to be necessary to give up a number of policy options. In its defense it could be argued that the upper regime also has a lower multiplier and a weaker accelerator, and a reduced propensity for inflation. (On the other hand, it does have higher risk.)

To see this, first, consider the effects on rE and IF of the normal policy instruments. Then we will turn to the alternative policy approaches. (It will be assumed that a well-trained, independent civil service is available to carry out policy, and that governments are capable of acting decisively on a unified program. Obviously, this is unrealistic, but politics are not at issue here.)

[14] Fiscal Keynesianism proposes a pattern of spending adjustments to counter the cycle (which would shift the rE), but it does nothing to control shifts of the IF or swings of the rE. Monetarists, of course, wish to target monetary growth directly. This misunderstands the market forces that govern the growth of the money supply, and it fails to address the causes of the cycle. Such a policy is likely to worsen unemployment and increase volatility, as, indeed, it has whenever put into practice (Godley, 1994; Nell, 1984, 1992). (It may be useful in establishing the regime of i > g, but that is not what it claims to do.) Real business cycle theorists and New Classicals miss important variables altogether, and mistake effects for causes among the ones they do exmaine. Their policy proposals amount to surrender.

Policy instruments

Taxes: A tax on profits, t_p, will tend to cause the IF to shift upwards. For any given level of I less profit is available – by the amount of the tax – to underwrite the spending on the investment and meet fixed obligations. (A progressive tax would cause it to shift upward and to increase in curvature.) A tax on wages, t_w, will reduce the multiplier, and thus the amount of aggregate income generated by a given level of investment spending. It will now be the case that instead of $I = P$, we have $I = P + t_w W$, which implies that $P < I$. So a tax on wages will reduce the slope of the rE line (the reduction in P for each level of I will be greater the higher the level of I, since this implies a higher W). A progressive tax on wages will lower the RE line proportionately more at higher levels of I, since t_w will be higher. Hence this will cause the slope of the rE curve to decline.

A tax on profits will reduce employment through its effect on the IF/rE intersection, but a tax on wages will reduce employment not only by affecting that intersection, but in addition through its impact on the multiplier. For a tax on wages acts as a withdrawal; in regard to respending it is equivalent to saving out of wages. Hence a tax on wages that raises a given amount of revenue, has a greater negative impact on employment than a tax on profits raising that same amount, that is, where $t_w W = t_p P$.

If taxes and wages and profits just balance government spending at full capacity, then if investment spending should drop, tax collections will fall, while welfare spending, compensating for unemployment, will rise, the two effects creating a deficit. (On the other hand, a fixed amount of government spending shifts the rE line up – a higher r, for each level of investment spending – with the shift being proportionately more at lower levels of investment spending.)

Deficits: By deficits we mean net government stimulus. Suppose deficit spending takes place, when capacity is underutilized and labor is available. The effect of deficit spending is to create demand for goods and/or services, without reducing either investment spending or household spending. The government spending in question can be purchases of capital goods, in which case the spending will travel through the secondary circuit, or it could be the hiring of labor, with the wages or salaries spent on consumer goods. In the first case, government spending affects both sectors equally, maintaining the relationship between them, whereas in the second case it impacts only on the consumer sector, thus tending to unbalance them. Net government stimulus will shift the rE; if the effect is unbalanced, it may trigger an accelerator response, depending on capacity utilization and the phase of the cycle.

Government spending will normally be financed by money creation; modern governments cannot wait for tax collection in order to spend. The Central Bank creates deposits for the Treasury, which writes checks buying goods and services. (Technically – in the U.S. – the Treasury borrows from the Fed and pays interest; but the net earnings of the Fed are paid to the Treasury, so from the latter's point of view the transaction is essentially costless.) The funds thrown into circulation then return as taxes; small deficits may be useful in providing currency for precautionary balances. But larger deficits will leave funds in circulation which will accumulate as excess reserves. Two cases are possible:

1. The funds corresponding to the deficit are left in circulation: These funds circulate, either through the capital goods sector in the secondary circulation, being paid out as wages, and thus flowing to consumer goods as well; or they may be paid only to workers as wages, and so flow directly to consumer goods. Either way, the funds end up in the hands of firms in the consumer goods sector, where they will increase the realized profit. Profit will be in excess of the current spending on investment; the surplus profit will be placed in securities, driving up prices and lowering yields. Moreover, when deposited, the funds corresponding to surplus profit will accumulate as excess reserves, which banks will try to lend, leading to a tendency for interest rates to drift down (Mosler, 1994, 1997). Moreover, money creation by the Central Bank means that private commercial banks do not earn profits from financing these activities. Hence their capital, growing through reinvested earnings, will not keep pace with the expansion of output. In the long run this will raise banking system risk and tend to intensify the curvature of iBC.

2. The Treasury borrows back the funds corresponding to the deficit: In this case the government borrows from the private sector, mediated by the banking system. The Treasury issues bonds; these are handled by investment banks, and taken up by commercial banks, pension funds, insurance companies, and others. These bonds will absorb the surplus profit and reduce the downward pressure on interest rates. Investment banks earn commissions for handling, underwriting, etc., and interest is earned by those who buy the bonds. This has two further effects:

> It enables the institutions of the financial system to earn profits, and reinvesting them, to grow at the pace at which output is expanding. Hence these institutions will be able to finance lending (or in the case of pensions, payments) at an appropriate capitalization ratio. To put it differently, if output expands at a faster rate than the capital of the commercial banking system, the latter will only be able to finance the former, in the absence

of further government/Central Bank money creation, at a higher level of risk, that is, at a lower capitalization ratio. Hence borrowing to finance the deficit ensures that the capital of the banking system expands at the rate at which output grows.

It puts a supply of safe assets in the portfolios of financial institutions, enabling them to design their portfolios around a core of safe assets growing at the same pace as output, which for equilibrium, must be the same pace at which the holdings of financial institutions are growing. If the supply of safe assets were to grow more slowly than financial holdings, then portfolio earnings would have to be higher to compensate for greater risk. Wealth holders would demand higher interest rates, and iBC would shift up.

Notice, however, that open-market operations operate on *asset* holdings of money. They do not necessarily affect the quantity of money in circulation. Changes in bank portfolios *may* induce banks to change their lending patterns, which would affect money in circulation, but the incentive effects of portfolio changes will have to contend with other, perhaps often more powerful, pressures on bank leading.

The Golden Rule and a wealth tax: So the growth of government debt serves useful functions; moreover, its cost – the "burden of the debt on future generations" – need cause no sleepless nights. Let us define a tax on wealth of all kinds and forms. The tax rate will be τ, where W is income-earning wealth, and D is the total debt. Then the tax will be given by

$$iD = \tau W, \text{ or } \tau = i\frac{D}{W}$$

That is, the interest cost of the debt will be exactly covered by the tax on wealth of all kinds. Note the implication when the extended golden rule holds, so that $i = g$, and the total debt grows at the same pace as the economy. In that case,

$$i = g = \frac{\text{Deficit}}{\text{Debt}}$$

which implies that

$$\text{Deficit} = \tau W = \text{interest on the debt}$$

Of course, by the same token, if $i < g$, the deficit will be falling in relation to the debt, so that debt will be falling in relation to output, and if $i > g$, both will be rising.

Wages and incomes policies: By setting the minimum wage and establishing wage and salary levels in government service, and through regulation of pensions and working conditions, and through imposing standards in government contracts, government policy can raise or lower the general level of money wages. By itself, this could tend to be inflationary, although it will also provide a stimulus to business to raise productivity. This stimulus can be promoted further by offering tax advantages, subsidies and preference in government contracting to those companies that innovate and improve productivity. Government can also provide incentives to keep prices down, through hard bargaining in government contracts, tax penalties on firms that raise prices and selective controls.

A significant rise in real wages, if sufficiently widespread, can have another important effect: it can create a new or expanded market for consumer goods, especially durables. In this way it can raise the "normal" rate of growth, or help to form one, if market growth is diverse. And even a small increase in real wages will tend to strengthen consumer demand and raise the level of capacity utilization.

On the other hand under conditions of prolonged full employment wage demands will tend to put pressure on profits, and may create excessive inflation. Incomes policies will help to keep them in line with what can be expected in productivity growth.

Controls: During and after World War II direct controls on prices (including money wages and interest rates), imports – and sometimes exports, too – and capital flows were an important part of economic policy, and, indeed, played an important role in maintaining the i < g regime. Indeed, dismantling these controls was a necessary precondition for moving to stagnation and financial expansion. Price controls provided a policy tool for controlling and preventing inflation. Interest rate controls could help to keep interest down, encouraging investment and construction. They could also prevent banks from *paying* interest on demand deposits, preventing bidding wars for deposits. Wage controls lowered the pressures on business in boom times. Import controls can keep the balance of payments from getting out of hand; and export controls can prevent shortages developing in goods needed by a weak domestic economy where foreigners can outbid locals. Controls on capital movements can stem capital flight, and can sometimes prevent or weaken speculation against a currency.

The case against controls is twofold. It is claimed that controls "distort" the market. No doubt, but so what? If we reject the idea that the market "allocates" optimally, this is no problem. The question is whether controls direct the forces of the market in more desirable channels – which may sometimes be a difficult matter to determine. Market forces are complex and may not always be obvious, particularly the incentives to innovate. Second, it is claimed that controls don't work. Agents come to find ways around them. They merely increase costs. Again, so what? If controls increase the costs of certain lines of activity, they will tend to reduce the quantity of it – which is usually the point. No doubt agents will find ways around the controls; the controllers then need to find better ways of controlling. It is a developing system, what evolutionary theorists sometimes call an "arms race" (the gazelles get faster, and then the leopards catch up).

In many instances a form of indirect controls may be superior. Instead of regulating prices or wages bureaucratically, government can use its own market position to enforce a competitive price or a wage norm. Government is a substantial producer in many markets, e.g. public power. In others it is a (sometimes the) major purchaser, as in markets for military and space equipment. And it has an important position as an employer. It could use its market position to establish "yardstick" prices and wage norms – as the Tennessee Valley Authority did for decades. In addition Government could make use of its taxing powers, for example, to impose penalties on inflationary price or wage increase – recall the "tax-based incomes policies" which the Carter Administration toyed with. Finally, a truly activist government could use its substantial holdings of assets (and its power to issue money) to take ownership positions in major firms, on the basis of which it could appoint public interest directors, audit the books and bring their activities in line with its economic program, at the very least ensuring that the firms not violate the public interest while pursuing profitable activities. The Government would not need to remain in any given firm for a long period; it could take a position in a set of firms for a time, then sell out and move into a different sector of the economy, so that over the course of a decade or so the Government would have held a position in most of the major firms in the economy (Nell, 1996, Chapter 5). To work this out in practical terms would require extensive study; there would have to be safeguards for each firm's sensitive plans and information. Nor would it be easy to implement politically. The point, however, in respect to all of the above, is that Government's extensive powers, and the many ways it could use these powers to direct the course of the economy, have not been explored fully. Indeed, the *laissez faire* mentality would prefer to pretend that these powers do not even exist!

Policy approaches

Broadly speaking, a strong and sustained Austerity package will be needed to shift the system towards the $i > g$ regime, while a similarly persistent Keynesian Expansionist package would tend to establish $i < g$.

Temporarily raising i above g will not be sufficient to establish the austerity regime; a high level of fixed costs must be put in place, with a corresponding high pattern of spending based on wealth. The markup must be raised and the multiplier brought down, weakening the accelerator. This requires Austerity to rein in wages, and support increases in managerial salaries. Inflation must be curbed and inflationary expectations eradicated. High interest rates, and high debt are required to increase risk, intensifying the curvature of the IF.

In the same way, to establish i < g, fixed costs must be lowered, and wealth-based spending reduced. This can be accomplished by leveling the income distribution, which will also lower the markup, raising the multiplier and strengthening the accelerator. Low-level inflation will become endemic, productivity growth will improve, risk will be lowered, and labor's position in wage bargaining will strengthen.

Furthermore, policy works differently in the two regimes. Monetary policy affects i directly, by shifting the iBC curve, but it affects g indirectly, through the i,g interaction and the effect on IF. Fiscal policy, on the other hand, affects g directly, shifting the rE, and i indirectly, through the effect of g on dP/P. When $g - i < 0$ this effect will be negligible, but when $g - i > 0$, it will be important.

Given the regime i > g, monetary policy will tend to be effective in adjusting g, because the changes due to shifts in the iBC and IF will be relatively large, and because changes in g have only small effects on inflation. Fiscal policy, however, will tend to be less effective; since corporate debt will be growing faster than corporate income, risk will tend to be high, so that the IF will be steep. Hence an upward shift of the rE will raise r, but bring only a small rise in g. Indeed if the iBC is upward sloping because of risk and corporate indebtedness, fiscal policy designed to shift rE may also drive up i, an effect which would have to be counteracted by monetary policy.

Given the regime i < g, monetary policy will tend to be ineffective in adjusting g, because changes in g will tend to bring changes in inflation in the same direction, undermining the policy change in the real interest rate. Fiscal policy, however, will lead to changes in real interest rates that will tend to support the policy direction – thus a policy-driven rise in g will raise inflation, lowering real interest rates.

In the long run, when i < g bank capital will grow more slowly than activity, and will have to be supplemented. When i > g it will grow faster, leading to a relative expansion of the financial sector and, very likely, to speculative excesses.

Next, consider the cycle itself. Under either regime, a cycle of this kind could easily have a wide and hard-to-predict amplitude. The relationships between the variables and their directions of movement may be reasonably stable in the upswing and downswing, but the magnitudes may be influenced by a wide variety of factors. As for the turning points it is likely to be very difficult to predict the timing. In short, this cycle is itself somewhat unstable.

To bring it under control, the amplitude must be narrowed, the booms dampened, and slumps raised. But the policy tools available for this, used in the ways that have proved their effectiveness, are likely to favor expansion. Automatic stabilizers, such as unemployment compensation, antipoverty spending, "bracket creep," will help to reduce the amplitude, but they will also tend to promote growth. A large and stable government sector will put a floor under slumps. The spending of this sector, if financed by taxes on wages, of course, will not add to demand, though it may stabilize it. But if financed by taxes on profits, or by borrowing, it makes a net addition to aggregate demand. Provided there is a significant corporate income tax, offering investment tax credits can help to encourage investment in the slump, while an excess profits tax can help to dampen the boom. Ideally, these should be set to kick in automatically at given points. A wages and incomes policy can keep low-level inflation under control. Tax penalties and incentives can encourage productivity growth and keep rE from swinging up as the downturn begins. International or even national buffer stocks could be developed to prevent demand-sensitive primary products from rising during the boom. Finally, pegging the interest rate, as suggested earlier, will help to slow down shifts in the IF, while taxing speculative funds can aid in preventing outbreaks of euphoria. By curbing overlending and over-leveraging this will dampen the increase of risk. But all these measures tend to promote growth and curb financial capital.

The government, in short, can stabilize and diminish the cycle. To do this, it must put a floor under the downswing by means of government spending, narrow the amplitude of the cycle by means of automatic stabilizers, and soften both the upswing and downswing through direct fiscal policy. In addition, control over credit could be used to further reduce the amplitude, and prevent speculative crashes. But these policies will tend to hold interest rates down while encouraging growth, in short, will tend to establish the regime in which $i < g$.

Medium and long-term interest rates are affected by Central Bank and/or Treasury purchases and sales of securities; short-term rates tend to be dominated by the Central Bank's rediscount rate. These provide policy tools. Using them, discretionary monetary policy has chiefly tried to keep interest rates moving with the cycle, rather than lagging behind it. Yet this is not likely to provide much help in controlling the cycle. To raise interest rates in the boom and lower them in the slump simply reinforces the pattern of the cycle. Artful timing might help to dampen strong movements and prevent overshooting, but the effects can hardly be very great, since policy is following what the market would do anyway. Raising interest rates at a time when profits are rising will not dampen investment, any more than lowering them when profits are falling will encourage it. The important point is that a procyclical adjustment of interest rates does nothing to prevent funds from shifting to, or staying in, financial markets when speculation is rising; certainly it does nothing to curb speculative excesses. But this is where credit controls and bank regulations could have a beneficial impact.

Conclusions

With the development of Mass Production, real wage changes are no longer necessary to induce changes in employment, when aggregate demand varies. The system is designed to adapt to changes in the rate of sales. In the short run employment is governed by the multiplier; the labor market reflects this, and is weakly unstable, under normal conditions generating steady but mild inflationary pressures. The monetary system accommodates. Financial markets play a complex role; they tend to turn around destabilizing movements of the goods and labor markets. On the other hand, they are themselves unstable, and the system runs in a cycle that loosely follows financial expansions and crashes, with the real wage tending to move procyclically. The cycle arises from the interaction of the financial and real sides of the system.

Following Keynes does lead, through arbitrage in capital assets, to a tendency toward a uniform average rate of return, although it is the expected rate in money terms, on both real and financial assets, not the Ricardian notion. When the rate of interest equals the rate of growth, the ability of the banking system to supply credit for working capital will rise at the same rate that the need for such funds increases – bank capital will grow through compound interest at the same rate that real capital accumulates. Long-term debt will remain stable. Government debt will grow pari passu with the economy, and the deficit will equal the interest on the debt. The long-term expected rate of growth of markets will be associated with benchmark prices – but while these are stable and provide an appropriate foundation for forming aggregates, they are not centers of gravitation, and do not imply a long-period equilibrium of any sort. Keynesian arbitrage, moreover, provides the basis for a powerful internal critique of the neo-Classical synthesis. Surprisingly, it also enables us to show the inadequacy of a monetary theory based solely on money as an asset.

The Mass Production model bears on a prime objective for many post-Keynesians, developing an account of the business cycle in which instability depends on the financial system. However, a closer look at real-financial interactions does suggest that this case has been overstated when it is claimed that finance destabilizes an otherwise stable system. That claim applies to the Craft economy. The instability of Mass Production has roots in the real system; indeed, the unstable real system is partially *stabilized* by the financial system. By constraining and turning around the unstable movement it, in effect, creates the basic pattern of the cycle. But the fluctuations arise from the interaction of all markets – goods, labor, and financial. The cycle is deeply embedded, and substantial policy interventions are needed to control it.

But the cycle comes in two varieties: one austerity-based, the other expansionist. Governments must choose between them, and their ability to control the cycle will depend on this choice. If they choose the austerity-based regime, i > g, they sacrifice control over the cycle, and must hope that the weaker multiplier and accelerator, and reduced inflationary tendency will suffice. However, this regime provides lower output growth, lower productivity growth, greater inequality and higher unemployment. The only real beneficiaries are rentiers and financial capital. The expansionist choice offers benefits to a far wider constituency, and it encourages the active use of policy instruments to control the volatility of the cycle.

But the choice is not so easily made. For the dilemma posed to business by the Mass Production cycle has generated a set of responses. Innovations have made it possible to reorganize production on a world scale; in doing so the power of organized labor in the advanced countries has been seriously undermined. This reorganization has created powerful pressures to remove the restrictions on trade. But freer trade has led to stagnation. Interest rates are high not just as a matter of policy, but because the opening of markets to capital flows – partly the result of new communications technologies – has created pressures driving them up. In short, the responses of business to the adjustment process of Mass Production have tended to bring about stagnation, high interest rates, and booming financial markets.

Before policy can return the economy to expansion, it seems that control over world financial markets will have to be established. At present there is no institutional framework for this.

Conclusions

This book began with a complaint and a vision. The complaint charged that much of what passes for modern economics is actually the theory of the working of markets in an earlier age, propped up and kept alive by infusions of rational choice, prescriptions masquerading as descriptions. The result is a set of models that fails to give us much insight into how our world actually works, presented in rancorous debates over issues far removed from practical concerns.

The vision put History in place of Equilibrium, and saw the development of markets as a succession of patterns of adjustment, generated by technological change, itself driven by incentives created by markets. Market adjustment creates problems; innovation solves them, but changes the character of costs, which, in turn, changes the way markets adjust. Markets work through money, and the forms of money change as markets change. Money and credit are inseparably linked; as money develops, so does credit, and with credit, financial markets. These markets interact with those for commodities and labor in varying ways. Our stylized economic history is a parade of forms, marching through time. In fact, of course, reality will always be a mixture – the new developments will crowd in before the old have departed the stage. Only in theory can we see the unadulterated systems.

CHAPTER 14

Keynesian themes on Classical grounds

It's been a long journey. We've covered a lot of ground, but, hopefully, we've climbed a peak, giving us some perspective to see around us, and assess what we've accomplished.

We began with History. Economics based on rational choice does not have much of a place for history. If rationality is the same always and everywhere, all History can do is shift the constraints and incidental circumstances. Markets, fundamentally, will always be the same. Not much room for Evolution.

However, basing the theory of market equilibrium on rational choice involves a philosophical mistake, a confusion of prescription with description. Rational choice – optimizing – has a place in economics, but not in the service of equilibrium. It is a tool to be used in pursuit of innovation. To situate innovations, to understand the forces generating them, it is necessary to have a picture of the structure of the system, at the basis of which lies technology. Technology, in turn, responds to incentives, which the market generates plentifully. Hence we have evolutionary pressures driving the development of markets, and market institutions – Transformational Growth.

These pressures will be expressed in monetary terms. To understand them we will need to know how the monetary system works. And this brings us face to face with our title: just what do we mean, calling our theory *Keynes after Sraffa*?

Keynes after Sraffa?

Keynes argued that capitalism had to be understood as a theory of monetary production. That is the way we approached the Classical system; the result was the theory of monetary circulation, in which the movement of money reflects the interdependence of production.

This theory requires stable monetary values, so that when distribution changes, for example, funds can be shifted from one part of the circuit – paying wages, for instance – to another – profits – without a change in the value of the funds. But this relies on a condition very close to the "Standard Commodity," namely the linear wage-profit trade-off. This, we showed, could be derived on the basis of the "Golden Rule," which, in

717

turn, was implied by capital arbitrage – capital in monetary form, seeking the highest return.

So the theory of monetary circulation implies capitalist arbitrage – market pressures pulling the rate of interest, the rate of profits, and the rate of growth together. But the implications of such arbitrage indicate that the theory of liquidity preference, treating money simply as an asset, is inadequate. In Keynes's own system capitalist arbitrage recreates the "Classical Dichotomy," and it shows that the "neo-Classical synthesis" is inconsistent.

Keynes's theory of effective demand, on the other hand, can be generalized to earlier eras. To do this, however, we have to show how the Classical Price Equations apply, and how they are to be understood in different eras. The theory of effective demand is a theory of behavior; the Classical Equations present a picture of the structure of capitalism. In the earliest period, with employment and output virtually fixed in the short period, Sraffa's interpretation applies – the symbols representing inputs are *fixed amounts*, not coefficients in the usual sense. Under Mass Production, output and employment are flexible, but the coefficients are fixed, since constant returns can be assumed. In the middle period, the coefficients of production must be understood to represent input requirements at the *minimum* points of the average cost curves. Given this and an account of prices, variations in demand can be shown to have different effects in different periods.

In the earlier periods the price mechanism acted in a moderately stabilizing way, so that variations in investment and consumption tended to be offsetting, rather than reinforcing, as they are in the modern era. The early monetary and financial systems tended to be unstable, and the interaction of the two led to periodic crises. The price behavior that brought about stabilizing "real" responses, however, was costly to firms, and tended to encourage the development of more flexible technology. This, in turn, led to unstable commodity and labor markets, which interacted with a more fully developed monetary system in a cyclical pattern that can be considered mildly stabilizing. But behind the very price flexibility that brought such stability to the system loomed the spectre of bankruptcy for firms. Price flexibility implied fluctuating profits. Firms sought and found ways to stabilize prices and profits by making employment flexible, and in the process moved the entire economy into a new age. But the resulting system, in its turn, burdened firms with unwanted capacity for too much of the cycle and threatened them with profit squeezes in good times – leading to the innovations that we now call high-tech.

These changing patterns of market behavior and the pressures that brought them about make up the centerpiece of Transformational Growth. This is a very different conception of markets and the principles of economics than that *shared* by conventional theory and neo-Ricardian thinking – and this difference – History in place of Equilibrium – is what has made it possible to develop a theory of effective demand on the basis provided by the Classical Equations.

Approaches to theory

Like neo-Ricardians, Transformational Growth places the Classical Equations in a central position, but it sees a different role for them in the redevelopment of economic theory. The former school tends to deploy the Classical Equations in a critique of neo-Classical theory, grounded in the choice of technique framework. From a positive perspective they are considered to provide a general explanation of prices in a long-period, comparatively static framework, and this can be used to unravel a number of puzzles in the History of Economic Thought. The explanation of prices, however, has not been closely connected to pricing behavior.

Joan Robinson started the capital theory/production function controversies in the 1950s. After Sraffa's book in 1960 the next decades saw major battles in the journals, battles which resulted in conclusions widely held today: to wit, the technical errors are conceded, but their significance is contested. This has a practical meaning: open any major journal at random today, and there will be marginal products, aggregate production functions, *et hoc genus omnia* – with no hint that any technical error is involved. The critique is simply ignored. It can't be answered, but it is held to be unimportant.

The neo-Ricardian project initially aimed at reviving the Classical approach. The idea, it seemed, was to develop an *alternative* economics, a science of economic phenomena grounded on different principles. Such a science would have to cover most of the same topics, even though it might break up the subject-matter differently. Yet since Sraffa's book was published virtually no progress has been made in this. The *only* area of economics in which the neo-Ricardians are major players is the History of Thought. There is no neo-Ricardian macroeconomics, there are no studies of inflation and unemployment, no (or very weak) monetary theory, no financial market theory, no theory of demand, no labor market theory (some historical and history of thought work on wages), no theory of the firm, no theories of other institutions, or of the state or of economic policy, not much on agriculture, one good book on the environment, some largely critical work on foreign trade, some work on technological development. Finally, a few good books on prices and the theory of production, which, however, are largely amplifications and developments of Sraffa, *not extensions of the approach to the rest of economics*. Not much to show for a generation of work.

Why should anyone give up neo-Classical theory, any more than Swiss cheese, just because it has a few holes in it? The Sraffian equations are fine, but the neo-Ricardian criticism is a sponge that proposes to wipe away the entire horizon! What is going to replace it? How can we answer the questions that every economist faces – what makes the economy grow, why do people have to pay for some goods and not others, and why are they willing to, why are corporations growing and merging, why are some people rich and others poor and how is this changing, will the developing nations develop and what will happen to the environment if they do, will the banks crash and if so when, why is there so much unemployment/ inflation/poverty/etc., poverty-in-the-midst-of-plenty, and so on. The neo-Ricardians have not developed answers to these and other obvious questions; in fact, for the most part, no one has tried. What has been done, on the other hand, is a comprehensive reanalysis of the History of Economic Thought.

The original idea was to move toward a complete reconstruction of economics, on a revived *and revised* form of the Classical approach, *not* merely criticism of neo-Classical arguments, nor clarification of Classical arguments. The approach would be different: it would be sound theory, but theory based on a realistic account of institutions and history. Furthermore, such analyses could be expected to lead to new, useful, and progressive formulations of policy. That was also the hope of the summer school in Trieste.

What has actually emerged must be considered disappointing. A Classical "general equilibrium" theory has been worked out, together with a critique of neo-Classical – but there has been no development of a new *economics*. To be sure, there are a few scattered articles on a number of the above topics. But besides the critical work and the development of price theory, the important and widely recognized work has centered on the History of Economic Thought.[1]

Contrast the contributions, apart from capital theory, with other contemporary and related approaches. Unlike the Old and New Keynesians, neo-Ricardians have had little to say about modern labor markets (as opposed to the history of wage concepts), or poverty, or urban decay, or inflation and unemployment, or the growth and change in the corporation, or balance of payments problems on current account. Unlike the post-Keynesians, neo-Ricardians have little to say about money and finance,

---

[1] Indeed, the major controversies in that field appear to focus on the work of Sraffians – especially with respect to Ricardo, Malthus, Smith, and Marx. Few major articles appear in which neo-Ricardian writings are not cited. In Marxian economics, considered as a separate field, Sraffians have also made a major impact.

except for some ideas that no one else finds convincing. Nor have they related production prices to corporate pricing. In spite of a theoretical framework of great mathematical elegance and power, very little has been done to develop the theory of growth – and even less to deal with dynamics. Post-Keynesians may sing only one uncertain note, but people of widely different persuasions are interested in hearing it. Post-Keynesians, old-fashioned Keynesians, and even New Keynesians may all be weak or misguided theorists – but they speak to the issues of the day. For the most part neo-Ricardians don't, and, even more revealingly, when they (we) do, *they draw on conventional techniques having little or no connection to their basic theory.*

It can easily be retorted – the same is true of the neo-Classicals. Their applied economics has nothing to do with their general equilibrium theory. But this leads to a further point: the positive development of neo-Ricardian theory has followed a course through the same valley in which the mainstream flows, meandering along just a little to the left of that stream. As a result, in spite of their obvious differences, the two theories are very similar in important respects:

> Both develop their analysis in barter (nonmonetary) terms
> Both concentrate chiefly on the stationary case
> Both represent production abstractly, paying little attention to the details of the technologies
> Both treat prices as signals (NC as signals to offer or demand goods, NR as signals to make a choice of technique)
> Both consider prices to be flexible (in NR prices must be flexible to converge to the long-period position)
> Both accept the same "choice of technique" problematic – although institutional and historical analysis would suggest that no technique was ever "chosen" in such a manner
> Both argue that their systems will converge to a stable point (although in both cases difficulties have emerged with respect to such claims)

Small wonder such theories cannot shed light on real problems! Capitalism has never operated in barter terms. Money is fundamental. It has never been stationary; it always grows. Technologies differ, and the differences matter for the way markets adjust; in particular, Craft technologies are rigid in ways that Mass Production is flexible – and the new Information technologies are flexible where Mass Production was rigid. Prices are either not signals, or are not important signals. Quantities and productivities are. Prices were only flexible prior to World War I, or in underdeveloped economies; they have not been flexible in the postwar period (in the sense required by theory). No technique was ever chosen in the sense or manner required by the approach; but *competition* has been an important force driving the development of technology. Technology develops as part of competitive dynamics. Finally, instability is and has been a major problem; different kinds of instability need to be analyzed and classified – in particular the frequently unstable relationships between commodity and labor markets, on the one hand, and money and financial markets, on the other. The applicability of the Classical Equations *cannot* depend upon their being the point upon which a process can be expected to converge. In the postwar period market processes are as likely – or more likely – to diverge as to converge.

So, should we abandon theory and become institutionalists? Not at all; we need to develop theory more appropriately, with particular attention to dynamics. Should we then abandon the work of the last thirty years? Not at all; that work is excellent. It's not what has been done that is the problem; it's what *hasn't*.

The first problem – from the point of view developed here – is that the modeling of the Classical system has largely proceeded in barter terms. Inflation, wages and labor (money wages), aggregate demand, corporate pricing, interest and finance, balance of payments (among others!), all arguably involve money in some sort of essential way. These issues have to be analyzed in monetary terms, which precludes using the Classical Equations unless they can be expressed in monetary terms. This has required, first, a theory of monetary circulation, second, a theory of money as an asset – a store of value – and finally, a theory of the interaction between the two.

The second problem is *dynamics*: to take account of the historical development of institutions, and to note how this interacts with the changing character of markets, and how, through their relation to competition, both institutions and markets affect and are affected by the development of technology.

Both of these called for a rethinking of the Classical Equations.

The Classical system: accumulation and profitability

The Classical Equations, as developed here, present the central core of capitalism. But this core has to be understood as the model of a working economy, involving *both sides* of the system, at the same time. Profits and prices comprise one side, accumulation and quantities, the other. Monetary transactions only take place when both are involved. Capital arbitrage means shifting capital, which is a sum of quantities times prices, expressed – and conveyed – in money.

So we began from the equations describing wages, the rate of profits and prices, on the one hand, and consumption, the rate of growth and quantities, on the other. Taking these together, and assuming effective capital arbitrage, we described the relationships that remain invariant during the process of Transformational Growth, and which identify the basic structural relationships in capital maintenance and accumulation. In particular, the wage-bill/capital requirements relationship provides the foundations for both monetary circulation and the multiplier. Moreover, this relationship turns out to be equivalent to – to imply and be implied by – the Golden Rule, when the Classical Equations are written in the correct dual form. In turn, the Golden Rule, and its extension to cover interest and financial capital, are fundamental to market adjustment in capitalism. The relationships between i and r, and between i and g, underlie the basic mechanisms by which the monetary system and financial markets interact with commodity and labor markets. These are different in different eras, to be sure, but in all eras adjustment depends on capital arbitrage. Finally, the Golden Rule is not only equivalent to the wage-bill/capital requirements condition; it also implies and is implied by a linear wage-profit trade-off. This has the fortunate property that changes in the wage bill are exactly offset by opposite changes in profit payments, a property that ensures that changes in distribution need not disrupt banking practices. Moreover, it ensures that simple calculations can be made of the gains or losses from real wage changes, and it supports the view that under conditions of labor shortage there is an unambiguous incentive to labor-saving, capital-using technical change.

The Classical Equations give us the mathematical relationships of the system, but markets are organized by and embedded in institutions, which develop and change as the technology develops.

Institutions

Institutions play an important role in structuring the incentives that drive the process of Transformational Growth. In summary, starting with households, going on to production and firms, then to families and the state, we find that at each stage institutions and the way markets adjust mutually influence one another.

The household and the market

From its earliest beginnings market-driven industry has been guided by a central principle: appropriate the activities or products of the household, make them better or cheaper, and sell them back. The household economy of several hundred years ago, or even of the last century on the American frontier, bought or grew or traded its own products for raw materials – raw produce, dry goods, skins, grain – and produced its own finished goods, for its own use or for trading in a local exchange system.

Traditional society grew up around settled cultivation, organized through ties of kinship, which bound the peasant – and his masters – to the soil. These bonds were fixed at birth, remained through life, and were passed on from generation to generation. The seasons rolled, the weather varied, the cycle turned, but nothing changed. The world went on forever, under the gaze of heaven. The emergence of the market shattered the fixed and rigid peace of the countryside. Market pressures created unparalleled opportunites for geographic and social mobility. They also forced the development of new skills, opened opportunities for new talents, and in general broke apart the frozen stupor of rural life, bringing openings for initiative, and chances for adventure. In the process, of course, a lot was lost: the safety of custom, the wisdom of tradition, the ease of settled ways, and the closeness and good fit, the caring relationship, between human society and the earth (Laslett, 1984).

But the gains were enormous. Individuals had the opportunity to choose their own paths of development, to choose careers, to try to become whatever they wished. Of course, it wasn't really possible except in a relatively few cases, but even the purely formal establishment of equality of opportunity changed thinking. People were no longer what they were born; birth no longer defined the horizon of the possible. There were choices, and there were chances.

The very technological dynamism of capitalism has required changes in the way children are socialized. The system of fathers passing along their skills to their sons, mothers to their daughters, could not be continued, once innovation became intense and widespread; the skills of the parents would be outmoded, out-of-date. Children needed to be taught things their parents never knew, skills their parents never could have learned. In traditional society the work of the world was the same generation after generation. But in capitalism the jobs of tomorrow may be nothing like the jobs of yesterday, and the skills and knowledge required may be altogether different. Under these circumstances vocational education cannot take place in the home. The consequence has been an enormous growth in the activities of the State.

Yet market pressures have also had terrible consequences for the development of people. The socialization institutions of the state and private sector provide nothing comparable to the family in regard to nurturing. Love can't be bought, or doled out by a bureaucracy. The destruction of traditions and the withering of the extended family have left people lonely, isolated, afraid, subject to anomie, and with rootless, apparently unfocused anger. There is an evident spiritual malaise throughout Western society, a loss of values, of rootedness, of purpose in life, which no amount of entrepreneurship or calculated self-interest can overcome. Yet none of the "commune" movements of the last century or so, whether religious or political, seems to have succeeded either.

Before the rise of the modern economy the world was the world we found; now it is the world we have made. In traditional society the household fitted into the rhythms of the earth; the land, the seasons and the cycle of animal life provided a setting that had to be respected and cared for, and to which human life had to adapt. But with the growth of industry and technology, the world can be changed in whatever directions the market calls for.

Yet all is not well. The pace of innovation and improvement has probably never been greater, yet today at least in the United States, and perhaps in many advanced economies, there is a sense of malaise and disillusion, a feeling that greater and greater efforts are required to achieve the most modest improvements – even to stand still! Innovations are dramatic and far-reaching; so, however, are their consequences, often in unwanted ways. The overall impact of economic growth on the quality of life is uncertain, bordering on the undesirable.

Our theme has been the role of the market in generating innovation and productivity growth. Let us take stock of where we have arrived.

The transformation of consumption

To begin with consumption: a century ago, the management of consumption was a household project. The basic necessities of food, clothing, and shelter, together with light and heat, made up the bulk of consumer demand. These and many other elements in consumption were bought or brought into the household in unfinished form and turned into final products by domestic labor. Foods would be bought in the market or grown; but meals were prepared, preserves put up, jams and jellies made, and food packed for winter storage in the household. Yarn and cloth would be bought at the dry-goods store; clothes would be sewn and spun at home. Candles, furniture, wood for heating, and many other necessities and conveniences were produced at home.

Modern capitalism has transformed these household acitivities into industries, in which products serving the same functions, usually better, are turned out more cheaply by mass production. Canning replaced home preserves, modern textiles home sewing, prepared foods home cooking, radio and TV "home entertainment"; the old craft skills survive only as hobbies, which, in turn, have become an industry as well, so that the old craft tools are now made by modern precision methods. The household has been deskilled, stripped of its traditional functions. Women were thereby set free, but they may also have been left at a loss as to what to do next.

By the last part of the twentieth century it was clear that there was very little left in the way of household activities for the industrial system to appropriate. Birth and death, hospitals and funeral homes, nursing homes for care of the aged, craft activities of all sorts, the manufacture of household items in ordinary use, all have long since disappeared into the market, to return polished, improved, and advertised in color. The household itself, stripped of its traditional functions, appears to be changing in nature. Moreover, technological developments have been intensifying these changes, as consumption has taken on an increasingly social character. Households used to buy food and cook meals, buy cloth and make clothes; food and clothing next became increasingly prepared or ready-made. Now more and more people eat out and wear whatever is in style, both leaving households less to do and making them less the focus of activity. And the newest developments are in communications, transportation, and data processing – all social processes impacting on households in ways that tend to reduce their importance as centers of activity and decision.

In the process of moving the final stage of production – the finishing – of consumer goods out of the household and into industry, the nature and relative importance of these goods changed. This went through several stages. At first, under Craft economy conditions, a great deal of both production and consumption took place in the household, in ways that involved only household members. Much artisan production was organized by the household; the shop and the farm were household affairs, the home was also the workplace. And the most important goods, food and clothes and other household basics, could be considered "individual" goods in the sense that the textbook Robinson Crusoe could in principle both produce and consume them in isolation. Growing food and making clothes involve traditional skills that a household could master and carry out – even if a larger number might do the job more efficiently. Households often grew part of their food, butchered their own animals, laid up preserves, bought bolts of cloth and made their clothes. Certainly they were under pressure to live up to the appropriate social standards, but to a very large extent their consumption pattern was up to them, and depended on their own skills (Larkin, 1988, chs. 1–3).

But with the development of the factory system, even in its early stages, production became separated from the household, so that its organization could adapt to functional needs, rather than reflecting kinship and family relationships. Consumption, however, remained as before. But all this changed with Mass Production and the emergence of standardized products.

Mass production changed the supply side, redesigning and standardizing the goods, replacing Robinson by Henry Ford. Production requires an organized collective effort, independent of the family system, and mass production requires a mass market. The goods are finished in the factory;

they do not have to be completed at home, and they will not be produced at all unless there is enough of a market to justify mass production. The act of consumption remains up to the household, but the social dimension is implicit. Advertising and marketing, interacting with product design soon lead to further changes, adding a social character to personal consumption. Clothes, furniture, household appliances, and food acquire symbolic status, and come to express the style and social position of the household or individual.[2]

Transformational Growth thus takes both production and consumption from activities performed in the household, under the aegis of the kinship system, to collective and socially defined projects, practiced according to public criteria and governed by the rules of the market place. But besides this, there is another dimension to the change in consumption.

Some economic activities cannot be individual, in principle. That's why even the textbook Robinson needs Friday. Services, for example are performed by one party for another. There are many different kinds: a telephone call requires someone else to answer; a game requires the other players – and the other team – to show up. A Broadway show needs a cast and an audience. In other words, these acts of consumption are inherently social; they cannot in principle be undertaken by one person alone. At least one other person is involved, and because the act establishes a relationship, the timing has to be coordinated.

So this is where a second transformation can be found. Besides the evolution in the character of basic consumption from domestic finishing to industrial Mass Production, there can be seen another development, in the changing composition of household consumption. In the earlier stages of capitalism, the basic items, whose consumption remains an individual act even when it is mass produced for a mass market – for example, food, clothing, and shelter – made up the major part of the household budget.

---

[2] We referred earlier, in Chapters 7 and 10, to the complementary character of the principal categories of consumer spending. The different kinds of consumption together make up a lifestyle, and have to be increased or decreased in proportion. There is little room for substitution between them – we cannot make up for a deficient diet with more clothes or better transportation. (We can substitute potatoes for rice, but we can't get a balanced diet by substituting potatoes for meat – tofu for meat, perhaps, but that is a different lifestyle!) Given the categories that make up a lifestyle, we can't increase one without at the same time increasing others – we can't have more appliances without a larger house, or better appliances without more power. We don't need a better washer-drier unless we have better clothes; we don't need better cooking utensils unless we plan a more gourmet menu – but we do need all of these if we plan to live in a certain style, a style, which in turn, goes with a certain level of education. These complementarities make it difficult for price flexibility to bring about adjustment, and they also require that the production of the various consumer goods grow in pace.

Later the proportions shift, as collectively consumed goods, and goods with a social aspect, make up a progressively larger fraction of household expenditure. Transportation, communications, entertainment, and education all increase with industrial development. The rise of services, and the changes in the character of services, has been a persistent feature of both CBF/MA and MP/CA.

This increasingly social character of consumption has been matched by a corresponding change in the nature of expenditure. In the past consumption goods were purchased by means of direct payments of cash. Households handled their money, often keeping it in a safe or strongbox, or under the mattress. Very little credit was involved. Today cash is now largely confined to incidentals and illegal goods. Major purchases and regular household expenses have long been paid by check and are increasingly managed through the servicing of credit – charge accounts, debit or credit cards, often handled through electronic funds transfer. At the end of the month, as income is received, accounts are balanced. Household expenditure is managed through banks and credit card companies.

The shift from largely personal to largely social consumption seems now to be accelerating as we move into the Information Economy. This poses new problems for the market and may also make generating productive innovations more difficult. For the market works best when consumer preferences are independent, not only from each other, but especially from those of producers. Costs and benefits must be clearly defined, and must be internal to the market. Everything relevant should have a price. Collective consumption, however, necessarily involves externalities; think of transportation systems, media networks, education, health care, and the Internet. Social costs and benefits will differ from private. Nor can preferences be taken as "given"; they are the result of complex interactions, in which the power of money to influence preferences and to direct learning is not to be underestimated. Consequently, the line between what is "internal" and what is "external" to the market may not be clearly drawn; but if costs and benefits are not both private and independent of each other, the market will not be able to balance them.

Traditional markets were composed of private producers and private consumers, where each agent was assumed to be small and to act independently, taking the world as given. In the simple "private agent" models, the external environment could be treated as given and largely unaffected by innovations and technical progress. This is what provides the basis for the pressure that market relations put on producers to innovate. Taylorism – cost-cutting and deskilling jobs – works most effectively in this context. When it is applied to complex services with collective dimensions, like health care or education, the quality of service is reduced along with the costs. When consumers are interdependent both with each other and with producers, and if large groups of agents tend to react together, we are in a different world. The traditional claims for the market have to be modified, and the impacts on the environment and on other agents – externalities – must be set against the gains from innovation.

The transformation of production

The changes in producer technology have run along parallel lines. Just as households have been deskilled by the development of mass production, so have the traditional crafts. Entire crafts have been embodied in machinery, as mechanical energy has displaced human, mechanical skills manual ones. Crafts were first embodied in factories, driven by water and steam power, then by electricity. Scale expanded, and science was applied directly to the development of products and processes. This has been explored earlier – there's no point in repeating the argument here.

The result of all these changes has been an enormous increase in the productivity of labor, and a reduction in both work hours and in the intensity and effort required. But now those processes which hitherto helped the market to develop may be undermining its ability to function.

Processes initially producing a single product have increasingly developed into joint production, as uses have been found for by-products and wastes. (A recent example: lumber mills used to give away scraps and ends; later they began to split them to sell for home heating; now they process all scrap, to market wood chips.) Formerly separate processes have been merged or run in tandem to take advantage of synergy or mutually advantageous externalities. (Think of the construction of modular homes.) The result is that various goods are produced together; the proportions in which they are produced are fixed, or at any rate can only vary within limits. This may limit the flexibility of the market. Just as consumers have become increasingly interdependent, so have producers; just as different consuming activities have become more closely complementary, so have different outputs.

Products have become systems, rather than single units. Companies sell services along with goods – repairs, and insurance, for example. But they also provide upgrades, replacements, and complementary goods, together with installation. What is sold is a *system*, rather than a single item, binding the customer to the company. This tends to change the nature of competition.

Production has become heavily dependent on social and financial infrastructure. Not just roads and bridges, but the information highway and bridging loans, are essential to modern production. Production, more than ever, is a collective activity, resting on the structure of the whole society. This puts a premium on coordination and a burden on governments.

At the same time capital has become extremely mobile. Financial capital has always been mobile; it can move faster than ever with electronic funds transfers. Financial markets have become closely linked, forming a world market. But real capital has also become highly mobile. Firms can move their production facilities around the globe with unprecedented ease; plants can be closed down one place and reopened in another in weeks, in some cases even days. Computers and modern communications enable headquarters to monitor processes in full detail, regardless of location.

Production processes have become so complex that neither they nor most products can be understood without special training. Consumers cannot evaluate products adequately, and workers cannot know in advance if their workplace is dangerous or safe. Because of large-scale production and mass consumption, the dangers of a mistake may be very great, yet the effects might also take a long time to show up.

The effects of production on workers, of products on consumers, and of both on the environment are technically complex and can only be evaluated by scientific studies. Pollutants, however, often interact in complex ways, multiplicatively, so that the effects of increased pollution grow exponentially. Under these circumstances, studies of a single pollution problem, even very good ones, may miss a large part of the effect.

These changes imply that proposed innovations to increase productivity have become harder to evaluate. The mobility of capital also makes them harder to regulate. An innovation or a cost-cutting improvement may appear to be worthwhile, in terms of its immediate effects, but its more distant ramifications, both in regard to private and to collective or external impacts, may be difficult to forecast. Even separating internal effects from external may be difficult. If long-run exposure to a new chemical injures workers or consumers, lawsuits as well as loss of markets may be the result. On the other hand, governments find that the increased mobility of capital makes it harder to impose regulations, controls, and taxes.

Hierarchy, ownership, and the family

As technology has changed, so have the organizations that administer it. These have grown both larger and more complex, and at the same time, they have become less rooted in the kinship system. Growth in size follows from the general growth of the system in conjunction with economies of scale; growth in complexity from the "synergy" which leads to the combination and joint operation of processes, and from the development of joint production through finding uses for by-products and waste. The problem of coordinating different kinds of activities makes the hierarchy more complex.

At one time most businesses were controlled through the family and were handed down from father to son (Marshall, 1961, ch. 12; Skidelsky, 1992, vol. 2, ch. 8). Many family firms still exist, though most are small and, regrettably, short-lived. But the major form of contemporary business, the modern corporation, has become independent of the kinship system, and represents a different principle of social organization.

The family firm determines control and fixes succession through a patriarchal hierarchy. The father who runs the family runs the business, and the oldest son inherits, subject to various qualifications and special circumstances. But a corporation is an independent entity – a legal "person" – completely separate from its "owners," who do not and cannot own it directly. Instead they own shares, entitling them to vote for directors and officers; a controlling interest is therefore 51 percent of the shares. But a controlling interest does not own the corporation or its assets. An individual or a family may own a controlling interest, and the shares may be passed on from generation to generation. But here again the market intervenes: it will seldom if ever be in an individual's interest to keep his funds tied up in the assets of a single corporation. If a portfolio is to be managed to maximize its growth or earnings, it will have to be diversified. Hence the family's interests over time will very likely lead to the loss or dilution of its controlling position. But in any case, stock positions are determined by portfolio management, on the basis of professional criteria for performance.

A corporation is controlled by directors and officers who are elected to represent certain interests; the actual directors are often selected by the interests on the basis of their knowledge and expertise. Corporations are actually run by managers who are appointed and promoted on the basis of their qualifications and experience. In contrast to the family, where appointment and succession are determined by birth, in the corporation it is determined functionally, by performance.

Interestingly, exactly the same can be said about the managers of portfolios. They are chosen on the basis of qualifications and performance – professional competence. Thus both the management and the control of the modern corporation are determined on the basis of functional, as opposed to kinship, criteria, even when the assets are still in the hands of a family. Further, when the assets are professionally managed, it will usually be advantageous to convert them into a trust, making the former owners into beneficiaries. In other words, the ownership itself becomes an independent corporate person, controlled by trustees, who are elected by the beneficiaries and some overseeing institutions according to some formula. This is a far cry from determining control by birth, or through family.

The picture can be taken a step further. Consider a corporation, all of whose stock is held by other corporations, banks, insurance companies, each of whose stock, in turn, is held only by other institutions. The set of these institutions, taken together, constitutes a unit of capital completely independent of ownership by any individuals or families. These institutions own each other and, taken together, own themselves. They form a self-perpetuating, self-subsistent independent entity, operating in the market in pursuit of profit. (Self-perpetuating so long as raiders can be fended off! But raiders just represent other blocs of capital.) Do such blocs of capital owned or controlled only by other capital exist? Evidently. Institutional holdings and bank trusts together make up a solid and growing majority of all stock holdings. In leading countries families and individuals have generally been a minority of total stock holdings since the 1960s.

Ownership and control

The development of such hierarchical systems led, of course, to the well-known divorce between ownership and control. To rest with this, however, is to miss the point. Looking more closely, we see, first, that control has become professionalized and embodied in a *self-renewing* management, operating according to (supposedly) objective standards. But secondly, exactly the same has happened with ownership; professionalism and hierarchical management have taken over there, too, for ownership has itself become a process requiring skills and experience. Funds must be held in portfolios to spread risks and hedge against inflation; such portfolios, in turn, must be managed by competent and knowledgeable professionals. Control over funds therefore passes into the hands of managers, and again, the organizational form will be the self-renewing hierarchy.

But it is not just that control over portfolios has passed from owners to professional managers; the very core of ownership itself has become institutionalized. The family holdings are put in trust; the members of the family, formerly owners, now become the designated beneficiaries. Not only do they no longer manage the firm bearing their name; not only do they no longer own it, having diversified; they have now, first, given up control of their funds to the bank's trust department, and then have given over the ownership itself to a newly created entity, the family foundation or trust, in return for the (safer, more sensible, tax-saving) status of beneficiary, entitled to income. Thus the entire circuit of capital, $M–C–P–C'–M'$, including both the commodity circuit and the money circuit, has become institutionalized, which is to say, embedded in, and operated by, self-renewing systems of professional management, run according to objective principles. These principles, by which day to day practice is guided and evaluated, are, of course, those of the balance sheet.

At the same time that family holdings were being restructured, the financing of business changed. Instead of borrowing, businesses increasingly sought to finance their investments out of profits. When this posed problems they turned to nonthreatening institutional sources, such as the pension funds of their own workers, or going further afield, to insurance companies, to sources, in short, that would keep their distance and pose no threat to management.

Implications for the family

The institutionalization of capital separates the latter's self-renewing management from the kinship system. In the case of the true family firm, both changes in ownership and changes in management – the transfer of property from one generation to the next – were mediated through, and sometimes wholly determined by, kinship arrangements and the extended family structure. (In Anglo-Saxon societies, this meant male primogeniture, or some modification thereof.) But once the institutionalization of capital is complete (which, of course, it is not even today) capital – both ownership and control, that is to say, the management of finance and the management of production and sales – will be fully embedded in institutions, capable of renewing their personnel without directly depending on the family or kinship structure. Instead of owning and controlling capital, formerly capitalist families have simply become entitled to an income, a status which, magnitudes apart, may not differentiate them much from other groups entitled to incomes for other reasons (for example, pensioners, veterans, recipients of disability awards, and lottery winners). And their certificates of ownership, their equity in companies, are little more than betting slips in the casino known as the stock exchange.

Of course, nuclear families remain fundamental in the reproduction of the population. But they are less basic to socialization, education, and training than they were in the era of family farms and firms. These processes are being taken over by the state or by private nonprofit, professional orgnaizations. First through universal compulsory education, but now through child care, welfare, and social work, vocational training and vocational guidance, and the provision of various medical services, the state, employers, and the nonprofit sector together have become the major forces shaping the statistical profile of the population in terms of health, general education, and specific skills. The extended family, the church and the neighborhood, once central to socialization, no longer have practical roles commensurate with their mythic status.

To say that "family" is no longer so important does not imply, however, that *birth* is any less significant in determining life's chances. Rather, we should see status and position at birth as a complex of social coordinates, including along with family, educational prospects, proximity to other opportunities for socialization, and occupational opportunities. Of course, these largely depend on the parents, connections, but not only on their *kinship* connections.

Transformational Growth and the state

As production and consumption have developed in a manner that combines both quantitative and qualititative changes – economies of scale, the growth of externalities, and joint production – so have economic organizations. The family firm became the modern corporation, which then grew into the transnational conglomerate. The Night Watchman State became the Welfare State; the standing army became the Military-Industrial Complex.

As a beginning, we should note that this expansion coincided with the shift from craft methods to Mass Production in industry, and with the correlative changes in the economic status and role of the family. At present, another vast change in technology appears to be taking shape – a shift from Mass Production to Information Technology, which may have implications for the public sector, also, though any such implications are hard to discern as yet.

The shift from the Craft economy to Mass Production brought with it many important changes in the structure of the economy, for the most part causally connected, even if only loosely. For example, the change was accompanied by a decline in employment in agriculture, a rise in urbanization, the shrinking of the extended family to the nuclear (and now a weakening and breaking apart of the nuclear), a decline in the economic functions of the family, and a rise in the educational requirements for citizenship and employment. All these contributed to the pressures that led to the increased role of the state.

Each of these changes was partly caused by, and in turn contributed to, the technological shift from the Craft economy to Mass Production. And each, in turn, led to reasons for the state to adopt an expanded role. Taking them in turn: the decline in employment in agriculture was brought about by the labor-displacing effects first, of agricultural machinery, and then by the rise in productivity which followed on the development of fertilizers and pesticides.

The exodus of labor from agriculture contributed strongly to the development of the urban labor force, and the growth of cities. Mass Production needed a large labor force concentrated in one place. So a symbiosis developed: as Mass Production grew, it produced machinery and gene-rated innovations that increased agricultural productivity, bringing new waves of labor to the cities, where new forms of transport, built by Mass Production (streetcars and railroads, at first, then automobiles and buses on roads built by earth-moving machinery), could convey them between home and work. The concentration of labor, in turn, made it easier to build large-scale industries further increasing productivity.

The growth of cities required an expanded role for the state. Concen-trations of people require more intensive police work and more regulation of traffic and also of business. (Arguably, transactions between economic agents in an area increase with the square of the number of agents, and the cost of regulation rises linearly with the number of transactions.) In addition, life moves at a more rapid pace, and there is less time to acquire knowledge of suppliers and products. Cities also require public utilities – public lighting, maintenance of roads, water, sewage, and garbage dis-posal, and so on. In a rural or small town setting, water will come from wells, sewage will be handled by a septic system, and garbage can be composted or buried. Concentration of people in one area raises public costs more than proportionally to the numbers, both by adding new functions and by raising costs exponentially.

A separate cause of increased public spending can be found in the decline of rural and small-town life, and the increased horizontal mobility of the labor force, which put stresses on the traditional extended family. As labor moved from the farm to the city, the close connections arising from proximity were lost. Moreover, the old-fashioned extended kinship system provided the basis for the division of labor in domestic production. (Within the extended family essentially barter exchanges of goods and services take place – preserves for help with harvesting, dresses for shoeing the horses.) Such a division of labor was possible in rural and small town cir-cumstances, where many people were partially or wholly self-employed, and where the women worked in farmhouses with large kitchens and capacious cellars and storerooms. But it is much more difficult to maintain in the urban setting. With the decline of rural life there had to come a replacement of domestic production, and here too, a symbiosis can be found. For Mass Production replaces domestic produc-tion – the canning industry displaces home canning, the textile industry displaces home dressmaking and sewing, the furniture industry, home carpentry, pharma-ceuticals, home remedies, and so on.

The decline of the extended family has created a need for "human" services. The extended family cared for the aged and the infirm, supplied built-in day care and babysitting, and provided preschool training for the very young. The nuclear family cannot do this, yet needs these services.

They must either be provided by the state, or if provided for profit, must be regulated by the state, since those who receive these services will normally be different from those who pay (and may very well be helpless) so that the opportunities for abuse are very great. Such services are needed even more with the weakening of the nuclear family.

The nuclear family has come under strain, first, from the development of horizontal mobility and the career structure of Mass Production, and more recently from the changing nature of the corporation and the job market as information technology develops. As the nuclear family disintegrates, the regulation of child development must devolve increasingly on the state – or some other representative of the general community. For children must be socialized, and both equity and the good of the community require that all children be given at least a minimum set of social skills. Divorce and single parenting lay heavy burdens on children.

With the development of Mass Production came the rationalization – and increased mechanization – of work. Technical progress and reorganization of tasks became a regular feature of the economy. Traditional skills were displaced; and as a direct result, fathers could no longer pass on a trade to their sons. The position of fathers vis-à-vis sons changed dramatically. Sons no longer needed follow in their father's footsteps; nor, in many cases, could they, for in an industrial system Transformational Growth can be expected to redesign the methods of production, eliminating or changing beyond recognition not only jobs, but also entire crafts and trades. To learn one of the new trades it was necessary for the sons to go to school, to learn things their parents never knew. Education had to be moved out of the home.

Such a transformation of the system, especially the elimination of the family farm, requires management by the state, to ensure an adequate supply of appropriately trained and socialized labor for the new *kinds* of occupations. Moreover, speaking generally, we should expect the institutionalization of capital to increase the proportion of office jobs relative to factory ones, while the substitution of bureaucratic control for technical control in the factory led to changes in the required skill mix.

These changes generated social tensions, because they undermined the traditional authority of the family. Yet this was absolutely necessary. If the occupational destiny of children were determined by the class position of parents, the system would have been hard put to accommodate these changes. If status lines were unbreachable, the sizes of the classes would change only slowly, with population. How, then, would the need for professionals and managers be met? Some children, perhaps many, had to rise above the station of their parents – a move that their parents could not prepare them for, and which was likely to create a rift in manners and mores. If socialization and education were to remain in the family, the children would only learn what their parents could teach them: each generation would have to reflect the previous one. But where would the new generation of handloom weavers find work? The family-based socialization system had to be destroyed and replaced. (And this did mean that children had to learn not to respect their parents, or the old ways, too much – a real stress was created here.)

Mass production technology brings economies of scale, and to manage large-scale operations requires a professional bureaucracy. Thus the family firm grew into the modern corporation. But a corporation has different legal responsibilities, and must be regulated differently. Also a giant corporation can easily dominate a locality or region; its customers and suppliers may become dependent on it, very possibly including government or the military. Defective products, or failure to fulfill contracts, may cause widespread disruption. The failure of a large corporation can bring distress to a whole region. Regulating and overseeing such companies is necessary to protect the public interest.

With the rise of Mass Production significant changes in the working of the economy took place. In particular the system became more unstable. The Craft economy had operated in such a way that price changes ensured that short-run fluctuations in aggregate demand would be offset – a collapse of investment or of exports would be at least partially offset by a rise in consumption spending. With Mass Production, a fluctuation would be transmitted by the multiplier-accelerator, ensuring that it was amplified. Hence the system needed a different kind of government presence than before.

New responsibilities for the state

So this leads to new roles for the state, and for the nonprofit sector, replacing some of the functions formerly performed by the family system. There also developed new demands on the state to regulate and contain conflicts, to resolve economic disputes, and to provide legitimacy. Moreover, in the absense of a central steering or self-adjusting mechanism in the economy, such as existed, however imperfectly, in the craft-based world, the state has had to assume responsibility, in the Keynesian era, for managing the overall level of economic activity.

These new roles, however, cannot be performed unless the state itself develops appropriate permanent institutions for doing so. Such institutions must be professionally managed and self-renewing; they will therefore tend to adopt the same hierarchical form that capital itself has. Moreover, where the institutions of the state and private capital interlock – vocational education and certification of training, regulation of business, military supply, antitrust – they will be drawing on the same pool of professionally trained personnel, some of whom will tend to circulate, moving back and forth from state offices to private.

Since the late nineteenth century observers have noted the tendency of state expenditures to rise relative to other categories of spending (Wagner, 1883; Musgrave, 1967; Gemmell, 1993). The increase has been irregular, but marked. This is as we should expect. State expenditures rise relative to others and to GNP partly because the state must take on new functions as the economy moves from craft technologies to Mass Production, and partly because the state, like the market, is taking over func-tions formerly performed by the family.[3]

Capital and the state will thus be drawn increasingly into a symbiotic relationship, each dependent on the other, each limiting and defining the other. Capital will depend on the state for trained personnel, for regulation, for the management and limiting of competition, for subsidies, for demand management, for planning, for managing the monetary system, and for legitimation. The state will depend on capital for some of its trained personnel, for supplies and equipment, for revenues, for cooperation in directing the economy, in protecting the environment, and in developing and exploiting natural resources. The state must work *through* the existing system to achieve its goals; to put it another way, unless the economy is functioning properly, the state cannot carry out its functions. Its *first* objective therefore must be to see that the system of production for private profit is working smoothly. Policies which come into conflict with this cannot be carried out, except in a limited way.

This provides a definite constraint on policy, but one which nevertheless allows a great deal of leeway, since there are many ways in which the system can be encouraged, and in which business can be rendered profitable. Moreover, different ways of stimulating, regulating, or managing the economy will usually benefit some groups more than others, or benefit some at the expense of others, giving rise to the possibilities of complicated political trade-offs. However, the system is facing new problems that may be less amenable to traditional forms of regulation.

Increasing returns and the possibilities of "lock-in" – implying that the winner takes all – have led to intensified competition, and pressures to cut

---

[3] A long-standing and inconclusive debate has raged over whether or not "government goods" are income-elastic. Government size increases because the "demand" for government rises faster, with income, than the demand for other goods. The "demand" is metaphorical; the goods are paid for by taxes (or deficits). Part of the difficulty has been defining the goods at issue; another problem concerns whose income is relevant. But another difficulty arises from the irregular and discontinuous way the increases in the relative size of government have come about (cf. Peacock and Wiseman, 1961).

back on regulation. The ability to shift production about the globe has weakened domestic labor, and has also reduced the loyalty of companies to localities. National states have been weakened, but supranational agencies are not strong enough to step in. Globalization seems to have brought stagnation; it might yet bring us worse.

Stages of Transformational Growth

The development of technology interacts with the changing character of market adjustment. Both are driven by the forces of competition, which itself is channeled through economic institutions – the family, the firm, and the state. Economics is not a "separate and inexact science," in Mill's words, adopted by Hausman. It is just the opposite. It is not separate, because, as just seen, markets, institutions, and technology all interact, and to be fully understood, must be studied together, although aspects can be usefully treated in isolation. When understood historically, economics appears to be reasonably exact, compared to other social sciences, as is shown by the table below, where the theories are grouped according to the stages they fit best. One reason that it has been thought to be "inexact" is that theories have been taken to embody universal laws, on the analogy of physical science, and the "laws" which accurately describe the working of markets in one era may fit poorly or not even be applicable in another.

Three stages of the development of capitalism have been identified, and the movement through them is the subject of the theory of Transformational Growth. To sharpen the picture, at the risk of further oversimplifying what is already a "stylized" story, we can present the essentials in a table. The left-hand column lists the different aspects of the system to be compared. The first row shows the three basic periods defined according to "Technology." The next two rows show the organization of production in the sectors of the economy, with Manufacturing, Retail Trade, and Services all grouped together, with Agriculture below. Family firms changed first to family-dominated companies and then developed into modern corporations. A similar evolution took place more slowly in Agriculture. Households and the kinship system are listed next, showing the gradual change of the family from extended to nuclear. In the case of government, the historical form during most of the AS/TA period was actually slowly decaying Mercantilism; but arguably this was due to institutional inertia and the resistance of the aristocracy. If Adam Smith and Bentham were right, the form of government *should* have been the Night Watchman State, developing somewhat in CBF/MA into a source of subsidies for desirable large-scale projects, and finally into the Welfare State under Mass Production. Following government we have the economic characteristics, the patterns of Market Adjustment, the different

Monetary Systems, and the forms of Credit, developing from commercial paper to bonds to the market for corporate control. In AS/TA the Quantity of Money affects its value and is regulated by the Mint. A new system begins in CBF/MA and develops fully in MP/CA in which it affects the rate of interest and is regulated by the Central Bank. Of course, in these, what we are listing is the *dominant* form. Flex price markets remain, for example, especially in the primary sector, even when Mass Production has become the dominant form. In each later period the earlier forms remain, still functioning, and often still important, but altered. Metallic money is present in Mass Production – but it is used for small transactions and change. Commercial paper develops significantly in CBF/MA as new forms emerge. The bond market is enormously important in Mass Production, but bonds are far different from those of CBF/MA; the market and the paper are more complex and the market has developed altogether new instruments.

Characteristics of the stages of development

Technology:	AS/TA	CBF/MA	MP/CA
Sectors:			
Manufacturing, Trade, Services	Family firms	Family firms, companies	Corporations
Agriculture	Family farms	Family farms	Corporate agriculture
Households	Embedded nuclear family	Embedded (weak) nuclear family	Nuclear family
Government	Night Watchman	Night Watchman with subsidies	Welfare State
Economic relations:			
Market adjustment	Flex price/ flex labor	Flex price/ flex labor	Fix price/ flex labor
Monetary system	Metallic M → prices	Paper Mixed	Bank deposits M → interest
Financial markets	Bills of exchange	Bonds	Stock market

It is worth spelling out the market systems more fully. The left-hand column shows the three types of market determination: the long-run, the short-run, and the cycle. The long-run position means "normal" or natural prices, and normal or planned outputs, which as we argued, are those on the basis of which capital decisions are made. The short-run position is that determined by current market forces, and the cycle shows the characteristic pattern of fluctuations.

Characteristics of market adjustment in different periods

	AS/TA	CBF/MA	MP/CA
Long-run	Growth/real wage	Growth/real wage	Price/Investment
Short-run	Classical gravitation	Marshallian	Multiplier
Cycle	C/I cobweb (SR)	Financial (SR)	Financial-real
	Marx-Goodwin	Marx-Goodwin	(Modified
	(LR)	(LR)	multi.-accel.)

An interesting exercise might be to consider some well-known models, and ask, to what periods do they apply? The traditional "theory of the firm" would seem to apply to CBF/MA, while models of "administered pricing," or theories of the markup would seem to fit the corporate world of MP/CA. Models that assume "outside money," for example, clearly appear to fit under AS/TA. Models such as the IS-LM that assume an asset market with only money and bonds would seem to fit under CBF/MA, although the multiplier also fits MP/CA. Well-behaved aggregate production functions likewise appear to belong to CBF/MA. By contrast, multiplier-accelerator models, Kaleckian, and most post-Keynesian work seem to be appropriate to MP/CA. In international trade, Ricardian comparative advantage presumably fits the Artisan era – and perhaps describes quite a lot in later periods as well – while "absorption" certainly seems to fit the post-World War II relations between the ACEs.[4] Unfortunately, many models mix up aspects of the adjustment mechanisms from different eras and end up fitting none of them very well. (The popular "Aggregate Demand–Aggregate Supply" model might be an example – "real balance effects," if they belong anywhere, belong to the era of metallic money and limited credit, AS/TA, the multiplier belongs to MP/CA, as do any assumptions about sophisticated expectations, and the liquidity preference equation to CBF/MA.)

Suppose, just for argument, that we tried to fill in a fourth column, for the era we seem to be moving toward – Computerized Production and Bio-Tech Agriculture. Running down the column we might have something like this:

Transnational corporations
Transnational corporations
Serial monogamy, single-parent and varied families
Weak supranational, fragmented national governments
World multiplier, world capital mobility – real and financial

---

[4] Unfortunately, Heckscher-Ohlin, which might seem the appropriate candidate for the era of Craft-based factories, appears not to fit *any* period empirically (Leontief, 1953; Bowen, Leamer, and Sveikaukas, 1987; Baldwin, 1971).

Electronic funds transfer system, credit and debit cards
World derivatives, world securities, foreign exchange

And how might we expect the working of markets to change? For one thing, the cost structure of business appears to be changing. Labor costs are a lower proportion of total costs, and a higher proportion of labor costs are fixed. Research and development and setup costs are rising in relation to other costs. But prices cannot easily become flexible, since many are locked into contracts, and others are established in market in which expectations are significant for marketing. Companies appear more and more to market *systems* rather than single commodities or services. Prices tend to change only with technical progress in the form of product or service improvement, and even then many prices and money wages will tend to stay fixed or move less. Automation will substitute energy and equipment use for skilled and expensive labor; with computerization and flexibility in production, components can be contracted out, all over the world. It will no longer be necessary to keep large capacity and skilled production teams on hand, so large production systems will be broken down. But high setup and development costs will keep firm sizes large, even if production units can be smaller. Fluctuating sales in the face of high monetary fixed costs will lead to financial difficulties; borrowing will rise, and reorganizations will become more frequent. One response has been to create variable interest rates. Profits can be protected to some extent by outsourcing and contracting abroad, in arrangements that can quickly be varied with changes in the state of markets, and by variable interest rates. Nevertheless with low variable labor costs there will be less of a cushion, hence there will be downward pressure on wages and costs to create a margin of safety in profits. This continues the tendency to stagnation. One possible result is that the multiplier may come to work differently.

Globalization will increase outsourcing and intensify pressure on unionized labor and high wages. It will also make it more difficult for governments to tax, or to keep interest rates low – capital and business can always move or threaten to move. Very broadly, globalization and worldwide capital mobility appear to tend to promote austerity, and to weaken the ability of nation-states to regulate their economies – at a time when technological innovation is creating instability and generating new products and processes with large, and largely unknown, externalities.

To explain and try to defend these summary judgments would be an entirely new project. They are presented here only as a list of ways one might extend this analysis. Nevertheless, the list is suggestive, and the suggestions unsettling.

Prospects: the future of the market

For five centuries the market has been the chief engine driving technological innovation and productivity growth in the West. Along with innovation has come institutional and social change, including changes in the political system. The process is extremely powerful, but has been largely uncontrolled. It has always generated dangerous and costly by-products. In this century we have seen the shift from Craft technologies to Mass Production, and with it the development of the modern corporation out of the family firm. It has begun to alter and destroy parts of the environment. Under the influence of social and economic pressures, the Night Watchman assumed new duties and became the provider of welfare.

Transformational Growth has seen the decline of agriculture, the rise and then stagnation, even decline of manufacturing, and the rise of new service sector. These changes have led to a transformation in the character of consumption, from predominantly private to substantially collective. But the incentives generated by the market only encourage the development of private goods and private aspects of collective goods.

With science-based technology innovations are not improvements of well-known and well-tried products and processes; they are often completely new. No one knows what their long-term effects will be, on people or on the environment. Regulation and comprehensive testing are therefore more important than ever. But market processes are undermining the powers of the state. It is time for another "revision of the agenda of Government," in Keynes's memorable phrase, but this time it is the terms of the political debate that most need revision.

Effects of externalities

The dynamics of capital accumulation under Mass Production have set up powerful incentives to cost-cutting and product development; innovation is central to modern competition. But the form this techno-logical development takes appears to generate increasingly burdensome externalities. These changes have been going on for a long time. At first they simply reinforced the effect of Mass Production technology, in which the market began to adjust to variations in demand through adapting output and employment, rather than by adjusting prices. In other words they reinforced the multiplier. But before long other consequences began to emerge.[5] The most important is the increase in "external" effects in production and consumption, for example, the production of polluting waste products by industrial processes, and the generation of garbage,

---

[5] These have become sufficiently important, apparently, to slow the growth of Net Economic Welfare relative to GNP (cf. Samuelson and Nordhaus, 1990).

especially solid waste, as a by-product of consumption (part of this being discarded packaging, designed to increase sales and consolidate markets). These external effects can be both negative and positive, but both kinds undermine the ability of markets to adjust.

The positive effects, as we have just seen, link preferences of different consumers and also interlock consumption activities with the prospective behavior and preferences of producers, making it difficult for price adjustments to work. The negative effects either reduce productivity (smog increases sick days, lowers efficiency on the job) or undermine well-being by destroying beauty or ruining health. But the market attaches no price to either the positive or the negative effects; no one has to pay to get the benefits, nor do they have to pay for inflicting the costs. The community bears the burdens, but no one pays directly. From the point of view of economic incentives, this situation is a disaster. Positive externalities should be encouraged, but since no price is paid directly, there is no incentive to maximize them; negative externalities should be discouraged, but since the producer faces no costs, there is no reason to minimize them.

The traditional approach, of course, is to cover the external costs with taxes, and provide subsidies to indicate the external benefits. However effective this may seem in theory, in practice it falls foul of the efforts of lobbyists. A more effective route to containing social costs may lie in empowering the potential victims.

The growing interdependence between consumers, on the one hand, and between consumers and producers on the other, creates another set of coordination problems which the market cannot handle. These are problems of scheduling. Each economic agent chooses the best time for his or her activity; but because the activities are interdependent, the result is a rush hour or a traffic jam, or a crowded beach – often with underutilized facilities at other times.

To paraphrase the poet: The market's a fine and private place, though none, I think, do there embrace. Indeed, the market tends to corrupt anything immaterial or spiritual that comes into contact with it. The incentives in the market are Machiavellian; the object, the attainment of wealth and power through whatever means are efficient and arguably legal. Hence when the market comes to deal with public goods, collective activities, or social consumption, it will attempt to privatize them, in order to make money. Ideas are patented, works of art copyrighted – for once the language is perfectly clear, so they can be exploited. To privatize a public good or collective activity means to establish boundaries, fence it off, charge admission or user fees – in short to limit access to it. This may encourage development and innovation. But gains must be weighed against costs. Inevitably, it sets up two classes, those who can afford it, and those who can't. Exclusion tends to breed envy and frustration, which in turn breed anger and resentment, feelings very much in evidence today.

Growing dysfunctionality

In the past the pressures of the market generated innovations, growing productivity, greater control over nature and the environment, all of which contributed to growing affluence, for many, perhaps at times for all, households. Much of this remains true today, but the negative side effects and externalities may now be growing faster than the positive benefits, and the imbalance may be further accelerated as we move into CP/BA. As an experiment, we can apply the market's own rules of calculation to this problem, in a rough and ready way, to see if and to what extent the pressures of the market will continue to benefit something like the "average household." The prospects may not be good.

Alarmingly, it seems that we cannot expect capital to expand indefinitely, by continually modernizing and marketing new products. And the reason lies in the way the system is developing: the cost externalities associated with the kinds of innovation generated by incentives thrown up by the "demand-scarce" market appear to be rising faster than the benefits from such innovation. As a result, capital accumulation, at the margin, may actually be making us worse off! Any such judgment, at this point, of course, must remain speculative – but the signs are pointing this way.

Consider on the one hand, the amenities and benefits that flow from the forces and pressures of the market, and let us suppose that we can aggregate them in monetary terms. These amenities will increase as the per capita wealth of the society expands, but certainly after a point, they will begin to increase less rapidly. That is, in the language of traditional economics, diminishing returns will set in, since material incentives can be expected to weaken as people become richer (marginal utility declines), and in any case such incentives will work less effectively as consumption becomes increasingly collective, and therefore increasingly subject to social pressures other than monetary ones.

On the other hand, now consider the disamenities and negative effects of market pressures and forces, disamenities such as those we've just discussed. These can also be aggregated in monetary terms, and they can be expected not only to increase, but very likely to increase at an increasing rate, as per capita wealth rises. This will be particularly obvious in the case of pollution, the costs of which may well rise exponentially, if pollutants combine multiplicatively, but it may also apply to the growth of additional social costs in market situations – for example misinformation and corruption associated with marketing and selling, the costs of dangerous or unsafe but profitable goods (cigarettes, alchohol, cars "unsafe at any speed") and so on.

Thus as wealth per person increases, the benefits of market pressures rise, but at a diminishing rate while the costs of such pressures also rise, but at an increasing rate. It should be inituitively clear that at some point these must balance – and at this point the market system will no longer be socially advantageous from the point of view of households, although it may continue to be profitable according to market calculations. Even earlier, however, that is, at an even lower level of wealth, there will be a point at which the *additional* (rising) costs of the system are no longer offset by the additional (falling) gains from it. So at that level of wealth per capita it would be desirable to cease to rely on market incentives as guidelines for further investment.

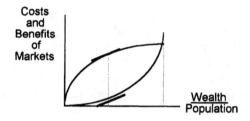

Markets are a little like guns: they are powerful and deadly, can accomplish great things, and in a sense civilization may be said to rest on them. They are becoming more powerful and more dangerous all the time. If they are handled carefully, they can be a boon to progress. But they are extremely dangerous and should not be allowed to spread through the world unregulated. If they do, a lot of people will suffer and die unnecessarily, without much good being accomplished.

Yet this may not be our most serious potential problem. For five hundred years markets have been the driving force of Western development. Now they have taken the whole world for their stage. During this time markets have wreaked havoc on traditional societies, unhinged manners and morals, and brought terrible disasters on masses of uprooted people – but they have delivered the goods, in unprecedented quantity. Living standards not only rose beyond belief, things that had barely been imagined in our great-grandfathers' wildest dreams have become everyday realities. Who could doubt that the light was worth the candle?

But in the foreseeable future this certainty may no longer be justified. If, as seems possible, the market's negative impact is rising faster than the benefits it delivers, the market may begin to impoverish us with the same vigor and drama – and speed – with which it once enriched us!

Bibliography

Abraham-Frois, G., and E. Berrebi. 1979. *Theory of Value, Prices and Accumulation.* Cambridge University Press.

Aftalion, A. 1913. *Les crises periodiques de surproduction.* 2 vols. Paris: Riviere.

Ahmad, S. 1991. *Capital in Economic Theory.* London: Edward Elgar.

Akerlof, G., and J. Yellin. 1985a. "Can Small Deviations from Rationality Make Significant Differences to Economic Equilibrium?" *American Economic Review* 75 (Sept.): 708–21.

——— 1985b. "A Near-Rational Model of the Business Cycle with Wage and Price Inertia." *Quarterly Journal of Economics* (Supplement) 100: 823–38.

Alchian, A. 1950. "Uncertainty, Evolution and Economic Theory." *Journal of Political Economy* 58: 211–21.

Aldrich, N. 1892. *Wholesale Prices, Wages and Transportation.* Washington, D.C.: U.S. Government Printing Office.

Altman, M. 1992 "Business Cycle Volatility in Developed Market Economies, 1870–1986: Revisions and Conclusions." *Eastern Economic Journal* 18 (3): 259–75.

Andrews, P. W. S. 1949. *Manufacturing Business.* London: Macmillan.

——— 1964. *On Competition in Economic Theory.* London: Macmillan.

Anscombe, G. E. M. 1958. *Intentions.* Oxford: Basil Blackwell.

Argyrous, G. 1991. "Investment, Demand and Technological Change: Transformational Growth and the State in America during World War II." Ph.D. dissertation, New School for Social Research, New York.

Arthur, W. B. 1989. "Competing Technologies, Increasing Returns, and Lock-In by Historical Small Events." *Economic Journal* 99 (March): 116–31.

Aschauer, D. 1990. *Public Investment and Private Sector Growth.* Washington, D.C.: Economic Policy Institute.

Asimakopulos, A. 1978. *Microeconomics.* Oxford: Oxford University Press.

——— 1988. *Investment, Employment and Income Distribution.* Oxford: Polity Press.

——— 1991. *Keynes's General Theory and Accumulation.* Cambridge: Cambridge University Press.

——— 1992. "The Determinants of Profits: United States, 1950–88." In Papadimitriou 1992.

Austin, J. L. 1960. In J. O. Urmson and G. J. Warnock (eds). 1960. *Philosophical Papers.* Oxford: Clarendon Press.

Axelrod, R. 1984. *The Evolution of Cooperation.* New York: Basic Books.

Baldwin, R. 1971. "Determinants of the Commodity Structure of U.S. Trade." *American Economic Review* 61 (March): 126–45.

——— 1988. "Hysteresis in Import Prices: The Beachhead Effect." *American Economic Review* (Sep.): 773–85.

Balke, N. S., and R. J. Gordon. 1989. "The Estimation of Prewar National Product: Methodology and New Evidence." *Journal of Political Economy* 97: 38–92.

Ball, L. and D. Romer. 1990. "Real Rigidities and the Non-Neutrality of Money," *Review of Economic Studies*, 57 (April): 183–203. Reprinted in Mankiw and Romer, 1993, *New Keynesian Economics*, vol 1. Cambridge, MA: MIT Press.

Baran, P., and P. Sweezy. 1996. *Monopoly Capital*. New York: Monthly Review Press.

Baranzini, M., and G. C. Harcourt. 1993. *The Dynamics of the Wealth of Nations*. London: Macmillan.

Barnes, R. M. 1952. *Motion and Time Studies*. 6th ed. New York: Wiley and Sons.

Baumol, W. 1970. *Economic Dynamics, 3rd edn*. New York: Macmillan.

Basalla, G. 1988. *The Evolution of Technology*. Cambridge: Cambridge University Press.

Benetti, C. 1996. "The Ambiguity of General Equilibrium with Zero-Price for Money." In Delaplace and Nell 1996, 366–76.

Bernanke, B. S., and A. S. Blinder. 1993. "Credit, Money and Aggregate Demand." In Mankiw and Romer 1993, 325–33.

Bernstein, R. J. 1978. *The Restructuring of Social and Political Theory*. Philadelphia: University of Pennsylvania Press.

Bhaduri, A. 1996. "Implications of Globalization for Macroeconomic Theory and Policy." Manuscript.

Bhardwaj, K. 1989. *Themes in Value and Distribution: Classical Theory Reappraised*. London: Unwin Hyman.

Bharadwaj, K., and B. Schefold. (eds.). 1990. *Essays on Piero Sraffa: Critical Perspectives on the Revival of Classical Theory*. London: Unwin Hyman.

Bhaskar, R. 1993. *Reclaiming Reality*. London: Verso.

Bienenfeld, Mel. 1988. "Regularity in Price Changes as an Effect of Changes in Distribution," *Cambridge Journal of Economics*, 12: 247–255.

Black, F. 1970. "Banking and Interest Rates in a World without Money: The Effects of Uncontrolled Banking." *Journal of Banking Research*. Autumn: 9–20.

Blackwell, R., J. Chatha, and E. J. Nell. 1993. *Economics as Worldly Philosophy*. New York: St. Martin's Press.

Blanshard, O. J., and D. H. Fischer. 1989. *Lectures on Macroeconomics*. Cambridge, Mass.: MIT Press.

Blaug, M. 1980. *The Methodology of Economics*. Cambridge: Cambridge University Press.

Blaug, M. *Economic Theories: True or False?* London: Edward Elgar.

Blinder, A. 1988. "The Rise and Fall of Keynesian Economics." Mimeo. Australian National University.

Bloomfield, A. I. 1959. *Monetary Policy under the International Gold Standard: 1880–1914*. New York: Federal Reserve Bank of New York.

Boggio, L. 1990. "The Dynamic Stability of Production Prices: A Synthetic Discussion." *Special Issue: Convergence to Long-Period Positions. Political Economy, Studies in the Surplus Approach, vol. 6, nos. 1–2* [Hereafter: *Political Economy: Special Issue*].

Bowen, H., E. Leamer, and L. Sveikauskas. 1987. "Multicountry, Multifactor Tests of the Factor Abundance Theory." *American Economic Review* 7 (5): 791–809.

Bowles, S., and M. Gintis. 1977. "The Marxian Theory of Value and Heterogeneous Labor: A Critique and Reformulation." *Cambridge Journal of Economics* 1 (2): 173–92.

Boyd, I., and J. M. Blatt. 1985. "Investment Confidence and Trade Cycles." Mimeo. University of New South Wales.

Boyer, R. 1989. *Regulation Theory: A Critical Assessment*, New York: Columbia University Press.

Bradford, F. A. 1932. *Banking*. New York: Longmans, Green.

Braudel, F. 1979. *Civilization and Capitalism*, 3 vols. New York: Harper and Row.

Broehl, W. D. 1959. *Precision Valley: The Machine Tool Companies of Springfield, VT.* Engelwood Cliffs, NJ: Prentice Hall.

Brunner, K., and A. H. Meltzer. 1977. *Stabilization of the Domestic and International Economy.* Carnegie Conference on Public Policy, 5. Amsterdam: North Holland.

Bryant, R. 1988. *Empirical Macroeconomics for Interdependent Economies.* Wash. D.C.: Brookings Institution.

Bryce, R. B. 1973. "An Introduction to a Monetary Theory of Employment." In D. Moggridge (ed.). 1979. *The Collected Writings of John Maynard Keynes – The General Theory and after,* vol. 29, 132–50. London: Macmillan and Cambridge University Press.

Buchanan, R. A. 1994. *The Power of the Machine.* Penguin.

Bukharin, N. 1930. *Empirialism and World Economy.* London: Martin Lawrence.

Bureau of Labor Statistics, United States. 1931. "Index Numbers of Estimates in Output, Employment and Productivity in the United States After 1800." *Studies in Income and Wealth,* vol. 30.

Burns, A. R. 1927. *Money and Monetary Policy in Early Times.* New York: Knopf.

Burns, A. and W. C. Mitchell. 1946. *Measuring Business Cycles.* New York: NBER.

Caldwell, B. 1991. "Clarifying Popper." *Journal of Economic Literature* 29: 1–33.

Caminati, M. 1990. "Gravitation: An Introduction." *Political Economy: Special Issue.*

Caminati, M., and F. Petri. 1990. "Preface." *Political Economy: Special Issue.*

Canova, F. 1991. *Detrending and Business Cycle Facts.* Badia Fiesolana: European University Institute.

Cartelier, J. 1990. "The Stability Problem in Capitalism: Are Long-term Positions the Problem?" *Political Economy: Special Issue.*

―――― 1995. In Deleplace and Nell 1996, 20–38.

Carter, A. 1970. *Structural Change in the American Economy.* Cambridge, MA: Harvard Unviresity Press.

Carus-Wilson, E. M. 1952. "The Woolen Industry." *Essays in Economic History,* ed. Carm. Wilsom 1952. *Economic History of Europe* – Vol. II.

―――― 1954a. "An Industrial Revolution of the Thirteenth Century." Macmillian. Vol. I *Essays in Economic History.*

―――― 1954b. "Evidences of Industrial Growth on Some Fifteenth Century Manors." *Essays in Economic History.*

Carvalho, F. J. 1992. *Mr. Keynes and the Post Keynesians.* Aldershot: Edward Elgar.

Caskey, J., and S. Fazzari. 1992. "Rising Debt in the Private Sector: A Cause for Concern?" In Papadimitriou 1992, 202–18.

Cesarotto, S. 1995. "Long-Period Method and Analysis of Technological Change: Is There Any Inconsistency?" *Review of Poltical Economy* 7 (3): yy. Vol. 7, No. 3.

Chamberlin, E. 1993. *The Theory of Monopolistic Competition.* Cambridge, Mass.: Harvard University Press.

Chandler, A. D. 1977. *The Visible Hand: The Managerial Revolution in American Business.* Cambridge, Mass.: Belknap Press.

―――― 1990. *Scale and Scope: The Dynamics of Industrial Capitalism.* Cambridge, Mass.: Belknap Press.

Chenery, H. B., and L. Taylor. 1968. "Development Patterns among Countries and over Time." *Review of Economics and Statistics* 50 (November): 391–416.

Cherry, R., P. Clawson, and J. Dean. 1984. "The Micro Foundations of the Short Run Phillips Curve." in Nell 1984, 119–33.

Chick, V. 1983. *Macroeconomics after Keynes.* Oxford: Philip Allan.

Christofides, N. 1975. *Graph Theory: An Algorithmic Approach.* London: Academic Press.

Clark, J. B. 1895. *The Distribution of Wealth.*

—— 1899. *Distribution of Wealth.*

Coase, R. 1937. "The Nature of the Firm." *Economica* 4 (November): 386–405.

—— 1938. "The Problem of Social Cost." *Journal of Law and Economics* 3 (October): 1–44.

Commons, J. R. 1968 (1924). *Legal Foundations of Capitalism.* Madison: University of Wisconsin Press.

Coontz, S. 1966. *Productive Labor and Effective Demand.* New York: Augustus Kelley.

Costabile, L., and R. E. Rowthorn. 1985. "Malthus's Theory of Wages and Growth," *Economic Journal* 95 (378) (June): 418–35.

Coulton, G. G. 1945. *Medieval Panorama.* New York: Macmillan.

Coutts, K., W. Godley, and W. Nordhaus. 1978. *Industrial Pricing in the United Kingdom* Cambridge: Cambridge University Press.

Currie, L. 1968. *The Supply and Control of Money in the U.S.* Ed. Karl Brunner. New York: Russell and Russell.

Davidson, D. 1984. "Reality without Reference." In *Inquiries into Truth and Interpretation.* Oxford: Oxford University Press.

Davidson, P. 1978. *Money and the Real World.* 2nd ed. London: Macmillan.

—— 1991. "Money: Cause or Effect? Exogenous or Endogenous?" In Nell and Semmler 1991.

—— 1996. "What Are the Essential Elements of Post Keynesian Monetary Theory." In Deleplace and Nell 1996.

Davidson, P and E. Smolensky. 1964. *Aggregate Supply and Demand Analysis.* New York: Harper and Row.

Davis, C., and W. Charemza. (eds.). 1989. *Models of Disequilibrium and Shortage in Centrally Planned Economies.* London: Chapman and Hall.

Deaton, A. 1992. *Understanding Consumption.* Oxford: Clarendon Press.

De Brunhoff, S. 1973. *Marx on Money.* New York: Urizen Books.

Deleplace, G., and E. J. Nell. (eds.). 1996. *Money in Motion: The Circulation and Post Keynesian Approaches.* London: Macmillan.

Denison, E. 1962. *The Sources of Economic Growth in the US.* New York: Committee for Economic Development.

—— 1985. *Trends in American Economic Growth: 1929–82.* Washington, D.C.: Brookings. Institute.

Dennett, D. 1995. *Darwin's Dangerous Idea.* New York: Simon and Schuster.

Dickinson, H. D. 1956–7. "The Falling Rate of Profit in Marxiam Economics," *Review of Economics Studies.*

Dobb, M. 1973. *Theories of Value and Distribution since Adam Smith.* Cambridge: Cambridge University Press.

Domar, E. 1957. *Essays in the Theory of Economic Growth.* New York: Oxford University Press.

Dornbusch, R. 1987. "Exchange Rates and Prices." *American Economic Review* 77 (1) (March): 93–106.

Dosi, G. 1988. "Sources, Procedures and Microeconomic Effects of Innovation." *Journal of Economic Literature* 36: 1126–71.

Dosi, G., C. Freeman, R. Silverman, and L. Soete (eds.). 1988. *Technical Change and Economic Theory.* London: Frances Printer.

Dumenil, G., and D. Levy. 1987. "The Dynamics of Competition: A Restoration of the Classical Analysis." *Cambridge Journal of Economics* 11 (2): 133–64

—— 1990a. "Stability in Capitalism: Are Long-term Positions the Problem?" *Political Economy: Special Issue.*

———— 1990b. "Convergence to Long-Period Positions: An Addendum." *Political Economy: Special Issue.*

Dunlop, J. T. 1938. "The Movement of Real and Money Wage Rates." *Economic Journal* 48 (Sept.): 413–34.

Dusenberry, J. 1958. *Business Cycles and Economic Growth.* New York: McGraw-Hill.

Dutt, A. K., and E. J. Amadeo. 1990. *Keynes's Third Alternative.* Aldershot: Edward Elgar.

Dymski, G., and R. Pollin. 1994. *New Perspectives in Monetary Macroeconomics.* Ann Arbor: University of Michigan Press.

Eatwell, J. 1975. "Mr Sraffa's Standard Commodity and the Rate of Exploitation." *Quarterly Journal of Economics* 89: 543–55.

———— 1983b. "Theories of Value, Output and Employment." In Eatwell and Milgate 1983a, 93–128.

Eatwell, J., and M. Milgate (eds.) 1983a. *Keynes's Economics and the Theory of Value and Distribution.* New York: Oxford University Press.

Eatwell, J., M. Milgate., and P. Newman. (eds.). 1987. *The New Palgrave.*

Edwards, R. 1979. *Contested Terrain: The Transformation of the Workplace in the Twentieth Century.* New York: Basic Books.

Ehrbar, H., and M. Glick. 1988a. "Micro Advancement toward Long-Term Equilibrium." *Journal of Economic Theory* 53 (2): 369–95.

———— 1988b. "Structural Change in Profit Rate Differentials: The Post World War II U.S. Economy." *Boston Review of Economic Issues* 10 (22): 81–102.

Eichengreen, B. 1985. *The Gold Standard in Theory and History.* New York and London: Methuen.

Eichner, A. 1976. *The Mega-corp and Oligopoly.* Cambridge: Cambridge University Press.

———— 1987. *The Macrodynamics of Advanced Market Economies.* Armonk, N.Y.: M. E. Sharpe.

Eichner, A., and P. Arestis. 1988. "The Post-Keynesian and Institutionalist Theory of Money and Credit." *Journal of Economic Issues* 22 (4).

Einzig, P. 1966. *Primative Money.* 2nd ed. Oxford: Pergamon Press.

Ellis, H. S. 1937. "Some Fundamentals in the Theory of Velocity," *Quarterly Journal of Economics,* 52: 431–72. Reprinted in Lutz, F. and L Mints, *Readings in Monetary Theory,* London: Allen and Unwin.

Enke, S. 1951. "A Distinction between Chamberlin and Robinson." *American Economic Review* 61: 566–78.

Erlich, A. 1960. *The Soviet Industrialization Debate.* Cambridge, MA: Harvard.

Evans, M. 1969. *Macroeconomics: Theory, Forecasting and Control.* New York: Harper and Row.

Fair, R. 1984. Specification, Estimation and Analysis of Macroeconometric Models. Cambridge. Harwad University Press.

Fama, E. 1980. "Banking in the Theory of Finance." *Journal of Monetary Economic* 6: 39–57.

Feyerabend, P. 1975. *Against Method.* London: Verso Press.

Fischer, D. H. 1996. *The Great Wave: Price Revolutions and the Rhythm of History.* New York: Oxford University Press.

Flaschel, P. 1987. "Classical and Neoclassical Competitive Adjustment Processes." *Manchester School* 55: 13–37.

———— 1990. "Cross-Dual Dynamics, Derivative Control and Global Stability: A Neoclassical Presentation of a Classical Theme." *Political Economy: Special Issue.*

Flaschel, P., and W. Semmler. 1991. "Classical Competitive Dynamics and Technical Change." In Halevi, Laibman, and Nell 1992, 198–222.

Florence, P. S. 1953. *The Logic of British and American Industry*. London: Routledge and Kegan Paul.

Foley, D. 1982. "Realization and Accumulation in a Marxian Model of the Circuit of Capital." *Journal of Economic Theory* 28.

—— 1983. "Money and Effective Demand in Marx's Scheme of Expanded Reproduction." In P. Desai (eds.). 1983. *Marxism, Central Planning and the Soviet Economy: Essay in Honor of Alexander Ehrlich*. Cambridge, Mass.: MIT Press.

—— 1986a. *Money, Accumulation and Crisis*. New York: Harwood.

—— 1986b. *Understanding Capital: Marx's Economic Theory*. Cambridge, Mass.: Harvard University Press.

Foley, D. and A. Marquetti. 1997. "Economic Growth from a Classical Perspective," in Teixeira, J., *Money, Growth, Distribution and Structural Change*, University of Brasilia Press.

Freeman, C. 1994. "Economics of Technical Change." *Cambridge Journal of Economics* 18: 463–514.

Friedman, B. M. 1986. "Money, Credit and Interest Rates in the Business Cycle." In Gordon 1986, 395–458.

Friedman, M. 1953. *Essay in Positive Economics*. Chicago: University of Chicago Press.

—— 1959. A Proposal for Monetary Stability. New York: Fordham University Press.

—— 1970. "The New Monetarism: Comment." *Lloyd's Bank Review* (October): xx.

Gallman, R. E. 1966. "Gross National Product in the United States, 1834–1909." *Studies in Income and Wealth* 30: NBER.

Gandolfo, E. 1983. *Economic Dynamics: Methods and Models*. New York: North Holland.

Garegnani, P. 1976. "On a Change in the Notion of Equilibrium in Recent Work on Capital Theory." In M. Brown, K. Sato, and T. Zarembka (eds.). 1976. *Essays in Modern Capital Theory*. Amsterdam: North Holland. Reprinted in Eatwell and Milgate 1983a.

—— 1983. "Notes on Consumption, Investment and Effective Demand" and "Reply to Joan Robinson." In Eatwell and Milgate 1983a.

—— 1984. "Value and Distribution in the Classical Economists and Marx." *Oxford Economic Papers* 36: 291–325.

—— 1990. "On Some Supposed Obstacles to the Tendency of Market Prices towards Natural Prices." *Political Economy: Special Issue*.

—— 1992. "Some Notes for an Analysis of Accumulation." In Halevi, Laibman, and Nell 1992, 47–72.

Geertz, C. 1973. *The Interpretation of Cultures*. New York: Basic Books.

Gemmel, N. 1990. "Comments on Asimakopolus." In Bhardwaj and Schefold, 1990, 345–52.

—— 1993. *The Growth of the Public Sector: Theories and International Evidence*. London: Edward Elgar.

Georgescu-Roegen, N. 1971. *The Entrophy Law and Economic Process*. Cambridge, Mass.: Harvard University Press.

Gilboy, E. 1967. "Demand as a Factor in the Industrial Revolution." In R. M. Hartwell (ed.) 1967. *The Causes of the Industrial Revolution*. London: Methuen.

Goldfeld, S. M. 1976. "The Case of the Missing Money." *Brookings Papers on Economic Activity*, 683–730. Washington, D.C.: Brookings Institute.

Goodwin, R. 1953. "The Static and Dynamic Linear General Equilibrium Models." In *Input-Output Relations, Procedings of a Conference at Driebergen*. Driebergen, Holland: Netherlands Economic Institute. Reprinted in Goodwin, 1983. *Linear Economic Structures*. London: Macmillan.

—— 1966. "A Growth Cycle." In C. H. Feinstein (ed.) 1966. *Socialism, Capitalism and Economic Growth*. Cambridge: Cambridge University Press.

—— 1970. *Elementary Economics from the Higher Standpoint*. Cambridge: Cambridge University Press.

Goodwin, R., and L. Punzo. 1986. *The Dynamics of a Capitalist Economy*. Cambridge:: Polity Press.

Gordon, R. 1983. "A Century of Evidence on Wage and Price Stickiness in the United States, the United Kingdom, and Japan." In Tobin 1983, 1185–34.

—— (ed.). 1986. *The American Business Cycle: Continuity and Change*. Chicago: University of Chicago Press.

—— 1990. "What Is New Keynesian Economics." *Journal of Economics Literature* 28 (3): 1115–71.

Graziani, A. 1996. In Deleplace and Nell, 1996, 139–54.

Greenfield, R. L., and L. B. Yeager. 1983. "A Laissez-Faire Approach to Monetary Stability." *Journal of Money and Credit* 15 (3).

Greenwald, B. C., and J. Stiglitz. 1989. "Toward as Theory of Rigidities." NBER Working Paper no. 2938 (April).

Grossman, G., and E. Helpman. 1994. Innovation and Growth in Global Economy, Cambridge, Mass.: MIT Press.

Haavelmo, T. 1960. *The Pure Theory of Investment*. Chicago: Chicago University Press.

Hagemann, H. 1992. "Traverse Analysis in a Post-Classical Model." In Halevi, Laibman, and Nell 1992, 235–63.

Hahn, F. 1973. *On the Notion of Equilibrium in Economics: An Inaugural Lecture*. Cambridge: Cambridge University Press.

Hahn, F. 1985. *Money, Growth and Stability*. Cambridge, MA: MIT Press.

—— 1988. "The Neo-Ricardians,"*Cambridge Journal of Economics*.

Halevi, J., D. Laibman, and E. J. Nell (eds.). 1992. *Beyond the Steady State*. London: Macmillan.

Hall, R. 1982. "Explorations in the Gold Standard and Related Policies for Stabilizing the Dollar." In R. Hall (ed.). 1982. *Cause and Effects*, 111–22. Chicago: University of Chicago Press.

Hands, D. W. 1993. "Popper and Lakatos in Economic Methodology." In U. Maki, B. Gundafsson, and C. Knudson (eds.). 1993. *Rationality, Institutions, and Economic Methodology*. London: Routledge.

Hansen, A. 1948. *Monetary Theory and Fiscal Policy*. New York: McGraw-Hill.

Harcourt, A. 1980. "A Post-Keynesian Development of the 'Keynesian Model.' " In E. J. Nell 1980, 151–64.

—— 1982. *The Social Science Imperialists*. Ed. Prue Kerr. London: Routledge.

—— 1983. "The Sraffian Contribution: An Evaluation." In I. Bradley and M. Howard (eds.). 1983. *Classical and Marxian Political Economy*. London: Macmillan.

—— 1995. *Capitalism, Socialism and Post-Keynesianism*. Aldershot: Edward Elgar.

Harcourt, G., and P. Kenyon. 1976. "Price Theory and the Investment Decision." *Kyklos* 29: 449–77.

Hargreaves-Heap, S. 1989. *New Keynesian Economics*. Aldershot: Edward Elgar.

Harvey, D. 1989. *The Condition of Post Modernity*. Oxford: Basil Blackwell.

Harris, D. 1978. *Capital Accumulation and Income Distribution*. Stanford: Stanford University Press.

Hausman, D. 1992. *The Separate and Inexact Science of Economics*. Cambridge: Cambridge University Press.

Hayek, F. 1932. *Monetary Theory and the Trade Cycle*. Trans. N. Kaldor and H. Croome. New York: Harcourt Brace.

Hayek, F. A. 1937. *Monetary Nationalism and International Stability*, London: Longmans, Green. Reprinted, New York: Augustus Kelley, 1964.

——— 1941. *The Pure Theory of Capital*. London: Routledge and Kegan Paul.

Hayek, F. A. 1976. *Choice in Currency: A Way to Stop Inflation*, Occ. Paper 48. London: Institute of Economic Affairs

Herman, E. S. 1981. *Corporate Control, Corporate Power*. Cambridge University Press.

Hicks, J. R. 1939. *Value and Capital*. Oxford: Clarendon Press.

——— 1950. *A Contribution to the Theory of the Trade Cycle*. Oxford: Oxford University Press.

——— 1963. *The Theory of Wages*. 2nd ed. London: Macmillan.

——— 1965. *Capital and Growth*. Oxford: Clarendon Press.

——— 1967. *Critical Essays in Monetary Theory*. Oxford: Clarendon Press.

Hicks, J. R. 1969. *A Theory of Economic History*. Oxford: Clarendon Press.

——— 1977. *Economic Perspectives: Further Essays on Monetary Growth*. Oxford: Clarendon Press.

——— 1979. *Causality in Economics*. New York: Basic Books.

——— 1989. *A Market Theory of Money*. London: Macmillan.

Hirsch, F., and H. Goldthorpe. 1978. *The Political Economy of Inflation*. Oxford: Martin Robertson.

Hirschman, A. 1959. *The Strategy of Economic Development*. New Haven: Yale University Press.

Hodgson, G. 1994. "Hayek, Evolution and Spontaneous Order." In Mirowski 1994, 408–50.

Hoffman, W. G. 1958. *The Growth of Industrial Economics*. Manchester: Manchester University Press.

——— 1965. *British Industry 1700–1950*. Oxford: Basil Blackwell.

Hollis, M., and E. J. Nell. 1975. *Rational Economic Man*. Cambridge: Cambridge University Press.

——— 1995. *Philosophy of Social Science*. Cambridge: Cambridge University Press.

Hoover, X. 1990. *The Elements of Social Scientific Thinking*. NY. St. Martin's Press.

Hoover, K. D. 1990. *The New Classical Macroeconomics*. Oxford: Basil Blackwell.

Horsman, G. 1988. *Inflation in the Twentieth Century*. New York: St. Martin's Press.

Hounshell, D. 1984. *From the American System to Mass Production: The Development of Manufacturing Technology in the U.S.* Baltimore: Johns Hopkins University Press.

Howell, D. 1993. "Stages of Technical Advance, Industrial Segmentation and Employment." In Nell 1993c.

Hume, D. 1888. *A Treatise of Human Nature*. Oxford: Oxford University Press.

Hunter, L. C. 1979. *A History of Industrial Power in the U.S. 1780–1930. I: Waterpower.* Charlottesville: University of Virginia Press.

——— 1985. *A History of Industrial Power in the U.S. 1780–1930. II: Steam Power.* Charlottesville: University of Virginia Press.

——— 1991. *A History of Industrial Power in the U.S. 1780–1930. III: The Transmission of Power.* Cambridge, Mass.: MIT Press.

Hymer, S. 1980. "Robinson Crusoe and the Secret of Primitive Accumulation," in Nell, ed., *Growth, Profits and Property*, pp. 29–40.

Ironmonger, D. S. 1972. *New Commodities and Causal Behavior*. Cambridge University Press, Cambridge.

Johnson, W. E. 1933. *Logic*. Oxford: Oxford University Press.

Johnston, J. 1970. *Bishop Berkeley's Querist in Historical Perspective*. Dundalk, Ireland: Dundalgan Press.

Julca, A. 1997. *Peruvian Migration: Countryside to City to Overseas*. New School for Social Research: Dissertation.

Kahn, R. 1905. *Selected Essays on Employment and Growth*. Cambridge: Cambridge University Press.

Kaldor, N. 1940. "A Model of the Trade Cycle." *Economic Journal* 5 (March): 78–89.

——— 1950. "The Economic Aspects of Advertising." *Review of Economic Studies* 18: 1–27.

——— 1956. "Alternative Theories of Distribution." *Review of Economic Studies* 23: 83–100.

——— 1970. "The New Monetarism." *Lloyd's Bank Review* (July).

——— 1982. *The Scourge of Monetarism*. New York: Oxford University Press.

——— 1985. *Economics without Equilibrium*. Armonk, N.Y.: M.E. Sharpe.

Kalecki, M. 1955. *Economic Dynamics*. London: George Allen and Unwin.

——— 1971a. *Selected Essays on the Dynamics of the Capitalist Economy*. Cambridge: Cambridge University Press.

——— 1971b. "Real Wages and Employment." In Kalecki 1971a.

——— 1990. *The Collected Works of Michal Kalecki*. Ed. J. Osiatinski, Oxford: Clarendon Press.

Kauffman, S. 1995. *At Home in the Universe: The Search for Laws of Self-Organization and Complexity*. New York: Oxford University Press.

Kelly, K. 1994. *Out of Control*: The New Biology of Machines, Social Systems, and the Economic World Leading, MA Addison-Wesley.

Keynes J. M. 1930. *Treatise on Money*. London: Macmillan.

——— 1936. *The General Theory of Employment, Interest and Money*. London: Macmillan.

——— 1972–79. *The Collected Works of John Maynard Keynes*. Ed. D. Middridge. Vol. 13, 1972; vol. 14, 1973.

Khalil, E. 1993. "New Classical Economics and Neo-Darwinism: Clearing the Way for Historical Thinking." In Blackwell, Chathu, and Nell 1993, 22–72.

Kindleberger, C. 1978. *Manias, Panics and Crashes: A History of Financial Crises*. New York: Basic Books.

Kirwan, A. 1993. "Representative Agents." *Economic Perspectives*.

Kitson, M., and J. Michie. 1995. "Trade and Growth: A Historical Perspective." In Michie and Smith 1995.

Klein, L., and R. F. Kosobud. 1961. "Some Econometres of Growth: Great Ratios of Economics." *Quarterly Journal of Economics* 75: 173–98.

Koopmans, T. 1953. *Three Essays on the State of Economic Science*. New Haven: Basil Blackwell.

Kornai, J. 1986. *Contradictions and Dilemmas: Studies on the Socialist Economy and Society*. Cambridge, Mass.: MIT Press.

Kozul-Wright, R. 1995. "Transnational Corporations and the Nation-State," in Michie, J. and Smith, J., *Managing the Global Economy*, Oxford: Oxford University Press.

Krause, U. 1971. *Money and Abstract Labour*. London: Verso.

Kregel, J. (ed.) 1983. *Distribution, Effective Demand and International Economic Relations*. London: Macmillan.

——— 1987. "Natural and Warranted Rates of Growth." In Eatwell et al. 19xx.

Krugman, P., and M. Obsfeld. 1994. *International Economics*. 3rd ed. New York: Harper Collins.

Kubin, I. 1990. "Market Prices and Natural Prices: A Model with a Value Effectual Demand." *Political Economy: Special Issue.*

Kurdas, C. 1993. "A Classical Perspective on Investment: Exegesis of Behavioral Assumptions." In Nell 1993c, 333–55.

Kurihara, K. 1957. *Post-Keynesian Economics.* New Brunswick: Rutgers University Press.

Kurihara, K. K. 1951. *Monetary Theory and Public Policy.* London: George Allen and Unwin.

Kurz, H. 1990. *Capital, Distribution and Effective Demand.* Cambridge: Polity Press and Basil Blackwell.

—— 1991. "Technological Change, Growth and Distribution: A Steady-state Approach to Unsteady Growth on Kaldorian Lines." In Nell and Semmler 1991.

Kurz, H., and N. Salvadori. 1993. "The 'Standard Commodity' and Ricardo's Search for an 'Invariable Measure of Value.'" In M. Baranzini and G. C. Harcourt (eds.). 1993. *The Dynamics of the Wealth of Nations.* London: Macmillan.

—— 1995. *The Theory of Production.* Cambridge: Cambridge University Press.

Kuznets, S. 1961. *Capital in the American Economy: Its Formation and Financing.* New York: National Bureau of Economic Research.

Laibman, D. 1992. *Value, Technical Change and Crisis.* Armonk, N.Y. M. E. Sharpe.

Laibman, D., and E. J. Nell. 1977. "Reswitching, Wicksell Effects, and the Neoclassical Production Function." *American Economic Review* 63: 100–13.

Lakatos, I. 1970. "Falsification and the Methodology of Scientific Research Programmes." In I. Lakatos and A. Musgrave (eds.). 1970. *Criticism and the Growth of Knowledge.* Cambridge: Cambridge University Press.

—— 1978. "The Methodology of Scientific Research Programmes." In J. Worral and G. Currie (eds.). 1978. *Philosophical Papers,* 1. Cambridge: Cambridge University Press.

Lancaster, K. 1966. "A New Approach to Consumer Theory." *Journal of Political Economy* 74: 132–57.

—— 1979. *Variety, Equity and Efficiency.* New York: Columbia University Press.

Larkin, J. 1988. *The Reshaping of Everyday Life: 1780–1840.* New York: Harper and Row.

Laslett, P. 1984. *The World We Have Lost: England before the Industrial Age.* 3rd ed. New York: Scribner's.

Latouche, Robert. 1967. *The Birth of Western Economy: Economic Aspects of the Dark Ages.* London: Methuen.

Latsis, S. (ed.). *Method and Appraisal Economics.* Cambridge: Cambridge University Press.

Lavoie, M. 1992. *Foundations of Post-Keynesian Analysis.* Aldershot: Edward Elgar.

Lebergott, S. 1964. *Manpower in Economic Growth.* New York: McGraw-Hill.

—— 1986. "Discussion," *Journal of Economic History.* 46(2): 367–71.

Leijonhufvud, A. 1985. "Capitalism and the Factory System." In R. Langloys (ed.). 1985. *Economics as a Process: Essays in the New Industrial Economics.* Cambridge: Cambridge University Press.

Leontief, W. 1951. *Structure of the American Economy.* New York: Oxford University Press.

—— 1953. "Domestic Production and Foreign Trade: The American Capital Position Reexamined." *Proceedings of the American Philosophical Society* 97: 331–49.

—— 1966. *Input-Output Analysis.* New York: Oxford University Press.

Leontief, W. 1986. *Input-Output Analysis, 2nd edn.* New York: Oxford University Press.

Lloyd, C. 1964. "The Real Balance Effect and the Slutsky Equation." *Journal of Poltical Economy* 72 (June).

Lowe, A. 1955. "Structural Analysis of Real Capital Formation." In Abramovitz M. (eds.) 1955, *Capital Formation and Economic Growth.* Princeton: Princeton University Press.

—— 1965. *On Economic Knowledge.* Armonk, N.Y.: M. E. Sharpe.

—— 1976. *The Path of Economic Growth.* Cambridge: Cambridge University Press.

—— 1987. In A. Oakley (ed.). 1987. *Essays in Political Economics: Public Control in a Democratic Society.* Brighton: Wheatsheaf Books.

Lowe, P. 1970. *The Study of Production.* London: Macmillan.

Lucas, R. 1977. "Understanding Business Cycles." In Brunner and Meltzer, 1977.

—— 1987. *Models of Business Cycles.* London: Basil Blackwell.

Luxemburg, R. 1963 (1951). *The Accumulation of Capital.* trans. Schwarzchild. London: Routledge and Kegan Paul.

Maddison, A. 1982. *Phases of Capitalist Development.* New York: Oxford University Press.

—— 1984. "Origins and Impacts of the Welfare State." *Banca Nazional del Lavoro* x: 55–87.

Madrick, J. 1995. *The End of Affluence.* New York: Random House.

Majewski, R. F. 1994. "Elasticity to Effective Demand and Institutions of Exchange." Mimeo. New School For Social Research.

Malthus, T. R. 1967 (1823). *The Measure of Value Stated and Illustrated, with an Application of It to the Alternations in the Value of the English Currency Since 1790.* New York: Augustus M. Kelley.

Mankiw, N. G. 1990. "A Quick Refresher Course in Macroeconomics." *Journal of Economic Literature* 28: 1645–60.

Mankiw, N. G., and D. Romer. 1993. *New Keynesian Economics.* 2 vols. Cambridge Mass.: MIT Press.

Mansfield, E. 1978. *Monopoly Power and Economic Performance.* New York: Norton.

Mantoux, E. 1961. *The Industrial Revolution in the Eighteenth Century.* London: Methuen.

Marcuzzo, C., and A. Rosselli. 1986. "The Theory of the Gold Standard and Ricardo's Standard Commodity." Discussion Paper 13. Department of Political Economy, University of Modena.

Marglin, S., and J. Schor. 1990. *The Golden Age of Capitalism.* Oxford: Clarendon Press.

Marshall, A. 1961. *Principles of Economics,* Ed. C. W. Guillebaud. 9th ed. London: Macmillan.

Marx, K. 1967. *Capital,* 3 vols. New York: International Publishers.

Matthews, R. C. O. 1959. *The Trade Cycle.* Cambridge: Cambridge University Press.

Mayer, T. 1993. *Truth vs. Precision.* Aldershot: Edward Elgar.

McCloskey, D. 1985. *The Rhetoric of Economics,* Madison: University of Wisconsin Press.

Michie, J. 1987. *Wages in the Business Cycle.* London: Frances Pinter.

Michie, J., and J. G. Smith (eds.). 1995. *Managing the Global Economy,* Oxford: Oxford University Press.

Michl, T. 1992. "Why Is the Rate of Profit Still So Low?" In Papadimitriou 1992, 40–59.

Milgate, M. 1986. *Capital and Employment: A Study of Keynes' Economics.* London: Academic Press.

Mill, J. S. 1848 (1987). *Principles of Political Economy.* Book III. Fairfield, N.J.: Augustus Kelley.

Minsky, H. 1975. *John Maynard Keynes.* New York: Columbia University Press.

—— 1990. "Sraffa and Keynes: Effective Demand in the Long Run." In Bharadwaj and Schefold 1990.

—— 1986. *Stabilizing an Unstable Economy.* New Haven: Yale University Press.

Mirowski, P. 1992. *"What Could Replication Mean in Econometrics?"* Mimeo. University of Notre Dame.

—— (ed.). 1994. *Natural Images in Economic Thought.* Cambridge University Press.

Mishkin, F. 1981. "The Real Interest Rate: An Empirical Investigation." *Carnegie-Rochester Conference Series on Public Policy* 15: 151–200.

—— 1992. *Money, Banking and Financial Markets*. 3rd edition. New York: Harper Collins.

Modigliani, F. 1992. "Liquidity Preference and the Theory of Interest and Money." *Econometrica* 12: 45–88.

Mommsen, T. 1911 (1854–6). *History of Rome, 4 vols*. London and New York: Everyman.

Moore, B. J. 1988. *Horizontalists and Verticalists: The Macroeconomics of Credit Money*. Cambridge: Cambridge University Press.

Moore, G. H., and J. P. Cullity. 1988. "Trends and Cycles in Productivity, Unit Costs, and Prices: An International Perspective." In G. H. Moore (ed.). 1988. *Business Cycles, Inflation and Forecasting*, 2nd ed. NBER. Cambridge: Cambridge University Press.

Morishima, M. 1975. *Theory of Economic Growth*. Oxford University Press.

Mosler, W. 1995. *Soft Currency Economics*. 3rd edn. West Palm Beach, FL: III Finance.

—— 1997–8. "Full Employment AND Price Stability," *Journal of Post Keynesian Economics*, Fall-Winter.

Moss, S. 1980. "The End of Orthodox Capital Theory," in Nell, *Growth, Profits and Property*, Cambrdige: Cambridge University Press.

Munnell, A. 1990. "Why has productivity growth declined? Productivity and Public Investment," *New England Economic Review*, Jan.–Feb., pp. 3–22.

Musgrave, R. A. 1969. *Fiscal Systems*. New Haven: Princeton University Press.

——, and G. Catephores. 1985. *Value, Capital and Exploitation*. xx.

Nayyar, D. 1995. "Globalization: The Past in Our Present." Presidential Address to the Indian Economic Association.

Neisser, H. 1928. *Der Tauschwert des Geldes*. Jena.

Nell, E. J. 1967. "Wicksell's Theory of Circulation." *Journal of Poltical Economy* xx: yy.

—— 1968. "Advantages of Money over Barter." *Australian Economic Papers* xx: yy.

—— 1970. "A Note on Cambridge Controversies in Capital Theory." *Journal of Economic Literature* 8.

Nell, E. J. 1973. "The Fall of the House of Efficiency," *The Annal of the American Academy of Political and Social Science*, Special Issue on *Income Inequality*, pp. 102–112.

—— 1975. "The Simple Theory of Effective Demand." In E. J. Nell (ed.). 1975. *Political Economy at the New School*. New York: New School.

—— 1976. "An Alternative Presentation of Lowe's Basic Model." In A. Lowe, 1976, 289–329.

—— 1976a. "No Statement is Immune to Revision," *Social Research*.

—— 1977. "Credito, Circulacao, e trocas na transformacao do sociedade agricola" in Garegnani, Steindl, et al., *Progresso Technico e Teoria Economia*, Editora Hucitec: Universidade Estadual de Campinas.

—— 1980. *Growth, Profits and Property*. Cambridge: Cambridge University Press.

—— 1982. "Growth, Distribution and Inflation." Journal of Post Keynesian Economics 5(1): 104–13.

—— 1983. "Keynes after Sraffa." In Kregel 1983.

—— 1988a. "Does the Rate of Interest Determine the Rate of Profit?" *Political Economy: Studies in the Surplus Approach* 4(2).

—— 1988b. *Prosperity and Public Spending*. Boston: Unwin Hyman.

—— 1988c. "On Monetary Circulation and the Rate of Exploitation." In P. Arestis (ed.), 1988. *Post-Keynesian Monetary Economics*. London: Edward Elgar.

—— 1991a. "Capitalism, Socialism and Effective Demand." In Nell and Semmler 1991.

—— 1992a. "Demand Equilibrium." In Halevi, Laibman, and Nell 1992, 96–128.

—— 1992b. "Transformation Growth and the Multiplier." In Halevi, Laibman, and Nell 1992, 131–74.

—— 1992c. *Transformational Growth and Effective Demand.* London: Macmillan, New York: New York University Press.

—— 1993a. "Transformational Growth and Learning: Developing Craft Technology into Scientific Mass Production." In R. Thomson (ed.), 1993. *Learning and Technical Change.* London: Macmillan.

—— 1993b. "Demand, Pricing and Investment." In W. Milberg (ed.), 1993. *The Megacorp and Macrodynamics*, Armonk, N.Y.: M.E. Sharpe.

—— 1993c. *Economics and Worldly Philosophy.* London: Macmillan.

—— 1994. "Minsky, Keynes and Sraffa: Investment and the Long Period." In G. Dymski and R. Pollin (eds.), 1994. *New Perspectives in Monetary Macroeconomics.* Ann Arbor: University of Michigan Press.

—— 1995b. "The Circuit of Money in a Production Model." In Deleplace and Nell 1996, 245–304.

—— 1996. *Making Sense of a Changing Economy.* New York and London: Routledge.

—— 1997. *Transformational Growth and the Business Cycle.* New York and London: Routledge.

—— 1997a. "Wicksell after Sraffa." In G. Mongiovi and F. Petri (eds.), 1997. *Essays in Honor of Piero Garegnani*, Elwarl Elgar.

Nell, E. J., and G. Deleplace (eds.). 1995c. *Money in Motion: The Post-Keynesian and Circulation Approaches.* London: Macmillan.

Nell, E. J., and E. Delamonica. 1995d. "Oil Shocks and Missing Money." Mimeo. New School for Social Research.

Nell, E. J., and W. Semmler. 1991b. *Nicholas Kaldor and Mainstream Economics.* London: Macmillan.

Nell, E. J., and T. F. Phillips. 1995a. "Transformational Growth and the Business Cycle." *Eastern Economic Journal* 21 (2): 125–42.

Nelson, R., and S. Winter, 1982. *An Evolutionary Theory of Economic Change.* Cambridge, Mass.: Belknap Press.

Nickell, S. J. 1978. *The Investment Decisions of Forms.* Cambridge: Cambridge University Press.

Nield, R. R. 1963. "Pricing and Employment in the Trade-Cycle: A Study of British Manufacturing Industry." National Institute of Economic and Social Research Occasional Paper 21.

Niehaus, J. 1988. *The Theory of Money.* Baltimore: John Hopkins University Press.

Nikaido, H. 1985. "Dynamics of Growth and Capital Mobility in Marx s Schemes of Reproduction," *Zeitschrift fur Nationalokonomie.*

Nurske, R. 1953. *Problems of Capital Formation in Underdeveloped Countries.* Oxford: Oxford University Press.

Ochoa, E. 1984. "Labor Values and Prices of Production: An Interindustry Study of the U.S. Economy, 1947–72." Ph.D. dissertation. New York: New School for Social Research.

—— 1986. "Is Reswitching Empirically Relevant: U.S. Wage, Profit Rate Frontiers, 1947–1972." *Economic Forum* 16: 45–67.

O'Neill, O. 1994. *Constructions of Reason.* Cambridge: Cambridge University Press.

Organization for Economic Cooperation and Development. 1994. *Economic Survey.* Paris: OECD.

Okun, A. 1981. *Prices and Quantities.* Washington: Bookings Institute.

O'Neill, O. 1986. *Faces of Hunger.* London: Allen and Unwin.

Osiatynski, J. 1990. *The Collected Works of Michal Kalecki. Vol. 1, Capitalism: Business Cycles and Full Employment*, Oxford: Clarendon Press.

Ostroy, J., and S. Starr. 1990. "The Transactions Role of Money." In B. Friedman and F. Hahn (eds.) 1990. *Handbook of Monetary Economics. I.* North Holland/Elsevier Science.

Pacey, A. 1976. *The Maze of Ingenuity: Ideas and Idealism in the Development of Technology.* Cambridge Mass.: MIT Press.

Panico, C. 1988. *Interest and Profit in the Theory of Value and Distribution.* London: Macmillan.

Pap, A. 1959. *Semantics and Necessary Truth.* New Haven: Yale University Press.

Papadimitriou, D. (ed.). 1992. *Profits, Deficits and Instability,* London: Macmillan.

Parguez, A. 1996. "Beyond Scarcity: A Reappraisal of the Theory of the Circuit of Money." In Deleplace and Nell 1996, 155-99.

Park, M. C. 1994. "Normal Values and Average Values." School of Business and Economic Studies. University of Leeds.

Parrinello, S. 1990. "Some Reflexions on Classical Equilibrium, Expectations and Random Distrubances." In *Political Economy: Special Issue.*

Pasinetti, L. 1960. "A Mathematical Formulation of the Ricardian System." *Review of Economic Studies* 27: 78–98.

——— 1974. *Growth and Income Distribution.* Cambridge: Cambridge University Press.

——— 1977. *Lectures on the Theory of Production.* New York: Columbia University Press.

——— 1981. *Structural Change and Economic Growth: A Theoretical Essay on the Dynamics of the Wealth of Nations.* Cambridge University Press.

Patinkin, D. 1965. *Money, Interest and Prices*, 2nd ed. New York: Harper & Row.

——— 1991. "Introduction." *Money, Interest and Prices*, 3rd edn.

Peacock, A. T., and J. Wiseman. 1961. *The Growth of Public Expenditure in the United Kingdom.* Princeton: Princeton University Press.

Pedersen, J., and O. Petersen. 1938. *An Analysis of Price Behaviour, 1855–1913.* London: Humphrey Milford, Oxford University Press.

Peitgen, H. O., H. Jurgens, and D. Soupe. 1992. *Chaos and Fractals: New Frontiers in Science.* New York and Berlin: Springer-Verlag.

Penrose, E. 1954. *A Theory of the Growth of the Firm.* Oxford: Basil Blackwell.

——— 1974. *The Large International Firm.* Oxford: Oxford University Press.

Perez, C. 1983. "Structural Change and Assimilation of New Technologies in Economic Social Systems." *Futures* (October): 357–75.

——— 1985. "Microelectronics, Long Waves and World Structural Change: New Perspectives for Developing Economies." *World Development* 13(3): 441–63.

Phelps Brown, H., and S. Hopkins. 1981. *A Perspective of Wages and Prices.* London: Methuen.

Pickering, J. F. 1971. "The Prices and Incomes Board and Private Sector Prices: A Survey." *Economic Journal* 81 (June): 225–41.

Pigou, A. C. 1927. *Industrial Fluctuations.* London: Macmillan; reprinted, New York: Augustus Kelley, 1967.

——— 1944. *Employment and Equilibrium.* London: Macmillan.

Pivetti, M. 1988. "On the Monetary Explanation of Distribution." In Bharadwaj and Schefold 1990, 443–53.

——— 1991. *An Essay on Money and Distribution.* London: Macmillan.

Polanyi, K. 1944. *The Great Transformation: The Political and Economic Origins of Our Time.* Boston: Beacon Press.

Portes, A., M. Castells, and L. Benton. 1989. *The Informal Economy: Studies in Advanced and Less Developed Countries*. Baltimore: Johns Hopkins University Press.

Popper, K. 1959. *The Logic of Scientific Discovery*. London: Hutchinson.

———— 1963. *Conjectures and Refutations*. London: Routledge.

———— 1985. "The Rationality Principle." In D. Miller (ed.), 1985. *Popper Selections*, Princeton: Princeton University Press.

Poulon, F. 1982. *Economie Generale*. Paris: Dunod

Quine, W. V. O. 1953. *From a Logical Point of View*. Cambridge, Mass.: Harvard University Press.

———— 1960. *Word and Object*. New York: Wiley and Sons.

Radcliffe-Brown, A. R. 1963. *Structure and Function in Primitive Society*. New York: Free Press.

Ricardo, D. 1951. *On the Principles Economy and Taxation*. In P. Sraffa (ed.). 1951. Cambridge: Cambridge University Press.

Rizvi, S. 1991. "Specialization and the Existence Problem in General Equilibrium Theory." *Contributions to Poltical Economy* 10: 1–20.

Robbins, L. 1930. "The Conception of Stationary Equilibrium." *Economic Journal* 40: 194–214.

———— 1935. *An Essay on the Nature and Significance of Economic Science*, 2nd ed. London: Macmillan.

Robertson, D. 1931. "Wage-Grumbles." In *Economic Fragments*, 42–57.

———— 1957. *Lectures on Economic Principles*. London: Fontana.

———— 1962. *Money*. Chicago: University of Chicago Press.

Robinson, E. A. G. 1931. *The Structure of Competitive Industry*. Cambridge: Cambridge University Press.

Robinson, J. 1962. *Essays in the Theory of Economic Growth*. London: Macmillan.

———— 1965. *Exercises in Economic Analysis*. London: Macmillan.

———— 1971. *Economic Heresies*. New York: Basic Books.

———— 1983. "Garegnani on Effective Demand." In Eatwell and Milgate 1983a.

Robinson, J., and J. Eatwell. 1973. *An Introduction to Modern Economics*. London: McGraw-Hill.

Romer, C. 1986. "New Estimates of Prewar Gross National Product and Unemployment." *Journal of Economic History* 46 (2): 341–52.

———— 1989. "The Prewar Business Cycle Reconsidered: New Estimates of Gross National Product, 1869–1908." *Journal of Poltical Economy* 97(1): 1–37.

Romer, D. 1984. "The Theory of Social Custom," *Quarterly Journal of Economics*, 99: 717–27.

Roncaglia, A. 1985. "The Neo-Ricardian Approach and the Distribution of Income." In A. Asimakopolus (ed.), 1985. *Theories of Income Distribution*. Boston: Kluwer-Nijhoff.

———— 1978. *Sraffa and the Theory of Prices*, New York: John Wiley.

———— 1990. "Is the Notion of Long-Period Positions Compatible with Classical Political Economy?" *Political Economy: Studies in the Surplus Approach* 6: 103–11.

Rorty, R. 1991. *Objectively, Relativism and Truth*. Cambridge University Press.

Rosenberg, A. 1992. *Economics: Mathematical Politics or Science of Diminishing Returns*. Chicago: University of Chicago Press.

———— 1994. "Does Evolutionary Theory Give Inspiration to Economics?" In Mirowski 1994, 384–407.

Rostow, W. W. 1990a. *The Stages of Economic Growth*. 3rd ed. Cambridge: Cambridge University Press.

762   Bibliography

——— 1990b. *Theories of Economic Growth from David Hume to the Present.* New York: Oxford University Press.

Rowthorn, R. 1982. "Demand, Real Wages, and Economic Growth." *Studi Economici* 18: 2–53.

Rymes, T. K. (ed.). 1989. *Keynes's Lectures, 1932–35: Notes of a Representative Student.* Ann Arbor: University of Michigan Press.

Samuelson, P. 1966. "Wages and Interest: A Modern Dissection of Marxian Economic Models," reprinted in *Collected Scientific Papers of Paul A. Samuelson,* vol. 1, Cambridge, MA: MIT Press.

——— 1979. "Insight and Detour in the Theory of Exploitation: A Reply to Baumol," reprinted in *Collected Scientific Papers of Paul A. Samuelson,* vol. 4. Cambridge, MA: MIT Press.

Samuelson, P. 1989. "Revisionist Findings on Sraffa." In Bharadwaj and Schefold (1990).

Samuelson, P., and W. Nordhaus. 1990. *Economics.* 13th ed. New York: McGraw-Hill.

Sargent, T. 1987. *Macroeconomic Theory.* 2nd ed. New York: Academic Press.

Sayers, R. 1957. *Central Banking after Bagehot.* Oxford: Oxford University Press.

Schefold, B. 1986. "The Standard Commodity as a Tool of Economic Analysis: A Comment on Flaschel." *Journal of Institutional and Theoretical Economics* 142: 603–22.

——— 1989. *Mr Sraffa on Joint Production and Other Essays.* London: Unwin Hyman.

Schefold, B. 1997. *Normal Prices, Technical Change and Accumulation.* London: Macmillan.

Schelling, T. 1960. *Strategy of Conflict.* Cambridge Mass.: Harvard University Press.

Schoeffler, S. 1995. *The Failures of Economics: A Diagnostic Study.* Cambridge, Mass.: Harvard University Press.

Schumpeter, J. 1934. *The Theory of Economic Development.* New York: Harper and Bros.

——— 1942. *Capitalism, Socialism and Democracy.* New York: Harper and Bros.

Semmler, W. 1942. *Competition, Monopoly and Differential Profit Rates.* New York: Columbia University Press.

Sammler, W. 1982. *Competition, Monopoly and Differential Profit Rates.* New York: Columbia University Press

——— 1990. "On Composite Market Dynamics: Simultaneous Microeconomic Price and Quantity Adjustments." *Poltical Economy: Special Issue.*

Semmler, W., and R. Franke. 1989. "Debt-Financing of Firms, Stability and Cycles in A Dynamical Macro Model." In W. Semmler (ed.). 1989. *Financial Dynamics and Business Cycles.* Armonk, N.Y.: M. E. Sharpe.

——— 1996. "The Financial-Real Interaction, and Investment in the Business Cycle: Theories and Empirical Evidence." In Deleplace and Nell 1996, 606–34.

Sen, A. 1981. *Poverty and Famines.* Oxford: Clarendon Press.

Sethi, R. 1991. "Notes on Debt, Growth and the 'Cambridge' Equation." Mimeo. New School for Social Research.

Sheffrin, S. M. 1989. *The Making of Economic Policy.* Oxford: Basil Blackwell.

Skidelsky, R. 1992. *John Maynard Keynes.* Vols. 1 and 2. New York: Viking.

Smith, A. 1961 (1776). *An Inquiry into the Nature and Causes of the Wealth of Nations.* Cannan ed. London: Methuen.

Synder, C. 1924. "New Measures in the Equation of Exchange." *American Economic Review* 14: 698–713.

Solomou, S. 1987. *Phases of Economic Growth 1850–1973: Kondratieff Waves and Kuznets Swings.* Cambridge: Cambridge University Press.

Spaventa, L. 1970. "Rate of Profit, Rate of Growth, and Capital Intensity in a Simple Production Model." *Oxford Economic Papers* 22: 129–47.

Sraffa, P. 1926. "The Laws of Returns under Competitive Conditions." *Economic Journal* (Dec.): 535–50.

——— 1960. *Production of Commodities by Means of Commodities*. Cambridge: Cambridge University Press.

——— 1990. "Questions and Suggestions re Convergence." *Poltical Economy: Special Issue.*

Steedman, I. 1977. *Max after Sraffa*. London: Verso.

——— 1984. "Natural Prices, Differential Profit Rates and the Classical Competitive Process." *Manchester School* 52: 123–39.

Steedman, I. 1991. "Questions for Kaleckians," mimeo.

Steindl, J. 1976. *Maturity and Stagnation in American Capitalism*. New York: Monthly Review Press.

Stonier, A. and D. C. Hague. 1961. *A Textbook of Economic Theory*. New York: John Wiley.

Strawson, P. 1959. *Individuals*. London: Methuen.

Summers, L. 1991. *Understanding Unemployment*. Cambridge, MA, MIT Press.

Sutcliffe, R. B. 1971. *Industry and Underdevelopment*. London: Addison-Wesley.

Sylos-Labini, P. 1969. *Obligology and Technical Progress*. Cambridge, Mass.: Harvard University Press.

——— 1984. *The Forces of Economic Growth and Decline*. Cambridge Mass.: MIT Press.

——— 1989. "Changing Character of the So-called Business Cycle." *Atlantic Economic Journal.*

——— 1993. "Long-Run Changes in the Wage and Price Mechanisms and the Processes of Growth." In Baranzini and Harcourt 1993.

Szostak, R. 1994. *Technological Innovation and the Great Depression*. Chicago: Westview Press.

Tambiah, S. J. 1990. *Magic, Science, Religion, and the Scope of Rationality*. Cambridge: Cambridge University Press.

Targetti, F. 1989. *Nicholas Kaldor.* Oxford: Oxford University Press.

Tarshis, L. 1939. "Changes in Real and Money Wages." *Economic Journal* 49 (March): 150–4.

Taylor, F. W. 1991. *The Principles of Scientific Management*. New York: Harper and Bros.

Taylor, J. B. 1986. "Improvements in Macroeconomic Stability: The Role of Wages and Prices." In Gordon 1986, 639–79.

Taylor, L., and D. Weiser. 1972. "Advertising and the Aggregate Consumption Function." *American Economic Review* 62: 642–55.

Tobin, J. 1980. *Asset Accumulation and Economic Activity*. Chicago: University of Chicago Press.

——— (ed.). 1983. *Macroeconomics, Prices and Quantities*. Oxford: Basil Blackwell.

Trautwein, H.-M. 1990. "Money Matters in Post-Keynesian Theories and New Monetary Economics." Mimeo. Lunenburg University.

Tylecote, A. 1991. *The Long Wave in the World*. London: Routledge.

Uimonen, P. 1993. *Public Debt, Public Investment and Productivity Growth*. New School For Social Research: Dissertation.

Urquhart, M. C. 1986. "New Estimates of Gross National Product, Canada, 1870–1926: Some Implications for Canadian Development." In S. L. Engerman and R. E. Gallman (eds.). 1986. Long-term Factors in American Economic Growth. NBER Studies in Income and Wealth, vol. 51, Chicago: University of Chicago Press.

U.S. Department of Commerce. 1966. *Long Term Economic Growth: 1860–1965.* Washington, D.C.: U.S. Dept. of Commerce.

Vega, Garcilaso de la. 1871 (1609). *Royal Commentary on the Origin of the Incas.* London (Lisbon).

Vianello, F. 1986. "The Pace of Accumulation." *Political Economy: Studies in the Surplus Approach.*

——— 1989. "Natural (or Normal) Prices: Some Pointers." *Poltical Economy: Studies in the Surplus Approach* 2: 89–105.

——— 1995. "Joan Robinson's Theory of Accumulation." Mimeo. University of Rome.

Vilar, P. 1969. *A History of Gold and Money,* 1450–1920. London: New Left Books.

Von Neumann, J. 1945–6. "A Model of General Economic Equilibrium." *Review of Economics and Statistics.*

Von Weizacker, C. 1971. *Steady State Capital Theory.* New York: Springer-Verlag.

Wagner, A. 1883. *Finanzwissenschaft* 2nd ed. (3rd ed. 1890). Leipzig. Translated and reprinted in R. A. Musgrave and A. T. Peacock (eds.). 1958. *Classics in the Theory of Public Finance.* London: Macmillan.

Waldrop, M. 1992. *Complexity: The Emerging Science at the Edge of Order and Chaos.* New York: Simon and Schuster.

Walras, L. 1926. [W. Jaffe (ed.). 1954.] *Elements of Pure Economics or The Theory of Social Wealth.* London: Allen and Unwin.

Walsh, V., and H. Gram. 1980. *Classical and Neoclassical Theories of General Equilibrium.* Oxford: Oxford University Press.

Weber, M. 1949. "Critical Studies in the Logic of the Cultural Sciences." In E. A. Shils and H. A. Finch (eds.), 1949. *Max Weber on the Methodology of the Social Sciences*, 113–63. Glencoe, Ill.: Free Press.

Weintraub, S. 1978. *Capitalism's Inflation and Unemplyment Crises.* Reading, Mass.: Addison-Wesley.

Weisberg, D. 1967. *Guild Structure and Political Allegiance in Early Achaemenid Mesopotamia.* New Haven: Yale University Press.

Wells, P. 1960. "Keynes' Aggregate Supply Function: A Suggested Interpretation," *Economic Journal*, Dec.

——— 1962. "Aggregate Supply and Demand: An Explanation of Ch. 3 of the General Theory," *Canadian Journal of Economics*, Nov.

Wicksell, K. 1898 [1936]. *Interest and Prices.* Trnas. R. F. Kahn, London: Macmillan.

——— 1935. *Lectures on Political Economy.* Vols. 1 and 2. Trans. E. Classen. Ed. Lionel Robbins. New York: Macmillan.

——— 1958. *Selected Papers on Economic Theory.* London: Allen and Unwin.

Wiggins, D. 1980. *Sameness and Substance.* Oxford: Basil Blackwell.

Williamson, O. E. 1980. "The Organization of Work: A Comparative Institutional Assessment." *Journal of Economic Behavior and Organization* 1: 5–38.

Wilson, T., and P. W. S. Andrews. 1951. *Oxford Studies in the Price Mechanism.* Oxford: Clarendon Press.

Winch, P. 1958. *The Idea of a Social Science and its Relation to Philsophy.* London: Routledge.

Withers, H. 1937. *The Meaning of Money.* London: John Murray.

Wittgenstein, L. 1956. *Philosophical Investigations.* Oxford: Basil Blackwell.

Wolfson, M. 1987. "Science and History: Economics and Thermodynamics." Paper presented at the Fourteenth History of Economics Meeting, Boston.

Wood, A. 1978. *A Theory of Profit.* Cambridge: Cambridge University Press.

Woods, J. E. 1978. *Mathematical Economics.* New York: Longman.

Wray, L. R. 1990. *Endogenous Money.* Aldershot: Edward Elgar.

Wright, I. 1926. *Readings in Money, Credit and Banking Principles.* New York: Harper and Bros.

Zarnowitz, V. 1985. "Recent Work on Business Cycles in Historical Perspective: A Review of Theories and Evidence." *Journal of Economic Literature* 23: 523–80.

——— 1986. "Major Changes in Cyclical Behavior." In Gordon 1986, 529–82.

——— 1992. *Business Cycles: Theory, History, Indicators, and Forecasting.* Chicago: University of Chicago Press.

INDEX

accelerator, *see* multiplier
ACEs, *see* advanced capitalist economies
adjustment process
  capitalism and scualigm, 604–08
  convergence to long-period positions, 389
  in Craft economy, 390
  Marshallian, 410, 419
administrative hierarchy, *see* bureaucracy
advanced capitalist economies (ACEs), xxiv–xv
Aftalion, A., 43
agents
  abstract neo-Classical rational, 124–6
  with active and passive minds, 71–3
  in behavioral model situated in structure, 126–8
  fieldwork related to, 112–13
  in models explaining and predicting behavior, 126–8
  in models of neo-Classical economics, 71, 73, 115–18, 125–6
  in neo-Classical economics, 105, 124–5
  participation in Craft-based economy, 564, 654–55
  participation in Mass Production economy, 565
  in prediction models, 138
  in programming models, 128
  rational, 76, 129
  in rational choice model, 128–9
  uncertainty for, 135–7
  in Walrasian system, 166
agents, economic
  activity of, 92–4
  identity of, 96
  rationality of, 100
  using fieldwork to understand, 103
aggregate theory, xxii
agriculture
  in AS/TA period, 22–4
  bio-tech, 28–9
  in CBF/MA period, 24–6
  corporate, 26–8
  in CP/BA period, 28–9

  in MP/CA period, 26–8
  traditional, 22–24
  transition from traditional to mechanized, 24–6, 687
Akerlof, G., 63
allocation
  by organization, 134
  of stimulus-response models, 149
Altman, M., 68
Amadeo, E. J., 647
analytic-synthetic distinction (ASD), 82–4, 86
Ancombe, G. E. M., 114
Andrews, P., 101
arbitrage
  of capital, 266–83
  with changes in value of money, 250–1
  in Craft economy, 267
  in growth stocks, 264–5
  interaction between i and g, 687
  in Mass Production, 267
  in monetary system, 281
  see also capital arbitrage
artisan shops
  in early industry, 22–24
artisan shops/traditional agriculture (AS/TA) period
  characteristics of, 739–41
  pressure for mechanization, 429
  short-run labor demand, 566
  transition to CBF/MA, 412, 415, 429
  trends in sector relationships, 410–12, 740
ASD, *see* analytic-synthetic distinction
Asimakopoulos, A., 343, 524n31, 619–20
AS/TA period, *see* artisan shops/traditional agriculture period
auctioneer (Walras), 152
Austerity Dilemma, 698
Austin, J. L., 108

Baldwin, R., 538n34, 741n4
Balke, N. S., 68
bank capital, 225
bank deposits, 227

784   Index